Complete Business Statistics

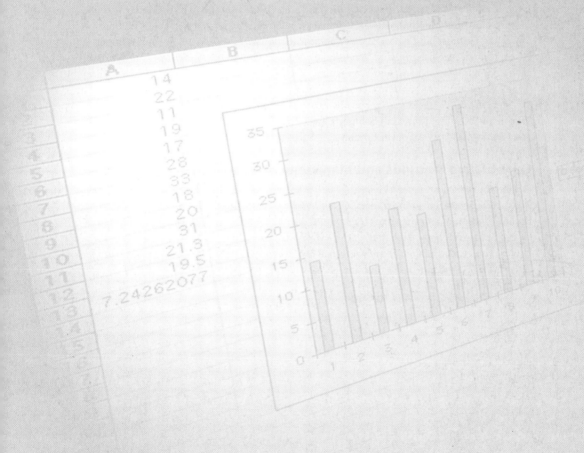

CD-ROM ON BACK PAGE

The McGraw-Hill/Irwin Series:
Operations and Decision Sciences

Business Statistics

Alwan, *Statistical Process Analysis,* First Edition

Aczel and Sounderpandian, *Complete Business Statistics,* Fifth Edition

Bowerman, O'Connell, and Hand, *Business Statistics in Practice,* Second Edition

Bryant and Smith, *Practical Data Analysis: Case Studies in Business Statistics, Volumes I and II,* Second Edition; *Volume III,* First Edition

Cooper and Schindler, *Business Research Methods,* Seventh Edition

Delurgio, *Forecasting Principles and Applications,* First Edition

Doane, Mathieson, and Tracy, *Visual Statistics,* Second Edition, 2.0

Doane, Mathieson, and Tracy, *Visual Statistics: Statistical Process Control,* First Edition, 1.0

Gitlow, Oppenheim, and Oppenheim, *Quality Management: Tools and Methods for Improvement,* Second Edition

Lind, Mason, and Marchal, *Basic Statistics for Business and Economics,* Third Edition

Mason, Lind, and Marchal, *Statistical Techniques in Business and Economics,* Tenth Edition

Merchant, Goffinet, and Koehler, *Basic Statistics Using Excel for Office 97,* Second Edition

Neter, Kutner, Nachtsheim, and Wasserman, *Applied Linear Statistical Models,* Fourth Edition

Neter, Kutner, Nachtsheim, and Wasserman, *Applied Linear Regression Models,* Third Edition

Siegel, *Practical Business Statistics,* Fourth Edition

Webster, *Applied Statistics for Business and Economics: An Essentials Version,* Third Edition

Wilson and Keating, John Galt Solutions, Inc., *Business Forecasting with Accompanying Excel-based ForecastX Software,* Fourth Edition

Quantitative Methods and Management Science

Bodily, Carraway, Frey, and Pfeifer, *Quantitative Business Analysis: Casebook,* First Edition

Bodily, Carraway, Frey, and Pfeifer, *Quantitative Business Analysis: Text and Cases,* First Edition

Bonini, Hausman, and Bierman, *Quantitative Analysis for Business Decisions,* Ninth Edition

Hesse, *Managerial Spreadsheet Modeling and Analysis,* First Edition

Hillier, Hillier, and Lieberman, *Introduction to Management Science: A Modeling and Case Studies Approach with Spreadsheets,* First Edition

Complete Business Statistics

FIFTH EDITION

Amir D. Aczel
Bentley College

Jayavel Sounderpandian
University of Wisconsin–Parkside

McGraw-Hill
Irwin

Boston   Burr Ridge, IL   Dubuque, IA   Madison, WI   New York   San Francisco   St. Louis
Bangkok   Bogotá   Caracas   Kuala Lumpur   Lisbon   London   Madrid   Mexico City
Milan   Montreal   New Delhi   Santiago   Seoul   Singapore   Sydney   Taipei   Toronto

McGraw-Hill Higher Education

A Division of The **McGraw-Hill** *Companies*

COMPLETE BUSINESS STATISTICS

Published by McGraw-Hill/Irwin, an imprint of The McGraw-Hill Companies, Inc. 1221 Avenue of the Americas, New York, NY, 10020. Copyright © 2002, 1999, 1996, 1993, 1989 by The McGraw-Hill Companies, Inc. All rights reserved. No part of this publication may be reproduced or distributed in any form or by any means, or stored in a database or retrieval system, without the prior written consent of The McGraw-Hill Companies, Inc., including, but not limited to, in any network or other electronic storage or transmission, or broadcast for distance learning.
Some ancillaries, including electronic and print components, may not be available to customers outside the United States.

This book is printed on acid-free paper.

2 3 4 5 6 7 8 9 0 VNH/VNH 0 9 8 7 6 5 4

ISBN 0-07-112290-7

www.mhhe.com

To my father, Captain E. L. Aczel, and to the memory of my mother, Miriam Aczel

Amir D. Aczel

To the memory of my father,
V. M. Sounderpandian

Jayavel Sounderpandian

Amir D. Aczel teaches business statistics at Bentley College in Waltham, Massachusetts, and is an occasional business commentator on CNN. He earned his B.A. in mathematics and M.S. in operations research from the University of California–Berkeley, and he was awarded his Ph.D. in statistics from the University of Oregon. Prior to assuming his current position, Aczel lived and taught in California, Alaska, Europe, the Middle East, and the Far East.

Besides his two textbooks with McGraw-Hill/Irwin, Aczel has written four trade books, each drawing on his mathematical and statistical background to examine a real-world problem. *Fermat's Last Theorem* discusses the quest to prove a heretofore unprovable mathematical theorem, *How to Beat the IRS at Its Own Game* suggests measures readers can use to avoid an IRS audit, *Probability 1* examines whether life exists in the universe beyond our own planet, and the recently published *The Riddle of the Compass* discusses the invention of the compass.

Aczel's wide-ranging interests, which are evident in *Complete Business Statistics*, reflect the international scope of the author's experience, the authority of his educational background, and the practical bent he has displayed in his other writing. Aczel lives with his wife and daughter in the greater Boston area.

Jayavel Sounderpandian is a professor of quantitative methods at the University of Wisconsin–Parkside, where he has been teaching business statistics and operations management for 18 years. He earned his doctoral and master's degrees in business from Kent State University and received his bachelor's degree in mechanical engineering from the Indian Institute of Technology, Madras. Before turning to academia, Sounderpandian worked as an engineer for seven years at an aircraft manufacturing company in India.

Besides this textbook, Sounderpandian has written four supplemental textbooks that focus on the use of spreadsheets in business statistics, market research, and operations management. He has several years of experience in using spreadsheet templates in the classroom, and he has applied that expertise in designing the more than 100 templates in this textbook. Sounderpandian's research interests are in decision and risk analysis, and he has published widely in leading journals such as *Operations Research, Abacus, Journal of Risk and Uncertainty, International Journal of Production Economics,* and *Interfaces.*

In his spare time, Sounderpandian composes and solves mathematical puzzles. His mathematical creativity and industrial experience are behind many of the case studies in this textbook. Sounderpandian and his wife live in Racine, Wisconsin. They have three daughters.

There are two significant changes to the fifth edition of *Complete Business Statistics* that we are pleased to announce. The first is the addition of a new co-author, Professor Jayavel Sounderpandian of the University of Wisconsin–Parkside. A longtime adopter of *Complete Business Statistics* and a major contributor to the fourth edition of the book, Professor Sounderpandian possesses the intimate knowledge of and respect for the book that you would expect from a dedicated user, and he brings a fresh perspective to this revision of it. He has helped to improve the book by incorporating a number of new features.

One of these features constitutes the second significant change—the thorough integration of Microsoft Excel into the fifth edition. Although Excel was introduced in the previous edition of the book, Professor Sounderpandian has fully woven it into the fabric of the text and the pedagogy. Thus, Excel is now integral to the text rather than overlaid on it.

The motivation for this change is the increasing use of Microsoft Excel in the marketplace. Certainly statistical software packages continue to be used in business statistics courses, and few would disagree that for some statistical applications these packages are superior to Excel. Despite the prominent treatment of Excel in the fifth edition, the use of alternative statistical packages remains a viable option for those who prefer them. Today, however, an overwhelming majority of managers and other practitioners in business have immediate access to Excel. Our increased emphasis on Excel in the fifth edition of *Complete Business Statistics* is simply a response to the realities of the business environment, the worldwide computing environment, and the students we teach who are preparing to join these communities.

A specific improvement in the treatment of Excel in the fifth edition is the incorporation into the text of spreadsheet templates for maximum flexibility and utility in solving problems. Although templates are not without disadvantages—most serious is the possibility that students will use templates without understanding the concepts behind them—we believe that the benefits of using templates outweigh their liabilities. We also thoroughly discuss the use and misuse of templates in Chapter 0, so that students will be aware of both their advantages and disadvantages prior to tackling the topical material in *Complete Business Statistics*. It should also be noted that although we believe templates are particularly useful and efficient aids to problem solving, illustrations for manual calculations have been retained so that students can manually calculate any result found in a template.

Besides these changes, the fifth edition offers the following improvements:

1. Many new problems and examples.
2. New cases, incorporating the use of Excel and Excel templates, are found in every chapter.
3. Chapter 7 on hypothesis testing has been almost completely rewritten for greater clarity and accessibility.
4. Chapter 3 on random variables contains new materials facilitated by the use of templates.
5. In the chapters covering simple and multiple regression, the addition of templates enables students to completely solve problems rather than just interpret computer printouts. Cost considerations have been brought into focus wherever appropriate—a feature that business students should find especially useful and appealing.

So far we have focused on the changes to the fifth edition. Just as important as what has been added to the book, however, is what has been retained. As with previous editions, all the important subjects within business statistics are covered, though they are emphasized in a new light for this edition to reflect the evolution of business practices. Topics that are not widely taught at business schools have been abridged, and some statistical methods with less-than-universal applicability in today's business world have been omitted. The seldom-used chapter on multivariate statistics (Chapter 16 in the previous edition) has been relocated to the student CD-ROM that is packaged at the back of the text–making for a leaner text, which many have requested, yet without sacrificing material that may still appeal to some longtime users of the book.

The fifth edition, like its predecessors, retains its global emphasis, maintaining its position of being at the vanguard of international issues in business. The economies of countries around the world are becoming increasingly intertwined. Events in Asia have direct impact on Wall Street, and the Russian economy's move toward capitalism has immediate effects on Europe as well as on the United States. The publishing industry, in which large international conglomerates have acquired entire companies; the financial industry, in which stocks are now traded around the clock at markets all over the world; and the retail industry, which now offers consumer products that have been manufactured at a multitude of different locations throughout the world–all testify to the ubiquitous globalization of the world economy. A large proportion of the problems and examples in this new edition are concerned with international issues. We hope that instructors welcome this approach as it increasingly reflects the context of almost all business issues.

Many people have contributed greatly to the development of this fifth edition and we are grateful to all of them. They are:

Steen Andersen, The Aarhus School of Business
Theodore Bos, University of Alabama at Birmingham
Margaret Capen, East Carolina University
Bradford R. Crain, Portland State University
Nicholas R. Farnum, California State University–Fullerton
Frank Kelly, University of New Mexico
Duk Lee, Indiana Wesleyan University
Gary Lynch, Indiana University–Northwest
Jerrold H. May, University of Pittsburgh
Constance McLaren, Indiana State University
Tom Page, Michigan State University
Rakesh Sarin, University of California–Los Angeles
Axel Schultz-Nielsen, University of Southern Denmark
Walter Simmons, John Carroll University
Robert K. Smidt, California Polytechnic University
Cindy van Es, Cornell University
Charles W. Williams, Troy State University

We would like to thank the authors of the supplements that have been developed to accompany the text. Lou Patille, Keller Graduate School of Management, updated the Instructor's Manual and the Student Problem Solving Guide. Jeff Jung, University of Phoenix, updated the Test Bank, and Lloyd Jaisingh, Morehead State University, created data files and updated the PowerPoint Presentation Software. Don Robinson, Illinois State University, provided an accuracy check of the page proofs. Also, a special thanks to David Doane, Ronald Tracy, and Kieran Mathieson, all of Oakland University, who have permitted us to include their statistical graphics package, *Visual Statistics,* on the CD-ROM that accompanies this text.

We are indebted to the dedicated personnel at McGraw-Hill/Irwin. Scott Isenberg, sponsoring editor, helped pave the way for the inclusion of templates in the fifth edition. We owe him a lot for that. Wanda Zeman, development editor, was a constant source of support, enthusiasm, and cheer. She made sure the project was carried out on schedule and budget. Kimberly Hooker managed the production and Carole Schwager did the copyediting. We thank all the technical staff for their help and hard work.

Amir D. Aczel   *Jayavel Sounderpandian*
Bentley College   *University of Wisconsin–Parkside*

CONTENTS IN BRIEF

Chapter 3   Random Variables   **114**

Chapter 7   Hypothesis Testing   282

Chapter 8   The Comparison of Two Populations   328

These templates are included on the student CD-ROM in the back of the text.

Workbook	Sheet	Short Description	Page
		Chapter 1	
Basic Statistics.xls	Sheet 1	Calculates basic statistics such as mean, variance, percentiles, for sample, and population data.	55
Histogram.xls	Raw Data	Creates a histogram from raw data.	56
	Grouped Data	Creates a histogram, a frequency polygon, and an ogive from grouped data.	57
Frequency Polygon 2.xls	Sheet 1	Superposes two frequency polygons from two sets of grouped data.	58
Pie Chart.xls	Pie Chart	Creates a pie chart.	58
Bar Chart.xls	Sheet 1	Creates a vertical bar chart.	59
Box Plot.xls	Sheet 1	Creates a box plot from raw data.	59
Box Plot 2.xls	Sheet 1	Superposes two box plots from two sets of raw data.	60
Time Plot.xls	Time Plot	Creates a time plot from raw data.	60
Time Plot 2.xls	Time Plot 2	Superposes two time plots from two sets of raw data.	61
		Chapter 2	
Permutation and Combination.xls	Sheet 1	Computes $_nC_r$ and $_nP_r$ values.	96
Probability of at least 1.xls	Sheet 1	Computes the probability of at least one success in n different attempts.	104
Contingency Table.xls	Sheet 1	Computes joint, conditional, and marginal probabilities from a contingency table.	104
		Chapter 3	
Sum of Random Variables.xls	Sheet 1	Computes the mean, standard deviation, and variance of the sum of up to five independent random variables.	135
Random Variable.xls	Sheet	Computes the basic statistics of X and $h(X)$.	135
Binomial.xls	Binomial	Computes binomial exact, cumulative, and decumulative probabilities.	141
Negative Binomial.xls	Negative Binomial	Computes negative binomial exact, cumulative, and decumulative probabilities.	146
Geometric.xls	Geometric	Computes geometric exact, cumulative, and decumulative probabilities.	147
Hypergeometric.xls	Hypergeometric	Computes hypergeometric exact, cumulative, and decumulative probabilities.	150
Poisson.xls	Poisson	Computes Poisson exact, cumulative, and decumulative probabilities.	152
Uniform.xls	Uniform	Computes uniform probabilities.	158

Complete
Business
Statistics

0 Working with Templates

0–1 The Idea of Templates

The idea of templates can be traced back to the Greek mathematician Heron of Alexandria, who lived 2,000 years ago. Heron wrote several books, including a few meant for mechanical engineers. In these books, he presented solutions to practical engineering problems in such an illustrative manner that readers could solve similar problems simply by substituting their data in clearly designated places in his calculations.[1] In effect, his solutions were **templates** that others could use to solve similar problems with ease and confidence.

Heron's templates helped engineers by removing the tedium required to find the right formulas and the right sequence of calculations to solve a given problem. But the tedium of hand calculations endured. Over the years, abacuses, slide rules, electromechanical calculators, and electronic calculators lessened it. But even electronic calculators have a lot of buttons to be pressed, each presenting the chance for error. With the advent of computers and spreadsheets, even that tedium has been overcome.

A **spreadsheet template** is a specially designed workbook that carries out a particular computation on any data, requiring little or no effort beyond entering the data in designated places. Spreadsheet templates completely remove the tedium of computation and thus enable the user to concentrate on other aspects of the problem, such as sensitivity analysis or decision analysis. **Sensitivity analysis** refers to the examination of how the solution changes when the data change. **Decision analysis** refers to the evaluation of the available alternatives of a decision problem in order to find the best alternative. Sensitivity analyses are useful when we are not sure about the exact value of the data, and decision analyses are useful when we have many decision alternatives. In most practical problems, there will be uncertainty about the data; and in all decision problems, there will be two or more decision alternatives. The templates can therefore be very useful to students who wish to become practical problem solvers and decision makers.

Another kind of tedium is the task of drawing charts and graphs. Here too spreadsheet templates can completely remove the tedium by automatically drawing necessary charts and graphs.

The templates provided with this book are designed to solve statistical problems using the techniques discussed in the book and can be used to conduct sensitivity analyses and decision analyses. To conduct these analyses, one can use powerful features of Excel—the Data|Table command, the Goal Seek command, and the Solver macro—are explained in this chapter. Many of the templates contain charts and graphs that are automatically created.

The Dangers of Templates and How to Avoid Them

As with any other powerful tool, there are some dangers with templates. The worst danger is the **black box** issue: the use of a template by someone who does not know the concepts behind what the template does. This can result in continued ignorance about those concepts as well as the application of the template to a problem to which it should not be applied. Clearly, students should learn the concepts behind a template before using it, so this textbook explains all the concepts before presenting the templates. Additionally, to avoid the misuse of templates, wherever possible, the necessary conditions for using a template are displayed on the template itself.

---

[1]Morris Klein, *Mathematics in Western Culture* (New York: Oxford University Press, 1953), pp. 62–63.

FIGURE 0–1   **A Sample Template**
[Testing Population Mean.xls; Sheet: Power]

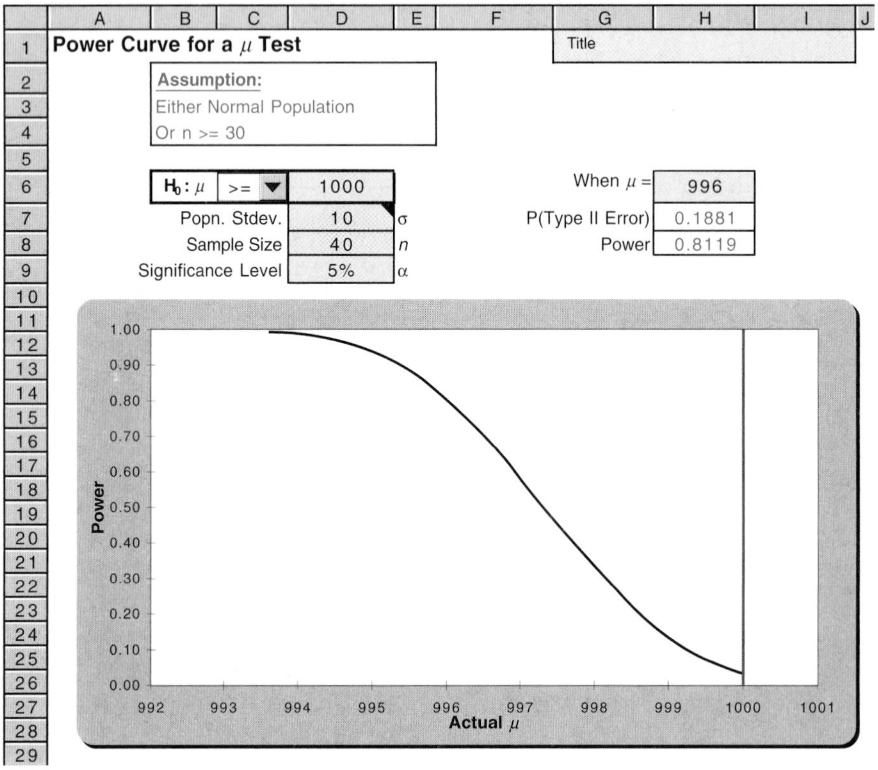

Another danger is that a template may contain errors that the user is unaware of. In the case of hand calculations, there is a good chance that the same error will not be made twice. But an error in a template is going to recur every time it is used and with every user. Thus template errors are quite serious. The templates provided with this book have been tested for errors over a period of several years. Many errors were indeed found, often by students, and have been corrected. But there is no guarantee that the templates are error-free. If you find an error, please communicate it to the authors or the publisher. That would be a very good service to a lot of people.

A step that has been taken to minimize the dangers is the avoidance of macros. *No macros* have been used in any of the templates. The user can view the formula in any cell by clicking on that cell and looking at the formula bar. By viewing all the formulas, the user can get a good understanding of the calculations performed by the template. An added advantage is that one can detect and correct mistakes or make modifications more easily in formulas rather than in macros.

Conventions Employed in the Templates

Figure 0–1 is a sample template that computes the power of a hypothesis test (from Chapter 7).[2] The first thing to note is the name of the workbook and the name of the

---

[2]The concepts of power of a hypothesis test have not been discussed yet. Our aim here is only to describe the features of a template.

sheet where you can locate this template. These details are given in square brackets immediately following the caption. This particular template is in the workbook named "Testing Population Mean.xls" and within that workbook this template is on the sheet named "Power." If you wish, you may open the template right now and look at it.

Several conventions have been employed to facilitate proper use of the templates. *The areas designated for data entry are shaded in blue.* In Figure 0–1, the cell H6[3] and the range D6:D9[4] appear shaded in blue, and therefore are meant for data entry. The range G1:I1, also shaded in blue, can be used for entering a title for the problem solved on the template.

Important results appear in red fonts; in the present case the values in the cells H7 and H8 are results and they appear in red (on the computer screen). Intermediate results appear in black. In the present case, there is no such result.

Instructions and necessary assumptions for the use of a template appear in magenta. On this template, the assumptions appear in the range B2:E4.[5] The user should make sure that the assumptions are satisfied before using the template. To avoid crowding of the template with all kinds of instructions, some instructions are placed in **comments** behind cells. A red marker at the upper right corner of a cell indicates the presence of a comment behind that cell. Placing the pointer on that cell will pop up the comment. Cell D7 in the figure has a comment. Such comments are usually instructions that pertain to the content of that cell.

A template may have a **drop down box,** in which the user will need to make a selection. There is a drop down box in the figure in the location of cell C6. Drop down boxes are used when the choice has to be one of a few possibilities. In the present case, the choices are the symbols =, <=, and >=. In the figure >= has been chosen. The user makes the choice based on the problem to be solved.

A template may have a **chart** embedded in it. The charts are very useful to visualize how one variable changes with respect to another. In the present example, the chart depicts how the power of the test changes with the actual population mean μ. An advantage with templates is that such charts are automatically created, and they are automatically updated when data change.

0–2   Working with Templates

Protecting and Unprotecting a Sheet

The computations in a template are carried out by **formulas** already entered into many of its cells. To protect these formulas from accidental erasure, all the cells except the (blue-shaded) data cells are "**locked.**" The user can change the contents of only the unlocked data cells. If for some reason, such as having to correct an error, you want to change the contents of a locked cell, you must first **unprotect** the sheet. To unprotect a sheet, use the Unprotect Sheet command under the Tools|Protection menu. Once you have made the necessary changes, make it a habit to reprotect the sheet using the Protect Sheet command under the Tools|Protection menu. While protecting a sheet in this manner, you will be asked for a **password.** *It is better not to use any password.* Just leave the password area blank. If you use a password, you will need it to unprotect the sheet. If you forget the password, you cannot unprotect the sheet.

---

[3]H6 refers to the cell at the intersection of Row 6 and Column H.

[4]Range D6:D9 refers to the rectangular area from cell D6 to cell D9.

[5]Range B2:E4 refers to the rectangular area with cell B2 at top left and cell E4 at bottom right.

FIGURE 0–2   **Dialog Box of Paste Special Command**

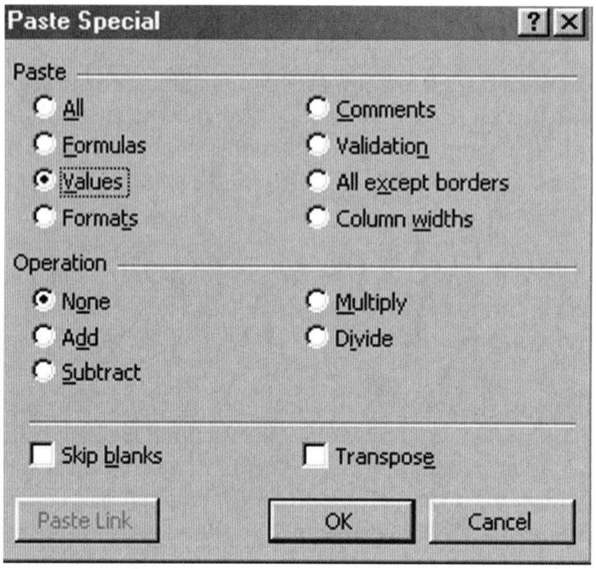

Entering Data into the Templates

It is a good habit to *erase all old data* before entering new data into a template. To erase the old data, select the range containing the data and press the Delete key on the keyboard. *Do not type a space to remove a data.* The computer will treat the space character as data and will try to make sense out of it rather than ignore it. This can give rise to error messages or, worse, erroneous results. Always use the Delete key to delete any data. Also make sure that you erase only the old data and nothing else.

At times your new data may be already there on another spreadsheet. If so, copy that data using the Edit|Copy command. Select the area where you want it pasted and use the Paste Special command. In the dialog box that appears (see Figure 0–2), select Values under Paste and None under Operation and then click the OK button. This will avoid any unwanted formulas or formats in the copied data getting pasted into the template. Sometimes the copied data may be in a *row,* but you may want to paste it into a *column,* or vice versa. In this case, the data need to be **transposed.** When you are in the Paste Special dialog box, additionally, select the Transpose box under Operation.

0–3   The AutoCalculate Command

The bar at the bottom of a spreadsheet screen image is known as the **status bar.** In the status bar there is an area, known as the **AutoCalculate** area, that can be used to quickly calculate certain statistics such as the sum or the average of several numbers. Figure 0–3 shows a spreadsheet in which a range of cells has been selected by dragging the mouse over them. In the AutoCalculate area, the sum of the numbers in the selected range appears. If the numbers you want to add are not in a single range, use the CTRL+click method to select more than one range of cells.

The AutoCalculate area can be used to calculate other statistics, such as the average. You can right-click the mouse on the area to see other statistics that are available,

FIGURE 0–3   **The AutoCalculate Area in the Status Bar**

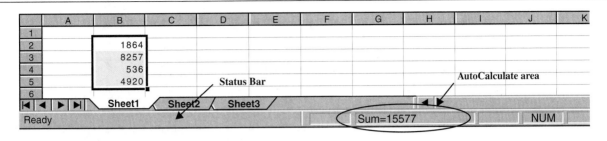

FIGURE 0–4   **Creating a Table**

	A	B	C	D	E	F
1						
2	Annual growth rate		2%			
3						
4	Year	2001	2002	2003	2004	2005
5	Sales	316	322	329	335	342
6						
7				Growth Rate	2004 Sales	2005 Sales
8				2%	335	342
9				3%	345	356
10				4%	355	370
11				5%	366	384
12				6%	376	399
13				7%	387	414

and select the one you want. When Excel is closed and opened again, the statistic will revert back to Sum.

0–4   The Data|Table Command

When a situation calls for comparing many alternatives at once, a tabulation of the results makes the comparison easy. Many of the templates have built-in comparison tables. In others, an exercise may ask you to create one yourself. You can create such tables using the Data|Table command.

In Figure 0–4, sales figures of a company have been calculated for years 2001 to 2005 using an annual growth rate of 2%, starting with 316 in year 2001. [In cell C5, the formula =B5*(1+C2) has been entered and copied to the right.] Suppose we are not sure about the growth rate and believe it may be anywhere between 2% and 7%. Suppose further we are interested in knowing what the sales figures would be in the years 2004 and 2005 at different growth rates. In other words, we want to input many different values for the growth rate in cell C2 and see its effect on cells E5 and F5. It is best seen as a table shown in the range D8:F13. To create this table,

- Enter the growth rates 2%, 3%, etc., in the range D8:D13.
- Enter the formula =E5 in cell E8 and =F5 in cell F8.
- Select the range D8:F13.
- Under the Data menu, select the Table command.
- In the dialog box that appears, in the Column Input Cell box, type C2 and press Enter. (We use the Column Input Cell rather than Row Input Cell because the input values are in a column, in the range D8:D13.)
- The desired table now appears in the range D8:F13. It is worth noting here that this table is "live," meaning that if any of the input values in the range D8:D13 is changed, the table is immediately updated. Also, the input values could have been calculated using formulas or could have been a part of another table to the left.

In general, the effect of changing the value in *one* cell on the values of one or more other cells can be tabulated using the Data|Table command. At times, we may want to tabulate the effect of changing *two* cells. In this case we can tabulate the effect on only one other cell. In the previous example, suppose that in addition to the growth rate we are not sure about the starting sales figure of 316 in year 2001, and we believe it could be anywhere between 316 and 324. Suppose further that we are interested only in the sales figure for year 2005. A table varying both the growth rate and 2001 sales has been calculated in Figure 0–5.

To create the table in Figure 0–5,

- Enter the input values for growth rate in the range B9:B14.
- Enter the input values for 2001 sales in the range C8:G8.
- Enter the formula =F5 in cell B8. (This is because we are tabulating what happens to cell F5.)

FIGURE 0–5   **Creating a Two-Dimensional Table**

	A	B	C	D	E	F	G
1							
2	Annual growth rate		2%				
3							
4	Year	2001	2002	2003	2004	2005	
5	Sales	316	322	329	335	342	
6							
7			2001 Sales				
8		342	316	318	320	322	324
9		2%	342	344	346	349	351
10		3%	356	358	360	362	365
11	Growth	4%	370	372	374	377	379
12	Rate	5%	384	387	389	391	394
13		6%	399	401	404	407	409
14		7%	414	417	419	422	425

- Select the range B8:G14.
- Under the Data menu select the Table command.
- In the dialog box that appears enter B5 in the Row Input Cell box and C2 in the Column Input Cell box and press Enter.

The table appears in the range B8:G14. This table is "live" in that when any of the input values is changed the table updates automatically. The appearance of 342 in cell B8 is distracting. It can be hidden by either changing the text color to white or formatting the cell with ;;. Suitable borders also improve the appearance of the table.

0–5   The Goal Seek Command

The Goal Seek command can be used to change a numerical value in any cell, called the **changing cell,** to make the numerical value in another cell, called the **target cell,** reach a "goal." Clearly, the value in the target cell must depend on the value in the changing cell for this scheme to work. In the previous example, suppose we are interested in finding the growth rate that would attain the goal of a sales value of 400 in year 2005. (We assume the sales in 2001 to be 316.) One way to do it is to manually change the growth rate in cell C2 up or down until we see 400 in cell F5. But that would be tedious, so we automate it using the Goal Seek command as follows:

- Under the Tools menu select the Goal Seek command. A dialog box appears.
- In the Set Cell box enter F5.
- In the To Value box enter 400.
- In the By Changing Cell box enter C2.
- Click OK.

The computer makes numerous trials and stops when the value in cell F5 equals 400 accurate to several decimal places. The value in cell C2 is the desired growth rate, and it shows up as 6.07%.

0–6   The Solver Macro

The Solver tool is a giant leap forward from the Goal Seek command. It can be used to make the value of a target cell equal a predetermined value or, more commonly, reach its maximum or minimum possible value, by changing the values in *many* other changing cells. In addition, some constraints can be imposed on the values of selected cells, called **constrained cells,** such as restricting a constrained cell's value to be, say, between 10 and 20. Note that the Solver can accommodate many changing cells and many constrained cells, and is thus a very powerful tool.

Solver Installation

Since the Solver macro is very large, it will not be installed during the installation of Excel or Office software unless you specifically ask for it to be installed. Before you can use the Solver, you therefore have to determine if it has been installed in your computer and "added in." If it hasn't been, then you need to install it and add it in.

To check if it has already been installed and added in, click on the Tools menu (and make sure the menu is opened out fully). If the command "Solver…" appears in the menu, you have nothing more to do. If the command does not appear in the

FIGURE 0–6   **Solver Application**

	A	B	C	D
1				
2		Batch size	10	
3		Cost	50	=4*C2+100/C2

menu, then the Solver has not been installed or perhaps has been installed but not added in. Select the Add-Ins command under the Tools menu. Look for Solver Add-In in the list that appears. If you find it, select it and click OK. It will be added in and after that you will see the Solver command under the Tools menu. If you don't see Solver Add-In in the list, see if the file named Solver.xla is available in your hard disk by checking the pathname

c:\Program files\Microsoft Office\office\library\Solver\Solver.xla

If the file is present, open it. This action will add in the Solver and after that the "Solver…" command will appear under the Tools menu. If the Solver.xla file is not available, it means the Solver has not been installed in the computer. You have to get the original Excel or Office CD, go through the setup process, and install the Solver files. You will then find the above Solver.xla file and you should open it. After that, the "Solver…" command will appear under the Tools menu. If you are using Excel in your workplace and the Solver is not installed, then you may have to seek the help of your Information Systems department to install it.

In all the templates that use the Solver, the necessary settings for the Solver have already been entered. The user merely needs to press Solve in the Solver dialog box (see Figure 0–7), and when the problem is solved, the user needs to click Keep Solver Solution in a message box that appears.

Just to make you a little more comfortable with the use of the Solver, let's consider an example. A production manager finds that the cost of manufacturing 100 units of a product in batches of x units at a time is $4x + 100/x$. She wants to find the most economic batch size. (This is a typical economic batch size problem, famous in production planning.) In solving this problem, we note that there is a constraint, namely, x must be between 0 and 100. Mathematically, we can express the problem as

$$\text{Minimize} \quad 4x + 100/x$$
$$\text{Subject to} \quad x >= 0$$
$$x <= 100$$

We set up the problem as shown in Figure 0–6:

- In cell C3 the formula =4*C2+100/C2 has been entered. The batch size in cell C2 can be manually changed and corresponding cost can be read off from cell C3. A batch size of 20, for instance, yields a cost of 85; a batch size of 2 yields a cost of 58. To find the batch quantity that has the least cost, we use the Solver as follows.
- Under the Tools menu select the Solver… command.
- In the Solver dialog box, enter C3 in the Set Cell box.
- Click Minimum (because we want to minimize the cost).
- Enter C2 in the Changing Cells box.

FIGURE 0–7　The Solver Dialog Box

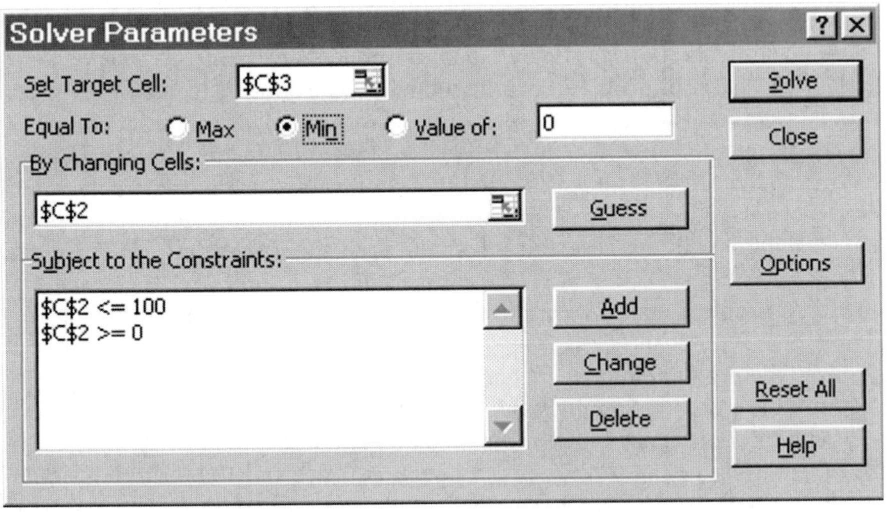

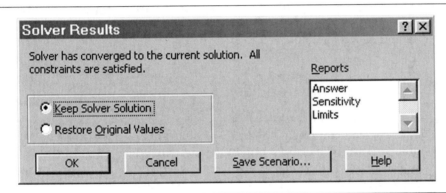

FIGURE 0–8　Solver Solution Dialog Box

- Click Add to add a constraint.
- In the dialog box that appears, click the left-hand-side box and enter C2.
- In the middle drop down box, select >=.
- Click the right-hand-side box, enter 0, and click Add. (We click on the Add button because we have one more constraint to add.)
- In the new dialog box that appears click the left-hand-side box and enter C2.
- In the middle drop down box, select <=.
- Click the right-hand-side box, enter 100, and click OK. (The Solver dialog box should reappear as in Figure 0–7.)
- Click Solve.

The Solver carries out a sophisticated computation process internally and finds the solution, if one exists. When the solution is found, the Solver Results dialog box appears (see Figure 0–8). It asks if you want to keep the solution or retain the original values for batch quantity and cost. Select Keep Solver Solution and click OK. The solution is a batch size of 5, which has the least cost of 40. The full solution appears on the spreadsheet.

Some comments about the Solver are appropriate here. First, it is a very powerful tool. A great variety of problems can be modeled on a spreadsheet and solved using this tool.

Second, not all problems may be solvable, particularly when the constraints are so restrictive that there is no feasible solution. In this case, some constraints must be removed or relaxed. Another possibility is that the solution may diverge to positive or negative infinity. In this case, the Solver will flash a message about divergence and abort the calculation.

Third, a problem may have more than one solution. In this case, the Solver will find only one of them and stop.

Fourth, the problem may be too large for the Solver. Although the manuals claim that a problem with 200 variables and 200 constraints can be solved, it may be safer to restrict the problem size to not more than 50 variables and 50 constraints.

Last, entering the constraints has some syntax rules and some shortcuts. A constraint line may read A1:A20 <= B1:B20. This is a shortcut for A1 <= B1; B1 <= B2; and so on. In effect, 20 constraints have been entered in one line. Also, A1:A20 <= 100 would imply A1 <= 100; A2 <= 100; and so on. One of the syntax rules is that the left-hand side of a constraint, as entered into the Solver, cannot be a number but must be a reference to a cell or a range of cells. For instance, C2 <= 100 cannot be entered as 100 >= C2, although they mean the same thing.

For further details about the Solver tool, you may consult online help or the Excel manual.

0–7 Some Formatting Tips

If you find a cell filled with ########, you know that the cell is not wide enough to display its contents. To see the contents, you should unprotect the sheet and widen the column. (It is a good habit to reprotect the sheet after that.)

Excel displays very large numbers and very small numbers in **scientific format.** For example, the number 1,234,500,000 will be displayed as 1.2345E+09. The "E+09" at the end means that the decimal point needs to be moved 9 places to the right. In the case of very small numbers Excel once again uses the scientific format. For example, the number 0.0000012345 will be displayed as 1.2345E-06 where the "E-06" signifies that the decimal point is to be moved 6 places to the left.

If you do not wish to see the numbers in scientific format, widening the column might help. In the case of very small numbers you may format the cell, using the Format|Cells command to display the number in decimal form with any desired number of decimal places. For all probability values that appear in the templates, four decimal places are recommended.

Many templates contain graphs. The axes of the graphs are set to rescale automatically when the plotted values change. Yet, at times, the scale on an axis may have to be adjusted. To adjust the scale, you have to first unprotect the sheet. Then double-click on the axis and format the scale as needed. (Once again, reprotect the sheet when you are done.)

0–8   Saving the Templates

All the templates discussed in this textbook are available in the CD attached. They are stored in the folder named Templates. It is recommended that you save these templates in your computer's hard drive in a suitably named folder, say, c:\Stat Templates.

0–1.   While working with a template, you find that a cell displays ######## instead of a numerical result. What should you do to see the result?

0–2.   While working with a template a probability value in a cell appears as $4.839E - 04$. What should you do to see it as a number with four decimal places?

0–3.   The scale on the axis of a chart in a template starts at 0 and ends at 100. What should you do to make it start at 50 and end at 90?

0–4.   You want to copy a *row* of data from a spreadsheet into a *column* in the data area of a template. How should this be done?

0–5.   Why are the nondata cells in a template locked? How can they be unlocked? Mention a possible reason for unlocking them.

0–6.   How can you detect if a cell has a comment attached to it? How can you view the comment?

0–7.   Suppose an error has been found in the formula contained in a particular cell of a template, and a corrected formula has been announced. Give a step-by-step description of how to incorporate the correction on the template.

0–8.   How many target cells, changing cells, and constrained cells can be there when the Solver is used?

0–9.   How many target cells, changing cells, and constrained cells can be there when the **Goal Seek** command is used?

0–10.   What is the "black box" issue in the use of templates? How will you, as a student, avoid it?

0–11.   Find the average of the numbers 78, 109, 44, 38, 50, 11, 136, 203, 117, and 34 using the AutoCalculate feature of Excel.

1 Introduction and Descriptive Statistics

It is better to be roughly right
than precisely wrong.
            —*John Maynard Keynes*

You all have probably heard the story about Malcolm Forbes, who once got lost floating for miles in one of his famous balloons and finally landed in the middle of a cornfield. He spotted a man coming toward him and asked, "Sir, can you tell me where I am?" The man said. "Certainly, you are in a basket in a field of corn." Forbes said, "You must be a statistician." The man said, "That's amazing, how did you know that?" "Easy," said Forbes, "your information is concise, precise, and absolutely useless!"[1]

The purpose of this book is to convince you that information resulting from a good statistical analysis is always concise, often precise, and never useless! The spirit of statistics is, in fact, very well captured by the quotation above from Keynes. This book should teach you how to be at least roughly right a high percentage of the time.

The word *statistics* is derived from the Italian word *stato,* which means "state" and *statista* refers to a person involved with the affairs of state. Therefore, *statistics* originally meant the collection of facts useful to the *statista.* Statistics in this sense was used in 16th-century Italy and then spread to France, Holland, and Germany. We note, however, that surveys of people and property actually began in ancient times.[2] Today, statistics is not restricted to information about the state but extends to almost every realm of human endeavor. Neither do we restrict ourselves to merely collecting numerical information, called *data.* Our data are summarized, displayed in meaningful ways, and analyzed. Statistical analysis often involves an attempt to generalize from the data. Statistics is a science—the science of information. Information may be *qualitative* or *quantitative.* To illustrate the difference between these two types of information, let's consider an example.

Realtors who help sell condominiums in the Boston area provide prospective buyers with the information given in Table 1–1. Which of the variables in the table are quantitative and which are qualitative?

EXAMPLE 1–1

The asking price is a *quantitative* variable: it conveys a quantity—the asking price in dollars. The number of rooms is also a quantitative variable. The direction the apartment faces is a *qualitative* variable since it conveys a quality (east, west, north, south). Whether a condominium has a washer and dryer in the unit (yes or no) and whether heat is provided as part of the condo fee are also qualitative variables.

Solution

A **quantitative variable** can be described by a number for which arithmetic operations such as averaging make sense. A **qualitative (or categorical) variable**

---

[1]From an address by R. Gnanadesikan to the American Statistical Association, reprinted in *American Statistician* 44, no. 2 (May 1990), p. 122.

[2]See Anders Hald, *A History of Probability and Statistics and Their Applications before 1750* (New York: Wiley, 1990), pp. 81–82.

TABLE 1–1   Boston Condominium Data

Asking Price	Number of Rooms	Direction Facing	Washer/Dryer?	Heated?
$168,000	2	E	Y	Y
152,000	2	N	N	Y
187,000	3	N	Y	Y
142,500	1	W	N	N
166,800	2	W	Y	N

simply records a quality. If a number is used for distinguishing members of different categories of a qualitative variable, the number assignment is arbitrary.

The field of statistics deals with **measurements**—some quantitative and others qualitative. The measurements are the actual numerical values of a variable. (Qualitative variables could be described by numbers, although such a description might be arbitrary; for example, $N = 1$, $E = 2$, $S = 3$, $W = 4$, $Y = 1$, $N = 0$.)

There are four generally used **scales of measurement,** listed here from weakest to strongest.

Nominal Scale.   In the **nominal scale** of measurement, numbers are used simply as labels for groups or classes. If our data set consists of blue, green, and red items, we may designate blue as 1, green as 2, and red as 3. In this case, the numbers 1, 2, and 3 stand only for the category to which a data point belongs. "Nominal" stands for "name" of category. The nominal scale of measurement is used for qualitative rather than quantitative data: blue, green, red; male, female; professional classification; geographic classification; and so on.

Ordinal Scale.   In the **ordinal scale** of measurement, data elements may be ordered according to their relative size or quality. Four products ranked by a consumer may be ranked as 1, 2, 3, and 4, where 4 is the best and 1 is the worst. In this scale of measurement we do not know how much better one product is than others, only that it is better.

Interval Scale.   In the **interval scale** of measurement the value of zero is assigned arbitrarily and therefore we cannot take ratios of two measurements. But *we can take ratios of intervals.* A good example is how we measure time of day, which is in interval scale. We cannot say 10:00 A.M. is twice as long as 5:00 A.M. But we can say that the interval between 0:00 A.M. (midnight) and 10:00 A.M., which is a duration of 10 hours, is twice as long as the interval between 0:00 A.M. and 5:00 A.M., which is a duration of 5 hours. This is because 0:00 A.M. does not mean absence of any time. Another example is temperature. When we say 0°F, we do not mean zero heat. A temperature of 100°F is not twice as hot as 50°F. But it is true that to heat an object from 0°F to 100°F we will need twice as much heat as we would need to heat the same object from 0°F to 50°F.

Ratio Scale.   If two measurements are in **ratio scale,** then we can take ratios of those measurements. The zero in this scale is an absolute zero. Money, for example, is measured in ratio scale. A sum of $100 is twice as large as $50. A sum of $0 means absence of any money and is thus an absolute zero. We have already seen that measurement of duration (but not time of day) is in ratio scale. In general, the interval between two interval scale measurements will be in ratio scale. Other examples of ratio scale are measurements of weight, volume, area, or length.

Samples and Populations

In statistics we make a distinction between two concepts: a population and a sample.

> The **population** consists of the set of all measurements in which the investigator is interested. The population is also called the **universe**.
>
> A **sample** is a subset of measurements selected from the population. Sampling from the population is often done randomly, such that every possible sample of n elements will have an equal chance of being selected. A sample selected in this way is called a **simple random sample**, or just a **random sample**. A random sample allows chance to determine its elements.

For example, Farmer Jane owns 1,264 sheep. These sheep constitute her entire *population* of sheep. If 15 sheep are selected to be sheared, then these 15 represent a *sample* from Jane's population of sheep. Further, if the 15 sheep were selected at *random* from Jane's population of 1,264 sheep, then they would constitute a *random sample* of sheep.

The definitions of *sample* and *population* are relative to what we want to consider. If Jane's sheep are all we care about, then they constitute a population. If, however, we are interested in all the sheep in the county, then all Jane's 1,264 sheep are a sample of that larger population (although this sample would not be random).

The distinction between a sample and a population is very important in statistics.

Data and Data Collection

A set of measurements obtained on some variable is called a **data set.** For example, heart rate measurements for 10 patients may constitute a data set. The variable we're interested in is heart rate, and the scale of measurement here is a ratio scale. (A heart that beats 80 times per minute is twice as fast as a heart that beats 40 times per minute.) Our actual observations of the patients' heart rates, the data set, might be 60, 70, 64, 55, 70, 80, 70, 74, 51, 80.

Data are collected by various methods. Sometimes our data set consists of the entire population we're interested in. If we have the actual point spread for five football games, and if we are interested only in these five games, then our data set of five measurements is the entire population of interest. (In this case, our data are on a ratio scale. Why? Suppose the data set for the five games told only whether the home or visiting team won. What would be our measurement scale in this case?)

In other situations data may constitute a sample from some population. If the data are to be used to draw some conclusions about the larger population they were drawn from, then we must collect the data with great care. A conclusion drawn about a population based on the information in a sample from the population is called a **statistical inference.** Statistical inference is an important topic of this book. To ensure the accuracy of statistical inference, data must be drawn randomly from the population of interest, and we must make sure that every segment of the population is adequately and proportionally represented in the sample.

Statistical inference may be based on data collected in surveys or experiments, which must be carefully constructed. For example, when we want to obtain information from people, we may use a mailed questionnaire, or a telephone interview, as a convenient instrument. In such surveys, however, we want to minimize any **nonresponse bias.** This is the biasing of the results that occurs when we disregard the fact that some people will simply not respond to the survey. The bias distorts the findings, because the people who do not respond may belong more to one segment of the population than to another. In social research some questions may be sensitive—for example, "Have you ever been arrested?" This may easily result in a nonresponse bias, because people who have indeed been arrested may be less likely to answer the question (unless they can be perfectly certain of remaining anonymous). Surveys conducted by popular magazines often suffer from nonresponse bias, especially when their questions are provocative. What makes good magazine reading often makes bad statistics.

Suppose we want to measure the speed performance or gas mileage of an automobile. Here the data will come from experimentation. In this case we want to make sure that a variety of road conditions, weather conditions, and other factors are represented. Pharmaceutical testing is also an example where data may come from experimentation. Drugs are usually tested against a placebo as well as against no treatment at all. When an experiment is designed to test the effectiveness of a sleeping pill, the variable of interest may be the time, in minutes, that elapses between taking the pill and falling asleep.

In experiments, as in surveys, it is important to **randomize** if inferences are indeed to be drawn. People should be randomly chosen as subjects for the experiment if an inference is to be drawn to the entire population. Randomization should also be used in assigning people to the three groups: pill, no pill, or placebo. Such a design will minimize potential biasing of the results.

In other situations data may come from published sources, such as statistical abstracts of various kinds or government publications. The published unemployment rate over a number of months is one example. Here, data are "given" to us without our having any control over how they are obtained. Again, caution must be exercised. The unemployment rate over a given period is not a random sample of any *future* unemployment rates, and making statistical inferences in such cases may be complex and difficult. If, however, we are interested only in the period we have data for, then our data do constitute an entire population, which may be described. In any case, however, we must also be careful to note any missing data or incomplete observations.

In this chapter, we will concentrate on the processing, summarization, and display of data—the first step in statistical analysis. In the next chapter, we will explore the theory of probability, the connection between the random sample and the population. Later chapters build on the concepts of probability and develop a system that allows us to draw a logical, consistent inference from our sample to the underlying population.

Why worry about inference and about a population? Why not just look at our data and interpret them? Mere inspection of the data will suffice when interest centers on the particular observations you have. If, however, you want to draw meaningful conclusions with implications extending beyond your limited data, statistical inference is the way to do it.

In marketing research, we are often interested in the relationship between advertising and sales. A data set of randomly chosen sales and advertising figures for a given firm may be of some interest in itself, but the information in it is much more useful if it leads to implications about the underlying process—the relationship between the firm's level of advertising and the resulting level of sales. An understanding of the true relationship between advertising and sales—the relationship in the population of advertising and sales possibilities for the firm—would allow us to predict sales for any level of advertising and thus to set advertising at a level that maximizes profits.

A pharmaceutical manufacturer interested in marketing a new drug may be required by the Food and Drug Administration to prove that the drug does not cause serious side effects. The results of tests of the drug on a random sample of people may then be used in a statistical inference about the entire population of people who may use the drug if it is introduced.

A bank may be interested in assessing the popularity of automatic teller machines. The machines may be tried on a randomly chosen group of bank customers. The conclusions of the study could then be generalized by statistical inference to the entire population of the bank's customers.

A quality control engineer at a plant making disk drives for computers needs to make sure that no more than 3% of the drives produced are defective. The engineer may routinely collect random samples of drives and check their quality. Based on the random samples, the engineer may then draw a conclusion about the proportion of defective items in the entire population of drives.

These are just a few examples illustrating the use of statistical inference in business situations. In the rest of this chapter, we will introduce the descriptive statistics needed to carry out basic statistical analyses. The next chapters will develop the elements of the inference from samples to populations.

PROBLEMS

1–1. A survey by an electric company contains questions on the following:
1. Age of household head.
2. Sex of household head.
3. Number of people in household.
4. Use of electric heating (yes or no).
5. Number of large appliances used daily.
6. Thermostat setting in winter.
7. Average number of hours heating is on.
8. Average number of heating days.
9. Household income.
10. Average monthly electric bill.
11. Ranking of this electric company as compared with two previous electricity suppliers.

Describe the variables implicit in these 11 items as quantitative or qualitative, and describe the scales of measurement.

1–2. Discuss the various data collection methods described in this section.

1–3. Discuss and compare the various scales of measurement.

1–4. Describe each of the following variables as qualitative or quantitative.[3]

Name	Investing Style	% Gain to July 1, 1997			Minimum Initial Investment	% Sales Charge
		6 Months	3 Years[1]	5 Years[1]		
LARGE- AND MIDCAP STOCK						
Legg Mason Value	Lg/Bl	22.1	35.6	24.4	$1,000	None
Harbor Capital Appreciation	Lg/Gro	18.1	29.7	21.8	2,000	None
Vanguard/Primecap	Lg/Bl	21.1	29.1	23.6	3,000	None
Westwood Equity	Lg/Bl	18.5	28.8	22.4	1,000	None
Vanguard Index 500	Lg/Bl	20.5	28.7	19.6	3,000	None
Standard & Poor's 500-stock index	—	20.6	28.8	19.8	—	—

[3]From *Money*, August 1997, p. 84.

Name	Investing Style	% Gain to July 1, 1997			Minimum Initial Investment	% Sales Charge
		6 Months	3 Years[1]	5 Years[1]		
SMALL-CAP STOCK						
Franklin Small Cap Growth I	Sm/Bl	7.9	31.6	25.3	$ 100	4.50
Wasatch Growth	Sm/Val	15.5	29.1	19.5	2,000	None
Baron Asset	Md/Bl	15.2	28.3	24.2	2,000	None
Colonial Small Cap Value A	Sm/Val	11.5	26.0	22.0	1,000	5.75
Fidelity Low-Priced Stock	Sm/Val	12.2	23.2	21.1	2,500	3.00
Russell 2000 stock index	—	10.2	20.1	17.9	—	—
FOREIGN STOCK						
Hotchkis & Wiley International	Lg/Val	10.3	17.0	15.0	10,000	None
Vanguard International Growth	Lg/Bl	16.6	15.6	15.7	3,000	None
AIM International Equity A	Lg/Gro	11.3	15.6	15.7	500	5.50
Templeton Foreign Smaller Companies	Sm/Val	9.0	14.3	14.1	100	4.50
MS EAFE index	—	11.4	9.4	13.2	—	—
EMERGING MARKETS						
Templeton Developing Markets I	Md/Val	20.1	14.3	15.9	100	5.75
SSGA Emerging Markets	Md/Val	19.4	14.1	N.A.	1,000	None
Vanguard International Equity Emerg. Mkt.	Md/Bl	12.1	10.7	N.A.	3,000	None
MS emerging markets index	—	7.5	12.2	13.4	—	—
INTERMEDIATE-TERM BOND						
Strong Corporate Bond	L/Med	4.4	12.4	10.7	2,500	None
Vanguard Bond Index Total	Int/Hi	2.9	8.4	7.0	3,000	None
Vanguard Muni Intermediate-Term	Int/Hi	2.7	6.7	6.8	3,000	None
SHORT-TERM BOND						
Vanguard F/I Short-Term Corporate	Sh/Med	2.8	7.0	6.1	3,000	None
Fidelity Spartan Ltd. Mat. Gov.	Int/Hi	2.7	7.1	5.8	10,000	None
USAA Tax-Exempt Short-Term	Sh/Med	2.4	5.3	4.8	3,000	None
Vanguard Muni Limited-Term	Sh/Hi	2.1	5.0	4.9	3,000	None
Lehman Bros. aggregate bond index	—	3.1	8.5	7.1	—	—

[1]Annualized: **Stock fund styles: Gro**—Buys companies with accelerating earnings; **Val**—Buys stocks that are inexpensive relative to their earnings or assets; **Bl**—Buys stocks that blend growth and value characteristics; **Lg**—Buys stocks with total market values of more than $5 billion; **Md**—Buys stocks with total market values of $1 billion to $5 billion; **Sm**—Buys stocks with total market values of less than $1 billion. **Bond fund styles: Hi**—Buys bonds rated AA or better; **Med**—Buys bonds rated BBB or better; **L**—Buys bonds with maturities over 10 years; **Int**—Buys bonds with maturities between 4 and 10 years; **Sh**—Buys bonds with maturities of less than 4 years. N.A. Not applicable. **Source:** Morningstar.

1–5. Five ice cream flavors are rank-ordered by preference. What is the scale of measurement?

1–6. What is the difference between a qualitative and a quantitative variable?

1–7. A town has 15 neighborhoods. If you interviewed everyone living in one particular neighborhood, would you be interviewing a population or a sample from the town? Would this be a random sample? If you had a list of everyone living in the town, called a **frame,** and you randomly selected 100 people from all the neighborhoods, would this be a random sample?

1–8.   What is the difference between a sample and a population?

1–9.   What is a random sample?

1–10.   For each tourist entering the United States, the U.S. Immigration and Naturalization Service computer is fed the tourist's nationality and length of intended stay. Characterize each variable as quantitative or qualitative.

1–11.   What is the scale of measurement for the color of a karate belt?

1–12.   An individual federal tax return form asks, among other things, for the following information: income (in dollars and cents), number of dependents, whether filing singly or jointly with a spouse, whether or not deductions are itemized, amount paid in local taxes. Describe the scale of measurement of each variable, and state whether the variable is qualitative or quantitative.

1–2   Percentiles and Quartiles

Given a set of numerical observations, we may order them according to magnitude. Once we have done this, it is possible to define the boundaries of the set. Any student who has taken a nationally administered test, such as the Scholastic Aptitude Test (SAT), is familiar with *percentiles*. Your score on such a test is compared with the scores of all people who took the test at the same time, and your position within this group is defined in terms of a percentile. If you are in the 90th percentile, 90% of the people who took the test received a score lower than yours. We define a percentile as follows.

> The *P*th **percentile** of a group of numbers is that value below which lie *P*% (*P* percent) of the numbers in the group. The position of the *P*th percentile is given by $(n + 1)P/100$, where n is the number of data points.

Let's look at an example.

A large department store collects data on sales made by each of its salespeople. The data, number of sales made on a given day by each of 20 salespeople, are as follows: **EXAMPLE 1–2**

9, 6, 12, 10, 13, 15, 16, 14, 14, 16, 17, 16, 24, 21, 22, 18, 19, 18, 20, 17

Find the 50th, 80th, and 90th percentiles of this data set.

First, let's order the data from smallest to largest: *Solution*

6, 9, 10, 12, 13, 14, 14, 15, 16, 16, 16, 17, 17, 18, 18, 19, 20, 21, 22, 24

To find the 50th percentile, we need to determine the data point in position $(n + 1)P/100 = (20 + 1)(50/100) = (21)(0.5) = 10.5$. Thus, we need the data point in position 10.5. Counting the observations from smallest to largest, we find that the 10th observation is 16, and so is the 11th. Therefore, the observation that would lie in position 10.5 (halfway between the 10th and 11th observations) is 16. Thus, the 50th percentile is 16.

Similarly, we find the 80th percentile of the data set as the observation lying in position $(n + 1)P/100 = (21)(80/100) = 16.8$. The 16th observation is 19, and the 17th

FIGURE 1–1 Spreadsheet Template Results for Data in Example 1–2
[Basic Stastics.xls]

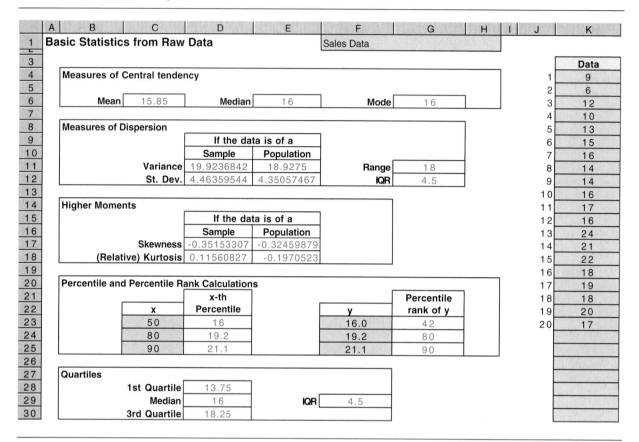

is 20; therefore, the 80th percentile is a point lying 0.8 of the way from 19 to 20, that is, 19.8.

The 90th percentile is found as the observation in position $(n + 1)P/100 = (20 + 1)(90/100) = (21)(0.9) = 18.9$, which is 21.9.

Figure 1–1 shows the basic statistics calculated on data from Example 1–2 using a template that is described in section 1–10 of this chapter. Figure 1–2 shows the bottom portion of the same template. This portion contains additional statistics and an area that can be used to apply two concepts discussed later in this chapter: Chebyshev's theorem and the empirical rule.

Certain percentiles have greater importance than others because they break down the **distribution** of the data (the way the data points are distributed along the number line) into four groups. These are the quartiles. **Quartiles** are the percentage points that break down the data set into quarters—first quarter, second quarter, third quarter, and fourth quarter.

The **first quartile** is the 25th percentile. It is that point below which lie one-fourth of the data.

FIGURE 1–2 **The Lower Part of the Template in Figure 1–1**
[Basic Stastics.xls]

Other Statistics							
Sum	317						
Size	20						
Maximum	24						
Minimum	6						

Chebyshev's Theorem observation
Data points within 1.5 Std. Devns from mean   17
out of   20
which is 85.00%
Minimum predicted by Chebyshev's Theorem   55.56%
Minimum predicted by Empirical Rule   86.64%

Similarly, the second quartile is the 50th percentile, as we computed in Example 1–2. This is a most important point and has a special name—the *median*.

The **median** is the point below which lie half the data. It is the 50th percentile.

We define the third quartile correspondingly:

The **third quartile** is the 75th percentile point. It is that point below which lie 75 percent of the data.

The 25th percentile is often called the **lower quartile;** the 50th percentile point, the median, is called the **middle quartile;** and the 75th percentile is called the **upper quartile.**

Find the lower, middle, and upper quartiles of the data set in Example 1–2.   **EXAMPLE 1–3**

Based on the procedure we used in computing the 80th and 90th percentiles, we find   *Solution*
that the lower quartile is the observation in position $(21)(0.25) = 5.25$, which is 13.25. The middle quartile was already computed (it is the 50th percentile, the median, which is 16). The upper quartile is the observation in position $(21)(75/100) = 15.75$, which is 18.75.

We define the **interquartile range** as the difference between the first and third quartiles.

The interquartile range is a measure of the spread of the data. In Example 1–2, the interquartile range is equal to Third quartile − First quartile = $18.75 - 13.25 = 5.5$.

PROBLEMS

1–13. The following data are numbers of passengers on flights of Delta Air Lines between San Francisco and Seattle over 33 days in April and early May.

128, 121, 134, 136, 136, 118, 123, 109, 120, 116, 125, 128, 121, 129, 130, 131, 127, 119, 114, 134, 110, 136, 134, 125, 128, 123, 128, 133, 132, 136, 134, 129, 132

Find the lower, middle, and upper quartiles of this data set. Also find the 10th, 15th, and 65th percentiles. What is the interquartile range?

1–14.   The following data are annualized returns on a group of 15 stocks.

   12.5, 13, 14.8, 11, 16.7, 9, 8.3, −1.2, 3.9, 15.5, 16.2, 18, 11.6, 10, 9.5

Find the median, the first and third quartiles, and the 55th and 85th percentiles for these data.

1–15.   The following data are scores on a management examination taken by a group of 22 people.

   88, 56, 64, 45, 52, 76, 54, 79, 38, 98, 69, 77, 71, 45, 60, 78, 90, 81, 87, 44, 80, 41

Find the median and the 20th, 30th, 60th, and 90th percentiles.

1–16.   Following are the numbers of daily bids received by the government of a developing country from firms interested in winning a contract for the construction of a new port facility.

   2, 3, 2, 4, 3, 5, 1, 1, 6, 4, 7, 2, 5, 1, 6

Find the quartiles and the interquartile range. Also find the 60th percentile.

1–17.   Find the median, the interquartile range, and the 45th percentile of the following data.

   23, 26, 29, 30, 32, 34, 37, 45, 57, 80, 102, 147, 210, 355, 782, 1,209

1–3   Measures of Central Tendency

Percentiles, and in particular quartiles, are measures of the relative positions of points within a data set or a population (when our data set constitutes the entire population). The median is a special point, since it lies in the center of the data in the sense that half the data lie below it and half above it. The median is thus a measure of the *location* or *centrality* of the observations.

In addition to the median, there are two other commonly used measures of central tendency. One is the *mode* (or modes—there may be several of them), and the other is the *arithmetic mean,* or just the *mean.* We define the mode as follows.

> The **mode** of the data set is the value that occurs most frequently.

Let us look at the frequencies of occurrence of the data values in Example 1–2, shown in Table 1–2. We see that the value 16 occurs most frequently. There are three data points with this value—more points than for any other value in the data set. Therefore, the mode is equal to 16.

The most commonly used measure of central tendency of a set of observations is the mean of the observations.

> The **mean** of a set of observations is their **average.** It is equal to the sum of all observations divided by the number of observations in the set.

Let us denote the observations by $x_1, x_2, \ldots x_n$. That is, the first observation denoted by x_1, the second by x_2, and so on to the nth observation, x_n. (In Example 1–2, $x_1 = 6$, $x_2 = 9, \ldots$, and $x_n = x_{20} = 24$.) The sample mean is denoted by $\bar{x}$.

TABLE 1–2 **Frequencies of Occurrence of Data Values in Example 1–2**

Value	Frequency
6	1
9	1
10	1
12	1
13	1
14	2
15	1
16	3
17	2
18	2
19	1
20	1
21	1
22	1
24	1

Mean of a sample:

$$\bar{x} = \frac{\sum_{i=1}^{n} x_i}{n} = \frac{x_1 + x_2 + \cdots + x_n}{n} \tag{1–1}$$

where Σ is summation notation. The summation extends over all data points.

When our observation set constitutes an entire population, instead of denoting the mean by $\bar{x}$ we use the symbol μ (the Greek letter mu). For a population, we use N as the number of elements instead of n. The population mean is defined as follows.

Mean of a population:

$$\mu = \frac{\sum_{i=1}^{N} x_i}{N} \tag{1–2}$$

The mean of the observations in Example 1–2 is found as

$$\bar{x} = (x_1 + x_2 + \cdots + x_{20})/20 = (6 + 9 + 10 + 12 + 13 + 14 + 14 + 15$$
$$+ 16 + 16 + 16 + 17 + 17 + 18 + 18 + 19 + 20 + 21 + 22 + 24)/20$$

$$= 317/20$$

$$= 15.85$$

The mean of the observations of Example 1–2, their average, is 15.85.

Figure 1–3 shows the data of Example 1–2 drawn on the number line along with the mean, median, and mode of the observations. If you think of the data points as

FIGURE 1–3   **Mean, Median, and Mode for Example 1–2**

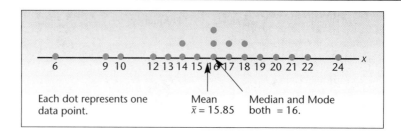

Each dot represents one   Mean   Median and Mode
data point. $\bar{x} = 15.85$ both = 16.

little balls of equal weight located at the appropriate places on the number line, the mean is that point where all the weights balance. It is the *fulcrum* of the point-weights, as shown in Figure 1–3.

What characterizes the three measures of centrality, and what are the relative merits of each? The mean summarizes all the information in the data. It is the average of all observations. The mean is a single point that can be viewed as the point where all the mass—the weight—of the observations is concentrated. It is the center of mass of the data. If all the observations in our data set were the same size, then (assuming the total is the same) each would be equal to the mean.

The median, on the other hand, is an observation (or a point between two observations) in the center of the data set. One-half of the data lie above this observation, and one-half of the data lie below it. When we compute the median, we do not consider the exact location of each data point on the number line; we only consider whether or not it falls in the half lying above the median or in the half lying below the median.

What does this mean? If you look at the picture of the data set of Example 1–2, Figure 1–3, you will note that the observation $x_{20} = 24$ lies to the far right. If we shift this particular observation (or any other observation to the right of 16) to the right, say, move it from 24 to 100, what will happen to the median? The answer is: absolutely *nothing* (prove this to yourself by calculating the new median). The exact location of any data point is not considered in the computation of the median, only its relative standing with respect to the central observation. *The median is resistant to extreme observations.*

The mean, on the other hand, is sensitive to extreme observations. Let us see what happens to the mean if we change x_{20} from 24 to 100. The new mean is

$$\bar{x} = (6 + 9 + 10 + 12 + 13 + 14 + 14 + 15 + 16 + 16 + 16 + 17 + 17$$
$$+\ 18 + 18 + 19 + 20 + 21 + 22 + 100)/20$$
$$=\ 19.65$$

We see that the mean has shifted almost 4 units to the right to accommodate the change in the single data point x_{20}.

The mean, however, does have strong advantages as a measure of central tendency. *The mean is based on information contained in all the observations in the data set,* rather than being an observation lying "in the middle" of the set. The mean also has some desirable mathematical properties that make it useful in many contexts of statistical inference. In cases where we want to guard against the influence of a few outlying observations (called *outliers*), however, we may prefer to use the median.

To continue with the condominium prices from Example 1–1, a larger sample of asking prices for two-bedroom units in Boston (numbers in thousand dollars, rounded to the nearest thousand) is

EXAMPLE 1–4

   168, 152, 167, 155, 171, 165, 150, 177, 295

What are the mean and the median? Interpret their meaning in this case.

Solution

Arranging the data from smallest to largest, we get

   150, 152, 155, 165, 167, 168, 171, 177, 295

There are nine observations, so the median is the value in the middle, that is in the fifth position. That value is 167 thousand dollars.

   To compute the mean, we add all data values and divide by 9, giving 177.778 thousand dollars—that is, $177,778. Now notice some interesting facts. The value 295 is clearly an *outlier*. It lies far to the right, away from the rest of the data bunched together in the 150–177 range.

   In this case, the median is a very descriptive measure of this data set: it tells us where our data (with the exception of the outlier) are located. The mean, on the other hand, pays so much attention to the large observation 295 that it locates itself at 177.8, a value larger than our largest observation, except for the outlier. If our outlier had been more like the rest of the data, say, 175 instead of 295, the mean would have been 164.4. Notice that the median does not change and is still 167. This is so because 175 is on the same side of the median as 295.

   Sometimes an outlier is due to an error in recording the data. In such a case it should be removed. Other times there is a good reason for its being "out in left field" (actually, right field in this case).

   As it turned out, the condominium with asking price of $295,000 was quite different from the rest of the two-bedroom units of roughly equal square footage and location. This unit was located in a prestigious part of town (away from the other units, geographically as well). It had a large whirlpool bath adjoining the master bedroom; its floors were marble from the Greek island of Paros; all light fixtures and faucets were gold-plated; the chandelier was Murano crystal. "This is not your average condominium," the realtor said, inadvertently reflecting a purely statistical fact in addition to the intended meaning of the expression.

   The mode tells us our data set's most frequently occurring value. There may be several modes. In Example 1–2, our data would possess two modes if we had another data point equal to 18, for example. Of the three measures of central tendency, we are most interested in the mean.

   If a data set or population is *symmetric* (i.e., if one side of the distribution of the observations is a mirror image of the other) and if the distribution of the observations has only one mode, then the mode, the median, and the mean are all equal. Such a situation is demonstrated in Figure 1–4. Generally, when the data distribution is not symmetric, then the mean, median, and mode will not all be equal. The relative positions of the three measures of centrality in such situations will be discussed in section 1–6.

   In the next section, we discuss measures of variability of a data set or population.

FIGURE 1–4   **A Symmetrically Distributed Data Set**

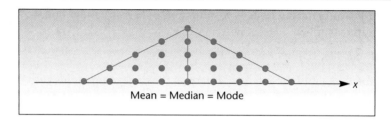

Mean = Median = Mode

PROBLEMS

1–18. Discuss the differences among the three measures of centrality.

1–19. Find the mean, median, and mode(s) of the observations in problem 1–13.

1–20. Do the same as problem 1–19, using the data of problem 1–14.

1–21. Do the same as problem 1–19, using the data of problem 1–15.

1–22. Do the same as problem 1–19, using the data of problem 1–16.

1–23. Do the same as problem 1–19, using the observation set in problem 1–17.

1–24. Do the same as problem 1–19 for the data in Example 1–1.

1–25. Find the mean and median for the 6-month percentage gain for all funds in problem 1–4.

1–26. For the following company price-earnings ratios, plot the data and identify any outliers. Find the mean and median.[4]

Arbor Software	39
Ascent	35
Compaq	17
EFI	25
LSI	30
Sierra Semiconductor	27
Teradyne	33
Texas Instruments	26

1–27. The following data are the price, in French francs, of a 14-milliliter bottle of Chanel No. 5 perfume at duty-free shops at various airports around the world.[5] Find the mean, median, and outliers.

Abu Dhabi	399
Dubai	570
Bangkok	616
Seoul	642
Hong Kong	616
Singapore	940

---

[4]G. Smith, "How to Play This Market," *Business Week,* June 16, 1997, p. 50.

[5]L. Citrinot, "Voyages d'Affaires," *Le Monde,* June 14, 1997, p. IX.

New York	515
Amsterdam	540
Frankfurt	554
Zurich	562
Paris	560
Copenhagen	548
London	627
Rome	612

1–4   Measures of Variability

Consider the following two data sets.

>   Set I: 1, 2, 3, 4, 5, 6, 6, 7, 8, 9, 10, 11
>   Set II: 4, 5, 5, 5, 6, 6, 6, 6, 7, 7, 7, 8

Compute the mean, median, and mode of each of the two data sets. As you see from your results, the two data sets have the same mean, the same median, and the same mode, all equal to 6. The two data sets also happen to have the same number of observations, $n = 12$. But the two data sets are different. What is the main difference between them?

Figure 1–5 shows data sets I and II. The two data sets have the same central tendency (as measured by any of the three measures of centrality), but they have a different *variability*. In particular, we see that data set I is more variable than data set II. The values in set I are more spread out: they lie farther away from their mean than do those of set II.

There are several measures of **variability,** or **dispersion.** We have already discussed one such measure—the interquartile range. (Recall that the interquartile range is defined as the difference between the upper quartile and the lower quartile.) The

FIGURE 1–5   **Comparison of Data Sets I and II**

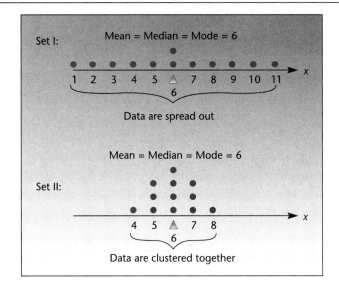

interquartile range for data set I is 5.5, and the interquartile range of data set II is 2 (show this). The interquartile range is one measure of the dispersion or variability of a set of observations. Another such measure is the *range*.

> The **range** of a set of observations is the difference between the largest observation and the smallest observation.

The range of the observations in Example 1–2 is Largest number − Smallest number = 24 − 6 = 18. The range of the data in set I is 11 − 1 = 10, and the range of the data in set II is 8 − 4 = 4. We see that, conforming with what we expect from looking at the two data sets, the range of set I is greater than the range of set II. Set I is more variable.

The range and the interquartile range are measures of the dispersion of a set of observations, the interquartile range being more resistant to extreme observations. There are also two other, more commonly used measures of dispersion. These are the *variance* and the square root of the variance—the *standard deviation*.

The variance and the standard deviation are more useful than the range and the interquartile range because, like the mean, they use the information contained in all the observations in the data set or population. (The range contains information only on the distance between the largest and smallest observations, and the interquartile range contains information only about the difference between upper and lower quartiles.) We define the variance as follows.

> The **variance** of a set of observations is the average squared deviation of the data points from their mean.

When our data constitute a sample, the variance is denoted by s^2, and the averaging is done by dividing the sum of the squared deviations from the mean by $n - 1$. (The reason for this will become clear in Chapter 5.) When our observations constitute an entire population, the variance is denoted by σ^2, and the averaging is done by dividing by N. (And σ is the Greek letter sigma; we call the variance *sigma squared*. The capital sigma is known to you as the symbol we use for summation, Σ.)

Sample variance:

$$s^2 = \frac{\sum_{i=1}^{n}(x_1 - \bar{x})^2}{n - 1} \tag{1-3}$$

Recall that $\bar{x}$ is the sample mean, the average of all the observations in the sample. Thus, the numerator in equation 1–3 is equal to the sum of the squared differences of the data points x_i (where $i = 1, 2, \ldots, n$) from their mean $\bar{x}$. When we divide the numerator by the denominator $n - 1$, we get a kind of average of the items summed in the numerator. This average is based on the assumption that there are only $n - 1$ data points. (Note, however, that the summation in the numerator extends over all n data points, not just $n - 1$ of them.) This will be explained in section 5–5.

When we have an entire population at hand, we denote the total number of observations in the population by N. We define the population variance as follows.

Population variance:

$$\sigma^2 = \frac{\sum\limits_{i=1}^{N} (x_1 - \mu)^2}{N} \qquad (1\text{--}4)$$

where μ is the population mean.

Unless noted otherwise, we will assume that all our data sets are samples and do not constitute entire populations; thus, we will use equation 1–3 for the variance, and not equation 1–4. We now define the standard deviation.

The **standard deviation** of a set of observations is the (positive) square root of the variance of the set.

The standard deviation of a sample is the square root of the sample variance, and the standard deviation of a population is the square root of the variance of the population.[6]

Sample standard deviation:

$$s = \sqrt{s^2} = \sqrt{\frac{\sum\limits_{i=1}^{n} (x_i - \bar{x})^2}{n - 1}} \qquad (1\text{--}5)$$

Population standard deviation:

$$\sigma = \sqrt{\sigma^2} = \sqrt{\frac{\sum\limits_{i=1}^{N} (x_i - \mu)^2}{N}} \qquad (1\text{--}6)$$

Why would we use the standard deviation when we already have its square, the variance? The standard deviation is a more meaningful measure. The variance is the average squared deviation from the mean. It is squared because if we just compute the deviations from the mean and then averaged them, we get zero (prove this with any of the data sets). Therefore, when seeking a measure of the variation in a set of observations, we square the deviations from the mean; this removes the negative signs, and thus the measure is not equal to zero. The measure we obtain—the variance—is still a *squared* quantity; it is an average of squared numbers. By taking its square root, we "unsquare" the units and get a quantity denoted in the original units of the problem (e.g., dollars instead of dollars squared, which would have little meaning in most applications). The variance tends to be large because it is in squared units.

---

[6]A note about calculators: If your calculator is designed to compute means and standard deviations, find the key for the standard deviation. Typically, there will be two such keys. Consult your owner's handbook to be sure you are using the key that will produce the correct computation for a sample (division by $n - 1$) versus a population (division by N).

TABLE 1–3   Calculations Leading to the Sample Variance in Example 1–2

x	$x - \bar{x}$	$(x - \bar{x})^2$
6	$6 - 15.85 = -9.85$	97.0225
9	$9 - 15.85 = -6.85$	46.9225
10	$10 - 15.85 = -5.85$	34.2225
12	$12 - 15.85 = -3.85$	14.8225
13	$13 - 15.85 = -2.85$	8.1225
14	$14 - 15.85 = -1.85$	3.4225
14	$14 - 15.85 = -1.85$	3.4225
15	$15 - 15.85 = -0.85$	0.7225
16	$16 - 15.85 = 0.15$	0.0225
16	$16 - 15.85 = 0.15$	0.0225
16	$16 - 15.85 = 0.15$	0.0225
17	$17 - 15.85 = 1.15$	1.3225
17	$17 - 15.85 = 1.15$	1.3225
18	$18 - 15.85 = 2.15$	4.6225
18	$18 - 15.85 = 2.15$	4.6225
19	$19 - 15.85 = 3.15$	9.9225
20	$20 - 15.85 = 4.15$	17.2225
21	$21 - 15.85 = 5.15$	26.5225
22	$22 - 15.85 = 6.15$	37.8225
24	$24 - 15.85 = 8.15$	66.4225
	0	378.5500

Statisticians like to work with the variance because its mathematical properties simplify computations. People applying statistics prefer to work with the standard deviation because it is more easily interpreted.

Let us find the variance and the standard deviation of the data in Example 1–2. It is convenient to carry out hand computations of the variance by use of a table. After doing the computation using equation 1–3, we will show a shortcut that will help in the calculation. Table 1–3 shows how the mean $\bar{x}$ is subtracted from each of the values and the results are squared and added. At the bottom of the last column we find the sum of all squared deviations from the mean. Finally, the sum is divided by $n - 1$, giving s^2, the sample variance. Taking the square root gives us s, the sample standard deviation.

By equation 1–3, the variance of the sample is equal to the sum of the third column in the table, 378.55, divided by $n - 1$: $s^2 = 378.55/19 = 19.923684$. The standard deviation is the square root of the variance: $s = \sqrt{19.923684} = 4.4635954$, or, using two-decimal accuracy,[7] $s = 4.46$.

If you have a calculator with statistical capabilities, you may avoid having to use a table such as Table 1–3. If you need to compute by hand, there is a shortcut formula for computing the variance and the standard deviation.

---

[7]In quantitative fields such as statistics, there is always the problem of decimal accuracy. How many digits after the decimal point should we carry? This question has no easy answer; everything depends on the required level of accuracy. As a rule, we will use only two decimals, since this suffices in most applications in this book. In some procedures, such as regression analysis, it is recommended that more digits be used in computations (these computations, however, are usually done by computer).

Shortcut formula for the sample variance:

$$s^2 = \frac{\sum\limits_{i=1}^{n} x_i^2 - \left(\sum\limits_{i=1}^{n} x_i\right)^2 / n}{n-1} \qquad (1\text{--}7)$$

Again, the standard deviation is just the square root of the quantity in equation 1–7. We will now demonstrate the use of this computationally simpler formula with the data of Example 1–2. We will then use this simpler formula and compute the variance and the standard deviation of the two data sets we are comparing: set I and set II.

As before, a table will be useful in carrying out the computations. The table for finding the variance using equation 1–7 will have a column for the data points x and a column for the squared data points x^2. Table 1–4 shows the computations for the variance of the data in Example 1–2.

Using equation 1–7, we find

$$s^2 = \frac{\sum\limits_{i=1}^{n} x_i^2 - \left(\sum\limits_{i=1}^{n} x_i\right)^2 / n}{n-1} = \frac{5{,}403 - (317)^2/20}{19} = \frac{5{,}403 - 100{,}489/20}{19}$$
$$= 19.923684$$

TABLE 1–4   **Shortcut Computations for the Variance in Example 1–2**

x	x^2
6	36
9	81
10	100
12	144
13	169
14	196
14	196
15	225
16	256
16	256
16	256
17	289
17	289
18	324
18	324
19	361
20	400
21	441
22	484
24	576
317	5,403

The standard deviation is obtained as before: $s = \sqrt{19.923684} = 4.46$. Using the same procedure demonstrated with Table 1–4, we find the following quantities leading to the variance and the standard deviation of set I and of set II. Both are assumed to be samples, not populations.

Set I: $\Sigma x = 72, \Sigma x^2 = 542, s^2 = 10,$ and $s = \sqrt{10} = 3.16$
Set II: $\Sigma x = 72, \Sigma x^2 = 446, s^2 = 1.27,$ and $s = \sqrt{1.27} = 1.13$

As expected, we see that the variance and the standard deviation of set II are smaller than those of set I. While each has a mean of 6, set I is more variable. That is, the values in set I vary more about their mean than do those of set II, which are clustered more closely together.

The sample standard deviation and the sample mean are very important statistics used in inference about populations.

EXAMPLE 1–5 In financial analysis, the standard deviation is often used as a measure of *volatility* and of the *risk* associated with financial variables. The data below are exchange rate values of the British pound, given as the value of one U.S. dollar's worth in pounds. The first column of 10 numbers is for a period in the beginning of 1993, and the second column of 10 numbers is for a similar period in the beginning of 1995. During which period, of these two precise sets of 10 days each, was the value of the pound more volatile?

1993	1995
0.6666	0.6332
0.6464	0.6254
0.6520	0.6286
0.6522	0.6359
0.6510	0.6336
0.6437	0.6427
0.6477	0.6209
0.6473	0.6214
0.6507	0.6204
0.6536	0.6325

Solution We are looking at two *populations* of 10 specific days at the start of each year (rather than a random sample of days), so we will use the formula for the population standard deviation. For the 1993 period we get $\sigma = 0.005929$, while for the 1995 period we get $\sigma = 0.007033$. We conclude that during the 1995 ten-day period the British pound was more volatile than in the same period in 1993. Notice that if these had been random samples of days, we would have used the sample standard deviation. In such cases we might have been interested in statistical inference to some population.

EXAMPLE 1–6[8] The data for the second quarter 1997 earnings per share (EPS) for major banks in the Northeast are tabulated below. Compute the mean, the variance, and the standard deviation of the data.

---

[8] *Business Week,* August 25, 1997, p. 162.

Name	EPS
Bank of New York	$2.53
BankBoston	4.38
Banker's Trust/New York	7.53
Chase Manhattan	7.53
Citicorp	7.96
Fleet	4.35
MBNA	1.50
Mellon	2.75
Morgan JP	7.25
PNC Bank	3.11
Republic	7.44
State Street	2.04
Summit	3.25

The results are tabulated below. [They were obtained from the template described in section 1–10.]   *Solution*

Sum	$61.62
Mean	$ 4.74
(Sample) Variance	5.94
(Sample) Standard Deviation	$ 2.44

PROBLEMS

1–28. Explain why we need measures of variability and what information these measures convey.

1–29. What is the most important measure of variability and why?

1–30. What is the computational difference between the variance of a sample and the variance of a population?

1–31. Find the range, the variance, and the standard deviation of the data set in problem 1–13 (assumed to be a sample).

1–32. Do the same as problem 1–31, using the data in problem 1–14.

1–33. Do the same as problem 1–31, using the data in problem 1–15.

1–34. Do the same as problem 1–31, using the data in problem 1–16.

1–35. Do the same as problem 1–31, using the data in problem 1–17.

1–5   Grouped Data and the Histogram

It often happens that data are grouped. This happened naturally in Example 1–2, where we had a group of three points with a value of 16 and three groups of two points (14s, 17s, and 18s). In other cases, especially when we have a large data set, the collector of the data may break the data into groups even if the points in each group are not equal in value. The data collector may set some (often arbitrary) group boundaries for ease of recording the data. When the salaries of 5,000 executives are considered, for example, the data may be reported in the form: 1,548 executives in the salary range $60,000 to $65,000; 2,365 executives in the salary range $65,001 to

$70,000; and so on. In this case, the data collector or analyst has processed all the salaries and put them into groups with defined boundaries. In such cases, there is a loss of information. We are unable to find the mean, variance, and other measures because we do not know the actual values. (There are, however, formulas that allow us to find the approximate mean, variance, and standard deviation. The formulas assume that all data points in a group are placed in the midpoint of the interval.) In this example, we assume that all 1,548 executives in the $60,000–$65,000 *class* make exactly ($60,000 + $65,000)/2 = $62,500; we estimate similarly for executives in the other groups.

We define a group of data values within specified group boundaries as a **class**.

When data are grouped into classes, we may also plot a frequency distribution of the data. Such a frequency plot is called a *histogram*.

A **histogram** is a chart made of bars of different heights. The height of each bar represents the **frequency** of values in the class represented by the bar. Adjacent bars share sides.

We demonstrate the use of histograms in the following example. Note that a histogram is used only for measured, or ordinal, data.

EXAMPLE 1–7   Management of an appliance store recorded the amounts spent at the store by the 184 customers who came in during the last day of the big sale. The data, amounts spent, were grouped into categories as follows: $0 to less than $100, $100 to less than $200, and so on up to $600, a bound higher than the amount spent by any single buyer. The classes and the frequency of each class are shown in Table 1–5. The frequencies, denoted by $f(x)$, are shown in a histogram in Figure 1–6.

As you can see from Figure 1–6, a histogram is just a convenient way of plotting the frequencies of grouped data. Here the frequencies are *absolute frequencies,* or **counts** of data points. It is also possible to plot *relative frequencies.*

The **relative frequency** of a class is the count of data points in the class divided by the total number of data points.

TABLE 1–5   **Classes and Frequencies, Example 1–7**

x Spending Class ($)	*f(x)* Frequency (Number of Customers)
0 to less than 100	30
100 to less than 200	38
200 to less than 300	50
300 to less than 400	31
400 to less than 500	22
500 to less than 600	13
	184

FIGURE 1–6   **A Histogram of the Data in Example 1–7**

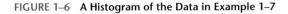

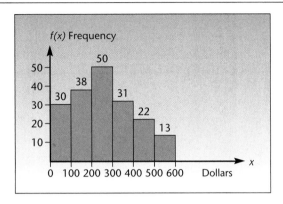

TABLE 1–6   **Relative Frequencies, Example 1–7**

x Class ($)	f(x) Relative Frequency
0 to less than 100	0.163
100 to less than 200	0.207
200 to less than 300	0.272
300 to less than 400	0.168
400 to less than 500	0.120
500 to less than 600	0.070
	1.000

The relative frequency in the first class, $0 to less than $100, is equal to count/total = $30/184 = 0.163$. We can similarly compute the relative frequencies for the other classes. The advantage of relative frequencies is that they are standardized: They add to 1.00. The relative frequency in each class represents the proportion of the total sample in the class. Table 1–6 gives the relative frequencies of the classes.

Figure 1–7, on the following page, is a histogram of the relative frequencies of the data in this example. Note that the shape of the histogram of the relative frequencies is the same as that of the absolute frequencies, the counts. The shape of the histogram does not change; only the labeling of the $f(x)$ axis is different.

Relative frequencies—proportions that add to 1.00—may be viewed as probabilities, as we will see in the next chapter. Hence, such frequencies are very useful in statistics, and so are their histograms.

1–6   Skewness and Kurtosis

In addition to measures of location, such as the mean or median, and measures of variation, such as the variance or standard deviation, there are two attributes of a frequency distribution of a data set that may be of interest to us. These are *skewness* and *kurtosis*.

FIGURE 1–7   A Histogram of the Relative Frequencies in Example 1–7

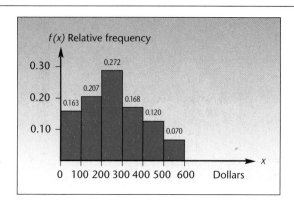

FIGURE 1–8   Skewness of Distributions

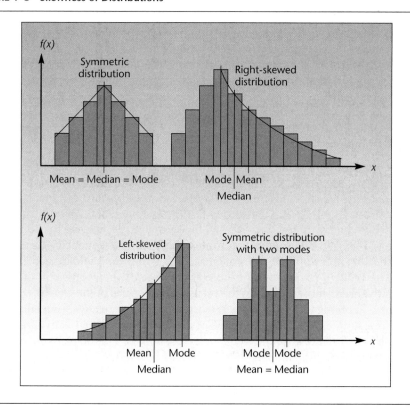

Skewness is a measure of the degree of asymmetry of a frequency distribution.

When the distribution stretches to the right more than it does to the left, we say that the distribution is *right skewed*. Similarly, a *left-skewed* distribution is one that stretches asymmetrically to the left. Four graphs are shown in Figure 1–8: a symmetric distri-

FIGURE 1–9   Kurtosis of Distributions

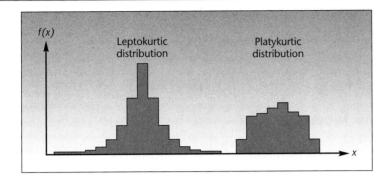

bution, a right-skewed distribution, a left-skewed distribution, and a symmetrical distribution with two modes.

Recall that a symmetric distribution with a single mode has mode = mean = median. Generally, for a right-skewed distribution, the mean is to the right of the median, which in turn lies to the right of the mode (assuming a single mode). The opposite is true for left-skewed distributions.

Skewness is calculated[9] and reported as a number that may be positive, negative, or zero. *Zero skewness* implies a symmetric distribution. A *positive skewness* implies a right-skewed distribution, and a *negative skewness* implies a left-skewed distribution.

Two distributions that have the same mean, variance, and skewness could still be significantly different in their shape. We may then look at their kurtosis.

Kurtosis is a measure of the peakedness of a distribution.

The larger the kurtosis, the more peaked will be the distribution. The kurtosis is calculated[10] and reported either as an absolute or a relative value. *Absolute kurtosis* is always a positive number. The absolute kurtosis of a *normal distribution,* a famous distribution about which we will learn in Chapter 4, is 3. This value of 3 is taken as the datum to calculate the *relative kurtosis.* The two are related by the equation

$$\text{Relative kurtosis} = \text{Absolute kurtosis} - 3$$

The relative kurtosis can be negative. We will always work with relative kurtosis. As a result, in this book, "kurtosis" means "relative kurtosis."

A negative kurtosis implies a flatter distribution than the normal distribution, and it is called *platykurtic.* A positive kurtosis implies a more peaked distribution than the normal distribution, and it is called *leptokurtic.* Figure 1–9 shows these examples.

---

[9]The formula used for calculating the skewness of a population is $\sum_{i=1}^{N} \left[\dfrac{x_i - \mu}{\sigma} \right]^3 / N$.

[10]The formula used for calculating the absolute kurtosis of a population is $\sum_{i=1}^{N} \left[\dfrac{x_i - \mu}{\sigma} \right]^4 / N$.

1–7   Relations between the Mean and the Standard Deviation

The mean is a measure of the centrality of a set of observations, and the standard deviation is a measure of their spread. There are two general rules that establish a relation between these measures and the set of observations. The first is called Chebyshev's theorem, and the second is the empirical rule.

Chebyshev's Theorem

A mathematical theorem called **Chebyshev's theorem** establishes the following rules:

1. At least three-quarters of the observations in a set will lie within 2 standard deviations of the mean.
2. At least eight-ninths of the observations in a set will lie within 3 standard deviations of the mean.

In general, the rule states that at least $1 - 1/k^2$ of the observations will lie within k standard deviations of the mean. (We note that k does not have to be an integer.) In Example 1–2 we found that the mean was 15.85 and the standard deviation was 4.46. According to rule 1 above, at least three-quarters of the observations should fall in the interval Mean $\pm 2s = 15.85 \pm 2(4.46)$, which is defined by the points 6.93 and 24.77. From the data set itself, we see that only one observation, 6, lies outside this range of values. Since there are 20 observations in the set, nineteen-twentieths are within the specified range, so the rule that at least three-quarters will be within the range is satisfied.

The Empirical Rule

If the distribution of the data is mound-shaped—that is, if the histogram of the data is more or less symmetric with a single mode or high point—then tighter rules will apply. This is the **empirical rule:**

1. Approximately 68% of the observations will be within 1 standard deviation of the mean.
2. Approximately 95% of the observations will be within 2 standard deviations of the mean.
3. A vast majority of the observations (all, or almost all) will be within 3 standard deviations of the mean.

For the data set in Example 1–2, even though the distribution of the data set is not perfectly symmetric, the empirical rule holds approximately (especially its last two parts). The mean is 15.85 and the standard deviation is 4.46. The two points that are 1 standard deviation on either side of the mean are $15.85 - 4.46 = 11.39$ and $15.85 + 4.46 = 20.31$. We see that 14 out of our 20 data points lie within this range. This is 70% of our observations. The two points that lie 2 standard deviations on either side of the mean are $15.85 - 2(4.46) = 6.93$ and $15.85 + 2(4.46) = 24.77$. We see that 19 out of our 20 data points lie within this range, or 95.83% of our observations. The 3 standard deviation limits around the mean are 2.47 and 29.23. One hundred percent of our observations lie within this range.

1–36.   Check the applicability of Chebyshev's theorem and the empirical rule for the data set in problem 1–13.

1–37.   Check the applicability of Chebyshev's theorem and the empirical rule for the data set in problem 1–14.

1–38.   Check the applicability of Chebyshev's theorem and the empirical rule for the data set in problem 1–15.

1–39.   Check the applicability of Chebyshev's theorem and the empirical rule for the data set in problem 1–16.

1–40.   Check the applicability of Chebyshev's theorem and the empirical rule for the data set in problem 1–17.

1–8   Methods of Displaying Data

In section 1–5, we saw how a histogram is used to display frequencies of occurrence of values in a data set. In this section, we will see a few other ways of displaying data, some of which are descriptive only. We will introduce frequency polygons, cumulative frequency plots (called *ogives*), pie charts, and bar charts. We will also see examples of how descriptive graphs can sometimes be misleading. We will start with pie charts.

Pie Charts

A **pie chart** is a simple descriptive display of data that sum to a given total. A pie chart is probably the most illustrative way of displaying quantities as percentages of a given total. The total area of the pie represents 100% of the quantity of interest (the sum of the variable values in all categories), and the size of each slice is the percentage of the total represented by the category the slice denotes. Pie charts are used to present frequencies for categorical data. The scale of measurement may be nominal or ordinal. Figure 1–10 is a pie chart of the geographical locations of the world's largest telecommunications companies.

FIGURE 1–10   Headquarters of the World's Largest Telecommunications Companies

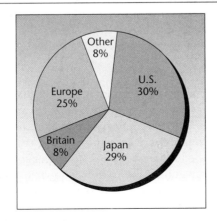

FIGURE 1–11   **Airline Operating Expenses and Revenues**

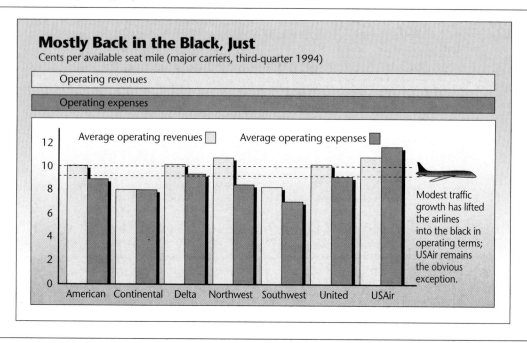

Reprinted by permission of *Forbes Magazine,* March 27, 1995, p. 79, with data from Avitas Aviation, based on carrier Form 41 filings with the Department of Transportation. © Forbes, Inc., 1995.

Bar Charts

Bar charts (which use horizontal or vertical rectangles) are often used to display categorical data where there is no emphasis on the percentage of a total represented by each category. The scale of measurement is nominal or ordinal.

Charts using horizontal bars and those using vertical bars are essentially the same. In some cases, it may be more convenient for the purpose at hand to use one versus the other. For example, if we want to write the name of each category inside the rectangle that represents that category, then a horizontal bar chart may be more convenient. If we want to stress the height of the different columns as measures of the quantity of interest, we use a vertical bar chart. Figure 1–11 is an example of how a bar chart can be used effectively to display and interpret information.

Frequency Polygons and Ogives

A **frequency polygon** is similar to a histogram except that there are no rectangles, only a point in the midpoint of each interval at a height proportional to the frequency or relative frequency (in a relative-frequency polygon) of the category of the interval. The rightmost and leftmost points are zero. Table 1–7 gives the relative frequency of sales volume, in thousands of dollars per week, for pizza at a local establishment.

A relative-frequency polygon for these data is shown in Figure 1–12. Note that the frequency is located in the middle of the interval as a point with height equal to the relative frequency of the interval. Note also that the point zero is added at the left boundary and the right boundary of the data set: The polygon starts at zero and ends at zero relative frequency.

TABLE 1–7   Pizza Sales

Sales ($000)	Relative Frequency
6–14	0.20
15–22	0.30
23–30	0.25
31–38	0.15
39–46	0.07
47–54	0.03

FIGURE 1–12   Relative-Frequency Polygon for Pizza Sales

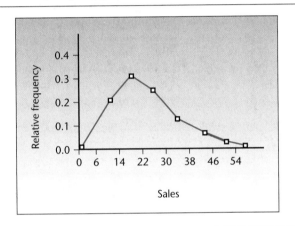

Figure 1–13 shows the Northeast banks' data from Example 1–6 displayed as a column chart. This is done using Excel's Chart Wizard.

An **ogive** is a cumulative-frequency (or cumulative relative-frequency) graph. An ogive starts at 0 and goes to 1.00 (for a relative-frequency ogive) or to the maximum cumulative frequency. The point with height corresponding to the cumulative frequency is located at the right endpoint of each interval. An ogive for the data in Table 1–7 is shown in Figure 1–14. While the ogive shown is for the cumulative *relative* frequency, an ogive can also be used for the cumulative absolute frequency.

A Caution about Graphs

A picture is indeed worth a thousand words, but pictures can sometimes be deceiving. Often, this is where "lying with statistics" comes in: presenting data graphically on a stretched or compressed scale of numbers with the aim of making the data show whatever you want them to show. This is one important argument against a merely descriptive approach to data analysis and an argument for statistical *inference*. Statistical tests tend to be more objective than our eyes and are less prone to deception as long as our assumptions (random sampling and other assumptions) hold. As we will see, statistical inference gives us tools that allow us to objectively evaluate what we see in the data.

FIGURE 1–13   Excel-Produced Graph of the Data in Example 1–6

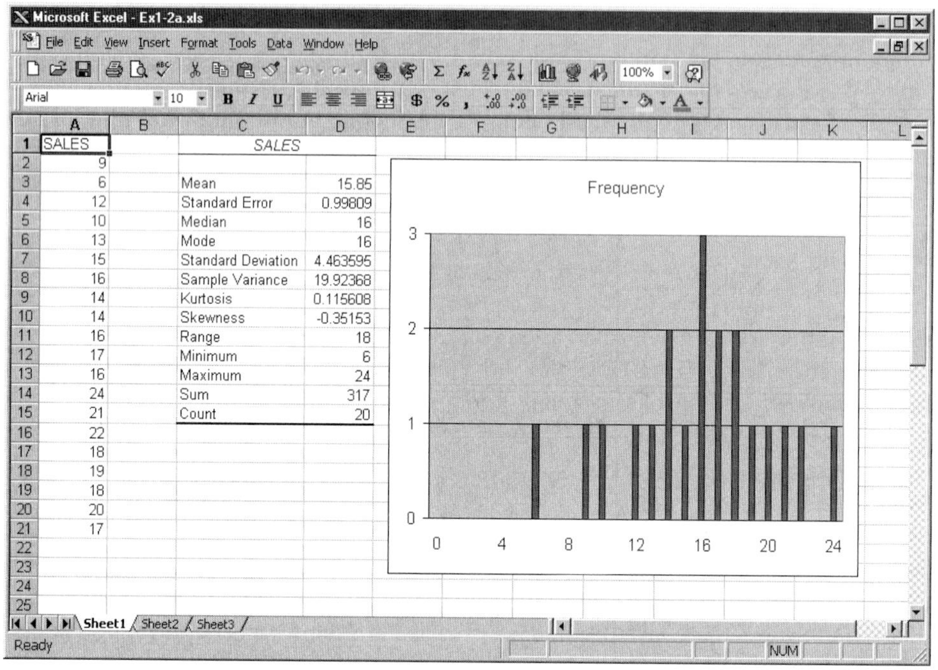

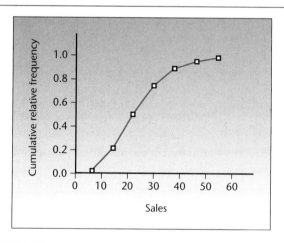

	A	B	C	D
1	SALES		*SALES*	
2	9			
3	6		Mean	15.85
4	12		Standard Error	0.99809
5	10		Median	16
6	13		Mode	16
7	15		Standard Deviation	4.463595
8	16		Sample Variance	19.92368
9	14		Kurtosis	0.115608
10	14		Skewness	-0.35153
11	16		Range	18
12	17		Minimum	6
13	16		Maximum	24
14	24		Sum	317
15	21		Count	20
16	22			
17	18			
18	19			
19	18			
20	20			
21	17			

FIGURE 1–14   Ogive of Pizza Sales

Pictures are sometimes deceptive even though there is no intention to deceive. When someone shows you a graph of a set of numbers, there may really be no particular scale of numbers that is "right" for the data.

Figure 1–15 is reprinted exactly as it appeared on the front page of *Investor's Business Daily*, February 14, 1995. Notice that there is *no scale* for this graph, and only one

FIGURE 1–15   **Median Income for Families**

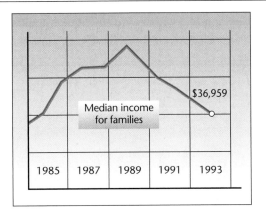

FIGURE 1–16   **The Dow-Jones Average**

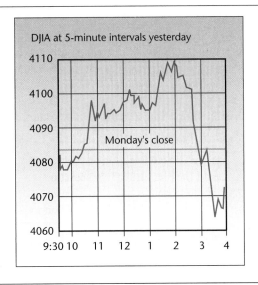

number, $36,959, is shown. What are the figures for the other years? For 1989, was the median income $50,000, or $36,970? This would make a big difference in how we interpret the graph.

Time Plots

Often we want to graph changes in a variable over time. An example is given in Figure 1–16.

PROBLEMS

1-41. The following data are estimated worldwide appliance sales (in millions of dollars). Use the data to construct a pie chart for the worldwide appliance sales of the listed manufacturers.

Electrolux	$5,100
General Electric	4,350
Matsushita Electric	4,180
Whirlpool	3,950
Bosch-Siemens	2,200
Philips	2,000
Maytag	1,580

1-42. Draw a bar graph for the data on the first five funds in problem 1–14. Is any one of the three kinds of plot more appropriate than the others for these data? If so, why?

1-43. Draw a bar graph for the endowments (stated in billions of dollars) of each of the universities specified in the following list.

Harvard	$3.4
Texas	2.5
Princeton	1.9
Yale	1.7
Stanford	1.4
Columbia	1.3
Texas A&M	1.1

1-44. Countries sending the most visitors to the United States in 1997 are as follows:[11]

Canada	15.9 million
Mexico	8.9 million
Japan	5.5 million
United Kingdom	3.3 million
Germany	2.1 million
France	1.1 million
Brazil	1.0 million
South Korea	1.0 million
Italy	0.6 million
Australia	0.5 million

Find the mean, median, and standard deviation. Draw a bar graph.

1-45. Foreign airlines with the most consumer complaints to the Transportation Department in 1996 were as follows:[12]

British Airways	98
Air France	48
KLM	39
Alitalia	36
AerLingus	26

---

[11] *USA Today*, June 2, 1997, p. 1.

[12] *USA Today*, June 2, 1997, p. 10B.

Draw a pie chart. Find the mean and median.

1–46. The following are the amounts from the sales slips of a department store (in dollars): 3.45, 4.52, 5.41, 6.00, 5.97, 7.18, 1.12, 5.39, 7.03, 10.25, 11.45, 13.21, 12.00, 14.05, 2.99, 3.28, 17.10, 19.28, 21.09, 12.11, 5.88, 4.65, 3.99, 10.10, 23.00, 15.16, 20.16. Draw a frequency polygon for these data (start by defining intervals of the data and counting the data points in each interval). Also draw an ogive and a column graph.

1–9 Exploratory Data Analysis

Exploratory data analysis (EDA) is the name given to a large body of statistical and graphical techniques. These techniques provide ways of looking at data to determine relationships and trends, identify outliers and influential observations, and quickly describe or summarize data sets. Pioneering methods in this field, as well as the name *exploratory data analysis,* derive from the work of John W. Tukey [John W. Tukey, *Exploratory Data Analysis* (Reading, Massachusetts: Addison-Wesley, 1977)].

Stem-and-Leaf Displays

A **stem-and-leaf display** is a quick way of looking at a data set. It contains some of the features of a histogram but avoids the loss of information in a histogram that results from aggregating the data into intervals. The stem-and-leaf display is based on the tallying principle: | || ||| |||| |||| ; but it also uses the decimal base of our number system. In a stem-and-leaf display, the *stem* is the number without its rightmost digit (the *leaf*). The stem is written to the left of a vertical line separating the stem from the leaf. For example, suppose we have the numbers 105, 106, 107, 107, 109. We display them as

$$10 \mid 56779$$

With a more complete data set with different stem values, the last digit of each number is displayed at the appropriate place to the right of its stem digit(s). Stem-and-leaf displays help us identify, at a glance, numbers in our data set that have high frequency. Let's look at an example.

Virtual reality is the name given to a system of simulating real situations on a computer in a way that gives people the feeling that what they see on the computer screen is a real situation. Flight simulators were the forerunners of virtual reality programs. A particular virtual reality program has been designed to give production engineers experience in real processes. Engineers are supposed to complete certain tasks as responses to what they see on the screen. The following data are the time, in seconds, it took a group of 42 engineers to perform a given task:

EXAMPLE 1–8

11, 12, 12, 13, 15, 15, 15, 16, 17, 20, 21, 21, 21, 22, 22, 22, 23, 24, 26, 27, 27, 27, 28, 29, 29, 30, 31, 32, 34, 35, 37, 41, 41, 42, 45, 47, 50, 52, 53, 56, 60, 62

Use a stem-and-leaf display to analyze these data.

The data are already arranged in increasing order. We see that the data are in the 10s, 20s, 30s, 40s, 50s, and 60s. We will use the first digit as the stem and the second digit of each number as the leaf. The stem-and-leaf display of our data is shown in Figure 1–17.

Solution

FIGURE 1–17   **Stem-and-Leaf Display of the Task Performance Times of Example 1–8**

```
1 122355567
2 0111222346777899
3 012457
4 11257
5 0236
6 02
```

FIGURE 1–18   **Refined Stem-and-Leaf Display for Data of Example 1–8**

```
1*      1223
1.      55567
2*      011122234
2.      6777899
3*      0124
3.      57
4*      112
4.      57
5*      023
5.      6
6*      02
```

As you can see, the stem-and-leaf display is a very quick way of arranging the data in a kind of a histogram (turned sideways) that allows us to see what the data look like. Here, we note that the data do not seem to be symmetrically distributed; rather, they are skewed to the right.

We may feel that this display does not convey very much information because there are too many values with first digit 2. To solve this problem, we may split the groups into two subgroups. We will denote the stem part as 1* for the possible numbers 10, 11, 12, 13, 14 and as 1. for the possible numbers 15, 16, 17, 18, 19. Similarly, the stem 2* will be used for the possible numbers 20, 21, 22, 23, and 24; stem 2. will be used for the numbers 25, 26, 27, 28, and 29; and so on for the other numbers. Our stem-and-leaf diagram for the data of Example 1–8 using this convention is shown in Figure 1–18. As you can see from the figure, we now have a more spread-out histogram of the data. The data still seem skewed to the right.

If desired, a further refinement of the display is possible by using the symbol * for a stem followed by the leaf values 0 and 1; the symbol t for leaf values 2 and 3; the symbol f for leaf values 4 and 5; s for 6 and 7; and . for 8 and 9. Also, the class containing the median observation is often denoted with its stem value in parentheses. We demonstrate this version of the display for the data of Example 1–8 in Figure 1–19. Note that the median is 27 (why?).

Note that for the data set of this example, the refinement offered in Figure 1–19 may be too much: We may have lost the general picture of the data. In cases where

FIGURE 1–19   **Further Refined Stem-and-Leaf Display of Data of Example 1–8**

	1*	1
	t	223
	f	555
	s	67
	.	
	2*	0111
	t	2223
	f	4
(Median in this class)	(s)	6777
	.	899
	3*	01
	t	2
	f	45
	s	7
	.	
	4*	11
	t	2
	f	5
	s	7
	.	
	5*	0
	t	23
	f	
	s	6
	.	
	6*	0
	t	2

there are many observations with the same value (for example, 22, 22, 22, 22, 22, 22, 22, . . .), it may be necessary to use a more stretched-out display in order to get a good picture of the way our data are clustered.

Box Plots

A *box plot* (also called a *box-and-whisker plot*) is another way of looking at a data set in an effort to determine its central tendency, spread, skewness, and the existence of outliers.

A **box plot** is a set of five summary measures of the distribution of the data:

1. The median of the data
2. The lower quartile
3. The upper quartile
4. The smallest observation
5. The largest observation

FIGURE 1–20   The Box Plot

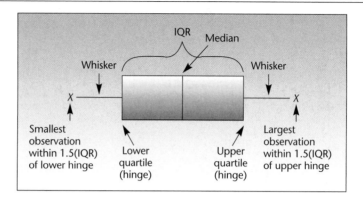

There are two qualifications to the preceding statements. First, we will assume that the *hinges* of the box plot are essentially the quartiles of the data set. (We will define hinges shortly.) The median is a line inside the box.

Second, the **whiskers** of the box plot are made by extending a line from the upper quartile to the largest observation and from the lower quartile to the smallest observation, only if the largest and smallest observations are within a distance of 1.5 times the interquartile range from the appropriate hinge (quartile). If one or more observations are farther away than that distance, they are marked as suspected outliers. If these observations are at a distance of over 3 times the interquartile range from the appropriate hinge, they are marked as outliers. The whisker then extends to the largest or smallest observation that is at a distance less than or equal to 1.5 times the interquartile range from the hinge.

Let us make these definitions clearer by using a picture. Figure 1–20 shows the parts of a box plot and how they are defined. The median is marked as a vertical line across the box. The **hinges** of the box are the upper and lower quartiles (the rightmost and leftmost sides of the box). The interquartile range (IQR) is the distance from the upper quartile to the lower quartile (the length of the box from hinge to hinge): $IQR = Q_U - Q_L$. We define the **inner fence** as a point at a distance of 1.5(IQR) above the upper quartile; similarly, the lower inner fence is $Q_L - 1.5(IQR)$. The **outer fences** are defined similarly but are at a distance of 3(IQR) above or below the appropriate hinge. Figure 1–21 shows the fences (these are not shown on the actual box plot; they are only guidelines for defining the whiskers, suspected outliers, and outliers) and demonstrates how we mark outliers.

Box plots are very useful for the following purposes.

1. To identify the location of a data set based on the median.
2. To identify the spread of the data based on the length of the box, hinge to hinge (the interquartile range), and the length of the whiskers (the range of the data without extreme observations: outliers or suspected outliers).
3. To identify possible skewness of the distribution of the data set. If the portion of the box to the right of the median is longer than the portion to the left of the median, and/or the right whisker is longer than the left whisker, the data are right-skewed. Similarly, a longer left side of the box and/or left whisker

FIGURE 1–21　**The Elements of a Box Plot**

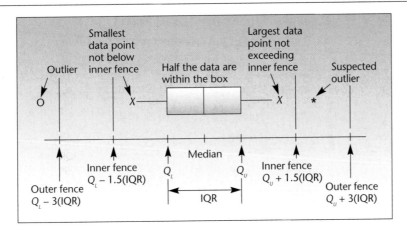

implies a left-skewed data set. If the box and whiskers are symmetric, the data are symmetrically distributed with no skewness.

4. To identify suspected outliers (observations beyond the inner fences but within the outer fences) and outliers (points beyond the outer fences).

5. To compare two or more data sets. By drawing a box plot for each data set and displaying the box plots on the same scale, we can compare several data sets.

It has even been suggested that a special form of a box plot may be used for conducting a test of the equality of two population medians. The various uses of a box plot are demonstrated in Figure 1–22.

Let us now construct a box plot for the data of Example 1–8. For this data set, the median is 27, and we find that the lower quartile is 20.75 and the upper quartile is 41. The interquartile range is $IQR = 41 - 20.75 = 20.25$. One and one-half times this distance is 30.38; hence, the inner fences are -9.63 and 71.38. Since no observation lies beyond either point, there are no suspected outliers and no outliers, so the whiskers extend to the extreme values in the data: 11 on the left side and 62 on the right side. The box plot for the data of Example 1–8 is shown in Figure 1–29 (on page 59). The plot in the figure was created using a template, described later.

As you can see from the figure, there are no outliers or suspected outliers in this data set. The data set is skewed to the right. This confirms our observation of the skewness from consideration of the stem-and-leaf diagrams of the same data set, in Figures 1–17 to 1–19.

PROBLEMS

1–47. The following data are monthly steel production figures, in millions of tons.

7.0, 6.9, 8.2, 7.8, 7.7, 7.3, 6.8, 6.7, 8.2, 8.4, 7.0, 6.7, 7.5, 7.2, 7.9, 7.6, 6.7, 6.6, 6.3, 5.6, 7.8, 5.5, 6.2, 5.8, 5.8, 6.1, 6.0, 7.3, 7.3, 7.5, 7.2, 7.2, 7.4, 7.6.

Draw a stem-and-leaf display of these data.

1–48. Draw a box plot for the data in problem 1–47. Are there any outliers? Is the distribution of the data symmetric or skewed? If it is skewed, to what side?

FIGURE 1–22   **Box Plots and Their Uses**

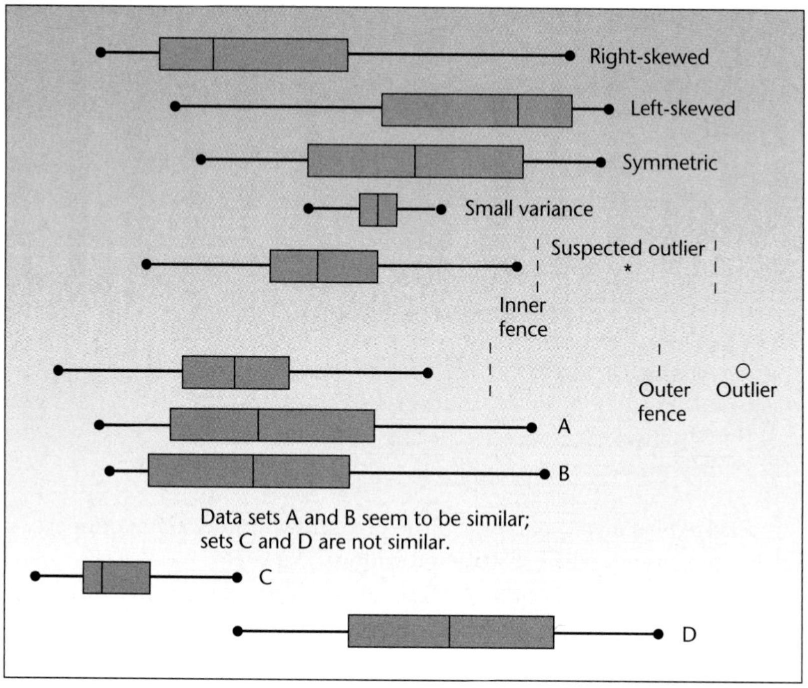

1–49.   What are the uses of a stem-and-leaf display? What are the uses of a box plot?

1–50.   Worker participation in management is a new concept that involves employees in corporate decision making. The following data are the percentages of employees involved in worker participation programs in a sample of firms. Draw a stem-and-leaf display of the data.

    5, 32, 33, 35, 42, 43, 42, 45, 46, 44, 47, 48, 48, 48, 49, 49, 50, 37, 38, 34, 51, 52, 52, 47, 53, 55, 56, 57, 58, 63, 78

1–51.   Draw a box plot of the data in problem 1–50, and draw conclusions about the data set based on the box plot.

1–52.   Consider the two box plots in Figure 1–30 (on page 60), and draw conclusions about the data sets.

1–53.   Refer to the following data on distances between seats in business class for various airlines.[13] Find μ, σ, σ^2, draw a box plot, and find the mode and any outliers.

Characteristics of Business-Class Carriers

	Distance between Rows (in cm)
Europe	
Air France	122
Alitalia	140

---

[13]From Luc Citrinot, "Voyages d'Affaires," *Le Monde,* June 14, 1997, p. 3.

Characteristics of Business-Class Carriers

	Distance between Rows (in cm)
British Airways	127
Iberia	107
KLM/Northwest	120
Lufthansa	101
Sabena	122
SAS	132
SwissAir	120
Asia	
All Nippon Airw	127
Cathay Pacific	127
JAL	127
Korean Air	127
Malaysia Air	116
Singapore Airl	120
Thai Airways	128
Vietnam Airl	140
North America	
Air Canada	140
American Airl	127
Continental	140
Delta Airlines	130
TWA	157
United	124

1–54. The following data are the daily price quotations for a certain stock over a period of 45 days. Construct a stem-and-leaf display for these data. What can you conclude about the distribution of daily stock prices over the period under study?

10, 11, 10, 11, 11, 12, 12, 13, 14, 16, 15, 11, 18, 19, 20, 15, 14, 14, 22, 25, 27, 23, 22, 26, 27, 29, 28, 31, 32, 30, 32, 34, 33, 38, 41, 40, 42, 53, 52, 47, 37, 23, 11, 32, 23

1–55. Discuss ways of dealing with outliers—their detection and what to do about them once they are detected. Can you always discard an outlier? Why or why not?

1–56. Define the inner fences and the outer fences of a box plot; also define the whiskers and the hinges. What portion of the data is represented by the box? By the whiskers?

1–57. The following data are the number of ounces of silver per ton of ore for two mines.

Mine A: 34, 32, 35, 37, 41, 42, 43, 45, 46, 45, 48, 49, 51, 52, 53, 60, 73, 76, 85
Mine B: 23, 24, 28, 29, 32, 34, 35, 37, 38, 40, 43, 44, 47, 48, 49, 50, 51, 52, 59

Construct a stem-and-leaf display for each data set and a box plot for each data set. Compare the two displays and the two box plots. Draw conclusions about the data.

1–58. Can you compare two *populations* by looking at box plots or stem-and-leaf displays of random samples from the two populations? Explain.

1–59. Refer to the following figure.[14] Draw a box plot of business rents in all cities listed.

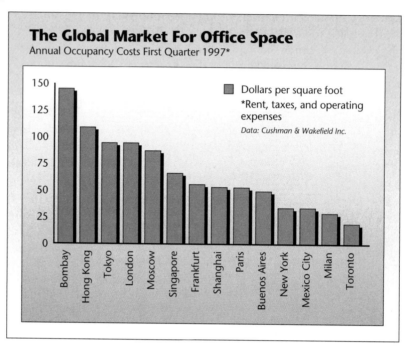

1–60. Consult the following data in "Plenty to Show Off About."[15] Find the mean unemployment rate and the median. Compare the two.

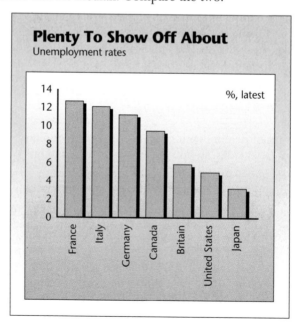

---

[14]G. Koretz, "Less of a Gulf in Office Rents," *Business Week*, June 16, 1997, p. 12. Reprinted by permission of *Business Week*. All rights reserved.

[15]*The Economist*, June 21, 1997, p. 27–28. © 1997 The Economist Newspaper Group, Inc. Reprinted with permission. Further production prohibited.

FIGURE 1–23   Template for Calculating Basic Statistics
[Basic Statistics.xls]

	A	B	C	D	E	F	G	H	I	J	K
1	**Basic Statistics from Raw Data**					Sales Data					
2											Data
3										1	9
4		Measures of Central tendency								2	6
5										3	12
6		Mean	15.85		Median	16		Mode	16	4	10
7										5	13
8		Measures of Dispersion								6	15
9				If the data is of a						7	16
10				Sample	Population					8	14
11			Variance	19.923684	18.9275		Range	18		9	14
12			St. Dev.	4.4635954	4.3505747		IQR	4.5		10	16
13										11	17
14		Higher Moments								12	16
15				If the data is of a						13	24
16				Sample	Population					14	21
17			Skewness	-0.351533	-0.324599					15	22
18			(Relative) Kurtosis	0.1156083	-0.197052					16	18
19										17	19
20		Percentile and Percentile Rank Calculations						Percentile		18	18
21					x-th			rank of y		19	20
22				x	Percentile		y			20	17
23				50	16		16.0	42			
24				80	19.2		19.2	80			
25				90	21.1		21.1	90			
26											
27		Quartiles									
28			1st Quartile	13.75							
29			Median	16		IQR	4.5				
30			3rd Quartile	18.25							
31											

1–10   Using the Computer

In this section, we shall discuss the use of templates for the computations and charts covered in the chapter. If you have not yet read the general instructions about using templates in Chapter 0, read them now.

Figure 1–23 shows the template that can be used for calculating basic statistics of a data set. As soon as the data are entered in the shaded area in column K, all the statistics are automatically calculated and displayed. All the statistics have been explained in this chapter, but some aspects of this template will be discussed next.

Percentile and Percentile Rank Computation

The percentile and percentile rank computations are done slightly differently in Excel. Do not be alarmed if your manual calculation differs (slightly) from the result you see in the template. These discrepancies in percentile and percentile rank computations occur because of approximation and rounding off. In Figure 1–23, notice that the 50th percentile is 16, but the percentile rank of 16 is 42. Such discrepancies will get smaller as the size of the data set increases. For large data sets, the discrepancy will be negligible or absent.

Histograms

A histogram can be drawn either from raw data or from grouped data, so the workbook contains one sheet for each case. Figure 1–24 shows the template that used raw

**FIGURE 1–24   Template for Histograms
[Histogram.xls; Sheet: Raw Data]**

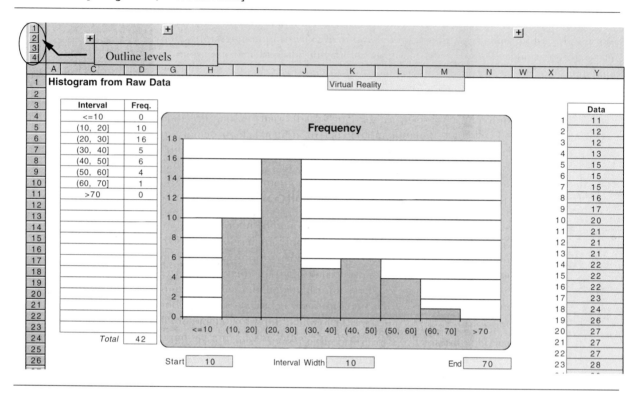

data. After entering the data in the shaded area in column Y, select appropriate values for the start, interval width, and end values for the histogram in cells H26, K26, and N26 respectively. When selecting the start and end values, make sure that the first bar and the last bar of the chart have zero frequencies. This will ensure that no value in the data has been omitted. The interval width should be selected to make the histogram a good representation of the distribution of the data.

An additional feature of this template is the outline feature. As shown in the figure, the template outline is organized in four different levels. By increasing the level you can see more details of the template. Currently, Level 2 has been selected. If Level 1 is chosen, columns C and D will be hidden, showing only the histogram. If Level 3 is chosen, you can see a Relative Frequency histogram as well. At Level 4, some intermediate calculations can also be seen.

Figure 1–25 shows the template used with grouped data. For a frequency polygon and/or an ogive, enter the interval and frequency data on the right. These data are the same as the results seen in columns C and D of the raw data sheet. You can copy and paste the *values only* into the grouped data sheet, using the Paste Special command (see section 0–2).

At times, you may have grouped data rather than raw data to start with. In this case, go directly to the grouped data sheet shown in Figure 1–25 and enter the data in the shaded area on the right. This sheet contains a total of five charts. If any of these is not needed, unprotect the sheet and delete it before printing.

**FIGURE 1–25   Histograms from Grouped Data
[Histogram.xls; Sheet: Grouped Data]**

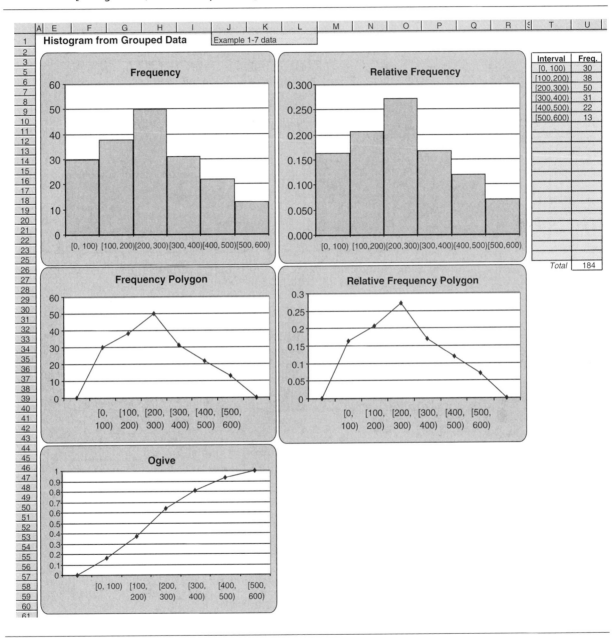

An advantage of frequency polygons is that unlike histograms, we can superpose two or more polygons to compare the distributions. Figure 1–26 shows the template that can be used to compare two distributions.

Pie Charts

Figure 1–27 shows the pie chart template. To use this template, enter the data in the shaded range, B4:C23. The entries in column C need not be percentages, and even if

FIGURE 1–26   Comparing Two Frequency Polygons
[Frequency Polygon 2.xls]

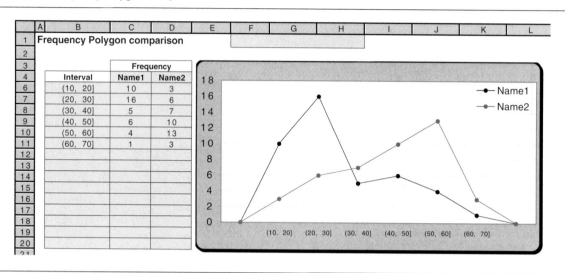

FIGURE 1–27   Pie Chart Template.
[Pie Chart.xls]

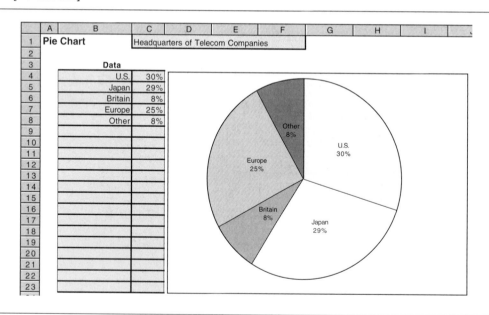

they are percentages, they need not add up to 100%, since the spreadsheet recalculates the proportions.

If you wish to modify the format of the chart, for example, by changing the colors of the slices or the location of legends, unprotect the sheet and use the Chart Wizard.

FIGURE 1–28 **The Bar Chart Template**
[Bar Chart.xls]

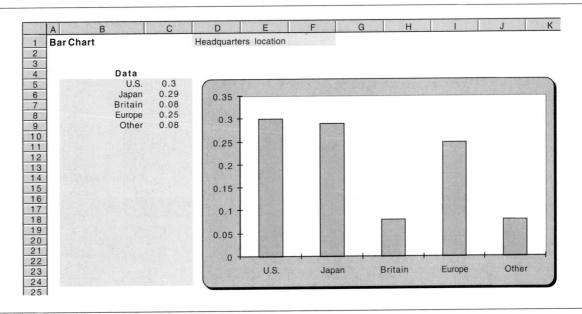

FIGURE 1–29 **Box Plot Template**
[Box Plot.xls]

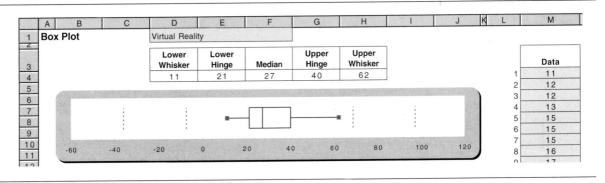

To use the Chart Wizard, click on the icon that looks like this: 📊 . Protect the sheet after you are done.

Bar Charts

Figure 1–28 shows the template that can be used to draw bar charts. Many refinements are possible on the bar charts, such as making it a 3-D chart. You can unprotect the sheet and use the Chart Wizard to make the refinements.

Box Plots

Figure 1–29 shows the template that can be used to create box plots. Figure 1–30 shows the template that draws two box plots of two different data sets. Thus it can be used to compare two data sets. Cells N3 and O3 are used to enter the name for each

FIGURE 1–30   Box Plot Template to Compare Two Data Sets
               [Box Plot 2.xls]

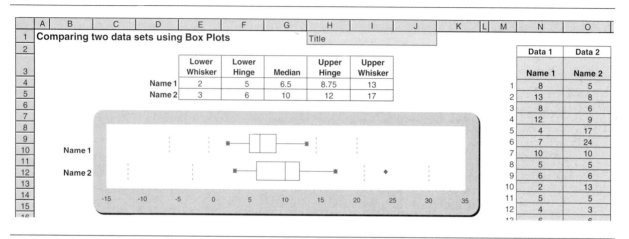

FIGURE 1–31   Time Plot Template
               [Time Plot.xls]

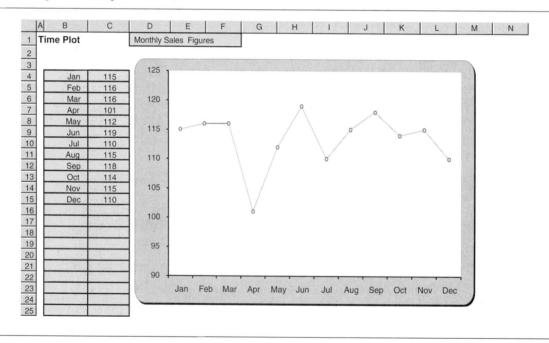

data set. The comparison shows that the second data set is more varied and contains relatively larger numbers than the first set.

Time Plots

Figure 1–31 shows the template that can be used to create time plots. The plot in the figure reveals that something extraordinary must have happened in the month of April. Whatever happened must be worth investigating.

FIGURE 1–32   **Time Plot Comparison**
[Time Plot 2.xls]

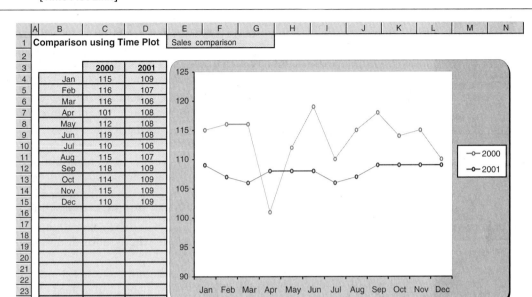

To compare two data sets, use the template shown in Figure 1–32. Comparing sales in years 2000 and 2001, the figure shows that Year 2001 sales were consistently below those of Year 2000, except in April. Moreover, the Year 2001 sales show less variance than those of Year 2001. Reasons for both facts may be worth investigating.

1–11   Summary and Review of Terms

In this chapter we introduced many terms and concepts. We defined a **population** as the set of all measurements in which we are interested. We defined a **sample** as a smaller group of measurements chosen from the larger population (the concept of random sampling will be discussed in detail in Chapter 4). We defined the process of using the sample for drawing conclusions about the population as **statistical inference.**

We discussed **descriptive statistics** as quantities computed from our data. We also defined the following statistics: **percentile,** a point below which lie a specified percentage of the data, and **quartile,** a percentile point in multiples of 25. The first quartile, the 25th percentile point, is also called the **lower quartile.** The 50th percentile point is the second quartile, also called the middle quartile, or the **median.** The 75th percentile is the **third quartile,** or the upper quartile. We defined the **interquartile range** as the difference between the upper and lower quartiles. We said that the median is a measure of central tendency, and we defined two other measures of central tendency: the **mode,** which is a *most frequent* value, and the **mean.** We called the mean the most important measure of central tendency, or location, of the data set. We said that the mean is the average of all the data points and is the point where the entire distribution of data points balances.

We defined measures of variability: the **range,** the **variance,** and the **standard deviation.** We defined the range as the difference between the largest and smallest data points. The variance was defined as the average squared deviation of the data points from their mean. For a sample (rather than a population), we saw that this averaging is done by dividing the sum of the squared deviations from the mean by $n - 1$ instead of by n. We defined the standard deviation as the square root of the variance.

We discussed grouped data and **frequencies** of occurrence of data points in **classes** defined by intervals of numbers. We defined **relative frequencies** as the absolute frequencies, or counts, divided by the total number of data points. We saw how to construct a **histogram** of a data set: a graph of the frequencies of the data. We mentioned **skewness,** a measure of the asymmetry of the histogram of the data set. We also mentioned **kurtosis,** a measure of the flatness of the distribution. We introduced **Chebyshev's theorem** and the **empirical rule** as ways of determining the proportions of data lying within several standard deviations of the mean.

We defined four scales of measurement of data: **nominal**—name only; **ordinal**— data that can be ordered as greater than or less than; **interval**—with meaningful distances as intervals of numbers; and **ratio**—a scale where ratios of distances are also meaningful.

The next topic we discussed was graphical techniques. These extended the idea of a histogram. We saw how a **frequency polygon** may be used instead of a histogram. We also saw how to construct an **ogive:** a cumulative frequency graph of a data set. We also talked about **bar charts** and **pie charts,** which are types of charts for displaying data, both categorical and numerical.

Then we discussed **exploratory data analysis,** a statistical area devoted to analyzing data using graphical techniques and other techniques that do not make restrictive assumptions about the structure of the data. Here we encountered two useful techniques for plotting data in a way that sheds light on their structure: **stem-and-leaf displays** and **box plots.** We saw that a stem-and-leaf display, which can be drawn quickly, is a type of histogram that makes use of the decimal structure of our number system. We saw how a box plot is made out of five quantities: the median, the two **hinges,** and the two **whiskers.** And we saw how the whiskers, as well as outliers and suspected outliers, are determined by the **inner fences** and **outer fences;** the first lies at a distance of 1.5 times the interquartile range from the hinges, and the second is found at 3 times the interquartile range from the hinges.

ADDITIONAL PROBLEMS

1–61. Open the workbook named Problem 1–61.xls. Study the statistics that have been calculated in the worksheet. Of special interest to this exercise are the two cells marked Mult and Add. If you enter 2 under Mult, all the data points will be multiplied by 2, as seen in the modified data column. Entering 1 under Mult leaves the data unchanged, since multiplying a number by 1 does not affect it. Similarly, entering 5 under Add will add 5 to all the data points. Entering 0 under Add will leave the data unchanged.

    1. Set Mult = 1 and Add = 5, which corresponds to adding 5 to all data points. Observe how the statistics have changed in the modified statistics column. Keeping Mult = 1 and changing Add to different values, observe how the statistics change. Then make a formal statement such as

"If we add x to all the data points, then the average would increase by x," for each of the statistics, starting with average.

2. Add an explanation for each statement made in part 1 above. For the average, this will be "If we add x to all the data points, then the sum of all the numbers will increase by $x*n$ where n is the number of data points. The sum is divided by n to get the average. So the average will increase by x."

3. Repeat part 1 for multiplying all the data points by some number. This would require setting Mult equal to desired values and Add = 0.

4. Repeat part 1 for multiplying and adding at once. This would require setting both Mult and Add to desired values.

1–62. Twenty randomly chosen people are shown a television commercial and asked to rank it as to overall appeal on a scale of 0 to 100. The results are given below.

89, 75, 59, 96, 88, 71, 43, 62, 80, 92, 76, 72, 67, 60, 79, 85, 77, 83, 87, 53

Find the mean, variance, and standard deviation of the sample of ratings.

1–63. The following data are the number of tons shipped weekly across the Pacific by a shipping company.

398, 412, 560, 476, 544, 690, 587, 600, 613, 457, 504, 477, 530, 641, 359, 566, 452, 633, 474, 499, 580, 606, 344, 455, 505, 396, 347, 441, 390, 632, 400, 582

Assume these data represent an entire population. Find the population mean and the population standard deviation.

1–64. Group the data in problem 1–63 into classes, and draw a histogram of the frequency distribution.

1–65. Find the 90th percentile, the quartiles, and the range of the data in problem 1–63.

1–66. The following data are numbers of color television sets manufactured per day at a given plant: 15, 16, 18, 19, 14, 12, 22, 23, 25, 20, 32, 17, 34, 25, 40, 41. Draw a frequency polygon and an ogive for these data.

1–67. Construct a stem-and-leaf display for the data in problem 1–66.

1–68. Construct a box plot for the data in problem 1–66. What can you say about the data?

1–69. The following data are the number of cars passing a point on a highway per minute: 10, 12, 11, 19, 22, 21, 23, 22, 24, 25, 23, 21, 28, 26, 27, 27, 29, 26, 22, 28, 30, 32, 25, 37, 34, 35, 62. Construct a stem-and-leaf display of these data. What does the display tell you about the data?

1–70. For the data problem 1–69, construct a box plot. What does the box plot tell you about these data?

1–71. Consult the following data on Swiss banks.[16] Plot the data and find outliers, the mean, and the standard deviation of return on equity (ROE).

---

[16]"The Best Banks," *Bilan, The Swiss Economic Magazine,* July–August 1997, p. 23.

The Best Banks

Rank	Institution	ROE 1996 in %
24	LGT	18.1
28	Vontobel	15.8
31	LLB	15.1
36	Sarasin	13.8
48	Rothschild	12.1
58	Bär	11.0
65	VP Bank	10.5
69	Credit Suisse Group	10.0
76	KB Zug	8.7
77	SBS	8.7

1–72. The following are P/E ratios (price of stock/projected earnings per share) of firms *Fortune* calls "Small Stocks, Big Promise":[17]

28, 33, 11, 12, 14, 15, 15, 16, 20, 23, 14, 17, 21, 26, 15, 22

Find the mean and the variance, plot the data, determine outliers, and do a box plot.

1–73. Consult the corporate data shown in "It's Not Just Silicon Valley."[18] Plot data; find μ, σ, σ^2; and identify outliers.

It's Not Just Silicon Valley

Of the 200 largest U.S. companies, 15 had at least 24% of their shares set aside for options and other stock awards as of 1996.

Morgan Stanley	91.36%
Merrill Lynch	40.26
Travelers	39.42
Warner-Lambert	35.00
Microsoft	32.95
J.P. Morgan & Co.	29.62
Lehman Brothers	28.25
US Airways	26.71
Sun Microsystems	25.99
Marriott	25.81
Bankers Trust	25.53
General Mills	25.41
MCI	24.39
AlliedSignal	24.23
ITT Industries	24.14

1–74. The following are quoted interest rates (%) on Italian bonds.[19]

---

[17]"Small Stocks, Big Promise," *Fortune,* July 7, 1997, p. 97.

[18]J. Fox, "It's Not Just Silicon Valley," *Fortune,* July 7, 1997, p. 32, with data from Pearl Meyer and Partners. Fortune © 1997 Time, Inc. All rights reserved.

[19]"Obligazioni," *Corriere Della Sera,* June 24, 1997, p. 18.

2.95, 4.25, 3.55, 1.90, 2.05, 1.78, 2.90, 1.85, 3.45, 1.75, 3.50, 1.69, 2.85, 4.10, 3.80, 3.85, 2.85, 8.70, 1.80, 2.87, 3.95, 3.50, 2.90, 3.45, 3.40, 3.55, 4.25, 1.85, 2.95

Plot the data; find μ, σ, and σ^2. Identify outliers (one is private, the rest are banks and government).

1–75. Refer to the box plot below to answer the questions.

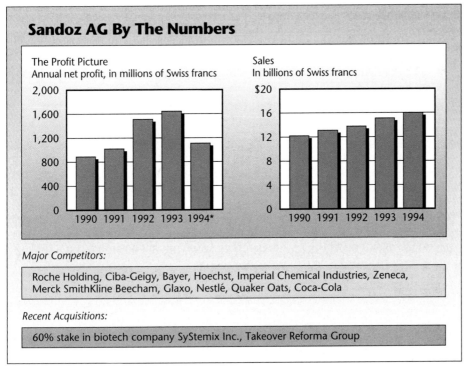

1. What is the interquartile range for this data set?
2. What can you say about the skewness of this data set?
3. For this data set, the value of 9.5 is more likely to be
   a. The first quartile rather than the median.
   b. The median rather than the first quartile.
   c. The mean rather than the mode.
   d. The mode rather than the mean.
4. If a data point that was originally 13 is changed to 14, how would the box plot be affected?

1–76. Interpret the graph "Sandoz AG by the Numbers."

Sandoz AG By The Numbers

The Profit Picture
Annual net profit, in millions of Swiss francs

Sales
In billions of Swiss francs

Major Competitors:

Roche Holding, Ciba-Geigy, Bayer, Hoechst, Imperial Chemical Industries, Zeneca, Merck SmithKline Beecham, Glaxo, Nestlé, Quaker Oats, Coca-Cola

Recent Acquisitions:

60% stake in biotech company SyStemix Inc., Takeover Reforma Group

Reprinted from *The Wall Street Journal,* February 21, 1995, p. B4, with data from Sandoz AG, Merrill Lynch & Co.

1–77. Consult the following figure.[20] What are the two averages? How do they differ in computation? In meaning? At least for the United States, verify that the per capita average is derived correctly from the overall U.S. average.

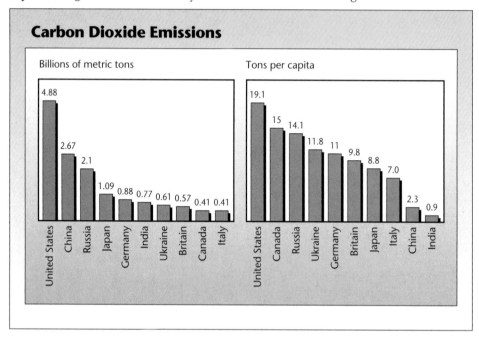

Carbon Dioxide Emissions

1–78. Profits from the Big Three[21]—GM, Chrysler, and Ford—in billions of dollars for 1990 to 1997 are as follows:

2.9, −7.5, −30.2*, 2.2, 13.9, 13.0, 12.9, 14.8†

Plot the data. Find μ and σ^2.

1–79. The following[22] are values of the Dow Jones Industrial Average at *5-minute* intervals on Friday, June 13, 1997. Their variance would be a good estimate for the *instantaneous* (limiting, as the time interval decreases to zero) variance to the process on that day. Compute the mean, variance, and standard deviation.

7,765, 7,760, 7,790, 7,780, 7,766, 7,768, 7,759, 7,772, 7,773, 7,767, 7,781, 7,774, 7,775, 7,780, 7,776, 7,771, 7,772, 7,794, 7,792, 7,793, 7,791, 7,792, 7,790, 7,795, 7,789, 7,788, 7,780, 7,787, 7,788, 7,778, 7,779, 7,790, 7,788, 7,782, 7,787, 7,795, 7,790, 7,784, 7,789, 7,781, 7,795, 7,791, 7,788, 7,792, 7,780, 7,772, 7,759, 7,783

1–80. The future Euroyen is the price of the Japanese yen as traded in the European futures market. The following are 30-day Euroyen prices on an index from 0 to 100%: 99.24, 99.37, 98.33, 98.91, 98.51, 99.38, 99.71, 99.21, 98.63, 99.10. Find μ, σ, σ^2, and the median.

---

[20]"Ecoefficiency: Business and the Environment," *International Herald Tribune,* June 23, 1997, p. 17, with data from Carbon Dioxide Information Center.

[21]L. Schiff, "The Outlook for the Big Three Is Getting Worse," *Fortune,* June 23, 1997, p. 14.

*Reflects accounting changes.

†1997 estimate.

[22]"The Dow's Performance," *The Wall Street Journal,* June 16, 1997, p. 18.

1–81. The daily price of Brent crude oil in dollars per barrel in summer 1997 was as follows: 17.5, 17.6, 18.3, 17.9, 17.4, 16.9, 17.1, 17.1, 18.0, 17.2, 18.3, 17.8, 17.1, 18.3, 17.5, 17.4. Find the mean, standard deviation, and variance.

1–82. The number of issues traded[23] on the American Stock Exchange over six trading days (Tuesday, Wednesday, Thursday, Friday, Monday, and Tuesday on June 10, 9, 8, 7, 4, and 3) are as follows:

June 10	749
June 9	760
June 8	719
June 7	715
June 4	730
June 3	721

Find the mean, median, standard deviation, and variance of these numbers.

1–83. The following is a random sample of 90-day futures prices in dollars for 1 troy oz. of silver from *The Wall Street Journal* issues in May and June of 1997:

4.74, 4.77, 4.87, 4.91, 4.83, 4.72, 4.92, 4.86, 4.97, 4.71, 4.90, 4.93, 4.75, 4.88, 4.79, 4.83, 4.89

Find the mean, median, and standard deviation of the 90-day future price of silver data.

1–84. Find the daily stock price of Wal-Mart for the last three months. (A good source for the data is http://moneycentral.msn.com. You can ask for the three-month chart and export the data to a spreadsheet.)

1. Calculate the mean and the standard deviation of the stock prices.
2. Get the corresponding data for Kmart and calculate the mean and the standard deviation.
3. The coefficient of variation (CV) is defined as the ratio of the standard deviation over the mean. Calculate the CV of Wal-Mart and Kmart stock prices.
4. If the CV of the daily stock prices is taken as an indicator of risk of the stock, how do Wal-Mart and Kmart stocks compare in terms of risk? (There are better measures of risk, but we will use CV in this exercise.)
5. Get the corresponding data of the Dow Jones Industrial Average (DJIA) and compute its CV. How do Wal-Mart and Kmart stocks compare with the DJIA in terms of risk?
6. Suppose you bought 100 shares of Wal-Mart stock three months ago and held it. What are the mean and the standard deviation of the daily market price of your holding for the three months?

1–85. To calculate variance and standard deviation, we take the deviations from the mean. At times, we need to consider the deviations from a target value rather than the mean. Consider the case of a machine that bottles cola into 2-liter ($2,000\text{-cm}^3$) bottles. The target is thus $2,000 \text{ cm}^3$. The machine, however, may be bottling $2,004 \text{ cm}^3$ on average into every bottle. Call this $2,004 \text{ cm}^3$ the *process mean*. The damage from process errors is determined by the deviations from the target rather than from the

---

[23]From *The Wall Street Journal*, June 11, 1997, p. 16.

process mean. The variance, though, is calculated with deviations from the process mean, and therefore is not a measure of the damage. Suppose we want to calculate a new variance using deviations from the target value. Let "SSD(Target)" denote the sum of the squared deviations from the target. [For example, SSD(2,000) denotes the sum of squared deviations when the deviations are taken from 2,000.] Dividing the SSD by the number of data points gives the Average SSD(Target).

The following spreadsheet is set up to calculate the deviations from the target, SSD(Target), and the Average SSD(Target). Column B contains the data, showing a process mean of 2,004. (Strictly speaking, this would be sample data. But to simplify matters, let us assume that this is population data.) Note that the population variance (VARP) is 3.5 and the Average SSD(2,000) is 19.5.

In the range G5:H13, a table has been created to see the effect of changing the target on Average SSD(Target). The offset refers to the difference between the target and the process mean.

1. Study the table and find an equation that relates the Average SSD to VARP and the Offset. [Hint: Note that while calculating SSD, the deviations are squared, so think in squares.]

2. Using the equation you found in part 1, prove that the Average SSD(Target) is minimized when the target equals the process mean.

Working with Deviations from a Target
[Problem 1–85.xls]

	A	B	C	D	E	F	G	H	I
1	Deviations from a Target								
2									
3		Target	2000						
4									
5		Data	Deviation from Target	Squared Deviation		Offset	Target	Average SSd	
6		2003	3	9					
7		2002	2	4		-4	2000	19.5	
8		2005	5	25		-3	2001	12.5	
9		2004	4	16		-2	2002	7.5	
10		2006	6	36		-1	2003	4.5	
11		2001	1	1		0	2004	3.5	<- VARP
12		2004	4	16		1	2005	4.5	
13		2007	7	49		2	2006	7.5	
14						3	2007	12.5	
15		Mean	2004	SSd	156	4	2008	19.5	
16		VARP	3.5	Avg. SSd	19.5				

1–86. The Consumer Price Index (CPI) is an important indicator of the general level of prices of essential commodities. It is widely used in making cost of living adjustments to salaries, for example.

1. Log on to the Consumer Price Index (CPI) home page of the Bureau of Labor Statistics website (stats.bls.gov/cpihome.htm). Get a table of the last 48 months' CPI for U.S. urban consumers with 1982–1984 as the base. Make a time plot of the data. Discuss any seasonal pattern you see in the data.

2. Go to the Average Price Data area and get a table of the last 48 months' average price of unleaded regular gasoline. Make a comparison time plot of the CPI data in part 1 and the gasoline price data. Comment on the gasoline prices.

1–87. Log on to the Center for Disease Control website and go to the HIV statistics page (www.cdc.gov/hiv/stats.htm).

1. Download the data on the cumulative number of AIDS cases reported in the United States and its age-range breakdown. Draw a pie chart of the data.

2. Download the race/ethnicity breakdown of the data. Draw a pie chart of the data.

1–88. Log on to the CBS Sportsline website and go to the Major League Baseball (MLB) player salaries page (cbs.sportsline.com/u/baseball/mlb/salaries_index.html).

1. Get the Chicago Cubs players' salaries for the current year. Draw a box plot of the data. (Enter the data in thousands of dollars to make the numbers smaller.) Are there any outliers?

2. Get the Chicago White Sox players' salaries for the current year. Make a comparison box plot of the two data. Describe your comparison based on the plot.

CASE 1   NASDAQ Volatility

The NASDAQ Combined Composite Index is a measure of the aggregate value of technological stocks. During the year 2000, the index moved up and down considerably, indicating the rapid changes in e-business that took place in that year and the high uncertainty in the profitability of technology-oriented companies. Historical data of the index is available at many websites, including **Finance. Yahoo.com.**

1. Download the monthly data of the index for the calendar year 2000 and make a time plot of the data. Comment on the volatility of the index, looking at the plot. Report the standard deviation of the data.

2. Download the monthly data of the index for the calendar year 1999 and compare the data for 1999 and 2000 on a single plot. Which year has been more volatile? Calculate the standard deviations of the two sets of data. Do they confirm your answer about the relative volatility of the two years?

3. Download the monthly data of the S&P 500 index for the year 2000. Compare this index with the NASDAQ index for the same year on a single plot. Which index has been more volatile? Calculate and report the standard deviations of the two sets of data.

4. Download the monthly data of the Dow Jones Industrial Average for the year 2000. Compare this index with the NASDAQ index for the same year on a single plot. Which index has been more volatile? Calculate and report the standard deviations of the two sets of data.

5. Repeat part 1 with the monthly data for the latest 12 full months.

2 Probability

2–1 Using Statistics

Probability is everywhere.
It affects all aspects of our lives.

Giants Win Is No Surprise—They Had a 65.635738% Chance

BY STEVE RUBENSTEIN

The San Francisco Giants had a 65.6 percent chance of winning yesterday's game, said a statistics professor who worked it out.

"It wasn't that difficult a calculation," said statistics professor Bill Kaigh.

Today, statistics say, the Giants have a 100 percent chance of winning yesterday's game. Numbers don't lie.

"The estimates are derived from principal components and linear and logistic regression analyses of historical data," said Kaigh, a professor at the University of Texas–El Paso, shortly before Will Clark hit a triple to center field to give the Giants the lead.

Five thousand statisticians are in town this week for the meeting of the American Statistical Association. Kaigh, addressing a morning seminar on sports statistics, passed out a 24-page research paper showing exactly why the Giants would probably win yesterday's game, which they did, 10–7.

The formula involves multiplying the Giants' home-game record by the Cincinnati Reds' away-game record subtracted from one and dividing the whole thing by something else with four variables in it.

The answer is actually 65.635738 percent. That explains why Kurt Manwaring hit a two-run homer in the sixth inning.

A **probability** is a quantitative measure of uncertainty—a number that conveys the strength of our belief in the occurrence of an uncertain event. Since life is full of uncertainty, people have always been interested in evaluating probabilities. The statistician I. J. Good suggests that "the theory of probability is much older than the human species," since the assessment of uncertainty incorporates the idea of learning from experience, which most creatures do.[1]

The theory of probability as we know it today was largely developed by European mathematicians such as Galileo Galilei (1564–1642), Blaise Pascal (1623–1662), Pierre de Fermat (1601–1665), Abraham de Moivre (1667–1754), and others.

As in India, the development of probability theory in Europe is often associated with gamblers, who pursued their interests in the famous European casinos, such as the one at Monte Carlo. Many books on probability and statistics tell the story of the

---

[1] I. J. Good, "Kinds of Probability," *Science,* no. 129 (February 20, 1959), pp. 443–47.

Chevalier de Mère, a French gambler who enlisted the help of Pascal in an effort to obtain the probabilities of winning at certain games of chance, leading to much of the European development of probability.

Today, the theory of probability is an indispensable tool in the analysis of situations involving uncertainty. It forms the basis for inferential statistics as well as for other fields that require quantitative assessments of chance occurrences, such as quality control, management decision analysis, and areas in physics, biology, engineering, and economics.

While most analyses using the theory of probability have nothing to do with games of chance, gambling models provide the clearest examples of probability and its assessment. The reason is that games of chance usually involve dice, cards, or roulette wheels—mechanical devices. If we assume there is no cheating, these mechanical devices tend to produce sets of outcomes that are *equally likely,* and this allows us to compute probabilities of winning at these games.

Suppose that a single die is rolled and that you win a dollar if the number 1 or 2 appears. What are your chances of winning a dollar? Since there are six equally likely numbers (assuming the die is fair) and you win as a result of either of two numbers appearing, the probability that you win is 2/6, or 1/3.

As another example, consider the following situation. An analyst follows the price movements of IBM stock for a time and wants to assess the probability that the stock will go up in price in the next week. This is a different type of situation. The analyst does not have the luxury of a known set of equally likely outcomes, where "IBM stock goes up next week" is one of a given number of these equally likely possibilities. Therefore, the analyst's assessment of the probability of the event will be a *subjective* one. The analyst will base her or his assessment of this probability on knowledge of the situation, guesses, or intuition. Different people may assign different probabilities to this event depending on their experience and knowledge, hence the name *subjective* probability.

Objective probability is probability based on symmetry of games of chance or similar situations. It is also called *classical probability*. This probability is based on the idea that certain occurrences are equally likely (the term *equally likely* is intuitively clear and will be used as a starting point for our definitions): The numbers 1, 2, 3, 4, 5, and 6 on a fair die are each equally likely to occur. Another type of objective probability is long-term *relative-frequency* probability. If, in the long run, 20 out of 1,000 consumers given a taste test for a new soup like the taste, then we say that the probability that a given consumer will like the soup is 20/1,000 = 0.02. If the probability that a head will appear on any one toss of a coin is 1/2, then if the coin is tossed a large number of times, the proportion of heads will approach 1/2. Like the probability in games of chance and other symmetric situations, relative-frequency probability is objective in the sense that no personal judgment is involved.

Subjective probability, on the other hand, involves personal judgment, information, intuition, and other subjective evaluation criteria. The area of subjective probability—which is relatively new, having been first developed in the 1930s—is somewhat controversial.[2] A physician assessing the probability of a patient's recovery

---

[2]The earliest published works on subjective probability are Frank Ramsey's *The Foundation of Mathematics and Other Logical Essays* (London: Kegan Paul, 1931) and the Italian statistician Bruno de Finetti's "La Prévision: Ses Lois Logiques, Ses Sources Subjectives," *Annales de L'Institut Henri Poincaré* 7, no. 1 (1937).

and an expert assessing the probability of success of a merger offer are both making a personal judgment based on what they know and feel about the situation. Subjective probability is also called *personal probability*. One person's subjective probability may very well be different from another person's subjective probability of the same event.

Whatever the kind of probability involved, the same set of mathematical rules holds for manipulating and analyzing probability. We now give the general rules for probability as well as formal definitions. Some of our definitions will involve counting the number of ways in which some event may occur. The counting idea is implementable only in the case of objective probability, although conceptually this idea may apply to subjective probability as well, if we can imagine a kind of lottery with a known probability of occurrence for the event of interest.

2–2 Basic Definitions: Events, Sample Space, and Probabilities

In order to understand probability, it is useful to have some familiarity with sets and with operations involving sets.

> A **set** is a collection of elements.

The elements of a set may be people, horses, desks, cars, files in a cabinet, or even numbers. We may define our set as the collection of all horses in a given pasture, all people in a room, all cars in a given parking lot at a given time, all the numbers between 0 and 1, or all integers. The number of elements in a set may be infinite, as in the last two examples.

A set may also have no elements.

> The **empty set** is the set containing *no elements*. It is denoted by Ø.

We now define the universal set.

> The **universal set** is the set containing *everything* in a given context. We denote the universal set by S.

Given a set A, we may define its *complement*.

> The **complement** of set A is the set containing all the elements in the universal set S that are *not* members of set A. We denote the complement of A by Ā. The set Ā is often called "not A."

A **Venn diagram** is a schematic drawing of sets that demonstrates the relationships between different sets. In a Venn diagram, sets are shown as circles, or other closed figures, within a rectangle corresponding to the universal set, S. Figure 2–1 is a Venn diagram demonstrating the relationship between a set A and its complement Ā.

As an example of a set and its complement, consider the following. Let the universal set S be the set of all students at a given university. Define A as the set of all students who own a car (at least one car). The complement of A, or Ā, is thus the set of all students at the university who do *not* own a car.

FIGURE 2–1   A Set A and Its Complement Ā

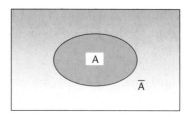

FIGURE 2–2   Sets A and B and Their Intersection

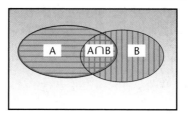

Sets may be related in a number of ways. Consider two sets A and B within the context of the same universal set S. (We say that A and B are *subsets* of the universal set S.) If A and B have some elements in common, we say they *intersect*.

> The **intersection** of A and B, denoted A ∩ B, is the set containing all elements that are members of *both* A and B.

When we want to consider all the elements of two sets A and B, we look at their *union*.

> The **union** of A and B, denoted A ∪ B, is the set containing all elements that are members of *either* A *or* B or *both*.

As you can see from these definitions, the union of two sets contains the intersection of the two sets. Figure 2–2 is a Venn diagram showing two sets A and B and their intersection A ∩ B. Figure 2–3 is a Venn diagram showing the union of the same sets.

As an example of the union and intersection of sets, consider again the set of all students at a university who own a car. This is set A. Now define set B as the set of all students at the university who own a bicycle. The universal set S is, as before, the set of all students at the university. And A ∩ B is the intersection of A and B—it is the set of all students at the university who own *both* a car and a bicycle. And A ∪ B is the union of A and B—it is the set of all students at the university who own either a car or a bicycle or both.

Two sets may have no intersection: They may be **disjoint.** In such a case, we say that the intersection of the two sets is the empty set Ø. In symbols, when A and B are disjoint, A ∩ B = Ø. As an example of two disjoint sets, consider the set of all

FIGURE 2–3   **The Union of A and B**

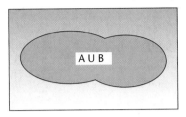

FIGURE 2–4   **Two Disjoint Sets**

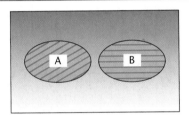

students enrolled in a business program at a particular university and all the students at the university who are enrolled in an art program. (Assume no student is enrolled in both programs.) A Venn diagram of two disjoint sets is shown in Figure 2–4.

In probability theory we make use of the idea of a set and of operations involving sets. We will now provide some basic definitions of terms relevant to the computation of probability. These are an *experiment,* a *sample space,* and an *event.*

> An **experiment** is a process that leads to one of several possible **outcomes**.
> An **outcome** of an experiment is some observation or measurement.

Drawing a card out of a deck of 52 cards is an experiment. One outcome of the experiment may be that the queen of diamonds is drawn.

A single outcome of an experiment is called a *basic outcome* or an *elementary event.* Any particular card drawn from a deck is a basic outcome.

> The **sample space** is the universal set S pertinent to a given experiment. The sample space is the set of all possible outcomes of an experiment.

The sample space for the experiment of drawing a card out of a deck is the set of all cards in the deck. The sample space for an experiment of reading the temperature is the set of all numbers in the range of temperatures.

> An **event** is a subset of a sample space. It is a set of basic outcomes. We say that the event *occurs* if the experiment gives rise to a basic outcome belonging to the event.

FIGURE 2–5   Sample Space for Drawing a Card

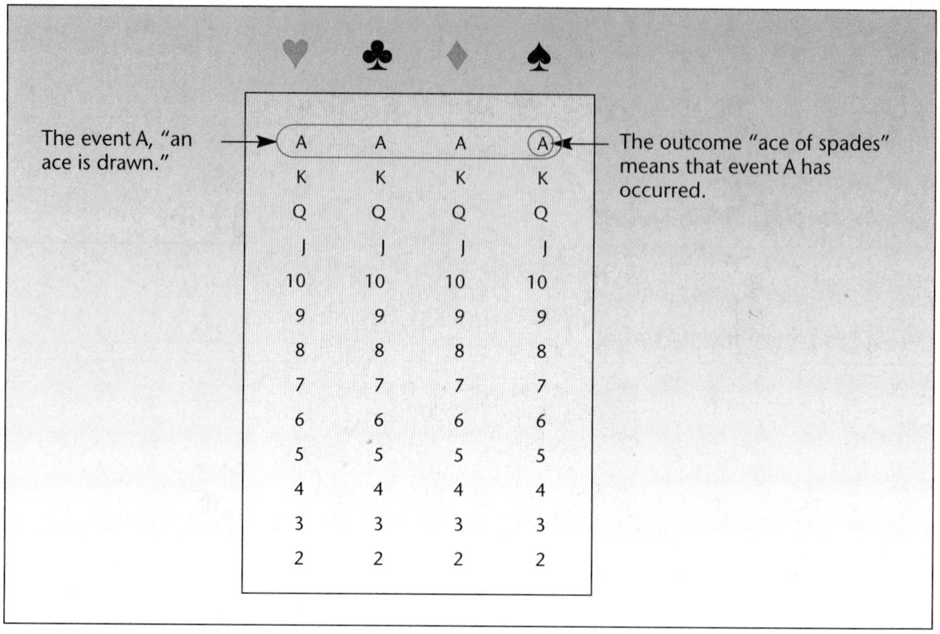

The event A, "an ace is drawn." — The outcome "ace of spades" means that event A has occurred.

For example, the event "an ace is drawn out of a deck of cards" is the set of the four aces within the sample space consisting of all 52 cards. This event occurs whenever one of the four aces (the basic outcomes) is drawn.

The sample space for the experiment of drawing a card out of a deck of 52 cards is shown in Figure 2–5. The figure also shows event A, the event that an ace is drawn.

In this context, for a given experiment we have a sample space with equally likely basic outcomes. When a card is drawn out of a well-shuffled deck, every one of the cards (the basic outcomes) is as likely to occur as any other. In such situations, it seems reasonable to define the probability of an event as the *relative size* of the event with respect to the size of the sample space. Since there are 4 aces and there are 52 cards, the size of A is 4 and the size of the sample space is 52. Therefore, the probability of A is equal to 4/52.

The rule we use in computing probabilities, assuming equal likelihood of all basic outcomes, is as follows:

Probability of event A:

$$P(A) = \frac{n(A)}{n(S)} \tag{2–1}$$

where

$n(A)$ = the number of elements in the set of the event A

$n(S)$ = the number of elements in the sample space S

FIGURE 2–6   The Events A and ♥ and Their Union and Intersection

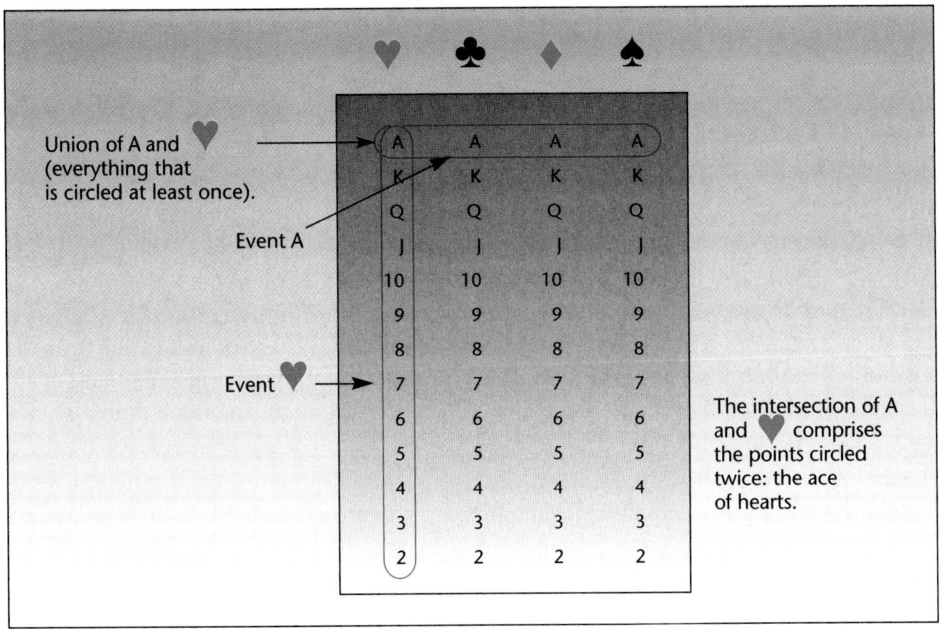

The probability of drawing an ace is $P(A) = n(A)/n(S) = 4/52$.

EXAMPLE 2–1

Roulette is a popular casino game. As the game is played in Las Vegas or Atlantic City, the roulette wheel has 36 numbers, 1 through 36, and the number 0 as well as the number 00 (double zero). What is the probability of winning on a single number that you bet?

Solution

The sample space S in this example consists of 38 numbers (0, 00, 1, 2, 3,..., 36), each of which is equally likely to come up. Using our counting rule P(any one given number) = 1/38.

Let's now demonstrate the meaning of union and intersection with the example of drawing a card from a deck. Let A be the event that an ace is drawn and ♥ the event that a heart is drawn. The sample space is shown in Figure 2–6. Note that the event A ∩ ♥ is the event that the card drawn is both an ace and a heart (i.e., the ace of hearts). The event A ∪ ♥ is the event that the card drawn is either an ace or a heart or both.

PROBLEMS

2–1.   What are the two main types of probability?

2–2.   What is an event? What is the union of two events? What is the intersection of two events?

2–3.   Define a sample space.

2–4.   Define the probability of an event.

2–5.   Let G be the event that a girl is born. Let F be the event that a baby over 5 pounds is born. Characterize the union and the intersection of the two events.

2–6.   Consider the event that a player scores a point in a game against team A and the event that the same player scores a point in a game against team B. What is the union of the two events? What is the intersection of the two events?

2–7.   A die is tossed twice and the two outcomes are noted. Draw the Venn diagram of the sample space and indicate the event "the second toss is greater than the first." Calculate the probability of the event.

2–8.   Ford Motor Company advertises its cars on radio and on television. The company is interested in assessing the probability that a randomly chosen person is exposed to at least one of these two modes of advertising. If we define event R as the event that a randomly chosen person was exposed to a radio advertisement and event T as the event that the person was exposed to a television commercial, define R ∪ T and R ∩ T in this context.

2–9.   A brokerage firm deals in stocks and bonds. An analyst for the firm is interested in assessing the probability that a person who inquires about the firm will eventually purchase stock (event S) or bonds (event B). Define the union and the intersection of these two events.

2–10.   The European version of roulette is different from the U.S. version in that the European roulette wheel doesn't have 00. How does this change the probability of winning when you bet on a single number? European casinos charge a small admission fee, which is not the case in U.S. casinos. Does this make sense to you, based on your answer to the earlier question?

2–11.   A fair coin is tossed twice and the two outcomes are noted. Consider the question: What is the probability that both outcomes are heads? Given below are three versions of the sample space leading to three different answers. Only one of them is correct. Which answer is correct? Explain why it is correct and why the others are incorrect.

• Both heads
• Not both heads

Answer: 1/2

• No heads
• 1 head
• 2 heads

Answer: 1/3

• H H
• H T
• T H
• T T

Answer: 1/4

2–12.   An e-commerce website gets 2,385 visitors on a particular day. Among these, 1,790 visitors explore the products by looking at more pages at the site. Among these 1,790 visitors who explore the products, 387 make a purchase.

a. If a visitor is chosen at random from all those who visited the site, what is the probability that the visitor explored the products?

b. If a visitor is chosen at random from all those who visited the site, what is the probability that the visitor made a purchase?

c. If a visitor is chosen at random from all those who explored the products, what is the probability that the visitor made a purchase?

d. Which of the preceding three probabilities is relevant to the design of the home page that leads to product pages?

2–13. You draw a card at random from a standard deck of 52 cards. Neither you nor anyone else looks at the card you picked. You keep it face down. Your friend then picks a card at random from the remaining 51 cards.

a. What is the probability that your card is the ace of spades?

b. What is the probability that your friend's card is the ace of spades? (*Hint:* Construct the sample space for what your friend's card can be.)

c. You turn over your card and it is the 10 of diamonds. Now what is the probability that your friend's card is the ace of spades?

2–3   Basic Rules for Probability

We have explored probability on a somewhat intuitive level and have seen rules that help us evaluate probabilities in special cases when we have a known sample space with equally likely basic outcomes. We will now look at some general probability rules that hold regardless of the particular situation or kind of probability (objective or subjective). First, let us give a general definition of probability.

> Probability is a measure of uncertainty. The **probability** of event A is a numerical measure of the likelihood of the event's occurring.

The Range of Values

Probability obeys certain rules. The first rule sets the range of values that the probability measure may take.

For any event A, the probability $P(A)$ satisfies
$$0 \leq P(A) \leq 1 \qquad (2\text{–}2)$$

When an event cannot occur, its probability is zero. The probability of the empty set is zero: $P(\emptyset) = 0$. In a deck where half the cards are red and half are black, the probability of drawing a green card is zero because the set corresponding to that event is the empty set: There are no green cards.

Events that are certain to occur have probability 1.00. The probability of the entire sample space S is equal to 1.00: $P(S) = 1.00$. If we draw a card out of a deck, 1 of the 52 cards in the deck will certainly be drawn, and so the probability of the sample space, the set of all 52 cards, is equal to 1.00.

FIGURE 2–7   **Interpretation of a Probability**

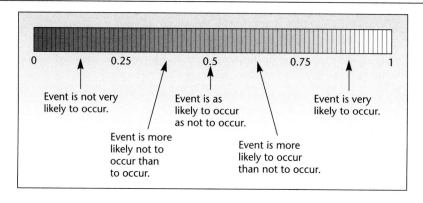

Within the range of values 0 to 1, the greater the probability, the more confidence we have in the occurrence of the event in question. A probability of 0.95 implies a very high confidence in the occurrence of the event. A probability of 0.80 implies a high confidence. When the probability is 0.5, the event is as likely to occur as it is not to occur. When the probability is 0.2, the event is not very likely to occur. When we assign a probability of 0.05, we believe the event is unlikely to occur, and so on. Figure 2–7 is an informal aid in interpreting probability.

Note that probability is a measure that goes from 0 to 1. In everyday conversation we often describe probability in less formal terms. For example, people sometimes talk about **odds.** If the odds are 1 to 1, the probability is 1/2; if the odds are 1 to 2, the probability is 1/3; and so on. Also, people sometimes say, "The probability is 80 percent." Mathematically, this probability is 0.80.

The Rule of Complements

Our second rule for probability defines the probability of the complement of an event in terms of the probability of the original event. Recall that the complement of set A is denoted by $\overline{A}$.

Probability of the complement:

$$P(\overline{A}) = 1 - P(A) \tag{2–3}$$

As a simple example, if the probability of rain tomorrow is 0.3, then the probability of no rain tomorrow must be $1 - 0.3 = 0.7$. If the probability of drawing an ace is 4/52, then the probability of the drawn card's not being an ace is $1 - 4/52 = 48/52$.

The Rule of Unions.   We now state a very important rule, the **rule of unions.** The rule of unions allows us to write the probability of the union of two events in terms of the probabilities of the two events and the probability of their intersection:[3]

---

[3]The rule can be extended to more than two events. In the case of three events, we have $P(A \cup B \cup C) = P(A) + P(B) + P(C) - P(A \cap B) - P(A \cap C) - P(B \cap C) + P(A \cap B \cap C)$. With more events, this becomes even more complicated.

The rule of unions:

$$P(A \cup B) = P(A) + P(B) - P(A \cap B) \qquad (2\text{--}4)$$

[The probability of the intersection of two events $P(A \cap B)$ is called their **joint probability**.] The meaning of this rule is very simple and intuitive: When we add the probabilities of A and B, we are measuring, or counting, the probability of their intersection *twice*—once when measuring the relative size of A within the sample space and once when doing this with B. Since the relative size, or probability, of the intersection of the two sets is counted twice, we subtract it once so that we are left with the true probability of the union of the two events (refer to Figure 2–6). Note that instead of finding the probability of A $\cup$ B by direct counting, we can use the rule of unions: We know that the probability of an ace is 4/52, the probability of a heart is 13/52, and the probability of their intersection—the drawn card being the ace of hearts—is 1/52. Thus, $P(A \cup \heartsuit)$ = 4/52 + 13/52 − 1/52 = 16/52, which is exactly what we find from direct counting.

The rule of unions is especially useful when we do not have the sample space for the union of events but do have the separate probabilities. For example, suppose your chance of being offered a certain job is 0.4, your probability of getting another job is 0.5, and your probability of being offered both jobs (i.e., the intersection) is 0.3. By the rule of unions, your probability of being offered at least one of the two jobs (their union) is 0.4 + 0.5 − 0.3 = 0.6.

Mutually Exclusive Events

When the sets corresponding to two events are disjoint (i.e., have no intersection), the two events are called **mutually exclusive** (see Figure 2–4). For mutually exclusive events, the probability of the intersection of the events is zero. This is so because the intersection of the events is the empty set, and we know that the probability of the empty set $\varnothing$ is zero.

For mutually exclusive events A and B:

$$P(A \cap B) = 0 \qquad (2\text{--}5)$$

This fact gives us a special rule for unions of mutually exclusive events. Since the probability of the intersection of the two events is zero, there is no need to subtract $P(A \cap B)$ when the probability of the union of the two events is computed. Therefore,

For mutually exclusive events A and B:

$$P(A \cup B) = P(A) + P(B) \qquad (2\text{--}6)$$

This is not really a new rule since we can always use the rule of unions for the union of two events: If the events happen to be mutually exclusive, we subtract zero as the probability of the intersection.

To continue our cards example, what is the probability of drawing either a heart or a club? We have $P(\heartsuit \cup \clubsuit) = P(\heartsuit) + P(\clubsuit) = 13/52 + 13/52 = 26/52 = 1/2$. We need not subtract the probability of an intersection, since no card is both a club and a heart.

PROBLEMS

2–14. According to a report on CNN Business News in April 1995, the probability of being murdered (in the United States) in 1 year is 9 in 100,000. How might such a probability have been obtained?

2–15. Assign a reasonable numerical probability to the statement "Rain is very likely tonight."

2–16. How likely is an event that has a 0.65 probability? Describe the probability in words.

2–17. If a team has an 80% chance of winning a game, describe its chances in words.

2–18. ShopperTrak is a hidden electric eye designed to count the number of shoppers entering a store. When two shoppers enter a store together, one walking in front of the other, the following probabilities apply: There is a 0.98 probability that the first shopper will be detected, a 0.94 probability that the second shopper will be detected, and a 0.93 probability that both of them will be detected by the device. What is the probability that the device will detect at least one of two shoppers entering together?

2–19. A machine produces components for use in cellular phones. At any given time, the machine may be in one, and only one, of three states: operational, out of control, or down. From experience with this machine, a quality control engineer knows that the probability that the machine is out of control at any moment is 0.02, and the probability that it is down is 0.015.

    a. What is the relationship between the two events "machine is out of control" and "machine is down"?

    b. When the machine is either out of control or down, a repair person must be called. What is the probability that a repair person must be called right now?

    c. Unless the machine is down, it can be used to produce a single item. What is the probability that the machine can be used to produce a single component right now? What is the relationship between this event and the event "machine is down"?

2–20. Following are age and sex data for 20 midlevel managers at a service company: 34 F, 49 M, 27 M, 63 F, 33 F, 29 F, 45 M, 46 M, 30 F, 39 M, 42 M, 30 F, 48 M, 35 F, 32 F, 37 F, 48 F, 50 M, 48 F, 61 F. A manager must be chosen at random to serve on a companywide committee that deals with personnel problems. What is the probability that the chosen manager will be either a woman or over 50 years old or both? Solve both directly from the data and by using the law of unions. What is the probability that the chosen manager will be under 30?

2–21. Suppose that 25% of the population in a given area is exposed to a television commercial for Ford automobiles, and 34% is exposed to Ford's radio advertisements. Also, it is known that 10% of the population is exposed to both means of advertising. If a person is randomly chosen out of the entire population in this area, what is the probability that he or she was exposed to at least one of the two modes of advertising?

2–22. Suppose it is known that 85% of the people who inquire about investment opportunities at a brokerage house end up purchasing stock, and 33% end up purchasing bonds. It is also known that 28% of the inquirers end up getting a portfolio with both stocks and bonds. If a person is just making an inquiry, what is the probability that she or he will get stock or bonds or both (i.e., open any portfolio)?

2–23. A firm has 550 employees; 380 of them have had at least some college education, and 412 of the employees underwent a vocational training program. Furthermore, 357 employees both are college-educated and have had the vocational training. If an employee is chosen at random, what is the probability that he or she is college-educated or has had the training or both?

2–24. A cross-tabulation of all the students at a small liberal arts college by gender and by school year is below. A student is selected at random. What is the probability that the selected student is

        a. a sophomore or male? (Express your answer as fraction.)

        b. a female senior?

        c. a female senior or a male junior?

        d. a sophomore given that the student is female?

        e. male given that the student is a junior?

        f. male and not a junior?

        g. male given that the student is not a junior?

	Freshman	Sophomore	Junior	Senior
Male	310	291	189	172
Female	338	306	238	230

2–25. As part of a student project for the 1994 Science Fair in Orange, Massachusetts, 28 horses were made to listen to Mozart and heavy-metal music. The results were as follows: 11 of the 28 horses exhibited some head movements when Mozart was played; 8 exhibited some head movements when the heavy metal was played; and 5 moved their heads when both were played. If a horse is chosen at random, what is the probability the horse exhibited head movements to Mozart or to heavy metal or to both?

2–4   Conditional Probability

As a measure of uncertainty, probability depends on information. Thus, the probability you would give the event "Xerox stock price will go up tomorrow" depends on what you know about the company and its performance; the probability is *conditional* upon your information set. If you know much about the company, you may assign a different probability to the event than if you know little about the company. We may define the probability of event A *conditional* upon the occurrence of event B. In this example, event A may be the event that the stock will go up tomorrow, and event B may be a favorable quarterly report.

> The **conditional probability** of event A given the occurrence of event B is
>
> $$P(A \mid B) = \frac{P(A \cap B)}{P(B)} \qquad (2\text{--}7)$$
>
> assuming $P(B) \neq 0$.

The vertical line in $P(A \mid B)$ is read *given,* or *conditional upon.* The probability of event A given the occurrence of event B is defined as the probability of the intersection of A and B, divided by the probability of event B.

EXAMPLE 2–2   As part of a drive to modernize the economy, the government of an eastern European country is pushing for starting 100 new projects in computer development and telecommunications. Two U.S. giants, IBM and AT&T, have signed contracts for these projects: 40 projects for IBM and 60 for AT&T. Of the IBM projects, 30 are in the computer area and 10 are in telecommunications; of the AT&T projects, 40 are in telecommunications and 20 are in computer areas. Given that a randomly chosen project is in telecommunications, what is the probability that it is undertaken by IBM?

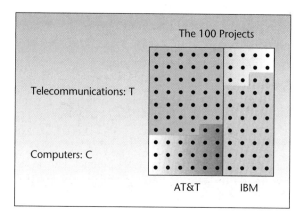

Solution

$$P(\text{IBM} \mid \text{T}) = \frac{P(\text{IBM} \cap \text{T})}{P(\text{T})} = \frac{10/100}{50/100} = 0.2$$

But we see this directly from the fact that there are 50 telecommunications projects and 10 of them are by IBM. This confirms the definition of conditional probability in an intuitive sense.

When two events and their complements are of interest, it may be convenient to arrange the information in a **contingency table.** In Example 2–2 the table would be set up as follows:

	AT&T	IBM	Total
Telecommunications	40	10	50
Computers	20	30	50
Total	60	40	100

Contingency tables help us visualize information and solve problems. There are two other useful forms of the definition of conditional probability (equation 2–7).

Variation of the conditional probability formula:

$$P(A \cap B) = P(A \mid B)P(B)$$

and

$$P(A \cup B) = P(B \mid A)P(A) \qquad (2\text{–}8)$$

These are illustrated in Example 2–3.

EXAMPLE 2–3

A consulting firm is bidding for two jobs, one with each of two large multinational corporations. The company executives estimate that the probability of obtaining the consulting job with firm A, event A, is 0.45. The executives also feel that if the company should get the job with firm A, then there is a 0.90 probability that firm B will also give the company the consulting job. What are the company's chances of getting *both* jobs?

Solution

We are given $P(A) = 0.45$. We also know that $P(B \mid A) = 0.90$, and we are looking for $P(A \cap B)$, which is the probability that both A and B will occur. From the equation we have $P(A \cap B) = P(B \mid A)P(A) = 0.90 \times 0.45 = 0.405$.

EXAMPLE 2–4

Twenty-one percent of the executives in a large advertising firm are at the top salary level. It is further known that 40% of all the executives at the firm are women. Also, 6.4% of all executives are women *and* are at the top salary level. Recently, a question arose among executives at the firm as to whether there is any evidence of salary inequity. Assuming that some statistical considerations (explained in later chapters) are met, do the percentages reported above provide any evidence of salary inequity?

Solution

To solve this problem, we pose it in terms of probabilities and ask whether the probability that a randomly chosen executive will be at the top salary level is approximately equal to the probability that the executive will be at the top salary level given the executive is a woman. To answer, we need to compute the probability that the executive will be at the top level given the executive is a woman. Defining T as the event of a top salary and W as the event that an executive is a woman, we get

$$P(T \mid W) = \frac{P(T \cap W)}{P(W)} = \frac{0.064}{0.40} = 0.16$$

Since 0.16 is smaller than 0.21, we may conclude (subject to statistical considerations) that salary inequity does exist at the firm, because an executive is less likely to make a top salary if she is a woman.

Example 2–4 may incline us to think about the relations among different events. Are different events related, or are they independent of each other? In this example, we concluded that the two events, being a woman and being at the top salary level, are related in the sense that the event W made event T less likely. Section 2–5 quantifies the relations among events and defines the concept of independence.

2–26. SBC Warburg, Deutsche Morgan Grenfell, and UBS are foreign. Given that a security is foreign-underwritten, find the probability that it is by SBC Warburg (see the accompanying table).[4]

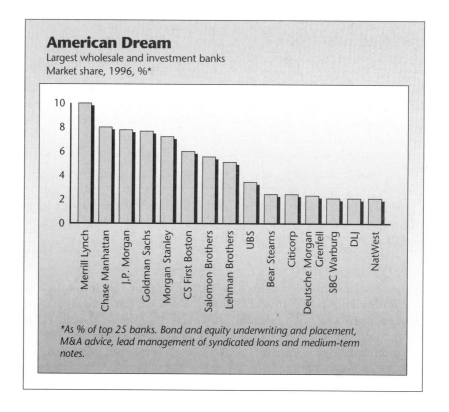

American Dream
Largest wholesale and investment banks
Market share, 1996, %*

As % of top 25 banks. Bond and equity underwriting and placement, M&A advice, lead management of syndicated loans and medium-term notes.

2–27. If a large competitor will buy a small firm, the firm's stock will rise with probability 0.85. The purchase of the company has a 0.40 probability. What is the probability that the purchase will take place and the firm's stock will rise?

2–28. A financial analyst believes that if interest rates decrease in a given period, then the probability that the stock market will go up is 0.80. The analyst further believes that interest rates have a 0.40 chance of decreasing during the period in question. Given the above information, what is the probability that the market will go up and interest rates will go down during the period in question?

2–29. A bank loan officer knows that 12% of the bank's mortgage holders lose their jobs and default on the loan in the course of 5 years. She also knows that 20% of the bank's mortgage holders lose their jobs during this period. Given that one of her mortgage holders just lost his job, what is the probability that he will now default on the loan?

2–30. An express delivery service promises overnight delivery of all packages checked in before 5 P.M. The delivery service is not perfect, however, and sometimes delays do occur. Management knows that if delays occur in the evening flight to a major city from which distribution is made, then a package will not arrive on time with probability 0.25. It is also known that 10% of the evening flights to the major city are delayed. What percentage of the packages arrive late? (Assume that all packages are sent out on the evening flight to the major city and that all packages arrive on time if the evening flight is not delayed.)

2–31. The following table gives numbers of claims at a large insurance company by kind and by geographic region.

	East	South	Midwest	West
Hospitalization	75	128	29	52
Physician's visit	233	514	104	251
Outpatient treatment	100	326	65	99

Compute column totals and row totals. What do they mean?

    a. If a bill is chosen at random, what is the probability that it is from the Midwest?

    b. What is the probability that a randomly chosen bill is from the East?

    c. What is the probability that a randomly chosen bill is either from the Midwest or from the South? What is the relation between these two events?

    d. What is the probability that a randomly chosen bill is for hospitalization?

    e. Given that a bill is for hospitalization, what is the probability that it is from the South?

    f. Given that a bill is from the East, what is the probability that it is for a physician's visit?

    g. Given that a bill is for outpatient treatment, what is the probability that it is from the West?

    h. What is the probability that a randomly chosen bill is either from the East or for outpatient treatment (or both)?

    i. What is the probability that a randomly selected bill is either for hospitalization or from the South (or both)?

2–32. One of the greatest problems in marketing research and other survey fields is the problem of nonresponse to surveys. In home interviews the problem arises when the respondent is not home at the time of the visit or, sometimes, simply refuses to answer questions. A market researcher believes that a respondent will answer all questions with probability 0.94 if found at home. He further believes that the probability that a given person will be found at home is 0.65. Given this information, what percentage of the interviews will be successfully completed?

2–33. An investment analyst collects data on stocks and notes whether or not dividends were paid and whether or not the stocks increased in price over a given period. Data are presented in the following table (top of page 88).

	Price Increase	No Price Increase	Total
Dividends paid	34	78	112
No dividends paid	85	49	134
Total	119	127	246

a. If a stock is selected at random out of the analyst's list of 246 stocks, what is the probability that it increased in price?

b. If a stock is selected at random, what is the probability that it paid dividends?

c. If a stock is randomly selected, what is the probability that it both increased in price and paid dividends?

d. What is the probability that a randomly selected stock neither paid dividends nor increased in price?

e. Given that a stock increased in price, what is the probability that it also paid dividends?

f. If a stock is known not to have paid dividends, what is the probability that it increased in price?

g. What is the probability that a randomly selected stock was worth holding during the period in question; that is, what is the probability that it increased in price or paid dividends or did both?

2–34. A recent cover article in *Business Week* dealt with the salaries of top executives at large corporations. The following table is compiled from data given in four tables in the article and lists the number of firms in the study where the top executive officer made over $1 million per year. The table also lists firms according to whether or not shareholder return was positive during the period in question.

	Top Executive Made More than $1 Million	Top Executive Made Less than $1 Million	Total
Shareholders made money	1	6	7
Shareholders lost money	2	1	3
Total	3	7	10

a. If a firm is randomly chosen from the list of 10 firms studied, what is the probability that its top executive made over $1 million per year?

b. If a firm is randomly chosen from the list, what is the probability that its shareholders lost money during the period studied?

c. Given that one of the firms in this group had negative shareholder return, what is the probability that its top executive made over $1 million?

d. Given that a firm's top executive made over $1 million, what is the probability that the firm's shareholder return was positive?

2–35. At the Paris airshow, the following aircraft sales results were announced:[5]

---

[5]From B. James, "As Air Show Ends, Paris Leaves Privatization Route Uncharted," *International Herald Tribune*, Business/Finance section, June 23, 1997, p. 11.

	Boeing	**Airbus**
Finnair		(A320) 12
TAM (Brazil)	(737-700) 8	(A330) 5
Northwest Airlines	(737-700) 12	(A319) 8
British Airways	(777) 5	

Given that a sale was by Boeing, find the probability that it was to British Airways; given the sale was to TAM, find the probability that the sale was by Airbus.

2–5   Independence of Events

In Example 2–4 we concluded that the probability that an executive made a top salary was lower when the executive was a woman, and we concluded that the two events T and W were *not* independent. We now give a formal definition of statistical independence of events.

Two events A and B are said to be *independent* of each other if and only if the following three conditions hold:

Conditions for the **independence of two events** A and B:

$$P(A \mid B) = P(A)$$
$$P(B \mid A) = P(B) \qquad\qquad (2\text{–}9)$$

and, most useful:

$$P(A \cap B) = P(A)P(B) \qquad\qquad (2\text{–}10)$$

The first two equations have a clear, intuitive appeal. The top equation says that when A and B are independent of each other, then the probability of A stays the same even when we know that B has occurred—it is a simple way of saying that knowledge of B tells us nothing about A when the two events are independent. Similarly, when A and B are independent, then knowledge that A has occurred gives us absolutely no information about B and its likelihood of occurring.

The third equation, however, is the most useful in applications. It tells us that when A and B are independent (and only when they are independent), we can obtain the probability of the joint occurrence of A and B (i.e., the probability of their intersection) simply by multiplying the two separate probabilities. This rule is thus called the **product rule** for independent events. (The rule is easily derived from the first rule, using the definition of conditional probability.)

As an example of independent events, consider the following: Suppose I roll a single die. What is the probability that the number 6 will turn up? The answer is 1/6. Now suppose that I told you that I just tossed a coin and it turned up heads. What is now the probability that the die will show the number 6? The answer is unchanged, 1/6, because events of the die and the coin are independent of each other. We see that $P(6 \mid H) = P(6)$, which is the first rule above.

In Example 2–2, we found that the probability that a project belongs to IBM given that it is in telecommunications is 0.2. We also knew that the probability that a project belongs to IBM was 0.4. Since these two numbers are not equal, the two events IBM and telecommunications are not independent.

When two events are not independent, neither are their complements. Therefore, AT&T and computers are not independent events (and neither are the other two possibilities).

EXAMPLE 2–5 The probability that a consumer will be exposed to an advertisement for a certain product by seeing a commercial on television is 0.04. The probability that the consumer will be exposed to the product by seeing an advertisement on a billboard is 0.06. The two events, being exposed to the commercial and being exposed to the billboard ad, are assumed to be independent. (*a*) What is the probability that the consumer will be exposed to both advertisements? (*b*) What is the probability that he or she will be exposed to at least one of the ads?

Solution (*a*) Since the two events are independent, the probability of the intersection of the two (i.e., being exposed to *both* ads) is $P(A \cap B) = P(A)P(B) = 0.04 \times 0.06 = 0.0024$. (*b*) We note that being exposed to at least one of the advertisements is, by definition, the union of the two events, and so the rule for union applies. The probability of the intersection was computed above, and we have $P(A \cup B) = P(A) + P(B) - P(A \cap B) = 0.04 + 0.06 - 0.0024 = 0.0976$. The computation of such probabilities is important in advertising research. Probabilities are meaningful also as proportions of the population exposed to different modes of advertising, and are thus important in the evaluation of advertising efforts.

Product Rules for Independent Events

The rules for the union and the intersection of two independent events extend nicely to sequences of more than two events. These rules are very useful in **random sampling.**

Much of statistics involves random sampling from some population. When we sample randomly from a large population, or when we sample randomly with replacement from a population of any size, the elements are independent of one another. For example, suppose that we have an urn containing 10 balls, 3 of them red and the rest blue. We randomly sample one ball, note that it is red, and return it to the urn (this is sampling *with* replacement). What is the probability that a second ball we choose at random will be red? The answer is still 3/10 because the second drawing does not "remember" that the first ball was red. Sampling with replacement in this way ensures independence of the elements. The same holds for random sampling without replacement (i.e., without returning each element to the population before the next draw) *if* the population is relatively large in comparison with the size of the sample. Unless otherwise specified, we will assume random sampling from a large population.

Random sampling from a large population implies independence.

Intersection Rule. The probability of the intersection of several independent events is just the product of the separate probabilities.

The rate of defects in corks of wine bottles is very high, 75%. Assuming independence, if four bottles are opened, what is the probability that all four corks are defective? Using this rule: P (all 4 are defective) $= P$ (first cork is defective) $\times P$ (second cork is defective) $\times P$ (third cork is defective) $\times P$ (fourth cork is defective) $= 0.75 \times 0.75 \times 0.75 \times 0.75 = 0.316$.

If these four bottles were randomly selected, then we would not have to specify independence—a random sample always implies independence.

Union Rule. The probability of the union of several independent events—$A_1, A_2, ..., A_n$—is given by the following equation:

$$P(A_1 \cup A_2 \cup \cdots \cup A_n) = 1 - P(\overline{A}_1)P(\overline{A}_2) \cdots P(\overline{A}_n) \qquad (2–11)$$

The union of several events is the event that at least one of the events happens. In the example of the wine corks, suppose we want to find the probability that at least one of the four corks is defective. We compute this probability as follows: P (at least one is defective) $= 1 - P$ (none are defective) $= 1 - 0.25 \times 0.25 \times 0.25 \times 0.25 = 0.99609$.

Read the accompanying article. Three women (assumed a random sample) in a developing country are pregnant. What is the probability that at least one will die?

EXAMPLE 2–6

Poor Nations' Mothers at Serious Health Risk

In the industrialized world, a woman's odds of dying from problems related to pregnancy are 1 in 1,687. But in the developing world the figure is 1 in 51. The World Bank also says that each year 7 million newborns die within a week of birth because of maternal health problems. The bank and the United Nations are in the midst of an initiative to cut maternal illnesses and deaths.

Edward Epstein, "Poor Nations' Mothers at Serious Health Risk," *World Insider*, *San Francisco Chronicle*, August 10, 1993, p. A9. © 1993 San Francisco Chronicle. Reprinted by permission.

P (at least 1 will die) $= 1 - P$ (all 3 will survive) $= 1 - (50/51)^3 = 0.0577$ *Solution*

EXAMPLE 2–7   A marketing research firm is interested in interviewing a consumer who fits certain qualifications, for example, use of a certain product. It is known that 10% of the public in a certain area use the product and would thus qualify to be interviewed. The company selects a random sample of 10 people from the population as a whole. What is the probability that at least 1 of these 10 people qualifies to be interviewed?

Solution   First, we note that if a sample is drawn at random, then the event that any one of the items in the sample fits the qualifications is independent of the other items in the sample. This is an important property in statistics. Let Q_i, where $i = 1, 2, \ldots, 10$, be the event that person i qualifies. Then the probability that at least 1 of the 10 people will qualify is the probability of the union of the 10 events Q_i ($i = 1, \ldots, 10$). We are thus looking for $P(Q_1 \cup Q_2 \cup \cdots \cup Q_{10})$.

Now, since 10% of the people qualify, the probability that person i does not qualify, or $P(\overline{Q}_i)$, is equal to 0.90 for each $i = 1, \ldots, 10$. Therefore, the required probability is equal to $1 - (0.9)(0.9) \cdots (0.9)$ (10 times), or $1 - (0.9)^{10}$. This is equal to 0.6513.

Be sure that you understand the difference between *independent* events and *mutually exclusive* events. Although these two concepts are very different, they often cause some confusion when introduced. When two events are mutually exclusive, they are *not* independent. In fact, they are dependent events in the sense that if one happens, the other one cannot happen. The probability of the intersection of two mutually exclusive events is equal to zero. The probability of the intersection of two independent events is *not* zero; it is equal to the product of the probabilities of the separate events.

PROBLEMS

2–36.   See the table in problem 2–26. If a random sample of six firms' securities is selected, find the probability that at least one is foreign-underwritten. SBC Warburg and Deutsche Morgan Grenfell and UBS are foreign (their market share = 10%, or 4% + 3% + 3%).[6]

2–37.   The chancellor of a state university is applying for a new position. At a certain point in his application process, he is being considered by seven universities. At three of the seven he is a finalist, which means that (at each of the three universities) he is in the final group of three applicants, one of which will be chosen for the position. At two of the seven universities he is a semifinalist, that is, one of six candidates (in each of the two universities). In two universities he is at an early stage of his application and believes there is a pool of about 20 candidates for each of the two positions. Assuming that there is no exchange of information, or influence, across universities as to their hiring decisions, and that the chancellor is as likely to be chosen as any other applicant, what is the chancellor's probability of getting at least one job offer?

2–38.   A package of documents needs to be sent to a given destination, and it is important that it arrive within one day. To maximize the chances of on-time delivery, three copies of the documents are sent via three different delivery services. Service A is known to have a 90% on-time delivery record, service B has an 88% on-time delivery record, and service C has a 91% on-time delivery record. What is the probability that at least one copy of the documents will arrive at its destination on time?

[6]From "Out of Their League?" *The Economist,* June 21, 1997, pp. 71–72. © 1997 The Economist Newspaper Group, Inc. Reprinted with permission. Further reproduction prohibited. www.economist.com.

2–39. The projected probability of increase in online holiday sales from 2001 to 2002 is 95% in the United States, 90% in Australia, and 85% in Japan. Assume these probabilities are independent. What is the probability that holiday sales will increase in all three countries from 2001 to 2002?

2–40. An electronic device is made up of two components A and B such that the device would work satisfactorily as long as at least one of the components works. The probability of failure of component A is 0.02 and that of B is 0.1 in some fixed period of time. If the components work independently, find the probability that the device will work satisfactorily during the period.

2–41. A recent survey conducted by Towers Perrin and published in the *Financial Times* showed that among 460 organizations in 13 European countries, 93% have bonus plans, 55% have cafeteria-style benefits, and 70% employ home-based workers. If the types of benefits are independent, what is the probability that an organization selected at random will have at least one of the three types of benefits?

2–42. It is reported that 31% of the population in a given area reads *The Wall Street Journal,* 20% read the *Financial Times,* and 9% read both. If a person is randomly selected from the area, what is the probability that he or she reads at least one of the newspapers?

2–43. Are the events "plane is bought by Northwest Airlines" and "plane is made by Boeing" independent of each other for the results given in problem 2–35?

2–44. In problem 2–31, are the events "hospitalization" and "the claim being from the Midwest" independent of each other?

2–45. In problem 2–33, are "dividends paid" and "price increase" independent events?

2–46. In problem 2–34, are the events "top executive made more than $1 million" and "shareholders lost money" independent of each other? If this is true for all firms, how would you interpret your finding?

2–47. The accompanying table shows the incidence of malaria and two other similar illnesses. If a person lives in an area affected by all three diseases, what is the probability that he or she will develop at least one of the three illnesses? (Assume that contracting one disease is an event independent from contracting any other disease.)

	Cases	Number at Risk (Millions)
Malaria	110 million per year	2,100
Schistosomiasis	200 million	600
Sleeping sickness	25,000 per year	50

2–48. A device has three components and works as long as at least one of the components is functional. The reliabilities of the components are 0.96, 0.91, and 0.80. What is the probability that the device will work when needed?

2–49. See the table in problem 2–26. Investment banks underwrite securities sold on the markets. According to *The Economist,* Merrill Lynch's market share is 10%.[7] If the Securities and Exchange Commission (SEC) selects a random sample of 10 securities issued by investment banks for scrutiny, what is the probability that at least 6 of them are underwritten by Merrill Lynch? Given this probability, would Merrill Lynch

[7] "Out of Their League," *The Economist,* June 21, 1997, p. 76.

have a case against the SEC, when 6 out of 10 scrutinized securities are underwritten by the firm, to claim that the choice targets Merrill Lynch and was not random? Explain. (Why do you use six or more?)

2–50.   The probabilities that three drivers will be able to drive home safely after drinking are 0.5, 0.25, and 0.2, respectively. If they set out to drive home after drinking at a party, what is the probability that at least one driver drives home safely?

2–51.   When one is randomly sampling four items from a population, what is the probability that all four elements will come from the top quartile of the population distribution? What is the probability that at least one of the four elements will come from the bottom quartile of the distribution?

2–6   Combinatorial Concepts

In this section we briefly discuss a few combinatorial concepts and give some formulas useful in the analysis. The interested reader may find more on combinatorial rules and their applications in the classic book by W. Feller or in other books on probability.[8]

> If there are n events and event i can occur in N_i possible ways, then the number of ways in which the sequence of n events may occur is $N_1 N_2 \cdots N_n$.

Suppose that a bank has two branches, each branch has two departments, and each department has three employees. Then there are $(2)(2)(3)$ choices of employees, and the probability that a particular one will be randomly selected is $1/(2)(2)(3) = 1/12$.

We may view the choice as done sequentially: First a branch is randomly chosen, then a department within the branch, and then the employee within the department. This is demonstrated in the tree diagram in Figure 2–8.

FIGURE 2–8   Tree Diagram for Computing the Total Number of Employees by Multiplication

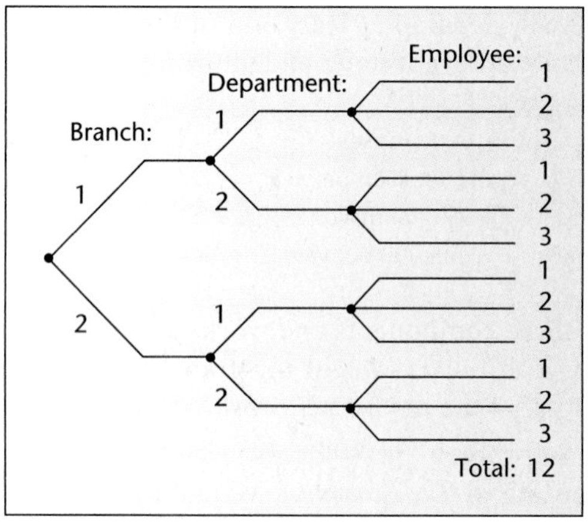

[8]William Feller, *An Introduction to Probability Theory and Its Applications,* vol. I, 3d ed. (New York: John Wiley & Sons, 1968).

For any positive integer *n*, we define *n* **factorial** as

$$n(n - 1)(n - 2) \cdots 1$$

We denote *n* factorial by *n*!. The number *n*! is the number of ways in which *n* objects can be ordered. By definition, 0! = 1.

For example, 6! is the number of possible arrangements of six objects. We have 6! = (6)(5)(4)(3)(2)(1) = 720. Suppose that six applications arrive at a center on the same day, all written at different times. What is the probability that they will be read in the order in which they were written? Since there are 720 ways to order six applications, the probability of a particular order (the order in which the applications were written) is 1/720.

Permutations are the possible ordered selections of *r* objects out of a total of *n* objects. The number of permutations of *n* objects taken *r* at a time is denoted *n***P***r*.

$$n\mathbf{P}r = \frac{n!}{(n - r)!} \tag{2-12}$$

Suppose that 4 people are to be randomly chosen out of 10 people who agreed to be interviewed in a market survey. The four people are to be assigned to four interviewers. How many possibilities are there? The first interviewer has 10 choices, the second 9 choices, the third 8, and the fourth 7. Thus, there are (10)(9)(8)(7) = 5,040 selections. You can see that this is equal to $n(n - 1)(n - 2) \cdots (n - r + 1)$, which is equal to $n!/(n - r)!$. If choices are made randomly, the probability of any predetermined assignment of 4 people out of a group of 10 is 1/5,040.

Combinations are the possible selections of *r* items from a group of *n* items regardless of the order of selection. The number of combinations is denoted by $\binom{n}{r}$ and is read *n choose r*. An alternative notation is *n***C***r*. We define the number of combinations of *r* out of *n* elements as

$$\binom{n}{r} = \frac{n!}{r!(n - r)!} \tag{2-13}$$

This is the most important of the combinatorial rules given in this chapter and is the only one we will use extensively. This rule is basic to the formula of the binomial distribution presented in the next chapter and will find use also in other chapters.

Suppose that 3 out of the 10 members of the board of directors of a large corporation are to be randomly selected to serve on a particular task committee. How many possible selections are there? Using equation 2–13, we find that the number of combinations is $\binom{10}{3} = 10!/(3!7!) = 120$. If the committee is chosen in a truly random fashion, what is the probability that the three committee members chosen will be the three senior board members? This is 1 combination out of a total of 120, so the answer is 1/120 = 0.00833.

FIGURE 2–9   **Template for Calculating Permutations and Combinations [Permutation & Combination.xls]**

	A	B	C	D	E	F	G	H	I
1	**Permutation and Combination**								
2									
3		Permutation				Combination			
4		*n*	*r*	*nPr*		*n*	*r*	*nCr*	
5		10	3	720		10	3	120	
6									
7									
8									
9									
10									
11									
12									
13									
14									
15									
16									

Since permutation and combination calculations can get tedious, you may use the template shown in Figure 2–9. By unprotecting the sheet, you can use the rest of the spreadsheet for additional calculations such as probabilities.

EXAMPLE 2–8   A certain university held a meeting of administrators and faculty members to discuss some important issues of concern to both groups. Out of eight members, two were faculty, and both were missing from the meeting. If two members are absent, what is the probability that they should be the two faculty members?

Solution   By definition, there are $\binom{8}{2}$ ways of selecting two people out of a total of eight people, disregarding the order of selection. Only one of these ways corresponds to the pair's being the two faculty members. Hence, the probability is $1/\binom{8}{2} = 1/[8!/(2!6!)] = 1/28 = 0.0357$. This assumes randomness.

PROBLEMS

2–52.   A company has four departments: manufacturing, distribution, marketing, and management. The number of people in each department is 55, 30, 21, and 13, respectively. Each department is expected to send one representative to a meeting with the company president. How many possible sets of representatives are there?

2–53.   Nine sealed bids for oil drilling leases arrive at a regulatory agency in the morning mail. In how many different orders can the nine bids be opened?

2–54. Fifteen locations in a given area are believed likely to have oil. An oil company can only afford to drill at eight sites, sequentially chosen. How many possibilities are there, in order of selection?

2–55. A committee is evaluating six equally qualified candidates for a job. Only three of the six will be invited for an interview; among the chosen three, the order of invitation is of importance because the first candidate will have the best chance of being accepted, the second will be made an offer only if the committee rejects the first, and the third will be made an offer only if the committee should reject both the first and the second. How many possible ordered choices of three out of six candidates are there?

2–56. In the analysis of variance (discussed in Chapter 9) we compare several population means to see which is largest. After the primary analysis, pairwise comparisons are made. If we want to compare seven populations, each with all the others, how many pairs are there? (We are looking for the number of choices of seven items taken two at a time, regardless of order.)

2–57. In a shipment of 14 computer parts, 3 are faulty and the remaining 11 are in working order. Three elements are randomly chosen out of the shipment. What is the probability that all three faulty elements will be the ones chosen?

2–58. Megabucks is a lottery game played in Massachusetts with the following rules. A random drawing of 6 numbers out of all 36 numbers from 1 to 36 is made every Wednesday and every Saturday. The game costs $1 to play, and to win a person must have the correct six numbers drawn, regardless of their order. (The numbers are sequentially drawn from a bin and are arranged from smallest to largest. When a player buys a ticket prior to the drawing, the player must also arrange his or her chosen numbers in ascending order.) The jackpot depends on the number of players and is usually worth several million dollars. What is the probability of winning the jackpot?

2–59. In Megabucks, a player who correctly chooses five out of the six winning numbers gets $400. What is the probability of winning $400?

2–7   The Law of Total Probability and Bayes' Theorem

In this section we present two useful results of probability theory. The first one, **the law of total probability,** allows us at times to evaluate probabilities of events that are difficult to obtain alone, but become easy to calculate once we *condition* on the occurrence of a related event. First we assume that the related event occurs, and then we assume it does not occur. The resulting conditional probabilities help us compute the total probability of occurrence of the event of interest.

The second rule, the famous **Bayes' theorem,** is easily derived from the law of total probability and the definition of conditional probability. The rule, discovered in 1761 by the English clergyman Thomas Bayes, has had a profound impact on the development of statistics and is responsible for the emergence of a new philosophy of science. Bayes himself is said to have been unsure of his extraordinary result, which was presented to the Royal Society by a friend in 1763—after Bayes' death.

The Law of Total Probability

Consider two events A and B. Whatever may be the relation between the two events, we can *always* say that the probability of A is equal to the probability of the intersection of A and B, plus the probability of the intersection of A and the complement of B (event $\bar{B}$).

FIGURE 2–10
Partition of Set A into Its Intersections with the Two Sets B and B̄, and the Implied Law of Total Probability

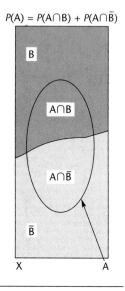

$P(A) = P(A \cap B) + P(A \cap \overline{B})$

The law of total probability:

$$P(A) = P(A \cap B) + P(A \cap \overline{B}) \qquad (2\text{–}14)$$

The sets B and B̄ form a **partition** of the sample space. A partition of a space is the division of the space into a set of events that are mutually exclusive (disjoint sets) and cover the whole space. Whatever event B may be, either B or B̄ must occur, but not both. Figure 2–10 demonstrates this situation and the law of total probability.

The law of total probability may be extended to more complex situations, where the sample space X is partitioned into more than two events. Say we partition the space into a collection of n sets $B_1, B_2, \ldots, B_n$. The law of total probability in this situation is

$$P(A) = \sum_{i=1}^{n} P(A \cap B_i) \qquad (2\text{–}15)$$

Figure 2–11 shows the partition of a sample space into the four events B_1, B_2, B_3, and B_4 and shows their intersections with set A.

We demonstrate the rule with a more specific example. Define A as the event that a picture card is drawn out of a deck of 52 cards (the picture cards are the aces, kings, queens, and jacks). Letting H, C, D, and S denote the events that the card drawn is a heart, club, diamond, or spade, respectively, we find that the probability of a picture card is $P(A) = P(A \cap H) + P(A \cap C) + P(A \cap D) + P(A \cap S) = 4/52 + 4/52 + 4/52 + 4/52 = 16/52$, which is what we know the probability of a picture card to be just by counting 16 picture cards out of a total of 52 cards in the deck. This demonstrates equation 2–15. The situation is shown in Figure 2–12. As can be seen from the figure, the event A is the set addition of the intersections of A with each of the four sets H, D, C, and S. Note that in these examples we denote the sample space X.

The law of total probability can be extended by using the definition of conditional probability. Recall that $P(A \cap B) = P(A \mid B)P(B)$ (equation 2–8) and, similarly, $P(A \cap \overline{B}) = P(A \mid \overline{B})P(\overline{B})$. Substituting these relationships into equation 2–14 gives us another form of the law of total probability. This law and its extension to a partition consisting of more than two sets are given in equations 2–16 and 2–17.

The law of total probability using conditional probabilities:

Two-set case:

$$P(A) = P(A \mid B)P(B) + P(A \mid \overline{B})P(\overline{B}) \qquad (2\text{–}16)$$

More than two sets in the partition:

$$P(A) = \sum_{i=1}^{n} P(A \mid B_i)P(B_i) \qquad (2\text{–}17)$$

where there are n sets in the partition: B_i, $i = 1, \ldots, n$.

EXAMPLE 2–9   An analyst believes the stock market has a 0.75 probability of going up in the next year if the economy should do well, and a 0.30 probability of going up if the econ-

FIGURE 2–11   The Partition of Set A into Its Intersection with Four Partition Sets

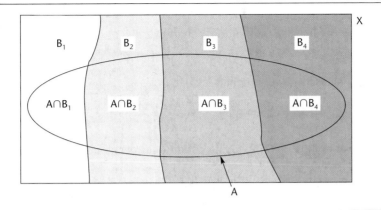

FIGURE 2–12   The Total Probability of Drawing a Picture Card as the Sum of the Probabilities of Drawing a Card in the Intersections of Picture and Suit

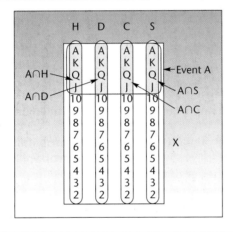

omy should not do well during the year. The analyst further believes there is a 0.80 probability that the economy will do well in the coming year. What is the probability that the stock market will go up next year (using the analyst's assessments)?

Solution

We define U as the event that the market will go up and W as the event the economy will do well. Using equation 2–16, we find $P(U) = P(U \mid W)P(W) + P(U \mid \overline{W})P(\overline{W}) = (0.75)(0.80) + (0.30)(0.20) = 0.66$.

Bayes' Theorem

We now develop the well-known Bayes' theorem. The theorem allows us to reverse the conditionality of events: we can obtain the probability of B given A from the probability of A given B (and other information).

By the definition of conditional probability, equation 2–7,

$$P(\text{B} \mid \text{A}) = \frac{P(\text{A} \cap \text{B})}{P(\text{A})} \qquad (2\text{--}18)$$

By another form of the same definition, equation 2–8,

$$P(\text{A} \cap \text{B}) = P(\text{A} \mid \text{B})P(\text{B}) \qquad (2\text{--}19)$$

Substituting equation 2–19 into equation 2–18 gives

$$P(\text{B} \mid \text{A}) = \frac{P(\text{A}|\text{B})P(\text{B})}{P(\text{A})} \qquad (2\text{--}20)$$

From the law of total probability using conditional probabilities, equation 2–16, we have

$$P(\text{A}) = P(\text{A} \mid \text{B})P(\text{B}) + P(\text{A} \mid \overline{\text{B}})P(\overline{\text{B}})$$

Substituting this expression for $P(\text{A})$ in the denominator of equation 2–20 gives us Bayes' theorem.

Bayes' theorem:

$$P(\text{B} \mid \text{A}) = \frac{P(\text{A}|\text{B})P(\text{B})}{P(\text{A}|\text{B})P(\text{B}) + P(\text{A}|\overline{\text{B}})P(\overline{\text{B}})} \qquad (2\text{--}21)$$

As we see from the theorem, the probability of B given A is obtained from the probabilities of B and $\overline{\text{B}}$ and from the conditional probabilities of A given B and A given $\overline{\text{B}}$.

The probabilities $P(\text{B})$ and $P(\overline{\text{B}})$ are called **prior probabilities** of the events B and $\overline{\text{B}}$; the probability $P(\text{B} \mid \text{A})$ is called the **posterior probability** of B. It is possible to write Bayes' theorem in terms of $\overline{\text{B}}$ and A, thus giving the posterior probability of $\overline{\text{B}}$, $P(\overline{\text{B}} \mid \text{A})$. Bayes' theorem may be viewed as a means of transforming our prior probability of an event B into a posterior probability of the event B—posterior to the known occurrence of event A.

The use of prior probabilities in conjunction with other information—often obtained from experimentation—has been questioned. The controversy arises in more involved statistical situations where Bayes' theorem is used in mixing the objective information obtained from sampling with prior information that could be subjective. We will explore this topic in greater detail in Chapter 15. We now give some examples of the use of the theorem.

EXAMPLE 2–10   Consider a test for an illness. The test has a known reliability:

1. When administered to an ill person, the test will indicate so with probability 0.92.
2. When administered to a person who is not ill, the test will erroneously give a positive result with probability 0.04.

Suppose the illness is rare and is known to affect only 0.1% of the entire population. If a person is randomly selected from the entire population and is given the test and the result is positive, what is the posterior probability (posterior to the test result) that the person is ill?

Let Z denote the event that the test result is positive and I the event that the person tested is ill. The preceding information gives us the following probabilities of events:

$$P(I) = 0.001 \qquad P(\bar{I}) = 0.999 \qquad P(Z \mid I) = 0.92 \qquad P(Z \mid \bar{I}) = 0.04$$

We are looking for the probability that the person is ill given a positive test result; that is, we need $P(I \mid Z)$. Since we have the probability with the reversed conditionality, $P(Z \mid I)$, we know that Bayes' theorem is the rule to be used here. Applying the rule, equation 2–21, to the events Z, I, and $\bar{I}$, we get

$$P(I \mid Z) = \frac{P(Z|I)P(I)}{P(Z|I)P(I) + P(Z|\bar{I})P(\bar{I})} = \frac{(0.92)(0.001)}{(0.92)(0.001) + (0.04)(0.999)}$$

$$= 0.0225$$

This result may surprise you. A test with a relatively high reliability (92% correct diagnosis when a person is ill and 96% correct identification of people who are not ill) is administered to a person, the result is positive, and yet the probability that the person is actually ill is only 0.0225!

The reason for the low probability is that we have used two sources of information here: the reliability of the test and the very small probability (0.001) that a randomly selected person is ill. The two pieces of information were mixed by Bayes' theorem, and the posterior probability reflects the mixing of the high reliability of the test with the fact that the illness is rare. The result is perfectly correct as long as the information we have used is accurate. Indeed, subject to the accuracy of our information, if the test were administered to a large number of people selected randomly from the entire population, it would be found that about 2.25% of the people in the sample who test positive are indeed ill.

Problems with Bayes' theorem arise when we are not careful with the use of prior information. In this example, suppose the test is administered to people in a hospital. Since people in a hospital are more likely to be ill than people in the population as a whole, the overall population probability that a person is ill, 0.001, no longer applies. If we applied this low probability in the hospital, our results would not be correct. This caution extends to all situations where prior probabilities are used: We must always examine the appropriateness of the prior probabilities.

Bayes' theorem may be extended to a partition of more than two sets. This is done using equation 2–17, the law of total probability involving a partition of sets B_1, $B_2, \ldots, B_n$. The resulting extended form of Bayes' theorem is given in equation 2–22. The theorem gives the probability of one of the sets in partition B_1 given the occurrence of event A. A similar expression holds for any of the events B_i.

Extended Bayes' theorem:

$$P(B_1 \mid A) = \frac{P(A|B_1)P(B_1)}{\displaystyle\sum_{i=1}^{n} P(A|B_i)P(B_i)} \qquad (2\text{--}22)$$

We demonstrate the use of equation 2–22 with the following example. In the solution, we use a table format to facilitate computations. We also demonstrate the computations using a tree diagram.

EXAMPLE 2–11   An economist believes that during periods of high economic growth, the U.S. dollar appreciates with probability 0.70; in periods of moderate economic growth, the dollar appreciates with probability 0.40; and during periods of low economic growth, the dollar appreciates with probability 0.20. During any period of time, the probability of high economic growth is 0.30, the probability of moderate growth is 0.50, and the probability of low economic growth is 0.20. Suppose the dollar has been appreciating during the present period. What is the probability we are experiencing a period of high economic growth?

Solution   Our partition consists of three events: high economic growth (event H), moderate economic growth (event M), and low economic growth (event L). The prior probabilities of the three states are $P(H) = 0.30$, $P(M) = 0.50$, and $P(L) = 0.20$. Let A denote the event that the dollar appreciates. We have the following conditional probabilities: $P(A \mid H) = 0.70$, $P(A \mid M) = 0.40$, and $P(A \mid L) = 0.20$. Applying equation 2–22 while using three sets $(n = 3)$, we get

$$P(H \mid A) = \frac{P(A \mid H)P(H)}{P(A \mid H)P(H) + P(A \mid M)P(M) + P(A \mid L)P(L)}$$

$$= \frac{(0.70)(0.30)}{(0.70)(0.30) + (0.40)(0.50) + (0.20)(0.20)} = 0.467$$

We can obtain this answer, along with the posterior probabilities of the other two states, M and L, by using a table. In the first column of the table we write the prior probabilities of the three states H, M, and L. In the second column we write the three conditional probabilities $P(A \mid H)$, $P(A \mid M)$, and $P(A \mid L)$. In the third column we write the joint probabilities $P(A \cap H)$, $P(A \cap M)$, and $P(A \cap L)$. The joint probabilities are obtained by multiplying across in each of the three rows (these operations make use of equation 2–8). The sum of the entries in the third column is the total probability of event A (by equation 2–15). Finally, the posterior probabilities $P(H \mid A)$, $P(M \mid A)$, and $P(L \mid A)$ are obtained by dividing the appropriate joint probability by the total probability of A at the bottom of the third column. For example, $P(H \mid A)$ is obtained by dividing $P(H \cap A)$ by the probability $P(A)$. The operations and the results are given in Table 2–1 and demonstrated in Figure 2–13.

Note that both the prior probabilities and the posterior probabilities of the three states add to 1.00, as required for probabilities of all the possibilities in a given situation. We conclude that, given that the dollar has been appreciating, the probability that our period is one of high economic growth is 0.467, the probability that it is one of moderate growth is 0.444, and the probability that our period is one of low economic growth is 0.089. The advantage of using a table is that we can obtain all posterior probabilities at once. If we use the formula directly, we need to apply it once for the posterior probability of each state.

TABLE 2–1 Bayesian Revision of Probabilities, Example 2–11

Event	Prior Probability	Conditional Probability	Joint Probability	Posterior Probability
H	$P(H) = 0.30$	$P(A \mid H) = 0.70$	$P(A \cap H) = 0.21$	$P(H \mid A) = \dfrac{0.21}{0.45} = 0.467$
M	$P(M) = 0.50$	$P(A \mid M) = 0.40$	$P(A \cap M) = 0.20$	$P(M \mid A) = \dfrac{0.20}{0.45} = 0.444$
L	$P(L) = 0.20$	$P(A \mid L) = 0.20$	$P(A \cap L) = 0.04$	$P(L \mid A) = \dfrac{0.04}{0.45} = 0.089$
	Sum = 1.00		$P(A) = 0.45$	Sum = 1.000

FIGURE 2–13 Tree Diagram for Example 2–11

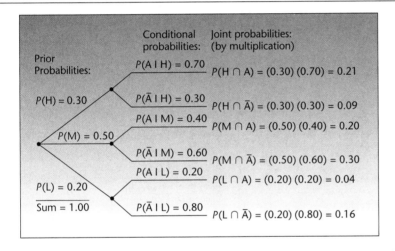

2–8 Using the Computer

Two templates are discussed in this section. The first is used for calculating the probability of at least one success in many independent trials, each with its own probability of success.

Figure 2–14 shows the template. The probability of success in each trial is entered under the Success Probs column. At times, the probabilities are related by a formula, in which case the user must figure out the formulas to be entered.

The second template, seen in Figure 2–15, is used for calculating joint, marginal, and conditional probabilities starting from a contingency table. If the starting point is a joint probability table, rather than a contingency table, this template can still be used. Enter the joint probability table in place of the contingency table.

The user needs to know how to read off the conditional probabilities from this template. The conditional probability of 0.6667 in cell C23 is $P(\text{Telecom} \mid \text{AT\&T})$, which is the row label in cell B23 and the column label in cell C22 put together. Similarly, the conditional probability of 0.8000 in cell K14 is $P(\text{AT\&T} \mid \text{Telecom})$.

FIGURE 2–14   Template for Calculating the Probability of At Least One Success
[Probability of at least 1.xls]

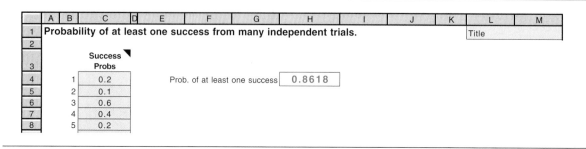

FIGURE 2–15   Template for Calculating Probabilities from a Contingency Table
[Contingency Table.xls]

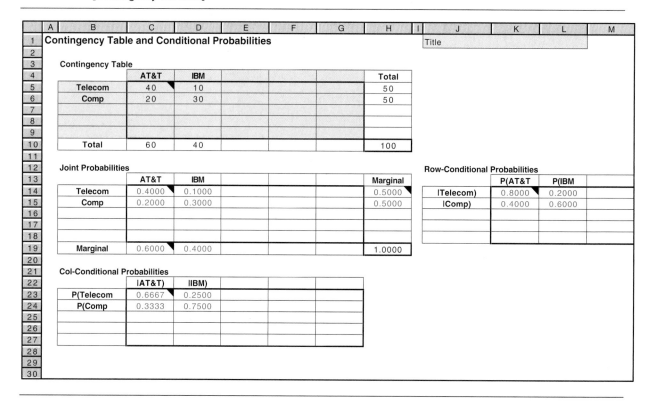

PROBLEMS

2–60. In a takeover bid for a certain company, management of the raiding firm believes that the takeover has a 0.65 probability of success if a member of the board of the raided firm resigns, and a 0.30 chance of success if she does not resign. Management of the raiding firm further believes that the chances for a resignation of the member in question are 0.70. What is the probability of a successful takeover?

2–61.   A drug manufacturer believes there is a 0.95 chance that the Food and Drug Administration (FDA) will approve a new drug the company plans to distribute if the results of current testing show that the drug causes no side effects. The manufacturer further believes there is a 0.50 probability that the FDA will approve the drug if the test shows that the drug does cause side effects. A physician working for the drug manufacturer believes there is a 0.20 probability that tests will show that the drug causes side effects. What is the probability that the drug will be approved by the FDA?

2–62.   An import–export firm has a 0.45 chance of concluding a deal to export agricultural equipment to a developing nation if a major competitor does not bid for the contract, and a 0.25 probability of concluding the deal if the competitor does bid for it. It is estimated that the competitor will submit a bid for the contract with probability 0.40. What is the probability of getting the deal?

2–63.   A realtor is trying to sell a large piece of property. She believes there is a 0.90 probability that the property will be sold in the next 6 months if the local economy continues to improve throughout the period, and a 0.50 probability the property will be sold if the local economy does not continue its improvement during the period. A state economist consulted by the realtor believes there is a 0.70 chance the economy will continue its improvement during the next 6 months. What is the probability that the piece of property will be sold during the period?

2–64.   Holland America Cruise Lines has three luxury cruise ships that sail to Alaska during the summer months. Since the business is very competitive, it is important that the ships run full during the summer if the company is to turn a profit on this line. A tourism expert hired by Holland America believes there is a 0.92 chance the ships will sail full during the coming summer if the dollar does not appreciate against European currencies, and a 0.75 chance they will sail full if the dollar does appreciate in Europe (appreciation of the dollar in Europe draws U.S. tourists there, away from U.S. destinations). Economists believe the dollar has a 0.23 chance of appreciating against European currencies soon. What is the probability the ships will sail full?

2–65.   Saflok is an electronic door lock system made in Troy, Michigan, and used in modern hotels and other establishments. To open a door, you must insert the electronic card into the lock slip. Then a green light indicates that you can turn the handle and enter; a yellow light indicates that the door is locked from inside, and you cannot enter. Suppose that 90% of the time when the card is inserted, the door should open because it is not locked from inside. When the door should open, a green light will appear with probability 0.98. When the door should not open, a green light may still appear (an electronic error) 5% of the time. Suppose that you just inserted the card and the light is green. What is the probability that the door will actually open?

2–66.   A chemical plant has an emergency alarm system. When an emergency situation exists, the alarm sounds with probability 0.95. When an emergency situation does not exist, the alarm system sounds with probability 0.02. A real emergency situation is a rare event, with probability 0.004. Given that the alarm has just sounded, what is the probability that a real emergency situation exists?

2–67.   When the economic situation is "high," a certain economic indicator rises with probability 0.6. When the economic situation is "medium," the economic indicator rises with probability 0.3. When the economic situation is "low," the indicator rises with probability 0.1. The economy is high 15% of the time, it is medium 70% of the time, and it is low 15% of the time. Given that the indicator has just gone up, what is the probability that the economic situation is high?

2–68.   An oil explorer orders seismic tests to determine whether oil is likely to be found in a certain drilling area. The seismic tests have a known reliability: When oil does exist in the testing area, the test will indicate so 85% of the time; when oil does not exist in the test area, 10% of the time the test will erroneously indicate that it does exist. The explorer believes that the probability of existence of an oil deposit in the test area is 0.4. If a test is conducted and indicates the presence of oil, what is the probability that an oil deposit really exists?

2–69.   Before marketing new products nationally, companies often test them on samples of potential customers. Such tests have a known reliability. For a particular product type, it is known that a test will indicate success of the product 75% of the time if the product is indeed successful and 15% of the time when the product is not successful. From past experience with similar products, a company knows that a new product has a 0.60 chance of success on the national market. If the test indicates that the product will be successful, what is the probability that it really will be successful?

2–70.   A market research field worker needs to interview married couples about use of a certain product. The researcher arrives at a residential building with three apartments. From the names on the mailboxes downstairs, the interviewer infers that a married couple lives in one apartment, two men live in another, and two women live in the third apartment. The researcher goes upstairs and finds that there are no names or numbers on the three doors, so that it is impossible to tell in which of the three apartments the married couple lives. The researcher chooses a door at random and knocks. A woman answers the door. Having seen a woman at the door, what *now* is the probability of having reached the married couple? Make the (possibly unrealistic) assumptions that if the two men's apartment was reached, a woman cannot answer the door; if the two women's apartment was reached, then only a woman can answer; and that if the married couple was reached, then the probability of a woman at the door is 1/2. Also assume a 1/3 prior probability of reaching the married couple. Are you surprised by the numerical answer you obtained?

2–9   Summary and Review of Terms

In this chapter, we discussed the basic ideas of probability. We defined **probability** as a relative measure of our belief in the occurrence of an **event**. We defined a **sample space** as the set of all possible outcomes in a given situation and saw that an event is a set within the sample space. We set some rules for handling probabilities: the **rule of unions**, the definition of **conditional probability**, the **law of total probability**, and **Bayes' theorem**. We also defined **mutually exclusive events** and **independence of events.** We saw how certain computations are possible in the case of independent events, and we saw how we may test whether events are independent.

In the next chapter, we will extend the ideas of probability and discuss random variables and probability distributions. These will bring us closer to statistical inference, the main subject of this book.

PROBLEMS

2–71.   AT&T was running commercials in 1990 aimed at luring back customers who may have switched to one of the other long-distance phone service providers. One such commercial shows a businessman trying to reach Phoenix and mistakenly getting Fiji, where a half-naked native on a beach responds incomprehensibly in Polynesian. When asked about this advertisement, AT&T admitted that the portrayed incident did not actually take place but added that this was an enactment of something that "could happen."[9] Suppose that one in 200 long-distance telephone calls is misdirected. What is the probability that at least one in five attempted telephone calls reaches the wrong number? (Assume independence of attempts.)

2–72.   Refer to the information in the previous problem. Given that your long-distance telephone call is misdirected, there is a 2% chance that you will reach a foreign country (such as Fiji). Suppose that I am now going to dial a single long-distance number. What is the probability that I will erroneously reach a foreign country?

2–73.   The probability that a builder of airport terminals will win a contract for construction of terminals in country A is 0.40, and the probability that it will win a contract in country B is 0.30. The company has a 0.10 chance of winning the contracts in both countries. What is the probability that the company will win at least one of these two prospective contracts?

2–74.   See the figure below.[10]

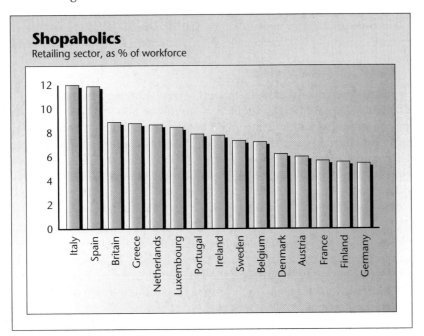

a. An Italian, a Spaniard, a Briton, and a Greek are independently selected from the workforce in these four countries. What is the probability that all work in retailing? What is the probability that at least one of them works in retailing?

b. A person is selected from the workforce in each and every one of the countries on the list. Find the probability that at least one works in the retailing sector.

[9]While this may seem virtually impossible due to the different dialing procedure for foreign countries, AT&T argues that erroneously dialing the prefix 679 instead of 617, for example, would get you Fiji instead of Massachusetts.

[10]From "More in Store," *The Economist,* June 21, 1997, p. 59, with data from Corporate Intelligence on Retailing; European Commission. © 1997 The Economist Newspaper Group, Inc. Reprinted with permission. Further reproduction prohibited. www.economist.com.

2–75. The probability that a consumer entering a retail outlet for microcomputers and software packages will buy a computer of a certain type is 0.15. The probability that the consumer will buy a particular software package is 0.10. There is a 0.05 probability that the consumer will buy both the computer and the software package. What is the probability that the consumer will buy the computer or the software package or both?

2–76. The probability that a graduating senior will pass the certified public accountant (CPA) examination is 0.60. The probability that the graduating senior will both pass the CPA examination and get a job offer is 0.40. Suppose that the student just found out that she passed the CPA examination. What is the probability that she will be offered a job?

2–77. Two stocks A and B are known to be related in that both are in the same industry. The probability that stock A will go up in price tomorrow is 0.20, and the probability that both stocks A and B will go up tomorrow is 0.12. Suppose that tomorrow you find that stock A did go up in price. What is the probability that stock B went up as well?

2–78. The probability that production will increase if interest rates decline more than 0.5 percentage point for a given period is 0.72. The probability that interest rates will decline by more than 0.5 percentage point in the period in question is 0.25. What is the probability that, for the period in question, both the interest rate will decline and production will increase?

2–79. A large foreign automaker is interested in identifying its target market in the United States. The automaker conducts a survey of potential buyers of its high-performance sports car and finds that 35% of the potential buyers consider engineering quality among the car's most desirable features and that 50% of the people surveyed consider sporty design to be among the car's most desirable features. Out of the people surveyed, 25% consider *both* engineering quality and sporty design to be among the car's most desirable features. Based on this information, do you believe that potential buyers' perceptions of the two features are independent? Explain.

2–80. Consider the situation in problem 2–79. Three consumers are chosen randomly from among a group of potential buyers of the high-performance automobile. What is the probability that all three of them consider engineering quality to be among the most important features of the car? What is the probability that at least one of them considers this quality to be among the most important ones? How do you justify your computations?

2–81. A financial service company advertises its services in magazines, runs billboard ads on major highways, and advertises its services on the radio. The company estimates that there is a 0.10 probability that a given individual will see the billboard ad during the week, a 0.15 chance that he or she will see the ad in a magazine, and a 0.20 chance that she or he will hear the advertisement on the radio during the week. What is the probability that a randomly chosen member of the population in the area will be exposed to at least one method of advertising during a given week? (Assume independence.)

2–82. An accounting firm carries an advertisement in *The Wall Street Journal*. The firm estimates that 60% of the people in the potential market read *The Wall Street Journal;* research further shows that 85% of the people who read the *Journal* remember seeing the advertisement when questioned about it afterward. What percentage of the people in the firm's potential market see and remember the advertisement?

2–83. A quality control engineer knows that 10% of the microprocessor chips produced by a machine are defective. Out of a large shipment, five chips are chosen at random. What is the probability that none of them is defective? What is the probability that at least one is defective? Explain.

2–84. A fashion designer has been working with the colors green, black, and red in preparing for the coming season's fashions. The designer estimates that there is a 0.3 chance that the color green will be "in" during the coming season, a 0.2 chance that black will be among the season's colors, and a 0.15 chance that red will be popular. Assuming that colors are chosen independently of each other for inclusion in new fashions, what is the probability that the designer will be successful with at least one of her colors?

2–85. A company president always invites one of her three vice presidents to attend business meetings and claims that her choice of the accompanying vice president is random. One of the three has not been invited even once in five meetings. What is the probability of such an occurrence if the choice is indeed random? What conclusion would you reach based on your answer?

2–86. A multinational corporation is considering starting a subsidiary in an Asian country. Management realizes that the success of the new subsidiary depends, in part, on the ensuing political climate in the target country. Management estimates that the probability of success (in terms of resulting revenues of the subsidiary during its first year of operation) is 0.55 if the prevailing political situation is favorable, 0.30 if the political situation is neutral, and 0.10 if the political situation during the year is unfavorable. Management further believes that the probabilities of favorable, neutral, and unfavorable political situations are 0.6, 0.2, and 0.2, respectively. What is the success probability of the new subsidiary?

2–87. The probability that a shipping company will obtain authorization to include a certain port of call in its shipping route is dependent on whether certain legislation is passed. The company believes there is a 0.5 chance that both the relevant legislation will pass and it will get the required authorization to visit the port. The company further estimates that the probability that the legislation will pass is 0.75. If the company should find that the relevant legislation just passed, what is the probability that authorization to visit the port will be granted?

2–88. The probability that a bank customer will default on a loan is 0.04 if the economy is high and 0.13 if the economy is not high. Suppose the probability that the economy will be high is 0.65. What is the probability that the person will default on the loan?

2–89. Researchers at Kurume University in Japan surveyed 225 workers aged 41 to 60 years and found that 30% of them were skilled workers and 70% were unskilled. At the time of survey, 15% of skilled workers and 30% of unskilled workers were on an assembly line. A worker is selected at random from the age group 41 to 60.

    a. What is the probability that the worker is on an assembly line?

    b. Given that the worker is on an assembly line, what is the probability that the worker is unskilled?

2–90. SwissAir maintains a mailing list of people who have taken trips to Europe in the last three years. The airline knows that 8% of the people on the mailing list will make arrangements to fly SwissAir during the period following their being mailed a brochure. In an experimental mailing, 20 people are mailed a brochure. What is the probability that at least one of them will book a flight with SwissAir during the coming season?

2–91.   A company's internal accounting standards are set to ensure that no more than 5% of the accounts are in error. From time to time, the company collects a random sample of accounts and checks to see how many are in error. If the error rate is indeed 5% and 10 accounts are chosen at random, what is the probability that none will be in error?

2–92.   At a certain university, 30% of the students who take basic statistics are first-year students, 35% are sophomores, 20% are juniors, and 15% are seniors. From records of the statistics department it is found that out of the first-year students who take the basic statistics course 20% get As; out of the sophomores who take the course 30% get As; out of the juniors 35% get As; and out of the seniors who take the course 40% get As. Given that a student got an A in basic statistics, what is the probability that she or he is a senior?

2–93.   The probability that a new product will be successful if a competitor does not come up with a similar product is 0.67. The probability that the new product will be successful in the presence of a competitor's new product is 0.42. The probability that the competing firm will come out with a new product during the period in question is 0.35. What is the probability that the product will be a success?

2–94.   In an effort to increase productivity, Motorola Corporation has instituted new book-balancing procedures. These procedures are reported to ensure that credit or debit entries in the general ledger are correct 99.92% of the time.[11] If a random sample of 1,000 book entries are selected (out of the company's more than 1.3 million monthly entries), what is the probability that at least one error will be found?

2–95.   Blackjack is a popular casino game in which the objective is to reach a card count greater than the dealer's without exceeding 21. One version of the game is referred to as the "hole card" version. Here, the dealer starts by drawing a card for himself or herself and putting it aside, face down, without the player's seeing what it is. This is the dealer's *hole card* (and the origin of the expression "an ace in the hole"). At the end of the game, the dealer has the option of turning this additional card face up if it may help him or her win the game. The no-hole-card version of the game is exactly the same, except that at the end of the game the dealer has the option of drawing the additional card from the deck for the same purpose (assume that the deck is shuffled prior to this draw). Conceptually, what is the difference between the two versions of the game? Is there any practical difference between the two versions as far as a player is concerned?

2–96.   Diamonds may soon become useful in making semiconductors for use in communication satellites. Theory predicts that chips from diamond films would run faster than the currently used silicon chips, would tolerate higher temperatures than silicon chips, and would be immune to radiation damage (encountered in space).[12] Suppose that the probabilities of these three events are 0.9, 0.9, and 0.95, respectively. The three events are believed to be independent. Cost considerations suggest that diamond semiconductors should be developed only if we can be at least 70% sure that all three properties described above hold. Should the project be considered?

---

[11]"Make Your Office More Productive," *Fortune,* February 25, 1991, pp. 72–76.

[12]"Will Diamond Transistors Leave Silicon in the Dust?" *Business Week,* February 11, 1991, p. 80.

2–97.   Recall from Chapter 1 that the median is that number such that one-half the observations lie above it and one-half the observations lie below it. If a random sample of two items is to be drawn from some population, what is the probability that the population median will lie between these two data points?

2–98.   Extend your result from the previous problem to a general case as follows. A random sample of n elements is to be drawn from some population and arranged according to their value, from smallest to largest. What is the probability that the population median will lie somewhere between the smallest and the largest values of the drawn data?

2–99.   A research journal states: "Rejection rate for submitted manuscripts: 86%." A prospective author believes that the editor's statement reflects the probability of acceptance of any author's *first* submission to the journal. The author further believes that for any subsequent submission, an author's acceptance probability is 10% lower than the probability he or she had for acceptance of the preceding submission. Thus, the author believes that the probability of acceptance of a first submission to the journal is $1 - 0.86 = 0.14$, the probability of acceptance of the second submission is 10% lower, that is, $(0.14)(0.90) = 0.126$, and so on for the third submission, fourth submission, etc. Suppose the author plans to continue submitting papers to the journal indefinitely until one is accepted. What is the probability that at least one paper will eventually be accepted by the journal?[13]

2–100.   (*The Von Neumann device*) Suppose that one of two people is to be randomly chosen, with equal probability, to attend an important meeting. One of them claims that using a coin to make the choice is not fair because the probability that it will land on a head or a tail is not exactly 0.50. How can the coin still be used for making the choice? (*Hint:* Toss the coin *twice,* basing your decision on two possible outcomes.) Explain your answer.

2–101.   At the same time as new hires were taking place, many retailers were cutting back. Out of 1,000 Kwik Save stores in Britain, 107 were to be closed. Out of 424 Somerfield stores, 424 were to be closed.[14] Given that a store is closing, what is the probability that it is a Kwik Save? What is the probability that a randomly chosen store is either closing or Kwik Save? Find the probability that a randomly selected store is not closing given that it is a Somerfield.

2–102.   Major hirings in retail in Britain, according to "More in Store,"[15] are as follows: 9,000 at Safeway; 5,000 at Boots; 3,400 at Debenhams; and 1,700 at Marks and Spencer. What is the probability that a randomly selected new hire from these was hired by Marks and Spencer?

2–103.   The House Ways and Means Committee is considering lowering airline taxes.[16] The committee has 38 members and needs a simple majority to pass the new legislation. If the probability that each member votes yes is 0.25, find the probability that the legislation will pass. (Assume independence.)

---

[13]Since its appearance in the first edition of the book, this interesting problem has been generalized. See N. H. Josephy and A. D. Aczel, "A Note on a Journal Selection Problem," *ZOR-Methods and Models of Operations Research* 34 (1990), pp. 469–76.

[14]From "More in Store," *The Economist,* June 21, 1997, p. 39.

[15]*The Economist,* June 21, 1997, p. 39.

[16]G. Hitt, "Airlines Keep Lobbying to Reverse U.S. Tax Plan," *The Wall Street Journal,* June 16, 1997, p. 11.

Given that taxes are reduced, the probability that Northwest Airlines will compete successfully is 0.7. If the resolution does not pass, Northwest cannot compete successfully. Find the probability that Northwest can compete successfully.

2–104. Hong Kong's Mai Po marsh is an important migratory stopover for more than 100,000 birds per year from Siberia to Australia. Many of the bird species that stop in the marsh are endangered, and there are no other suitable wetlands to replace Mai Po. Currently the Chinese government is considering building a large housing project at the marsh's boundary, which could adversely affect the birds.[17] Environmentalists estimate that if the project goes through, there will be a 60% chance that the black-faced spoonbill (current world population = 450) will not survive. It is estimated that there is a 70% chance the Chinese government will go ahead with the building project. What is the probability of the species' survival (assuming no danger if the project doesn't go through)?

CASE 2   Job Applications

A business graduate very much wants to get a job in any one of the top 10 accounting firms. Applying to any of these companies requires a lot of effort and paperwork and is therefore costly. She estimates the cost of applying to each of the 10 companies and the probability of getting a job offer there. These data are tabulated below. The tabulation is in the decreasing order of cost.

1. If the graduate applies to all 10 companies, what is the probability that she will get at least one offer?

2. If she can apply to only one company, based on cost and success probability criteria alone, should she apply to company 5? Why or why not?

3. If she applies to companies 2, 5, 8, and 9, what is the total cost? What is the probability that she will get at least one offer?

4. If she wants to be at least 75% confident of getting at least one offer, to which companies should she apply to minimize the total cost? *(This is a trial-and-error problem.)*

5. If she is willing to spend $1,500, to which companies should she apply to maximize her chances of getting at least one job? *(This is a trial-and-error problem.)*

Company	1	2	3	4	5	6	7	8	9	10
Cost	$870	$600	$540	$500	$400	$320	$300	$230	$200	$170
Probability	0.38	0.35	0.28	0.20	0.18	0.18	0.17	0.14	0.14	0.08

---

[17]S. Dallas, "Hong Kong's Birders Are Worried, Too," *Business Week,* European edition, June 16, 1997. p. 4EU2.

3 Random Variables

Marilyn Vos Savant was interested in the matter of the probabilities of the number of babies of each gender in a given number of births. Consider the sample space made up of the 16 equally likely points:

BBBB	BBBG	BGGB	GBGG
GBBB	GGBB	BGBG	GGBG
BGBB	GBGB	BBGG	GGGB
BBGB	GBBG	BGGG	GGGG

All these 16 points are equally likely because when four children are born, the gender of each child is assumed to be independent of those of the other three. Hence the probability of each quadruple (e.g., GBBG) is equal to the product of the probabilities of the four separate, single outcomes—G, B, B, and G—and is thus equal to $(1/2)(1/2)(1/2)(1/2) = 1/16$.

Now, let's look at the variable "the number of girls out of four births." This number *varies* among points in the sample space, and it is *random*—given to chance. That's why we call such a number a **random variable.**

> A **random variable** is an uncertain quantity whose value depends on chance.

A random variable has a probability law—a rule that assigns probabilities to the different values of the random variable. The probability law, the probability assignment, is called the **probability distribution** of the random variable. We usually denote the random variable by a capital letter, often X. The probability distribution will then be denoted by $P(X)$.

Look again at the sample space for the sexes of four babies, and remember that our variable is the number of girls out of four births. The first point in the sample space is BBBB; because the number of girls is zero here, $X = 0$. The next four points in the sample space all have one girl (and three boys). Hence, each one leads to the value $X = 1$. Similarly, the next six points in the sample space all lead to $X = 2$; the next four points to $X = 3$; and, finally, the last point in our sample space gives $X = 4$. The correspondence of points in the sample space with values of the random variable is as follows:

Sample Space	Random Variable
BBBB}	$X = 0$
GBBB⎫	
BGBB⎪	
BBGB⎬	$X = 1$
BBBG⎭	
GGBB⎫	
GBGB⎪	
GBBG⎪	
BGGB⎬	$X = 2$
BGBG⎪	
BBGG⎭	

Sample Space	Random Variable
BGGG⎫	
GBGG⎪	
GGBG⎬	$X = 3$
GGGB⎭	
GGGG}	$X = 4$

This correspondence, when a sample space clearly exists, allows us to define a random variable as follows.

A **random variable** is a function of the sample space.

What is this function? The correspondence between points in the sample space and values of the random variable allows us to determine the probability distribution of X as follows: Notice that 1 of the 16 equally likely points of the sample space leads to $X = 0$. Hence, the probability that $X = 0$ is 1/16. Because 4 of the 16 equally likely points lead to a value $X = 1$, the probability that $X = 1$ is 4/16, and so forth. Thus, looking at the sample space and counting the number of points leading to each value of X, we find the following probabilities:

$$P(X = 0) = 1/16 = 0.0625$$
$$P(X = 1) = 4/16 = 0.2500$$
$$P(X = 2) = 6/16 = 0.3750$$
$$P(X = 3) = 4/16 = 0.2500$$
$$P(X = 4) = 1/16 = 0.0625$$

The probability statements above constitute the probability distribution of the random variable $X =$ the number of girls in four births. Notice how this probability law was obtained simply by associating values of X with sets in the sample space. (For example, the set GBBB, BGBB, BBGB, BBBG leads to $X = 1$.) It is useful to write down the probability distribution of X in a table format, but first let's make a small, simplifying notational distinction so that we do not have to write complete probability statements such as $P(X = 1)$.

As stated earlier, we use a capital letter, such as X, to denote the random variable. But we use a lowercase letter to denote a particular value that the random variable can take. For example, $x = 3$ means that some particular set of four births resulted in three girls. Think of X as random and x as known. Before a coin is tossed, the number of heads (in one toss) is an unknown, X. Once the coin lands, we have $x = 0$ or $x = 1$.

Now let's return to the number of girls in four births. We can write the probability distribution of this random variable in a table format, as shown in Table 3–1.

Note an important fact: The sum of the probabilities of all the values of the random variable X must be 1.00. A picture of the probability distribution of the random variable X is given in Figure 3–1. Such a picture is a **probability histogram** for the random variable.

Marilyn is interested in the number of girls (or boys) in any fixed number of births, not necessarily four. Thus her discussion extends beyond this case. In fact, the random variable she describes, which in general counts the number of "successes" (here, a girl is a success) in a fixed number n of trials, is called a *binomial random variable*. We will study this particular important random variable in section 3–3.

TABLE 3–1 Probability Distribution of the Number of Girls in Four Births

Number of Girls x	Probability P(x)
0	1/16
1	4/16
2	6/16
3	4/16
4	1/16
	16/16 = 1.00

FIGURE 3–1 Probability Distribution of the Number of Girls in Four Births

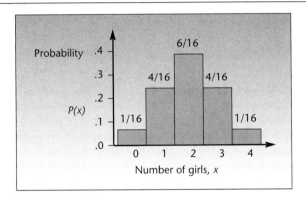

Figure 3–2 shows the sample space for the experiment of rolling two dice. As can be seen from the sample space, the probability of every pair of outcomes is 1/36. This can be seen from the fact that, by the independence of the two dice, for example, $P(6$ on red die $\cap$ 5 on green die$) = P(6$ on red die$) \times P(5$ on green die$) = (1/6)(1/6) = 1/36$, and that this holds for all 36 pairs of outcomes. Let $X =$ the sum of the dots on the two dice. What is the distribution of x?

EXAMPLE 3–1

Solution

Figure 3–3 shows the correspondence between sets in our sample space and the values of X. The probability distribution of X is given in Table 3–2. The probability distribution allows us to answer various questions about the random variable of interest. Draw a picture of this probability distribution. Such a graph need not be a histogram, used earlier, but can also be a bar graph or column chart of the probabilities of the different values of the random variable. Note from the graph you produced that the distribution of the random variable "the sum of two dice" is symmetric. The central value is $x = 7$, which has the highest probability, $P(7) = 6/36 = 1/6$. This is the *mode*, the most likely value. Thus, if you were to bet on one sum of two dice, the best bet is that the sum will be 7.

FIGURE 3–2   Sample Space for Two Dice

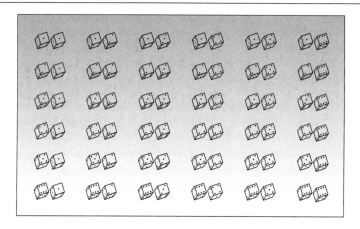

FIGURE 3–3   Correspondence between Sets and Values of X

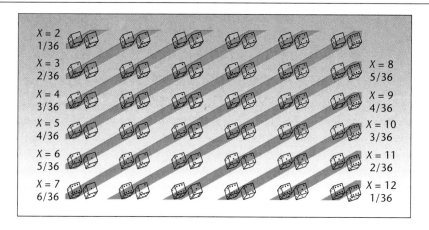

We can answer other probability questions, such as: What is the probability that the sum will be at most 5? This is $P(X \leq 5)$. Notice that to answer this question, we require the sum of all the probabilities of the values that are less than or equal to 5:

$$P(2) + P(3) + P(4) + P(5) = 1/36 + 2/36 + 3/36 + 4/36 = 10/36$$

Similarly, we may want to know the probability that the sum is greater than 9. This is calculated as follows:

$$P(X > 9) = P(10) + P(11) + P(12) = 3/36 + 2/36 + 1/36 = 6/36 = 1/6$$

Most often, unless we are dealing with games of chance, there is no evident sample space. In such situations the probability distribution is often obtained from lists or other data that give us the relative frequency in the recorded past of the various values of the random variable. This is demonstrated in Example 3–2.

TABLE 3–2   Probability Distribution of the Sum of Two Dice

x	P(x)
2	1/36
3	2/36
4	3/36
5	4/36
6	5/36
7	6/36
8	5/36
9	4/36
10	3/36
11	2/36
12	1/36
	36/36 = 1.00

EXAMPLE 3–2

800, 900, and Now: the 500 Telephone Numbers

The new code 500 is for busy, affluent people who travel a lot: It can work with a cellular phone, your home phone, office phone, second-home phone, up to five additional phones besides your regular one. The computer technology behind this service is astounding—the new phone service can find you wherever you may be on the planet at a given moment (assuming one of the phones you specify is cellular and you keep it with you when you are not near one of your stationary telephones). What the computer does is to first ring you up at the telephone number you specify as your primary one (your office phone, for example). If there is no answer, the computer switches to search for you at your second-specified phone number (say, home); if you do not answer there, it will switch to your third phone (maybe the phone at a close companion's home, or your car phone, or a portable cellular phone); and so on up to five allowable switches. The switches are the expensive part of this service (besides arrangements to have your cellular phone reachable overseas), and the service provider wants to get information on these switches. From data available on an experimental run of the 500 program, the following probability distribution is constructed for the number of dialing switches that are necessary before a person is reached. When $X = 0$, it means that the person was reached on her or his primary phone (no switching was necessary); when $X = 1$, it means that a dialing switch was necessary, and the person was found at the secondary phone; and so on up to five switches. Table 3–3 gives the probability distribution for this random variable.

A plot of the probability distribution of this random variable is given in Figure 3–4. When more than two switches occur on a given call, extra costs are incurred. What is the probability that for a given call there would be extra costs?

**TABLE 3–3
The Probability
Distribution of the
Number of Switches**

x	P(x)
0	0.1
1	0.2
2	0.3
3	0.2
4	0.1
5	0.1
	1.00

Solution

$$P(X > 2) = P(3) + P(4) + P(5) = 0.2 + 0.1 + 0.1 = 0.4$$

What is the probability that at least one switch will occur on a given call? $1 - P(0) = 0.9$, a high probability.

FIGURE 3–4   The Probability Distribution of the Number of Switches

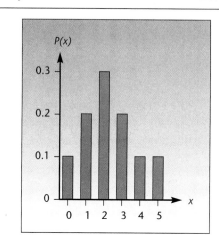

Discrete and Continuous Random Variables

Refer to Example 3–2. Notice that when switches occur, the number X jumps by 1. It is impossible to have one-half a switch or 0.13278 of one. The same is true for the number of dots on two dice (you cannot see 2.3 dots or 5.87 dots) and, of course, the number of girls in four births.

> A **discrete random variable** can assume at most a countable number of values.

The values of a discrete random variable do not have to be positive whole numbers; they just have to "jump" from one possible value to the next without being able to have any value in between. The price of a stock, for example, is a discrete random variable because it is reported to the nearest eighth of a dollar: 17 1/8, 17 2/8, 17 3/8, etc. As another example, the amount of money you make on an investment may be $500, or it may be a loss: −$200. At any rate, it can be measured at best to the nearest *cent,* so this variable is discrete.

What are continuous random variables, then?

> A **continuous random variable** may take on any value in an interval of numbers (i.e., its possible values are uncountably infinite).

The values of continuous random variables can be measured (at least in theory) to any degree of accuracy. They move continuously from one possible value to another, without having to jump. For example, temperature is a continuous random variable, since it can be measured as 72.00340981136 . . . °. Weight, height, and time are other examples of continuous random variables.

The difference between discrete and continuous random variables is illustrated in Figure 3–5. Is wind speed a discrete or a continuous random variable?

FIGURE 3–5   Discrete and Continuous Random Variables

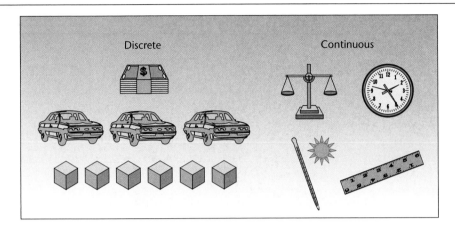

The probability distribution of a discrete random variable X must satisfy the following two conditions.

$$1. \ P(x) \geq 0 \text{ for all values } x \qquad\qquad (3\text{–}1)$$

$$2. \ \sum_{\text{all } x} P(x) = 1 \qquad\qquad\qquad (3\text{–}2)$$

These conditions must hold because the $P(x)$ values are probabilities. Equation 3–1 states that all probabilities must be greater than or equal to zero, as we know from Chapter 2. For the second rule, equation 3–2, note the following. For each value x, $P(x) = P(X = x)$ is the probability of the event that the random variable equals x. Since by definition *all x* means all the values the random variable X may take, and since X may take on only one value at a time, the occurrences of these values are mutually exclusive events, and one of them must take place. Therefore, the sum of all the probabilities $P(x)$ must be 1.00.

Cumulative Distribution Function

The probability distribution of a discrete random variable lists the probabilities of occurrence of different values of the random variable. We may be interested in *cumulative* probabilities of the random variable. That is, we may be interested in the probability that the value of the random variable is *at most* some value x. This is the sum of all the probabilities of the values i of X that are less than or equal to x. We define the *cumulative distribution function* (also called *cumulative probability function*) as follows.

The **cumulative distribution function,** $F(x)$, of a discrete random variable X is

$$F(x) = P(X \leq x) = \sum_{\text{all } i \leq x} P(i) \qquad\qquad (3\text{–}3)$$

TABLE 3–4   Cumulative Distribution Function of the Number of Switches (Example 3–2)

x	P(x)	F(x)
0	0.1	0.1
1	0.2	0.3
2	0.3	0.6
3	0.2	0.8
4	0.1	0.9
5	0.1	1.00
	1.00	

FIGURE 3–6   Cumulative Distribution Function of Number of Switches

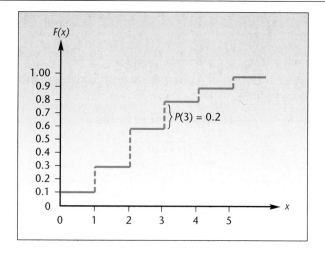

Table 3–4 gives the cumulative distribution function of the random variable of Example 3–2. Note that each entry of $F(x)$ is equal to the sum of the corresponding values of $P(i)$ for all values i less than or equal to x. For example, $F(3) = P(X \leq 3) = P(0) + P(1) + P(2) + P(3) = 0.1 + 0.2 + 0.3 + 0.2 = 0.8$. Of course, $F(5) = 1.00$ because $F(5)$ is the sum of the probabilities of all values that are less than or equal to 5, and 5 is the largest value of the random variable.

Figure 3–6 shows $F(x)$ for the number of switches on a given call. All cumulative distribution functions are nondecreasing and equal 1.00 at the largest possible value of the random variable.

Let us consider a few probabilities. The probability that the number of switches will be less than or equal to 3 is given by $F(3) = 0.8$. This is illustrated, using the probability distribution, in Figure 3–7.

The probability that *more than* one switch will occur, $P(X > 1)$, is equal to $1 - F(1) = 1 - 0.3 = 0.7$. This is so because $F(1) = P(X \leq 1)$, and $P(X \leq 1) + P(X > 1) = 1$ (the two events are complements of each other). This is demonstrated in Figure 3–8.

The probability that anywhere from one to three switches will occur is $P(1 \leq X \leq 3)$. From Figure 3–9 we see that this is equal to $F(3) - F(0) = 0.8 - 0.1 = 0.7$.

FIGURE 3–7   The Probability That at Most Three Switches Will Occur

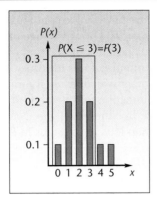

FIGURE 3–8   Probability That More than One Switch Will Occur

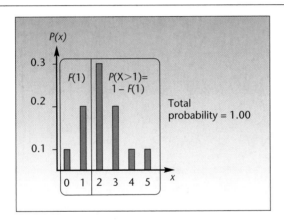

FIGURE 3–9   Probability That Anywhere from One to Three Switches Will Occur

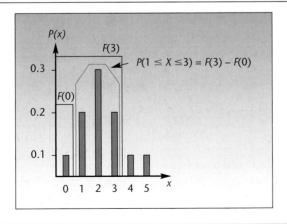

(This is the probability that the number of switches that occur will be less than or equal to 3 and greater than 0.) This, and other probability questions, could certainly be answered directly, without use of $F(x)$. We could just add the probabilities: $P(1) + P(2) + P(3) = 0.2 + 0.3 + 0.2 = 0.7$. The advantage of $F(x)$ is that probabilities may be computed by few operations [usually subtraction of two values of $F(x)$, as in this example], whereas use of $P(x)$ often requires lengthier computations.

If the probability distribution is available, use it directly. If, on the other hand, you have a cumulative distribution function for the random variable in question, you may use it as we have demonstrated. In either case, it is always helpful to draw a picture of the probability distribution. Then look at the signs in the probability statement, such as $P(X \le x)$ versus $P(X < x)$, to see which values to include and which ones to leave out of the probability computation.

PROBLEMS

3–1. The number of telephone calls arriving at an exchange during any given minute between noon and 1:00 P.M. on a weekday is a random variable with the following probability distribution.

x	P(x)
0	0.3
1	0.2
2	0.2
3	0.1
4	0.1
5	0.1

  a. Verify that $P(x)$ is a probability distribution.
  b. Find the cumulative distribution function of the random variable.
  c. Use the cumulative distribution function to find the probability that between 12:34 and 12:35 P.M. more than two calls will arrive at the exchange.

3–2. Typing errors per page for a certain typing pool are known to follow this probability distribution:

x	P(x)
0	0.01
1	0.09
2	0.30
3	0.20
4	0.20
5	0.10
6	0.10

  a. Verify that $P(x)$ is a probability distribution.
  b. Find the cumulative distribution function.
  c. Find the probability that at most four errors will be made on a page.
  d. Find the probability that at least two errors will be made on a page.

3–3. The percentage of people (to the nearest 10) responding to an advertisement is a random variable with the following probability distribution:

x (%)	P(x)
0	0.10
10	0.20
20	0.35
30	0.20
40	0.10
50	0.05

a. Show that $P(x)$ is a probability distribution.

b. Find the cumulative distribution function.

c. Find the probability that more than 20% will respond to the ad.

3–4. An automobile dealership records the number of cars sold each day. The data are used in calculating the following probability distribution of daily sales:

x	P(x)
0	0.1
1	0.1
2	0.2
3	0.2
4	0.3
5	0.1

a. Find the probability that the number of cars sold tomorrow will be between two and four (both inclusive).

b. Find the cumulative distribution function of the number of cars sold per day.

c. Show that $P(x)$ is a probability distribution.

3–5. Consider the roll of a pair of dice, and let X denote the sum of the two numbers appearing on the dice. Find the probability distribution of X, and find the cumulative distribution function. What is the most likely sum?

3–6. The number of intercity shipment orders arriving daily at a transportation company is a random variable X with the following probability distribution:

x	P(x)
0	0.1
1	0.2
2	0.4
3	0.1
4	0.1
5	0.1

a. Verify that $P(x)$ is a probability distribution.

b. Find the cumulative probability function of X.

c. Use the cumulative probability function computed in (b) to find the probability that anywhere from one to four shipment orders will arrive on a given day.

d. When more than three orders arrive on a given day, the company incurs additional costs due to the need to hire extra drivers and loaders. What is the probability that extra costs will be incurred on a given day?

e. Assuming that the numbers of orders arriving on different days are independent of each other, what is the probability that no orders will be received over a period of five working days?

f. Again assuming independence of orders on different days, what is the probability that extra costs will be incurred two days in a row?

3–7. The number of wooden sailboats constructed per month in a small shipyard is a random variable with the following probability distribution:

x	P(x)
2	0.2
3	0.2
4	0.3
5	0.1
6	0.1
7	0.05
8	0.05

a. Find the probability that the number of boats constructed next month will be between 4 and 7 (both inclusive).

b. Find the cumulative distribution function of X.

c. Use $F(x)$ computed in (b) to evaluate the probability that the number of boats constructed in a month will be less than or equal to 6.

d. Find the probability that the number of boats made will be greater than 3 and less than or equal to 6.

3–8. The number of defects in a machine-made product is a random variable X with the following probability distribution:

x	P(x)
0	0.1
1	0.2
2	0.3
3	0.3
4	0.1

a. Show that $P(x)$ is a probability distribution.

b. Find the probability $P(1 < X \le 3)$.

c. Find the probability $P(1 < X \le 4)$.

d. Find $F(x)$.

3–9. Returns on investments overseas, especially in Europe and the Pacific Rim, are expected to be higher than those of U.S. markets in the near term, and analysts are now recommending investments in international portfolios. An investment consultant believes that the probability distribution of returns (in percent per year) on one such portfolio is as follows:

x (%)	P(x)
9	0.05
10	0.15
11	0.30
12	0.20
13	0.15
14	0.10
15	0.05

a. Verify that $P(x)$ is a probability distribution.

b. What is the probability that returns will be at least 12%?

c. Find the cumulative distribution of returns.

3–10. The daily world price of sugar in cents per pound in June 1997 can be inferred to have the following distribution (from data in *The Wall Street Journal*[1]):

x	P(x)
78	0.05
79	0.10
80	0.25
81	0.40
82	0.15
83	0.05

a. Show that $P(x)$ is a probability distribution.

b. What is the probability that the price on a given day during this period will be at least 80 cents per pound?

c. What is the probability that the price on a given day during this period will be less than 82 cents per pound?

d. If daily prices are independent of one another, what is the probability that for two days in a row the price will be above 80 cents per pound?

3–2 Expected Values of Discrete Random Variables

In Chapter 1, we discussed summary measures of data sets. The most important summary measures discussed were the mean and the variance (also the square root of the variance, the standard deviation). We saw that the mean is a measure of *centrality,* or *location,* of the data or population, and that the variance and the standard deviation measure the *variability,* or *spread,* of our observations.

The mean of a probability distribution of a random variable is a measure of the centrality of the probability distribution. It is a measure that considers both the values of the random variable and their probabilities. The mean is a *weighted average* of the possible values of the random variable—the weights being the probabilities.

The mean of the probability distribution of a random variable is called the *expected value* of the random variable (sometimes called the *expectation* of the random variable). The reason for this name is that the mean is the (probability-weighted) average value of the random variable, and therefore it is the value we "expect" to occur.

[1]June 1, 1997, p. 21.

We denote the mean by two notations: μ for *mean* (as in Chapter 1 for a population) and $E(X)$ for *expected value of X*. In situations where no ambiguity is possible, we will often use μ. In cases where we want to stress the fact that we are talking about the expected value of a particular random variable (here, X), we will use the notation $E(X)$. The expected value of a discrete random variable is defined as follows.

The **expected value** of a discrete random variable X is equal to the sum of all values of the random variable, each value multiplied by its probability.

$$\mu = E(X) = \sum_{\text{all } x} xP(x) \tag{3-4}$$

Suppose a coin is tossed. If it lands heads, you win a dollar; but if it lands tails, you lose a dollar. What is the expected value of this game? Intuitively, you know you have an even chance of winning or losing the same amount, and so the average or expected value is zero. Your payoff from this game is a random variable, and we find its expected value from equation 3–4: $E(X) = (1)(1/2) + (-1)(1/2) = 0$. The definition of an expected value, or mean, of a random variable thus conforms with our intuition. Incidentally, games of chance with an expected value of zero are called *fair games*.

Let us now return to Example 3–2 and find the expected value of the random variable involved—the expected number of switches on a given call. It is convenient to compute the mean of a discrete random variable by using a table. In the first column of the table we write the values of the random variable. In the second column we write the probabilities of the different values, and in the third column we write the products $xP(x)$ for each value x. We then add the entries in the third column, giving us $E(X) = \Sigma xP(x)$, as required by equation 3–4. This is shown for Example 3–2 in Table 3–5.

As indicated in the table, $\mu = E(X) = 2.3$. We can say that, on the average, 2.3 switches occur per call. As this example shows, the mean does not have to be one of the values of the random variable. There are no calls with 2.3 switches, but 2.3 is the average number of switches. It is the *expected* number of switches per call, although here the exact expectation will not be realized on any call.

As the weighted average of the values of the random variable, with probabilities as weights, the mean is the *center of mass* of the probability distribution. This is demonstrated for Example 3–2 in Figure 3–10.

The Expected Value of a Function of a Random Variable

It is possible to compute the expected value of a *function* of a random variable. Let $h(X)$ be a function of the discrete random variable X.

The expected value of $h(X)$, a function of the discrete random variable X, is

$$E[h(X)] = \sum_{\text{all } x} h(x)P(x) \tag{3-5}$$

The function $h(X)$ could be X^2, $3X^4$, $\log X$, or any function. As we will see shortly, equation 3–5 is most useful for computing the expected value of the special function $h(X) =$

TABLE 3–5   Computing the Expected Number of Switches for Example 3–2

x	P(x)	xP(x)
0	0.1	0
1	0.2	0.2
2	0.3	0.6
3	0.2	0.6
4	0.1	0.4
5	0.1	0.5
	1.00	2.3 ← Mean, $E(X)$

FIGURE 3–10   The Mean of a Discrete Random Variable as a Center of Mass for Example 3–2

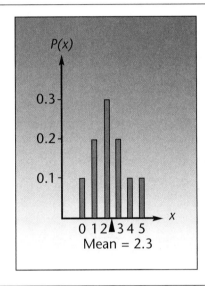

X^2. But let us first look at a simpler example, where $h(X)$ is a *linear* function of X. A linear function of X is a straight-line relation: $h(X) = a + bX$, where a and b are numbers.

Monthly sales of a certain product, recorded to the nearest thousand, are believed to follow the probability distribution given in Table 3–6. Suppose that the company has a fixed monthly production cost of $8,000 and that each item brings $2. Find the expected monthly profit from product sales.   **EXAMPLE 3–3**

The company's profit function from sales of the product is $h(X) = 2X - 8,000$. Equation 3–5 tells us that the expected value of $h(X)$ is the sum of the values of $h(X)$, each value multiplied by the probability of the particular value of X. We thus add two columns to Table 3–6: a column of values of $h(x)$ for all x and a column of the products $h(x)P(x)$. At the bottom of this column we find the required sum $E[h(X)] =$ *Solution*

TABLE 3–6   Probability Distribution of Monthly Product Sales for Example 3–3

Number of Items x	P(x)
5,000	0.2
6,000	0.3
7,000	0.2
8,000	0.2
9,000	0.1
	1.00

TABLE 3–7   Computing Expected Profit for Example 3–3

x	h(x)	P(x)	h(x)P(x)
5,000	2,000	0.2	400
6,000	4,000	0.3	1,200
7,000	6,000	0.2	1,200
8,000	8,000	0.2	1,600
9,000	10,000	0.1	1,000
		$E[h(X)] =$	5,400

$\Sigma_{\text{all }x}\, h(x)P(x)$. This is done in Table 3–7. As shown in the table, expected monthly profit from sales of the product is $5,400.

In the case of a linear function of a random variable, as in Example 3–3, there is a possible simplification of our calculation of the mean of $h(X)$. The simplified formula of the expected value of a linear function of a random variable is as follows:

> The expected value of a linear function of a random variable is
>
> $$E(aX + b) = aE(X) + b \qquad (3\text{–}6)$$
>
> where a and b are fixed numbers

Equation 3–6 holds for *any* random variable, discrete or continuous. Once you know the expected value of X, the expected value of $aX + b$ is just $aE(X) + b$. In Example 3–3 we could have obtained the expected profit by finding the mean of X first, and then multiplying the mean of X by 2 and subtracting from this the fixed cost of $8,000. The mean of X is 6,700 (prove this), and the expected profit is therefore $E[h(X)] = E(2X - 8,000) = 2E(X) - 8,000 = 2(6,700) - 8,000 = \$5,400$, as we obtained using Table 3–7.

As mentioned earlier, the most important expected value of a function of X is the expected value of $h(X) = X^2$. This is because this expected value helps us compute the *variance* of the random variable X and, through the variance, the standard deviation.

Variance and Standard Deviation of a Random Variable

The variance of a random variable is the expected squared deviation of the random variable from its mean. The idea is similar to that of the variance of a data set or a population, defined in Chapter 1. Probabilities of the values of the random variable are used as weights in the computation of the expected squared deviation from the mean of a discrete random variable. The definition of the variance follows. As with a population, we denote the variance of a random variable by σ^2. Another notation for the variance of X is $V(X)$.

The **variance** of a discrete random variable X is given by

$$\sigma^2 = V(X) = E[(X - \mu)^2] = \sum_{\text{all } x} (x - \mu)^2 P(x) \qquad (3\text{--}7)$$

Using equation 3–7, it is possible to compute the variance of a discrete random variable by subtracting the mean μ from each value x of the random variable, squaring the result, multiplying it by the probability $P(x)$, and finally adding the results for all x. Let us apply equation 3–7 and find the variance of the number of dialing switches in Example 3–2:

$$\sigma^2 = \Sigma(x - \mu)^2 P(x)$$
$$= (0 - 2.3)^2(0.1) + (1 - 2.3)^2(0.2) + (2 - 2.3)^2(0.3)$$
$$+ (3 - 2.3)^2(0.2) + (4 - 2.3)^2(0.1) + (5 - 2.3)^2(0.1)$$
$$= 2.01$$

There is, however, an easier way of computing the variance of a discrete random variable. It can be shown mathematically that equation 3–7 is equivalent to the following computational form of the variance.

Computational formula for the variance of a random variable:

$$\sigma^2 = V(X) = E(X^2) - [E(X)]^2 \qquad (3\text{--}8)$$

Equation 3–8 has the same relation to equation 3–7 as equation 1–7 has to equation 1–3 for the variance of a set of points.

Equation 3–8 states that the variance of X is equal to the expected value of X^2 minus the squared mean of X. In computing the variance using this equation, we use the definition of the expected value of a function of a discrete random variable, equation 3–5, in the special case $h(X) = X^2$. We compute x^2 for each x, multiply it by $P(x)$, and add for all x. This gives us $E(X^2)$. To get the variance, we subtract from $E(X^2)$ the mean of X, squared.

We now compute the variance of the random variable in Example 3–2, using this method. This is done in Table 3–8. The first column in the table gives the values of X, the second column gives the probabilities of these values, the third column gives the products of the values and their probabilities, and the fourth column is the product of the third column and the first [because we get $x^2 P(x)$ by just multiplying each entry

TABLE 3–8   Computations Leading to the Variance of the Number of Switches in Example 3–2 Using the Shortcut Formula (Equation 3–8)

x	$P(x)$	$xP(x)$	$x^2P(x)$
0	0.1	0	0
1	0.2	0.2	0.2
2	0.3	0.6	1.2
3	0.2	0.6	1.8
4	0.1	0.4	1.6
5	0.1	0.5	2.5
	1.00	2.3 ← Mean of X	7.3 ← Mean of X^2

$xP(x)$ by x from column 1]. At the bottom of the third column we find the mean of X, and at the bottom of the fourth column we find the mean of X^2. Finally, we perform the subtraction $E(X^2) - [E(X)]^2$ to get the variance of X:

$$V(X) = E(X^2) - [E(X)]^2 = 7.3 - (2.3)^2 = 2.01$$

This is the same value we found using the other formula for the variance, equation 3–7. Note that equation 3–8 holds for *all* random variables, discrete or otherwise. Once we obtain the expected value of X^2 and the expected value of X, we can compute the variance of the random variable using this equation.

For random variables, as for data sets or populations, the standard deviation is equal to the (positive) square root of the variance. We denote the standard deviation of a random variable X by σ or by $SD(X)$.

The **standard deviation** of a random variable:

$$\sigma = SD(X) = \sqrt{V(X)} \qquad\qquad (3\text{–}9)$$

In Example 3–2, the standard deviation is $\sigma = \sqrt{2.01} = 1.418$.

What are the variance and the standard deviation, and how do we interpret their meaning? By definition, the variance is the average squared deviation of the values of the random variable from their mean. Thus, it is a measure of the *dispersion* of the possible values of the random variable about the mean. The variance gives us an idea of the variation or uncertainty associated with the random variable: The larger the variance, the farther away from the mean are possible values of the random variable. Since the variance is a squared quantity, it is often more useful to consider its square root—the standard deviation of the random variable. When two random variables are compared, the one with the larger variance (standard deviation) is the more variable one. The risk associated with an investment is often measured by the standard deviation of investment returns. When comparing two investments with the same average (*expected*) return, the investment with the higher standard deviation is considered riskier (although a higher standard deviation implies that returns are expected to be more variable—both below and above the mean).

Variance of a Linear Function of a Random Variable

There is a formula, analogous to equation 3–6, that gives the variance of a linear function of a random variable. For a linear function of X given by $aX + b$, we have the following:

Variance of a linear function of a random variable:

$$V(aX + b) = a^2 V(X) = a^2 \sigma^2 \qquad (3\text{--}10)$$

where a and b are fixed numbers

Using equation 3–10, we will find the variance of the profit in Example 3–3. The profit is given by $2X - 8{,}000$. We need to find the variance of X in this example. We find

$$E(X^2) = (5{,}000)^2(0.2) + (6{,}000)^2(0.3) + (7{,}000)^2(0.2) + (8{,}000)^2(0.2)$$
$$+ (9{,}000)^2(0.1)$$
$$= 46{,}500{,}000$$

The expected value of X is $E(X) = 6{,}700$. The variance of X is thus

$$V(X) = E(X^2) - [E(X)]^2 = 46{,}500{,}000 - (6{,}700)^2 = 1{,}610{,}000$$

Finally, we find the variance of the profit, using equation 3–10, as $2^2(1{,}610{,}000) = 6{,}440{,}000$. The standard deviation of the profit is $\sqrt{6{,}440{,}000} = 2{,}537.72$.

Some Properties of Means and Variances

An investor has income from two sources. Investment A gives her a random amount X, per month, with a certain probability distribution and mean $E(X) = \$350$. Investment B also gives her a random amount Y, which has a different probability distribution, with mean $E(Y) = \$200$. What is the expected monthly income from both sources? The answer is very simple: $\$350 + \$200 = \$550$. The reason for this is a rule we have for means that says $E(X + Y) = E(X) + E(Y)$, which in this case is equal to $\$350 + \$200 = \$550$.

Our dice example demonstrates this property well. The expected value of the sum of the dots on two dice was 7. But let's look at each die separately. What is the expected number of a single die? Each of the six numbers 1 to 6 has probability 1/6. The mean, therefore, is found as follows:

$$(1)(1/6) + (2)(1/6) + (3)(1/6) + (4)(1/6) + (5)(1/6) + (6)(1/6) = 3.5$$

This should be intuitively clear since 3.5 is the center of the six equally likely points. Now, what is the mean of the sum of the two dice? By the rule above, it is just the sum of the two means: $3.5 + 3.5 = 7$.

The variance of the sum (or the difference) of two random variables is equal to the sum of the two variances if the two random variables are *independent*. If the variance of investment X is 84 and the variance of investment Y is 60, and the two investments are independent of each other, then the variance of the total income from both sources is $V(X) + V(Y) = 84 + 60 = 144$. This means that the standard deviation of the total income is the square root of this variance, which is $\$12$.

Chebyshev's Theorem

The standard deviation is useful in obtaining bounds on the possible values of the random variable with certain probability. The bounds are obtainable from a well-known theorem, *Chebyshev's theorem* (the name is sometimes spelled Tchebychev, Tchebysheff, or any of a number of variations). The theorem says that for any number k greater than 1.00, the probability that the value of a given random variable will be *within k standard deviations* of the mean is at least $1 - 1/k^2$. In Chapter 1, we listed some results for data sets that are derived from this theorem.

Chebyshev's Theorem

For a random variable X with mean μ and standard deviation σ, and for any number $k > 1$,

$$P(|X - \mu| < k\sigma) \geq 1 - 1/k^2 \qquad (3\text{–}11)$$

Let us see how the theorem is applied by selecting values of k. While k does not have to be an integer, we will use integers. When $k = 2$, we have $1 - 1/k^2 = 0.75$: The theorem says that there is at least a 0.75 probability that the value of the random variable will be within a distance of 2 standard deviations away from the mean. Letting $k = 3$, we find that there is at least a 0.89 probability that X will be within 3 standard deviations of its mean. We can similarly apply the rule for other values of k. The rule holds for data sets and populations in a similar way. When applied to a sample of observations, the rule says that at least 75% of the observations lie within 2 standard deviations of the sample mean $\bar{x}$. It says that at least 89% of the observations lie within 3 standard deviations of the mean, and so on. Applying the theorem to the random variable of Example 3–2, which has mean 2.3 and standard deviation 1.418, we find that there is at least a 0.75 probability that X will be anywhere from $2.3 - 2(1.418)$ to $2.3 + 2(1.418) = -0.536$ to 5.136. From the actual probability distribution in this example, Table 3–3, we know that the probability that X will be between 0 and 5 is 1.00.

Often, we will know the distribution of the random variable in question, in which case we will be able to use the distribution for obtaining actual probabilities rather than the bounds offered by Chebyshev's theorem. If the exact distribution of the random variable is not known, but we may assume an approximate distribution, the approximate probabilities may still be better than the general bounds offered by Chebyshev's theorem.

The Templates for Random Variables

The template shown in Figure 3–11 can be used to calculate the expected value and variance of the sum of up to five random variables. Note that the expected value of the sum of several random variables is simply the sum of the individual expected values. If the random variables are *independent*, then the variance of their sum is the sum of the individual variances. Once the variance is known, the standard deviation is calculated by taking the square root of the variance.

The template shown in Figure 3–12 can be used to calculate the descriptive statistics of a random variable and also those of a function $h(x)$ of that random variable. To calculate the statistics of $h(x)$, the Excel formula for the function must be entered in cell G12. For instance, if $h(x) = 5x^2 + 8$, enter the Excel formula =5*x^2+8 in cell G12.

FIGURE 3–11   Template for the Sum of Random Variables
              [Sum of Random Variables.xls]

	A	B	C	D	E	F	G	H	I	J
1	Sum of Independent Random Variables									
2										
3			X_1	X_2	X_3	X_4	X_5		Sum ▼	
4		Mean	12	15	8	11			46	Mean
5		Variance	2.57	5.4	0.88	1.07			9.92	Variance
6		Std Devn.	1.60312	2.32379	0.93808	1.03441			3.1496	Std Devn.

FIGURE 3–12   Descriptive Statistics of a Random Variable X and $h(x)$
              [Random Variable.xls]

	A	B	C	D	E	F	G	H	I
1	Descriptive Statistics of a Random Variable							Title	
2									
3		x	$P(x)$	$F(x)$					
4		0	0.1	0.1		Statistics of X			
5		1	0.2	0.3			Mean	2.3	
6		2	0.3	0.6			Variance	2.01	
7		3	0.2	0.8			Std. Devn.	1.41774	
8		4	0.1	0.9			Skewness	0.30319	
9		5	0.1	1			(Relative) Kurtosis	-0.6313	
10									
11						Definition of $h(x)$			x
12							$h(x) =$	8 ▼	0
13									
14						Statistics of $h(x)$			
15							Mean	44.5	
16							Variance	1400.25	
17							Std. Devn.	37.4199	
18							Skewness	1.24449	
19							(Relative) Kurtosis	0.51413	

PROBLEMS

3–11. Find the expected value of the random variable in problem 3–1. Also find the variance of the random variable and its standard deviation.

3–12. Find the mean, variance, and standard deviation of the random variable in problem 3–2.

3–13. What is the expected percentage of people responding to an advertisement when the probability distribution is the one given in problem 3–3? What is the variance of the percentage of people who respond to the advertisement?

3–14. Find the mean, variance, and standard deviation of the number of cars sold per day, using the probability distribution in problem 3–4.

3–15. What is the expected number of dots appearing on two dice? (Use the probability distribution you computed in your answer to problem 3–5.)

3–16. Use the probability distribution in problem 3–6 to find the expected number of shipment orders per day. What is the probability that on a given day there will be more orders than the average?

3–17. Find the mean, variance, and standard deviation of the number of wooden sailboats constructed in a month, using the probability distribution in problem 3–7.

3–18. According to Chebyshev's theorem, what is the minimum probability that a random variable will be within 4 standard deviations of its mean?

3–19. At least eight-ninths of a population lies within how many standard deviations of the population mean? Why?

3–20. The average annual return on a certain stock is 8.3%, and the variance of the returns on the stock is 2.3. Another stock has an average return of 8.4% per year and a variance of 6.4. Which stock is riskier? Why?

3–21. Returns on a certain business venture, to the nearest $1,000, are known to follow the probability distribution

x	P(x)
−2,000	0.1
−1,000	0.1
0	0.2
1,000	0.2
2,000	0.3
3,000	0.1

    a. What is the most likely monetary outcome of the business venture?

    b. Is the venture likely to be successful? Explain.

    c. What is the long-term average earning of business ventures of this kind? Explain.

    d. What is a good measure of the risk involved in a venture of this kind? Why? Compute this measure.

3–22. Management of an airline knows that 0.5% of the airline's passengers lose their luggage on domestic flights. Management also knows that the average value claimed for a lost piece of luggage on domestic flights is $600. The company is considering increasing fares by an appropriate amount to cover expected compensation to passengers who lose their luggage. By how much should the airline increase fares? Why? Explain, using the ideas of a random variable and its expectation.

3–23. Refer to problem 3–7. Suppose that the sailboat builders have fixed monthly costs of $25,000 and an additional construction cost of $5,000 per boat. Find the expected monthly cost of the operation. What property of expected values are you using?

3–24. Refer to problem 3–4. Suppose the car dealership's operation costs are well approximated by the square root of the number of cars sold, multiplied by $300. What is the expected daily cost of the operation? Explain.

3–25. In problem 3–2, suppose that a penalty is imposed on the typing pool of an amount equal to the square of the number of errors per page. What is the expected penalty per page? Explain.

3–26. All voters of Milwaukee County were asked a week before election day whether they would vote for a certain presidential candidate. Of these, 48% answered yes, 45% replied no, and the rest were undecided. If a yes answer is coded +1, a no answer is coded −1, and an undecided answer is coded 0, find the mean and the variance of the code.

3–27. Explain the meaning of the variance of a random variable. What are possible uses of the variance?

3–28. Why is the standard deviation of a random variable more meaningful than its variance for interpretation purposes?

3–29. Refer to problem 3–23. Find the variance and the standard deviation of the monthly production cost.

3–30. For problem 3–10, find the mean and the standard deviation of the price per pound.

3–31. Lobsters vary in sizes. The bigger the size, the more valuable the lobster per pound (a 6-pound lobster is more valuable than two 3-pound ones). Fishers will sell entire boatloads for a certain price. The boatload has a mixture of sizes. Suppose the distribution is as follows:

x (pound)	P(x)	v(x) ($)
½	0.1	2
¾	0.1	2.5
1	0.3	3.0
1¼	0.2	3.25
1½	0.2	3.40
1¾	0.05	3.60
2	0.05	5.00

What is a fair price for the shipload?

The Use of Standard Random Variables

When the probability distribution of a discrete random variable is known fully, the statistical calculations we have seen so far are all possible. In practice, however, the probability distribution of a random variable may not be known fully. Empirically assessing the probability of every possible value of a random variable can be difficult, even impossible, especially when the probabilities are very small. But we may be able to find out what *type* of random variable the one at hand is by examining the causes that make it random. Knowing the type, we can often approximate the random variable to a *standard* one for which convenient formulas are available to calculate expected value, variance, and other statistics. We shall study some popular standard random variables and look at related spreadsheet formulas. Side by side, we shall examine the conditions under which a random variable encountered in practice can be approximated to one of the standard ones.

3–3 Bernoulli Random Variable

The first standard random variable we shall study is the *Bernoulli random variable*, named in honor of the mathematician Jakob Bernoulli (1654–1705). It is the building block for other random variables in this chapter. The distribution of a Bernoulli random variable X is given in Table 3–9. As seen in the table, x is 1 with probability p and 0 with probability $(1 - p)$. The case where $x = 1$ is called "success" and the case where $x = 0$ is called "failure."

Observe that

TABLE 3–9
Bernoulli Distribution

x	P(x)
1	p
0	$1 - p$

$$E(X) = 1*p + 0*(1 - p) = p$$
$$E(X^2) = 1^{2*}p + 0^{2*}(1 - p) = p$$
$$V(X) = E(X^2) - [E(X)]^2 = p - p^2 = p(1 - p).$$

Often the quantity $(1 - p)$, which is the probability of failure, is denoted by the symbol q, and thus $V(X) = pq$. If X is a Bernoulli random variable with probability of success p, then we write $X \sim \text{BER}(p)$, where the symbol "$\sim$" is read "is distributed as" and BER stands for Bernoulli. The characteristics of a Bernoulli random variable are summarized in the following box.

Bernoulli Distribution

If $X \sim \text{BER}(p)$, then

$$P(1) = p; \qquad P(0) = 1 - p$$
$$E[X] = p$$
$$V(X) = p(1 - p)$$

For example, if $p = 0.8$, then

$$E[X] = 0.8$$
$$V(X) = 0.8*0.2 = 0.16$$

Let us look at a practical instance of a Bernoulli random variable. Suppose an operator uses a lathe to produce pins, and the lathe is not perfect in that it does not always produce a good pin. Rather, it has a probability p of producing a good pin and $(1 - p)$ of producing a defective one.

Just after the operator produces one pin, let X denote the "number of good pins produced." Clearly, X is 1 if the pin is good and 0 if it is defective. Thus, X follows exactly the distribution in Table 3–9, and therefore $X \sim \text{BER}(p)$.

If the outcome of a trial can only be either a success or a failure, then the trial is a **Bernoulli trial**.

The number of successes X in one Bernoulli trial, which can be 1 or 0, is a **Bernoulli random variable**.

Another example is tossing a coin. If we take heads as 1 and tails as 0, then the outcome of a toss is a Bernoulli random variable.

A Bernoulli random variable is too simple to be of immediate practical use. But it forms the building block of the binomial random variable, which is quite useful in practice. The binomial random variable in turn is the basis for many other useful cases.

3–4   The Binomial Random Variable

In the real world we often make several trials, not just one, to achieve one or more successes. Since we now have a handle on Bernoulli-type trials, let us consider cases where there are n number of Bernoulli trials. A condition we need to impose on these trials is that the outcome of any trial be independent of the outcome of any other trial. Very often this independence condition is true. For example, when we toss a coin several times, the outcome of one toss is not affected by the outcome of any other toss.

Consider n number of *identically and independently distributed* Bernoulli random variables $X_1, X_2, \ldots, X_n$. Here, identically means that they all have the same p, and independently means that the value of one X does not in any way affect the value of another. For example, the value of X_2 does not affect the value of X_3 or X_5, and so on. Such a *sequence* of identically and independently distributed Bernoulli variables is called a **Bernoulli process.**

Suppose an operator produces n pins, one by one, on a lathe that has probability p of making a good pin at each trial. If this p remains constant throughout, then independence is guaranteed and the sequence of numbers (1 or 0) denoting the good and bad pins produced in each of the n trials is a Bernoulli process. For example, in the sequence of eight trials denoted by

$$0\ 0\ 1\ 0\ 1\ 1\ 0\ 0$$

the third, fifth, and sixth are good pins, or successes. The rest are failures.

In practice, we are usually interested in the total number of good pins rather than the sequence of 1's and 0's. In the example above, three out of eight are good. In the general case, let X denote the total number of good pins produced in n trials. We then have

$$X = X_1 + X_2 + \cdots + X_n$$

where all $X_i \sim \text{BER}(p)$ and are independent.

An X that counts the number of successes in many independent, identical Bernoulli trials is called a **binomial random variable.**

Conditions for a Binomial Random Variable

Note the conditions that need to be satisfied for a binomial random variable:

1. The trials must be Bernoulli trials in that the outcomes can only be either success or failure.
2. The outcomes of the trials must be independent.
3. The probability of success in each trial must be constant.

The first condition is easy to understand. Coming to the second condition, we already saw that the outcomes of coin tosses will be independent. As an example of dependent outcomes, consider the following experiment. We toss a fair coin and if it is heads we record the outcome as success, or 1, and if it is tails we record it as failure, or 0. For the second outcome, we do not toss the coin but we record the opposite of the previous outcome. For the third outcome, we toss the coin again and repeat the process of writing the opposite result for every other outcome. Thus in the sequence of all outcomes, every other outcome will be the opposite of the previous outcome. We stop after recording 20 outcomes. In this experiment, all outcomes are random and of Bernoulli type with success probability 0.5. But they are not independent in that every other outcome is the opposite of, and thus dependent on, the previous outcome. And for this reason, the number of successes in such an experiment will not be binomially distributed. (In fact, the number is not even random. Can you guess what that number will be?)

The third condition of constant probability of success is important and can be easily violated. Tossing two different coins with differing probabilities of success will violate the third condition (but not the other two). Another case that is relevant to the third condition, which we need to be aware of, is *sampling with and without replacement.* Consider an urn that contains 10 green marbles (successes) and 10 red marbles (failures). We pick a marble from the urn at random and record the outcome. The probability of success is $10/20 = 0.5$. For the second outcome, suppose *we replace the first marble drawn* and then pick one at random. In this case the probability of success remains at $10/20 = 0.5$, and the third condition is satisfied. But if *we do not replace the first marble* before picking the second, then the probability of the second outcome being a success is $9/19$ if the first was a success and $10/19$ if the first was a failure. Thus the probability of success does not remain constant (and is also dependent of the previous outcomes). Therefore, the third condition is violated (as is the second condition). This means that sampling with replacement will follow a binomial distribution, but sampling without replacement will not. Later we will see that sampling without replacement will follow a hypergeometric distribution.

Binomial Distribution Formulas

Consider the case where five trials are made, and in each trial the probability of success is 0.6. To get to the formula for calculating binomial probabilities, let us analyze the probability that the number of successes in the five trials is exactly three.

First, we note that there are $\binom{5}{3}$ ways of getting three successes out of five trials. Next we observe that each of these $\binom{5}{3}$ possibilities has $0.6^3 * 0.4^2$ probability of occurrence corresponding to 3 successes and 2 failures. Therefore,

$$P(X = 3) = \binom{5}{3} * 0.6^3 * 0.4^2 = 0.3456$$

We can generalize this equation with n denoting the number of trials and p the probability of success:

$$P(X = x) = \binom{n}{x} p^x (1 - p)^{(n-x)} \qquad \text{for } x = 0, 1, 2, \ldots, n \qquad (3\text{--}12)$$

Equation 3–12 is the famous binomial probability formula.

To describe a binomial random variable we need two parameters, n and p. We write $X \sim B(n, p)$ to indicate that X is binomially distributed with n number of trials and p probability of success in each trial. The letter B in the expression stands for binomial.

With any random variable, we will be interested in its expected value and its variance. Let us consider the expected value of a binomial random variable X. We note that X is the sum of n number of Bernoulli random variables, each of which has an expected value of p. Hence the expected value of X must be np, that is, $E(X) = np$. Furthermore, the variance of each Bernoulli random variable is $p(1 - p)$, and they are all independent. Therefore variance of X is $np(1 - p)$, that is, $V(X) = np(1 - p)$. The formulas for the binomial distribution are summarized in the next box, which also presents sample calculations that use these formulas.

Binomial Distribution

If $X \sim B(n, p)$, then

$$P(X = x) = \binom{n}{x} p^x (1 - p)^{(n-x)} \qquad x = 0, 1, 2, \ldots, n$$

$$E(X) = np$$

$$V(X) = np(1 - p)$$

For example, if $n = 5$ and $p = 0.6$, then

$$P(X=3) = 10 * 0.6^3 * 0.4^2 = 0.3456$$

$$E(X) = 5 * 0.6 = 3$$

$$V(X) = 5 * 0.6 * 0.4 = 1.2$$

The Template

The calculation of binomial probabilities, especially the cumulative probabilities, can be tedious. Hence we shall use a spreadsheet template. The template that can be used to calculate binomial probabilities is shown in Figure 3–13. When we enter the values for n and p, the template automatically tabulates the probability of "Exactly x," "At most x," and "At least x" number of successes. This tabulation can be used to solve many kinds of problems involving binomial probabilities, as explained in the next section. Besides the tabulation, a histogram is also created on the right. The histogram helps the user to visualize the shape of the distribution.

FIGURE 3–13   **Binomial Distribution Template**
                [Binomial.xls]

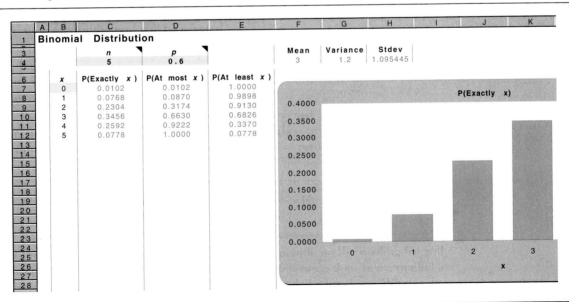

Problem Solving with the Template

Suppose an operator wants to produce *at least* two good pins. (In practice, one would want *at least* some number of *good* things, or *at most* some number of *bad* things. Rarely would one want *exactly* some number of good or bad things.) He produces the pins using a lathe that has 0.6 probability of making a good pin in each trial, and this probability stays constant throughout. Suppose he produces five pins. What is the probability that he would have made at least two good ones?

Let us see how we can answer the question using the template. After making sure that *n* is filled in as 5 and *p* as 0.6, the answer is read off as 0.9130 (in cell E9). That is, the operator can be 91.3% confident that he would have at least two good pins.

Let us go further with the problem. Suppose it is critical that the operator have at least two good pins, and therefore he wants to be at least 99% confident that he would have at least two good pins. (In this type of situation, the phrases "at least" and "at most" occur often. You should read carefully.) With five trials, we just found that he can be only 91.3% confident. To increase the confidence level, one thing he can do is increase the number of trials. How many more trials? Using the spreadsheet template, we can answer this question by progressively increasing the value of *n* and stopping when P(At least 2) in cell E9 just exceeds 99%. On doing this, we find that eight trials will do and seven will not. Hence the operator should make at least eight trials.

Increasing *n* is not the only way to increase confidence. We can increase *p*, if that is possible in practice. To see it, we pose another question.

Suppose the operator has enough time to produce only five pins, but he still wants to have at least 99% confidence of producing at least two good pins by improving the lathe and thus increasing *p*. How much should *p* be increased? To answer this, we can keep increasing *p* and stop when P(At least 2) just exceeds 99%. But this process could get tiresome if we need, say, four decimal place accuracy for *p*. This is where the Goal seek . . . command (see Chapter 0) in the spreadsheet comes in handy. The Goal seek command yields 0.7777. That is, *p* must be increased to at least 0.7777 in order to be 99% confident of getting at least two good pins in five trials.

We will complete this section by pointing out the use of the AutoCalculate command. We first note that the probability of at most *x* number of successes is the same as the cumulative probability $F(x)$. Certain types of probabilities are easily calculated using $F(x)$ values. For example, in our operator's problem, consider the probability that the number of successes will be between 1 and 3, both inclusive. We know that

$$P(1 \le x \le 3) = F(3) - F(0)$$

Looking at the template in Figure 3–13, we calculate this as $0.6630 - 0.0102 = 0.6528$. A quicker way is to use the AutoCalculate facility. When the range of cells containing $P(1)$ to $P(3)$ is selected, the sum of the probabilities appears in the Auto-Calculate area as 0.6528.

PROBLEMS

3–32. Three of the 10 airplane tires at a hangar are faulty. Four tires are selected at random for a plane; let F be the number of faulty tires found. Is F a binomial random variable? Explain.

3–33. A salesperson finds that, in the long run, two out of three sales calls are successful. Twelve sales calls are to be made; let X be the number of concluded sales. Is X a binomial random variable? Explain.

3–34. A large shipment of computer chips is known to contain 10% defective chips. If 100 chips are randomly selected, what is the expected number of defective ones? What is the standard deviation of the number of defective chips? Use Chebyshev's theorem to give bounds such that there is at least a 0.75 chance that the number of defective chips will be within the two bounds.

3–35. A new treatment for baldness is known to be effective in 70% of the cases treated. Four bald members of the same family are treated; let X be the number of successfully treated members of the family. Is X a binomial random variable? Explain.

3–36. What are Bernoulli trials? What is the relationship between Bernoulli trials and the binomial random variable?

3–37. Look at the histogram of probabilities in the binomial distribution template [Binomial.xls] for the case $n = 5$ and $p = 0.6$.

   a. Is this distribution symmetric or skewed? Now, increase the value of n to 10, 15, 20, . . . Is the distribution becoming more symmetric or more skewed? Make a formal statement about what happens to the distribution's shape when n increases.

   b. With $n = 5$, change the p value to 0.1, 0.2, . . . Observe particularly the case of $p = 0.5$. Make a formal statement about how the skewness of the distribution changes with p.

3–38. A salesperson goes door-to-door in a residential area to demonstrate the use of a new household appliance to potential customers. At the end of a demonstration, there is a constant 0.2107 probability that the potential customer would place an order for the product. To perform satisfactorily on the job, the salesperson needs at least four orders. Assume that each demonstration is a Bernoulli trial.

   a. If the salesperson makes 15 demonstrations, what is the probability that there would be exactly 4 orders?

   b. If the salesperson makes 16 demonstrations, what is the probability that there would be at most 4 orders?

   c. If the salesperson makes 17 demonstrations, what is the probability that there would be at least 4 orders?

   d. If the salesperson makes 18 demonstrations, what is the probability that there would be anywhere from 4 to 8 (both inclusive) orders?

   e. If the salesperson wants to be at least 90% confident of getting at least 4 orders, at least how many demonstrations should she make?

   f. The salesperson has time to make only 22 demonstrations, and she still wants to be at least 90% confident of getting at least 4 orders. She intends to gain this confidence by improving the quality of her demonstration and thereby improving the chances of getting an order at the end of a demonstration. At least to what value should this probability be increased in order to gain the desired confidence? Your answer should be accurate to four decimal places.

3–39. An MBA graduate is applying for nine jobs, and believes that she has in each of the nine cases a constant and independent 0.48 probability of getting an offer.

   a. What is the probability that she will have at least three offers?

   b. If she wants to be 95% confident of having at least three offers, how many more jobs should she apply for? (Assume each of these additional applications will also have the same probability of success.)

    c. If there are no more than the original nine jobs that she can apply for, what value of probability of success would give her 95% confidence of at least three offers?

3–40. A computer laboratory in a school has 33 computers. Each of the 33 computers has 90% reliability. Allowing for 10% of the computers to be down, an instructor specifies an enrollment ceiling of 30 for his class. Assume that a class of 30 students is taken into the lab.

    a. What is the probability that each of the 30 students will get a computer in working condition?

    b. The instructor is surprised to see the low value of the answer to (*a*) and decides to improve it to at least 95% by doing one of the following:

        i. Decreasing the enrollment ceiling.

        ii. Increasing the number of computers in the lab.

        iii. Increasing the reliability of all the computers.

To help the instructor, find out what the increase or decrease should be for each of the three alternatives.

3–41. A commercial jet aircraft has four engines. For an aircraft in flight to land safely, at least two engines should be in working condition. Each engine has an independent reliability of $p = 92\%$.

    a. What is the probability that an aircraft in flight can land safely?

    b. If the probability of landing safely must be at least 99.5%, what is the minimum value for *p?* Repeat the question for probability of landing safely to be 99.9%.

    c. If the reliability cannot be improved beyond 92% but the number of engines in a plane can be increased, what is the minimum number of engines that would achieve at least 99.5% probability of landing safely? Repeat for 99.9% probability.

    d. One would certainly desire 99.9% probability of landing safely. Looking at the answers to (*b*) and (*c*), what would you say is a better approach to safety, increasing the number of engines or increasing the reliability of each engine?

3–5 Negative Binomial Distribution

Consider again the case of the operator who wants to produce two good pins using a lathe that has 0.6 probability of making one good pin in each trial. Under binomial distribution, we assumed that he produces five pins and calculated the probability of getting at least two good ones. In practice, though, if two pins are all that are needed, the operator would produce the pins one by one and stop when he gets two good ones. For instance, if the first two are good, then he would stop right there; if the first and the third are good, then he would stop with the third; and so on. Notice that in this scenario, the number of successes is held constant at 2, and the number of trials is random. The number of trials could be 2, 3, 4, ... (Contrast this with the binomial distribution where the number of trials is fixed and the number of successes is random.)

    The number of trials made in this scenario is said to follow a **negative binomial distribution.** Let *s* denote the exact number of successes desired and *p* the probability of success in each trial. Let X denote the number of trials made until the desired number of successes is achieved. Then X will follow a negative binomial distribution and we shall write $X \sim NB(s, p)$ where NB denotes negative binomial.

Negative Binomial Distribution Formulas

What is the formula for $P(X = x)$ when $X \sim NB(s, p)$? We know that the very last trial must be a success; otherwise, we would have already had the desired number of successes with $x - 1$ trials, and we should have stopped right there. The last trial being a success, the first $x - 1$ trials should have had $s - 1$ successes. Thus the formula should be

$$P(X = x) = \binom{x - 1}{s - 1} p^s (1 - p)^{(x-s)}$$

The formula for the mean can be arrived at intuitively. For instance, if $p = 0.3$, and 3 successes are desired, then the expected number of trials to achieve 3 successes is 10. Thus the mean should have the formula $\mu = s/p$. The variance is given by the formula $\sigma^2 = s(1 - p)/p^2$.

Negative Binomial Distribution

If $X \sim NB(s, p)$, then

$$P(X = x) = \binom{x - 1}{s - 1} p^s (1 - p)^{(x-s)} \qquad x = s,\ s + 1,\ s + 2,\ \ldots$$

$$E(X) = s/p$$

$$V(X) = s(1 - p)/p^2$$

For example, if $s = 2$ and $p = 0.6$, then

$$P(X = 5) = \binom{4}{1} * 0.6^2 * 0.4^3 = 0.0922$$

$$E(X) = 2/0.6 = 3.3333$$

$$V(X) = 2 * 0.4 / 0.6^2 = 2.2222$$

Problem Solving with the Template

Figure 3–14 shows the negative binomial distribution template. When we enter the s and p values, the template updates the probability tabulation and draws a histogram on the right.

Let us return to the operator who wants to keep producing pins until he has two good ones. The probability of getting a good one at any trial is 0.6. What is the probability that he would produce exactly five? Looking at the template, we see that the answer is 0.0922, which agrees with the calculation in the preceding box. We can, in addition, see that the probability of producing at most five is 0.9130 and at least five is 0.1792.

Suppose the operator has enough time to produce only four pins. How confident can he be that he would have two good ones within the available time? Looking at the template, we see that the probability of needing at most four trials is 0.8208 and hence he can be about 82% confident.

FIGURE 3–14   **Negative Binomial Distribution Template**
              [Negative Binomial.xls]

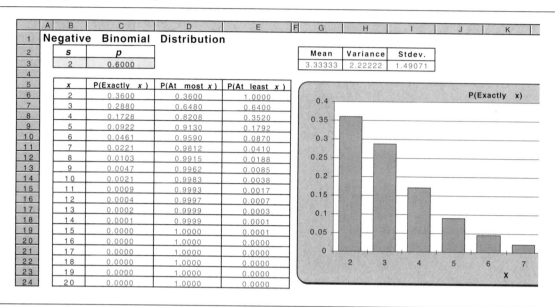

If he wants to be at least 95% confident, at least how many trials should he be prepared for? Looking at the template in the "At most" column, we infer that he should be prepared for at least six trials, since five trials yield only 91.30% confidence and six trials yield 95.90%.

Suppose the operator has enough time to produce only four pins and still wants to be at least 95% confident of getting two good pins within the available time. Suppose, further, he wants to achieve this by increasing the value of p. What is the minimum p that would achieve this? Using the Goal Seek command, this can be answered as 0.7514. Specifically, you set cell D8 to 0.95 by changing cell C3.

3–6   The Geometric Distribution

In a negative binomial distribution, the number of desired successes s can be any number. But in some practical situations, the number of successes desired is just one. For instance, if you are attempting to pass a test or get some information, it is enough to have just one success. Let X be the (random) number of Bernoulli trials, each having p probability of success, required to achieve just one success. Then X follows a **geometric distribution,** and we shall write $X \sim G(p)$. Note that the geometric distribution is a special case of the negative binomial distribution where $s = 1$. The reason for the name "geometric distribution" is that the sequence of probabilities $P(X = 1)$, $P(X = 2)$, ..., follows a *geometric progression.*

Geometric Distribution Formulas

Because the geometric distribution is a special case of the negative binomial distribution where $s = 1$, the formulas for the negative binomial distribution with s fixed as 1 can be used for the geometric distribution.

FIGURE 3–15 **Geometric Distribution Template [Geometric.xls]**

Geometric Distribution

p
0.6

Mean	Variance	Stdev.
1.66666	1.11111	1.05409

x	P(Exactly x)	P(At most x)	P(At least x)
1	0.6000	0.6000	1.0000
2	0.2400	0.8400	0.4000
3	0.0960	0.9360	0.1600
4	0.0384	0.9744	0.0640
5	0.0154	0.9898	0.0256
6	0.0061	0.9959	0.0102
7	0.0025	0.9984	0.0041
8	0.0010	0.9993	0.0016
9	0.0004	0.9997	0.0007
10	0.0002	0.9999	0.0003
11	0.0001	1.0000	0.0001
12	0.0000	1.0000	0.0000
13	0.0000	1.0000	0.0000
14	0.0000	1.0000	0.0000
15	0.0000	1.0000	0.0000
16	0.0000	1.0000	0.0000
17	0.0000	1.0000	0.0000
18	0.0000	1.0000	0.0000
19	0.0000	1.0000	0.0000
20	0.0000	1.0000	0.0000
21	0.0000	1.0000	0.0000
22	0.0000	1.0000	0.0000

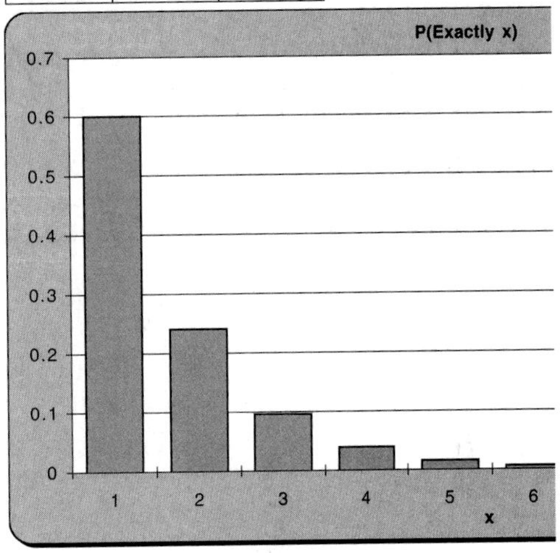

Geometric Distribution Formulas

If $X \sim G(p)$, then

$$P(X = x) = p(1 - p)^{(x-1)} \qquad x = 1, 2, \ldots$$
$$E(X) = 1/p$$
$$V(X) = (1 - p)/p^2$$

For example, if $p = 0.6$, then

$$P(X = 5) = 0.6{*}0.4^4 = 0.0154$$
$$E(X) = 1/0.6 = 1.6667$$
$$V(X) = 0.4/0.6^2 = 1.1111$$

Problem Solving with the Template

Consider the operator who produces pins one by one on a lathe that has 0.6 probability of producing a good pin at each trial. Suppose he wants only one good pin and stops as soon as he gets one. What is the probability that he would produce exactly five pins? The template that can be used to answer this and related questions is shown in Figure 3–15. On that template, we enter the value 0.6 for p. The answer can now be read off as 0.0154, which agrees with the example calculation in the preceding box. Further, we can read on the template that the probability of at most five is 0.9898 and at least five is 0.0256. Also note that the probability of exactly 1, 2, 3, . . . , trials follows the sequence 0.6, 0.24, 0.096, 0.0384, . . . , which is indeed a geometric progression with common ratio 0.4.

Now suppose the operator has time enough for at most two pins; how confident can he be of getting a good one within the available time? From the template, the answer is 0.8400, or 84%. What if he wants to be at least 95% confident? Again from the template, he must have enough time for four pins, because three would yield only 93.6% confidence and four yields 97.44%.

Suppose the operator wants to be 95% confident of getting a good pin by producing at most two pins. What value of p will achieve this? Using the Goal Seek command the answer is found to be 0.7761.

3–7   The Hypergeometric Distribution

Assume that a box contains 10 pins of which 6 are good and the rest defective. An operator picks 5 pins at random from the 10, and is interested in the number of good pins picked. Let X denote the number of good pins picked. We should first note that this is a case of sampling without replacement and therefore X is *not* a binomial random variable. The probability of success p, which is the probability of picking a good pin, is neither constant nor independent from trial to trial. The first pin picked has 0.6 probability of being good; the second has either 5/9 or 6/9 probability, depending on whether or not the first was good. Therefore, X does not follow a binomial distribution, but follows what is called a **hypergeometric distribution**. In general, when a pool of size N contains S successes and $(N - S)$ failures, and a random sample of size n is drawn from the pool, the number of successes X in the sample follows a hypergeometric distribution. We shall then write $X \sim \text{HG}(n, S, N)$. The situation is depicted in Figure 3–16.

Hypergeometric Distribution Formulas

Let us derive the formula for $P(X = x)$ when X is hypergeometrically distributed. The x number of successes have to come from the S successes in the pool, which can happen in $\binom{S}{x}$ ways. The $(n - x)$ failures have to come from the $(N - S)$ failures in the pool, which can happen in $\binom{N - S}{n - x}$ ways. Together, the x successes and $(n - x)$ failures can happen in $\binom{S}{x}\binom{N - S}{n - x}$ ways. Finally, there are $\binom{N}{n}$ ways of selecting a sample of size n. Putting them all together,

FIGURE 3–16   **Schematic for Hypergeometric Distribution**

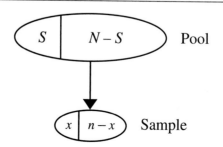

$$P(X = x) = \frac{\binom{S}{x}\binom{N-S}{n-x}}{\binom{N}{n}}$$

In this formula n cannot exceed N since the sample size cannot exceed the pool size. There is also a minimum possible value and a maximum possible value for x, depending on the values of n, S, and N. For instance, if $n = 9$, $S = 5$, and $N = 12$, you may verify that there would be at least two successes and at most five. In general, the minimum possible value for x is $\text{Max}(0, n - N + S)$ and the maximum possible value is $\text{Min}(n, S)$.

Hypergeometric Distribution Formulas

If $X \sim HG(n, S, N)$, then

$$P(X = x) = \frac{\binom{S}{x}\binom{N-S}{n-x}}{\binom{N}{n}} \qquad \text{Max}(0, n - N + S) \le x \le \text{Min}(n, S)$$

$$E(X) = np \qquad \text{where } p = S/N$$

$$V(X) = np(1 - p)\left[\frac{N-n}{N-1}\right]$$

For example, if $n = 5$, $S = 6$, and $N = 10$, then

$$P(X = 2) = \frac{\binom{6}{2}\binom{10-6}{5-2}}{\binom{10}{5}} = 0.2381$$

$$E(X) = 5*(6/10) = 3.00$$
$$V(X) = 5*0.6*(1 - 0.6)*(10 - 5)/(10 - 1) = 0.6667$$

The proportion of successes in the pool, which is the ratio S/N, is the probability of the first trial being a success. This ratio is denoted by the symbol p since it resembles the p used in the binomial distribution. The expected value and variance of X are expressed using p as

$$E(X) = np$$
$$V(X) = np(1 - p)\left[\frac{N-n}{N-1}\right]$$

Notice that the formula for $E(X)$ is the same as for the binomial case. The formula for $V(X)$ is similar to but not the same as the binomial case. The difference is the additional factor in square brackets. This additional factor approaches 1 as N becomes larger and larger compared to n and may be dropped when N is, say, 100 times as large as n. We can then approximate the hypergeometric distribution as a binomial distribution.

FIGURE 3–17 **The Template for the Hypergeometric Distribution [Hypergeometric.xls]**

Hypergeometric Distribution

n	*S*	*N*
5	6	10

Mean	Variance	Stdev.	Min *x*	Max *x*
3	0.666667	0.816497	1	5

x	P(exactly *x*)	P(at most *x*)	P(al least *x*)
1	0.0238	0.0238	1.0000
2	0.2381	0.2619	0.9762
3	0.4762	0.7381	0.7381
4	0.2381	0.9762	0.2619
5	0.0238	1.0000	0.0238

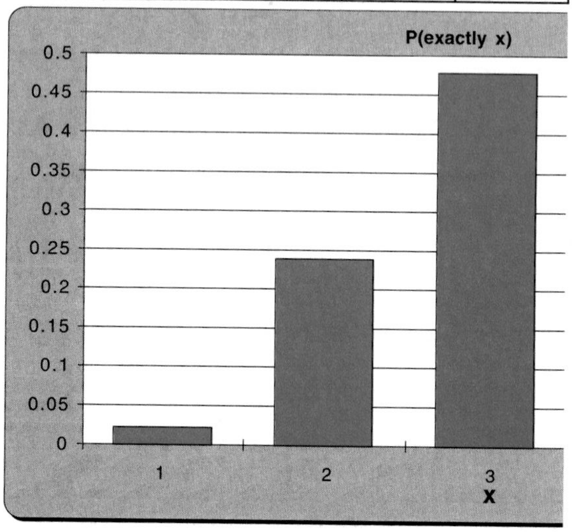

Problem Solving with the Template

Figure 3–17 shows the template used for the hypergeometric distribution. Let us consider the case where a box contains 10 pins out of which 6 are good, and the operator picks 5 at random. What is the probability that exactly 2 good pins are picked? The answer is 0.2381 (Cell C8). Additionally, the probabilities that at most two and at least two good ones are picked are, respectively, 0.2619 and 0.9762.

Suppose the operator needs at least three good pins. How confident can he be of getting at least three good pins? The answer is 0.7381 (Cell E9). Suppose the operator wants to increase this confidence to 90% by adding some good pins to the pool. How many good pins should be added to the pool? This question, unfortunately, cannot be answered using the Goal Seek command for three reasons. First, the Goal Seek command works on a continuous scale, whereas S and N must be integers. Second, *when* n, S, *or* N *is changed the tabulation may shift and* P*(at least 3) may not be in cell E9!* Third, the Goal Seek command can change only one cell at a time. But in many problems, two cells (S and N) may have to change. Hence do not use the Goal Seek or the Solver on this template. Also, be careful to read the probabilities from the correct cells.

Let us solve this problem without using the Goal Seek command. If a good pin is added to the pool, what happens to S and N? They *both* increase by 1. Thus we should enter 7 for S and 11 for N. When we do, P(at least 3) $= 0.8030$, which is less than the desired 90% confidence. So we add one more good pin to the pool. Continuing in this fashion, we find that at least four good pins must be added to the pool.

Another way to increase P(at least 3) is to remove a bad pin from the pool. What happens to S and N when a bad pin is removed? S will remain the same and N will decrease by one. Suppose the operator wants to be 80% confident that at least three good pins will be selected. How many bad pins must be removed from the pool? Decreasing N one by one, we find that it is enough if one bad pin is removed.

3–8   The Poisson Distribution

Imagine an automatic lathe that mass produces pins. On rare occasions, let us assume that the lathe produces a gem of a pin which is so perfect that it can be used for a very special purpose. To make the case specific, let us assume the lathe produces 20,000 pins and has 1/10,000 chance of producing a perfect one. Suppose we are interested in the number of perfect pins produced. We could try to calculate this number by using the binomial distribution with $n = 20,000$ and $p = 1/10,000$. But the calculation would be almost impossible because n is so large, p is so small, and the binomial formula calls for $n!$ and p^{n-x}, which are hard to calculate even on a computer. However, the expected number of defectives produced is $np = 20,000*(1/10,000) = 2$, which is neither too large nor too small. It turns out that as long as the expected value $\mu = np$ is neither too large nor too small, say, lies between 0.01 and 50, the binomial formula for $P(X = x)$ can be approximated as

$$P(X = x) = \frac{e^{-\mu}\mu^x}{x!} \qquad x = 0, 1, 2, \ldots$$

where e is the natural base of logarithms, equal to 2.71828. . . . This formula is known as the **Poisson formula,** and the distribution is called the **Poisson distribution.** In general, if we count the number of times a rare event occurs during a fixed interval, then that number would follow a Poisson distribution. We know the mean $\mu = np$.

Considering the variance of a Poisson distribution, we note that the binomial variance is $np(1 - p)$. But since p is very small, $(1 - p)$ is close to 1 and therefore can be omitted. Thus the variance of a Poisson random variable is np, which happens to be the same as its mean. The Poisson formula needs only μ, and not n or p.

We suddenly realize that we need not know n and p separately. All we need to know is their product, μ, which is the mean and the variance of the distribution. Just one number, μ, is enough to describe the whole distribution, and in this sense, the Poisson distribution is a simple one, even simpler than the binomial. If X follows a Poisson distribution, we shall write $X \sim P(\mu)$ where μ is the expected value of the distribution. The following box summarizes the Poisson distribution.

Poisson Distribution Formulas

If $X \sim P(\mu)$, then

$$P(X = x) = \frac{e^{-\mu}\mu^x}{x!} \qquad x = 0, 1, 2, \ldots$$

$$E(X) = np = \mu$$
$$V(X) = np = \mu$$

For example, if $\mu = 2$, then

$$P(X = 3) = \frac{e^{-2}2^3}{3!} = 0.1804$$

$$E(X) = \mu = 2.00$$
$$V(X) = \mu = 2.00$$

The Poisson template is shown in Figure 3–18. The only input needed is the mean μ in cell C4. The starting value of *x* in cell B7 is usually zero, but it can be changed as desired.

Problem Solving with the Template

Let us return to the case of the automatic lathe that produces perfect pins on rare occasions. Assume that the lathe produces on the average two perfect pins a day, and an operator wants at least three perfect pins. What is the probability that it will produce at least three perfect pins on a given day? Looking at the template, we find the answer to be 0.3233. Suppose the operator waits for two days. In two days the lathe will produce on average four perfect pins. We should therefore change the mean in cell C4 to 4. What is the probability that the lathe will produce at least three perfect pins in two days? Using the template, we find the answer to be 0.7619. If the operator wants to be at least 95% confident of producing at least three perfect pins, how many days should he be prepared to wait? Again, using the template, we find that the operator should be prepared to wait at least four days.

A Poisson distribution also occurs in other types of situations leading to other forms of analysis. Consider an emergency call center. The number of distress calls received within a specific period, being a count of rare events, is usually Poisson-distributed. In this context, suppose the call center receives on average two calls per hour. In addition, suppose the crew at the center can handle up to three calls in an hour. What is the probability that the crew can handle all the calls received in a given hour? Since

FIGURE 3–18   Poisson Distribution Template [Poisson.xls]

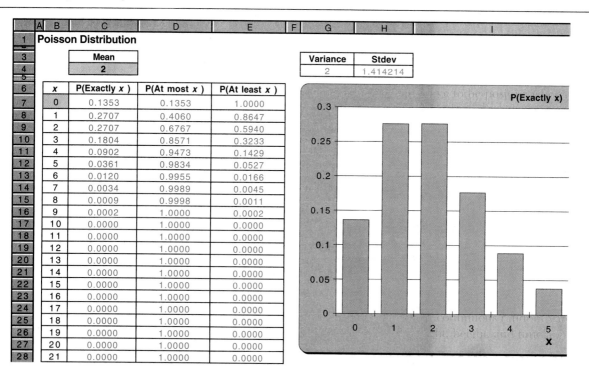

x	P(Exactly x)	P(At most x)	P(At least x)
0	0.1353	0.1353	1.0000
1	0.2707	0.4060	0.8647
2	0.2707	0.6767	0.5940
3	0.1804	0.8571	0.3233
4	0.0902	0.9473	0.1429
5	0.0361	0.9834	0.0527
6	0.0120	0.9955	0.0166
7	0.0034	0.9989	0.0045
8	0.0009	0.9998	0.0011
9	0.0002	1.0000	0.0002
10	0.0000	1.0000	0.0000
11	0.0000	1.0000	0.0000
12	0.0000	1.0000	0.0000
13	0.0000	1.0000	0.0000
14	0.0000	1.0000	0.0000
15	0.0000	1.0000	0.0000
16	0.0000	1.0000	0.0000
17	0.0000	1.0000	0.0000
18	0.0000	1.0000	0.0000
19	0.0000	1.0000	0.0000
20	0.0000	1.0000	0.0000
21	0.0000	1.0000	0.0000

Mean: 2 — Variance: 2 — Stdev: 1.414214

the crew can handle up to three calls, we look for the probability of at most three calls. From the template, the answer is 0.8571. If the crew wanted to be at least 95% confident of handling all the calls received during a given hour, how many calls should it be prepared to handle? Again, from the template, the answer is five, because the probability of at most four calls is less than 95% and of at most five calls is more than 95%.

3–9 Continuous Random Variables

Instead of depicting probability distributions by simple graphs, where the height of the line above each value represents the probability of that value of the random variable, let us use a histogram. We will associate the *area* of each rectangle of the histogram with the probability of the particular value represented. Let us look at a simple example. Let X be the time, measured in minutes, it takes to complete a given task. A histogram of the probability distribution of X is shown in Figure 3–19.

The probability of each value is the area of the rectangle over the value and is written on top of the rectangle. Since the rectangles all have the same base, the height of each rectangle is proportional to the probability. Note that the probabilities add to 1.00, as required. Now suppose that X can be measured more accurately. The distribution of X, with time now measured to the nearest half-minute, is shown in Figure 3–20.

FIGURE 3–19 **Histogram of the Probability Distribution of Time to Complete a Task, with Time Measured to the Nearest Minute**

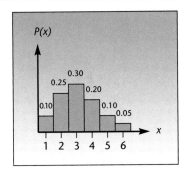

FIGURE 3–20 **Histogram of the Probability Distribution of Time to Complete a Task, with Time Measured to the Nearest Half-Minute**

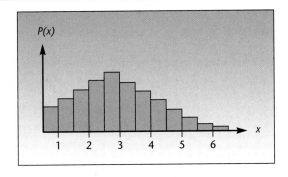

Let us continue the process. Time is a continuous random variable; it can take on any value measured on an interval of numbers. We may, therefore, refine our measurement to the nearest quarter-minute, the nearest 5 seconds, or the nearest second, or we can use even more finely divided units. As we refine the measurement scale, the number of rectangles in the histogram increases and the width of each rectangle decreases. The probability of each value is still measured by the area of the rectangle above it, and the total area of all rectangles remains 1.00, as required of all probability distributions. As we keep refining our measurement scale, the discrete distribution of X tends to a continuous probability distribution. The steplike surface formed by the tops of the rectangles in the histogram tends to a smooth function. This function is denoted by $f(x)$ and is called the **probability density function** of the continuous random variable X. Probabilities are still measured as areas under the curve. The probability that the task will be completed in 2 to 3 minutes is the area under $f(x)$ between the points $x = 2$ and $x = 3$. Histograms of the probability distribution of X with our measurement scale refined further and further are shown in Figure 3–21. Also shown is the density function $f(x)$ of the limiting continuous random variable X. The density function is the limit of the histograms as the number of rectangles approaches infinity and the width of each rectangle approaches zero.

Now that we have developed an intuitive feel for continuous random variables, and for probabilities of intervals of values as areas under a density function, we make some formal definitions.

A **continuous random variable** is a random variable that can take on any value in an interval of numbers.

The probabilities associated with a continuous random variable X are determined by the **probability density function** of the random variable. The function, denoted $f(x)$, has the following properties.

1. $f(x) \geq 0$ for all x.
2. The probability that X will be between two numbers a and b is equal to the area under $f(x)$ between a and b.
3. The total area under the entire curve of $f(x)$ is equal to 1.00.

When the sample space is continuous, the probability of any single given value is zero. For a continuous random variable, therefore, the probability of occurrence of any given value is zero. We see this from property 2, noting that the area under a curve between a point and itself is the area of a line, which is zero. *For a continuous random variable, nonzero probabilities are associated only with intervals of numbers.*

We define the cumulative distribution function $F(x)$ for a continuous random variable similarly to the way we defined it for a discrete random variable: $F(x)$ is the probability that X is less than (or equal to) x.

The **cumulative distribution function** of a continuous random variable:[2]

$$F(x) = P(X \leq x) = \text{area under } f(x) \text{ between the } \textit{smallest} \text{ possible value of } X \text{ (often } -\infty) \text{ and point } x$$

---

[2]If you are familiar with calculus, you know that the area under a curve of a function is given by the integral of the function. The probability that X will be between a and b is the definite integral of $f(x)$ between these two points: $P(a < X < b) = \int_a^b f(x)\, dx$. In calculus notation, we define the cumulative distributive function as $F(x) = \int_{-\infty}^x f(y)\, dy$.

FIGURE 3–21   **Histograms of the Distribution of Time to Complete a Task as Measurement Is Refined to Smaller and Smaller Intervals of Time, and the Limiting Density Function $f(x)$**

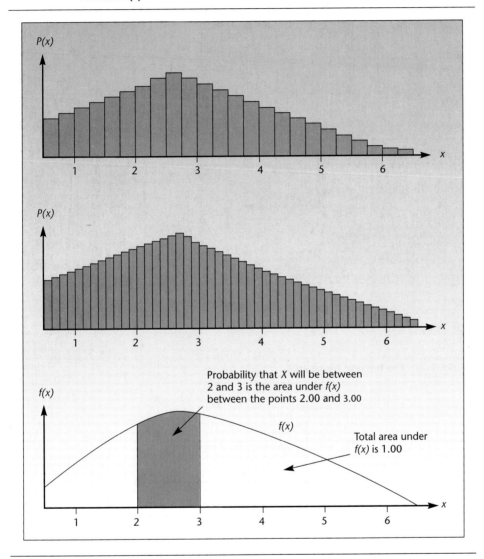

The cumulative distribution function $F(x)$ is a smooth, nondecreasing function that increases from 0 to 1.00. The connection between $f(x)$ and $F(x)$ is demonstrated in Figure 3–22.

The expected value of a continuous random variable X, denoted by $E(X)$, and its variance, denoted by $V(X)$, require the use of calculus for their computation.[3]

---

[3] $E(X) = \int_{-\infty}^{\infty} x f(x)\,dx;\ V(X) = \int_{-\infty}^{\infty} [x - E(X)]^2 f(x)\,dx.$

FIGURE 3–22   **Probability Density Function and Cumulative Distribution Function of a Continuous Random Variable**

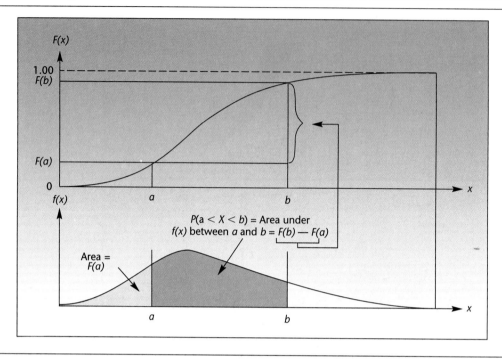

FIGURE 3–23   **The Uniform Distribution**

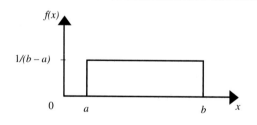

3–10   Uniform Distribution

The uniform distribution is the simplest of continuous distributions. The probability density function is

$$f(x) = 1/(b - a) \qquad a \leq x \leq b$$
$$= 0 \qquad \text{all other } x$$

where a is the minimum possible value and b is the maximum possible value of X. The graph of $f(x)$ is shown in Figure 3–23. Because the curve of $f(x)$ is a flat line, the area under it between any two points x_1 and x_2 will be a rectangle with height $1/(b - a)$ and width $(x_2 - x_1)$. Thus $P(x_1 \leq X \leq x_2) = (x_2 - x_1)/(b - a)$. If X is uniformly distributed between a and b, we shall write $X \sim U(a, b)$.

The mean of the distribution is the midpoint between a and b, which is $(a + b)/2$. By using integration, it can be shown that the variance is $(b - a)^2/12$. Because the shape of a uniform distribution is always a rectangle, the skewness and kurtosis are the same for all uniform distributions. The skewness is zero. (Why?) Because the shape is flat, the (relative) kurtosis is negative, always equal to -1.2.

The formulas for uniform distribution are summarized in the following box. Because the probability calculation is simple, there is no special spreadsheet function for uniform distribution. The box contains some sample calculations.

Uniform Distribution Formulas

If $X \sim U(a, b)$, then

$$f(x) = 1/(b - a) \qquad a \leq x \leq b$$
$$= 0 \qquad \text{all other } x$$
$$P(x_1 \leq X \leq x_2) = (x_2 - x_1)/(b - a) \qquad a \leq x_1 < x_2 \leq b$$
$$E(X) = (a + b)/2$$
$$V(X) = (b - a)^2/12$$

For example, if $a = 10$ and $b = 20$, then

$$P(12 \leq X \leq 18) = (18 - 12)/(20 - 10) = 0.6$$
$$E(X) = (10 + 20)/2 = 15$$
$$V(X) = (20 - 10)^2/12 = 8.3333$$

A common instance of uniform distribution is waiting time for a facility that goes in cycles. Two good examples are a shuttle bus and an elevator, which move, roughly, in cycles with some cycle time. If a user comes to a stop at a random time and waits till the facility arrives, the waiting time will be uniformly distributed between a minimum of zero and a maximum equal to the cycle time. In other words, if a shuttle bus has a cycle time of 20 minutes, the waiting time would be uniformly distributed between 0 and 20 minutes.

Problem Solving with the Template

Figure 3–24 shows the template for the uniform distributions. If $X \sim U(10, 20)$, what is $P(12 \leq X \leq 18)$? In the template, make sure the Min and Max are set to 10 and 20 in cells B4 and C4. Enter 12 and 18 in cells H10 and J10. The answer of 0.6 appears in cell I10.

What is the probability $P(X < 12)$? To answer this, enter 12 in cell C10. The answer 0.2 appears in cell B10. What is $P(X > 12)$? To answer this, enter 12 in cell E10. The answer 0.8 appears in F10.

Inverse calculations are possible in the bottom area of the template. Suppose you want to find x such that $P(X < x) = 0.2$. Enter 0.2 in cell B20. The answer, 12, appears in cell C20. To find x such that $P(X > x) = 0.3$, enter 0.3 in cell F20. The answer, 17, appears in cell E20.

As usual, you may also use facilities such as the **Goal Seek** command or the **Solver** tool in conjunction with this template.

FIGURE 3–24   **Template for the Uniform Distribution**
              [Uniform.xls]

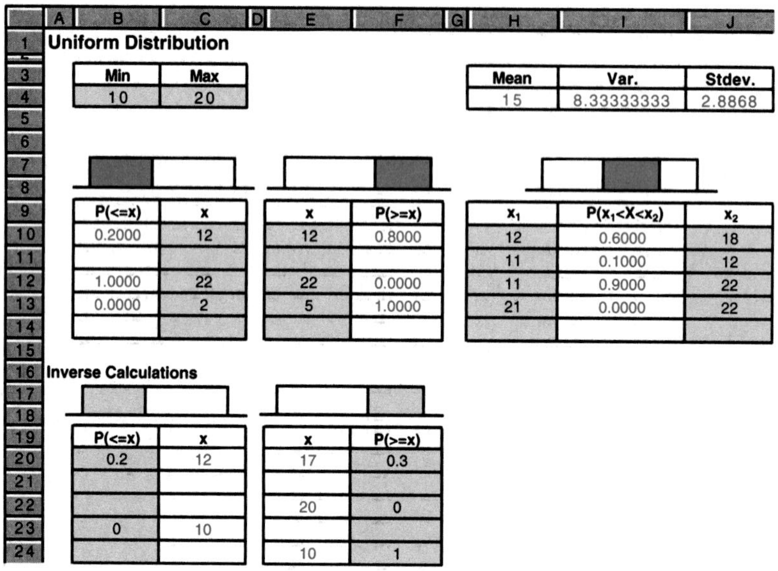

3–11   The Exponential Distribution

Suppose an event occurs with an average frequency of λ occurrences per hour and this average frequency is constant in that the probability that the event will occur during any tiny duration t is λt. Suppose further we arrive at the scene at any given time and wait till the event occurs. The waiting time will then follow an **exponential distribution,** which is the continuous limit of the geometric distribution. Suppose our waiting time was x. For the event (or success) to occur at time x, every tiny duration t from time 0 to time x should be a failure and the interval x to $x + t$ must be a success. This is nothing but a geometric distribution. To get the continuous version, we take the limit of this process as t approaches zero.

The exponential distribution is fairly common in practice. Here are some examples.

1. The time between two successive breakdowns of a machine will be exponentially distributed. This information is relevant to maintenance engineers. The mean μ in this case is known as the **mean time between failures** or **MTBF.**

2. The life of a product that fails by accident rather than by wear-and-tear follows an exponential distribution. Electronic components are good examples. This information is relevant to warranty policies.

3. The time gap between two successive arrivals to a waiting line, known as the **interarrival time,** will be exponentially distributed. This information is relevant to waiting line management.

When X is exponentially distributed with frequency λ, we shall write $X \sim E(\lambda)$. The probability density function $f(x)$ of the exponential distribution has the form

$$f(x) = \lambda e^{-\lambda x}$$

where λ is the frequency with which the event occurs. The frequency λ is expressed as so many times per unit time, such as 1.2 times per month. The mean of the distribution is $1/\lambda$ and the variance is $(1/\lambda)^2$. Just like the geometric distribution, the exponential distribution is positively skewed.

A Remarkable Property

The exponential distribution has a remarkable property. Suppose the time between two successive breakdowns of a machine is exponentially distributed with an MTBF of 100 hours, and we have just witnessed one breakdown. If we start a stopwatch as soon as it is repaired and put back into service so as to measure the time until the next failure then that time will, of course, be exponentially distributed with a μ of 100 hours. What is remarkable is the following. Suppose we arrive at the scene at some random time and start the stopwatch (instead of starting it immediately after a breakdown); the time until next breakdown will still be exponentially distributed with the same μ of 100 hours. In other words, it is immaterial when the event occurred last and how much later we start the stopwatch. For this reason, an exponential process is known as a *memoryless process*. It does not depend on the past at all.

The Template

The template for this distribution is seen in Figure 3–25. The following box summarizes the formulas and provides example calculations.

Exponential Distribution Formulas

If $X \sim E(\lambda)$, then

$$f(x) = \lambda e^{-\lambda x} \qquad x \geq 0$$
$$P(X \leq x) = 1 - e^{-\lambda x} \qquad \text{for } x \geq 0$$
$$P(X \geq x) = e^{-\lambda x} \qquad \text{for } x \geq 0$$
$$P(x_1 \leq X \leq x_2) = e^{-\lambda x_1} - e^{-\lambda x_2} \qquad 0 \leq x_1 < x_2$$
$$E(X) = 1/\lambda$$
$$V(X) = 1/\lambda^2$$

For example, if $\lambda = 1.2$, then

$$P(X \geq 0.5) = e^{-1.2*0.5} = 0.5488$$
$$P(1 \leq X \leq 2) = e^{-1.2*1} - e^{-1.2*2} = 0.2105$$
$$E(X) = 1/1.2 = 0.8333$$
$$V(X) = 1/1.2^2 = 0.6944$$

To use the exponential distribution template seen in Figure 3–25 (page 160), the value of λ must be entered in cell B4. At times, the mean μ rather than λ may be known, in which case its reciprocal $1/\mu$ is what should be entered as λ in cell B4. Note that λ is the average number of occurrences of a rare event in unit time and μ is the average time gap between two successive occurrences. The shaded cells are the input

FIGURE 3–25   **Exponential Distribution Template**
[Exponential.xls]

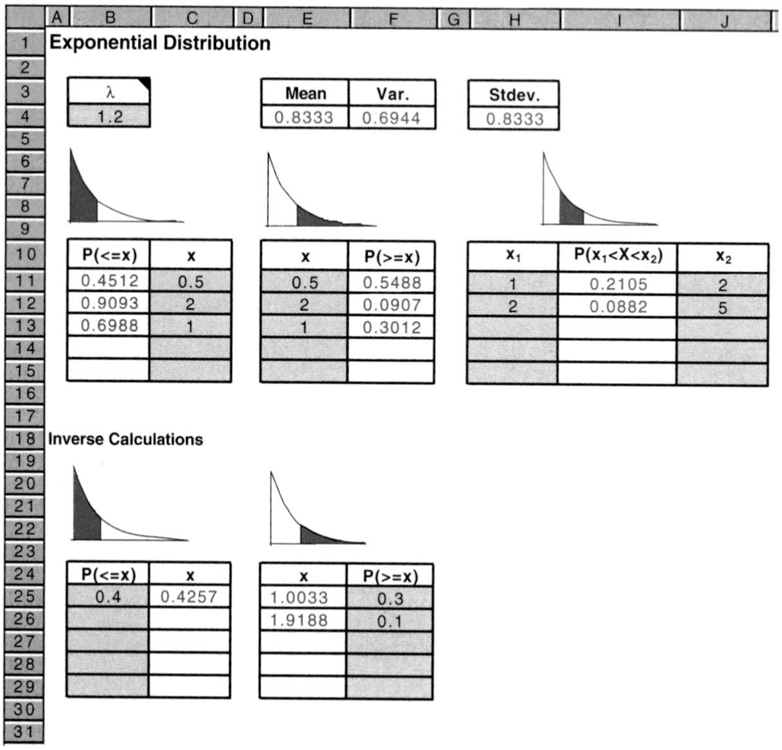

cells and the rest are protected. As usual, the Goal Seek command and the Solver tool can be used in conjunction with this template to solve problems.

EXAMPLE 3–4   A particular brand of handheld computers fails following an exponential distribution with a μ of 54.82 months. The company gives a warranty for 6 months.

a. What percent of the computers will fail within the warranty period?
b. If the manufacturer wants only 8% of the computers to fail during the warranty period, what should be the average life?

Solution   a. Enter the reciprocal of $54.82 = 0.0182$ as λ in the template. (You may enter the formula "$=1/54.82$" in the cell. But then you will not be able to use the Goal Seek command to change this entry. The Goal Seek command requires that the changing cell contain a number rather than a formula.) The answer we are looking for is the area to the left of 6. Therefore, enter 6 in cell C11. The area to the left, 0.1037, appears in cell B11. Thus 10.37% of the computers will fail within the warranty period.

b. Enter 0.08 in cell B25. Invoke the Goal Seek command to set cell C25 to the value of 6 by changing cell B4. The λ value in cell B4 reaches 0.0139 which corresponds to a μ value of 71.96 months, as seen in cell E4. Therefore, the average life of the computers must be 71.96 months.

3–12   Summary and Review of Terms

In this chapter we described several important standard random variables, the associated formulas, and problem solving with spreadsheets. In order to use a spreadsheet template, you need to know *which* template to use, but first you need to know the kind of random variable at hand. This summary concentrates on this question.

A **discrete random variable** X will follow a **binomial distribution** if it is the number of successes in n independent **Bernoulli trials.** Make sure that the probability of success, p, remains constant in all trials. X will follow a **negative binomial distribution** if it is the number of Bernoulli trials made to achieve a desired number of successes. It will follow a **geometric distribution** when the desired number of successes is one. X will follow a **hypergeometric distribution** if it is the number of successes in a random sample drawn from a finite pool of successes and failures. X will follow a **Poisson distribution** if it is the number of occurrences of a rare event during a finite period.

Waiting time for an event that occurs periodically is **uniformly distributed.** Waiting time for a rare event is **exponentially distributed.**

PROBLEMS

3–42. A random variable X is described in each of the following items. Find out which standard distribution X is likely to follow in each case.

    a. X is the number of rescue calls received at a 911 call center on a given day.

    b. Your friend is equally likely to be found at any one of a number of places during coffee break. X is the number of places you have to look to find him.

    c. A drawer contains 8 red socks and 12 green socks. You pick 6 socks at random from the drawer. X is the number of red socks you picked.

    d. An assembly line breaks down and stops about once in six months. X is the number of times it breaks down in a given year.

    e. X is the time gap between two successive crashes of a mainframe computer.

    f. A company buys bolts in bulk and rarely finds defective ones. X is the number of defective bolts in a shipment that just arrived.

    g. Twenty students take a test in which each of them has a 30% chance of getting an A grade. X is the number of students who get A grades.

    h. You need six volunteers for a task and you ask your friends one by one until you get six volunteers. Each friend has a 60% chance of agreeing to volunteer. X is the number of friends you have to ask.

    i. Fifty guests are sent invitations to a party, and each guest has an 80% chance of attending the party. X is the number of guests who actually show up.

    j. A factory has a large number of machines that occasionally break down. X is the number of machines that break down on a given day.

    k. A machine breaks down occasionally. *X* is the time gap between two successive breakdowns.

    l. *Headline News* on TV starts every half hour. You switch on the TV at a random time and wait till the start. *X* is the time you have to wait.

    m. A committee is formed by picking members at random from the U.S. Senate. *X* is the number of Democrats picked.

    n. An airline has sold 132 tickets for a flight. Each ticket buyer has a 96% chance of showing up. *X* is the actual number of passengers who show up.

    o. A company sends out tens of thousands of free samples to prospective customers, each of whom has a very small chance of buying the product. *X* is the actual number of customers who buy the product.

    p. A machine breaks down occasionally. You visit the machine and wait till it breaks down. *X* is the time you have to wait.

    q. A shuttle bus leaves an airport every 45 minutes. You arrive at a random time and wait for it. *X* is the time you have to wait.

3–43. A graduating student keeps applying for jobs until she has three offers. The probability of getting an offer at any trial is 0.48.

    a. What is the expected number of applications? What is the variance?

    b. If she has enough time to complete only six applications, how confident can she be of getting three offers within the available time?

    c. If she wants to be at least 95% confident of getting three offers, how many applications should she prepare?

    d. Suppose she has time for at most six applications. For what minimum value of *p* can she still have 95% confidence of getting three offers within the available time?

3–44. A real estate agent has four houses to sell before the end of the month by contacting prospective customers one by one. Each customer has an independent 0.24 probability of buying a house on being contacted by the agent.

    a. If the agent has enough time to contact only 15 customers, how confident can she be of selling all four houses within the available time?

    b. If the agent wants to be at least 70% confident of selling all the houses within the available time, at least how many customers should she contact? (If necessary extend the template downward to more rows.)

    c. What minimum value of *p* will yield 70% confidence of selling all four houses by contacting at most 15 customers?

    d. To answer (*c*) above more thoroughly, tabulate the confidence for *p* values ranging from 0.2 to 0.6 in steps of 0.05.

3–45. A graduating student keeps applying for jobs until she gets an offer. The probability of getting an offer at any trial is 0.35.

    a. What is the expected number of applications? What is the variance?

    b. If she has enough time to complete at most four applications, how confident can she be of getting an offer within the available time?

    c. If she wants to be at least 95% confident of getting an offer, how many applications should she prepare?

     d. Suppose she has time for at most four applications. For what minimum value of *p* can she have 95% confidence of getting an offer within the available time?

3–46. A shipment of pins contains 25 good ones and 2 defective ones. At the receiving department, an inspector picks three pins at random and tests them. If any defective pin is found among the three that are tested, the shipment would be rejected.

     a. What is the probability that the shipment would be accepted?

     b. To increase the probability of acceptance to at least 90%, it is decided to do one of the following:

       i. Add some good pins to the shipment.

       ii. Remove some defective pins in the shipment.

For each of the two options, find out exactly how many pins should be added or removed.

3–47. A committee of 7 members is to be formed by selecting members at random from a pool of 14 candidates consisting of 5 women and 9 men.

     a. What is the probability that there will be at least three women in the committee?

     b. It is desired to increase the chance that there are at least three women in the committee to 80% by doing one of the following:

       i. Adding more women to the pool.

       ii. Removing some men from the pool.

For each of the two options, find out how many should be added or removed.

3–48. A mainframe computer in a university crashes on the average 0.71 time in a semester.

     a. What is the probability that it will crash at least two times in a given semester?

     b. What is the probability that it will not crash at all in a given semester?

     c. It is desired to increase the probability of no crash at all in a semester to at least 90%. What is the largest μ that will achieve this goal?

3–49. The number of rescue calls received by a rescue squad in a city follows a Poisson distribution with $\mu = 2.83$ per day. The squad can handle at most four calls a day.

     a. What is the probability that the squad will be able to handle all the calls on a particular day?

     b. The squad wants to have at least 95% confidence of being able to handle all the calls received in a day. At least how many calls a day should the squad be prepared for?

     c. Assuming that the squad can handle at most four calls a day, what is the largest value of μ that would yield 95% confidence that the squad can handle all calls?

3–50. A student takes the campus shuttle bus to reach the classroom building. The shuttle bus arrives at his stop every 15 minutes but the actual arrival time at the stop is random. The student allows 10 minutes waiting time for the shuttle in his plan to make it in time to the class.

   a. What is the expected waiting time? What is the variance?

   b. What is the probability that the wait will be between four and six minutes?

   c. What is the probability that the student will be in time for the class?

   d. If he wants to be 95% confident of being on time for the class, how much time should he allow for waiting for the shuttle?

3–51. A hydraulic press breaks down at the rate of 0.1742 time per day.

   a. What is the MTBF?

   b. On a given day, what is the probability that it will break down?

   c. If four days have passed without a breakdown, what is the probability that it will break down on the fifth day?

   d. What is the probability that five consecutive days will pass without any breakdown?

3–52. Laptop computers produced by a company have an average life of 38.36 months. Assume that the life of a computer is exponentially distributed (which is a good assumption).

   a. What is the probability that a computer will fail within 12 months?

   b. If the company gives a warranty period of 12 months, what proportion of computers will fail during the warranty period?

   c. Based on the answer to (b), would you say the company can afford to give a warranty period of 12 months?

   d. If the company wants not more than 5% of the computers to fail during the warranty period, what should be the warranty period?

   e. If the company wants to give a warranty period of three months and still wants not more than 5% of the computers to fail during the warranty period, what should be the minimum average life of the computers?

3–53. In most statistics textbooks, you will find cumulative binomial probability tables in the format shown below. These can be created using spreadsheets using the Binomial template and DatalTable commands.

$n = 5$

						p				
		0.1	0.2	0.3	0.4	0.5	0.6	0.7	0.8	0.9
	0	0.5905	0.3277	0.1681	0.0778	0.0313	0.0102	0.0024	0.0003	0.0000
	1	0.9185	0.7373	0.5282	0.3370	0.1875	0.0870	0.0308	0.0067	0.0005
X	2	0.9914	0.9421	0.8369	0.6826	0.5000	0.3174	0.1631	0.0579	0.0086
	3	0.9995	0.9933	0.9692	0.9130	0.8125	0.6630	0.4718	0.2627	0.0815
	4	1.0000	0.9997	0.9976	0.9898	0.9688	0.9222	0.8319	0.6723	0.4095

   a. Create the above table.

   b. Create a similar table for $n = 7$.

3–54. Look at the shape of the binomial distribution for various combinations of n and p. Specifically, let $n = 5$ and try $p = 0.2, 0.5,$ and 0.8. Repeat the same for other values of n. Can you say something about how the skewness of the distribution is affected by p and n?

3–55. Try various values of s and p on the negative binomial distribution template and answer this question: How is the skewness of the negative binomial distribution affected by s and p values?

3–56. An MBA graduate keeps interviewing for jobs, one by one, and will stop interviewing on receiving an offer. In each interview he has an independent probability 0.2166 of getting the job.

    a. What is the expected number of interviews? What is the variance?

    b. If there is enough time for only six interviews, how confident can he be of getting a job within the available time?

    c. If he wants to be at least 95% confident of getting a job, how many interviews should he be prepared for?

    d. Suppose there is enough time for at most six interviews. For what minimum value of p can he have 95% confidence of getting a job within the available time?

    e. In order to answer (*d*) more thoroughly, tabulate the confidence level for p values ranging from 0.1 to 0.5 in steps of 0.05.

3–57. A shipment of thousands of pins contains some percentage of defectives. To decide whether to accept the shipment, the consumer follows a sampling plan where 80 items are chosen at random from the sample and tested. If the number of defectives in the sample is at most three, the shipment is accepted. (The number 3 is known as the *acceptance number* of the sampling plan.)

    a. Assuming that the shipment includes 3% defectives, what is the probability that the shipment will be accepted? (*Hint:* Use the binomial distribution.)

    b. Assuming that the shipment includes 6% defectives, what is the probability that the shipment will be accepted?

    c. Using the Data|Table command, tabulate the probability of acceptance for defective percentage ranging from 0% to 15% in steps of 1%.

    d. Plot a line graph of the table created in (*c*). (This graph is known as the *operating characteristic curve* of the sampling plan.)

3–58. A shipment of 100 pins contains some defectives. To decide whether to accept the shipment, the consumer follows a sampling plan where 15 items are chosen at random from the sample and tested. If the number of defectives in the sample is at most one, the shipment is accepted. (The number 1 is known as the *acceptance number* of the sampling plan.)

    a. Assuming that there are 5% defectives in the shipment, what is the probability that the shipment will be accepted? (*Hint:* Use the hypergeometric distribution.)

    b. Assuming that there are 8% defectives in the shipment, what is the probability that the shipment will be accepted?

    c. Using the Data|Table command, tabulate the probability of acceptance for defective percentage ranging from 0% to 15% in steps of 1%.

    d. Plot a line graph of the table created in part (*c*) above. (This graph is known as the *operating characteristic curve* of the sampling plan.)

3–59. A recent study published in the *Toronto Globe and Mail* reveals that 25% of mathematics degrees from Canadian universities and colleges are awarded to women.

If five recent graduates from Canadian universities and colleges are selected at random, what is the probability that

    a. At least one would be a woman.

    b. None of them would be a woman.

3–60. An article published in *Access* magazine states that according to a survey conducted by the American Management Association, 78% of major U.S. companies electronically monitor their employees. If five such companies are selected at random, find the probability that

    a. At most one company monitors its employees electronically.

    b. All of them monitor their employees electronically.

3–61. An article published in *Business Week* says that according to a survey by a leading organization 45% of managers change jobs for intellectual challenge, 35% for pay, and 20% for long-term impact on career. If nine managers who recently changed jobs are randomly chosen, what is the probability that

    a. Three changed for intellectual challenges.

    b. Three changed for pay reasons.

    c. Three changed for long-term impact.

3–62. Estimates published by the World Health Organization state that one out of every three workers may be toiling away in workplaces that make them sick. If seven workers are selected at random, what is the probability that a majority of them are made sick by their workplace?

3–63. Based on the survey conducted by a municipal administration in the Netherlands between October 1999 and February 2000, Monday appeared to be managements' preferred day for laying off workers. Of the total number of workers laid off in a given period, 30% were on Monday, 25% on Tuesday, 20% on Wednesday, 13% on Thursday, and 12% on Friday. If a random sample of 15 layoffs is taken, what is the probability that

    a. Five were laid off on Monday.

    b. Four were laid off on Tuesday.

    c. Three were laid off on Wednesday.

    d. Two were laid off on Thursday.

    e. One was laid off on Friday.

3–64. A recent survey published in *Business Week* (November 27, 2000) concludes that Gatorade commands an 83% of share of the sports drink market versus 11% for Coca-Cola's PowerAde and 3% for Pepsi's All Sport. A market research firm wants to conduct a new taste test for which it needs Gatorade drinkers. Potential participants for the test are selected by random screening of drink users to find Gatorade drinkers. What is the probability that

    a. The first randomly selected drinker qualifies.

    b. Three soft drink users will have to be interviewed to find the first Gatorade drinker?

3–65. The time between customer arrivals at a bank has an exponential distribution with a mean time between arrivals of three minutes. If a customer just arrived, what is the probability that another customer will not arrive for at least two minutes?

3–66. Lightbulbs manufactured by a particular company have an exponentially distributed life with mean 100 hours.

a. What is the probability that the lightbulb I am now putting in will last at least 65 hours?

b. What is the standard deviation of the lifetime of a lightbulb?

3–67. The Bombay Company offers reproductions of classic 18th- and 19th-century English furniture pieces, which have become popular in recent years. The following table gives the probability distribution of the number of Raffles tables sold per day at a particular Bombay store.

Number of Tables	Probability
0	0.05
1	0.05
2	0.10
3	0.15
4	0.20
5	0.15
6	0.15
7	0.10
8	0.05

a. Show that the probabilities above form a proper probability distribution.

b. Find the cumulative distribution function of the number of Raffles tables sold daily.

c. Using the cumulative distribution function, find the probability that the number of tables sold in a given day will be at least three and less than seven.

d. Find the probability that at most five tables will be sold tomorrow.

e. What is the expected number of tables sold per day?

f. Find the variance and the standard deviation of the number of tables sold per day.

g. Use Chebyshev's theorem to determine bounds of at least 0.75 probability on the number of tables sold daily. Compare with the actual probability for these bounds using the distribution itself.

3–68. Salvador Assael, a major competitor of Mikimoto, farms pearls in Tahiti. Five percent of these pearls are sold in Tahiti, and the rest are distributed worldwide.[4] Out of 20 randomly taken pearls on site, what is the probability that at most 3 will be sold in Tahiti?

3–69. The number of orders for installation of a computer information system arriving at an agency per week is a random variable X with the following probability distribution:

[4] Z. Moukheiber, "The Pearl King," *Forbes,* April 24, 1995, p. 106.

x	P(x)
0	0.10
1	0.20
2	0.30
3	0.15
4	0.15
5	0.05
6	0.05

a. Prove that $P(X)$ is a probability distribution.

b. Find the cumulative distribution function of X.

c. Use the cumulative distribution function to find probabilities $P(2 < X \le 5)$, $P(3 \le X \le 6)$, and $P(X > 4)$.

d. What is the probability that either four or five orders will arrive in a given week?

e. Assuming independence of weekly orders, what is the probability that three orders will arrive next week and the same number of orders the following week?

f. Find the mean and the standard deviation of the number of weekly orders.

3–70. Consider the situation in the previous problem, and assume that the distribution holds for all weeks throughout the year and that weekly orders are independent from week to week. Let Y denote the number of weeks in the year in which no orders are received (assume a year of 52 weeks).

a. What kind of random variable is Y? Explain.

b. What is the expected number of weeks with no orders?

3–71. An analyst kept track of the daily price quotation for a given stock. The frequency data led to the following probability distribution of daily stock price:

Price x in Dollars	P(x)
17	0.05
17 1/8	0.05
17 1/4	0.10
17 3/8	0.15
17 1/2	0.20
17 5/8	0.15
17 3/4	0.10
17 7/8	0.05
18	0.05
18 1/8	0.05
18 1/4	0.05

Assume that the stock price is independent from day to day.

a. If 100 shares are bought today at 17 1/4 and must be sold tomorrow, by prearranged order, what is the expected profit, disregarding transaction costs?

b. What is the standard deviation of the stock price? How useful is this information?

c. What are the limitations of the analysis in part (*a*)? Explain.

3–72. In problem 3–69, suppose that the company makes $1,200 on each order but has to pay a fixed weekly cost of $1,750. Find the expected weekly profit and the standard deviation of weekly profits.

3–73. The CD–ROM industry is growing fast. A major player in this expanding market is Multimedia, Inc. Out of 400 titles typically sold at an average store, 60 are made by Multimedia.[5]

    a. A customer buys 10 titles. Under what conditions is the number of those made by Multimedia a binomial random variable?

    b. Making the required assumptions above, find the probability that at least two are Multimedia's.

3–74. An advertisement claims that two out of five doctors recommend a certain pharmaceutical product. A random sample of 20 doctors is selected, and it is found that only 2 of them recommend the product.

    a. Assuming the advertising claim is true, what is the probability of the observed event?

    b. Assuming the claim is true, what is the probability of observing two or fewer successes?

    c. Given the sampling results, do you believe the advertisement? Explain.

    d. What is the expected number of successes in a sample of 20?

3–75. Five percent of the many cars produced at a plant are defective. Ten cars made at the plant are sent to a dealership. Let X be the number of defective cars in the shipment.

    a. Under what conditions can we assume that X is a binomial random variable?

    b. Making the required assumptions, write the probability distribution of X.

    c. What is the probability that two or more cars are defective?

    d. What is the expected number of defective cars?

3–76. Refer to the situation in the previous problem. Suppose that the cars at the plant are checked one by one, and let X be the number of cars checked until the first defective car is found. What type of probability distribution does X have?

3–77. Suppose that 5 of a total of 20 company accounts are in error. An auditor selects a random sample of 5 out of the 20 accounts. Let X be the number of accounts in the sample that are in error. Is X binomial? If not, what distribution does it have? Explain.

3–78. The time, in minutes, necessary to perform a certain task has the uniform [5, 9] distribution.

    a. Write the probability density function of this random variable.

    b. What is the probability that the task will be performed in less than 8 minutes? Explain.

    c. What is the expected time required to perform the task?

3–79. Suppose X has the following probability density function:

$$f(x) = \begin{cases} (1/8)(x - 3) & \text{for } 3 \le x \le 7 \\ 0 & \text{otherwise} \end{cases}$$

---

[5]R. Shaffer, "Shelf Battle," *Forbes*, April 24, 1995, p. 178.

a. Graph the density function.

b. Show that $f(x)$ is a density function.

c. What is the probability that X is greater than 5.00?

3–80.   Recently, the head of the Federal Deposit Insurance Corporation (FDIC) revealed that the agency maintains a secret list of banks suspected of being in financial trouble. The FDIC chief further stated that of the nation's 14,000 banks, 1,600 were on the list at the time. Suppose that, in an effort to diversify your savings, you randomly choose six banks and split your savings among them. What is the probability that no more than three of your banks are on the FDIC's suspect list?

3–81.   Corporate raider Asher Adelman, teaching a course at Columbia University's School of Business, made the following proposal to his students. He would pay $100,000 to any student who would give him the name of an undervalued company, which Adelman would then buy.[6] Suppose that Adelman has 15 students in his class and that 5% of all companies in this country are undervalued. Suppose also that due to liquidity problems, Adelman can give the award to at most three students. Finally, suppose each student chooses a single company at random without consulting others. What is the probability that Adelman would be able to make good on his promise?

3–82.   An applicant for a faculty position at a certain university is told by the department chair that she has a 0.95 probability of being invited for an interview. Once invited for an interview, the applicant must make a presentation and win the votes of a majority (at least 8) of the department's 14 current members. From previous meetings with four of these members, the candidate believes that three of them would certainly vote for her while one would not. She also feels that any member she has not yet met has a 0.50 probability of voting for her. Department members are expected to vote independently and with no prior consultation. What are the candidate's chances of getting the position?

3–83.   Nielsen Company rates television programs. The ratings for the three major networks during prime time on Friday, February 1, 1991, were as follows. Also shown is the proportion of viewers watching each program.

Program	Network	Rating	Proportion
20/20	ABC	13.8	0.44
George Burns 95th Birthday	CBS	10.4	0.33
Midnight Caller	NBC	7.5	0.23

a. What is the mean rating given a program that evening?

b. How many standard deviations above or below the mean is the rating for each one of the programs?

3–84.   A major ski resort in the eastern United States closes in late May. Closing day varies from year to year depending on when the weather becomes too warm for making and preserving snow.[7] The day in May and the number of years in which closing occurred that day are reported in the table:

---

[6]Columbia has since questioned this offer on ethical grounds, and the offer has been retracted.

[7]"Warm Weather Bargains," *Skiing,* March 1991, p. 72.

Day	Number of Years
21	2
22	5
23	1
24	3
25	3
26	1
27	2
28	1

a. Based only on this information, estimate the probability that you could ski at this resort after May 25 next year.

b. What is the average closing day based on history?

3–85. Ten percent of the items produced at a plant are defective. A random sample of 20 items is selected. What is the probability that more than three items in the sample are defective? If items are selected randomly until the first defective item is encountered, how many items, on average, will have to be sampled before the first defective item is found?

3–86. Lee Iacocca has volunteered to drive one of his Chryslers into a brick wall to demonstrate the effectiveness of airbags used in these cars.[8] Airbags are known to activate at random when the car decelerates anywhere from 9 to 14 miles per hour per second (mph/s). The probability distribution for the deceleration speed at which bags activate is given below.

mph/s	Probability
9	0.12
10	0.23
11	0.34
12	0.21
13	0.06
14	0.04

a. If the airbag activates at a deceleration of 12 mph/s or more, Iacocca would get hurt. What is the probability of his being hurt in this demonstration?

b. What is the mean deceleration at airbag activation moment?

c. What is the standard deviation of deceleration at airbag activation time?

3–87. In the previous problem, the time that it takes the airbag to completely fill up from the moment of activation has an exponential distribution with mean 1 second. What is the probability that the airbag will fill up in less than 1/2 second?

3–88. According to *The Economist,* 25% of all drugs sold in Japan are foreign, 20% are foreign drugs sold by Japanese companies, and the rest are Japanese-made.[9] If 12 drugs are randomly chosen at a pharmacy, what is the probability that 4 are foreign, 4 are foreign sold by Japanese companies, and 4 are Japanese?

---

[8]"Surviving the Crash," *The Economist,* March 2, 1991, pp. 65–66.

[9]"Too Small to Compete?" *The Economist,* April 22, 1995, p. 65.

3–89. The Dutch consumer-electronics giant, Philips, is protected against takeovers by a unique corporate voting structure that gives power only to a few trusted shareholders.[10] A decision of whether to sever Philips' links with the loss-producing German electronics firm Grundig had to be made. The decision required a simple majority of nine decision-making shareholders. If each is believed to have a 0.25 probability of voting yes on the issue, what is the probability that Grundig will be dumped?

3–90. Sales of appliances made by the Swedish firm Electrolux[11] are as follows: European Union, 52%; rest of Europe, 7%; Africa, 1%; Asia, 5.0%; Latin America, 7.0%; North America, 28.0%. If products are randomly selected, find the probability that at most 3 out of 10 products are sold within the European Union.

3–91. Plastic is now surpassing aluminum as the packaging material of choice for soft drinks. According to a recent article,[12] 80% of all new vending business is going to plastic. If a random sample of 10 new vending businesses is selected, what are $P(X > 2)$, $P(X < 5)$, and $P(X < 9)$ (using the binomial distribution)?

CASE 3   Microchip Contract

A company receives an order for five custom-made microchips at a price of $7,500 each. The company will produce the chips one by one using a complex process which has only a 67% chance of producing a defect-free chip at each trial. After five defect-free chips are produced the process will be stopped.

A cost accountant at the company has prepared the following cost report: The cost of production includes a $14,800 fixed cost and a $2,700 unit variable cost. Thus if X number of chips are produced, the total cost of production would be $14,800 + 2,700X$ dollars. The revenue minus the cost of production will be the profit.

After some analysis, the finance manager of the company says that the risk may be too high and thinks the order should not be accepted.

1. What distribution will the number of chips produced, X, follow?

2. What is the expected value and standard deviation of X?

3. What is the expected value and standard deviation of the profit?

4. What is the break-even X (allow fractional values for X)?

5. What is the probability that accepting the order will result in a loss?

6. A popular measure of risk in a venture is *value at risk*, which is the loss suffered at the 5th percentile of the return from the venture. In this problem, find an integer x such that $P(X \geq x)$ is approximately 5%.

7. For the x value found in part 6, calculate the loss, and thus the value at risk.

8. Express the value at risk as a percentage of the expected value of the profit.

9. What is your assessment of the risk and reward in the order? Should the company accept the order?

---

[10]From "The Dimmest Bulb of All," *The Economist,* June 7, 1997, p. 79.

[11]From B. Peltier, "Enterprises," *Le Monde,* June 14, 1997, p. 14.

[12]"Plastic Puts Pressure on Cans," *Financial Times,* June 11, 1997, p. 24.

The sales manager of the company says that the customer is very likely to agree to increase the order quantity from five to eight chips. But he is not sure whether the matter should be pursued with the customer.

10. "If accepting an order of five itself is risky, will it not be even more risky to accept an order for eight?" asks the sales manager. How would you answer him?

11. Calculate the expected value and standard deviation of the profit for an order quantity of eight.

12. What is the value at risk for an order quantity of eight, computed in a manner similar to parts 6 and 7 above? Express the value at risk as a percentage of the expected profit.

13. Looking at the answer to parts 3, 8, 11, and 12, would you say the risk and reward have become more favorable, compared to an order quantity of five?

14. Should the company pursue the matter of increasing the order quantity to eight with the customer?

4 The Normal Distribution

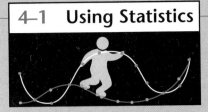

The **normal distribution** is an important continuous distribution because a good number of random variables occurring in practice can be approximated to it. *If a random variable is affected by many independent causes, and the effect of each cause is not overwhelmingly large compared to other effects, then the random variable will closely follow a normal distribution.* The lengths of pins made by an automatic machine, the times taken by an assembly worker to complete the assigned task repeatedly, the weights of baseballs, the tensile strengths of a batch of bolts, and the volumes of soup in a particular brand of canned soup are good examples of normally distributed random variables. All of these are affected by several independent causes where the effect of each cause is small. For example, the length of a pin is affected by many independent causes such as vibrations, temperature, wear and tear on the machine, and raw material properties.

Additionally, in the next chapter, on sampling theory, we shall see that many of the sample statistics are normally distributed.

For a normal distribution with mean μ and standard deviation σ, the probability density function $f(x)$ is given by the complicated formula

$$f(x) = \frac{1}{\sqrt{2\pi}\sigma} e^{-\frac{1}{2}\left(\frac{x-\mu}{\sigma}\right)^2} \qquad -\infty < x < +\infty \qquad (4\text{--}1)$$

In equation 4–1, e is the natural base logarithm, equal to 2.71828 . . . By substituting desired values for μ and σ, we can get any desired density function. For example, a distribution with mean 100 and standard deviation 5 will have the density function

$$f(x) = \frac{1}{\sqrt{2\pi}5} e^{-\frac{1}{2}\left(\frac{x-100}{5}\right)^2} \qquad -\infty < x < +\infty \qquad (4\text{--}2)$$

This function is plotted in Figure 4–1. This is the famous bell-shaped normal curve.

FIGURE 4–1 **A Normal Distribution with Mean 100 and Standard Deviation 5**

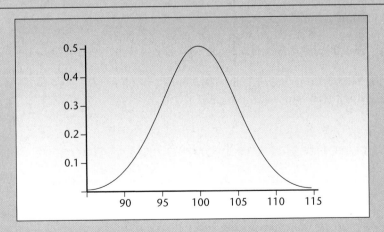

FIGURE 4–2 Three Normal Distributions

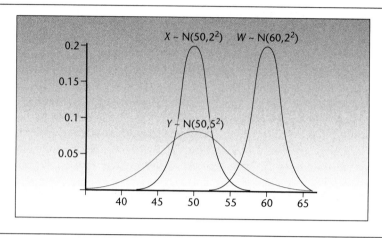

Over the years, many mathematicians have worked on the mathematics behind the normal distribution and have made many independent discoveries. The discovery of equation 4–1 for the normal density function is attributed to Carl Friedrich Gauss (1777–1855), who did much work with the formula. In science books, this distribution is often called the *Gaussian distribution*. But the formula was first discovered by the French-born English mathematician Abraham De Moivre (1667–1754). Unfortunately for him, his discovery was not discovered until 1924.

As seen in Figure 4–1, the normal distribution is symmetric about its mean. It has a (relative) kurtosis of 0, which means it has average peakedness. The curve reaches its peak at the mean of 100, and therefore its mode is 100. Due to symmetry, its median is 100 too. In the figure the curve seems to touch the horizontal axis at 85 on the left and at 115 on the right, these points are 3 standard deviations away from the center on either side. Theoretically, the curve never touches the horizontal axis and extends to infinity on both sides.

If X is normally distributed with mean μ and variance σ^2, we write $X \sim N(\mu, \sigma^2)$. If the mean is 100 and the variance is 9, we write $X \sim N(100, 3^2)$. Note how the variance is written. By writing 9 as 3^2, we explicitly show that the standard deviation is 3. Figure 4–2 shows three normal distributions: $X \sim N(50, 2^2)$; $Y \sim N(50, 5^2)$; $W \sim N(60, 2^2)$. Note their shapes and positions.

4–2 Properties of the Normal Distribution

There is a remarkable property possessed only by the normal distribution:

> If several *independent* random variables are normally distributed, then their sum will also be normally distributed. The mean of the sum will be the sum of all the individual means, and by virtue of the independence, the variance of the sum will be the sum of all the individual variances.

We can write this in algebraic form as

> If $X_1, X_2, \ldots, X_n$ are independent random variables that are normally distributed, then their sum S will also be normally distributed with
>
> $$E(S) = E(X_1) + E(X_2) + \cdots + E(X_n)$$
>
> and
>
> $$V(S) = V(X_1) + V(X_2) + \cdots + V(X_n)$$

It is important to note that it is the *variances* that can be added as in the preceding box, and *not the standard deviations*. We will never have an occasion to add standard deviations.

It is intuitive that the sum of many normal random variables will also be normally distributed, because the sum is affected by many independent individual causes, namely, those causes that affect each of the original random variables.

Let us see the application of this result through a few examples.

EXAMPLE 4–1

Let X_1, X_2, and X_3 be independent random variables that are normally distributed with means and variances as follows:

	Mean	Variance
X_1	10	1
X_2	20	2
X_3	30	3

Find the distribution of the sum $S = X_1 + X_2 + X_3$. Report the mean, variance, and standard deviation of S.

Solution

The sum S will be normally distributed with mean $10 + 20 + 30 = 60$ and variance $1 + 2 + 3 = 6$. The standard deviation of $S = \sqrt{6} = 2.45$.

EXAMPLE 4–2

The weight of a module used in a spacecraft is to be closely controlled. Since the module uses a bolt-nut-washer assembly in numerous places, a study was conducted to find the distribution of the weights of these parts. It was found that the three weights, in grams, are normally distributed with the following means and variances:

	Mean	Variance
Bolt	312.8	2.67
Nut	53.2	0.85
Washer	17.5	0.21

Find the distribution of the weight of the assembly. Report the mean, variance, and standard deviation of the weight.

Solution

The weight of the assembly is the sum of the weights of the three component parts, which are three normal random variables. Furthermore, the individual weights are independent since the weight of any one component part does not influence the weight of the other two. Therefore, the weight of the assembly will be normally distributed.

The mean weight of the assembly will be the sum of the mean weights of the individual parts: $312.8 + 53.2 + 17.5 = 383.5$ grams.

The variance will be the sum of the individual variances: $2.67 + 0.85 + 0.21 = 3.73$ gram^2.

The standard deviation $= \sqrt{3.73} = 1.93$ grams.

Another interesting property of the normal distribution is that if X is normally distributed, then $aX + b$ will also be normally distributed with mean $aE(X) + b$ and variance $a^2 V(X)$. For example, if X is normally distributed with mean 10 and variance 3, then $4X + 5$ will be normally distributed with mean $4*10 + 5 = 45$ and variance $4^2*3 = 48$.

We can combine the above two properties and make the following statement:

If $X_1, X_2, \ldots, X_n$ are independent random variables that are normally distributed, then the random variable Q defined as $Q = a_1X_1 + a_2X_2 + \cdots + a_nX_n + b$ will also be normally distributed with

$$E(Q) = a_1E(X_1) + a_2E(X_2) + \cdots + a_nE(X_n) + b$$

and

$$V(Q) = a_1^2V(X_1) + a_2^2V(X_2) + \cdots + a_n^2V(X_n)$$

The application of this result is illustrated in the following sample problems.

EXAMPLE 4-3   The four independent normal random variables $X_1, X_2, X_3,$ and X_4 have the following means and variances:

	Mean	Variance
X_1	12	4
X_2	−5	2
X_3	8	5
X_4	10	1

Find the mean and variance of $Q = X_1 - 2X_2 + 3X_3 - 4X_4 + 5$. Find also the standard deviation of Q.

Solution $E(Q) = 12 - 2(-5) + 3(8) - 4(10) + 5 = 12 + 10 + 24 - 40 + 5 = 11$

$V(Q) = 4 + (-2)^2 (2) + 3^2 (5) + (-4)^2 (1) = 4 + 8 + 45 + 16 = 73$

$SD(Q) = \sqrt{73} = 8.544$

EXAMPLE 4-4   A cost accountant needs to forecast the unit cost of a product for next year. He notes that each unit of the product requires 12 hours of labor and 5.8 pounds of raw material. In addition, each unit of the product is assigned an overhead cost of $184.50. He estimates that the cost of an hour of labor next year will be normally distributed with an expected value of $45.75 and a standard deviation of $1.80; the cost of the raw material will be normally distributed with an expected value of $62.35 and a standard deviation of $2.52. Find the distribution of the unit cost of the product. Report its expected value, variance, and standard deviation.

Let L be the cost of labor and M be the cost of the raw material. Denote the unit cost *Solution* of the product by Q. Then $Q = 12L + 5.8M + 184.50$. Since the cost of labor L may not influence the cost of raw material M, we can assume that the two are independent. This makes the unit cost of the product Q a normal random variable. Then

$$E(Q) = 12 \times 45.75 + 5.8 \times 62.35 + 184.50 = \$1095.13$$
$$V(Q) = 12^2 \times 1.80^2 + 5.8^2 \times 2.52^2 = 680.19$$
$$SD(Q) = \sqrt{680.19} = \$26.08$$

4–3 The Template

This normal distribution template is shown in Figure 4–3. As usual, it can be used in conjunction with the Goal Seek command and the Solver tool to solve many types of problems.

To use the template, make sure that the correct values are entered for the mean and the standard deviation in cells B4 and C4. Cell B11 gives the area to the left of the

FIGURE 4–3 **Normal Distribution Template**
[Normal Distribution.xls; Sheet: Normal]

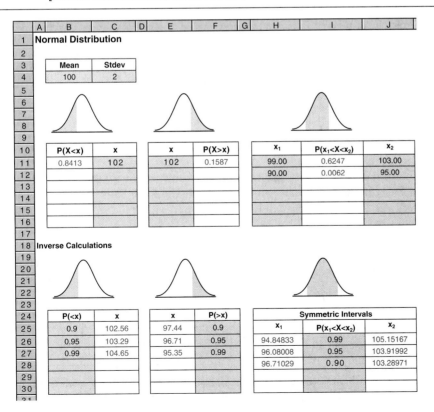

value entered in cell C11. The five cells below C11 can be similarly used. Cell F11 gives the area to the right of the value entered in cell E11. Cell I11 contains the area between the values entered in cells H11 and J11. In the area marked "Inverse Calculations," you can input areas (probabilities) and get x values corresponding to those areas. For example, on entering 0.9 in cell B25, we get the x value of 102.56 in cell C25. This implies that the area to the left of 102.56 is 0.9. Similarly, cell F25 has been used to get the x value that has 0.9 area to its right.

Sometimes we are interested in getting the *narrowest* interval that contains a desired amount of area. A little thought reveals that the narrowest interval has to be symmetric about the mean, because the distribution is symmetric and it peaks at the mean. In later chapters, we will study *confidence intervals,* many of which are also the narrowest intervals that contain a desired amount of area. Naturally, these confidence intervals are symmetric about the mean. For this reason, we have the "Symmetric Interval" area in the template. Once the desired area is entered in cell I26, the limits of the symmetric interval that contains that much area appear in cells H26 and J26. In the example shown in the Figure 4–3, the symmetric interval (94.85, 104.15) contains the desired area of 0.99.

Problem Solving with the Template

Most questions about normal random variables can be answered using the template in Figure 4–3. We will see a few problem-solving strategies through examples.

EXAMPLE 4–5   Suppose $X \sim N(100, 2^2)$. Find x_2 such that $P(99 \leq X \leq x_2) = 60\%$.

Solution   Fill in cell B4 with the mean 100 and cell C4 with standard deviation 2. Fill in cell H11 with 99. Select the Goal Seek command under the Tools menu. In the dialog box, ask to set cell I11 to value 0.6 by changing cell J11. Click OK when the computer finds the answer. The required value of 102.66 for x_2 appears in cell J11.

EXAMPLE 4–6   Suppose $X \sim N(\mu, 0.5^2)$; $P(X > 16.5) = 0.20$. What is μ?

Solution   Enter the σ of 0.5 in cell C4. Since we do not know μ, enter a *guessed* value of 15 in cell B4. Then enter 16.5 in cell E11. Now invoke the Goal Seek command to set cell F11 to value 0.20 by changing cell B4. The computer finds the value of μ in cell B4 to be 16.08.

The Goal Seek command can be used if there is only one unknown. With more than one unknown, the Solver tool has to be used. We shall illustrate the use of the Solver in the next example.

EXAMPLE 4–7   Suppose $X \sim N(\mu, \sigma^2)$; $P(X > 28) = 0.80$; $P(X > 32) = 0.40$. What are μ and σ?

Solution   *One way* to solve this problem is to use the Solver to find μ and σ with the objective of making $P(X > 28) = 0.80$ subject to the constraint $P(X > 32) = 0.40$. The following detailed steps will do just that:

- Fill in cell B4 with 30 (which is a guessed value for μ).
- Fill in cell C4 with 2 (which is a guessed value for σ).
- Fill in cell E11 with 28.
- Fill in cell E12 with 32.
- Under the Tools menu select the Solver.
- In the Set Cell box enter F11.
- In the To Value box enter 0.80 [which sets up the objective of $P(X > 28) = 0.80$].
- In the By Changing Cells box enter B4:C4.
- Click on the Constraints box and the Add button.
- In the dialog box on the left-hand side enter F12.
- Select the = sign in the middle drop down box.
- Enter 0.40 in the right-hand-side box [which sets up the constraint of $P(X > 32)$ = 0.40].
- Click the OK button.
- In the Solver dialog box that reappears, click the Solve button.
- In the dialog box that appears at the end, select the Keep Solver Solution option.

The Solver finds the correct values for the cells B4 and C4 as $\mu = 31.08$ and $\sigma = 3.67$.

EXAMPLE 4–8

A customer who has ordered 1-inch-diameter pins in bulk will buy only those pins with diameters in the interval 1 ± 0.003 inches. An automatic machine produces pins whose diameters are normally distributed with mean 1.002 inches and standard deviation 0.0011 inch.

1. What percentage of the pins made by the machine will be acceptable to the customer?
2. If the machine is adjusted so that the mean of the pins made by the machine is reset to 1.000 inch, what percentage of the pins will be acceptable to the customer?
3. Looking at the answer to parts 1 and 2, can we say that the machine must be reset?

Solution

1. Enter $\mu = 1.002$ and $\sigma = 0.0011$ into the template. From the template $P(0.997 < X < 1.003) = 0.8183$. Thus, 81.83% of the pins will be acceptable to the consumer.
2. Change μ to 1.000 in the template. Now, $P(0.997 < X < 1.003) = 0.9936$. Thus, 99.36% of the pins will be acceptable to the consumer.
3. Resetting the machine has considerably increased the percentage of pins acceptable to the consumer. It is therefore very much desirable that the machine be reset.

Finding Normal Probabilities Using the Tables

All kinds of normal probabilities are elegantly calculated using the templates. However, it is necessary to know how to calculate such probabilities using tables as well, because the normal distribution occurs frequently and you may not have access to the template all the time. To learn how to use the tables, we shall study the standard normal distribution and the details about the use of the tables in the next two sections.

4–4   The Standard Normal Distribution

Since, as noted earlier, there are infinitely many possible normal random variables, one is selected to serve as our *standard*. Probabilities associated with values of this standard normal random variable are tabulated. A special transformation then allows us to apply the tabulated probabilities to *any* normal random variable. The standard normal random variable has a special name, Z (rather than the general name X we use for other random variables).

> We define the **standard normal random variable** Z as the normal random variable with mean $\mu = 0$ and standard deviation $\sigma = 1$.

In the notation established in the previous section, we say

$$Z \sim N(0, 1^2) \tag{4–3}$$

Since $1^2 = 1$, we may drop the superscript 2 as no confusion of the standard deviation and the variance is possible. A graph of the standard normal density function is given in Figure 4–4.

Finding Probabilities of the Standard Normal Distribution

Probabilities of intervals are areas under the density $f(z)$ over the intervals in question. From the range of values in equation 4–1, $-\infty < x < \infty$, we see that any normal random variable is defined over the entire real line. Thus, the intervals in which we will be interested are sometimes *semi-infinite* intervals, such as a to ∞ or $-\infty$ to b (where a and b are numbers). While such intervals have infinite length, the probabilities associated with them are finite; they are, in fact, no greater than 1.00, as required of all probabilities. The reason for this is that the area in either of the "tails" of the distribution (the two narrow ends of the distribution, extending toward $-\infty$ and $+\infty$) becomes very small very quickly as we move away from the center of the distribution.

Tabulated areas under the standard normal density are probabilities of intervals extending from the mean $\mu = 0$ to points z to its right. Table 2 in Appendix C gives areas under the standard normal curve between 0 and points $z > 0$. The total area under the normal curve is equal to 1.00, and since the curve is symmetric, the area

FIGURE 4–4   **The Standard Normal Density Function**

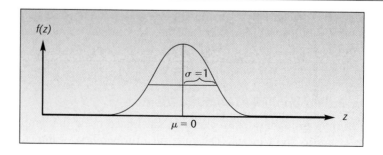

from 0 to $-\infty$ is equal to 0.5. The *table area* associated with a point z is thus equal to the value of the cumulative distribution function $F(z)$ minus 0.5.

We define the **table area** as

$$TA = F(z) - 0.5 \qquad (4\text{--}4)$$

The table area TA is shown in Figure 4–5. Part of Table 2 is reproduced here as Table 4–1. Let us see how the table is used in obtaining probabilities for the standard normal random variable. In the following examples, refer to Figure 4–5 and Table 4–1.

1. Let us find the probability that the value of the standard normal random variable will be between 0 and 1.56. That is, we want $P(0 < Z < 1.56)$. In Figure 4–5, substitute 1.56 for the point z on the graph. We are looking for the table area in the row labeled 1.5 and the column labeled 0.06. In the table, we find the probability 0.4406.

2. Let us find the probability that Z will be less than -2.47. Figure 4–6 shows the required area for the probability $P(Z < -2.47)$. By the symmetry of the normal curve, the area to the left of -2.47 is exactly equal to the area to the right of 2.47. We find

$$P(Z < -2.47) = P(Z > 2.47) = 0.5000 - 0.4932 = 0.0068$$

3. Find $P(1 < Z < 2)$. The required probability is the area under the curve between the two points 1 and 2. This area is shown in Figure 4–7. The table gives us the area under the curve between 0 and 1, and the area under the curve between 0 and 2. Areas are additive; therefore, $P(1 < Z < 2) = TA$(for 2.00) $- TA$(for 1.00) $= 0.4772 - 0.3413 = 0.1359$.

In cases where we need probabilities based on values with greater than second-decimal accuracy, we may use a linear interpolation between two probabilities obtained from the table. For example, $P(0 \le Z \le 1.645)$ is found as the midpoint between the two probabilities $P(0 \le Z \le 1.64)$ and $P(0 \le Z \le 1.65)$. This is found, using the table, as the midpoint of 0.4495 and 0.4505, which is 0.45. If even greater accuracy is required, we may use computer programs designed to produce standard normal probabilities.

FIGURE 4–5   The Table Area TA for a Point z of the Standard Normal Distribution

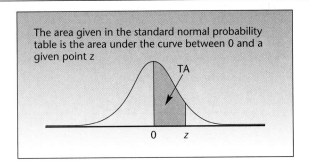

The area given in the standard normal probability table is the area under the curve between 0 and a given point z

TA

0   z

TABLE 4–1   Standard Normal Probabilities

z	.00	.01	.02	.03	.04	.05	.06	.07	.08	.09
0.0	.0000	.0040	.0080	.0120	.0160	.0199	.0239	.0279	.0319	.0359
0.1	.0398	.0438	.0478	.0517	.0557	.0596	.0636	.0675	.0714	.0753
0.2	.0793	.0832	.0871	.0910	.0948	.0987	.1026	.1064	.1103	.1141
0.3	.1179	.1217	.1255	.1293	.1331	.1368	.1406	.1443	.1480	.1517
0.4	.1554	.1591	.1628	.1664	.1700	.1736	.1772	.1808	.1844	.1879
0.5	.1915	.1950	.1985	.2019	.2054	.2088	.2123	.2157	.2190	.2224
0.6	.2257	.2291	.2324	.2357	.2389	.2422	.2454	.2486	.2517	.2549
0.7	.2580	.2611	.2642	.2673	.2704	.2734	.2764	.2794	.2823	.2852
0.8	.2881	.2910	.2939	.2967	.2995	.3023	.3051	.3078	.3106	.3133
0.9	.3159	.3186	.3212	.3238	.3264	.3289	.3315	.3340	.3365	.3389
1.0	.3413	.3438	.3461	.3485	.3508	.3531	.3554	.3577	.3599	.3621
1.1	.3643	.3665	.3686	.3708	.3729	.3749	.3770	.3790	.3810	.3830
1.2	.3849	.3869	.3888	.3907	.3925	.3944	.3962	.3980	.3997	.4015
1.3	.4032	.4049	.4066	.4082	.4099	.4115	.4131	.4147	.4162	.4177
1.4	.4192	.4207	.4222	.4236	.4251	.4265	.4279	.4292	.4306	.4319
1.5	.4332	.4345	.4357	.4370	.4382	.4394	.4406	.4418	.4429	.4441
1.6	.4452	.4463	.4474	.4484	.4495	.4505	.4515	.4525	.4535	.4545
1.7	.4554	.4564	.4573	.4582	.4591	.4599	.4608	.4616	.4625	.4633
1.8	.4641	.4649	.4656	.4664	.4671	.4678	.4686	.4693	.4699	.4706
1.9	.4713	.4719	.4726	.4732	.4738	.4744	.4750	.4756	.4761	.4767
2.0	.4772	.4778	.4783	.4788	.4793	.4798	.4803	.4808	.4812	.4817
2.1	.4821	.4826	.4830	.4834	.4838	.4842	.4846	.4850	.4854	.4857
2.2	.4861	.4864	.4868	.4871	.4875	.4878	.4881	.4884	.4887	.4890
2.3	.4893	.4896	.4898	.4901	.4904	.4906	.4909	.4911	.4913	.4916
2.4	.4918	.4920	.4922	.4925	.4927	.4929	.4931	.4932	.4934	.4936
2.5	.4938	.4940	.4941	.4943	.4945	.4946	.4948	.4949	.4951	.4952
2.6	.4953	.4955	.4956	.4957	.4959	.4960	.4961	.4962	.4963	.4964
2.7	.4965	.4966	.4967	.4968	.4969	.4970	.4971	.4972	.4973	.4974
2.8	.4974	.4975	.4976	.4977	.4977	.4978	.4979	.4979	.4980	.4981
2.9	.4981	.4982	.4982	.4983	.4984	.4984	.4985	.4985	.4986	.4986
3.0	.4987	.4987	.4987	.4988	.4988	.4989	.4989	.4989	.4990	.4990

FIGURE 4–6   Finding the Probability That Z Is Less than −2.47

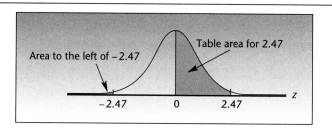

FIGURE 4–7   Finding the Probability That *Z* Is between 1 and 2

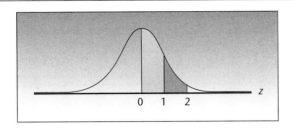

FIGURE 4–8   Using the Normal Table to Find a Value, Given a Probability

z	.00	.01	.02	.03	.04	.05	.06	.07	.08	.09
0.0	.0000	.0040	.0080	.0120	.0160	.0199	.0239	.0279	.0319	.0359
0.1	.0398	.0438	.0478	.0517	.0557	.0596	.0636	.0675	.0714	.0753
0.2	.0793	.0832	.0871	.0910	.0948	.0987	.1026	.1064	.1103	.1141
0.3	.1179	.1217	.1255	.1293	.1331	.1368	.1406	.1443	.1480	.1517
0.4	.1554	.1591	.1628	.1664	.1700	.1736	.1772	.1808	.1844	.1879
0.5	.1915	.1950	.1985	.2019	.2054	.2088	.2123	.2157	.2190	.2224
0.6	.2257	.2291	.2324	.2357	.2389	.2422	.2454	.2486	.2517	.2549
0.7	.2580	.2611	.2642	.2673	.2704	.2734	.2764	.2794	.2823	.2852
0.8	.2881	.2910	.2939	.2967	.2995	.3023	.3051	.3078	.3106	.3133
0.9	.3159	.3186	.3212	.3238	.3264	.3289	.3315	.3340	.3365	.3389
1.0	.3413	.3438	.3461	.3485	.3508	.3531	.3554	.3577	.3599	.3621
1.1	.3643	.3665	.3686	.3708	.3729	.3749	.3770	.3790	.3810	.3830
→ 1.2	.3849	.3869	.3888	.3907	.3925	.3944	.3962	.3980	.3997	.4015
1.3	.4032	.4049	.4066	.4082	.4099	.4115	.4131	.4147	.4162	.4177
1.4	.4192	.4207	.4222	.4236	.4251	.4265	.4279	.4292	.4306	.4319
1.5	.4332	.4345	.4357	.4370	.4382	.4394	.4406	.4418	.4429	.4441

Finding Values of Z Given a Probability

In many situations, instead of finding the probability that a standard normal random variable will be within a given interval, we may be interested in the reverse: finding an interval with a given probability. Consider the following examples.

1. Find a value z of the standard normal random variable such that the probability that the random variable will have a value between 0 and z is 0.40. We look *inside* the table for the value closest to 0.40; we do this by searching through the values inside the table, noting that they increase from 0 to numbers close to 0.5000 as we go down the columns and across the rows. The closest value we find to 0.40 is the table area .3997. This value corresponds to 1.28 (row 1.2 and column .08). This is illustrated in Figure 4–8.

2. Find the value of the standard normal random variable that cuts off an area of 0.90 to its left. Here, we reason as follows: Since the area to the left of the given point z is greater than 0.50, z *must be on the right side of 0*. Furthermore,

FIGURE 4–9 Finding *z* Such That $P(Z \leq z) = 0.9$

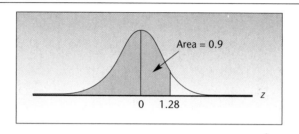

FIGURE 4–10 A Symmetric 0.99 Probability Interval about 0 for a
Standard Normal Random Variable

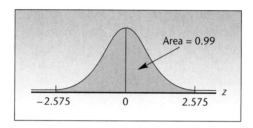

the area to the left of 0 all the way to $-\infty$ is equal to 0.5. Therefore, TA = $0.9 - 0.5 = 0.4$. We need to find the point *z* such that TA = 0.4. We know the answer from the preceding example: $z = 1.28$. This is shown in Figure 4–9.

3. Find a 0.99 probability interval, symmetric about 0, for the standard normal random variable. The required area between the two *z* values that are equidistant from 0 on either side is 0.99. Therefore, the area under the curve between 0 and the positive *z* value is TA = $0.99/2 = 0.495$. We now look in our normal probability table for the area closest to 0.495. The area 0.495 lies exactly between the two areas 0.4949 and 0.4951, corresponding to $z = 2.57$ and $z = 2.58$. Therefore, a simple linear interpolation between the two values gives us $z = 2.575$. This is correct to within the accuracy of the linear interpolation. The answer, therefore, is $z = \pm 2.575$. This is shown in Figure 4–10.

PROBLEMS

4–1.   Find the following probabilities: $P(-1 < Z < 1)$, $P(-1.96 < Z < 1.96)$, $P(-2.33 < Z < 2.33)$.

4–2.   What is the probability that a standard normal random variable will be between the values -2 and 1?

4–3.   Find the probability that a standard normal random variable will have a value between -0.89 and -2.50.

4–4. Find the probability that a standard normal random variable will have a value greater than 3.02.

4–5. Find the probability that a standard normal random variable will be between 2 and 3.

4–6. Find the probability that a standard normal random variable will have a value less than or equal to -2.5.

4–7. Find the probability that a standard normal random variable will be greater in value than -2.33.

4–8. Find the probability that a standard normal random variable will have a value between -2 and 300.

4–9. Find the probability that a standard normal variable will have a value less than -10.

4–10. Find the probability that a standard normal random variable will be between -0.01 and 0.05.

4–11. A sensitive measuring device is calibrated so that errors in the measurements it provides are normally distributed with mean 0 and variance 1.00. Find the probability that a given error will be between -2 and 2.

4–12. Find two values defining tails of the normal distribution with an area of 0.05 each.

4–13. Is it likely that a standard normal random variable will have a value less than -4? Explain.

4–14. Find a value such that the probability that the standard normal random variable will be above it is 0.85.

4–15. Find a value of the standard normal random variable cutting off an area of 0.685 to its left.

4–16. Find a value of the standard normal random variable cutting off an area of 0.50 to its right. (Do you need the table for this probability? Explain.)

4–17. Find z such that $P(Z > z) = 0.12$.

4–18. Find two values, equidistant from 0 on either side, such that the probability that a standard normal random variable will be between them is 0.40.

4–19. Find two values of the standard normal random variable, z and $-z$, such that $P(-z < Z < z) = 0.95$.

4–20. Find two values of the standard normal random variable, z and $-z$, such that the two corresponding "tail areas" of the distribution (the area to the right of z and the area to the left of $-z$) add to 0.01.

4–21. The deviation of a magnetic needle from the magnetic pole in a certain area in northern Canada is a normally distributed random variable with mean 0 and standard deviation 1.00. What is the probability that the absolute value of the deviation from the north pole at a given moment will be more than 2.4?

4–5 The Transformation of Normal Random Variables

The importance of the standard normal distribution derives from the fact that any normal random variable may be transformed to the standard normal random variable. We want to transform X, where $X \sim N(\mu, \sigma^2)$, into the standard normal random

FIGURE 4–11   Transforming a Normal Random Variable with Mean 50 and
               Standard Deviation 10 into the Standard Normal Random Variable

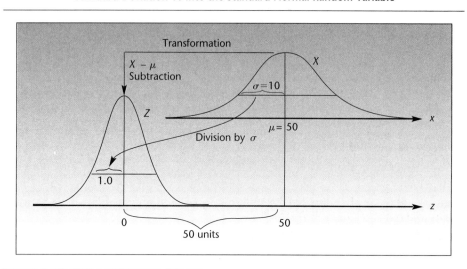

variable $Z \sim N(0, 1^2)$. Look at Figure 4–11. Here we have a normal random variable
X with mean $\mu = 50$ and standard deviation $\sigma = 10$. We want to transform this random variable to a normal random variable with $\mu = 0$ and $\sigma = 1$. How can we do
this?

   We move the distribution from its center of 50 to a center of 0. This is done by
subtracting 50 from all the values of X. Thus, we shift the distribution 50 units back so
that its new center is 0. The second thing we need to do is to make the width of the
distribution, its standard deviation, equal to 1. This is done by squeezing the width
down from 10 to 1. Because the total probability under the curve must remain 1.00,
the distribution must grow upward to maintain the same area. This is shown in Figure
4–11. Mathematically, squeezing the curve to make the width 1 is equivalent to dividing the random variable by its standard deviation. The area under the curve ad-
justs so that the total remains the same. *All probabilities* (areas under the curve) *adjust
accordingly.* The mathematical transformation from X to Z is thus achieved by first subtracting μ from X and then dividing the result by σ.

The transformation of X to Z:

$$Z = \frac{X - \mu}{\sigma}$$

(4–5)

The transformation of equation 4–5 takes us from a random variable X with mean μ
and standard deviation σ to the standard normal random variable. We also have an
opposite, or *inverse,* transformation, which takes us from the standard normal random
variable Z to the random variable X with mean μ and standard deviation σ. The inverse transformation is given by equation 4–6.

> The inverse transformation of Z to X:
>
> $$X = \mu + Z\sigma \qquad (4\text{–}6)$$

You can verify mathematically that equation 4–6 does the opposite of equation 4–5. Note that multiplying the random variable Z by the number σ increases the width of the curve from 1 to σ, thus making σ the new standard deviation. Adding μ makes μ the new mean of the random variable. The actions of multiplying and then adding are the opposite of subtracting and then dividing. We note that the two transformations, one an inverse of the other, transform a *normal* random variable into a *normal* random variable. If this transformation is carried out on a random variable that is not normal, the result will not be a normal random variable.

Using the Normal Transformation

Let us consider our random variable X with mean 50 and standard deviation 10, $X \sim N(50, 10^2)$. Suppose we want the probability that X is greater than 60. That is, we want to find $P(X > 60)$. We cannot evaluate this probability directly, but if we can transform X to Z, we will be able to find the probability in the Z table, Table 2 in Appendix C. Using equation 4–5, the required transformation is $Z = (X - \mu)/\sigma$. Let us carry out the transformation. In the probability statement $P(X > 60)$, we will substitute Z for X. If, however, we carry out the transformation on one side of the probability inequality, we must also do it on the other side. In other words, transforming X into Z requires us also to transform the value 60 into the appropriate value of the standard normal distribution. We transform the value 60 into the value $(60 - \mu)/\sigma$. The new probability statement is

$$P(X > 60) = P\left(\frac{X - \mu}{\sigma} > \frac{60 - \mu}{\sigma}\right) = P\left(Z > \frac{60 - \mu}{\sigma}\right)$$

$$= P\left(Z > \frac{60 - 50}{10}\right) = P(Z > 1)$$

Why does the inequality still hold? We subtracted a number from each side of an inequality; this does not change the inequality. In the next step we divide both sides of the inequality by the standard deviation σ. The inequality does not change because we can divide both sides of an inequality by a positive number, and a standard deviation is always a positive number. (Recall that dividing by 0 is not permissible; and dividing, or multiplying, by a negative value would reverse the direction of the inequality.) From the transformation, we find that the probability that a normal random variable with mean 50 and standard deviation 10 will have a value greater than 60 is exactly the probability that the standard normal random variable Z will be greater than 1. The latter probability can be found using Table 2 in Appendix C. We find: $P(X > 60) = P(Z > 1) = 0.5000 - 0.3413 = 0.1587$. Let us now look at a few examples of the use of equation 4–5.

EXAMPLE 4–9

Suppose that the time it takes the electronic device in the car to respond to the signal from the toll plaza is normally distributed with mean 160 microseconds and standard deviation 30 microseconds. What is the probability that the device in the car will respond to a given signal within 100 to 180 microseconds?

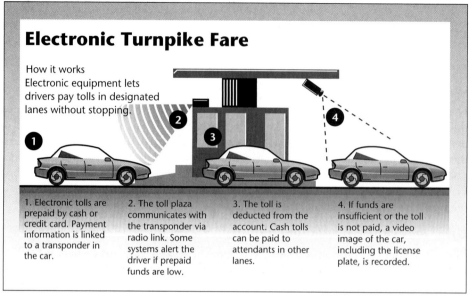

Electronic Turnpike Fare

How it works
Electronic equipment lets
drivers pay tolls in designated
lanes without stopping.

1. Electronic tolls are prepaid by cash or credit card. Payment information is linked to a transponder in the car.

2. The toll plaza communicates with the transponder via radio link. Some systems alert the driver if prepaid funds are low.

3. The toll is deducted from the account. Cash tolls can be paid to attendants in other lanes.

4. If funds are insufficient or the toll is not paid, a video image of the car, including the license plate, is recorded.

From *Boston Globe*, Tuesday, May 9, 1995, p. 1, with data from industry reports. Reprinted courtesy of the Boston Globe.

Solution   Figure 4–12 shows the normal distribution for $X \sim N(160, 30^2)$ and the required area on the scale of the original problem and on the transformed z scale. We have the following (where the probability statement inequality has three sides and we carry out the transformation of equation 4–5 on all three sides):

$$P(100 < X < 180) = P\left(\frac{100 - \mu}{\sigma} < \frac{X - \mu}{\sigma} < \frac{180 - \mu}{\sigma}\right)$$

$$= P\left(\frac{100 - 160}{30} < Z < \frac{180 - 160}{30}\right)$$

$$= P(-2 < Z < 0.6666) = 0.4772 + 0.2475 = 0.7247$$

FIGURE 4–12   Probability Computation for Example 4–9

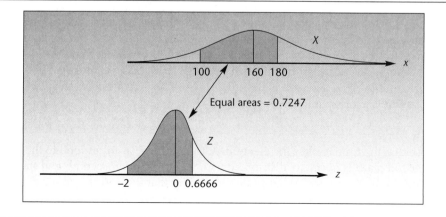

(Table area values were obtained by linear interpolation.) Thus, there is a 0.7247 chance that the device will respond within 100 to 180 microseconds.

EXAMPLE 4–10

The concentration of impurities in a semiconductor used in the production of microprocessors for computers is a normally distributed random variable with mean 127 parts per million and standard deviation 22. A semiconductor is acceptable only if its concentration of impurities is below 150 parts per million. What proportion of the semiconductors are acceptable for use?

Now $X \sim N(127, 22^2)$, and we need $P(X < 150)$. Using equation 4–5, we have

$$P(X < 150) = P\left(\frac{X - \mu}{\sigma} < \frac{150 - \mu}{\sigma}\right) = P\left(Z < \frac{150 - 127}{22}\right)$$
$$= P(Z < 1.045) = 0.5 + 0.3520 = 0.8520$$

(The TA of 0.3520 was obtained by interpolation.) Thus, 85.2% of the semiconductors are acceptable for use. This also means that the probability that a randomly chosen semiconductor will be acceptable for use is 0.8520. The solution of this example is illustrated in Figure 4–13.

EXAMPLE 4–11

Fluctuations in the prices of precious metals such as gold have been empirically shown to be well approximated by a normal distribution when observed over short intervals of time. In May 1995, the daily price of gold (1 troy ounce) was believed to have a mean of $383 and a standard deviation of $12. A broker, working under these assumptions, wanted to find the probability that the price of gold the next day would be between $394 and $399 per troy ounce. In this eventuality, the broker had an order from a client to sell the gold in the client's portfolio. What is the probability that the client's gold will be sold the next day?

Figure 4–14 shows the setup for this problem and the transformation of X, where $X \sim N(383, 12^2)$, into the standard normal random variable Z. Also shown are the required areas under the X curve and the transformed Z curve. We have

FIGURE 4–13 **Probability Computation for Example 4–10**

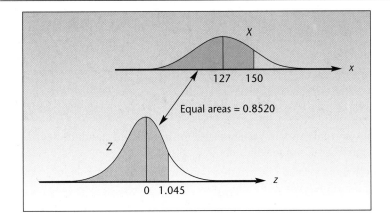

FIGURE 4–14   Probability Computation for Example 4–11

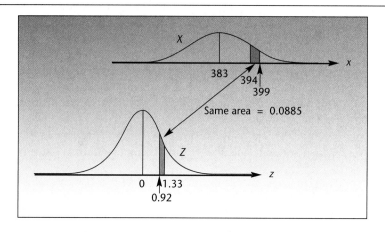

$$P(394 < X < 399) = P\left(\frac{394 - \mu}{\sigma} < \frac{X - \mu}{\sigma} < \frac{399 - \mu}{\sigma}\right)$$

$$= P\left(\frac{394 - 383}{12} < Z < \frac{399 - 383}{12}\right)$$

$$= P(0.9166 < Z < 1.3333) = 0.4088 - 0.3203 = 0.0885$$

(Both TA values were obtained by linear interpolation, although this is not necessary if less accuracy is acceptable.)

Let us summarize the transformation procedure used in computing probabilities of events associated with a normal random variable $X \sim N(\mu, \sigma^2)$.

Transformation formulas of X to Z, where a and b are numbers:

$$P(X < a) = P\left(Z < \frac{a - \mu}{\sigma}\right)$$

$$P(X > b) = P\left(Z > \frac{b - \mu}{\sigma}\right)$$

$$P(a < X < b) = P\left(\frac{a - \mu}{\sigma} < Z < \frac{b - \mu}{\sigma}\right)$$

PROBLEMS

4–22. For a normal random variable with mean 650 and standard deviation 40, find the probability that its value will be below 600.

4–23. Let X be a normally distributed random variable with mean 410 and standard deviation 2. Find the probability that X will be between 407 and 415.

4–24. If X is normally distributed with mean 500 and standard deviation 20, find the probability that X will be above 555.

4–25. For a normally distributed random variable with mean -44 and standard deviation 16, find the probability that the value of the random variable will be above 0.

4–26. A normal random variable has mean 0 and standard deviation 4. Find the probability that the random variable will be above 2.5.

4–27. Let X be a normally distributed random variable with mean $\mu = 16$ and standard deviation $\sigma = 3$. Find $P(11 < X < 20.)$ Also find $P(17 < X < 19)$ and $P(X > 15)$.

4–28. The time it takes an international telephone operator to place an overseas phone call is normally distributed with mean 45 seconds and standard deviation 10 seconds.

  a. What is the probability that my call will go through in less than 1 minute?

  b. What is the probability that I will get through in less than 40 seconds?

  c. What is the probability that I will have to wait more than 70 seconds for my call to go through?

4–29. The number of votes cast in favor of a controversial proposition is believed to be approximately normally distributed with mean 8,000 and standard deviation 1,000. The proposition needs at least 9,322 votes in order to pass. What is the probability that the proposition will pass? (Assume numbers are on a continuous scale.)

4–30. Under the system of floating exchange rates, the rate of foreign money to the U.S. dollar is affected by many random factors, and this leads to the assumption of a normal distribution of small daily fluctuations. The rate of German marks per U.S. dollar is believed in a certain period to have a mean of 2.06 and a standard deviation of 0.08. Find the following.

  a. The probability that tomorrow's rate will be above 2.10.

  b. The probability that tomorrow's rate will be below 1.90.

  c. The probability that tomorrow's exchange rate will be between 2.00 and 2.20.

4–31.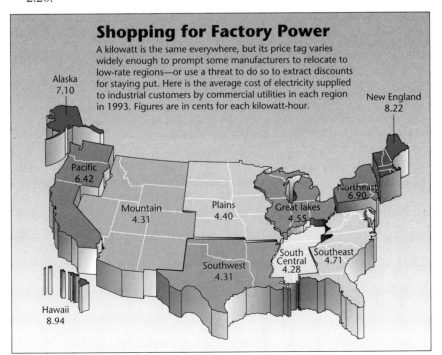

From *The New York Times*, May 18, 1995, p. D1, with data from Edison Electric Institute. Copyright 1995 by the New York Times Company. Reprinted by permission.

Assume that the rate charged in each region of the country is normally distributed with mean as shown on the map and standard deviation of 1.00.

    a. What is the probability that a randomly chosen customer in the Northeast will be charged over 8.00 cents per kilowatt-hour?

    b. What is the probability that a randomly chosen customer in the Mountain region will be charged less than 3.00 cents per kilowatt-hour?

4–32. Electricity rates at different regions of the country are independent of one another. Use the map given with problem 4–31 to find the following probabilities.

    a. What is the probability that a customer in New England and one in Alaska would both pay over 7.00 cents per kilowatt-hour?

    b. A company has one subsidiary in the Great Lakes, one in the Southeast, one in the Plains, and one in the Southwest. What is the probability that all of them pay less than 4.00 cents per kilowatt-hour?

4–33. Daily fluctuations[1] of the French CAC-40 stock index from March to June 1997 seem to follow a normal distribution with mean of 2,600 and standard deviation of 50. Find the probability that the CAC-40 will be between 2,520 and 2,670 on a random day in the period of study.

4–34. In May and June of 1997, the Greek drachma was fluctuating in its exchange rate to the U.S. dollar in a roughly normal distribution with mean of 271 drachmas per dollar and standard deviation of 2.5 drachmas per dollar. Fluctuations seemed to be independent of one another, each drawn from this distribution. Find the probability that on the next day during this period the dollar would have been worth anywhere between 265 and 275 drachmas.

4–35. Based on the research of Ibbotson Associates, a Chicago investment firm, and Prof. Jeremy Siegel of the Wharton School of the University of Pennsylvania,[2] the average return on large-company stocks since 1920 has been 10.5% per year and the standard deviation has been 4.75%. Assuming a normal distribution for stock returns (and that the trend will continue this year), what is the probability that a large-company stock you've just brought will make in 1 year at least 12%? Will lose money? Will make at least 5%?

4–36. A manufacturing company regularly consumes a special type of glue purchased from a foreign supplier. Because the supplier is foreign, the time gap between placing an order and receiving the shipment against that order is long and uncertain. This time gap is called "lead time." From past experience, the materials manager notes that the company's demand for glue during the uncertain lead time is normally distributed with a mean of 187.6 gallons and a standard deviation of 12.4 gallons. The company follows a policy of placing an order when the glue stock falls to a predetermined value called the "reorder point." Note that if the reorder point is *x* gallons and the demand during lead time exceeds *x* gallons, the glue would go "stock-out" and the production process would have to stop. Stock-out conditions are therefore serious.

    a. If the reorder point is kept at 187.6 gallons (equal to the mean demand during lead time) what is the probability that a stock-out condition would occur?

---

[1]From Data Stream International, reported in a chart in *The Wall Street Journal Europe,* June 11, 1997, p. 13.

[2]From J. Glassman, "Some Investing Essentials," The Money Report; *International Herald Tribune,* June 14–15, 1997, p. 17.

  b. If the reorder point is kept at 200 gallons what is the probability that a stock-out condition would occur?

  c. If the company wants to be 95% confident that the stock-out condition will not occur, what should be the reorder point? The reorder point minus the mean demand during lead time is known as the "safety stock." What is the safety stock in this case?

  d. If the company wants to be 99% confident that the stock-out condition will not occur, what should be the reorder point? What is the safety stock in this case?

4–37. The daily price of orange juice 30-day futures is normally distributed. In June through August 1997, the mean was 77.2 cents per pound, and standard deviation = 3.1 cents per pound. Assuming the price is independent from day to day, find $P(x < 80)$ on the next day.

4–6 The Inverse Transformation

Let us look more closely at the relationship between X, a normal random variable with mean μ and standard deviation σ, and the standard normal random variable. The fact that the standard normal random variable has mean 0 and standard deviation 1 has some important implications. When we say that Z is greater than 2, we are also saying that Z is more than 2 *standard deviations above its mean*. This is so because the mean of Z is 0 and the standard deviation is 1; hence, $Z > 2$ is the same event as $Z > [0 + 2(1)]$.

  Now consider a normal random variable X with mean 50 and standard deviation 10. Saying that X is greater than 70 is exactly the same as saying that X is 2 standard deviations above its mean. This is so because 70 is 20 units above the mean of 50, and 20 units = 2(10) units, or 2 standard deviations of X. Thus, the event $X > 70$ is the same as the event $X > (2$ standard deviations above the mean). This event is identical to the event $Z > 2$. Indeed, this is what results when we carry out the transformation of equation 4–5:

$$P(X > 70) = P\left(\frac{X - \mu}{\sigma} > \frac{70 - \mu}{\sigma}\right) = P\left(Z > \frac{70 - 50}{10}\right) = P(Z > 2)$$

  Normal random variables are related to one another by the fact that the probability that a normal random variable will be above (or below) its mean a certain number of standard deviations is exactly equal to the probability that any other normal random variable will be above (or below) its mean the same number of (its) standard deviations. In particular, this property holds for the standard normal random variable. The probability that a normal random variable will be greater than (or less than) z standard-deviation units above its mean is the same as the probability that the standard normal random variable will be greater than (less than) z. The change from a *z value* of the random variable Z to *z standard deviations* above the mean for a given normal random variable X should suggest to us the inverse transformation, equation 4–6:

$$x = \mu + z\sigma$$

That is, the value of the random variable X may be written in terms of the number z of standard deviations σ it is above or below the mean μ. Three examples are useful here. We know from the standard normal probability table that the probability that Z is greater than -1 and less than 1 is 0.6826 (show this). Similarly, we know that the probability that Z is greater that -2 and less than 2 is 0.9544. Also, the probability that Z is greater than -3 and less than 3 is 0.9974. These probabilities may be applied to *any* normal random variable as follows:[3]

1. The probability that a normal random variable will be within a distance of *1 standard deviation* from its mean (on either side) is 0.6826, or *approximately 0.68*.
2. The probability that a normal random variable will be within *2 standard deviations* of its mean is 0.9544, or *approximately 0.95*.
3. The probability that a normal random variable will be within *3 standard deviations* of its mean is 0.9974.

We use the inverse transformation, equation 4–6, when we want to get from a given probability to the value or values of a normal random variable X. We illustrate the procedure with a few examples.

EXAMPLE 4–12   PALCO Industries, Inc., is a leading manufacturer of cutting and welding products. One of the company's products is an acetylene gas cylinder used in welding. The amount of nitrogen gas in a cylinder is a normally distributed random variable with mean 124 units of volume and standard deviation 12. We want to find the amount of nitrogen x such that 10% of the cylinders contain more nitrogen than this amount.

Solution   We have $X \sim N(124, 12^2)$. We are looking for the value of the random variable X such that $P(X > x) = 0.10$. In order to find it, we look for the value of the standard normal random variable Z such that $P(Z > z) = 0.10$. Figure 4–15 illustrates how we find the value z and transform it to x. If the area to the right of z is equal to 0.10, the area

FIGURE 4–15   Solution of Example 4–12

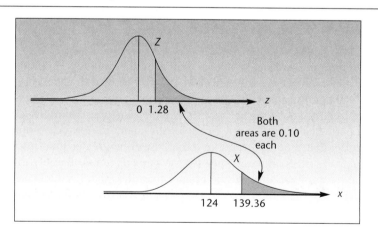

[3]This is the origin of the *empirical rule* (in Chapter 1) for mound-shaped data distributions. Mound-shaped data sets approximate the distribution of a normal random variable, and hence the proportions of observations within a given number of standard deviations away from the mean roughly equal those predicted by the normal distribution. Compare the empirical rule (section 1–7) with the numbers given here.

between 0 and z (the table area) is equal to $0.5 - 0.10 = 0.40$. We look inside the table for the z value corresponding to TA $= 0.40$ and find $z = 1.28$ (actually, TA $= 0.3997$, which is close enough to 0.4). We need to find the appropriate x value. Here we use equation 4–6:

$$x = \mu + z\sigma = 124 + (1.28)(12) = 139.36$$

Thus, 10% of the acetylene cylinders contain more than 139.36 units of nitrogen.

The amount of fuel consumed by the engines of a jetliner on a flight between two cities is a normally distributed random variable X with mean $\mu = 5.7$ tons and standard deviation $\sigma = 0.5$. Carrying too much fuel is inefficient as it slows the plane. If, however, too little fuel is loaded on the plane, an emergency landing may be necessary. The airline would like to determine the amount of fuel to load so that there will be a 0.99 probability that the plane will arrive at its destination.

EXAMPLE 4–13

We have $X \sim N(5.7, 0.5^2)$. First, we must find the value z such that $P(Z < z) = 0.99$. Following our methodology, we find that the required table area is TA $= 0.99 - 0.5 = 0.49$, and the corresponding z value is 2.33. Transforming the z value to an x value, we get $x = \mu + z\sigma = 5.7 + (2.33)(0.5) = 6.865$. Thus, the plane should be loaded with 6.865 tons of fuel to give a 0.99 probability that the fuel will last throughout the flight. The transformation is shown in Figure 4–16.

Solution

Weekly sales of Campbell's soup cans at a grocery store are believed to be approximately normally distributed with mean 2,450 and standard deviation 400. The store management wants to find two values, symmetrically on either side of the mean, such that there will be a 0.95 probability that sales of soup cans during the week will be between the two values. Such information is useful in determining levels of orders and stock.

EXAMPLE 4–14

Here $X \sim N(2,450, 400^2)$. From the section on the standard normal random variable, we know how to find two values of Z such that the area under the curve between

Solution

FIGURE 4–16   **Solution of Example 4–13**

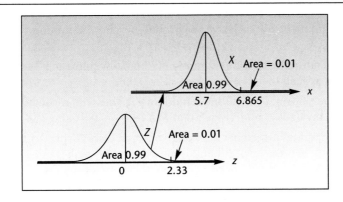

them is 0.95 (or any other area). We find that $z = 1.96$ and $z = -1.96$ are the required values. We now need to use equation 4–6. Since there are *two* values, one the negative of the other, we may combine them in a single transformation:

$$x = \mu \pm z\sigma \qquad\qquad (4\text{--}7)$$

Applying this special formula we get $x = 2{,}450 \pm (1.96)(400) = 1{,}666$ and $3{,}234$. Thus, management may be 95% sure that sales on any given week will be between 1,666 and 3,234 units.

The procedure of obtaining values of a normal random variable, given a probability, is summarized:

1. Draw a picture of the normal distribution in question and the standard normal distribution.
2. In the picture, shade in the area corresponding to the probability.
3. Use the table to find the z value (or values) that gives the required probability.
4. Use the transformation from Z to X to get the appropriate value (or values) of the original normal random variable.

PROBLEMS

4–38. If X is a normally distributed random variable with mean 120 and standard deviation 44, find a value x such that the probability that X will be less than x is 0.56.

4–39. For a normal random variable with mean 16.5 and standard deviation 0.8, find a point of the distribution such that there is a 0.85 probability that the value of the random variable will be above it.

4–40. For a normal random variable with mean 19,500 and standard deviation 400, find a point of the distribution such that the probability that the random variable will exceed this value is 0.02.

4–41. Find two values of the normal random variable with mean 88 and standard deviation 5 lying symmetrically on either side of the mean and covering an area of 0.98 between them.

4–42. For $X \sim N(32, 7^2)$, find two values x_1 and x_2, symmetrically lying on each side of the mean, with $P(x_1 < X < x_2) = 0.99$.

4–43. If X is a normally distributed random variable with mean -61 and standard deviation 22, find the value such that the probability that the random variable will be above it is 0.25.

4–44. If X is a normally distributed random variable with mean 97 and standard deviation 10, find x_2 such that $P(102 < X < x_2) = 0.05$.

4–45. Let X be a normally distributed random variable with mean 600 and variance 10,000. Find two values x_1 and x_2 such that $P(X > x_1) = 0.01$ and $P(X < x_2) = 0.05$.

4–46. Pierre operates a currency exchange office at Orly Airport in Paris. His office is open at night when the airport bank is closed, and he makes most of his business on returning U.S. tourists who need to change their remaining French francs back to U.S. dollars. From experience, Pierre knows that the demand for dollars on any given night during high season is approximately normally distributed with mean $25,000 and standard deviation $5,000. If Pierre carries too much cash in dollars overnight, he pays a penalty: interest on the cash. On the other hand, if he runs short of cash during the night, he needs to send a person downtown to an all-night financial agency to get the required cash. This, too, is costly to him. Therefore, Pierre would like to carry overnight an amount of money such that the demand on 85% of the nights will not exceed this amount. Can you help Pierre find the required amount of dollars to carry?

4–47. The demand for unleaded gasoline at a service station is normally distributed with mean 27,009 gallons per day and standard deviation 4,530. Find two values that will give a symmetric 0.95 probability interval for the amount of unleaded gasoline demanded daily.

4–48. The percentage of protein in a certain brand of dog food is a normally distributed random variable with mean 11.2% and standard deviation 0.6%. The manufacturer would like to state on the package that the product has a protein content of at least x_1% and no more than x_2%. It wants the statement to be true for 99% of the packages sold. Determine the values x_1 and x_2.

4–49. Debt as a percentage of equity (measured daily) for a firm tends to follow a normal distribution. A recent article reveals that for the Centex Corporation average debt as a percentage of equity in early 1995 was 113%.[4] If the standard deviation was 4%, find a value such that you can be 95% sure that debt as a percentage of equity was below that value on a given day.

4–50. The daily price of coffee is approximately normally distributed over a period of 15 days with a mean in June 1997 of $1.8781 per pound (on the wholesale market) and standard deviation of $0.15. Find a price such that the probability in the next 15 days that the price will go below it will be 0.90.

4–51. The daily price in dollars per metric ton of Ivory Coast cocoa, in summer 1997, was normally distributed with $\mu = \$1,617$ and $\sigma = \$12$. Find a price such that the probability that the actual price will be above it is 0.80.

4–7   Normal Approximation of Binomial Distributions

When the number of trials n in a binomial distribution is large ($>1,000$), the calculation of probabilities becomes difficult for the computer, because the calculation encounters some numbers that are too large and some that are too small to handle with needed accuracy. Fortunately, the binomial distribution approaches the normal distribution as n increases and therefore we can approximate it as a normal distribution. Note that the mean is np and the standard deviation is $\sqrt{np(1 - p)}$. The template is shown in Figure 4–17. When the values for n and p of the binomial distribution are entered in cells B4 and C4, the mean and the standard deviation of the corresponding normal distribution are calculated in cells E4 and F4. The rest of the template is similar to the normal distribution template we already saw.

Whenever a binomial distribution is approximated as a normal distribution, a **continuity correction** is required because a binomial is discrete and a normal is continuous. Thus, a column in the histogram of a binomial distribution for, say, $X = 10$ covers, in the continuous sense, the interval [9.5, 10.5]. Similarly, if we include the columns for $X = 10$, 11, and 12, then in the continuous case, the bars occupy the

---

[4]Eric S. Hardy, "Statistical Spotlight," *Forbes,* January 2, 1995, p. 119.

FIGURE 4–17   **The Template for Normal Approximation of Binomial Distribution**
[Normal Distribution.xls; Sheet: Normal Approximation]

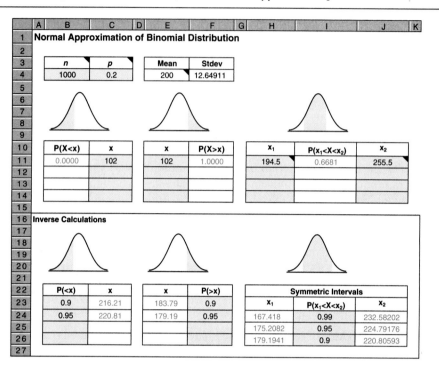

FIGURE 4–18   **Continuity Correction**

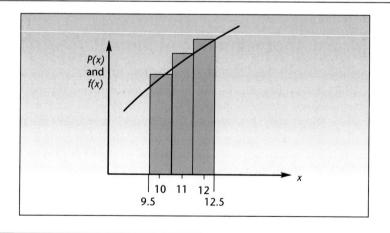

interval [9.5, 12.5], as seen in Figure 4–18. Therefore, when we calculate the binomial probability of an interval, say, $P(195 \leq X \leq 255)$, *we should subtract 0.5 from the left limit and add 0.5 to the right limit* to get the corresponding normal probability, namely, $P(194.5 < X < 255.5)$. Adding and subtracting 0.5 in this manner is known as the *con-*

tinuity correction. In Figure 4–17, this correction has been applied in cells H11 and J11. Cell I11 has the binomial probability of $P(195 \le X \le 255)$.

A total of 2,058 students take a difficult test. Each student has an independent 0.6205 probability of passing the test.

EXAMPLE 4–15

1. What is the probability that between 1,250 and 1,300 students, both numbers inclusive, will pass the test?
2. What is the probability that at least 1,300 students will pass the test?
3. If the probability of at least 1,300 students passing the test has to be at least 0.5, what is the minimum value for the probability of each student passing the test?

---

1. On the template for normal approximation, enter 2,058 for n and 0.6205 for p. Enter 1,249.5 in cell H11 and 1,300.5 in cell J11. The answer 0.7514 appears in cell I11.
2. Enter 1,299.5 in cell E11. The answer 0.1533 appears in cell F11.
3. Use the **Goal Seek** command to set cell F11 to value 0.5 by changing cell C4. The computer finds the answer as $p = 0.6314$.

Solution

PROBLEMS

In the following problems, use a normal distribution to compute the required probabilities. In each problem, also state the assumptions necessary for a binomial distribution, and indicate whether the assumptions are reasonable.

4–52. The manager of a restaurant knows from experience that 70% of the people who make reservations for the evening show up for dinner. The manager decides one evening to overbook and accept 20 reservations when only 15 tables are available. What is the probability that more than 15 parties will show up?

4–53. An advertising research study indicates that 40% of the viewers exposed to an advertisement try the product during the following four months. If 100 people are exposed to the ad, what is the probability that at least 20 of them will try the product in the following four months?

4–54. A computer system contains 45 identical microchips. The probability that any microchip will be in working order at a given time is 0.80. A certain operation requires that at least 30 of the chips be in working order. What is the probability that the operation will be carried out successfully?

4–55. Sixty percent of the managers who enroll in a special training program will successfully complete the program. If a large company sends 328 of its managers to enroll in the program, what is the probability that at least 200 of them will pass?

4–56. A large state university sends recruiters throughout the state in order to recruit graduating high school seniors to enroll in the university. From the university's records, it is known that 25% of the students who are interviewed by the recruiters actually enroll. If last spring the university recruiters interviewed 1,889 graduating seniors, what is the probability that at least 500 of them will enroll this fall?

4–57.

Formatting
Digital Esperanto

Electronic commerce is getting easier. Until now, corporations wanting to send documents to some courts and federal agencies had to convert them into a format called standard generalized markup language (SGML). This has required special editing software or tedious hand-coding. The conversion will be a breeze with an SGML version of WordPerfect coming from Novell Inc. Then, you can simply prepare documents in regular fashion; the program converts them when you save.

From *Business Week,* April 24, 1995, p. 19.

If 75% of executives shown this new product were impressed by its performance and indicated an intention to purchase it, what is the probability that at least 230 executives in a group of 300 would be interested in this product once it was demonstrated to them?

4–58.   According to a survey conducted by *Money* magazine, 64% of middle-class earners say it would be difficult to find a new job if they were fired.[5] If a group of 500 randomly chosen middle-class breadwinners is selected, what is the probability that at most 300 of them feel this way?

4–8   Summary and Review of Terms

In this chapter, we discussed the **normal probability distribution,** the most important probability distribution in statistics. We defined the **standard normal random variable** as the normal random variable with mean 0 and standard deviation 1. We saw how to use a table of probabilities for the standard normal random variable and how to transform a normal random variable with any mean and any standard deviation to the standard normal random variable by using the **normal transformation.**

We also saw how the standard normal random variable may, in turn, be transformed into any other normal random variable with a specified mean and standard deviation, and how this allows us to find values of a normal random variable that conform with some probability statement. We discussed a method of determining the mean and/or the standard deviation of a normal random variable from probability statements about the random variable. We saw how the normal distribution is used as a model in many real-world situations, both as the true distribution (a continuous one) and as an approximation to discrete distributions. In particular, we illustrated the use of the normal distribution as an approximation to the binomial distribution.

In the following chapters, we will make much use of the material presented here. Most statistical theory relies on the normal distribution and on distributions that are derived from it.

---

[5]J.S. Coyle, "How to Beat the Squeeze on the Middle Class," *Money,* May 1995, p. 108.

4–59. The time, in hours, that a copying machine may work without breaking down is a normally distributed random variable with mean 549 and standard deviation 68. Find the probability that the machine will work for at least 500 hours without breaking down.

4–60. The yield, in tons of ore per day, at a given coal mine is approximately normally distributed with mean 785 tons and standard deviation 60. Find the probability that at least 800 tons of ore will be mined on a given day. Find the proportion of working days in which anywhere from 750 to 850 tons is mined. Find the probability that on a given day, the yield will be below 665 tons.

4–61. Scores on a management aptitude examination are believed to be normally distributed with mean 650 (out of a total of 800 possible points) and standard deviation 50. What is the probability that a randomly chosen manager will achieve a score above 700? What is the probability that the score will be below 750?

4–62. Assume that the price of a share of TWA stock is normally distributed with mean 48 and standard deviation 6. What is the probability that on a randomly chosen day in the period for which our assumptions are made, the price of the stock will be more than $60 per share? Less than $60 per share? More than $40 per share? Between $40 and $50 per share? What are the limitations of your analysis?

4–63. The amount of oil pumped daily at Standard Oil's facilities in Prudhoe Bay is normally distributed with mean 800,000 barrels and standard deviation 10,000. In determining the amount of oil the company must report as its lower limit of daily production, the company wants to choose an amount such that for 80% of the days, at least the reported amount x is produced. Determine the value of the lower limit x.

4–64. Models of the pricing of stock options make the assumption of a normal distribution. An analyst believes that the price of an IBM stock option is a normally distributed random variable with mean $8.95 and variance 4. The analyst would like to determine a value such that there is a 0.90 probability that the price of the option will be greater than that value. Find the required value.

4–65. Weekly rates of return (on an annualized basis) for certain securities over a given period are believed to be normally distributed with mean 8.00% and variance 0.25. Give two values x_1 and x_2 such that you are 95% sure that annualized weekly returns will be between the two values.

4–66. The impact of a television commercial, measured in terms of excess sales volume over a given period, is believed to be approximately normally distributed with mean 50,000 and variance 9,000,000. Find 0.99 probability bounds on the volume of excess sales that would result from a given airing of the commercial.

4–67. A travel agency believes that the number of people who sign up for tours to Hawaii during the Christmas–New Year's holiday season is an approximately normally distributed random variable with mean 2,348 and standard deviation 762. For reservation purposes, the agency's management wants to find the number of people such that the probability is 0.85 that at least that many people will sign up. It also needs 0.80 probability bounds on the number of people who will sign up for the trip.

4–68. A loans manager at a large bank believes that the percentage of her customers who default on their loans during each quarter is an approximately normally distributed random variable with mean 12.1% and standard deviation 2.5%. Give a lower bound x with 0.75 probability that the percentage of people defaulting on their loans is at least x. Also give an upper bound x' with 0.75 probability that the percentage of loan defaulters is below x'.

4–69.   The power generated by a solar electric generator is normally distributed with mean 15.6 kilowatts and standard deviation of 4.1 kilowatts. We may be 95% sure that the generator will deliver at least how many kilowatts?

4–70.   Short-term rates fluctuate daily. It may be assumed that the yield for 90-day Treasury bills in 1995 was approximately normally distributed with mean 5.9% and standard deviation 0.8%.[6] Find a value such that 95% of the time during that year the yield of 90-day T-bills was below this value.

4–71.   In quality-control projects, engineers use charts where item values are plotted and compared with 3-standard-deviation bounds above and below the mean for the process. When items are found to fall outside the bounds, they are considered non-conforming, and the process is stopped when "too many" items are out of bounds. Assuming a normal distribution of item values, what percentage of values would you expect to be out of bounds when the process is in control? Accordingly, how would you define "too many"? What do you think is the rationale for this practice?

4–72.   Total annual textbook sales in a certain discipline are normally distributed. Forty-five percent of the time, sales are above 671,000 copies, and 10% of the time, sales are above 712,000 copies. Find the mean and the variance of annual sales.

4–73.   Typing speed on a new kind of keyboard for people at a certain stage in their training program is approximately normally distributed. The probability that the speed of a given trainee will be greater than 65 words per minute is 0.45. The probability that the speed will be more than 70 words per minute is 0.15. Find the mean and the standard deviation of typing speed.

4–74.   The number of people responding to a mailed information brochure on cruises of the Royal Viking Line through an agency in San Francisco is approximately normally distributed. The agency found that 10% of the time, over 1,000 people respond immediately after a mailing, and 50% of the time, at least 650 people respond right after the mailing. Find the mean and the standard deviation of the number of people who respond following a mailing.

4–75.   The Tourist Delivery Program was developed by several European automakers. In this program, a tourist from outside Europe—most are from the United States—may purchase an automobile in Europe and drive it in Europe for as long as six months, after which the manufacturer will ship the car to the tourist's home destination at no additional cost. In addition to the time limitations imposed, some countries impose mileage restrictions so that tourists will not misuse the privileges of the program. In setting the limitation, some countries use a normal distribution assumption. It is believed that the number of kilometers driven by a tourist in the program is normally distributed with mean 4,500 and standard deviation 1,800. If a country wants to set the mileage limit at a point such that 80% of the tourists in the program will want to drive fewer kilometers, what should the limit be?

---

[6]Investment Figures, *Business Week,* April 24, 1995, p. 145.

4–76. The number of newspapers demanded daily in a large metropolitan area is believed to be an approximately normally distributed random variable. If more newspapers are demanded than are printed, the paper suffers an opportunity loss, in that it could have sold more papers, and a loss of public goodwill. On the other hand, if more papers are printed than will be demanded, the unsold papers are returned to the newspaper office at a loss. Suppose that management believes it is most important to guard against the first type of error, unmet demand, and would like to set the number of papers printed at a level such that 75% of the time, demand for newspapers will be lower than that point. How many papers should be printed daily if the average demand is 34,750 papers and the standard deviation of demand is 3,560?

4–77. The Federal Funds rate in summer 1997 was approximately normal with $\mu = 5.5\%$ and $\sigma = 0.25\%$. Find the probability that the rate on a given day will be less than 5%.

4–78. The daily value of the European currency, ECU, in Belgian francs during the summer of 1997 was normally distributed with mean of 40.1 Belgian francs per ECU and standard deviation of 0.4 Belgian franc per ECU. Assume that every day the value of the ECU is independently and randomly drawn from this distribution. What is the probability that the next day's rate will be less than 42 Belgian francs per ECU?

4–79. A project consists of three phases to be completed one after the other. The duration of each phase, in days, is normally distributed as follows: Duration of Phase I $\sim N(84, 3^2)$; Duration of Phase II $\sim N(102, 4^2)$; Duration of Phase III $\sim N(62, 2^2)$.

    a. Find the distribution of the project duration. Report the mean and the standard deviation.

    b. If the project duration exceeds 250 days, a penalty will be assessed. What is the probability that the project will be completed within 250 days?

    c. If the project is completed within 240 days, a bonus will be earned. What is the probability that the project will be completed within 240 days?

4–80. The GMAT scores of students who are potential applicants to a university are normally distributed with a mean of 487 and a standard deviation of 98.

    a. What percentage of students will have scores exceeding 500?

    b. What percentage of students will have scores between 600 and 700?

    c. If the university wants only the top 75% of the students to be eligible to apply, what should be the minimum GMAT score specified for eligibility?

    d. Find the narrowest interval that will contain 75% of the students' scores.

    e. Find x such that the interval $[x, 2x]$ will contain 75% of the students' scores. (There are two answers. See if you can find them both.)

4–81. The profit (or loss) from an investment is normally distributed with a mean of $11,200 and a standard deviation of $8,250.

    a. What is the probability that there will be a loss rather than a profit?

    b. What is the probability that the profit will be between $10,000 and $20,000?

    c. Find x such that the probability that the profit will exceed x is 25%.

    d. If the loss exceeds $10,000 the company will have to borrow additional cash. What is the probability that the company will have to borrow additional cash?

    e. Comment on the risk in the investment.

CASE 4   Acceptable Pins

A company supplies pins in bulk to a customer. The company uses an automatic lathe to produce the pins. Due to many causes—vibration, temperature, wear and tear, and the like—the lengths of the pins made by the machine are normally distributed with a mean of 1.012 inches and a standard deviation of 0.018 inch. The customer will buy only those pins with lengths in the interval 1.00 ± 0.02 inch. In other words, the customer wants the length to be 1.00 inch but will accept up to 0.02 inch deviation on either side. This 0.02 inch is known as the *tolerance*.

1. What percentage of the pins will be acceptable to the consumer?

In order to improve percentage accepted, the production manager and the engineers discuss adjusting the population mean and standard deviation of the length of the pins.

2. If the lathe can be adjusted to have the mean of the lengths to any desired value, what should it be adjusted to? Why?

3. Suppose the mean cannot be adjusted, but the standard deviation can be reduced. What maximum value of the standard deviation would make 90% of the parts acceptable to the consumer? (Assume the mean to be 1.012.)

4. Repeat question 3, with 95% and 99% of the pins acceptable.

5. In practice, which one do you think is easier to adjust, the mean or the standard deviation? Why?

The production manager then considers the costs involved. The cost of resetting the machine to adjust the population mean involves the engineers' time and the cost of production time lost. The cost of reducing the population standard deviation involves, in addition to these costs, the cost of overhauling the machine and reengineering the process.

6. Assume it costs $\$150x^2$ to decrease the standard deviation by $(x/1000)$ inch. Find the cost of reducing the standard deviation to the values found in questions 3 and 4.

7. Now assume that the mean has been adjusted to the best value found in question 2 at a cost of $80. Calculate the reduction in standard deviation necessary to have 90%, 95%, and 99% of the parts acceptable. Calculate the respective costs, as in question 6.

8. Based on your answers to questions 6 and 7, what are your recommended mean and standard deviation?

CASE 5   Multicurrency Decision

A company sells precision grinding machines to four customers in four different countries. It has just signed a contract to sell, two months from now, a batch of these machines to each customer. The table below shows the number of machines (batch quantity) to be delivered to the four customers. The selling price of the machine is fixed in the local currency, and the company plans to convert the local currency at the exchange rate prevailing at the time of delivery. As usual, there is uncertainty in the exchange rates. The sales department estimates the exchange rate for each currency and its standard deviation, ex-

pected at the time of delivery, as shown in the table. Assume that the exchange rates are normally distributed and independent.

Customer	Batch Quantity	Selling Price	Exchange Rate Mean	Exchange Rate Standard Deviation
1	12	£ 57,810	$1.41/£	$0.041/£
2	8	¥ 8,640,540	$0.00904/¥	$0.00045/¥
3	5	€97,800	$0.824/€	$0.0342/€
4	2	R 4,015,000	$0.0211/R	$0.00083/R

1. Find the distribution of the uncertain revenue from the contract in U.S. dollars. Report the mean, the variance, and the standard deviation.

2. What is the probability that the revenue will exceed $2,250,000?

3. What is the probability that the revenue will be less than $2,150,000?

4. To remove the uncertainty in the revenue amount, the sales manager of the company looks for someone who would assume the risk. An international bank offers to pay a sure sum of $2,150,000 in return for the revenue in local currencies. What useful facts can you tell the sales manager about the offer, without involving any of your personal judgment?

5. What is your recommendation to the sales manager, based on your personal judgment?

6. If the sales manager is willing to accept the bank's offer, but the CEO of the company is not, who is more risk-averse?

7. Suppose the company accepts the bank's offer. Now consider the bank's risk, assuming that the bank will convert all currencies into U.S. dollars at the prevailing exchange rates. What is the probability that the bank will incur a loss?

8. The bank defines its value at risk as the loss that occurs at the 5th percentile of the uncertain revenue. What is the bank's value at risk?

9. What is the bank's expected profit?

10. Express the value at risk as a percentage of the expected profit. Based on this percentage, what is your evaluation of the risk faced by the bank?

11. Suppose the bank does not plan to convert all currencies into U.S. dollars, but plans to spend or save them as local currency or convert them into some other needed currency. Will this increase or decrease the risk faced by the bank?

12. Based on the answer to part 11, is the assumption (made in parts 7 to 10) that the bank will convert all currencies into U.S. dollars a good assumption?

Sampling and Sampling Distributions

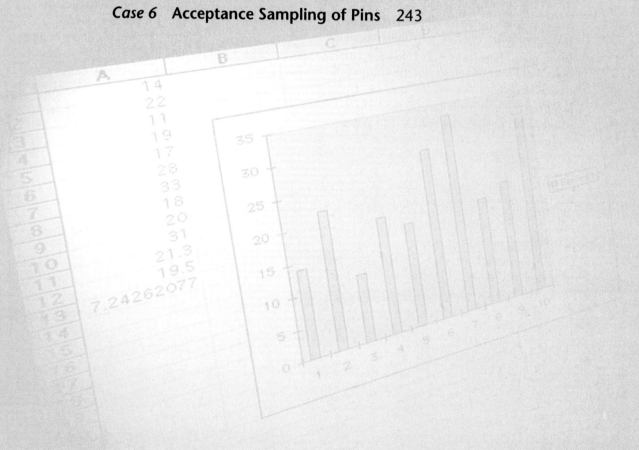

5–1   Using Statistics

Statistics is a science of *inference*. It is the science of generalization from a *part* (the randomly chosen sample) to the *whole* (the population).[1] Recall from Chapter 1 that the population is the entire collection of measurements in which we are interested, and the sample is a smaller set of measurements selected from the population. A random sample of *n* elements is a sample selected from the population in such a way that every set of *n* elements is as likely to be selected as any other set of *n* elements.[2] It is important that the sample be drawn randomly from the entire population under study. This makes it likely that our sample will be truly representative of the population of interest and minimizes the chance of errors. As we will see in this chapter, random sampling also allows us to compute the probabilities of sampling errors, thus providing us with knowledge of the degree of accuracy of our sampling results. The need to sample correctly is best illustrated by the well-known story of the *Literary Digest* (see page 210).

In 1936, the widely quoted *Literary Digest* embarked on the project of predicting the results of the presidential election to be held that year. The magazine boasted it would predict, to within a fraction of the percentage of the votes, the winner of the election—incumbent President Franklin Delano Roosevelt or the Republican governor of Kansas, Alfred M. Landon. The *Digest* tried to gather a sample of staggering proportion—10 million voters! One problem with the survey was that only a fraction of the people sampled, 2.3 million, actually provided the requested information. Should a link have existed between a person's inclination to answer the survey and his or her voting preference, the results of the survey would have been *biased*: slanted toward the voting preference of those who did answer. It is unknown whether such a link did exist in the case of the *Digest*. (This problem is known as *nonresponse bias* and is discussed in Chapter 16.) A very serious problem with the *Digest*'s poll, and one known to have affected the results, is the following.

The sample of voters chosen by the *Literary Digest* was obtained from lists of telephone numbers, automobile registrations, and names of *Digest* readers. Remember that this was 1936—not as many people owned phones or cars as today, and the ones who did tended to be wealthier and more likely to vote Republican (and the same goes for readers of the *Digest*). The selection procedure for the sample of voters was thus biased (slanted toward one kind of voter) because the sample was not randomly chosen from the entire population of voters. Figure 5–1 demonstrates a correct sampling procedure versus the sampling procedure used by the *Literary Digest*.

As a result of the *Digest* error, the magazine does not exist today; it went bankrupt soon after the 1936 election. Some say that hindsight is useful and that today we know more statistics, making it easy for us to deride mistakes made more than 50 years ago. Interestingly enough, however, the ideas of sampling bias were understood in 1936. A few weeks *before* the election, there appeared a small article in *The New York Times* criticizing the methodology of the *Digest* poll. Few paid it any attention.

---

[1]Not all of statistics concerns inferences about populations. One branch of statistics, called *descriptive statistics*, deals with describing data sets—possibly with no interest in an underlying population. The descriptive statistics of Chapter 1, when *not* used for inference, fall in this category.

[2]This is the definition of *simple random sampling*, and we will assume throughout that all our samples are simple random samples. Other methods of sampling are discussed in Chapter 6.

Digest Poll Gives Landon 32 States

Landon Leads 4–3 in Last Digest Poll

Final Tabulation Gives Him 370 Electoral Votes to 161 for President Roosevelt

Governor Landon will win the election by an electoral vote of 370 to 161, will carry thirty-two of the forty-eight States, and will lead President Roosevelt about four to three in their share of the popular vote, if the final figures in The Literary Digest poll, made public yesterday, are verified by the count of the ballots next Tuesday.

The New York Times, Friday, October 30, 1936. Copyright © 1936 by The New York Times Company. Reprinted by permission.

Roosevelt's Plurality Is 11,000,000

History's Largest Poll

46 States Won by President, Maine and Vermont by Landon

Many Phases To Victory

Democratic Landslide Looked Upon as Striking Personal Triumph for Roosevelt

BY ARTHUR KROCK

As the count of ballots cast Tuesday in the 1936 Presidential election moved toward completion yesterday, these facts appeared:

Franklin Delano Roosevelt was re-elected President, and John N. Garner Vice President, by the largest popular and electoral majority since the United States became a continental nation—a margin of approximately 11,000,000 plurality of all votes cast, and 523 votes in the electoral college to 8 won by the Republican Presidential candidate, Governor Alfred M. Landon of Kansas. The latter carried only Maine and Vermont of the forty-eight States of the Union . . . .

The New York Times, Thursday, November 5, 1936. Copyright © 1936 by The New York Times Company. Reprinted by permission.

Sampling is very useful in many situations besides political polling, including business and other areas where we need to obtain information about some population. Our information often leads to a *decision*. There are also situations, as demonstrated by the examples in the introduction to this book, where we are interested in a *process* rather than a single population. One such process is the relationship between advertising and sales. In these more involved situations, we still make the assumption of an underlying population—here, the population of *pairs* of possible advertising and sales values. Conclusions about the process are reached based on information in our data, which are assumed to constitute a random sample from the entire population. The ideas of a population and of a random sample drawn from the population are thus essential to all inferential statistics.

FIGURE 5–1   **A Good Sampling Procedure and the One Used by the** *Literary Digest*

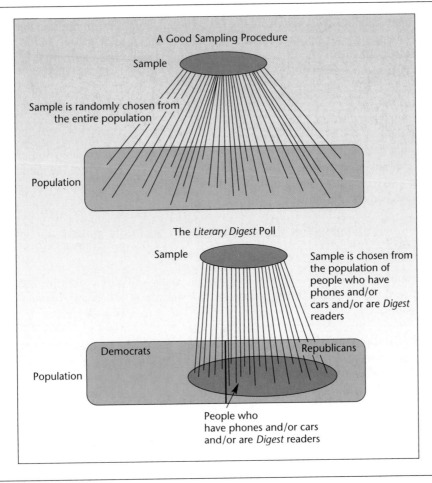

In statistical inference we are concerned with populations; the samples are of no interest to us in their own right. We wish to use our *known* random sample in the extraction of information about the *unknown* population from which it is drawn. The information we extract is in the form of summary statistics: a sample mean, a sample standard deviation, or other measures computed from the sample. A statistic such as the sample mean is considered an *estimator* of a population *parameter*—the population mean. In the next section, we discuss and define sample estimators and population parameters. Then we explore the relationship between statistics and parameters via the *sampling distribution*. Finally, we discuss desirable properties of statistical estimators.

5–2   Sample Statistics as Estimators of Population Parameters

A population may be a large, sometimes infinite, collection of elements. The population has a *frequency distribution*—the distribution of the frequencies of occurrence of its elements. The population distribution, when stated in relative frequencies, is also the probability distribution of the population. This is so because the relative frequency of a value in the population is also the probability of obtaining the particular value when an element is randomly drawn from the entire population. As with random variables, we may associate with a population its mean and its standard deviation. In the case of populations, the mean and the standard deviation are called *parameters*. They are denoted by μ and σ, respectively.

> A numerical measure of a population is called a **population parameter,** or simply a **parameter.**

Recall that in Chapter 4 we referred to the mean and the standard deviation of a normal probability distribution as the distribution parameters. Here we view parameters as descriptive measures of populations. Inference drawn about a population parameter is based on sample statistics.

> A numerical measure of the sample is called a **sample statistic,** or simply a **statistic.**

Population parameters are estimated by sample statistics. When a sample statistic is used to estimate a population parameter, the statistic is called an *estimator* of the parameter.

> An **estimator** of a population parameter is a sample statistic used to estimate the parameter. An **estimate** of the parameter is a *particular* numerical value of the estimator obtained by sampling. When a single value is used as an estimate, the estimate is called a **point estimate** of the population parameter.

The sample mean $\bar{X}$ is the sample statistic used as an estimator of the population mean μ. Once we sample from the population and obtain a value of $\bar{X}$ (using equation 1–1), we will have obtained a *particular* sample mean; we will denote this particular value by $\bar{x}$. We may have, for example, $\bar{x} = 12.53$. This value is our estimate of μ. The estimate is a point estimate because it constitutes a single number. In this chapter, every estimate will be a point estimate—a single number that, we hope, lies close to the population parameter it estimates. Chapter 6 is entirely devoted to the concept of an *interval estimate*—an estimate constituting an interval of numbers rather than a single number. An interval estimate is an interval believed likely to contain the unknown population parameter. It conveys more information than just the point estimate on which it is based.

In addition to the sample mean, which estimates the population mean, other statistics are useful. The sample variance S^2 is used as an estimator of the population variance σ^2. A particular estimate obtained will be denoted by s^2. (This estimate is computed from the data using equation 1–3 or an equivalent formula.)

As demonstrated by the political polling example with which we opened this chapter, interest often centers not on a mean or standard deviation of a population, but rather on a population *proportion*. The population proportion parameter is also called a binomial proportion parameter.

> The **population proportion** p is equal to the number of elements in the population belonging to the category of interest, divided by the total number of elements in the population.

The population proportion of voters for Governor Landon in 1936, for example, was the number of people who intended to vote for the candidate, divided by the total number of voters. The estimator of the population proportion p is the *sample proportion* $\hat{P}$, defined as the number of *binomial successes* in the sample (i.e., the number of elements in the sample that belong to the category of interest), divided by the sample size n. A particular estimate of the population proportion p is the sample proportion $\hat{p}$.

The **sample proportion** is

$$\hat{p} = \frac{x}{n} \qquad\qquad (5\text{--}1)$$

where x is the number of elements in the sample found to belong to the category of interest and n is the sample size.

Suppose that we want to estimate the proportion of consumers in a certain area who are users of a certain product. The (unknown) population proportion is p. We estimate p by the statistic $\hat{P}$, the sample proportion. Suppose a random sample of 100 consumers in the area reveals that 26 are users of the product. Our point estimate of p is then $\hat{p} = x/n = 26/100 = 0.26$.

In summary, we have the following estimation relationships:

Estimator (Sample Statistic)		Population Parameter
$\bar{X}$	estimates →	μ
S^2	estimates →	σ^2
$\hat{p}$	estimates →	p

Let us consider sampling to estimate the population mean, and let us try to visualize how this is done. Consider a population with a certain frequency distribution. The frequency distribution of the values of the population is the probability distribution of the value of an element in the population, drawn at random. Figure 5–2 shows a frequency distribution of some population and the population mean μ. If we knew the exact frequency distribution of the population, we would be able to determine μ directly in the same way we determine the mean of a random variable when we know its probability distribution. In reality, the frequency distribution of a population is not known; neither is the mean of the population. We try to estimate the population mean by the sample mean, computed from a random sample. Figure 5–2 shows the values of a random sample obtained from the population and the resulting sample mean $\bar{x}$, computed from the data.

In this example, $\bar{x}$ happens to lie close to μ, the population parameter it estimates, although this does not always happen. The sample statistic $\bar{X}$ is a *random variable* whose actual value depends on the particular random sample obtained. The random variable $\bar{X}$ has a relatively high probability of being close to the population mean it estimates, and it has decreasing probabilities of falling farther and farther from the population mean. Similarly, the sample statistic S is a random variable with a relatively high probability of being close to σ, the population parameter it estimates. Also, when sampling for a population proportion p, the estimator $\hat{P}$ has a relatively high probability of being close to p. How high a probability, and how close to the parameter? The answer to this question is the main topic of this chapter, presented in

FIGURE 5–2   **A Population Distribution, a Random Sample from the Population, and Their Respective Means**

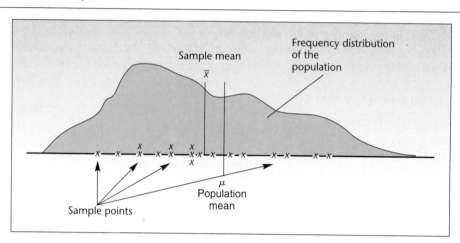

the next section. Before discussing this important topic, we will say a few things about the mechanics of obtaining random samples.

Obtaining a Random Sample

All along we have been referring to random samples. We have stressed the importance of the fact that our sample should always be drawn randomly from the entire population about which we wish to draw an inference. How do we draw a random sample?

To obtain a random sample from the entire population, we need a list of all the elements in the population of interest. Such a list is called a *frame*. The frame allows us to draw elements from the population by randomly generating the numbers of the elements to be included in the sample. Suppose we need a simple random sample of 100 people from a population of 7,000. We make a list of all 7,000 people and assign each person an identification number. This gives us a list of 7,000 numbers—our frame for the experiment. Then we generate by computer or by other means a set of 100 random numbers in the range of values from 1 to 7,000. This procedure gives every set of 100 people in the population an equal chance of being included in the sample.

As mentioned, a computer (or an advanced calculator) may be used for generating random numbers. We will demonstrate an alternative method of choosing random numbers—a random number table. Table 5–1 is a part of such a table. A random number table is given in Appendix C as Table 14. To use the table, we start at any point, pick a number from the table, and continue in the same row or the same column (it does not matter which), systematically picking out numbers with the number of digits appropriate for our needs. If a number is outside our range of required numbers, we ignore it. We also ignore any number already obtained.

For example, suppose that we need a random sample of 10 data points from a population with a total of 600 elements. This means that we need 10 random drawings of elements from our frame of 1 through 600. To do this, we note that the number 600 has three digits; therefore, we draw random numbers with three digits. Since our population has only 600 units, however, we ignore any number greater than 600

TABLE 5–1   Random Numbers

10480	15011	01536	02011	81647	91646	69179	14194
22368	46573	25595	85393	30995	89198	27982	53402
24130	48360	22527	97265	76393	64809	15179	24830
42167	93093	06243	61680	07856	16376	93440	53537
37570	39975	81837	16656	06121	91782	60468	81305
77921	06907	11008	42751	27756	53498	18602	70659

and take the next number, assuming it falls in our range. Let us decide arbitrarily to choose the first three digits in each set of five digits in Table 5–1; and we proceed by row, starting in the first row and moving to the second row, continuing until we have obtained our 10 required random numbers. We get the following random numbers: 104, 150, 15, 20, 816 (discard), 916 (discard), 691 (discard), 141, 223, 465, 255, 853 (discard), 309, 891 (discard), 279. Our random sample will, therefore, consist of the elements with serial numbers 104, 150, 15, 20, 141, 223, 465, 255, 309, and 279. A similar procedure would be used for obtaining the random sample of 100 people from the population of 7,000 mentioned earlier. Random number tables are included in books of statistical tables.

In many situations it is not possible to obtain a frame of the elements in the population. In such situations we may still randomize some aspect of the experiment and thus obtain a random sample. For example, we may randomize the location and the time and date of the collection of our observations, as well as other factors involved. In estimating the average miles-per-gallon rating of an automobile, for example, we may randomly choose the dates and times of our trial runs as well as the particular automobiles used, the drivers, the roads used, and so on. More will be said about sampling techniques in Chapter 16.

PROBLEMS

5–1. Discuss the concepts of a parameter, a sample statistic, an estimator, and an estimate. What are the relations among these entities?

5–2. An auditor selected a random sample of 12 accounts from all accounts receivable of a given firm. The amounts of the accounts, in dollars, are as follows: 87.50, 123.10, 45.30, 52.22, 213.00, 155.00, 39.00, 76.05, 49.80, 99.99, 132.00, 102.11. Compute an estimate of the mean amount of all accounts receivable. Give an estimate of the variance of all the amounts.

5–3. In problem 5–2, suppose the auditor wants to estimate the proportion of all the firm's accounts receivable with amounts over $100. Give a point estimate of this parameter.

5–4. Following is a random sample from the year 1986 of personal incomes of industry workers in the state of New York, in thousands of dollars per year: 14.5, 13.2, 15.4, 12.8, 19.3, 13.4, 16.5, 17.2, 17.8, 11.5, 13.6, 18.8. Compute point estimates of the mean and the standard deviation of the population of incomes of industry workers in the state.

5–5. Gasoline prices in the Midwest jumped between 25% and 50% during the summer of 2000. A random sample of prices on August 31, 2000, for regular unleaded gasoline was $1.59, 1.64, 1.69, 1.69, 1.57, 1.78, 1.65, 1.66, 1.69, 1.79, 1.77, 1.75, 1.69, 1.66, 1.80, 1.75, 1.66, 1.65, 1.59, 1.69, 1.79, 1.78, 1.59, 1.79, 1.65, 1.59.

    a. Calculate a point estimate of the average price.

    b. If the true mean is $1.64 and $\sigma = \$0.12$, what is the probability of getting a point estimate that is greater than or equal to the point estimate you calculated in part (*a*)?

5–6. A market research worker interviewed a random sample of 18 people about their use of a certain product. The results, in terms of Y or N (for Yes, a user of the product, or No, not a user of the product), are as follows: Y N N Y Y Y N Y N Y Y Y N Y N Y Y N. Estimate the population proportion of users of the product.

5–7. Use a random number table (you may use Table 5–1) to find identification numbers of elements to be used in a random sample of size $n = 25$ from a population of 950 elements.

5–8. Find five random numbers from 0 to 5,600.

5–9. Assume that you have a frame of 40 million voters (something the *Literary Digest* should have had for an unbiased polling). Randomly generate the numbers of five sampled voters.

5–10. Suppose you need to sample the concentration of a chemical in a production process that goes on continuously 24 hours per day, 7 days per week. You need to generate a random sample of six observations of the process over a period of one week. Use a computer, a calculator, or a random number table to generate the six observation times (to the nearest minute).

5–3 Sampling Distributions

> The **sampling distribution** of a statistic is the probability distribution of all possible values the statistic may take when computed from random samples of the same size, drawn from a specified population.

Let us first look at the sample mean $\overline{X}$. The sample mean is a random variable. The possible values of this random variable depend on the possible values of the elements in the random sample from which $\overline{X}$ is to be computed. The random sample, in turn, depends on the distribution of the population from which it is drawn. As a random variable, $\overline{X}$ has a *probability distribution*. This probability distribution is the sampling distribution of $\overline{X}$.

> The **sampling distribution of** $\overline{X}$ is the probability distribution of all possible values the random variable $\overline{X}$ may take when a sample of size *n* is taken from a specified population.

Let us derive the sampling distribution of $\overline{X}$ in the simple case of drawing a sample of size $n = 2$ items from a population uniformly distributed over the integers 1 through 8. That is, we have a large population consisting of equal proportions of the values 1 to 8. At each draw, there is a 1/8 probability of obtaining any of the values 1 through 8 (alternatively, we may assume there are only eight elements, 1 through 8, and that the sampling is done with replacement). The sample space of the values of the two sample points drawn from this population is given in Table 5–2. This is an example. In real situations, sample sizes are much larger.

TABLE 5–2   Possible Values of Two Sample Points from a Uniform Population of the Integers 1 through 8

Second Sample Point	First Sample Point							
	1	2	3	4	5	6	7	8
1	1,1	2,1	3,1	4,1	5,1	6,1	7,1	8,1
2	1,2	2,2	3,2	4,2	5,2	6,2	7,2	8,2
3	1,3	2,3	3,3	4,3	5,3	6,3	7,3	8,3
4	1,4	2,4	3,4	4,4	5,4	6,4	7,4	8,4
5	1,5	2,5	3,5	4,5	5,5	6,5	7,5	8,5
6	1,6	2,6	3,6	4,6	5,6	6,6	7,6	8,6
7	1,7	2,7	3,7	4,7	5,7	6,7	7,7	8,7
8	1,8	2,8	3,8	4,8	5,8	6,8	7,8	8,8

TABLE 5–3   The Sampling Distribution of $\bar{X}$ for a Sample of Size 2 from a Uniformly Distributed Population of the Integers 1 to 8

Particular Value $\bar{x}$	Probability of $\bar{x}$	Particular Value $\bar{x}$	Probability of $\bar{x}$
1	1/64	5	7/64
1.5	2/64	5.5	6/64
2	3/64	6	5/64
2.5	4/64	6.5	4/64
3	5/64	7	3/64
3.5	6/64	7.5	2/64
4	7/64	8	1/64
4.5	8/64		1.00

Using the sample space from the table, we will now find all possible values of the sample mean $\bar{X}$ and their probabilities. We compute these probabilities, using the fact that all 64 sample pairs shown are equally likely. This is so because the population is uniformly distributed and because in random sampling each drawing is independent of the other; therefore, the probability of a given pair of sample points is the product $(1/8)(1/8) = 1/64$. From Table 5–2, we compute the sample mean associated with each of the 64 pairs of numbers and find the probability of occurrence of each value of the sample mean. The values and their probabilities are given in Table 5–3. The table thus gives us the sampling distribution of $\bar{X}$ in this particular sampling situation. Verify the values in Table 5–3 using the sample space given in Table 5–2. Figure 5–3 shows the uniform distribution of the population and the sampling distribution of $\bar{X}$, as listed in Table 5–3.

Let us find the mean and the standard deviation of the *population*. We can do this by treating the population as a random variable (the random variable being the value of a single item randomly drawn from the population; each of the values 1 through 8 has a 1/8 probability of being drawn). Using the appropriate equations from Chapter 3, we find $\mu = 4.5$ and $\sigma = 2.29$ (verify these results).

FIGURE 5–3   The Population Distribution and the Sampling Distribution of the Sample Mean

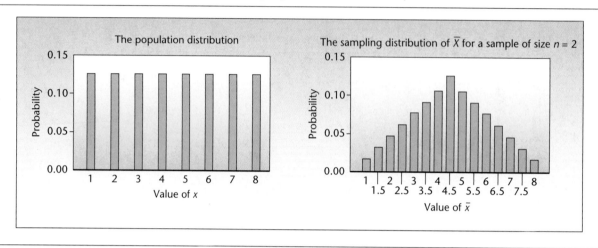

Now let us find the expected value and the standard deviation of the random variable $\bar{X}$. Using the sampling distribution listed in Table 5–3, we find $E(\bar{X}) = 4.5$ and $\sigma_{\bar{x}} = 1.62$ (verify these values by computation). Note that the expected value of $\bar{X}$ is equal to the mean of the population; each is equal to 4.5. The standard deviation of $\bar{X}$, denoted $\sigma_{\bar{x}}$, is equal to 1.62, and the population standard deviation σ is 2.29. But observe an interesting fact: $2.29/\sqrt{2} = 1.62$. The facts we have discovered in this example are not an accident—they hold in all cases. The expected value of the sample mean $\bar{X}$ is equal to the population mean μ and the standard deviation of $\bar{X}$ is equal to the population standard deviation divided by the square root of the sample size. Sometimes the estimated standard deviation of a statistic is called its *standard error*.

> The expected value of the sample mean is[3]
> $$E(\bar{X}) = \mu \qquad\qquad (5\text{–}2)$$
> The standard deviation of the sample mean is[4]
> $$\mathrm{SD}(\bar{X}) = \sigma_{\bar{x}} = \sigma/\sqrt{n} \qquad\qquad (5\text{–}3)$$

We know the two parameters of the sampling distribution of $\bar{X}$: We know the mean of the distribution (the expected value of $\bar{X}$) and we know its standard devia-

---

[3]The proof of equation 5–2 relies on the fact that the expected value of the sum of several random variables is equal to the sum of their expected values. Also, from equation 3–6 we know that the expected value of aX, where a is a number, is equal to a times the expected value of X. We also know that the expected value of each element X drawn from the population is equal to μ, the population mean. Using these facts, we find the following: $E(\bar{X}) = E(\Sigma X/n) = (1/n)E(\Sigma X) = (1/n)n\mu = \mu.$

[4]The proof of equation 5–3 relies on the fact that, when several random variables are *independent* (as happens in random sampling), the variance of the sum of the random variables is equal to the sum of their variances. Also, from equation 3–10, we know that the variance of aX is equal to $a^2 V(X)$. The variance of each X drawn from the population is equal to σ^2. Using these facts, we find $V(\bar{X}) = V(\Sigma X/n) = (1/n)^2(\Sigma\sigma^2) = (1/n)^2(n\sigma^2) = \sigma^2/n$. Hence, $\mathrm{SD}(\bar{X}) = \sigma/\sqrt{n}.$

FIGURE 5–4   **A Normally Distributed Population and the Sampling Distribution of the Sample Mean for Different Sample Sizes**

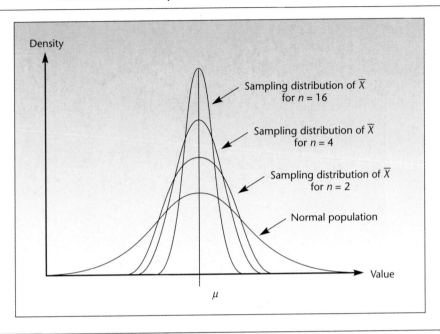

tion. What about the shape of the sampling distribution? If the population itself is *normally distributed,* the sampling distribution of $\overline{X}$ is also normal.

> When sampling is done from a *normal* distribution with mean μ and standard deviation σ, the sample mean $\overline{X}$ has a **normal sampling distribution:**
>
> $$\overline{X} \sim N(\mu, \sigma^2/n) \qquad\qquad (5\text{–}4)$$

Thus, when we sample from a normally distributed population with mean μ and standard deviation σ, the sample mean has a normal distribution with the same *center,* μ, as the population but with *width* (standard deviation) that is $1/\sqrt{n}$ the size of the width of the population distribution. This is demonstrated in Figure 5–4, which shows a normal population distribution and the sampling distribution of $\overline{X}$ for different sample sizes.

The fact that the sampling distribution of $\overline{X}$ has mean μ is very important. It means that, *on the average,* the sample mean is equal to the population mean. The distribution of the statistic is *centered* on the parameter to be estimated, and this makes the statistic $\overline{X}$ a good estimator of μ. This fact will become clearer in the next section, where we discuss estimators and their properties. The fact that the standard deviation of $\overline{X}$ is $\sigma/\sqrt{n}$ means that as the sample size *increases,* the standard deviation of $\overline{X}$ *decreases,* making $\overline{X}$ more likely to be close to μ. This is another desirable property of a good estimator, to be discussed later. Finally, when the sampling distribution of $\overline{X}$ is normal, this allows us

FIGURE 5–5 The Sampling Distribution of $\bar{X}$ as the Sample Size Increases

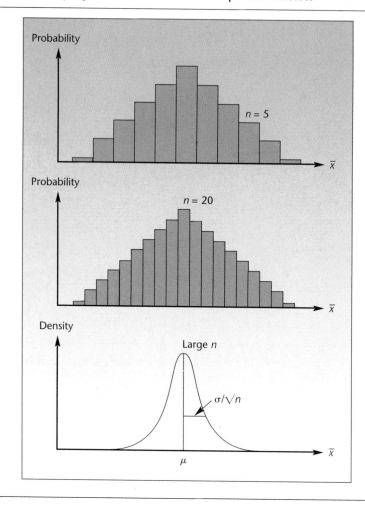

to compute probabilities that $\bar{X}$ will be within specified distances of μ. What happens in cases where the population itself is *not* normally distributed?

In Figure 5–3, we saw the sampling distribution of $\bar{X}$ when sampling is done from a uniformly distributed population and with a sample of size $n = 2$. Let us now see what happens as we increase the sample size. Figure 5–5 shows results of a simulation giving the sampling distribution of $\bar{X}$ when the sample size is $n = 5$, when the sample size is $n = 20$, and the *limiting* distribution of $\bar{X}$—the distribution of $\bar{X}$ as the sample size increases indefinitely. As can be seen from the figure, the limiting distribution of $\bar{X}$ is, again, the *normal distribution*.

The Central Limit Theorem

The result we just stated—that the distribution of the sample mean $\bar{X}$ tends to the normal distribution as the sample size increases—is one of the most important results in statistics. It is known as the *central limit theorem*.

The Central Limit Theorem

When sampling is done from a population with mean μ and finite standard deviation σ, the sampling distribution of the sample mean $\bar{X}$ will tend to a normal distribution with mean μ and standard deviation $\sigma/\sqrt{n}$ as the sample size n becomes large.

$$\text{For "large enough" } n \qquad \bar{X} \sim N(\mu, \sigma^2/n) \qquad (5-5)$$

The central limit theorem is remarkable because it states that the distribution of the sample mean $\bar{X}$ tends to a normal distribution *regardless* of the distribution of the population from which the random sample is drawn. The theorem allows us to make probability statements about the possible range of values the sample mean may take. It allows us to compute probabilities of how far away $\bar{X}$ may be from the population mean it estimates. For example, using our rule of thumb for the normal distribution, we know that the probability that the distance between $\bar{X}$ and μ will be less than $\sigma/\sqrt{n}$ is approximately 0.68. This is so because, as you remember, the probability that the value of a normal random variable will be within 1 standard deviation of its mean is 0.6826; here our normal random variable has mean μ and standard deviation $\sigma/\sqrt{n}$. Other probability statements can be made as well; we will see their use shortly. When is a sample size n "large enough" that we may apply the theorem?

The central limit theorem says that, *in the limit,* as n goes to infinity $(n \rightarrow \infty)$, the distribution of $\bar{X}$ becomes a normal distribution (regardless of the distribution of the population). The *rate* at which the distribution approaches a normal distribution does depend, however, on the shape of the distribution of the parent population. If the population itself is normally distributed, the distribution of $\bar{X}$ is normal for *any* sample size $n,$ as stated earlier. On the other hand, for population distributions that are very different from a normal distribution, a relatively large sample size is required to achieve a good normal approximation for the distribution of $\bar{X}$. Figure 5–6 shows several parent population distributions and the resulting sampling distributions of $\bar{X}$ for different sample sizes.

Since we often do not know the shape of the population distribution, it would be useful to have some general rule of thumb telling us when a sample is large enough that we may apply the central limit theorem.

In general, a sample of 30 or more elements is considered large enough for the central limit theorem to take effect.

We emphasize that this is a *general,* and somewhat arbitrary, rule. A larger minimum sample size may be required for a good normal approximation when the population distribution is very different from a normal distribution. By the same token, a smaller minimum sample size may suffice for a good normal approximation when the population distribution is close to a normal distribution.

Throughout this book, we will make reference to *small* samples versus *large* samples. By a small sample, we generally mean a sample of fewer than 30 elements. A large sample will generally mean a sample of 30 or more elements. The results we will discuss as applicable for large samples will be more meaningful, however, the larger the sample size. (By the central limit theorem, the larger the sample size, the better the approximation offered by the normal distribution.) The "30 rule" should, therefore, be applied with caution. Let us now look at an example of the use of the central limit theorem.

FIGURE 5–6   **The Effects of the Central Limit Theorem: The Distribution of $\bar{X}$ for Different Populations and Different Sample Sizes**

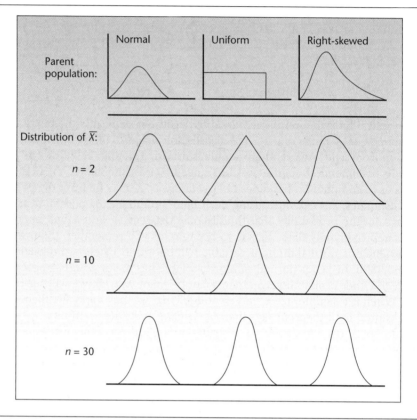

EXAMPLE 5–1   Mercury makes a 2.4-liter V-6 engine, the Laser XRi, used in speedboats. The company's engineers believe that the engine delivers an average power of 220 horsepower and that the standard deviation of power delivered is 15 horsepower. A potential buyer intends to sample 100 engines (each engine to be run a single time). What is the probability that the sample mean $\bar{X}$ will be less than 217 horsepower?

Solution   In solving problems such as this one, we use the techniques of Chapter 4. There we used μ as the mean of the normal random variable and σ as its standard deviation. Here our random variable $\bar{X}$ is normal (at least approximately so, by the central limit theorem because our sample size is large) and has mean μ. Note, however, that the standard deviation of our random variable $\bar{X}$ is $\sigma/\sqrt{n}$ and not just σ. We proceed as follows:

$$P(\bar{X} < 217) = P\left(Z - \frac{217 - \mu}{\sigma/\sqrt{n}}\right)$$

$$= P\left(Z < \frac{217 - 220}{15/\sqrt{100}}\right) = P(Z < -2) = 0.0228$$

Thus, if the population mean is indeed $\mu = 220$ horsepower and the standard deviation is $\sigma = 15$ horsepower, there is a rather small probability that the potential buyer's tests will result in a sample mean less than 217 horsepower.

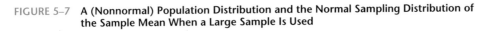

FIGURE 5–7   **A (Nonnormal) Population Distribution and the Normal Sampling Distribution of the Sample Mean When a Large Sample Is Used**

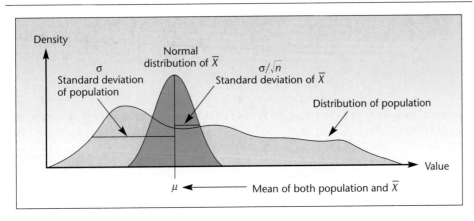

Figure 5–7 should help clarify the distinction between the population distribution and the sampling distribution of $\overline{X}$. The figure emphasizes the three aspects of the central limit theorem:

1. When the sample size is large enough, the sampling distribution of $\overline{X}$ is normal.
2. The expected value of $\overline{X}$ is μ.
3. The standard deviation of $\overline{X}$ is $\sigma/\sqrt{n}$.

The last statement is the key to the important fact that as the sample size increases, the variation of $\overline{X}$ about its mean μ decreases. Stated another way, as we buy *more information* (take a larger sample), our *uncertainty* (measured by the standard deviation) about the parameter being estimated *decreases*.

Eastern-Based Financial Institutions
Second-Quarter EPS and Statistical Summary

EXAMPLE 5–2

Corporation	EPS ($)	Summary	
Bank of New York	2.53	Sample size	13
Bank Boston	4.38	Mean EPS	4.7377
Banker's Trust NY	7.53	Median EPS	4.3500
Chase Manhattan	7.53	Standard deviation	2.4346
Citicorp	7.93		
Fleet	4.35		
MBNA	1.50		
Mellon	2.75		
JP Morgan	7.25		
PNC Bank	3.11		
Republic Bank	7.44		
State Street Bank	2.04		
Summit	3.25		

This example shows random samples from the data above. Here 100 random samples of five banks each are chosen. The mean for each sample is computed, and a frequency distribution is drawn. Note the shape of this distribution (Figure 5–8).

Data Set

2.53
4.38
7.53
7.53
7.93
4.35
1.50
2.75
7.25
3.11
7.44
2.04
3.25

	RS 1	RS 2	RS 3	RS 4	RS 5	RS 6	RS 7	RS 8	RS 9	RS 10	RS 11	RS 12	RS 13	RS 14	RS 15	RS 16	RS 17	RS 18	RS 19	RS 20
	2.53	2.04	2.53	3.25	7.53	4.35	7.93	2.04	7.53	2.04	7.53	3.11	3.25	2.75	7.44	7.44	2.04	2.53	7.93	7.53
	7.53	7.53	2.04	2.75	4.38	1.50	2.53	3.25	3.25	3.25	3.11	7.53	7.53	2.53	4.35	7.53	3.11	4.35	4.38	3.11
	7.93	7.44	7.93	7.93	2.04	7.93	7.53	4.35	2.04	7.53	2.75	1.50	3.25	3.25	7.44	7.53	7.53	4.35	7.25	3.25
	2.53	3.25	4.38	2.04	4.35	7.25	2.75	7.53	3.11	7.93	1.50	7.53	1.50	3.11	2.53	7.53	1.50	1.50	4.35	2.53
	2.75	4.38	4.38	2.53	7.93	3.11	3.25	4.35	2.04	7.53	7.53	2.75	7.53	7.44	3.11	3.25	3.11	4.38	3.25	3.11
Mean	4.65	4.93	4.25	3.70	5.25	4.83	4.80	4.30	3.59	5.66	4.48	4.48	4.61	3.82	4.97	6.66	3.46	3.42	5.43	3.91

	RS 21	RS 22	RS 23	RS 24	RS 25	RS 26	RS 27	RS 28	RS 29	RS 30	RS 31	RS 32	RS 33	RS 34	RS 35	RS 36	RS 37	RS 38	RS 39	RS 40
	3.11	1.50	2.75	7.53	7.44	7.93	2.53	7.93	7.53	4.38	7.93	7.93	7.44	4.35	7.53	7.93	4.38	4.35	7.44	2.53
	2.04	2.04	7.53	2.04	4.35	1.50	3.11	1.50	7.53	7.53	7.93	7.53	3.25	7.25	1.50	2.75	7.93	3.25	7.53	3.25
	3.25	1.50	2.04	4.38	2.75	7.53	3.25	3.11	4.38	2.53	2.75	4.35	4.38	7.25	4.35	1.50	7.93	3.11	4.35	2.53
	4.38	3.25	7.53	2.53	4.35	2.75	7.25	7.93	7.44	3.11	7.93	7.53	3.25	4.35	4.35	2.04	4.35	1.50	3.25	1.50
	2.75	2.75	7.93	2.75	2.04	2.75	1.50	1.50	3.11	7.44	3.11	3.11	7.44	7.53	7.93	2.04	4.38	2.04	2.53	7.53
Mean	3.11	2.21	5.56	3.85	4.19	4.49	3.53	4.39	6.00	5.00	5.93	6.09	5.15	6.15	5.13	3.25	5.79	2.85	5.02	3.47

	RS 41	RS 42	RS 43	RS 44	RS 45	RS 46	RS 47	RS 48	RS 49	RS 50	RS 51	RS 52	RS 53	RS 54	RS 55	RS 56	RS 57	RS 58	RS 59	RS 60
	1.50	1.50	2.75	2.75	4.35	7.53	7.44	7.53	4.35	7.44	3.25	2.53	2.53	7.53	7.25	2.75	7.53	1.50	2.75	2.75
	4.38	7.25	7.44	4.35	1.50	7.93	3.25	4.35	3.11	7.25	2.75	7.53	7.53	4.38	7.53	2.04	2.75	1.50	7.93	7.53
	4.38	7.25	1.50	4.35	3.25	3.25	7.25	7.53	7.44	3.11	4.35	2.75	1.50	4.38	1.50	7.53	3.11	2.04	3.11	7.53
	3.11	4.38	2.75	3.11	2.75	7.53	2.04	7.25	4.35	3.11	4.35	7.53	7.53	4.38	7.25	1.50	7.93	7.25	7.93	7.53
	3.25	7.53	2.04	4.38	7.44	2.04	3.11	4.38	3.25	7.53	4.35	1.50	2.04	7.53	3.25	7.93	2.75	2.75	7.25	3.11
Mean	3.32	5.58	3.30	3.79	3.86	5.66	4.62	6.21	4.50	5.69	3.81	4.37	4.23	5.64	5.36	4.35	4.81	3.01	5.79	5.69

	RS 61	RS 62	RS 63	RS 64	RS 65	RS 66	RS 67	RS 68	RS 69	RS 70	RS 71	RS 72	RS 73	RS 74	RS 75	RS 76	RS 77	RS 78	RS 79	RS 80
	4.38	7.93	3.25	7.53	3.25	2.53	7.25	3.11	7.25	7.53	2.04	7.44	7.25	7.25	7.44	3.25	7.53	7.44	2.53	3.25
	3.25	4.35	7.53	7.44	3.11	7.53	3.11	7.25	7.53	2.75	2.75	7.53	4.38	7.44	7.25	1.50	4.35	4.38	1.50	4.38
	7.93	7.53	3.25	4.35	3.11	7.25	7.25	7.44	7.53	7.53	7.44	4.38	7.25	7.53	2.75	7.25	3.11	1.50	7.53	3.25
	3.25	2.53	7.25	7.44	4.38	2.75	1.50	7.93	3.25	4.38	7.93	3.11	3.11	1.50	3.25	7.25	3.11	7.53	2.53	3.25
	4.35	4.38	3.25	3.25	7.53	4.38	4.38	2.75	7.93	7.25	7.53	7.53	2.04	2.75	3.11	2.04	2.75	2.53	3.25	2.75
Mean	4.63	5.34	4.91	6.00	4.28	4.89	4.70	5.70	6.70	5.89	5.54	6.00	4.81	5.29	4.76	4.26	4.17	4.68	3.47	3.38

	RS 81	RS 82	RS 83	RS 84	RS 85	RS 86	RS 87	RS 88	RS 89	RS 90	RS 91	RS 92	RS 93	RS 94	RS 95	RS 96	RS 97	RS 98	RS 99	RS 100
	7.53	3.25	7.44	7.93	2.04	7.53	2.75	7.93	7.53	7.25	7.93	7.53	7.53	3.25	2.75	7.93	7.44	2.04	4.35	7.53
	3.25	3.11	7.53	2.04	7.53	7.93	4.38	1.50	4.38	4.38	7.25	7.25	3.11	7.93	3.11	2.04	2.04	7.53	7.93	7.53
	7.25	7.25	7.25	7.93	7.93	3.11	2.75	7.93	4.38	2.75	2.04	7.93	1.50	2.75	2.04	3.25	4.38	7.53	2.75	7.25
	3.11	1.50	7.53	2.04	2.53	3.11	7.25	3.11	2.75	7.53	4.38	7.53	2.04	7.93	4.38	4.35	2.75	7.93	3.25	2.53
	4.38	7.53	2.53	1.50	7.25	4.35	7.44	4.35	7.53	1.50	1.50	7.53	7.25	4.38	7.25	2.75	4.35	7.53	2.53	7.53
Mean	5.10	4.53	6.46	4.29	5.46	5.21	4.91	4.96	5.31	4.68	4.62	7.55	4.29	5.25	3.91	4.06	4.19	6.51	4.16	6.47

FIGURE 5–8   EPS Mean Distribution—Excel Output

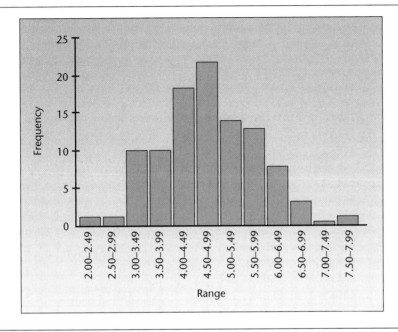

	A	B
1	Distribution	
2		
3	2.00–2.49	1
4	2.50–2.99	1
5	3.00–3.49	10
6	3.50–3.99	10
7	4.00–4.49	18
8	4.50–4.99	21
9	5.00–5.49	14
10	5.50–5.99	13
11	6.00–6.49	8
12	6.50–6.99	3
13	7.00–7.49	0
14	7.50–7.99	1

Figure 5–8 shows a graph of the means of the samples from the banks' data using Excel.

The History of the Central Limit Theorem

What we call the central limit theorem actually comprises several theorems developed over the years. The first such theorem was discussed at the beginning of Chapter 4 as the discovery of the normal curve by Abraham De Moivre in 1733. Recall that De Moivre discovered the normal distribution as the *limit* of the binomial distribution. The fact that the normal distribution appears as a limit of the binomial distribution as *n* increases is a form of the central limit theorem. Around the turn of the twentieth century, Liapunov gave a more general form of the central limit theorem, and in 1922 the final form we use in applied statistics was given by Lindeberg. The proof of the necessary condition of the theorem was given in 1935 by W. Feller [see W. Feller, *An Introduction to Probability Theory and Its Applications* (New York: Wiley, 1971), vol. 2]. A proof of the central limit theorem is beyond the scope of this book, but the interested reader is encouraged to read more about it in the given reference or in other books.

The Standardized Sampling Distribution of the Sample Mean When σ Is Not Known

To use the central limit theorem, we need to know the population standard deviation, σ. When σ is not known, we use its estimator, the sample standard deviation *S*, in its place. In such cases, the distribution of the standardized statistic

$$\frac{\overline{X} - \mu}{S/\sqrt{n}} \tag{5–6}$$

(where S is used in place of the unknown σ) is no longer the standard normal distribution. *If the population itself is normally distributed, the statistic in equation 5–6 has a* t *distribution with* n − *1 degrees of freedom.* The t distribution has wider tails than the standard normal distribution. Values and probabilities of t distributions with different degrees of freedom are given in Table 3 in Appendix C. The t distribution and its uses will be discussed in detail in Chapter 6. The idea of degrees of freedom is explained in the last section of this chapter.

The Sampling Distribution of the Sample Proportion $\hat{P}$

The sampling distribution of the sample proportion $\hat{P}$ is based on the binomial distribution with parameters n and p, where n is the sample size and p is the population proportion. Recall that the binomial random variable X counts the number of successes in n trials. Since $\hat{P} = X/n$ and n is fixed (determined before the sampling), the distribution of the number of successes X leads to the distribution of $\hat{P}$.

As the sample size n increases, the central limit theorem applies here as well. In the beginning of Chapter 4, we saw how the binomial distribution approaches the normal distribution when $p = 0.5$—the case of a symmetric distribution for all n. The symmetry of the distribution makes the convergence to a normal distribution relatively fast. Figure 5–9 shows the effects of the central limit theorem in the case where the binomial distribution is *not* symmetric. We use a distribution with $p = 0.3$ (a distribution skewed to the right, when n is small).

We now state the central limit theorem when sampling for the population proportion p.

> As the sample size n increases, the sampling distribution of $\hat{P}$ approaches a **normal distribution** with mean p and standard deviation $\sqrt{p(1 - p)/n}$.

(The estimated standard deviation of $\hat{P}$ is also called its *standard error*.) In order for us to use the normal approximation for the sampling distribution of $\hat{P}$, the sample size needs to be large. A commonly used rule of thumb says that the normal approximation to the distribution of $\hat{P}$ may be used only if *both np and n(1 − p) are greater than* 5. We demonstrate the use of the theorem with Example 5–3.

EXAMPLE 5–3   In recent years, convertible sport coupes have become very popular in Japan. Toyota is currently shipping Celicas to Los Angeles, where a customizer does a roof lift and ships them back to Japan. Suppose that 25% of all Japanese in a given income and lifestyle category are interested in buying Celica convertibles. A random sample of 100 Japanese consumers in the category of interest is to be selected. What is the probability that at least 20% of those in the sample will express an interest in a Celica convertible?

Solution   We need $P(\hat{P} \geq 0.20)$. Since $np = 100(0.25) = 25$ and $n(1 - p) = 100(0.75) = 75$, both numbers greater than 5, we may use the normal approximation to the distribution of $\hat{P}$. The mean of $\hat{P}$ is $p = 0.25$, and the standard deviation of $\hat{P}$ is $\sqrt{p(1 - p)/n} = 0.0433$. We have

$$P(\hat{P} \geq 0.20) = P\left(Z \geq \frac{0.20 - 0.25}{0.0433}\right) = P(Z \geq -1.15) = 0.8749$$

FIGURE 5–9   The Sampling Distribution of $\hat{P}$ When $p = 0.3$, as n Increases

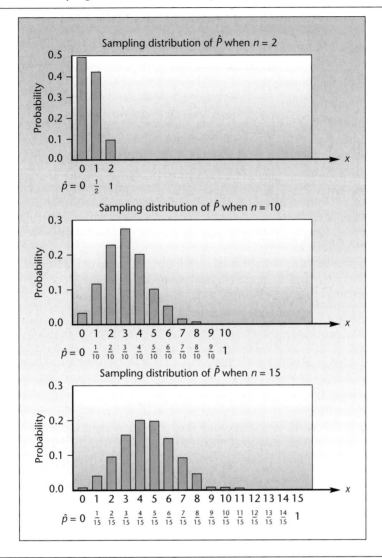

Sampling distributions are essential to statistics. In the following chapters, we will make much use of the distributions discussed in this section, as well as others that will be introduced as we go along. In the next section, we discuss properties of good estimators.

5–11. What is a sampling distribution, and what are the uses of sampling distributions?

5–12. A sample of size $n = 5$ is selected from a population. Under what conditions is the sampling distribution of $\overline{X}$ normal?

5–13.   In problem 5–12, suppose the population mean is $\mu = 125$ and the population standard deviation is 20. What are the expected value and the standard deviation of $\overline{X}$?

5–14.   What is the most significant aspect of the central limit theorem?

5–15.   Under what conditions is the central limit theorem most useful in sampling to estimate the population mean?

5–16.   What are the limitations of small samples?

5–17.   When sampling is done from a population with population proportion $p = 0.1$, using a sample size $n = 12$, what is the sampling distribution of $\hat{P}$? Is it reasonable to use a normal approximation for this sampling distribution? Explain.

5–18.   If the population mean is 1,247, the population variance is 10,000, and the sample size is 100, what is the probability that $\overline{X}$ will be less than 1,230?

5–19.   When sampling is from a population with standard deviation $\sigma = 55$, using a sample of size $n = 150$, what is the probability that $\overline{X}$ will be at least 8 units away from the population mean μ?

5–20.   The Colosseum, once the most popular monument in Rome, dates from about AD 70. Since then, earthquakes have caused considerable damage to the huge structure, and engineers are currently trying to make sure the building will survive future shocks. The Colosseum can be divided into several thousand small sections. Suppose that the average section can withstand a quake measuring 3.4 on the Richter scale with a standard deviation of 1.5. A random sample of 100 sections is selected and tested for the maximum earthquake force they can withstand. What is the probability that the average section in the sample can withstand an earthquake measuring at least 3.6 on the Richter scale?

5–21.   On June 10, 1997, the average price per share on the Big Board Composite Index in New York rose 15 cents.[5] Assume the population standard deviation that day was 5 cents. If a random sample of 50 stocks is selected that day, what is the probability that the average price change in this sample was a rise between 14 and 16 cents?

5–22.   An economist wishes to estimate the average family income in a certain population. The population standard deviation is known to be $4,500, and the economist uses a random sample of size $n = 225$. What is the probability that the sample mean will fall within $800 of the population mean?

5–23.   When sampling is done for the proportion of defective items in a large shipment, where the population proportion is 0.18 and the sample size is 200, what is the probability that the sample proportion will be at least 0.20?

5–24.   A study of the investment industry claims that 58% of all mutual funds outperformed the stock market as a whole last year. An analyst wants to test this claim and obtains a random sample of 250 mutual funds. The analyst finds that only 123 of the funds outperformed the market during the year. Determine the probability that another random sample would lead to a sample proportion as low as or lower than the one obtained by the analyst, assuming the proportion of all mutual funds that outperformed the market is indeed 0.58.

---

[5]R. Obrien, "Shares Rise Despite Earnings Worries," Abreast of the Market column, *The Wall Street Journal,* June 11, 1997, p. 16.

5–25 According to a recent article in *Forbes* magazine, the average price of a house in the Midwest is $119,600. Assume that the standard deviation of the prices is $35,000. A random sample of 75 houses is taken and the average price is computed.

    a. What is the probability that the sample mean exceeds $125,000?

    b. The population standard deviation of the prices is not known. It may be anywhere between $30,000 and $40,000. Using the DatalTable command, tabulate the probability of the sample mean exceeding $125,000 for population standard deviation values $30,000 to $40,000 in steps of $2,000.

5–26. It has been suggested that an investment portfolio selected randomly by throwing darts at the stock market page of *The Wall Street Journal* may be a sound (and certainly well-diversified) investment.[6] Suppose that you own such a portfolio of 16 stocks randomly selected from all stocks listed on the New York Stock Exchange (NYSE). On a certain day, you hear on the news that the average stock on the NYSE rose 1.5 points. Assuming that the standard deviation of stock price movements that day was 2 points and assuming stock price movements were normally distributed around their mean of 1.5, what is the probability that the average stock in your portfolio increased in price?

5–27. An advertisement for Citicorp Insurance Services, Inc., claims "one person in seven will be hospitalized this year." Suppose you keep track of a random sample of 180 people over an entire year. Assuming Citicorp's advertisement is correct, what is the probability that fewer than 10% of the people in your sample will be found to have been hospitalized (at least once) during the year? Explain.

5–28. Shimano mountain bikes are displayed in chic clothing boutiques in Milan, Italy, and the average price for the bike in the city is $700. Suppose that the standard deviation of bike prices is $100. If a random sample of 60 boutiques is selected, what is the probability that the average price for a Shimano mountain bike in this sample will be between $680 and $720?

5–29. A quality-control analyst wants to estimate the proportion of imperfect jeans in a large warehouse. The analyst plans to select a random sample of 500 pairs of jeans and note the proportion of imperfect pairs. If the actual proportion in the entire warehouse is 0.35, what is the probability that the sample proportion will deviate from the population proportion by more than 0.05?

5–4 Estimators and Their Properties[7]

The sample statistics we discussed—$\overline{X}$, S, and $\hat{P}$—as well as other sample statistics to be introduced later, are used as estimators of population parameters. In this section, we discuss some important properties of good statistical estimators: *unbiasedness, efficiency, consistency,* and *sufficiency.*

> An estimator is said to be **unbiased** if its expected value is equal to the population parameter it estimates.

---

[6]See the very readable book by Burton G. Malkiel, *A Random Walk Down Wall Street* (New York: W. W. Norton, 1995).

[7]An optional, but recommended, section.

Consider the sample mean $\overline{X}$. From equation 5–2, we know $E(\overline{X}) = \mu$. *The sample mean $\overline{X}$ is, therefore, an unbiased estimator of the population mean μ.* This means that if we sample repeatedly from the population and compute $\overline{X}$ for each of our samples, *in the long run,* the average value of $\overline{X}$ will be the parameter of interest μ. This is an important property of the estimator because it means that there is no systematic *bias* away from the parameter of interest.

If we view the gathering of a random sample and the calculating of its mean as shooting at a target—the target being the population parameter, say, μ—then the fact that $\overline{X}$ is an unbiased estimator of μ means that the device producing the estimates is aiming at the *center* of the target (the parameter of interest), with no systematic deviation away from it.

> Any *systematic* deviation of the estimator away from the parameter of interest is called a **bias**.

The concept of unbiasedness is demonstrated for the sample mean $\overline{X}$ in Figure 5–10.

Figure 5–11 demonstrates the idea of a biased estimator of μ. The hypothetical estimator we denote by Y is centered on some point M that lies away from the parameter μ. The distance between the expected value of Y (the point M) and μ is the *bias*.

It should be noted that, in reality, we usually sample *once* and obtain our estimate. The multiple estimates shown in Figures 5–10 and 5–11 serve only as an illustration of the expected value of an estimator as the center of a large collection of the actual estimates that would be obtained in repeated sampling. (Note also that, in reality, the "target" at which we are "shooting" is one-dimensional—on a straight line rather than on a plane.)

The next property of good estimators we discuss is *efficiency*.

> An estimator is **efficient** if it has a relatively small variance (and standard deviation).

Efficiency is a relative property. We say that one estimator is efficient *relative* to another. This means that the estimator has a smaller variance (also a smaller standard deviation) than the other. Figure 5–12 shows two hypothetical unbiased estimators of

FIGURE 5–10   The Sample Mean $\overline{X}$ as an Unbiased Estimator of the Population Mean μ

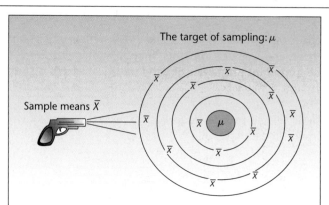

the population mean μ. The two estimators, which we denote by X and Z, are unbiased: Their distributions are centered at μ. The estimator X, however, is more efficient than the estimator Z because it has a smaller variance than that of Z. This is seen from the fact that repeated estimates produced by Z have a larger spread about their mean μ than repeated estimates produced by X.

Another desirable property of estimators is *consistency*.

> An estimator is said to be **consistent** if its probability of being close to the parameter it estimates increases as the sample size increases.

FIGURE 5–11   An Example of a Biased Estimator of the Population Mean μ

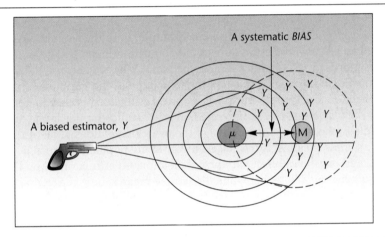

FIGURE 5–12   Two Unbiased Estimators of μ, Where the Estimator X Is Efficient Relative to the Estimator Z

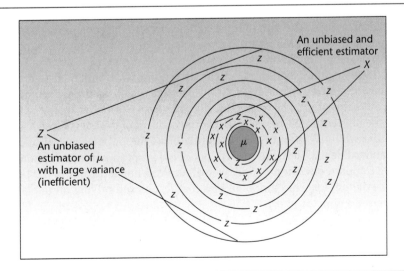

The sample mean $\overline{X}$ is a consistent estimator of μ. This is so because the standard deviation of $\overline{X}$ is $\sigma_{\overline{x}} = \sigma/\sqrt{n}$. As the sample size n increases, the standard deviation of $\overline{X}$ decreases and, hence, the probability that $\overline{X}$ will be close to its expected value μ increases.

We now define a fourth property of good estimators: *sufficiency*.

An estimator is said to be **sufficient** if it contains all the information in the data about the parameter it estimates.

Applying the Concepts of Unbiasedness, Efficiency, Consistency, and Sufficiency

We may evaluate possible estimators of population parameters based on whether they possess important properties of estimators and thus choose the best estimator to be used.

For a *normally distributed population,* for example, both the sample mean and the sample median are *unbiased* estimators of the population mean μ. The sample mean, however, is more *efficient* than the sample median. This is so because the variance of the sample median happens to be 1.57 times as large as the variance of the sample mean. In addition, the sample mean is a *sufficient* estimator because in computing it we use the *entire* data set. The sample median is not sufficient; it is found as the point in the middle of the data set, regardless of the exact magnitudes of all other data elements. The sample mean $\overline{X}$ is the *best* estimator of the population mean μ, because it is unbiased and has the smallest variance of all unbiased estimators of μ. The sample mean is also *consistent*. (Note that while the sample mean is best, the sample median is sometimes used because it is more resistant to extreme observations.)

The sample proportion $\hat{P}$ is the best estimator of the population proportion p. Since $E(\hat{P}) = p$, the estimator $\hat{P}$ is unbiased. It also has the smallest variance of all unbiased estimators of p.

What about the sample variance S^2? The sample variance, as defined in equation 1–3, is an unbiased estimator of the population variance σ^2. Recall equation 1–3:

$$S^2 = \frac{\Sigma(x_i - \overline{x})^2}{n - 1}$$

It seems logical to divide the sum of squared deviations in the equation by n rather than by $n - 1$ because we are seeking the *average* squared deviation from the sample mean. There are n deviations from the mean, so why not divide by n? It turns out that if we were to divide by n rather than by $n - 1$, our estimator of σ^2 would be biased. Although the bias becomes small as n increases, we will always use the statistic given in equation 1–3 as an estimator of σ^2. The reason for dividing by $n - 1$ rather than n will become clearer in the next section, when we discuss the concept of degrees of freedom.

Note that while S^2 is an unbiased estimator of the population variance σ^2, the sample standard deviation S (the square root of S^2) is *not* an unbiased estimator of the population standard deviation σ. Still, we will use S as our estimator of the population standard deviation, ignoring the small bias that results and relying on the fact that S^2 is the unbiased estimator of σ^2.

PROBLEMS

5–30. Suppose that you have two statistics A and B as possible estimators of the same population parameter. Estimator A is unbiased, but has a large variance. Estimator B has a small bias, but has only one-tenth the variance of estimator A. Which estimator is better? Explain.

5-31. Suppose that you have an estimator with a relatively large bias. The estimator is consistent and efficient, however. If you had a generous budget for your sampling survey, would you use this estimator? Explain.

5-32. Suppose that in a sampling survey to estimate the population variance, the biased estimator (with n instead of $n - 1$ in the denominator of equation 1–3) was used instead of the unbiased one. The sample size used was $n = 100$, and the estimate obtained was 1,287. Can you find the value of the unbiased estimate of the population variance?

5-33. What are the advantages of a sufficient statistic? Can you think of a possible disadvantage of sufficiency?

5-34. Suppose that you have two biased estimators of the same population parameter. Estimator A has a bias equal to $1/n$ (that is, the mean of the estimator is $1/n$ unit away from the parameter it estimates), where n is the sample size used. Estimator B has a bias equal to 0.01 (the mean of the estimator is 0.01 unit away from the parameter of interest). Under what conditions is estimator A better than B?

5-35. Why is consistency an important property?

5–5   Degrees of Freedom

The concept of degrees of freedom (df) is important for many statistical calculations and probability distributions. We have already seen one instance where the number of degrees of freedom enters the picture, and it is the calculation of sample variance. The formula for sample variance is

$$S^2 = \text{SSD}/(n - 1)$$

where n is the sample size and $\text{SSD} = \Sigma_{i=1}^{n} (x_i - \bar{x})^2$ is the sum of squared deviations. The reason given for $(n - 1)$ in the denominator of the formula is that the deviations have only $(n - 1)$ degrees of freedom. We shall see the details about why the degrees of freedom are $(n - 1)$.

We first note that in the calculation of SSD, the deviations are taken from the sample mean $\bar{x}$ and not from the population mean μ. The reason is simple: While sampling, almost always, the population mean μ is not known. Not knowing the population mean, we take the deviations from the sample mean. But this introduces a downward bias in the deviations. To see the bias, refer to Figure 5–13, which shows the deviation of a sample point x from the sample mean and from the population mean.

It can be seen from Figure 5–13 that for sample points that fall to the right of the midpoint between μ and $\bar{x}$, the deviation from the sample mean will be smaller than the deviation from the population mean. Since the sample mean is where the sample points gravitate, a majority of the sample points are expected to fall to the right of the midpoint. Thus, overall, there will be a downward bias in the deviations.

To compensate for the downward bias, we introduce the concept of **degrees of freedom.** Let the population be a uniform distribution of the values $\{1, 2, \ldots, 10\}$. The mean of this population is 5.5. Suppose a random sample of size 10 is taken from this population. Assume that we are told to take the deviations from this population mean. In Figure 5–14, the Sample column shows the sampled values. The calculation of SSD is shown taking deviations from the population mean of 5.5. The SSD works

FIGURE 5–13   **Deviations from the Population Mean and the Sample Mean**

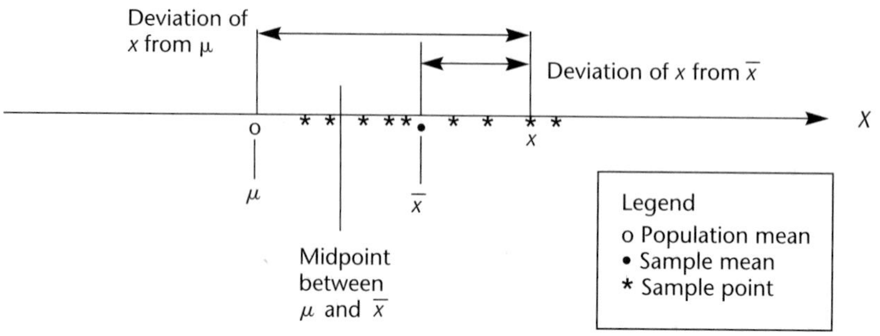

FIGURE 5–14   **SSD and df**

df = 10

	Sample	Deviation from	Deviation	Deviation Squared
1	10	5.5	4.5	20.25
2	3	5.5	−2.5	6.25
3	2	5.5	−3.5	12.25
4	6	5.5	0.5	0.25
5	1	5.5	−4.5	20.25
6	9	5.5	3.5	12.25
7	6	5.5	0.5	0.25
8	4	5.5	−1.5	2.25
9	10	5.5	4.5	20.25
10	7	5.5	1.5	2.25
			SSD	96.5

out to 96.5. *Since we had no freedom in taking the deviations, all the 10 deviations are completely left to chance.* Hence we say that the deviations have 10 degrees of freedom.

Suppose we do not know the population mean and are told that we can take the deviation from any number we choose. The best number to choose then is the sample mean, which will minimize the SSD (see problem 1–85). Figure 5–15a shows the calculation of SSD where the deviations are taken from the sample mean of 5.8. Because of the downward bias, the SSD has decreased to 95.6. The SSD would decrease further if we were allowed to select two different numbers from which the deviations are taken. Suppose we are allowed to use one number for the first five data points and another for the next five. Our best choices are the average of the first five numbers, 4.4, and the average of next five numbers, 7.2. Only these choices will minimize the SSD. The minimized SSD works out to 76, as seen in Figure 5–15b.

We can carry this process further. If we were allowed 10 different numbers from which the deviations are taken, then we could reduce the SSD all the way to zero.

FIGURE 5–15 **SSD and df** (*continued*)

df = 10−1 = 9

	Sample	Deviation from	Deviation	Deviation Squared
1	10	5.8	4.2	17.64
2	3	5.8	−2.8	7.84
3	2	5.8	−3.8	14.44
4	6	5.8	0.2	0.04
5	1	5.8	−4.8	23.04
6	9	5.8	3.2	10.24
7	6	5.8	0.2	0.04
8	4	5.8	−1.8	3.24
9	10	5.8	4.2	17.64
10	7	5.8	1.2	1.44
			SSD	95.6

(a)

df = 10−2 = 8

	Sample	Deviation from	Deviation	Deviation Squared
1	10	4.4	5.6	31.36
2	3	4.4	−1.4	1.96
3	2	4.4	−2.4	5.76
4	6	4.4	1.6	2.56
5	1	4.4	−3.4	11.56
6	9	7.2	1.8	3.24
7	6	7.2	−1.2	1.44
8	4	7.2	−3.2	10.24
9	10	7.2	2.8	7.84
10	7	7.2	−0.2	0.04
			SSD	76

(b)

FIGURE 5–16 **SSD and df** (*continued*)

df = 10−10 = 0

	Sample	Deviation from	Deviation	Deviation Squared
1	10	10	0	0
2	3	3	0	0
3	2	2	0	0
4	6	6	0	0
5	1	1	0	0
6	9	9	0	0
7	6	6	0	0
8	4	4	0	0
9	10	10	0	0
10	7	7	0	0
			SSD	0

How? See Figure 5–16. We choose the 10 numbers equal to the 10 sample points (which in effect are 10 means). In the case of Figure 5–15*a*, we had one choice, and this takes away 1 degree of freedom from the deviations. The df of SSD is then de-clared as $10 - 1 = 9$. In Figure 5–15*b*, we had two choices and this took away 2 degrees of freedom from the deviations. Thus the df of SSD is $10 - 2 = 8$. In Figure 5–16, the df of SSD is $10 - 10 = 0$.

In every one of these cases, *dividing the SSD by only its corresponding df will yield an unbiased estimate of the population variance* σ^2. Hence the concept of the degrees of freedom is important. This also explains the denominator of $(n - 1)$ in the formula for sample variance S^2. For the case in Figure 5–15a, SSD/df $= 95.6/9 = 10.62$, and this is an unbiased estimate of the population variance.

We can now summarize how the number of degrees of freedom is determined. If we take a sample of size n and take the deviations from the (known) population mean then the deviations, and therefore the SSD, will have df $= n$. But if we take the deviations from the sample mean, then the deviations, and therefore the SSD, will have df $= n - 1$. If we are allowed to take the deviations from $k (\leq n)$ different numbers that we choose, then the deviations, and therefore the SSD, will have df $= n - k$. While choosing each of the k numbers, we should choose the mean of the sample points to which that number applies. The case of $k > 1$ will be seen in Chapter 9, "Analysis of Variance."

EXAMPLE 5–4   A sample of size 10 is given below. We are to choose three different numbers from which the deviations are to be taken. The first number is to be used for the first five sample points; the second number is to be used for the next three sample points; and the third number is to be used for the last two sample points.

Sample

1	93
2	97
3	60
4	72
5	96
6	83
7	59
8	66
9	88
10	53

1. What three numbers should we choose to minimize the SSD?
2. Calculate the SSD with the chosen numbers.
3. What is the df for the calculated SSD?
4. Calculate an unbiased estimate of the population variance.

Solution

1. We choose the means of the corresponding sample points: 83.6, 69.33, 70.5.
2. SSD $= 2030.367$. See the spreadsheet calculation below.
3. df $= 10 - 3 = 7$.
4. An unbiased estimate of the population variance is SSD/df $= 2030.367/7 = 290.05$

	Sample	Mean	Deviation	Deviation Squared
1	93	83.6	9.4	88.36
2	97	83.6	13.4	179.56
3	60	83.6	−23.6	556.96
4	72	83.6	−11.6	134.56
5	96	83.6	12.4	153.76
6	83	69.33	13.6667	186.7778
7	59	69.33	−10.3333	106.7778
8	66	69.33	−3.33333	11.11111
9	88	70.5	17.5	306.25
10	53	70.5	−17.5	306.25
			SSD	2030.367
			SSD/df	290.0524

5–36. Three random samples of sizes, 30, 48, and 32, respectively, are collected, and the three sample means are computed. What is the total number of degrees of freedom for deviations from the means?

5–37. The data points in a sample of size 9 are 34, 51, 40, 38, 47, 50, 52, 44, 37.

a. If you can take the deviations of these data from any number you select, and you want to minimize the sum of the squared deviations (SSD), what number would you select? What is the minimized SSD? How many degrees of freedom are associated with this SSD? Calculate the mean squared deviation (MSD) by dividing the SSD by its degrees of freedom. (This MSD is an unbiased estimate of population variance.)

b. If you can take the deviations from three different numbers you select, and the first number is to be used with the first four data points to get the deviations, the second with the next three data points, and the third with the last two data points, what three numbers would you select? What is the minimized SSD? How many degrees of freedom are associated with this SSD? Calculate MSD.

c. If you can select nine different numbers to be used with each of the nine data points, what numbers would you select? What is the minimized SSD? How many degrees of freedom are associated with this SSD? Does MSD make sense in this case?

d. If you are told that the deviations are to be taken with respect to 50, what is the SSD? How many degrees of freedom are associated with this SSD? Calculate MSD.

5–38. Your bank sends you a summary statement, giving the average amount of all checks you wrote during the month. You have a record of the amounts of 17 out of the 19 checks you wrote during the month. Using this and the information provided by the bank, can you figure out the amounts of the two missing checks? Explain.

FIGURE 5–17   The Template for Sampling Distribution of a Sample Mean
[Sampling Distribution.xls; Sheet: X-bar]

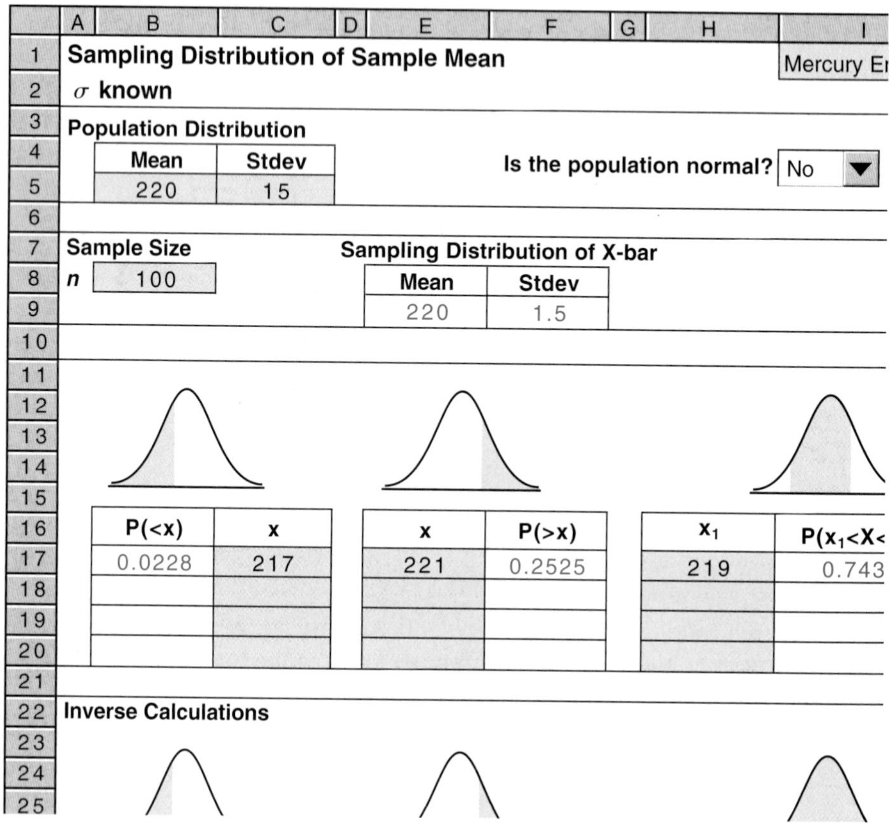

5–39. In problem 5–38, suppose you know the amounts of 18 of the 19 checks you wrote and the average of all the checks. Can you figure out the amount of the missing check? Explain.

5–40. You are allowed to take the deviations of the data points in a sample of size n, from k numbers you select, in order to calculate the sum of squared deviations (SSD). You select them to minimize SSD. How many degrees of freedom are associated with this SSD? As k increases, what happens to the degrees of freedom? What happens to SSD? What happens to MSD = SSD/df(SSD)?

5–6   The Template

Figure 5–17 shows the template that can be used to calculate the sampling distribution of a sample mean. It is largely the same as the normal distribution template. The additional items are the population distribution entries at the top. To use the template, enter the population mean and standard deviation in cells B5 and C5. Enter the sample size in cell B8. In the drop down box in cell I4, select Yes or No to answer the question "Is the population normally distributed?" The sample mean will follow a normal distribution if either the population is normally distributed or the sample size

FIGURE 5–18   The Template for Sampling Distribution of a Sample Proportion
[Sampling Distribution.xls; Sheet: P-hat]

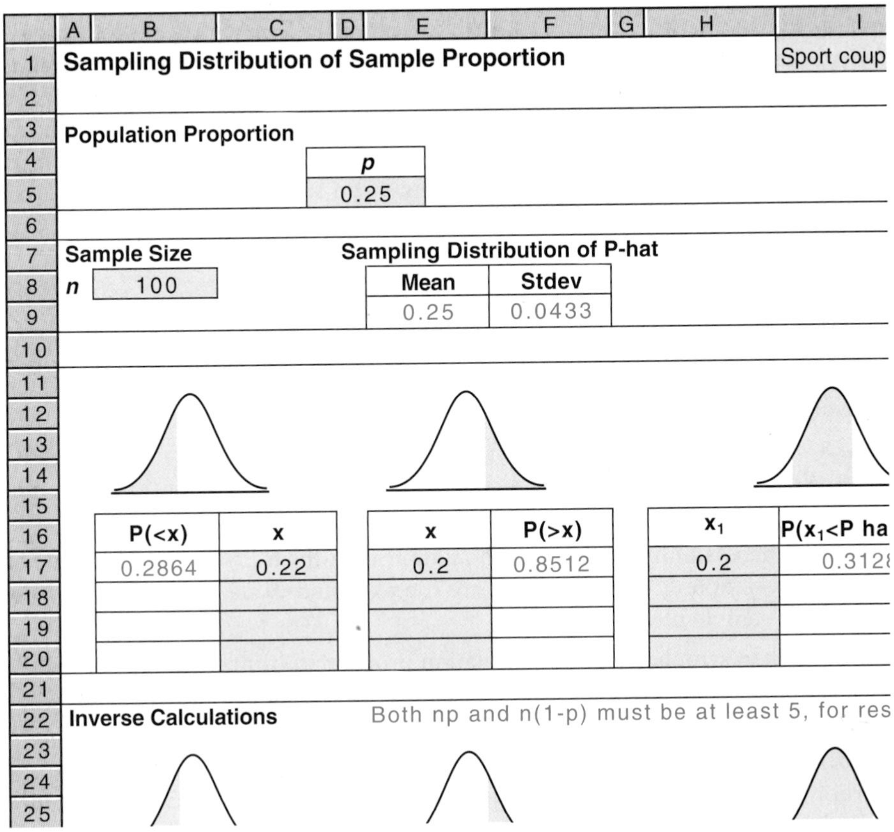

is at least 30. Only in such cases should this template be used. In other cases, a warning message—"Warning: The sampling distribution cannot be approximated as normal. Results appear anyway"—will appear in cell A10.

To solve Example 5–1, enter the population mean 220 in cell B5 and the population standard deviation 15 in cell C5. Enter the sample size 100 in cell B8. To find the probability that the sample mean will be less than 217, enter 217 in cell C17. The answer 0.0228 appears in cell B17.

Figure 5–18 shows the template that can be used to calculate the sampling distribution of a sample proportion. To use the template, enter the population proportion in cell E5 and the sample size in cell B8.

To solve Example 5–3, enter the population proportion 0.25 in cell E5 and the sample size 100 in cell B8. Enter the value 0.2 in cell E17 to get the probability of the sample proportion being more than 0.2 in cell F17. The answer is 0.8749.

5–7   Summary and Review of Terms

In this chapter, we saw how samples are randomly selected from populations for the purpose of drawing inferences about **population parameters**. We saw how **sample statistics** computed from the data—the sample mean, the sample standard deviation, and the sample proportion—are used as **estimators** of population parameters. We presented the important idea of a **sampling distribution** of a statistic, the probability distribution of the values the statistic may take. We saw how the **central limit theorem** implies that the sampling distributions of the sample mean and the sample proportion approach normal distributions as the sample size increases. Sampling distributions of estimators will prove to be the key to the construction of confidence intervals in the following chapter, as well as the key to the ideas presented in later chapters. We also presented important properties we would like our estimators to possess: **unbiasedness, efficiency, consistency,** and **sufficiency.** Finally, we discussed the idea of **degrees of freedom.**

ADDITIONAL PROBLEMS

5–41. Suppose you are sampling from a population with mean $\mu = 1,065$ and standard deviation $\sigma = 500$. The sample size is $n = 100$. What are the expected value and the variance of the sample mean $\overline{X}$?

5–42. Suppose you are sampling from a population with population variance $\sigma^2 = 1,000,000$. You want the standard deviation of the sample mean to be at most 25. What is the minimum sample size you should use?

5–43. When sampling is from a population with mean 53 and standard deviation 10, using a sample of size 400, what are the expected value and the standard deviation of the sample mean?

5–44. When sampling is for a population proportion from a population with actual proportion $p = 0.5$, using a sample of size $n = 120$, what is the standard deviation of our estimator $\hat{P}$?

5–45. What are the expected value and the standard deviation of the sample proportion $\hat{P}$ if the true population proportion is 0.2 and the sample size is $n = 90$?

5–46. For a fixed sample size, what is the value of the true population proportion p that maximizes the variance of the sample proportion $\hat{P}$? (*Hint:* Try several values of p on a grid between 0 and 1.)

5–47. According to an article in *The Economist,*[8] the average income per person in China is \$600 per year. If $\sigma = \$600$ and $n = 30$, find $P(500 < \overline{X} < 700)$.

5–48. In problem 5–41, what is the probability that the sample mean will be at least 1,000? Do you need to use the central limit theorem to answer this question? Explain.

5–49. In problem 5–43, what is the probability that the sample mean will be between 52 and 54?

5–50. In problem 5–44, what is the probability that the sample proportion will be at least 0.45?

5–51. "Searches at Switzerland's 406 commercial banks turned up only \$3.3 million in accounts belonging to Zaire's deposed president, Mubutu Sese Seko. The Swiss banks had been asked to look a little harder after finding nothing at all the first time round."[9]

> *a.* If President Mubutu's money was distributed in *all* 406 banks, how much was found, on average, per bank?

---

[8]"Car Jams in China," *The Economist,* June 21, 1997, p. 64.

[9]From "Business This Week," *The Economist,* June 7, 1997, p. 27.

b. If a random sample of 16 banks was first selected in a preliminary effort to estimate how much money was in all banks, then assuming that amounts were normally distributed with standard deviation of $2,000, what was the probability that the mean of this sample would have been less than $7,000?

5–52.   The proportion of defective microcomputer disks of a certain kind is believed to be anywhere from 0.06 to 0.10. The manufacturer wants to draw a random sample and estimate the proportion of all defective disks. How large should the sample be to ensure that the standard deviation of the estimator is *at most* 0.03?

5–53.   Explain why we need to draw random samples and how such samples are drawn. What are the properties of a (simple) random sample?

5–54.   Explain the idea of a bias and its ramifications.

5–55.   Is the sample median a biased estimator of the population mean? Why do we usually prefer the sample mean to the sample median as an estimator for the population mean? If we use the sample median, what must we assume about the population? Compare the two estimators.

5–56.   Explain why the sample variance is defined as the sum of squared deviations from the sample mean, divided by $n - 1$ and not by n.

5–57.   Average weekly family expenditure on entertainment for a population in a given region is $19.50, and the population standard deviation is $5.30. What is the probability that a random sample of size 100 will yield a sample mean greater than $20.00?

5–58.   In problem 5–57, give 0.95 probability bounds on the value of the sample mean that would be obtained. Also give 0.90 probability bounds on the value of the sample mean.

5–59.   An investment of $10,000 in January 1991 would be worth $25,000 if invested in the S&P 500. The same $10,000 would be worth $37,000 if invested in the S&P Technology Index, and $40,000 if invested in the S&P Financial Index.[10]

a. Suppose an "innocent" investor in 1991 chose a random sample of stocks from the S&P 500, $n = 25$ stocks, and assume normality of the return. Also assume $\sigma = \$1,500$. Find $P(9,500 < \bar{X} < 10,500)$.

b. A random sample of 25 technology stocks in 1991 was selected, with $\sigma = \$2,000$. Find $P(35,000 < \bar{X} < 39,000)$.

5–60.   Repeat problem 5–59 for financial stocks selected randomly from the S&P Financial Index population. Assume $n = 25$, normal distribution, and $\sigma = \$2,500$. Find $P(38,000 < \bar{X} < 42,000)$.

5–61.   Thirty-eight percent of all shoppers at a large department store are holders of the store's charge card. If a random sample of 100 shoppers is taken, what is the probability that at least 30 of them will be found to be holders of the card?

5–62.   When sampling is from a normal population with an unknown variance, is the sampling distribution of the sample mean normal? Explain.

5–63.   When sampling is from a normal population with a known variance, what is the smallest sample size required for applying a normal distribution for the sample mean?

---

[10]From N. Schwartz, "The Revolution in Financial Stocks," *Fortune*, June 23, 1997, p. 74.

5–64. Which of the following estimators are unbiased estimators of the appropriate population parameters: $\bar{X}$, $\hat{P}$, S^2, S? Explain.

5–65. Suppose a new estimator for the population mean is discovered. The new estimator is unbiased and has variance equal to σ^2/n^2. Discuss the merits of the new estimator compared with the sample mean.

5–66. Three independent random samples are collected, and three sample means are computed. The total size of the combined sample is 124. How many degrees of freedom are associated with the deviations from the sample means in the combined data set? Explain.

5–67. Discuss, in relative terms, the sample size needed for an application of a normal distribution for the sample mean when sampling is from each of the following populations. (Assume the population standard deviation is known in each case.)

   a. A normal population
   b. A mound-shaped population, close to normal
   c. A discrete population consisting of the values 1,006, 47, and 0, with equal frequencies
   d. A slightly skewed population
   e. A highly skewed population

5–68. When sampling is from a normally distributed population, is there an advantage to taking a large sample? Explain.

5–69. Suppose that you are given a new sample statistic to serve as an estimator of some population parameter. You are unable to assume any theoretical results such as the central limit theorem. Discuss how you would empirically determine the sampling distribution of the new statistic.

5–70. Recently, the federal government claimed that the state of Alaska had overpaid 20% of the Medicare recipients in the state. The director of the Alaska Department of Health and Social Services planned to check this claim by selecting a random sample of 250 recipients of Medicare checks in the state and determining the number of overpaid cases in the sample. Assuming the federal government's claim is correct, what is the probability that less than 15% of the people in the sample will be found to have been overpaid?

5–71. A new kind of alkaline battery is believed to last an average of 25 hours of continuous use (in a given kind of flashlight). Assume that the population standard deviation is 2 hours. If a random sample of 100 batteries is selected and tested, is it likely that the average battery in the sample will last less than 24 hours of continuous use? Explain.

5–72. Häagen-Dazs ice cream produces a frozen yogurt aimed at health-conscious ice cream lovers. Before marketing the product, the company wanted to estimate the proportion of grocery stores currently selling Häagen-Dazs ice cream that would sell the new product. If 80% of the grocery stores would sell the product and a random sample of 200 stores is selected, what is the probability that the percentage in the sample will deviate from the population percentage by no more than 7 percentage points?

5–73. Japan's birthrate is believed to be 1.57 per woman. Assume that the population standard deviation is 0.4. If a random sample of 200 women is selected, what is the probability that the sample mean will fall between 1.52 and 1.62?

5–74. The Toyota Prius uses both gasoline and electric power. Toyota claims its mileage per gallon is 52. A random sample of 40 cars is taken and each sampled car is tested for its fuel efficiency. Assuming that 52 miles per gallon is the population mean and 2.4 miles per gallon is the population standard deviation, answer the following questions.

    a. What is a point estimate for the sample mean?

    b. What is the probability that the sample mean will be between 52 and 53?

CASE 6   Acceptance Sampling of Pins

A company supplies pins in bulk to a customer. The company uses an automatic lathe to produce the pins. Factors such as vibration, temperature, and wear and tear affect the pins, so that the lengths of the pins made by the machine are normally distributed with a mean of 1.008 inches and a standard deviation of 0.045 inch. The company supplies the pins in large batches to a customer. The customer will take a random sample of 50 pins from the batch and compute the sample mean. If the sample mean is within the interval 1.000 inch ± 0.010 inch, then the customer will buy the whole batch.

1. What is the probability that a batch will be acceptable to the consumer? Is the probability large enough to be an acceptable level of performance?

To improve the probability of acceptance, the production manager and the engineers discuss adjusting the population mean and standard deviation of the lengths of the pins.

2. If the lathe can be adjusted to have the mean of the lengths at any desired value, what should it be adjusted to? Why?

3. Suppose the mean cannot be adjusted, but the standard deviation can be reduced. What maximum value of the standard deviation would make 90% of the parts acceptable to the consumer? (Assume the mean continues to be 1.008 inches.)

4. Repeat part 3 with 95% and 99% of the pins acceptable.

5. In practice, which one do you think is easier to adjust, the mean or the standard deviation? Why?

The production manager then considers the costs involved. The cost of resetting the machine to adjust the population mean involves the engineers' time and the cost of production time lost. The cost of reducing the population standard deviation involves, in addition to these costs, the cost of overhauling the machine and reengineering the process.

6. Assume it costs $150x^2$ to decrease the standard deviation by $(x/1,000)$ inch. Find the cost of reducing the standard deviation to the values found in parts 3 and 4.

7. Now assume that the mean has been adjusted to the best value found in part 2 at a cost of $80. Calculate the reduction in standard deviation necessary to have 90%, 95%, and 99% of the parts acceptable. Calculate the respective costs, as in part 6.

8. Based on your answers to parts 6 and 7, what are your recommended mean and standard deviation to which the machine should be adjusted?

6 Confidence Intervals

6-1 Using Statistics

Anthropologists, armed with knowledge of molecular biology and statistics, announced in 1988 that they had discovered Eve. A long and involved research study resulted in the implication that all human beings are genetically descended from *one* woman.[1] The scientists did not claim to have found the *first* woman, but rather a common ancestor of all humanity. The scientists' Eve was the most genetically successful woman living in our distant past, a woman whose progeny are all people living in the world today, regardless of race, ethnic origin, or native continent. This finding brings with it far-reaching implications, which have stirred up heated debates. The most controversial implication is that humans did not slowly evolve in different parts of the world, as many anthropologists believed. The evolution to modern *Homo sapiens,* according to the new theory, occurred only in *one* place (scientists are still uncertain where, although it is believed to have been somewhere in Asia or Africa). Sometime between 90,000 and 180,000 years ago, Eve's descendents left their homeland and began their slow migration to all parts of the world, replacing local populations as they went along and eventually settling the entire planet. What we call racial characteristics or external differences of appearance all developed *after* these migrations. How did the scientists develop and demonstrate these incredible assertions?

Scientists asked a random sample of 147 pregnant women of all races, ethnic origins, and continents of birth to donate their babies' placentas. The genetic matter in the placentas, strings of the protein DNA, was then carefully scrutinized with interest centered on a particular type of mitochondrial DNA inherited only from one's mother. The scientists found that the genetic matter in the placentas of all the sampled babies contained segments of DNA traceable to a single source—a woman they called Eve. Statistical estimation was then used in analyzing the number of mutations inherent in the genetic material. Based on their random sample, the scientists estimated the average percentage of mutated mitochondrial DNA in the entire population of living people at 2% to 4%. This is an **interval estimate** of a population parameter. Such an interval has associated with it a measure of **confidence** (usually stated as a percentage, often 95% or 99%). Since the scientists also knew the approximate rate of mutation over time, they were able to use the interval estimate of the average percentage of mutated mitochondrial DNA in the present-day population and extrapolate back in time toward the beginning of these mutations. This led them to conclude, with a high degree of confidence, that Eve lived anywhere between 140,000 and 290,000 years ago. In this chapter, we discuss the computation and interpretation of interval estimates of population parameters. An interval estimate with its associated measure of confidence is usually called a *confidence interval.*

In the last chapter, we saw how sample statistics are used as estimators of population parameters. We defined a point estimate of a parameter as a single value obtained from the estimator. We saw that an estimator, a sample statistic, is a random variable with a certain probability distribution—its sampling distribution. A given point estimate is a single realization of the random variable. The actual estimate may or may not be close to the parameter of interest. Therefore, if we only provide a point estimate of the parameter of interest, we are not giving any information about the

---

[1]See, for example, the cover article in *Newsweek,* January 11, 1988.

accuracy of the estimation procedure. For example, saying that the sample mean is 550 is giving a point estimate of the population mean. This estimate does not tell us how close μ may be to its estimate, 550. Suppose, on the other hand, that we also said: "We are *99% confident* that μ is in the interval [449, 551]." This conveys much more information about the possible value of μ. Now compare this interval with another one: "We are *90% confident* that μ is in the interval [400, 700]." This interval conveys less information about the possible value of μ, both because it is wider and because the level of confidence is lower. (When based on the same information, however, an interval of lower confidence level is narrower.)

> A **confidence interval** is a *range of numbers* believed to include an unknown population parameter. Associated with the interval is a measure of the *confidence* we have that the interval does indeed contain the parameter of interest.

The sampling distribution of the statistic gives a *probability* associated with a range of values the statistic may take. After the sampling has taken place and a *particular estimate* has been obtained, this probability is transformed to a *level of confidence* for a range of values that may contain the unknown parameter.

In the next section, we will see how to construct confidence intervals for the population mean μ, when the population standard deviation σ is known. Then we will alter this situation and see how a confidence interval for μ may be constructed without knowledge of σ. Other sections present confidence intervals in other situations.

6–2   Confidence Interval for the Population Mean When the Population Standard Deviation Is Known

The central limit theorem tells us that when we select a large random sample from any population with mean μ and standard deviation σ, the sample mean $\overline{X}$ is (at least approximately) normally distributed with mean μ and standard deviation $\sigma/\sqrt{n}$. If the population itself is normal, $\overline{X}$ is normally distributed for any sample size. Recall that the standard normal random variable Z has a 0.95 probability of being within the range of values -1.96 to 1.96 (you may check this using Table 2 in Appendix C). Transforming Z to the random variable $\overline{X}$ with mean μ and standard deviation $\sigma/\sqrt{n}$, we find that—*before the sampling*—there is a 0.95 probability that $\overline{X}$ will fall within the interval:

$$\mu \pm 1.96\frac{\sigma}{\sqrt{n}} \tag{6–1}$$

Once we have obtained our random sample, we have a particular value $\bar{x}$. This particular $\bar{x}$ either lies within the range of values specified by equation 6–1 or does not lie within this range. Since we do not know the (fixed) value of the population parameter μ, we have no way of knowing whether $\bar{x}$ is indeed within the range given in equation 6–1. Since the random sampling has already taken place and a particular $\bar{x}$ has been computed, we no longer have a random variable and may no longer talk about probabilities. We do know, however, that since the presampling probability that $\overline{X}$ will fall in the interval in equation 6–1 is 0.95, about 95% of the values of $\overline{X}$ obtained in a large number of repeated samplings will fall within the interval. Since we have a single value $\bar{x}$ that was obtained by this process, we may say that we are *95% confident that $\bar{x}$ lies within the interval*. This idea is demonstrated in Figure 6–1.

FIGURE 6–1   **Probability Distribution of $\bar{X}$ and Some Resulting Values of the Statistic in Repeated Samplings**

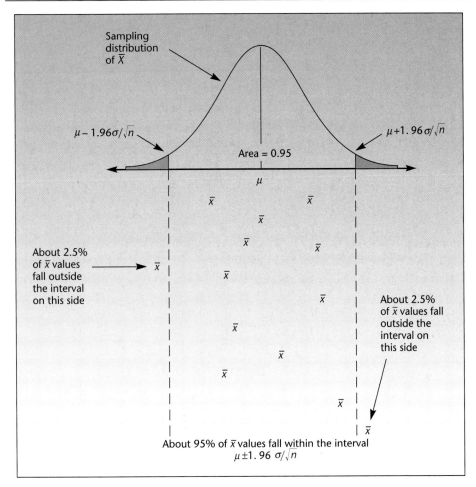

Consider a particular $\bar{x}$, and note that the distance between $\bar{x}$ and μ is the same as the distance between μ and $\bar{x}$. Thus, $\bar{x}$ falls inside the interval $\mu \pm 1.96\sigma/\sqrt{n}$ *if and only if* μ happens to be inside the interval $\bar{x} \pm 1.96\sigma/\sqrt{n}$. In a large number of re-peated trials, this would happen about 95% of the time. We therefore call the interval $\bar{x} \pm 1.96\sigma/\sqrt{n}$ a *95% confidence interval for the unknown population mean* μ. This is demonstrated in Figure 6–2.

Instead of measuring a distance of $1.96\sigma/\sqrt{n}$ on either side of μ (an impossible task since μ is unknown), we measure the same distance of $1.96\sigma/\sqrt{n}$ on either side of our *known* sample mean $\bar{x}$. Since, *before the sampling*, the random interval $\bar{X} \pm 1.96\sigma/\sqrt{n}$ had a 0.95 probability of capturing μ, *after the sampling* we may be 95% con-fident that our particular interval $\bar{x} \pm 1.96\sigma/\sqrt{n}$ indeed contains the population mean μ. We cannot say that there is a 0.95 *probability* that μ is inside the interval, because the interval $\bar{x} \pm 1.96\sigma/\sqrt{n}$ is not random, and neither is μ. The population mean μ is

FIGURE 6–2   Construction of a 95% Confidence Interval for the Population Mean μ

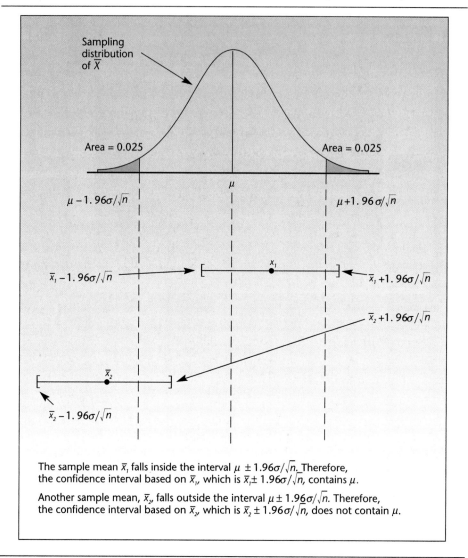

The sample mean $\bar{x}_1$, falls inside the interval $\mu \pm 1.96\sigma/\sqrt{n}$. Therefore, the confidence interval based on $\bar{x}_1$, which is $\bar{x}_1 \pm 1.96\sigma/\sqrt{n}$, contains μ.

Another sample mean, $\bar{x}_2$, falls outside the interval $\mu \pm 1.96\sigma/\sqrt{n}$. Therefore, the confidence interval based on $\bar{x}_2$, which is $\bar{x}_2 \pm 1.96\sigma/\sqrt{n}$, does not contain μ.

unknown to us but is a fixed quantity—not a random variable.[2] Either μ lies inside the confidence interval (in which case the probability of this event is 1.00), or it does not (in which case the probability of the event is 0). We do know, however, that 95% of all possible intervals constructed in this manner will contain μ. Therefore, we may say that we are *95% confident* that μ lies in the particular interval we have obtained.

---

[2]We are using what is called the *classical*, or *frequentist*, interpretation of confidence intervals. An alternative view, the Bayesian approach, will be discussed in Chapter 15. The Bayesian approach allows us to treat an unknown population parameter as a random variable. As such, the unknown population mean μ may be stated to have a 0.95 *probability* of being within an interval.

A 95% confidence interval for μ when σ is known and sampling is done from a normal population, or a large sample is used, is

$$\bar{x} \pm 1.96 \frac{\sigma}{\sqrt{n}} \qquad (6\text{--}2)$$

The quantity $1.96\sigma/\sqrt{n}$ is often called the *margin of error* or the *sampling error*. Its data-derived estimate (using s instead of the unknown σ) is commonly reported.

To compute a 95% confidence interval for μ, all we need to do is substitute the values of the required entities in equation 6–2. Suppose, for example, that we are sampling from a normal population, in which case the random variable $\bar{X}$ is normally distributed for any sample size. We use a sample of size $n = 25$, and we get a sample mean $\bar{x} = 122$. Suppose we also know that the population standard deviation is $\sigma = 20$. Let us compute a 95% confidence interval for the unknown population mean μ. Using equation 6–2, we get

$$\bar{x} \pm 1.96 \frac{\sigma}{\sqrt{n}} = 122 \pm 1.96 \frac{20}{\sqrt{25}} = 122 \pm 7.84 = [114.16, 129.84]$$

Thus, we may be 95% confident that the unknown population mean μ lies anywhere between the values 114.16 and 129.84.

In business and other applications, the 95% confidence interval is commonly used. There are, however, many other possible levels of confidence. You may choose any level of confidence you wish, find the appropriate z value from the standard normal table, and use it instead of 1.96 in equation 6–2 to get an interval of the chosen level of confidence. Using the standard normal table, we find, for example, that for a 90% confidence interval we use the z value 1.645, and for a 99% confidence interval we use $z = 2.58$ (or, using an accurate interpolation, 2.576). Let us formalize the procedure and make some definitions.

We define $z_{\alpha/2}$ as the z value that cuts off a right-tail area of $\alpha/2$ under the standard normal curve.

For example, 1.96 is $z_{\alpha/2}$ for $\alpha/2 = 0.025$ because $z = 1.96$ cuts off an area of 0.025 to its right. (We find from Table 2 that for $z = 1.96$, TA = 0.475; therefore, the right-tail area is $\alpha/2 = 0.025$.) Now consider the two points 1.96 and -1.96. Each of them cuts off a tail area of $\alpha/2 = 0.025$ in the respective direction of its tail. The area between the two values is therefore equal to $1 - \alpha = 1 - 2(0.025) = 0.95$. The area under the curve excluding the tails, $1 - \alpha$, is called the **confidence coefficient**. (And the combined area in both tails α is called the **error probability**. This probability will be important to us in the next chapter.) The confidence coefficient multiplied by 100, expressed as a percentage, is the **confidence level**.

A $(1 - \alpha)$ 100% confidence interval for μ when σ is known and sampling is done from a normal population, or with a large sample, is

$$\bar{x} \pm z_{\alpha/2} \frac{\sigma}{\sqrt{n}} \qquad (6\text{--}3)$$

FIGURE 6–3   **Construction of an 80% Confidence Interval for μ**

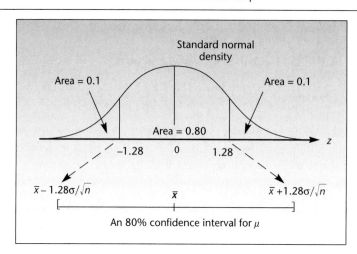

Thus, for a 95% confidence interval for μ we have

$$(1 - \alpha)100\% = 95\%$$
$$1 - \alpha = 0.95$$
$$\alpha = 0.05$$
$$\frac{\alpha}{2} = 0.025$$

From the normal table, we find $z_{\alpha/2} = 1.96$. This is the value we substitute for $z_{\alpha/2}$ in equation 6–3.

For example, suppose we want an 80% confidence interval for μ. We have $1 - \alpha = 0.80$ and $\alpha = 0.20$; therefore, $\alpha/2 = 0.10$. We now look in the standard normal table for the value of $z_{0.10}$, that is, the z value that cuts off an area of 0.10 to its right. We have TA $= 0.5 - 0.1 = 0.4$, and from the table we find $z_{0.10} = 1.28$. The confidence interval is therefore $\bar{x} \pm 1.28\sigma/\sqrt{n}$. This is demonstrated in Figure 6–3.

Let us compute an 80% confidence interval for μ using the information presented earlier. We have $n = 25$, and $\bar{x} = 122$. We also assume $\sigma = 20$. To compute an 80% confidence interval for the unknown population mean μ, we use equation 6–3 and get

$$\bar{x} \pm z_{\alpha/2} \frac{\sigma}{\sqrt{n}} = 1.22 \pm 1.28 \frac{20}{\sqrt{25}} = 122 \pm 5.12 = [116.88, 127.12]$$

Comparing this interval with the 95% confidence interval for μ we computed earlier, we note that the present interval is *narrower*. This is an important property of confidence intervals.

When sampling is from the same population, using a fixed sample size, *the higher the confidence level, the wider the interval.*

Intuitively, a wider interval has more of a presampling chance of "capturing" the unknown population parameter. If we want a 100% confidence interval for a parameter,

FIGURE 6–4   Width of a Confidence Interval as a Function of Sample Size

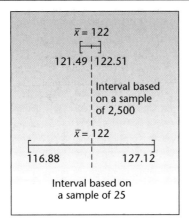

the interval must be $[-\infty, \infty]$. The reason for this is that 100% confidence is derived from a presampling probability of 1.00 of capturing the parameter, and the only way to get such a probability using the standard normal distribution is by allowing Z to be anywhere from $-\infty$ to ∞. If we are willing to be more realistic (nothing is *certain*) and accept, say, a 99% confidence interval, our interval will be finite and based on $z = 2.58$. The width of our interval will then be $2(2.58\sigma/\sqrt{n})$. If we further reduce our confidence requirement to 95%, the width of our interval will be $2(1.96\sigma/\sqrt{n})$. Since both σ and n are fixed, the 95% interval must be narrower. The more confidence you require, the more you need to sacrifice in terms of a wider interval.

If you want both a narrow interval *and* a high degree of confidence, you need to acquire a large amount of information—take a large sample. This is so because the larger the sample size n, the narrower the interval. This makes sense in that if you buy more information, you will have less uncertainty.

> When sampling is from the same population, using a fixed confidence level, *the larger the sample size* n, *the narrower the confidence interval.*

Suppose that the 80% confidence interval developed earlier was based on a sample size $n = 2,500$, instead of $n = 25$. Assuming that $\bar{x}$ and σ are the same, the new confidence interval should be 10 times as narrow as the previous one (because $\sqrt{2,500} = 50$, which is 10 times as large as $\sqrt{25}$). Indeed, the new interval is

$$\bar{x} \pm z_{\alpha/2} \frac{\sigma}{\sqrt{n}} = 122 \pm 1.28 \frac{20}{\sqrt{2,500}} = 122 \pm 0.512 = [121.49, 122.51]$$

This interval has width $2(0.512) = 1.024$, while the width of the interval based on a sample of size $n = 25$ is $2(5.12) = 10.24$. This demonstrates the value of information. The two confidence intervals are shown in Figure 6–4.

EXAMPLE 6–1

This example uses Excel to compute a 95% confidence interval for the population mean. The population consists of the Fortune 500 companies, as ranked by revenues. You are trying to estimate the average revenues for the companies on the list. The population standard deviation is $15,056.37. A random sample of 30 companies gives a sample mean of $10,672.87. Give a 95% and 90% confidence interval for average revenues.

Data Characteristics		
Number	30	
Population standard deviation	15,056.37	
Sample mean	10,672.87	
95% Confidence interval	5,387.85	Range is 5,285.02 to 16,060.72
90% Confidence interval	4,521.95	Range is 6,150.92 to 15,194.82

Note: The full list of 500 observations is obtainable from www.fortune.com.

The Template

In the workbook named Estimating Mean.xls there are sheets for computing confidence intervals for population means when

1. The sample statistics are known.
2. The sample data are known.

Figure 6–5 shows the first sheet. In this template, we enter the sample statistics in the top or bottom panel, depending on whether the population standard deviation σ is known or unknown. On the right, there is provision for finite population correction. The input needed in this panel is the population size N in cell P8. If the correction is not needed, it is a good idea to leave this cell blank to avoid creating a distraction.

FIGURE 6–5   **The Template for Estimating μ with Sample Statistics**
[Estimating Mean.xls; Sheet: Sample Stats]

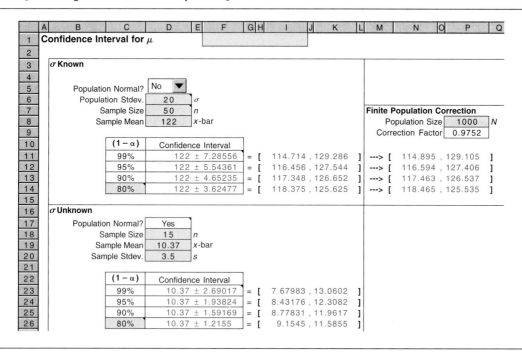

Since the population standard deviation σ may not be known for certain, on the extreme right there is a sensitivity analysis of the confidence interval with respect to σ. As can be seen in the plot below the panel, the half-width of the confidence interval is linearly related to σ.

Figure 6–6 shows the template to be used when the sample data are known. The data must be entered in column B. The sample size, the sample mean, and the sample standard deviation are automatically calculated and entered in cells F4, F5, F18, F19, and F20 as needed.

PROBLEMS

6–1. What is a confidence interval, and why is it useful? What is a confidence level?

6–2. Explain why in classical statistics it makes no sense to describe a confidence interval in terms of probability.

6–3. Explain how the postsampling confidence level is derived from a presampling probability.

6–4. Suppose that you computed a 95% confidence interval for a population mean. The user of the statistics claims your interval is too wide to have any meaning in the specific use for which it is intended. Discuss and compare two methods of solving this problem.

6–5. A real estate agent needs to estimate the average value of a residential property of a given size in a certain area. The real estate agent believes that the standard

FIGURE 6–6   **The Template for Estimating μ with Sample Data**
[Estimating Mean.xls; Sheet: Sample Data]

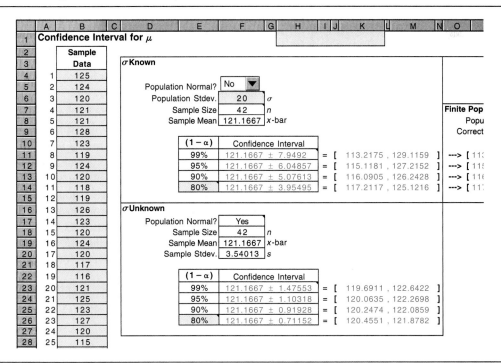

deviation of the property values is $\sigma = \$5,500.00$ and that property values are approximately normally distributed. A random sample of 16 units gives a sample mean of $89,673.12. Give a 95% confidence interval for the average value of all properties of this kind.

6–6.   In problem 6–5, suppose that a 99% confidence interval is required. Compute the new interval, and compare it with the 95% confidence interval you computed in problem 6–5.

6–7.   A car manufacturer wants to estimate the average miles-per-gallon highway rating for a new model. From experience with similar models, the manufacturer believes the miles-per-gallon standard deviation is 4.6. A random sample of 100 highway runs of the new model yields a sample mean of 32 miles per gallon. Give a 95% confidence interval for the population average miles-per-gallon highway rating.

6–8.   In problem 6–7, do we need to assume that the population of miles-per-gallon values is normally distributed? Explain.

6–9.   A wine importer needs to report the average percentage of alcohol in bottles of French wine. From experience with previous kinds of wine, the importer believes the population standard deviation is 1.2%. The importer randomly samples 60 bottles of the new wine and obtains a sample mean $\bar{x} = 9.3\%$. Give a 90% confidence interval for the average percentage of alcohol in all bottles of the new wine.

6–10.   A company is considering installing a fax machine at one of its offices. As part of the decision process as to whether to install the machine, the company's manager wants to estimate the average number of documents that would be transmitted daily if the machine were installed. From experience at other offices, the company manager believes the standard deviation of the number of documents sent daily is 32. The manager also believes the number of documents transmitted daily is a normally distributed random variable. The machine is tested over a random sample of 15 days, and the resulting sample mean is 267. Give a 99% confidence interval for the average number of documents that would be transmitted daily if the machine were installed.

6–11.   In problem 6–10, suppose that the manager would be interested in installing the machine if she could be fairly confident that the average number of documents transmitted daily would be above 245. Do the findings in problem 6–10 justify installing the machine? Explain.

6–12.   Beechcraft, Inc., wants to estimate the average time it takes the Beechjet corporate jet to climb from sea level to 41,000 feet. From previous experience, company engineers believe that the standard deviation of climbing times is 2 minutes. The model is tested in 100 random trials, and it is found that the sample mean is 22 minutes. Give a 95% confidence interval for the average climbing time from sea level to 41,000 feet.

6–13.   A mining company needs to estimate the average amount of copper ore per ton mined. A random sample of 50 tons gives a sample mean of 146.75 pounds. The population standard deviation is assumed to be 35.2 pounds. Give a 95% confidence interval for the average amount of copper in the "population" of tons mined. Also give a 90% confidence interval and a 99% confidence interval for the average amount of copper per ton.

6–14.   According to a newspaper article,[3] the average cost of information for firms looking for markets abroad is $166. If $\sigma = \$40$, $n = 16$, and it is a normal distribution, give the 95% confidence interval.

---

[3] B. Wall, "Small Firms Find Room for a Big View," *International Herald Tribune,* June 28, 1997, p. 17.

6–15. "Small-fry" funds trade at an average of 20% discount to net asset value.[4] If $\sigma = 8\%$ and $n = 36$, give the 95% confidence interval for average population percentage.

6–16. Suppose you have a confidence interval based on a sample of size n. Using the same level of confidence, how large a sample is required to produce an interval of one-half the width?

6–17. The width of a 95% confidence interval for μ is 10 units. If everything else stays the same, how wide would a 90% confidence interval be for μ?

6–3 Confidence Intervals for μ When σ Is Unknown— The *t* Distribution

In constructing confidence intervals for μ, we assume a normal population distribution or a large sample size (for normality via the central limit theorem). Until now, we have also assumed a known population standard deviation. This assumption was necessary for theoretical reasons so that we could use standard normal probabilities in constructing our intervals.

In real sampling situations, however, the population standard deviation σ is rarely known. The reason for this is that both μ and σ are population parameters. When we sample from a population with the aim of estimating its unknown mean, it is quite unlikely that the other parameter of the same population, the standard deviation, will be known.

The t Distribution

As we mentioned in Chapter 5, when the population standard deviation is not known, we may use the sample standard deviation S in its place. *If the population is normally distributed,* the standardized statistic

$$t = \frac{\overline{X} - \mu}{S/\sqrt{n}} \tag{6-4}$$

has a **t distribution** with $n - 1$ degrees of freedom. The degrees of freedom of the distribution are the degrees of freedom associated with the sample standard deviation S (as explained in the last chapter). The t distribution is also called *Student's distribution,* or *Student's t distribution*. What is the origin of the name *Student*?

W. S. Gossett was a scientist at the Guinness brewery in Dublin, Ireland. In 1908, Gossett discovered the distribution of the quantity in equation 6–4. He called the new distribution the t distribution. The Guinness brewery, however, did not allow its workers to publish findings under their own names. Therefore, Gossett published his findings under the pen name *Student*. As a result, the distribution became known also as Student's distribution.

The t distribution is characterized by its degrees-of-freedom parameter df. For any integer value df $= 1, 2, 3, \ldots$, there is a corresponding t distribution. The t distribution resembles the standard normal distribution Z: it is symmetric and bell-shaped. The t distribution, however, has wider tails than the Z distribution.

> The mean of a t distribution is zero. For df > 2, the variance of the t distribution is equal to df/(df $- 2$).

---

[4]From "A Dish of Small-Fry Funds," *International Herald Tribune,* June 28, 1997, p. 17.

FIGURE 6–7   **The *t*-Distribution Template**
[t.xls]

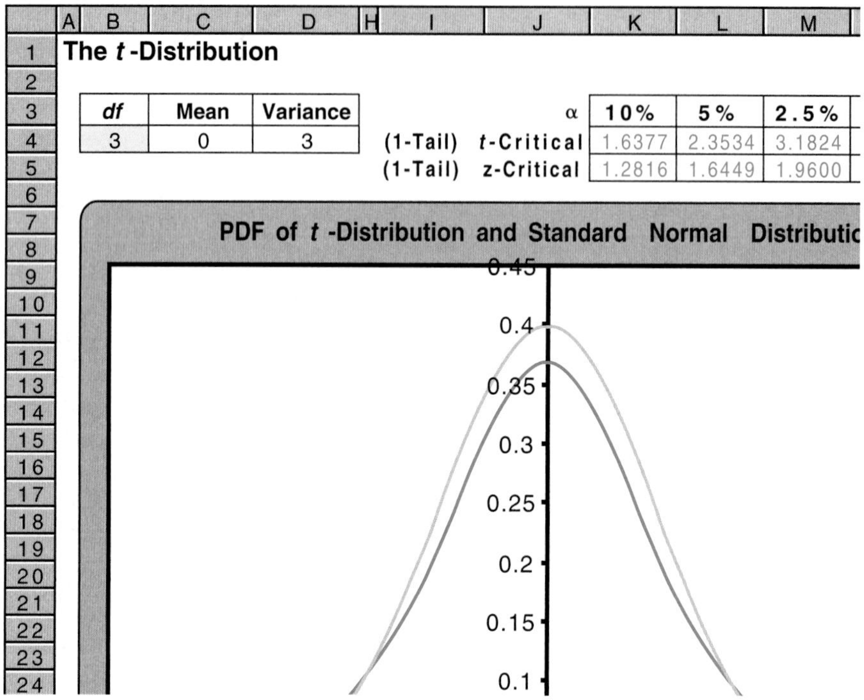

We see that the mean of *t* is the same as the mean of *Z*, but the variance of *t* is larger than the variance of *Z*. As df increases, the variance of *t* approaches 1.00, which is the variance of *Z*. Having wider tails and a larger variance than *Z* is a reflection of the fact that the *t* distribution applies to situations with a greater inherent *uncertainty*. The uncertainty comes from the fact that σ is unknown and is estimated by the *random variable S*. The *t* distribution thus reflects the uncertainty in *two* random variables, $\overline{X}$ and *S*, while *Z* reflects only an uncertainty due to $\overline{X}$. The greater uncertainty in *t* (which makes confidence intervals based on *t* wider than those based on *Z*) is the price we pay for not knowing σ and having to estimate it from our data. As df increases, the *t* distribution approaches the *Z* distribution.

Figure 6–7 shows the *t*-distribution template. We can enter any desired degrees of freedom in cell B4 and see how the distribution approaches the *Z* distribution, which is superimposed on the chart. In the range K3:O5 the template shows the critical values for all standard α values. The area to the right of the chart in this template can be used for calculating *p*-values, which we will learn in the next chapter.

Values of *t* distributions for selected tail probabilities are given in Table 3 in Appendix C (reproduced here as Table 6–1). Since there are infinitely many *t* distributions—one for every value of the degrees-of-freedom parameter—the table contains probabilities for only some of these distributions. For each distribution, the table gives values that cut off given areas under the curve to the *right*. The *t* table is thus a table of values corresponding to right-tail probabilities.

Let us consider an example. A random variable with a *t* distribution with 10 degrees of freedom has a 0.10 probability of exceeding the value 1.372. It has a 0.025

TABLE 6–1   Values and Probabilities of t Distributions

Degrees of Freedom	$t_{0.100}$	$t_{0.050}$	$t_{0.025}$	$t_{0.010}$	$t_{0.005}$
1	3.078	6.314	12.706	31.821	63.657
2	1.886	2.920	4.303	6.965	9.925
3	1.638	2.353	3.182	4.541	5.841
4	1.533	2.132	2.776	3.747	4.604
5	1.476	2.015	2.571	3.365	4.032
6	1.440	1.943	2.447	3.143	3.707
7	1.415	1.895	2.365	2.998	3.499
8	1.397	1.860	2.306	2.896	3.355
9	1.383	1.833	2.262	2.821	3.250
10	1.372	1.812	2.228	2.764	3.169
11	1.363	1.796	2.201	2.718	3.106
12	1.356	1.782	2.179	2.681	3.055
13	1.350	1.771	2.160	2.650	3.012
14	1.345	1.761	2.145	2.624	2.977
15	1.341	1.753	2.131	2.602	2.947
16	1.337	1.746	2.120	2.583	2.921
17	1.333	1.740	2.110	2.567	2.898
18	1.330	1.734	2.101	2.552	2.878
19	1.328	1.729	2.093	2.539	2.861
20	1.325	1.725	2.086	2.528	2.845
21	1.323	1.721	2.080	2.518	2.831
22	1.321	1.717	2.074	2.508	2.819
23	1.319	1.714	2.069	2.500	2.807
24	1.318	1.711	2.064	2.492	2.797
25	1.316	1.708	2.060	2.485	2.787
26	1.315	1.706	2.056	2.479	2.779
27	1.314	1.703	2.052	2.473	2.771
28	1.313	1.701	2.048	2.467	2.763
29	1.311	1.699	2.045	2.462	2.756
30	1.310	1.697	2.042	2.457	2.750
40	1.303	1.684	2.021	2.423	2.704
60	1.296	1.671	2.000	2.390	2.660
120	1.289	1.658	1.980	2.358	2.617
∞	1.282	1.645	1.960	2.326	2.576

probability of exceeding the value 2.228, and so on for the other values listed in the table. Since the t distributions are symmetric about zero, we also know, for example, that the probability that a random variable with a t distribution with 10 degrees of freedom will be less than -1.372 is 0.10. These facts are demonstrated in Figure 6–8.

As we noted earlier, the t distribution approaches the standard normal distribution as the df parameter approaches infinity. The t distribution with "infinite" degrees of freedom is defined as the standard normal distribution. The last row in Appendix C, Table 3 (Table 6–1) corresponds to df $= \infty$, the standard normal distribution. Note that the value corresponding to a right-tail area of 0.025 in that row is 1.96, which we

FIGURE 6–8   Table Probabilities for a Selected *t* Distribution (df = 10)

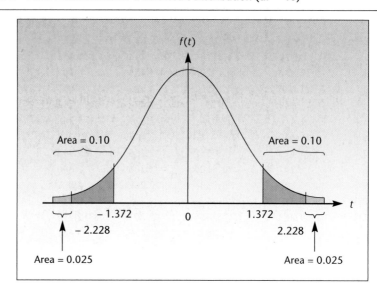

recognize as the appropriate *z* value. Similarly, the value corresponding to a right-tail area of 0.005 is 2.576, and the value corresponding to a right-tail area of 0.05 is 1.645. These, too, are values we recognize for the standard normal distribution. Look upward from the last row of the table to find cutoff values of the same right-tail probabilities for *t* distributions with different degrees of freedom. Suppose, for example, that we want to construct a 95% confidence interval for μ using the *t* distribution with 20 degrees of freedom. We may identify the value 1.96 in the last row (the appropriate *z* value for 95%) and then move up in the same column until we reach the row corresponding to df = 20. Here we find the required value $t_{\alpha/2} = t_{0.025} = 2.086$.

> A $(1 - \alpha)$ 100% confidence interval for μ when σ is not known (assuming a normally distributed population) is
>
> $$\bar{x} \pm t_{\alpha/2} \frac{s}{\sqrt{n}} \qquad (6\text{–}5)$$
>
> where $t_{\alpha/2}$ is the value of the *t* distribution with $n - 1$ degrees of freedom that cuts off a tail area of $\alpha/2$ to its right.

EXAMPLE 6–2   A stock market analyst wants to estimate the average return on a certain stock. A random sample of 15 days yields an average (annualized) return of $\bar{x} = 10.37\%$ and a standard deviation of $s = 3.5\%$. Assuming a normal population of returns, give a 95% confidence interval for the average return on this stock.

Solution   Since the sample size is $n = 15$, we need to use the *t* distribution with $n - 1 = 14$ degrees of freedom. In Table 3, in the row corresponding to 14 degrees of freedom and the column corresponding to a right-tail area of 0.025 (this is $\alpha/2$), we find $t_{0.025} = 2.145$. (We could also have found this value by moving upward from 1.96 in the last row.) Using this value, we construct the 95% confidence interval as follows:

$$\bar{x} \pm t_{\alpha/2} \frac{s}{\sqrt{n}} = 10.37 \pm 2.145 \frac{3.5}{\sqrt{15}} = [8.43, 12.31]$$

Thus, the analyst may be 95% sure that the average annualized return on the stock is anywhere from 8.43% to 12.31%.

Looking at the *t* table, we note the *convergence* of the *t* distributions to the *Z* distribution—the values in the rows preceding the last get closer and closer to the corresponding *z* values in the last row. Although the *t* distribution is the correct distribution to use whenever σ is not known (assuming the population is normal), when df is *large*, we may use the standard normal distribution as an adequate approximation to the *t* distribution. Thus, instead of using 1.98 in a confidence interval based on a sample of size 121 (df = 120), we will just use the *z* value 1.96.

We divide estimation problems into two kinds: small-sample problems and large-sample problems. Example 6–2 demonstrated the solution of a small-sample problem. In general, *large sample* will mean a sample of 30 items or more, and *small sample* will mean a sample of size less than 30. For small samples, we will use the *t* distribution as demonstrated above. For large samples, we will use the *Z* distribution as an adequate approximation. We note that the larger the sample size, the better the normal approximation. Remember, however, that this division of large and small samples is arbitrary.

> Whenever σ is not known (and the population is assumed normal), the correct distribution to use is the *t* distribution with $n - 1$ degrees of freedom. Note, however, that for large degrees of freedom, the *t* distribution is approximated well by the *Z* distribution.

If you wish, you may always use the more accurate values obtained from the *t* table (when such values can be found in the table) rather than the standard normal approximation. In this chapter and elsewhere (with the exception of some examples in Chapter 14), we will assume that the population satisfies, at least approximately, a normal distribution assumption. For large samples, this assumption is less crucial.

A large-sample $(1 - \alpha)$ 100% confidence interval for μ is

$$\bar{x} \pm z_{\alpha/2} \frac{s}{\sqrt{n}} \tag{6–6}$$

We demonstrate the use of equation 6–6 in Example 6–3.

EXAMPLE 6–3

An economist wants to estimate the average amount in checking accounts at banks in a given region. A random sample of 100 accounts gives $\bar{x} = \$357.60$ and $s = \$140.00$. Give a 95% confidence interval for μ, the average amount in any checking account at a bank in the given region.

Solution   We find the 95% confidence interval for μ as follows:

$$\bar{x} \pm z_{\alpha/2} \frac{s}{\sqrt{n}} = 357.60 \pm 1.96 \frac{140}{\sqrt{100}} = [330.16, 385.04]$$

Thus, based on the data and the assumption of random sampling, the economist may be 95% confident that the average amount in checking accounts in the area is anywhere from $330.16 to $385.04.

PROBLEMS

6–18.   A telephone company wants to estimate the average length of long-distance calls during weekends. A random sample of 50 calls gives a mean $\bar{x} = 14.5$ minutes and standard deviation $s = 5.6$ minutes. Give a 95% confidence interval and a 90% confidence interval for the average length of a long-distance phone call during weekends.

6–19.   An insurance company handling malpractice cases is interested in estimating the average amount of claims against physicians of a certain specialty. The company obtains a random sample of 165 claims and finds $\bar{x} = \$16,530$ and $s = \$5,542$. Give a 95% confidence interval and a 99% confidence interval for the average amount of a claim.

6–20.   The manufacturer of batteries used in small electric appliances wants to estimate the average life of a battery. A random sample of 12 batteries yields $\bar{x} = 34.2$ hours and $s = 5.9$ hours. Give a 95% confidence interval for the average life of a battery.

6–21.   A tire manufacturer wants to estimate the average number of miles that may be driven on a tire of a certain type before the tire wears out. A random sample of 32 tires is chosen; the tires are driven on until they wear out, and the number of miles driven on each tire is recorded. The data, in thousands of miles, are as follows:

32, 33, 28, 37, 29, 30, 25, 27, 39, 40, 26, 26, 27, 30, 25, 30, 31, 29, 24, 36, 25, 37, 37, 20, 22, 35, 23, 28, 30, 36, 40, 41

Give a 99% confidence interval for the average number of miles that may be driven on a tire of this kind.

6–22.   According to an article in *Fortune*,[5] the average hourly wages rose by an average of 3.9% based on a random sample of 1,000 firms, $s = 1.2\%$. Give the 90% confidence interval.

6–23.   Pier 1 Imports is a nationwide retail outlet selling imported furniture and other home items. From time to time, the company surveys its regular customers by obtaining random samples based on customer zip codes. In one mailing, customers were asked to rate a new table from Thailand on a scale of 0 to 100. The ratings of 25 randomly selected customers are as follows: 78, 85, 80, 89, 77, 50, 75, 90, 88, 100, 70, 99, 98, 55, 80, 45, 80, 76, 96, 100, 95, 90, 60, 85, 90. Give a 99% confidence interval for the rating of the table that would be given by an average member of the population of regular customers. Assume normality.

[5]From W. Woods, "Rising Wages," *Fortune,* July 7, 1997, p. 13.

6–24. An executive placement service needs to estimate the average salary of executives placed in a given industry. A random sample of 40 executives gives $\bar{x} =$ $42,539 and $s =$ $11,690. Give a 90% confidence interval for the average salary of an executive placed in this industry.

6–25. A bank is interested in extending its automated teller machine service to a new community. As part of the research prepared as an aid in making the decision, the bank's management undertakes an experiment to determine the average amount of a transaction, in dollars per person per day. A random sample of 10 experimental transactions on a trial run of the machine is collected. The following data (in dollars) are the result: 53, 40, 39, 10, 12, 60, 72, 65, 50, 45. Give a 95% confidence interval for the average amount of a transaction.

6–26. For advertising purposes, the Beef Industry Council needs to estimate the average caloric content of 3-ounce top loin steak cuts. A random sample of 400 pieces gives a sample mean of 212 calories and a sample standard deviation of 38 calories. Give a 95% confidence interval for the average caloric content of a 3-ounce cut of top loin steak. Also give a 98% confidence interval for the average caloric content of a cut.

6–27. A transportation company wants to estimate the average length of time goods are in transit across the country. A random sample of 20 shipments gives $\bar{x} = 2.6$ days and $s = 0.4$ day. Give a 99% confidence interval for the average transit time.

6–28. To aid in planning the development of a tourist shopping area, a state agency wants to estimate the average dollar amount spent by a tourist in an existing shopping area. A random sample of 56 tourists gives $\bar{x} =$ $258 and $s =$ $85. Give a 95% confidence interval for the average amount spent by a tourist at the shopping area.

6–29. A large drugstore wants to estimate average weekly sales for a brand of soap. A random sample of 13 weeks gives the following numbers: 123, 110, 95, 120, 87, 89, 100, 105, 98, 88, 75, 125, 101. Give a 90% confidence interval for average weekly sales.

6–30. Citibank Visa gives its cardholders "bonus dollars," which may be spent in partial payment for gifts purchased with the Visa card. The company wants to estimate the average amount of bonus dollars that will be spent by a cardholder enrolled in the program during a year. A trial run of the program with a random sample of 225 cardholders is carried out. The results are $\bar{x} =$ $259.60 and $s =$ $52.00. Give a 95% confidence interval for the average amount of bonus dollars that will be spent by a cardholder during the year.

6–31. An accountant wants to estimate the average amount of an account of a service company. A random sample of 46 accounts yields $\bar{x} =$ $16.50 and $s =$ $2.20. Give a 95% confidence interval for the average amount of an account.

6–32. An art dealer wants to estimate the average value of works of art of a certain period and type. A random sample of 20 works of art is appraised. The sample mean is found to be $5,139 and the sample standard deviation $640. Give a 95% confidence interval for the average value of all works of art of this kind.

6–33. A management consulting agency needs to estimate the average number of years of experience of executives in a given branch of management. A random sample of 28 executives gives $\bar{x} = 6.7$ years and $s = 2.4$ years. Give a 99% confidence interval for the average number of years of experience for all executives in this branch.

6–34. The Food and Drug Administration (FDA) needs to estimate the average content of an additive in a given food product. A random sample of 75 portions of the product gives $\bar{x} = 8.9$ units and $s = 0.5$ unit. Give a 95% confidence interval for the average number of units of additive in any portion of this food product.

6–35.   The management of a supermarket needs to make estimates of the average daily demand for milk. The following data are available (number of half-gallon containers sold per day): 48, 59, 45, 62, 50, 68, 57, 80, 65, 58, 79, 69. Assuming that this is a random sample of daily demand, give a 90% confidence interval for average daily demand for milk.

6–36.   According to A. Fisher,[6] the average refund Sun Television pays a customer who finds a competitor with a lower price is $3.90. If the sample size was 100 and sample standard deviation $2.00, construct a 95% confidence interval for the average refund.

6–37.   The data on the daily consumption of fuel by a delivery truck, in gallons, recorded during 25 randomly selected working days, are as follows:

9.7, 8.9, 9.7, 10.9, 10.3, 10.1, 10.7, 10.6, 10.4, 10.6, 11.6, 11.7, 9.7, 9.7, 9.7, 9.8, 12, 10.4, 8.8, 8.9, 8.4, 9.7, 10.3, 10, 9.2

Compute a 90% confidence interval for the daily fuel consumption.

6–38.   Yo-yos are back[7] and are so popular that the industry leader, Duncan Toys, cannot keep up with demand even with its Indiana plants running 24 hours per day, 7 days per week. One store in the San Francisco suburbs reported the following numbers of people on its daily waiting lists:

187, 195, 212, 180, 15, 217, 225, 169, 199, 208

Assuming this is a random sample from the population of waiting lists for the product at that time, give a 95% confidence interval for average waiting list size.

6–39.   Live Pictures Inc. markets a computer software package that allows pictures to be scanned into a personal computer (PC) to create a panoramic view.[8] The company needs to estimate the average maximum number of standard-size photographs that may be scaled to create a single computer picture. The following data are available from a random sample of trials:

15, 14.5, 16, 15.25, 14, 15.5, 16, 16, 16.25, 14.25, 15.75, 15.25, 17, 16.75, 14.25, 15, 14.5, 16.5, 15.75, 15.25

Plot the data, and find the mean. Construct a 95% confidence interval for μ.

6–4   Large-Sample Confidence Intervals for the Population Proportion p

Sometimes interest centers on a qualitative, rather than a quantitative, variable. We may be interested in the relative frequency of occurrence of some characteristic in a population. For example, we may be interested in the proportion of people in a population who are users of some product or the proportion of defective items produced by a machine. In such cases, we want to estimate the population proportion p.

---

[6]"The Latest Weapon in the Price Wars," *Fortune,* July 7, 1997, p. 91.

[7]H. Greenberg, "Around the World: Yo-Yo Shortages," *Fortune,* June 23, 1997, p. 18.

[8]P. Burrows, "A Panoramic View from Your PC," *Business Week,* June 16, 1997, p. 73.

The estimator of the population proportion p is the sample proportion $\hat{P}$. In Chapter 5, we saw that when the sample size is large, $\hat{P}$ has an approximately normal sampling distribution. The mean of the sampling distribution of $\hat{P}$ is the population proportion p, and the standard deviation of the distribution of $\hat{P}$ is $\sqrt{pq/n}$, where $q = 1 - p$. Since the standard deviation of the estimator depends on the unknown population parameter, its value is also unknown to us. It turns out, however, that for large samples we may use our actual estimate $\hat{P}$ instead of the unknown parameter p in the formula for the standard deviation. We will, therefore, use $\sqrt{\hat{p}\hat{q}/n}$ as our estimate of the standard deviation of $\hat{P}$. Recall our large-sample rule of thumb: For estimating p, a sample is considered large enough when both $n \cdot p$ and $n \cdot q$ are greater than 5. (We guess the value of p when determining whether the sample is large enough. As a check, we may also compute $n\hat{p}$ and $n\hat{q}$ once the sample is obtained.)

A large-sample $(1 - \alpha)$ 100% confidence interval for the population proportion p is

$$\hat{p} \pm z_{\alpha/2} \sqrt{\frac{\hat{p}\hat{q}}{n}} \qquad (6\text{--}7)$$

where the sample proportion $\hat{p}$ is equal to the number of successes in the sample x, divided by the number of trials (the sample size) n, and $\hat{q} = 1 - \hat{p}$.

We demonstrate the use of equation 6–7 in Example 6–4.

A market research firm wants to estimate the share that foreign companies have in the U.S. market for certain products. A random sample of 100 consumers is obtained, and it is found that 34 people in the sample are users of foreign-made products; the rest are users of domestic products. Give a 95% confidence interval for the share of foreign products in this market.

EXAMPLE 6–4

We have $x = 34$ and $n = 100$, so our sample estimate of the proportion is $\hat{p} = x/n = 34/100 = 0.34$. We now use equation 6–7 to obtain the confidence interval for the population proportion p. A 95% confidence interval for p is

Solution

$$\hat{p} \pm z_{\alpha/2} \sqrt{\frac{\hat{p}\hat{q}}{n}} = 0.34 \pm 1.96 \sqrt{\frac{(0.34)(0.66)}{100}}$$
$$= 0.34 \pm 1.96(0.04737) = 0.34 \pm 0.0928$$
$$= [0.2472, 0.4328]$$

Thus, the firm may be 95% confident that foreign manufacturers control anywhere from 24.72% to 43.28% of the market.

Suppose the firm is not happy with such a wide confidence interval. What can be done about it? This is a problem of *value of information,* and it applies to all estimation situations. As we stated earlier, for a fixed sample size, the higher the confidence you require, the wider will be the confidence interval. The sample size is in the denominator of the standard error term, as we saw in the case of estimating μ. If we should

FIGURE 6–9 **The Template for Estimating Population Proportions [Estimating Proportion.xls]**

Confidence Interval for Population Proportion

Sample size	100	n
Sample proportion	0.34	p-hat

$1-\alpha$	Confidence Interval		
99%	0.34 ± 0.1220	= [0.2180 , 0.4620]	------> [0.2210 , 0.4590]
95%	0.34 ± 0.0928	= [0.2472 , 0.4328]	------> [0.2495 , 0.4305]
90%	0.34 ± 0.0779	= [0.2621 , 0.4179]	------> [0.2640 , 0.4160]
80%	0.34 ± 0.0607	= [0.2793 , 0.4007]	------> [0.2808 , 0.3992]

Finite Population Correction

Population Size	2000	N
Correction Factor	0.9749	

increase n, the standard error of $\hat{P}$ will decrease, and there will be less uncertainty about the parameter being estimated. If the sample size cannot be increased but you still want a narrower confidence interval, you must reduce your confidence level. Thus, for example, if the firm agrees to reduce the confidence level to 90%, z will be reduced from 1.96 to 1.645, and the confidence interval will shrink to

$$0.34 \pm 1.645(0.04737) = 0.34 \pm 0.07792 = [0.2621, 0.4179]$$

The firm may be 90% confident that the market share of foreign products is anywhere from 26.21% to 41.79%. If the firm wanted a high confidence (say 95%) *and* a narrow confidence interval, it would have to take a larger sample. Suppose that a random sample of $n = 200$ customers gave us the same result; that is, $x = 68$, $n = 200$, and $\hat{p} = x/n = 0.34$. What would be a 95% confidence interval in this case? Using equation 6–7, we get

$$\hat{p} \pm z_{\alpha/2}\sqrt{\frac{\hat{p}\hat{q}}{n}} = 0.34 \pm 1.96\sqrt{\frac{(0.34)(0.66)}{200}} = [0.2743, 0.4057]$$

This interval is considerably narrower than our first 95% confidence interval, which was based on a sample of 100.

When proportions using small samples are estimated, the binomial distribution may be used in forming confidence intervals. Since the distribution is discrete, it may not be possible to construct an interval with an exact, prespecified confidence level such as 95% or 99%. We will not demonstrate the method here.

The Template

Figure 6–9 shows the template that can be used for computing confidence intervals for population proportions. The template also has provision for finite population correction. To make this correction, the population size N must be entered in cell N5. If the correction is not needed, it is a good idea to leave this cell blank to avoid creating a distraction.

PROBLEMS

6–40. A maker of portable exercise equipment, designed for health-conscious people who travel too frequently to use a regular athletic club, wants to estimate the proportion of traveling business people who may be interested in the product. A random sample of 120 traveling business people indicates that 28 may be interested in purchasing the portable fitness equipment. Give a 95% confidence interval for the proportion of all traveling business people who may be interested in the product.

6–41. The makers of a medicated facial skin cream are interested in determining the percentage of people in a given age group who may benefit from the ointment. A random sample of 68 people results in 42 successful treatments. Give a 99% confidence interval for the proportion of people in the given age group who may be successfully treated with the facial cream.

6–42. Of 1,383 U.S. workers randomly polled by *CareerPath.com*, 59% said they are underpaid. Give a 99% confidence interval for the proportion of employees in the United States who think they are underpaid.

6–43. A survey of financial executives at Fortune 100 companies showed that 60% were confident that U.S. economic growth would continue over the next two years.[9] If the survey included 984 executives, give a 95% confidence interval for the proportion of executives who are confident about U.S. economic growth.

6–44. A recent article describes the success of business schools in Europe and the demand on that continent for the MBA degree. The article reports that a survey of 280 European business positions resulted in the conclusion that only one-seventh of the positions for MBAs at European businesses are currently filled. Assuming that these numbers are exact and that the sample was randomly chosen from the entire population of interest, give a 90% confidence interval for the proportion of filled MBA positions in Europe.

6–45. A survey of 748 randomly selected employees of dot-com companies showed that 35% feel secure about their jobs.[10] Give a 90% confidence interval for the proportion of dot-com company employees who feel secure about their jobs.

6–46. Cramer J. Stiff is an undertaker in York, Pennsylvania; Bob Crooks, a used-car salesman in Illinois; Martin Lawyer, an attorney in Florida; Chris Roach, a pest-control inspector in Virginia; and John M. Hamburger (whose name predates the sandwich), head of a restaurant consulting firm in Minnesota. What's in a name? Research at Brown University tried to answer the question of whether people with unusual names are drawn to professions that suit them.[11] If 28 of 100 people randomly selected from those with unusual names indicated their name might have been a factor in career choice, give a 95% confidence interval.

6-47. A machine produces safety devices for use in helicopters. A quality-control engineer regularly checks samples of the devices produced by the machine, and if too many of the devices are defective, the production process is stopped and the machine is readjusted. If a random sample of 52 devices yields 8 defectives, give a 98% confidence interval for the proportion of defective devices made by this machine.

6–48. Before launching its Buyers' Assurance Program, American Express wanted to estimate the proportion of cardholders who would be interested in this automatic insurance coverage plan. A random sample of 250 American Express cardholders was selected and sent questionnaires. The results were that 121 people in the sample expressed interest in the plan. Give a 99% confidence interval for the proportion of all interested American Express cardholders.

[9]From "Fortune Business Confidence Index," *Fortune,* December 2000.

[10]From "Dot Comment," *Business 2.0,* January 2001.

[11]E. DeLisser, "Mr. Money Teaches Business, Dr. Bone Can Fix Your Back," *The Wall Street Journal,* June 16, 1997, p. 1.

6–49.   An airline wants to estimate the proportion of business passengers on a new route from New York to San Francisco. A random sample of 347 passengers on this route is selected, and 201 are found to be business travelers. Give a 90% confidence interval for the proportion of business travelers on the airline's new route.

6–50.   A medical researcher working for one of the firms in the forefront of fighting ulcers needs to estimate the percentage of ulcers that are caused by bacteria.[12] A random sample of 860 ulcer patients is selected, and it is found that 758 of these patients have bacteria-caused ulcers. Give a 95% confidence interval for the percentage of ulcers caused by bacteria.

6–51.   Italy's television went digital by an accord signed by the media giants RAI, Mediaset, Stet, and the French group Canal Plus for the TV of the Future (digital television).[13] In a poll of 2,500 consumers, 1,828 voted for the "TV of the Future." Give the 95% confidence interval.

6–5   Confidence Intervals for the Population Variance

In some situations, our interest centers on the population variance (or, equivalently, the population standard deviation). This happens in production processes, queuing (waiting line) processes, and other situations. As we know, the sample variance S^2 is the (unbiased) estimator of the population variance σ^2.

To compute confidence intervals for the population variance, we must learn to use a new probability distribution: the *chi-square distribution*. Chi (pronounced $k\bar{\imath}$) is one of two *X* letters in the Greek alphabet and is denoted by χ. Hence, we denote the chi-square distribution by χ^2.

The chi-square distribution, like the *t* distribution, has associated with it a degrees-of-freedom parameter df. In the application of the chi-square distribution to estimation of the population variance, df $= n - 1$ (as with the *t* distribution in its application to sampling for the population mean). Unlike the *t* and the normal distributions, however, the chi-square distribution is *not* symmetric.

> The **chi-square distribution** is the probability distribution of the sum of several independent, squared standard normal random variables.

As a sum of squares, the chi-square random variable cannot be negative and is therefore bounded on the left by zero. The resulting distribution is skewed to the right. Figure 6–10 shows several chi-square distributions with different numbers of degrees of freedom.

> The mean of a chi-square distribution is equal to the degrees of freedom parameter df. The variance of a chi-square distribution is equal to twice the number of degrees of freedom.

Note in Figure 6–10 that as df increases, the chi-square distribution looks more and more like a normal distribution. In fact, as df increases, the chi-square distribution approaches a normal distribution with mean df and variance 2(df).

---

[12]B. O'Reilly, "Why Doctors Aren't Curing Ulcers," *Fortune,* June 23, 1997, p. 44.

[13]R. Cotrono, "Accordo Fatto Sulla TV Digitale," "Economia," *Corriere Della Sera,* June 24, 1997, p. 17.

FIGURE 6–10   **Several Chi-Square Distributions with Different Values of the df Parameter**

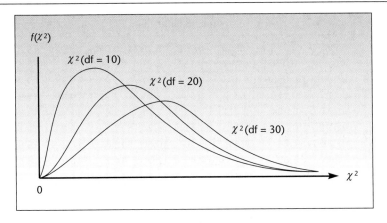

Table 4 in Appendix C gives values of the chi-square distribution with different degrees of freedom, for given tail probabilities. An abbreviated version of part of the table is given as Table 6–2. We apply the chi-square distribution to problems of estimation of the population variance, using the following property.

> In sampling from a normal population, the random variable
>
> $$\chi^2 = \frac{(n-1)S^2}{\sigma^2} \qquad (6\text{–}8)$$
>
> has a chi-square distribution with $n - 1$ degrees of freedom.

The distribution of the quantity in equation 6–8 leads to a confidence interval for σ^2. Since the χ^2 distribution is not symmetric, we cannot use equal values with opposite signs (such as ± 1.96 as we did with Z) and must construct the confidence interval using the two distinct tails of the distribution.

> A $(1 - \alpha)$ 100% confidence interval for the population variance σ^2 (where the population is assumed normal) is
>
> $$\left[\frac{(n-1)s^2}{\chi^2_{\alpha/2}}, \frac{(n-1)s^2}{\chi^2_{1-\alpha/2}} \right] \qquad (6\text{–}9)$$
>
> where $\chi^2_{\alpha/2}$ is the value of the chi-square distribution with $n - 1$ degrees of freedom that cuts off an area of $\alpha/2$ to its right and $\chi^2_{1-\alpha/2}$ is the value of the distribution that cuts off an area of $\alpha/2$ to its left (equivalently, an area of $1 - \alpha/2$ to its right).

We now demonstrate the use of equation 6–9 with an example.

TABLE 6–2 Values and Probabilities of Chi-Square Distributions

	Area in Right Tail									
df	0.995	0.990	0.975	0.950	0.900	0.100	0.050	0.025	0.010	0.005
1	0.0^4393	0.0^3157	0.0^3982	0.0^2393	0.158	2.71	3.84	5.02	6.63	7.88
2	0.0100	0.0201	0.0506	0.103	0.211	4.61	5.99	7.38	9.21	10.6
3	0.0717	0.115	0.216	0.352	0.584	6.25	7.81	9.35	11.3	12.8
4	0.207	0.297	0.484	0.711	1.06	7.78	9.49	11.1	13.3	14.9
5	0.412	0.554	0.831	1.15	1.61	9.24	11.1	12.8	15.1	16.7
6	0.676	0.872	1.24	1.64	2.20	10.6	12.6	14.4	16.8	18.5
7	0.989	1.24	1.69	2.17	2.83	12.0	14.1	16.0	18.5	20.3
8	1.34	1.65	2.18	2.73	3.49	13.4	15.5	17.5	20.1	22.0
9	1.73	2.09	2.70	3.33	4.17	14.7	16.9	19.0	21.7	23.6
10	2.16	2.56	3.25	3.94	4.87	16.0	18.3	20.5	23.2	25.2
11	2.60	3.05	3.82	4.57	5.58	17.3	19.7	21.9	24.7	26.8
12	3.07	3.57	4.40	5.23	6.30	18.5	21.0	23.3	26.2	28.3
13	3.57	4.11	5.01	5.89	7.04	19.8	22.4	24.7	27.7	29.8
14	4.07	4.66	5.63	6.57	7.79	21.1	23.7	26.1	29.1	31.3
15	4.60	5.23	6.26	7.26	8.55	22.3	25.0	27.5	30.6	32.8
16	5.14	5.81	6.91	7.96	9.31	23.5	26.3	28.8	32.0	34.3
17	5.70	6.41	7.56	8.67	10.1	24.8	27.6	30.2	33.4	35.7
18	6.26	7.01	8.23	9.39	10.9	26.0	28.9	31.5	34.8	37.2
19	6.84	7.63	8.91	10.1	11.7	27.2	30.1	32.9	36.2	38.6
20	7.43	8.26	9.59	10.9	12.4	28.4	31.4	34.2	37.6	40.0
21	8.03	8.90	10.3	11.6	13.2	29.6	32.7	35.5	38.9	41.4
22	8.64	9.54	11.0	12.3	14.0	30.8	33.9	36.8	40.3	42.8
23	9.26	10.2	11.7	13.1	14.8	32.0	35.2	38.1	41.6	44.2
24	9.89	10.9	12.4	13.8	15.7	33.2	36.4	39.4	43.0	45.6
25	10.5	11.5	13.1	14.6	16.5	34.4	37.7	40.6	44.3	46.9
26	11.2	12.2	13.8	15.4	17.3	35.6	38.9	41.9	45.6	48.3
27	11.8	12.9	14.6	16.2	18.1	36.7	40.1	43.2	47.0	49.6
28	12.5	13.6	15.3	16.9	18.9	37.9	41.3	44.5	48.3	51.0
29	13.1	14.3	16.0	17.7	19.8	39.1	42.6	45.7	49.6	52.3
30	13.8	15.0	16.8	18.5	20.6	40.3	43.8	47.0	50.9	53.7

EXAMPLE 6–5 In an automated process, a machine fills cans of coffee. If the average amount filled is different from what it should be, the machine may be adjusted to correct the mean. If the *variance* of the filling process is too high, however, the machine is out of control and needs to be repaired. Therefore, from time to time regular checks of the variance of the filling process are made. This is done by randomly sampling filled cans, measuring their amounts, and computing the sample variance. A random sample of 30 cans gives an estimate $s^2 = 18{,}540$. Give a 95% confidence interval for the population variance σ^2.

Figure 6–11 shows the appropriate chi-square distribution with $n - 1 = 29$ degrees of   *Solution*
freedom. From Table 6–2 we get, for df $= 29$, $\chi^2_{0.025} = 45.7$ and $\chi^2_{0.975} = 16.0$. Using
these values, we compute the confidence interval as follows:

$$\left[\frac{29(18,540)}{45.7}, \frac{29(18,540)}{16.0} \right] = [11,765, 33,604]$$

We can be 95% sure that the population variance is between 11,765 and 33,604.

The Template

In the workbook named Estimating Variance.xls there are sheets for computing confidence intervals for population variances when

1. The sample statistics are known.
2. The sample data are known.

Figure 6–12 shows the first sheet and Figure 6–13 shows the second. An assumption in both cases is that the population is normally distributed.

FIGURE 6–11   Values and Tail Areas of a Chi-Square Distribution with 29 Degrees of Freedom

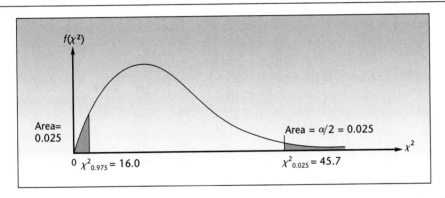

FIGURE 6–12   The Template for Estimating Population Variances [Estimating Variance.xls; Sheet: Sample Stats]

	A	B	C	D	E	F	G	H	I	J	K	L
1	Confidence Interval for Population Variance											
2						Assumption:						
3						The population is normally distributed						
4												
5		Sample Size	30		n		$1 - \alpha$		Confidence Interval			
6		Sample Variance	18.54		s^2		95%	[	11.7593 ,	33.5052	]	
7							95%	[	11.7593 ,	33.5052	]	
8							90%	[	12.6339 ,	30.3619	]	
9							80%	[	13.7553 ,	27.1989	]	
10												

FIGURE 6–13   **The Template for Estimating Population Variances**
[Estimating Variance.xls; Sheet: Sample Data]

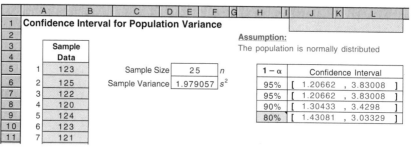

PROBLEMS

In the following problems, assume normal populations.

6–52. The service time in queues should not have a large variance; otherwise, the queue tends to build up. A bank regularly checks service time by its tellers to determine its variance. A random sample of 22 service times (in minutes) gives $s^2 = 8$. Give a 95% confidence interval for the variance of service time at the bank.

6–53. A sensitive measuring device should not have a large variance in the errors of measurements it makes. A random sample of 41 measurement errors gives $s^2 = 102$. Give a 99% confidence interval for the variance of measurement errors.

6–54. A random sample of 60 accounts gives a sample variance of 1,228. Give a 95% confidence interval for the variance of all accounts.

6–55. In problem 6–21, give a 99% confidence interval for the variance of the number of miles that may be driven on a tire.

6–56. In problem 6–25, give a 95% confidence interval for the variance of the population of transaction amounts.

6–57. In problem 6–26, give a 95% confidence interval for the variance of the caloric content of all 3-ounce cuts of top loin steak.

6–58. In problem 6–27, give a 95% confidence interval for the variance of the transit time for all goods.

6–6   Sample-Size Determination

One of the questions a statistician is most frequently asked before any actual sampling takes place is: "How large should my sample be?" From a *statistical* point of view, the best answer to this question is: "Get as large a sample as you can afford. If possible, 'sample' the entire population." If you need to know the mean or proportion of a population, and you can sample the entire population (i.e., carry out a census), you will have all the information and will know the parameter exactly. Clearly, this is better than any estimate. This, however, is unrealistic in most situations due to economic constraints, time constraints, and other limitations. "Get as large a sample as you can afford" is the best answer if we ignore all costs, because the larger the sample, the

FIGURE 6–14   Standard Error of a Statistic as a Function of Sample Size

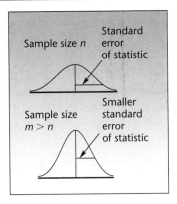

smaller the standard error of our statistic. The smaller the standard error, the less un-certainty with which we have to contend. This is demonstrated in Figure 6–14.

When the sampling budget is limited, the question often is how to find the *mini-mum* sample size that will satisfy some precision requirements. In such cases, you should explain to the designer of the study that he or she must first give you answers to the following three questions:

1. How close do you want your sample estimate to be to the unknown parameter? The answer to this question is denoted by B (for "bound").
2. What do you want the confidence level to be so that the distance between the estimate and the parameter is less than or equal to B?
3. The last, and often misunderstood, question that must be answered is: What is your estimate of the variance (or standard deviation) of the population in question?

Only after you have answers to all three questions can you specify the minimum re-quired sample size. Often the statistician is told: "How can I give you an estimate of the variance? I don't know. You are the statistician." In such cases, try to get from your client some idea about the variation in the population. If the population is approxi-mately normal and you can get 95% bounds on the values in the *population, divide the difference between the upper and lower bounds by* 4; this will give you a rough guess of σ. Or you may take a small, inexpensive *pilot* survey and estimate σ by the sample standard deviation. Once you have obtained the three required pieces of information, all you need to do is to substitute the answers into the appropriate formula that follows:

Minimum required sample size in estimating the population mean μ is

$$n = \frac{z_{\alpha/2}^2 \sigma^2}{B^2} \qquad\qquad (6\text{–}10)$$

> Minimum required sample size in estimating the population proportion p is
>
> $$n = \frac{z_{\alpha/2}^2 pq}{B^2} \qquad\qquad (6\text{--}11)$$

Equations 6–10 and 6–11 are derived from the formulas for the corresponding confidence intervals for these population parameters based on the normal distribution. In the case of the population mean, B is the half-width of a $(1 - \alpha)$ 100% confidence interval for μ, and therefore

$$B = z_{\alpha/2} \frac{\sigma}{\sqrt{n}} \qquad\qquad (6\text{--}12)$$

Equation 6–10 is the solution of equation 6–12 for the value of n. Note that B is the margin of error. We are solving for the minimum sample size for a given margin of error.

Equation 6–11, for the minimum required sample size in estimating the population proportion, is derived in a similar way. Note that the term pq in equation 6–11 acts as the population variance in equation 6–10. To use equation 6–11, we need a guess of p, the unknown population proportion. Any prior estimate of the parameter will do. When none is available, we may take a pilot sample, or—in the absence of any information—we use the value $p = 0.5$. This value maximizes pq and thus ensures us a minimum required sample size that will work for any value of p.

EXAMPLE 6–6 A market research firm wants to conduct a survey to estimate the average amount spent on entertainment by each person visiting a popular resort. The people who plan the survey would like to be able to determine the average amount spent by all people visiting the resort to within $120, with 95% confidence. From past operation of the resort, an estimate of the population standard deviation is $\sigma = \$400$. What is the minimum required sample size?

Solution Using equation 6–10, the minimum required sample size is

$$n = \frac{z_{\alpha/2}^2 \sigma^2}{B^2}$$

We know that $B = 120$, and σ^2 is estimated at $400^2 = 160{,}000$. Since we want 95% confidence, $z_{\alpha/2} = 1.96$. Using the equation, we get

$$n = \frac{(1.96)^2 160{,}000}{120^2} = 42.684$$

Therefore, the minimum required sample size is 43 people (we cannot sample 42.684 people, so we go to the next higher integer).

EXAMPLE 6–7 The manufacturer of a sports car wants to estimate the proportion of people in a given income bracket who are interested in the model. The company wants to know the population proportion p to within 0.10 with 99% confidence. Current company records indicate that the proportion p may be around 0.25. What is the minimum required sample size for this survey?

Using equation 6–11, we get

$$n = \frac{z_{\alpha/2}^2 pq}{B^2} = \frac{(2.576)^2(0.25)(0.75)}{0.10^2} = 124.42$$

The company should, therefore, obtain a random sample of at least 125 people. Note that a different guess of p would have resulted in a different sample size.

PROBLEMS

6–59. What is the required sample size for determining the proportion of defective items in a production process if the proportion is to be known to within 0.05 with 90% confidence? No guess as to the value of the population proportion is available.

6–60. How many test runs of an automobile are required for determining its average miles-per-gallon rating on the highway to within 2 miles per gallon with 95% confidence, if a guess is that the variance of the population of miles per gallon is about 100?

6–61. A company that conducts surveys of current jobs for executives wants to estimate the average salary of an executive at a given level to within $2,000 with 95% confidence. From previous surveys it is known that the variance of executive salaries is about 40,000,000. What is the minimum required sample size?

6–62. Find the minimum required sample size for estimating the average return on investments of a certain kind to within 0.5% per year with 95% confidence. The standard deviation of returns is believed to be 2% per year.

6–63. A company believes its market share is about 14%. Find the minimum required sample size for estimating the actual market share to within 5% with 90% confidence.

6–64. Find the minimum required sample size for estimating the average number of designer shirts sold per day to within 10 units with 90% confidence if the standard deviation of the number of shirts sold per day is about 50.

6–65. Find the minimum required sample size of accounts if the proportion of accounts in error is to be estimated to within 0.02 with 95% confidence. A rough guess of the proportion of accounts in error is 0.10.

6–7 The Templates

Optimizing Population Mean Estimates

Figure 6–15 shows the template that can be used for determining minimum sample size for estimating a population mean. Upon entering the three input data—confidence level desired, half-width (B) desired, and the population standard deviation—in the cells C5, C6, and C7, respectively, the minimum required sample size appears in cell C9.

Determining the Optimal Half-Width

Usually, the population standard deviation σ is not known for certain, and we would like to know how sensitive the minimum sample size is to changes in σ. In addition, there is no hard rule for deciding what the half-width B should be. Therefore, a

FIGURE 6–15   **The Template for Determining Minimum Sample Size**
[Sample Size.xls; Sheet: Population Mean]

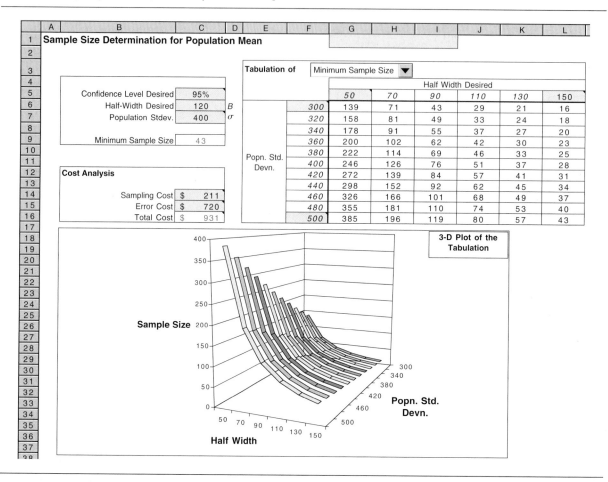

tabulation of the minimum sample size for various values of σ and *B* will help us to choose a suitable *B*. To create the table, enter the desired starting and ending values of σ in cells F6 and F16, respectively. Similarly, enter the desired starting and ending values for *B* in cells G5 and L5, respectively. The complete tabulation appears immediately. Note that the tabulation pertains to the confidence level entered in cell C5. If this confidence level is changed, the tabulation will be immediately updated.

To visualize the tabulated values a 3-D plot is created below the table. As can be seen from the plot, *the minimum sample size is more sensitive to the half-width* B *than the population standard deviation* σ. That is, when *B* decreases, the minimum sample size increases sharply. This sensitivity emphasizes the issue of how to choose *B*. The natural decision criterion is cost. When *B* is small, the possible error in the estimate is small and therefore the error cost is small. But a small *B* means large sample size, increasing the sampling cost. When *B* is large, the reverse is true. Thus *B* is a compromise between sampling cost and error cost. If cost data are available, an optimal *B* can be found. The template contains features that can be used to find optimal *B*.

The sampling cost formula is to be entered in cell C14. Usually, sampling cost involves a fixed cost that is independent of the sample size and a variable cost that increases linearly with the sample size. The fixed cost would include the cost of planning and organizing the sampling experiment. The variable cost would include the costs of selection, measurement, and recording of each sampled item. For example, if the fixed cost is \$125 and the variable cost is \$2 per sampled item, then the cost of sampling a total of 43 items will be \$125 + \$2*43 = \$211. In the template, we enter the formula "=125+2*C9" in cell C14, so that the sampling cost appears in cell C14. The instruction for entering the formula appears also in the comment attached to cell C14.

The error cost formula is entered in cell C15. In practice, error costs are more difficult to estimate than sampling costs. Usually, the error cost increases more rapidly than linearly with the amount of error. Often a quadratic formula is suitable for modeling the error cost. In the current case seen in the template, the formula "=0.05*C6^2" has been entered in cell C15. This means the error cost equals $0.05B^2$. Instructions for entering this formula appear in the comment attached to cell C15.

When proper cost formulas are entered in cells C14 and C15, cell C16 shows the total cost. It is possible to tabulate, in place of minimum sample size, the sampling cost, the error cost, or the total cost. You can select what you wish to tabulate using the drop down box in cell G3.

Using the Solver

By manually adjusting B, we can try to minimize the total cost. Another way to find the optimal B that minimizes the total cost is to use the Solver. Unprotect the sheet and select the Solver command under the Tools menu and click on the Solve button. If the formulas entered in cells C14 and C15 are realistic, the Solver will find a realistic optimal B, and a message saying that an optimal B has been found will appear in a dialog box. Select Keep Solver Solution and press the OK button. In the present case, the optimal B turns out to be 70.4.

For some combinations of sampling and error cost formulas, the Solver may not yield meaningful answers. For example, it may get stuck at a value of zero for B. In such cases, the manual method must be used. At times, it may be necessary to start the Solver with different initial values for B and then take the B value that yields the least total cost.

It must be noted that the total cost also depends on the confidence level (in cell C5) and the population standard deviation (in cell C7). If these values are changed then the total cost will change and so will the optimal B. The new optimum must be found once again, manually or using the Solver.

Optimizing Population Proportion Estimates

Figure 6–16 shows the template that can be used to determine the minimum sample size required for estimating a population proportion. This template is almost identical to the one in Figure 6–15, which is meant for estimating the population mean. The only difference is that instead of population standard deviation, we have population proportion in cell C7.

The tabulation shows that the minimum sample size increases with population proportion p until p reaches a value of 0.5 and then starts decreasing. Thus, the worst case occurs when p is 0.5.

The formula for error cost currently in cell C15 is "=40000*C6^2." This means the cost equals $40,000B^2$. Notice how different this formula looks compared to the formula $0.05B^2$ we saw in the case of estimating population mean. The coefficient

FIGURE 6–16   The Template for Determining Minimum Sample Size
[Sample Size.xls; Sheet: Population Proportion]

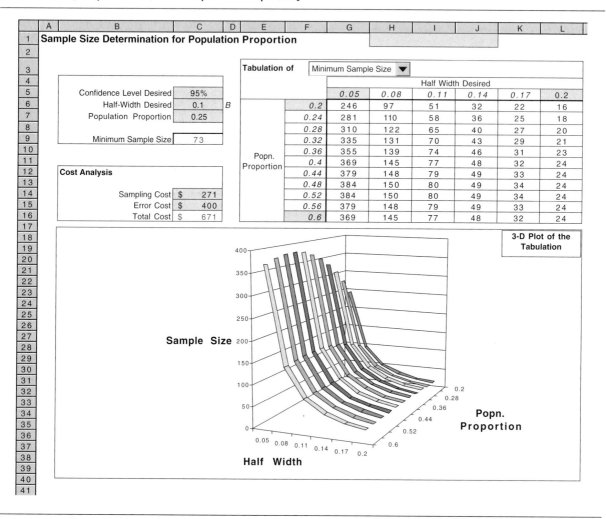

40,000 is much larger than 0.05. This difference arises because *B* is a much smaller number in the case of proportions. The formula for sampling cost currently entered in cell C14 is "=125+2*C9" (same as the previous case).

The optimal *B* that minimizes the total cost in cell C16 can be found using the Solver just as in the case of estimating population mean. Select the Solver command under the Tools menu and press the Solve button. In the current case, the Solver finds the optimal *B* to be 0.07472.

6–8   Summary and Review of Terms

In this chapter, we learned how to construct **confidence intervals** for population parameters. We saw how confidence intervals depend on the sampling distributions of the statistics used as estimators. We also encountered two new sampling distributions: the ***t* distribution,** used in estimating the population mean when the population standard deviation is unknown, and the **chi-square distribution,** used in estimating the population variance. The use of either distribution assumes a normal population. We saw how the new distributions, as well as the normal distribution, allow us to construct confidence intervals for population parameters. We saw how to determine the minimum required sample size for estimation.

ADDITIONAL PROBLEMS

6–66. Tradepoint, the electronic market set up to compete with the London Stock Exchange, is losing on average \$32,000 per day.[14] A potential long-term financial partner for Tradepoint needs a confidence interval for the actual (population) average daily loss, in order to decide whether future prospects for this venture may be profitable. In particular, this potential partner wants to be confident that the true average loss at this period is not over \$35,000 per day, which it would consider hopeless. Assume the \$32,000 figure given above is based on a random sample of 10 trading days, assume a normal distribution for daily loss, and assume a standard deviation of $s =$ \$6,000. Construct a 95% confidence interval for the average daily loss in this period. What decision should the potential partner make?

6–67. Landings and takeoffs[15] at Schiphol, Holland, per month are (in 1,000s) as follows:

   26, 19, 27, 30, 18, 17, 21, 28, 18, 26, 19, 20, 23, 18, 25, 29, 30, 26, 24, 22, 31, 18, 30, 19

Assume a random sample of months. Give a 95% confidence interval for the average monthly number of takeoffs and landings.

6–68. Thomas Stanley, who surveyed 200 millionaires in the United States for his book *The Millionaire Mind,* found that those in that bracket had an average net worth of \$9.2 million. The sample variance was \$1.3 million. Assuming that the surveyed subjects are a random sample of U.S. millionaires, give a 99% confidence interval for the average net worth of U.S. millionaires.

6–69. The Java computer language, developed by Sun Microsystems, has the advantage that its programs can run on types of hardware ranging from mainframe computers all the way down to handheld computing devices or even smart phones. A test of 100 randomly selected programmers revealed that 71 preferred Java to their other most used computer languages. Construct a 95% confidence interval for the proportion of all programmers in the population from which the sample was selected who prefer Java.

6–70. According to an article in *Fortune,* 39% of the companies in a survey of 1,000 now offer signing bonuses.[16] Give a 95% confidence interval.

6–71. Software companies in Silicon Valley now like to hire graduates of Stanford University's section leaders of the course CS 106, since the leaders know the best students.[17] Estimate the average number of students these "headhunter" hires can attract for a company from the following numbers per hire:

   12, 15, 9, 8, 17, 10, 11, 12, 16, 13, 10, 9, 13, 18, 8, 7, 10, 8, 5, 14, 3, 16, 20

---

[14]From "Business This Week," *The Economist,* June 7, 1997, p. 27.

[15]M. Dubois, "Holland's Schiphol Sees Room to Grow Out in North Sea," *The Wall Street Journal,* June 16, 1997, p. 1.

[16]From W. Woods, "Rising Wages," *Fortune,* July 7, 1997, p. 13.

[17]J. Aley, "The Heart of Silicon Valley," *Fortune,* July 7, 1997, p. 39.

6–72.   Finjan is a company that makes a new security product designed to protect software written in the Java programming language against hostile interference.[18] The market for this program materialized recently when Princeton University experts showed that a hacker could write misleading "applets" that fool Java's built-in security rules. If, in 430 trials, the system was fooled 47 times, give a 95% confidence interval for p = probability of successfully fooling the machine.

6–73.   Sony's new optical disk system prototype tested and claimed to be able to record an average of 1.2 hours of high-definition TV.[19] Assume $n = 10$ trials and $\sigma = 0.2$ hour. Give a 90% confidence interval.

6–74.   The average customer of the Halifax bank in Britain (of whom there are 7.6 million) received $3,600 when the institution changed from a building society to a bank.[20] If this is based on a random sample of 500 customers with standard deviation = $800, give a 95% confidence interval for the average amount paid to any of the 7.6 million bank customers.

6–75.   FinAid is a new, free website that helps people obtain information on 180,000 college tuition aid awards. A random sample of 500 such awards revealed that 368 were granted for reasons other than financial need. They were based on the applicant's qualifications, interests, and other variables.[21] Construct a 95% confidence interval for the proportion of all awards on this service made for reasons other than financial need.

6–76.   A Booz-Allen & Hamilton survey[22] found that

> The cost of an average payment transaction on the Internet was 13 cents. That compares with 26 cents for a proprietary personal computer banking service using the bank's own software, 54 cents for a telephone banking service, and $1.08 for a transaction conducted in a traditional bank branch. This means costs for Internet banking run at only 15–20 percent of income, compared with an average cost:income ratio for the banking industry of about 60 percent.

   a. If the Internet cost had $n = 100$ and standard deviation = 3 cents, give a 95% confidence interval.

   b. If the PC software cost had $n = 100$ and standard deviation = 8 cents, give a 95% confidence interval.

   c. If the telephone service had $n = 100$ and standard deviation = 10 cents, give a 95% confidence interval.

6–77.   A survey by Booz-Allen & Hamilton[23] found that the average bank spends $25,000 to set up a website. If the sample size was 80 banks, randomly chosen worldwide, and the standard deviation was $8,000, give a 95% confidence interval for μ.

6–78.   A small British computer-game firm, Eidos Interactive PLC, stunned the U.S.- and Japan-dominated market for computer games in June 1997, when it introduced

---

[18]E. Brown, "Finjan," *Fortune,* July 7, 1997, p. 52.

[19]S. Brull, "Digital Video's Next Generation," *Business Week,* June 16, 1997, p. 68.

[20]From "Feeling Too Good," *The Economist,* June 7, 1997, p. 39.

[21]M. Mangalindan, "Virtual Help with the Real College-Tuition Crunch," The Money Report, *International Herald Tribune,* June 14–16, 1997, p. B1.

[22]P. Taylor, "Banking without Branches," *Financial Times,* Business On-line, June 11, 1997, p. 7.

[23]Ibid.

Lara Croft, an Indiana Jones-like adventuress.[24] The successful product took two years to develop. One problem was whether Lara should have a swinging ponytail, which was decided after taking a poll. If in a random sample of 200 computer-game enthusiasts, 161 thought she should have a swinging ponytail (a computer programmer's nightmare to design), construct a 95% confidence interval for the proportion in this market. If the decision to incur the high additional programming cost was to be made if $p > 0.05$, was the right decision made (when Eidos went ahead with the ponytail)?

6–79. A company's book value is defined as the dollar value of its assets minus its liabilities. According to an article,[25] Wall Street acquisitions of investment banking firms have become very expensive recently. In particular, the article reports that the buyout of Robertson, Stephens & Co. by Bank America Corp. for $540 million amounts to five times the acquired firm's book value. In 10 recent acquisitions of similar firms, the average price-to-book-value ratio was 2.6, and the standard deviation of this ratio was 2.1. Construct a 90% confidence interval for the price-to-book-value ratio of such buyouts.

6–80. According to a survey published in the *Financial Times,* 56% of executives at Britain's top 500 companies are less willing than they had been five years ago to sacrifice their family lifestyle for their career. If the survey consisted of a random sample of 40 executives, give a 95% confidence interval for the proportion of executives less willing to sacrifice their family lifestyle.

6–81. Fifty years after the birth of duty-free shopping at international airports and border-crossing facilities, the European commission announced plans to end this form of business. A study by Cranfield University was carried out to estimate the average percentage rise in airline landing charges that would result as airlines try to make up for the loss of on-board duty-free shopping revenues. The study found the average increase to be 60%.[26] If this was based on a random sample of 22 international flights and the standard deviation of increase was 25%, give a 90% confidence interval for the average increase.

6–82. When NYSE, NASDAQ, and the British government bonds market were planning to change prices of shares and bonds from powers of 2, such as 1/2, 1/4, 1/8, 1/16, 1/32, to decimals (hence $1/32 = 0.03125$), they decided to run a test. If the test run of trading rooms using the new system revealed that 80% of the traders preferred the decimal system and the sample size was 200, give a 95% confidence interval for the percentage of all traders who will prefer the new system.

6–83. A survey of 5,250 business travelers worldwide conducted by OAG Business Travel Lifestyle indicated that 91% of business travelers consider legroom the most important in-flight feature. (Angle of seat recline and food service were second and third, respectively.)[27] Give a 95% confidence interval for the proportion of all business travelers who consider legroom the most important feature.

---

[24]From R. Strassel, "British Firms Face Struggle to Remain Video Game Force," *The Wall Street Journal* (Europe), June 11, 1997, p. 1.

[25]From A. Ragharan and P. McGeehan, "Heard on the Street," "Wall Street Takeovers Get Pricier," *The Wall Street Journal* (Europe), June 11, 1997, p. 16.

[26]From J. Brown, "Task Force to Fight Brussels Plan for End of Duty-Free Sales," *Financial Times,* June 11, 1997, p. 6.

[27]From *USA Today,* June 6, 1997, p. 12B, Money section.

6–84. In 1997, the *International Herald Tribune*, in collaboration with Nokia, launched the E-funds, a way of allowing investors to receive updates on international funds via e-mail.[28] If in a test run of the new service 550 international investors, randomly selected, revealed that 126 were seriously considering subscribing to E-funds, give a 95% confidence interval for the proportion of all investors in this market who will be interested.

6–85. The French company G C Tech introduced in 1997 the "virtual wallet" technology for making payments for goods and services by using a computer.[29] The technology was tested on 1,000 French Visa cardholders. If 86 of these people indicated complete satisfaction with the new technological system, construct a 99% confidence interval for the proportion of all French Visa cardholders who are satisfied.

6–86. An estimate of the average length of pins produced by an automatic lathe is wanted to within 0.002 inch with a 95% confidence level. σ is guessed to be 0.015 inch.

    a. What is the minimum sample size?

    b. If the value of σ may be anywhere between 0.010 and 0.020 inch, tabulate the minimum sample size required for σ values from 0.010 to 0.020 inch.

    c. If the cost of sampling and testing n pins is $(25 + 6n)$ dollars, tabulate the costs for σ values from 0.010 to 0.020 inch.

6–87. Wells Fargo Bank, based in San Francisco, offered the option of applying for a loan over the Internet in 1997. If a random sample of 200 test runs of the service reveal an average of 8 minutes to fill in the electronic application and standard deviation = 3 minutes, construct a 75% confidence interval for μ.

6–88. It is desired to estimate the percentage defective in a lot of pins supplied by a vendor to within 1% with a 90% confidence level. The actual percentage defective is guessed to be 4%.

    a. What is the minimum sample size?

    b. If the actual percentage defective may be anywhere between 3% and 6%, tabulate the minimum sample size required for actual percentage defective from 3% to 6%.

    c. If the cost of sampling and testing n pins is $(25 + 6n)$ dollars, tabulate the costs for the same percentage defective range as in part (b).

6–89. The lengths of pins produced by an automatic lathe are normally distributed. A random sample of 20 pins gives a sample mean of 0.992 inch and a sample standard deviation of 0.013 inch.

    a. Give a 95% confidence interval for the average lengths of all pins produced.

    b. Give a 99% confidence interval for the average lengths of all pins produced.

6–90. You take a random sample of 100 pins from the lot supplied by the vendor and test them. You find 8 of them defective. What is the 95% confidence interval for percentage defective in the lot?

---

[28] *International Herald Tribune,* June 3, 1997, p. 7.

[29] From G. Nairn, "French Pioneers Lead the Way," Doing Financial Business On-Line Guide, *Financial Times,* June 11, 1997, p. 23.

Presidential Polling

A company wants to conduct a telephone survey of randomly selected voters to estimate the proportion of voters who favor a particular candidate in a presidential election, to within 2% error with 95% confidence. It is guessed that the proportion is 53%.

1. What is the required minimum sample size?

2. The project manager assigned to the survey is not sure about the actual proportion or about the 2% error limit. The proportion may be anywhere from 40% to 60%. Construct a table for the minimum sample size required with half-width ranging from 1% to 3% and actual proportion ranging from 40% to 60%.

3. Inspect the table produced in question 2 above. Comment on the relative sensitivity of the minimum sample size to the actual proportion and to the desired half-width.

4. At what value of the actual proportion is the required sample size the maximum?

5. The cost of polling includes a fixed cost of $425 and a variable cost of $1.20 per person sampled, thus the cost of sampling n voters is $(425 + 1.20n)$. Tabulate the cost for range of values as in question 2 above.

6. A competitor of the company that had announced results to within ±3% with 95% confidence has started to announce results to within ±2% with 95% confidence. The project manager wants to go one better by improving the company's estimate to be within ±1% with 95% confidence. What would you tell the manager?

7

Hypothesis Testing

7–1 Using Statistics

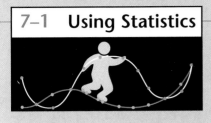

On June 18, 1964, a woman was robbed while walking home along an alley in San Pedro, California. Some time later, police arrested Janet Collins and charged her with the robbery. The interesting thing about this case of petty crime is that the prosecution had *no* direct evidence against the defendant. Janet Collins was convicted of robbery on purely statistical grounds.

The case, *People v. Collins,* drew much attention because of its use of probability—or, rather, what was perceived as a probability—in determining guilt. An instructor of mathematics at a local college was brought in by the prosecution and testified as an expert witness in the trial. The instructor "calculated the probability" that the defendant was a person *other* than the one who committed the crime as 1 in 12,000,000. This led the jury to convict the defendant.

The Supreme Court of California later reversed the guilty verdict against Janet Collins when it was shown that the method of calculating the probability was incorrect. The mathematics instructor had made some very serious errors.[1]

Despite the erroneous procedure used in deriving the probability, and the justified reversal of the conviction by the Supreme Court of California, the *Collins* case serves as an excellent analogy for statistical hypothesis testing. Under the U.S. legal system, the accused is assumed innocent until proved guilty "beyond a reasonable doubt." We will call this the *null hypothesis*—the hypothesis that the accused is *innocent*. We will hold the null hypothesis as true until a time when we can prove, beyond a reasonable doubt, that it is false and that the *alternative hypothesis*—the hypothesis that the accused is guilty—is true. We want to have a small probability (preferably *zero*) of convicting an innocent person, that is, of rejecting a null hypothesis when the null hypothesis is actually true.

In the *Collins* case, the prosecution claimed that the accused was guilty since, otherwise, an event with a very small probability had just been observed. The argument was that if Collins were *not* guilty, then another woman fitting her exact characteristics had committed the crime. According to the prosecution, the probability of this event was 1/12,000,000, and since the probability was so small, Collins was very likely the person who committed the robbery.

The *Collins* case illustrates **hypothesis testing,** an important application of statistics. A *thesis* is something that has been proven to be true. A *hypothesis* is something that has not yet been proven to be true. Hypothesis testing is the process of determining whether or not a given hypothesis is true. Most of the time, a hypothesis is tested through statistical means that use the concepts we learned in previous chapters.

The Null Hypothesis

The first step in a hypothesis test is to formalize it by specifying the *null hypothesis.*

---

[1]The instructor *multiplied* the probabilities of the separate events comprising the reported description of the robber: the event that a woman has blond hair, the event that she drives a yellow car, the event that she is seen with an African-American man, the event that the man has a beard. Recall that the probability of the intersection of several events is equal to the product of the probabilities of the separate events *only* if the events are independent. In this case, there was no reason to believe that the events were independent. There were also some questions about how the separate "probabilities" were actually derived since they were presented by the instructor with no apparent justification. See W. Fairley and F. Mosteller, "A Conversation about Collins," *University of Chicago Law Review* 41, no. 2 (Winter 1974), pp. 242–53.

A **null hypothesis** is an assertion about the value of a population parameter. It is an assertion that we hold as true unless we have sufficient statistical evidence to conclude otherwise.

For example, a null hypothesis might assert that the population mean is equal to 100. Unless we obtain sufficient evidence that it is not 100, we will accept it as 100. We write the null hypothesis compactly as

$$H_0: \mu = 100$$

where the symbol H_0 denotes the null hypothesis.

The **alternative hypothesis** is the *negation* of the null hypothesis.

For the null hypothesis $\mu = 100$, the alternative hypothesis is $\mu \neq 100$. We will write it as

$$H_1: \mu \neq 100$$

using the symbol H_1 to denote the alternative hypothesis.[2] Because the null and alternative hypotheses assert exactly opposite statements, only one of them can be true. Rejecting one is equivalent to accepting the other.

Hypotheses about other parameters such as population proportion or population variance are also possible. In addition, a hypothesis may assert that the parameter in question is at least or at most some value. For example, the null hypothesis may assert that the population proportion p is *at least* 40%. In this case, the null and alternative hypotheses are

$$H_0: p \geq 40\%$$
$$H_1: p < 40\%$$

Yet another example is where the null hypothesis asserts that the population variance is *at most* 50. In this case

$$H_0: \sigma^2 \leq 50$$
$$H_1: \sigma^2 > 50$$

Note that in all cases the equal to sign appears in the null hypothesis.

Although the idea of a null hypothesis is simple, determining what the null hypothesis should be in a given situation may be difficult. It is important to be clear about exactly what the null hypothesis is, or else the test is meaningless. Let us consider an example. Imagine an automatic bottling machine that fills two-liter bottles with cola. The amount of cola filled in every bottle on the average is expected to be, of course, 2 liters (or 2,000 cm³). Suppose a consumer advocate suspects that the average amount of cola is less than 2,000 cm³ and wants to test it. What should be the null hypothesis in this case?

The fact that the company sells cola in bottles labeled 2 liters implies a claim by the company that on the average each bottle contains *at least* 2,000 cm³. We have to accept this claim as true unless we have sufficient evidence against it. Thus the claim is the null hypothesis, and the consumer advocate's suspicion is the alternative hypothesis. We specify the hypotheses as

---

[2]In some books, the symbol H_a is used for alternative hypothesis.

$$H_0: \mu \geq 2{,}000 \text{ cm}^3$$
$$H_1: \mu < 2{,}000 \text{ cm}^3$$

As in the above example, the null hypothesis is often a claim made by someone. The alternative hypothesis is the suspicion that someone else has about that claim. But variations are possible. To illustrate another possibility, suppose the bottling company does not make any claim at all, but we simply wish to demonstrate that the average amount filled is less than $2{,}000 \text{ cm}^3$. In this case, *what we wish to demonstrate is what we write as the alternative hypothesis,* and its negation becomes the null hypothesis.

One way to ensure what the null hypothesis should be is to note that *if the null hypothesis is true, then no corrective action would be necessary. If the alternative hypothesis is true, then some corrective action would be necessary.* In the case of the consumer advocate, if the average amount of cola is greater than or equal to $2{,}000 \text{ cm}^3$, no corrective action is needed, and therefore "$\mu \geq 2{,}000 \text{ cm}^3$" should be the null hypothesis. If the average amount is less than $2{,}000 \text{ cm}^3$, the company has to halt the bottling operation and set right the error.

Now let us take another look at the same bottling operation. Assume that consumers are satisfied with the bottles. But suppose the owner of the bottling company suspects that the machine is filling more than $2{,}000 \text{ cm}^3$ on the average and is thus wasting cola. From the owner's point of view, no corrective action is necessary if the average is less than[3] or equal to $2{,}000 \text{ cm}^3$, and therefore that should be the null hypothesis. In this case we have

$$H_0: \mu \leq 2{,}000 \text{ cm}^3$$
$$H_1: \mu > 2{,}000 \text{ cm}^3$$

There is a third point of view regarding the bottling operation, and that is the engineering point of view. Suppose the engineer in charge of the accuracy of the machine wants to test the average amount filled. The engineer will have to take corrective action when the average is either more than or less than $2{,}000 \text{ cm}^3$. Only when the average equals $2{,}000 \text{ cm}^3$, no corrective action is necessary. In this case we have

$$H_0: \mu = 2{,}000 \text{ cm}^3$$
$$H_1: \mu \neq 2{,}000 \text{ cm}^3$$

As the bottling example indicates, there are three possible cases for the null hypothesis, involving $\geq$, $\leq$, and $=$ relationships. The exact null hypothesis should be finalized before any evidence is gathered, or the test will not be valid. Formulating the null and alternative hypotheses at one's convenience after collecting and looking at the evidence is unethical, and is referred to as **data snooping.**

A vendor claims that his company fills any accepted order, on the average, in at most six working days. You suspect that the average is greater than six working days and want to test the claim. How will you set up the null and alternative hypotheses?

EXAMPLE 7–1

The claim is the null hypothesis and the suspicion is the alternative hypothesis. Thus, with μ denoting the average time to fill an order,

Solution

---

[3]It is unethical and illegal to sell less than 2 liters as 2 liters. Just to make our point about the null hypothesis, let us overlook that fact.

$$H_0: \mu \le 6 \text{ days}$$
$$H_1: \mu > 6 \text{ days}$$

EXAMPLE 7–2   A manufacturer of golf balls claims that the variance of the weights of the company's golf balls is controlled to within 0.0028 oz². If you wish to test this claim, how will you set up the null and alternative hypotheses?

Solution   The claim is the null hypothesis. Thus, with σ^2 denoting the variance,

$$H_0: \sigma^2 \le 0.0028 \text{ oz}^2$$
$$H_1: \sigma^2 > 0.0028 \text{ oz}^2$$

EXAMPLE 7–3   It is claimed that at least 20% of the visitors to a particular commercial website where an electronic product is sold end up ordering the product. If you wish to test this claim, how will you set up the null and alternative hypotheses?

Solution   With p denoting the proportion of visitors ordering the product,

$$H_0: p \ge 0.20$$
$$H_1: p < 0.20$$

PROBLEMS

7–1.   A pharmaceutical company claims that four out of five doctors prescribe the pain medicine it produces. If you wish to test this claim, how would you set up the null and alternative hypotheses?

7–2.   A medicine is effective only if the concentration of a certain chemical in it is at least 200 parts per million (ppm). At the same time, the medicine would produce an undesirable side effect if the concentration of the same chemical exceeds 200 ppm. How would you set up the null and alternative hypotheses to test the concentration of the chemical in the medicine?

7–3.   It is found that Web surfers will lose interest in a Web page if downloading takes more than 12 seconds at 28K baud rate. If you wish to test the effectiveness of a newly designed Web page in regard to its download time, how will you set up the null and alternative hypotheses?

7–4.   It is claimed that the average cost of a traditional open-heart surgery is $49,160.[4] If you suspect that the claim exaggerates the cost, how would you set up the null and alternative hypotheses?

7–5.   During the sharp increase in gasoline prices in the summer of the year 2000, it was claimed by oil companies that the average price of unleaded gasoline with minimum octane rating of 89 in the Midwest was not more than $1.78. If you want to test this claim, how would you set up the null and alternative hypotheses?

---

[4] *Forbes,* December 11, 2000, p. 182.

7–2   The Concepts of Hypothesis Testing

We said that a null hypothesis is held as true unless there is sufficient evidence against it. When can we say that we have sufficient evidence against it and thus reject it? This is an important and difficult question. Before we can answer it we have to understand several preliminary concepts.

Evidence Gathering

After the null and alternative hypotheses are spelled out, the next step is to gather evidence. The best evidence is, of course, data that leave no uncertainty at all. If we could measure the whole population and calculate the exact value of the population parameter in question, we would have perfect evidence. Such evidence is perfect in that we can check the null hypothesis against it and be 100% confident in our conclusion that the null hypothesis is or is not true. For example, in the bottling example, if we can measure the contents of *all* the bottles coming out of the machine and calculate the population mean, then we can conclude that the null hypothesis is true and be 100% confident in our conclusion. Alas, it is not possible to get our hands on *all* the bottles coming out of the machine. For one thing, not all bottles have been produced yet. Even if they had been, the time, effort, and cost of measuring them all would be too large to be practicable. This is true of many situations. The population is either not fully produced yet or, if it is, it is too large for us to measure all of its elements. In all such cases, the evidence is gathered from a random sample of the population. In the rest of this chapter, unless otherwise specified, the evidence is from a random sample.

An important limitation of making inferences from sample data is that we cannot be 100% confident about it. How confident we can be depends on the sample size and parameters such as the population variance. In view of this fact, the sampling experiment for evidence gathering must be carefully designed. Among other considerations, the sample size needs to be large enough to yield a desired confidence level and small enough to contain the cost. We will see more details of sample size determination later in this chapter.

Type I and Type II Errors

In our professional and personal lives we often have to make an accept–reject type of decision based on incomplete data. An inspector has to accept or reject a batch of parts supplied by a vendor, usually based on test results of a random sample. A recruiter has to accept or reject a job applicant, usually based on evidence gathered from a résumé and interview. A bank manager has to accept or reject a loan application, usually based on financial data on the application. A person who is single has to accept or reject a suitor's proposal of marriage, perhaps based on the experiences with the suitor. A car buyer has to buy or not a car, usually based on a test drive. As long as such decisions are made based on evidence that does not provide 100% confidence, there will be chances for error. No error is committed when a good prospect is accepted or a bad one is rejected. But there is a small chance that a bad prospect is accepted or a good one is rejected. Of course, we would like to minimize the chances of such errors.

In the context of statistical hypothesis testing, rejecting a true null hypothesis is known as a **type I error** and accepting[5] a false null hypothesis is known as a **type II**

---

[5]Later we will see that "not rejecting" is a more accurate term than "accepting."

TABLE 7–1   Instances of Type I and Type II Errors

	H_0 True	H_0 False
Accept H_0	No error	Type II error
Reject H_0	Type I error	No error

error. (Unfortunately, these names are unimaginative and nondescriptive. Because they are nondescriptive, you have to memorize which is which.) Table 7–1 shows the instances of type I and type II errors.

Let us see how we can minimize the chances of type I and type II errors. Is it possible, even with imperfect sample evidence, to reduce the probability of type I error all the way down to zero? The answer is yes. Just accept the null hypothesis, no matter what the evidence is. Since you will never reject any null hypothesis, you will never reject a true null hypothesis and thus you will never commit a type I error! We immediately see that this would be foolish. Why? If we always accept a null hypothesis, then given a false null hypothesis, no matter how wrong it is, we are sure to accept it. In other words, our probability of committing a type II error will be 1. Similarly, it would be foolish to reduce the probability of type II error all the way to zero by always rejecting a null hypothesis, for we would then reject every true null hypothesis, no matter how right it is. Our probability of type I error will be 1.

The lesson is that we should not try to completely avoid either type of error. We should plan, organize, and settle for some small, optimal probability of each type of error. Before we can address this issue, we need to learn a few more concepts.

The p-*Value*

Suppose the null and alternative hypotheses are

$$H_0: \mu \geq 1,000$$
$$H_1: \mu < 1,000$$

A random sample of size 30 yields a sample mean of only 999. Because the sample mean is less than 1,000, the evidence goes against the null hypothesis (H_0). Can we reject H_0 based on this evidence? Immediately we realize the dilemma. If we reject it, there is some chance that we might be committing a type I error, and if we accept it, there is some chance that we might be committing a type II error. A natural question to ask at this situation is, What is the probability that H_0 can still be true despite the evidence? The question asks for the "credibility" of H_0 in light of unfavorable evidence. Unfortunately, due to mathematical complexities, it is not possible to compute the probability that H_0 is true. We therefore settle for a question that comes very close. Recall that $H_0: \mu \geq 1,000$. We ask,

> When the actual $\mu = 1,000$, and with sample size 30, what is the probability of getting a sample mean that is less than or equal to 999?

The answer to this question is then taken as the "credibility rating" of H_0. Study the question carefully. There are two aspects to note:

1. The question asks for the probability of the evidence being as unfavorable or more unfavorable to H_0. The reason is that in the case of continuous

distributions, probabilities can be calculated only for a range of values. Here we pick a range for the sample mean that disfavors H_0, namely, less than or equal to 999.

2. The condition assumed is $\mu = 1,000$, although H_0 states $\mu \geq 1,000$. The reason for assuming $\mu = 1,000$ is that *it gives the most benefit of doubt to H_0*. If we assume $\mu = 1,001$, for instance, the probability of the sample mean being less than or equal to 999 will only be smaller, and H_0 will only have less credibility. Thus the assumption $\mu = 1,000$ gives the maximum credibility to H_0.

Suppose the answer to the question is 26%. That is, there is a 26% chance for a sample of size 30 to yield a sample mean less than or equal to 999 when the actual $\mu = 1,000$. Statisticians call this 26% the **p-value.** As mentioned before, the p-value is a kind of "credibility rating" of H_0 in light of the evidence. A p-value of zero means H_0 is certainly false and a p-value of 1 means that H_0 is certainly true. A p-value of 26% means that there is *roughly*[6] 26% probability that H_0 is true, despite the evidence. Conversely, we can be roughly 74% confident that H_0 is false in light of the evidence. The implication is that if we reject H_0, then there is about a 74% chance that we are doing the right thing, and about a 26% chance that we are committing a type I error. The formal definition of the p-value follows:

> Given a null hypothesis and sample evidence with sample size *n*, the **p-value** is the probability of getting a sample evidence with the same *n* that is equally or more unfavorable to the null hypothesis while the null hypothesis is actually true. The p-value is calculated giving the null hypothesis the maximum benefit of doubt.

Most people in most circumstances would consider a 26% chance of committing a type I error to be too high and would not reject H_0. That is understandable. Now consider another scenario where the sample mean was 998 rather than 999. Here the evidence is more unfavorable to the null hypothesis. Hence there will be less credibility to H_0 and the p-value will be smaller. Suppose the new p-value is 2%, meaning that there is only about 2% chance that H_0 is true. Can we reject H_0 now? We clearly see a need for a *policy* for rejecting H_0 based on p-value. Let us see the most common policy.

The Significance Level

The most common policy in statistical hypothesis testing is to establish a **significance level,** denoted by α, and to reject H_0 when the p-value falls below it. When this policy is followed, one can be sure that the maximum probability of type I error is α.

Policy: When the p-value is less than α, reject H_0.

The standard values for α are 10%, 5%, and 1%. Suppose α is set at 5%. This means that whenever the p-value is less than 5%, H_0 will be rejected. In the preceding example, for a sample mean of 999 the p-value was 26%, and H_0 will not be rejected. For a sample mean of 998 the p-value was 2%, which has fallen below $\alpha = 5\%$. Hence H_0 will be rejected.

---

[6]As already mentioned, we don't know the exact probability that H_0 is true.

Let us see in more detail the implications of using a significance level α for rejecting a null hypothesis. The first thing to note is that *if we do not reject H_0, this does not prove that H_0 is true.* For example, if $\alpha = 5\%$ and the *p*-value $= 6\%$, we will not reject H_0. But there is only about 6% chance that H_0 is true, which is hardly proof that H_0 is true. It may very well be that H_0 is false and by not rejecting it, we are committing a type II error. For this reason, under these circumstances we should say "We cannot reject H_0 at an α of 5%" rather than "We accept H_0."

The second thing to note is that α is the maximum probability of type I error we set for ourselves. Since α is the maximum *p*-value at which we reject H_0, it is the maximum probability of committing a type I error. In other words, setting $\alpha = 5\%$ means that we are willing to put up with up to 5% chance of committing a type I error.

The third thing to note is that the selected value of α indirectly determines the probability of type II error as well. Consider the case of setting $\alpha = 0$. Although this may appear good because it reduces the probability of type I error to zero, this corresponds to the foolish case we already discussed: never rejecting H_0. Every H_0, no matter how wrong it is, is accepted and thus the probability of type II error becomes 1. To decrease the probability of type II error we have to increase α. In general, *other things remaining the same, increasing the value of α will decrease the probability of type II error.* This should be intuitively obvious. For example, increasing α from 5% to 10% means that in those instances with a *p*-value in the range 5% to 10% the H_0 that would not have been rejected before would now be rejected. Thus, some cases of false H_0 that escaped rejection before may not escape now. As a result, the probability of type II error will decrease.

Figure 7–1 is a graph of the probability of type II error versus α, for a case where H_0: $\mu \geq 1,000$, the evidence is from a sample of size 30, and the probability of type II error is calculated for the case $\mu = 994$. Notice how the probability of type II error decreases as α increases. That the probability of type II error decreases is good news.

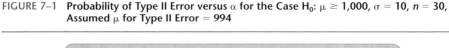

FIGURE 7–1   Probability of Type II Error versus α for the Case H_0: $\mu \geq 1,000$, $\sigma = 10$, $n = 30$, Assumed μ for Type II Error = 994

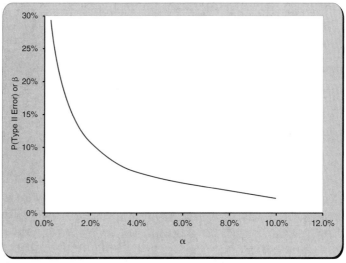

But as α increases, the probability of type I error increases. That is bad news. This brings out the important compromise between type I and type II errors. If we set a low value for α, we enjoy a low probability of type I error but suffer a high probability of type II error; if we set a high value for α, we will suffer a high probability of type I error but enjoy a low probability of type II error. Finding an optimal α is a difficult task. We will address the difficulties in the next subsection.

Our final note about α is the meaning of $(1 - \alpha)$. If we set $\alpha = 5\%$, then $(1 - \alpha) = 95\%$ is the minimum **confidence level** that we set in order to reject H_0. In other words, we want to be *at least* 95% confident that H_0 is false before we reject it. This concept of confidence level is the same that we saw in the previous chapter. It should explain why we use the symbol α for the significance level.

Optimal α and the Compromise between Type I and Type II Errors

Setting the value of α affects both type I and type II error probabilities as seen in Figure 7–1. But this figure is only one snapshot of a much bigger picture. In the figure, the type II error probability corresponds to the case where the actual $\mu = 994$. But the actual μ can be any one of an infinite number of possible values. For each one of those values, the graph will be different. In addition, the graph is only for a sample size of 30. When the sample size changes, so will the curve. This is the first difficulty in trying to find an optimal α.

Moreover, we note that selecting a value for α is a question of compromise between type I and type II error probabilities. To arrive at a fair compromise we should know the cost of each type of error. Most of the time the costs are difficult to estimate since they depend, among other things, on the unknown actual value of the parameter being tested. Thus, arriving at a "calculated" optimal value for α is impractical. Instead, we follow an intuitive approach of assigning one of the three standard values, 1%, 5%, and 10%, to α.

In the intuitive approach, we try to estimate the relative costs of the two types of errors. For example, suppose we are testing the average tensile strength of a large batch of bolts produced by a machine to see if it is above the minimum specified. Here type I error will result in rejecting a good batch of bolts and the cost of the error is roughly equal to the cost of the batch of bolts. Type II error will result in accepting a bad batch of bolts and its cost can be high or low depending on how the bolts are used. If the bolts are used to hold together a structure, then the cost is high because defective bolts can result in the collapse of the structure, causing great damage. In this case, we should strive to reduce the probability of type II error more than that of type I error. *In such cases where type II error is more costly, we keep a large value for α, namely, 10%.* On the other hand, if the bolts are used to secure the lids on trash cans, then the cost of type II error is not high and we should strive to reduce the probability of type I error more than that of type II error. *In such cases where type I error is more costly, we keep a small value for α, namely, 1%.*

Then there are cases where we are not able to determine which type of error is more costly. *If the costs are roughly equal, or if we have not much knowledge about the relative costs of the two types of errors, then we keep $\alpha = 5\%$.*

β and Power

The symbol used for the probability of type II error is β. It should be noted that β depends on the actual value of the parameter being tested, the sample size, and α. Let us see exactly how it depends. In the example plotted in Figure 7–1, if the actual μ is 993 rather than 994, it would make H_0 "even more wrong." That should make it easier to detect that it is wrong. Therefore, the probability of type II error, or β, will decrease. If the sample size increases, then the evidence becomes more reliable and the probability of any error, including β, will decrease. As Figure 7–1 depicts, as α increases, β decreases. Thus, β is affected by several factors.

The complement of β $(1 - \beta)$ is known as the *power* of the test.

> The **power** of a test is the probability that a false null hypothesis will be detected by the test.

You can see how α and β as well as $(1 - \alpha)$ and $(1 - \beta)$ are counterparts of each other and how they apply respectively to type I and type II errors. In a later section, we will see more about β and power.

Sample Size

Figure 7–1 depicts how α and β are related. In the discussion above we said that we can keep a low α or a low β depending on which type of error is more costly. What if both types of error are costly and we want to have low α as well as low β? The only way to do this is to make our evidence more reliable, which can be done only by increasing the sample size. Figure 7–2 shows the relationship between α and β for various values of the sample size n. As n increases, the curve shifts downward, reducing both α and β. Thus, when the costs of both types of error are high, the best policy is to have a large sample and a low α, such as 1%.

In this section, we have seen a number of important concepts about hypothesis testing. The mechanical details of computations, templates, and formulas remain. You must have a clear understanding of all the concepts discussed before proceeding further. If necessary, reread this entire section.

FIGURE 7–2 β versus α for Various Values of n
[*Taken from* **Testing Population Mean.xls; Sheet: Beta vs. Alpha**]

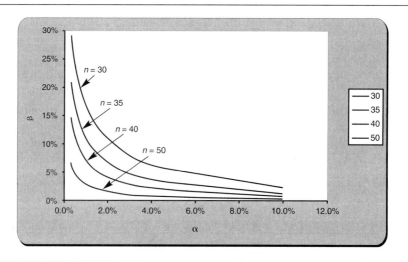

7–6. What is the power of a hypothesis test? Why is it important?

7–7. How is the power of a hypothesis test related to the significance level α?

7–8. How can the power of a hypothesis test be increased without increasing the sample size?

7–9. Consider the use of metal detectors in airports to test people for concealed weapons. In essence, this is a form of hypothesis testing.

    a. What are the null and alternative hypotheses?

    b. What are type I and type II errors in this case?

    c. Which type of error is more costly?

    d. Based on your answer to part (c), what value of α would you recommend for this test?

    e. If the sensitivity of the metal detector is increased, how would the probabilities of type I and type II errors be affected?

    f. If α is to be increased, should the sensitivity of the metal detector be increased or decreased?

7–10. When planning a hypothesis test, what should be done if the probabilities of both type I and type II errors are to be small?

7–3 Computing the *p*-Value

We will now examine the details of calculating the *p*-value. Recall that given a null hypothesis and sample evidence, the *p*-value is the probability of getting evidence that is equally or more unfavorable to H_0. Using what we have already learned in the previous two chapters, this probability can be calculated for hypotheses regarding population mean, proportion, and variance.

The Test Statistic

Consider the case

$$H_0: \mu \geq 1{,}000$$
$$H_1: \mu < 1{,}000$$

Suppose the population standard deviation σ is known and a random sample of size $n \geq 30$ is taken and the sample mean $\overline{X}$ is calculated. From sampling theory we know that when $\mu = 1{,}000$, $\overline{X}$ will be normally distributed with mean 1,000 and standard deviation $\sigma/\sqrt{n}$. This implies that $(\overline{X} - 1{,}000)/(\sigma/\sqrt{n})$ will follow a standard normal distribution, or Z distribution. Since we know the Z distribution well, we can calculate any probability and, in particular, the *p*-value. In other words, by calculating first

$$Z = \frac{\overline{X} - 1000}{\sigma/\sqrt{n}}$$

we can then calculate the *p*-value and decide whether or not to reject H_0. Since the test result boils down to checking just one value, the value of Z, we call Z the *test statistic* in this case.

A **test statistic** is a random variable calculated from the sample evidence, which follows a well-known distribution and thus can be used to calculate the *p*-value.

Most of the time, the test statistic we see in this book will be Z, t, χ^2, or F. The distributions of these random variables are well known and spreadsheet templates can be used to calculate the *p*-value.

p-*Value Calculations*

Once again consider the case

$$H_0: \mu \geq 1,000$$
$$H_1: \mu < 1,000$$

Suppose the population standard deviation σ is known and a random sample of size $n \geq 30$ is taken. This means $Z = (\overline{X} - 1,000)/(\sigma/\sqrt{n})$ is the test statistic. If the sample mean $\overline{X}$ is 1,000 or more, we have nothing against H_0 and we will not reject it. But if $\overline{X}$ is less than 1,000, say 999, then the evidence disfavors H_0 and we have reason to suspect that H_0 is false. If $\overline{X}$ decreases below 999, it becomes even more unfavorable to H_0. Thus the *p*-value when $\overline{X} = 999$ is the probability that $\overline{X} \leq 999$. This probability is the shaded area shown in Figure 7–3. But the usual practice is to calculate the probability using the distribution of the test statistic Z. So let us switch to the Z statistic.

Suppose the population standard deviation σ is 5 and the sample size, n, is 100. Then

$$Z = \frac{\overline{X} - 1000}{\sigma/\sqrt{n}} = \frac{999 - 1000}{5/\sqrt{100}} = -2.00$$

Thus the *p*-value $= P(Z < -2.00)$. See Figure 7–4, in which the probability is shaded. The figure also shows the direction in which $\overline{X}$ and Z decrease. The probability $P(Z < -2.00)$ can be calculated from the tables or using a spreadsheet template. We will see full details of the templates later. For now, let us use the tables. From the standard normal distribution table, the *p*-value is $0.5 - 0.4772 = 0.0228$, or 2.28%. This means H_0 will be rejected when α is 5% or 10% but will not be rejected when α is 1%.

FIGURE 7–3   The *p*-Value Shaded in the Distribution of $\overline{X}$

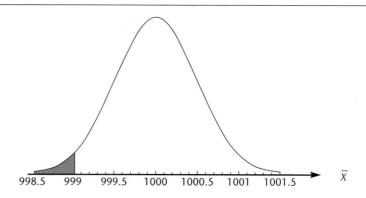

One-Tailed and Two-Tailed Tests

Let us repeat the null and alternative hypotheses for easy reference:

$$H_0: \mu \geq 1{,}000$$
$$H_1: \mu < 1{,}000$$

In this case, only when $\overline{X}$ is significantly less than 1,000 will we reject H_0, or only when Z falls significantly below zero will we reject H_0. Thus the rejection occurs only when Z takes a significantly low value in the *left tail* of its distribution. Such a case where rejection occurs in the left tail of the distribution of the test statistic is called a **left-tailed test,** as seen in Figure 7–5. At the bottom of the figure the direction in which Z, $\overline{X}$, and the p-value decrease is shown.

FIGURE 7–4   The p-Value Shaded in the Distribution of the Test Statistic Z where $H_0: \mu \geq 1{,}000$

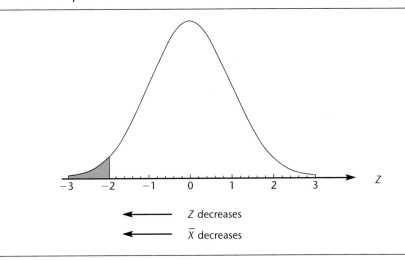

FIGURE 7–5   A Left-Tailed Test: The Rejection Region for $H_0: \mu \geq 1{,}000$; $\alpha = 5\%$

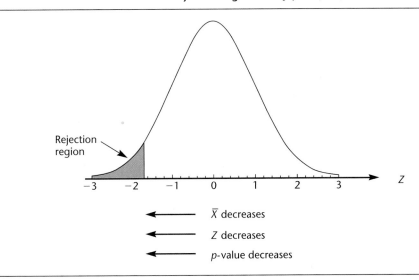

FIGURE 7–6   A Right-Tailed Test: The Rejection Region for H_0: $\mu \leq 1,000$; $\alpha = 5\%$

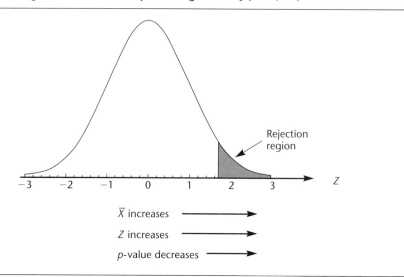

In the case of a left-tailed test, the p*-value is the area to the left of the calculated value of the test statistic.* The case we saw above is a good example. Suppose the calculated value of Z is -2.00. Then the area to the left of it, using tables, is $0.5 - 0.4772 = 0.0228$, or the p-value is 2.28%.

Now consider the case where H_0: $\mu \leq 1,000$. Here rejection occurs when $\overline{X}$ is significantly greater than 1,000 or Z is significantly greater than zero. In other words, rejection occurs on the right tail of the Z distribution. This case is therefore called a **right-tailed test,** as seen in Figure 7–6. At the bottom of the figure the direction in which the p-value decreases is shown.

In the case of a right-tailed test, the p*-value is the area to the right of the calculated value of the test statistic.* Suppose the calculated $z = +1.75$. Then the area to the right of it, using tables, is $0.5 - 0.4599 = 0.0401$, or the p-value is 4.01%.

In left-tailed and right-tailed tests, rejection occurs only on one tail. Hence each of them is called a **one-tailed test.**

Finally, consider the case H_0: $\mu = 1,000$. In this case, we have to reject H_0 in both cases, that is, whether $\overline{X}$ is significantly less than or greater than 1,000. Thus, rejection occurs when Z is significantly less than or greater than zero, which is to say that rejection occurs on both tails. Therefore, this case is called a **two-tailed test.** See Figure 7–7, where the shaded areas are the rejection regions. As shown at the bottom of the figure, the p-value decreases as the calculated value of test statistic moves away from the center in either direction.

In the case of a two-tailed test, the p*-value is twice the tail area. If the calculated value of the test statistic falls on the left tail, then we take the area to the left of the calculated value and multiply it by 2. If the calculated value of the test statistic falls on the right tail, then we take the area to the right of the calculated value and multiply it by 2.* For example, if the calculated $z = +1.75$, the area to the right of it is 0.0401. Multiplying that by 2, we get the p-value as 0.0802.

FIGURE 7–7   A Two-Tailed Test: Rejection Region for H_0: $\mu = 1{,}000$, $\alpha = 5\%$

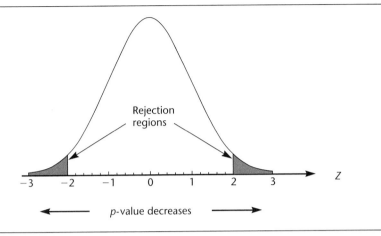

In a hypothesis test, the test statistic $Z = -1.86$.

EXAMPLE 7–4

1. Find the p-value if the test is (*a*) left-tailed, (*b*) right-tailed, and (*c*) two-tailed.
2. In which of these three cases will H_0 be rejected at an α of 5%?

Solution

1. (*a*) The area to the left of -1.86, from the tables, is $0.5 - 0.4686 = 0.0314$, or the p-value is 3.14%. (*b*) The area to the right of -1.86, from the tables, is $0.5 + 0.4686 = 0.9686$, or the p-value is 96.86%. (Such a large p-value means that the evidence greatly favors H_0, and there is no basis for rejecting H_0.) (*c*) The value -1.86 falls on the left tail. The area to the left of -1.86 is 3.14%. Multiplying that by 2, we get 6.28%, which is the p-value.
2. Only in the case of a left-tailed test does the p-value fall below the α of 5%. Hence that is the only case where H_0 will be rejected.

PROBLEMS

7–11. For each one of the following null hypotheses, determine if it is a left-tailed, a right-tailed, or a two-tailed test.

    a. $\mu \geq 10$.
    b. $p \leq 0.5$.
    c. μ is at least 100.
    d. $\mu \leq -20$.
    e. p is exactly 0.22.
    f. μ is at most 50.
    g. $\sigma^2 = 140$.

7–12. The calculated z for a hypothesis test is -1.75. What is the p-value if the test is (*a*) left-tailed, (*b*) right-tailed, and (*c*) two-tailed?

7–13. In which direction of $\bar{X}$ will the p-value decrease for the null hypotheses (*a*) $\mu \geq 10$, (*b*) $\mu \leq 10$, and (*c*) $\mu = 10$?

7–14.   What is a test statistic? Why do we have to know the distribution of the test statistic?

7–15.   The null hypothesis is $\mu \le 12$. The test statistic is Z. Assuming that other things remain the same, will the p-value increase or decrease when (a) $\bar{X}$ increases, (b) σ increases, and (c) n increases.

7–4   The Hypothesis Test

We now consider the three common types of hypothesis tests:

1. Tests of hypotheses about population means.
2. Tests of hypotheses about population proportions.
3. Tests of hypotheses about population variances.

Let us see the details of each type of test and the templates that can be used.

Testing Population Means

When the null hypothesis is about a population mean, the test statistic can be either Z or t. There are two cases in which it will be Z.

Cases in Which the Test Statistic Is Z
1. σ is known and the population is normal.
2. σ is known and the sample size is at least 30. (The population need not be normal.)

The normality of the population may be established by direct tests or the normality may be assumed based on the nature of the population. Recall that if a random variable is affected by many independent causes, then it can be assumed to be normally distributed.

The formula for calculating Z is

$$Z = \frac{\bar{X} - \mu}{\sigma/\sqrt{n}}$$

The value of μ in this equation is the claimed value that gives the maximum benefit of doubt to the null hypothesis. For example, if H_0: $\mu \ge 1{,}000$, we use the value of 1,000 in the equation. Once the Z value is known, the p-value is calculated using tables or the template described below.

Cases in Which the Test Statistic Is t
The population is normal and σ is unknown but the sample standard deviation S is known.

In this case, as we saw in the previous chapter, the quantity $(\bar{X} - \mu)/(S/\sqrt{n})$ will follow a t distribution with $(n - 1)$ degrees of freedom. Thus

$$t = \frac{\bar{X} - \mu}{S/\sqrt{n}}$$

becomes the test statistic. The value of μ used in this equation is the claimed value that gives the maximum benefit of doubt to the null hypothesis. For example, if H_0: $\mu \ge 1{,}000$, then we use the value of 1,000 for μ in the equation for calculating t.

A Note on t Tables and p-Values

Since the t table provides only the critical values, it cannot be used to find exact p-values. We have to use the templates described below or use other means of calculation. If we do not have access to the templates or other means, then the critical values found in the tables can be used to infer the *range* within which the p-value will fall. For example, if the calculated value of t is 2.000 and the degrees of freedom are 24, we see from the tables that $t_{0.05}$ is 1.711 and $t_{0.025}$ is 2.064. Thus, the p-value corresponding to $t = 2.000$ must be somewhere between 0.025 and 0.05, but we don't know its exact value. Since the exact p-value for a hypothesis test is generally desired, it is advisable to use the templates.

A careful examination of the cases covered above, in which Z or t is the test statistic, reveals that there are a few cases that do not fall under either category.

Cases Not Covered by the Z or t Test Statistic

1. The population is not normal and σ is unknown. (Many statisticians will be willing to accept a t test here, as long as the sample size is "large enough." The size is large enough if it is at least 30 in the case of populations believed to be not very skewed. If the population is known to be very skewed, then the size will have to be correspondingly larger.)
2. The population is not normal and the sample size is less than 30.
3. The population is normal and σ is unknown. Whoever did the sampling provided only the sample mean $\bar{X}$ but not the sample standard deviation S. The sample data are also not provided and thus S cannot be calculated. (Obviously, this case is rare.)

It is always better to use templates to solve hypothesis testing problems. But to understand the computation process, we shall do one example manually with Z as the test statistic.

EXAMPLE 7–5

An automatic bottling machine fills cola into 2-liter (2,000 cm^3) bottles. A consumer advocate wants to test the null hypothesis that the average amount filled by the machine into a bottle is at least 2,000 cm^3. A random sample of 40 bottles coming out of the machine was selected and the exact contents of the selected bottles are recorded. The sample mean was 1,999.6 cm^3. The population standard deviation is known from past experience to be 1.30 cm^3.

1. Test the null hypothesis at an α of 5%.
2. Assume that the population is normally distributed with the same σ of 1.30 cm^3. Assume that the sample size is only 20 but the sample mean is the same 1,999.6 cm^3. Conduct the test once again at an α of 5%.
3. If there is a difference in the two test results, explain the reason for the difference.

Solution

1.
$$H_0: \mu \geq 2,000$$
$$H_1: \mu < 2,000$$

Since σ is known and the sample size is more than 30, the test statistic is Z. Then

FIGURE 7–8　Testing Hypotheses about Population Means Using Sample Statistics
[Testing Population Mean.xls; Sheet: Sample Stats]

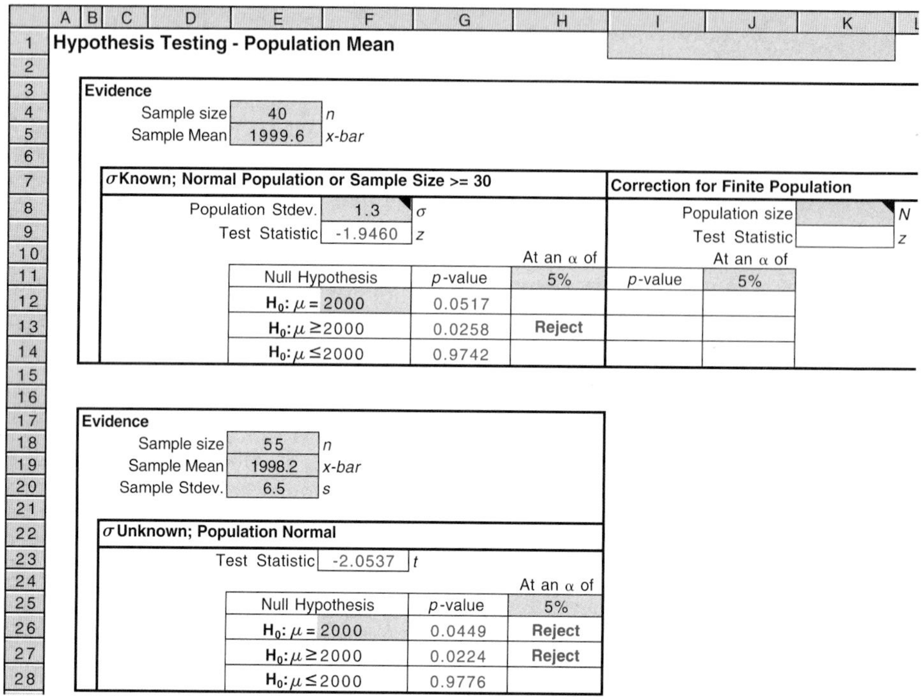

$$z = \frac{\bar{x} - \mu}{\sigma/\sqrt{n}} = \frac{1,999.6 - 2,000}{1.30/\sqrt{40}} = -1.95$$

Using the table for areas of Z distribution, the p-value $= 0.5000 - 0.4744 = 0.0256$, or 2.56%. Since this is less than the α of 5%, we reject the null hypothesis.

2. Since the population is normally distributed, the test statistic is once again Z:

$$z = \frac{\bar{x} - \mu}{\sigma/\sqrt{n}} = \frac{1,999.6 - 2,000}{1.30/\sqrt{20}} = -1.38$$

Using the table for areas of Z distribution, the p-value $= 0.5000 - 0.4162 = 0.0838$, or 8.38%. Since this is greater than the α of 5%, we do not reject the null hypothesis.

3. In the first case we could reject the null hypothesis but in the second we could not, although in both cases the sample mean was the same. The reason is that in the first case the sample size was larger and therefore the evidence against the null hypothesis was more reliable. This produced a smaller p-value in the first case.

The Templates

Figure 7–8 shows the template that can be used to test hypotheses about population means when sample statistics are known (rather than the raw sample data). The top

FIGURE 7–9   Testing Hypotheses about Population Means Using Sample Data
[Testing Population Mean.xls; Sheet: Sample Data]

	A	B	C D	E	F	G	H	I	J	
1	**Hypothesis Testing - Population Mean**								Example 7-6	
2										
3	**Sample**		Evidence							
4	**Data**			Sample size	37		n			
5	1	1998.41		Sample Mean	1999.54		x-bar			
6	2	2000.34								
7	3	2001.68		σ Known; Normal Population or Sample Size >= 30					Cor	
8	4	2000.98		Population Stdev.	1.8		σ			
9	5	2000.89		Test Statistic	-1.5472		z			
10	6	2001.07						At an α of		
11	7	1997.01			Null Hypothesis		p-value	5%		p
12	8	2000.34			$H_0: \mu = 2000$		0.1218			
13	9	1997.86			$H_0: \mu \geq 2000$		0.0609			
14	10	1998.43			$H_0: \mu \leq 2000$		0.9391			
15	11	1998.12								
16	12	1997.85								
17	13	2000.25		Evidence						
18	14	1997.65			Sample size	37		n		
19	15	2001.17			Sample Mean	1999.54		x-bar		
20	16	1997.44			Sample Stdev.	1.36884		s		
21	17	1998.7								
22	18	1998.67		σ Unknown; Population Normal						
23	19	1997.58			Test Statistic	-2.0345		t		
24	20	2000.28						At an α of		
25	21	1998.89			Null Hypothesis		p-value	5%		
26	22	2000.13			$H_0: \mu = 2000$		0.0493	Reject		
27	23	2000.1			$H_0: \mu \geq 2000$		0.0247	Reject		
28	24	2000.39			$H_0: \mu \leq 2000$		0.9753			

portion of the template is used when σ is known and the bottom portion when σ is unknown. On the top part, entries have been made to solve Example 7–5, part 1. The p-value of 0.0258 in cell G13 is read off as the answer to the problem. This answer is more accurate than the value of 0.0256 manually calculated using tables.

Correction for finite population is possible in the panel on the right. It is applied when $n/N > 1\%$. If no correction is needed, it is better to leave the cell K8, meant for the population size N, blank to avoid causing distraction.

Note that the hypothesized value entered in cell F12 is copied into cells F13 and F14. Only cell F12 is unlocked and therefore that is the only place where the hypothesized value of μ can be entered regardless of which null hypothesis we are interested in.

Once a value for α is entered in cell H11, the "Reject" message appears wherever the p-value is less than the α. All the templates on hypothesis testing work in this manner. In the case shown in Figure 7–8, the appearance of "Reject" in cell H13 means that the null hypothesis $\mu \geq 2,000$ is to be rejected at an α of 5%.

Figure 7–9 shows the template that can be used to test hypotheses about population means, when the sample data are known. Sample data are entered in column B. Correction for finite population is possible in the panel on the right.

A bottling machine is to be tested for accuracy of the amount it fills in 2-liter bottles. The null hypothesis is $\mu = 2,000$ cm^3. A random sample of 37 bottles is taken and the contents are measured. The data are shown below. Conduct the test at an α of 5%.

EXAMPLE 7–6

1. Assume $\sigma = 1.8$ cm^3. What is the test statistic and what is its value? What is the p-value?
2. Assume σ is not known and the population is normal. What is the test statistic and what is its value? What is the p-value?
3. Looking at the answers to parts 1 and 2, comment on any difference in the two results.

Sample Data

1998.41	1998.12	1998.89	2001.68
2000.34	1997.85	2000.13	2000.76
2001.68	2000.25	2000.1	1998.53
2000.98	1997.65	2000.39	1998.24
2000.89	2001.17	2001.27	1998.18
2001.07	1997.44	1998.98	2000.67
1997.01	1998.7	2000.21	2001.11
2000.34	1998.67	2000.36	
1997.86	1997.58	2000.17	
1998.43	2000.28	1998.67	

Solution   Open the template shown in Figure 7–9. Enter the data in column B. To answer part 1, use the top panel. Enter 1.8 for σ in cell H8, 2000 in cell H12, and 5% in cell J11. Since cell J12 is blank, it means the null hypothesis cannot be rejected. The test statistic is Z, and its value of -1.5472 appears in cell H9. The p-value is 0.1218, as seen in cell I12.

To answer part 2, use the bottom panel. Enter 2000 in cell H26 and 5% in cell J25. Since cell J26 says "Reject," we reject the null hypothesis. The test statistic is t, and its value of -2.0345 appears in cell H23. The p-value is 0.0493, as seen in cell I26.

The null hypothesis is not rejected in part 1, but is rejected in part 2. The main difference is that the sample standard deviation of 1.36884 (in cell G20) is less than the 1.8 used in part 1. This makes the value of the test statistic $t = -2.0345$ in part 2, significantly different from $Z = -1.5472$ in part 1. As a result, the p-value falls below 5% in part 2 and the null hypothesis is rejected.

Testing Population Proportions

Hypotheses about population proportions can be tested using the binomial distribution or normal approximation to calculate the p-value. The cases in which each approach is to be used are detailed below.

Cases in Which the Binomial Distribution Can Be Used

The binomial distribution can be used whenever we are able to calculate the necessary binomial probabilities. This means for calculations using tables, the sample size n and the population proportion p should have been tabulated. For calculations using spreadsheet templates, sample sizes up to 500 are feasible.

Cases in Which the Normal Approximation Is to Be Used

If the sample size n is too large (> 500) to calculate binomial probabilities, then the normal approximation method is to be used.

The advantage of using the binomial distribution, and therefore of this template, is that it is more accurate than the normal approximation. *When the binomial distribution is used, the number of successes* X *serves as the test statistic.* The *p*-value is the appropriate tail area, determined by *X*, of the binomial distribution defined by *n* and the hypothesized value of population proportion *p*. Note that *X* follows a *discrete* distribution, and recall that the *p*-value is the probability of the test statistic being *equally or more unfavorable to* H_0 *than* the value obtained from the evidence. As an example, consider a right-tailed test with H_0: $p \leq 0.5$. For this case, the *p*-value = $P(X \geq$ observed number of successes).

A coin is to be tested for fairness. It is tossed 25 times and only 8 heads are observed. Test if the coin is fair at $\alpha = 5\%$.   **EXAMPLE 7–7**

Let *p* denote the probability of getting a head, which must be 0.5 for a fair coin. Hence the null and alternative hypotheses are   *Solution*

$$H_0: p = 0.5$$
$$H_1: p \neq 0.5$$

Because this is a two-tailed test, the *p*-value = $2*P(X \leq 8)$. From the binomial distribution table (Appendix C, Table 1), this value is $2*0.054 = 0.108$. Since this value is more than the α of 5%, we cannot reject the null hypothesis. (For the use of the template to solve this problem, see Figure 7–10.)

Figure 7–10 shows the template that can be used to test hypotheses regarding population proportions using the binomial distribution. *This template will work only for sample sizes up to approximately 500.* Beyond that, the template that uses normal approximation (shown in Figure 7–11) should be used. The data entered in Figure 7–10 correspond to Example 7–7.

Figure 7–11 shows the template that can be used to test hypotheses regarding population means using normal distribution. The test statistic is *Z* defined by

$$Z = \frac{\hat{p} - p_0}{\sqrt{p_0(1 - p_0)/n}}$$

FIGURE 7–10 **Testing Population Proportion Using the Binomial Distribution**
[Testing Population Proportion.xls; Sheet: Binomial]

	A	B	C	D	E	F	G
1	Testing Population Proportion						
2							
3		Evidence				Assumption	
4			Sample size	25	*n*	Large Population	
5			#Successes	8	*x*		
6			Sample Proportion	0.3200	*p-hat*		
7							
8						At an α of	
9			Null Hypothesis		*p-value*	5%	
10			H_0: p = 0.5		0.1078		
11			H_0: p >= 0.5		0.0539		
12			H_0: p <= 0.5		0.9461		

FIGURE 7–11   A Normal Distribution Template for Testing Population Proportion
[Testing Population Proportion.xls; Sheet: Normal]

	A	B	C	D	E	F	G	H	I	J	K
1		z-Test for Population Proportion									
2											
3		Evidence					Assumption				
4							Both np and $n(1-p) >= 5$				
5			Sample size	210	n						
6			#Successes	132	x		Correction for Finite Population				
7			Sample Proportion	0.6286	p-hat						
8			Test statistic	-2.2588	z			Population Size	2000	N	
9								Test statistic	-2.3870	z	
10						At an α of		At an α of			
11			Null Hypothesis		p-value	5%	p-value	5%			
12			H_0: p = 0.7		0.0239	Reject	0.0170	Reject			
13			H_0: p >= 0.7		0.0119	Reject	0.0085	Reject			
14			H_0: p <= 0.7		0.9881		0.9915				

FIGURE 7–12   The Template for Testing Population Variances
[Testing Population Variance.xls; Sheet: Sample Stats]

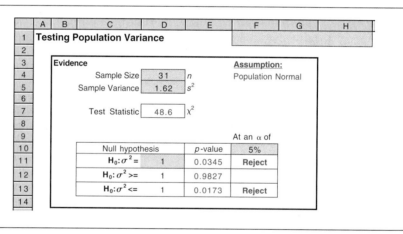

	A	B	C	D	E	F	G	H
1		Testing Population Variance						
2								
3		Evidence				Assumption:		
4			Sample Size	31	n	Population Normal		
5			Sample Variance	1.62	s^2			
6								
7			Test Statistic	48.6	χ^2			
8								
9						At an α of		
10			Null hypothesis		p-value	5%		
11			H_0: σ^2 =	1	0.0345	Reject		
12			H_0: σ^2 >=	1	0.9827			
13			H_0: σ^2 <=	1	0.0173	Reject		
14								

where p_0 is the hypothesized value for the proportion, $\hat{p}$ is the sample proportion, and n is the sample size. A correction for finite population can also be applied in this case. The correction is based on the hypergeometric distribution, and is applied if the sample size is more than 1% of the population size. If a correction is not needed, it is better to leave the cell J8, meant for population size N, blank to avoid any distraction.

Testing Population Variance

For testing hypotheses about population variances, the test statistic is $\chi^2 = (n-1)S^2/\sigma_0^2$. Here σ_0 is the claimed value of population variance in the null hypothesis. The degrees of freedom for this χ^2 is $(n-1)$. Since the χ^2 table provides only the critical values, it cannot be used to calculate exact p-values. As in the case of t tables, only a range of possible values can be inferred. It is therefore better to use a spreadsheet template for this test. Figure 7–12 shows the template that can be used for testing hypotheses regarding population variances when sample statistics are known.

FIGURE 7–13   The Template for Testing Population Variances with Raw Sample Data
[Testing Population Variance.xls; Sheet: Sample Data]

	A	B	C	D	E	F	G	H	I
1	Testing Population Variance								
2									
3		Data							
4	1	154		Evidence				Assumption:	
5	2	135			Sample Size	15	n	Population Normal	
6	3	187			Sample Variance	1702.6	s^2		
7	4	198							
8	5	133			Test Statistic	23.837	χ^2		
9	6	126							
10	7	200						At an α of	
11	8	149			Null hypothesis		p-value	5%	
12	9	187			$H_0: \sigma^2 =$ 1000		0.0959		
13	10	214			$H_0: \sigma^2 >=$ 1000		0.9521		
14	11	156			$H_0: \sigma^2 <=$ 1000		0.0479	Reject	
15	12	257							

A manufacturer of golf balls claims that the company controls the weights of the golf balls accurately so that the variance of the weights is not more than 1 mg^2. A random sample of 31 golf balls yields a sample variance of 1.62 mg^2. Is that sufficient evidence to reject the claim at an α of 5%?

EXAMPLE 7–8

The null and alternative hypotheses are

Solution

$$H_0: \sigma^2 \le 1$$
$$H_1: \sigma^2 > 1$$

In the template (see Figure 7–12), enter 31 for sample size and 1.62 for sample variance. Enter the hypothesized value of 1 in cell D11. The p-value of 0.0173 appears in cell E13. Since this value is less than the α of 5%, we reject the null hypothesis. This conclusion is also confirmed by the "Reject" message appearing in cell F13 with 5% entered in cell F10.

Figure 7–13 shows the template that can be used to test hypotheses regarding population variances when the sample data are known. The sample data are entered in column B.

PROBLEMS

7–16.   An automobile manufacturer substitutes a different engine in cars that were known to have an average miles-per-gallon rating of 31.5 on the highway. The manufacturer wants to test whether the new engine changes the miles-per-gallon rating of the automobile model. A random sample of 100 trial runs gives $\bar{x} = 29.8$ miles per gallon and $s = 6.6$ miles per gallon. Using the 0.05 level of significance, is the average miles-per-gallon rating on the highway for cars using the new engine different from the rating for cars using the old engine?

7–17.   A certain prescription medicine is supposed to contain an average of 247 parts per million (ppm) of a certain chemical. If the concentration is higher than 247 ppm, the drug may cause some side effects; and if the concentration is below 247 ppm, the drug may be ineffective. The manufacturer wants to check whether the average concentration in a large shipment is the required 247 ppm or not. A random sample of 60 portions is tested, and it is found that the sample mean is 250 ppm and the sample standard deviation is 12 ppm. Test the null hypothesis that the average concentration in the entire large shipment is 247 ppm versus the alternative hypothesis that it is not 247 ppm using a level of significance $\alpha = 0.05$. Do the same using $\alpha = 0.01$. What is your conclusion? What is your decision about the shipment? If the shipment were guaranteed to contain an average concentration of 247 ppm, what would your decision be, based on the statistical hypothesis test? Explain.

7–18.   A metropolitan transit authority wants to determine whether there is any need for changes in the frequency of service over certain bus routes. The transit authority needs to know whether the frequency of service should increase, decrease, or remain the same. It is determined that if the average number of miles traveled by bus over the routes in question by all residents of a given area is about 5 per day, then no change will be necessary. If the average number of miles traveled per person per day is either more than 5 or less than 5, then changes in service may be necessary. The authority wants, therefore, to test the null hypothesis that the average number of miles traveled per person per day is 5.0 versus the alternative hypothesis that the average is not 5.0 miles. The required level of significance for this test is $\alpha = 0.05$. A random sample of 120 residents of the area is taken, and it is found that the sample mean is 2.3 miles per resident per day and the sample standard deviation is 1.5 miles. Advise the authority on what should be done. Explain your recommendation. Could you state the same result at different levels of significance? Explain.

7–19.   A study was undertaken to determine customer satisfaction in Canadian automobile markets following certain changes in customer service. Suppose that before the changes, the average customer satisfaction rating, on a scale of 0 to 100, was 77. A survey questionnaire was sent to a random sample of 350 residents who bought new cars after the changes in customer service were instituted, and the average satisfaction rating for this sample was found to be $\bar{x} = 84$; the sample standard deviation was found to be $s = 28$. Use an α of your choice, and determine whether there is statistical evidence of a change in customer satisfaction. If you determine that a change did occur, state whether you believe customer satisfaction has improved or deteriorated.

7–20.   An investment services company claims that the average annual return on stocks within a certain industry is 11.5%. An investor wants to test whether this claim is true and collects a random sample of 50 stocks in the industry of interest. He finds that the sample average annual return is 10.8% and that the sample standard deviation is 3.4%. Does the investor have enough evidence to reject the investment company's claim? (Use $\alpha = 0.05$.)

7–21.   A certain commodity is known to have a price that is stable through time and does not change according to any known trend. Price, however, does change from day to day in a random fashion. If the price is at a certain level one day, it is as likely to be at any level the next day within some probability bounds approximately given by a normal distribution. The mean daily price is believed to be $14.25. To test the hypothesis that the average price is $14.25 versus the alternative hypothesis that it is not $14.25, a random sample of 16 daily prices is collected. The results are $\bar{x} = 16.50 and $s = 5.8. Using $\alpha = 0.05$, can you reject the null hypothesis?

7–22.   Average total daily sales at a small food store are known to be $452.80. The store's management recently implemented some changes in displays of goods, order within aisles, and other changes, and it now wants to know whether average sales volume has changed. A random sample of 12 days shows $\bar{x} = \$501.90$ and $s = \$65.00$. Using $\alpha = 0.05$, is the sampling result significant? Explain.

7–23.   An article[7] on new software companies that create programs for World Wide Web applications implies that average staff age at these companies is 27. To test this two-tailed hypothesis, a random sample is collected:

    41, 18, 25, 36, 26, 35, 24, 30, 28, 19, 22, 22, 26, 23, 24, 31, 22, 22, 23, 26, 27, 26, 29, 28, 23,
    19, 18, 18, 24, 24, 24, 25, 24, 23, 20, 21, 21, 21, 21, 32, 23, 21, 20

Test, using $\alpha = 0.05$.

7–24.   A study was undertaken to evaluate how stocks are affected by being listed in the Standard & Poor's 500 Index. The aim of the study was to assess average excess returns for these stocks, above returns on the market as a whole. The average excess return on *any* stock is zero because the "average" stock moves with the market as a whole. As part of the study, a random sample of 13 stocks newly included in the S&P 500 Index was selected. Before the sampling takes place, we allow that average "excess return" for stocks newly listed in the Standard & Poor's 500 Index may be either positive or negative; therefore, we want to test the null hypothesis that average excess return is equal to zero versus the alternative that it is not zero. If the excess return on the sample of 13 stocks averaged 3.1% and had a standard deviation of 1%, do you believe that inclusion in the Standard & Poor's 500 Index changes a stock's excess return on investment, and if so, in which direction? Explain. Use $\alpha = 0.05$.

7–25.   A new chemical process is introduced in the production of nickel-cadmium batteries. For batteries produced by the old process, it is known that the average life of a battery is 102.5 hours. To determine whether the new process affects the average life of the batteries, the manufacturer collects a random sample of 25 batteries produced by the new process and uses them until they run out. The sample mean life is found to be 107 hours, and the sample standard deviation is found to be 10 hours. Are these results significant at the $\alpha = 0.05$ level? Are they significant at the $\alpha = 0.01$ level? Explain. Draw your conclusion.

7–26.   Average soap consumption in a certain country is believed to be 2.5 bars per person per month. *The standard deviation of the population is known to be* $\sigma = 0.8$. While the standard deviation is not believed to have changed (and this may be substantiated by several studies), it is believed that the mean consumption may have changed either upward or downward. A survey is therefore undertaken to test the null hypothesis that average soap consumption is still 2.5 bars per person per month versus the alternative that it is not. A sample of size $n = 20$ is collected and gives $\bar{x} = 2.3$. The population is assumed to be normally distributed. What is the appropriate test statistic in this case? Conduct the test and state your conclusion. Use $\alpha = 0.05$. Does the choice of level of significance change your conclusion? Explain.

7–27.   A survey of schools of business indicates that 16% of the faculty positions in schools of business are currently vacant. A placement service working for a renowned university wants to test whether the claim is true and collects information on a random sample of 300 business faculty positions chosen from universities around the

---

[7]E. Brown, "Cool Companies," *Fortune*, September 7, 1997, p. 53.

country. The results indicate that 51 out of the 300 positions surveyed are vacant. Use $\alpha = 0.05$ to conduct the test.

7–28.   Suppose that the Goodyear Tire Company has historically held 42% of the market for automobile tires in the United States. Recent changes in company operations, especially its diversification to other areas of business, as well as changes in competing firms' operations, prompt the firm to test the validity of the assumption that it still controls 42% of the market. A random sample of 550 automobiles on the road shows that 219 of them have Goodyear tires. Conduct the test at $\alpha = 0.01$.

7–29.   The manufacturer of electronic components needs to inform its buyers of the proportion of defective components in its shipments. The company has been stating that the percentage of defectives is 12%. The company wants to test whether the proportion of all components that are defective is as claimed. A random sample of 100 items indicates 17 defectives. Use $\alpha = 0.05$ to test the hypothesis that the percentage of defective components is 12%.

7–30.   The percentage of farmers using fertilizers in an African country was known to be 35%. The drought and other events of the last few years are believed to have had a potential impact on the proportion of farmers using fertilizers. An international aid program wants to test whether the percentage is still around 35% and gathers a random sample of 150 farmers. The findings reveal that 68 of the farmers use fertilizers. Conduct the test at $\alpha = 0.05$ and $\alpha = 0.01$. State your conclusion.

7–31.   A company's market share is very sensitive to both its level of advertising and the levels of its competitors' advertising. A firm known to have a 56% market share wants to test whether this value is still valid in view of recent advertising campaigns of its competitors and its own increased level of advertising. A random sample of 500 consumers reveals that 298 use the company's product. Is there evidence to conclude that the company's market share is no longer 56%, at the 0.01 level of significance?

7–32.   According to financial planner E. Kra, individuals should in theory save 7% to 10% of their income over their working life, if they desire a reasonably comfortable retirement.[8] An agency wants to test whether this actually happens with people in the United States, suspecting the overall savings rate may be lower than this range. A random sample of 41 individuals revealed the following savings rates per year:

4, 0, 1.5, 6, 3.1, 10, 7.2, 1.2, 0, 1.9, 0, 1.0, 0.5, 1.7, 8.5, 0, 0, 0.4, 0, 1.6, 0.9, 10.5, 0, 1.2, 2.8, 0, 2.3, 3.9, 5.6, 3.2, 0, 1, 2.6, 2.2, 0.1, 0.6, 6.1, 0, 0.2, 0, 6.8

Conduct the test and state your conclusions. Use the lower value, 7%, in the null hypothesis. Use $\alpha = 0.01$. Interpret.

7–33.   The theory of finance allows for the computation of "excess" returns, either above or below the current stock market average. An analyst wants to determine whether stocks in a certain industry group earn either above or below the market average at a certain time period. The null hypothesis is that there are no excess returns, on the average, in the industry in question. "No average excess returns" means that the population excess return for the industry is zero. A random sample of 24 stocks in the industry reveals a sample average excess return of 0.12 and sample standard deviation of 0.2. State the null and alternative hypotheses, and carry out the test at the $\alpha = 0.05$ level of significance.

---

[8]A. Brooklehurst, "Investment Options for Slow Starters," The Money Report, *International Herald Tribune,* June 14–15, 1997, p. 17.

7–34. The average weekly earnings for all full-time-equivalent employees are reported to be $344. Suppose that you want to check this claim since you believe it is too low. You want to prove that average weekly earnings of all employees are higher than the amount stated. You collect a random sample of 1,200 employees in all areas and find that the sample mean is $361 and the sample standard deviation $110. Can you disprove the claim?

7–35. According to *Money,* the average amount of money that a typical person in the United States would need to make him or her feel rich is $1.5 million. A researcher wants to test this claim. A random sample of 100 people in the United States reveals that their mean "amount to feel rich" is $2.3 million and the standard deviation is $0.5 million. Conduct the test.

7–36. The U.S. Department of Commerce estimates that 17% of all automobiles on the road in the United States at a certain time are made in Japan. An organization that wants to limit imports believes that the proportion of Japanese cars on the road during the period in question is higher than 17% and wants to prove this. A random sample of 2,000 cars is observed, 381 of which are made in Japan. Conduct the hypothesis test at $\alpha = 0.01$, and state whether you believe the reported figure.

7–37. Airplane tires are sensitive to the heat produced when the plane taxis along runways. A certain type of airplane tire used by Boeing is guaranteed to perform well at temperatures as high as 125°F. From time to time, Boeing performs quality control checks to determine whether the average maximum temperature for adequate performance is as stated, or whether the average maximum temperature is lower than 125°F, in which case the company must replace all tires. Suppose that a random sample of 100 tires is checked. It is found that the average maximum temperature for adequate performance in the sample is 121°F and the sample standard deviation is 2°F. Conduct the hypothesis test, and conclude whether the company should take action to replace its tires.

7–38. S. Zesiger, in an article entitled "Vetting the American Dream,"[9] claims that 85% of Corvette buyers are men. A study aimed at proving that $p < 0.85$ looks at a random sample of 300 Corvette owners and finds that 238 are men. Conduct the appropriate test, using $\alpha = 0.05$.

7–39. A study of top executives' midlife crises indicates that 45% of all top executives suffer from some form of mental crisis in the years following corporate success. An executive who had undergone a midlife crisis opened a clinic providing counseling for top executives in the hope of reducing the number of executives who might suffer from this problem. A random sample of 125 executives who went through the program indicated that only 49 eventually showed signs of a midlife crisis. Do you believe that the program is beneficial and indeed reduces the proportion of executives who show signs of the crisis?

7–40. The unemployment rate in Britain during a certain period was believed to have been 11%. At the end of the period in question, the government embarked on a series of projects to reduce unemployment. It was of interest to determine whether the average unemployment rate in the country had decreased as a result of these projects, or whether previously employed people were the ones hired for the project jobs, while the unemployed remained unemployed. A random sample of 3,500 people was

---

[9]*Fortune,* July 7, 1997, p. 106.

chosen, and 421 were found to be unemployed. Do you believe that the government projects reduced the unemployment rate?

7–41. Certain eggs are stated to have reduced cholesterol content, with an average of only 2.5% cholesterol. A concerned health group wants to test whether the claim is true. The group believes that more cholesterol may be found, on the average, in the eggs. A random sample of 100 eggs reveals a sample average content of 5.2% cholesterol, and a sample standard deviation of 2.8%. Does the health group have cause for action?

7–42. Spurred by the demand of business travelers using battery-operated notebook computers, which often run out of power before the flight is over, several airlines decided in 1997 to experiment with making direct-current power available at passenger seats.[10] If on a test run of 120 flights, the average percentage of passengers in business class who took advantage of the option was 18% with a standard deviation of 4%, test the hypothesis that 15% or fewer of business passengers will use the new service initially if it is introduced versus the alternative that more than 15% of business passengers will do so. Use $\alpha = 0.05$.

7–43. Under U.S. government fuel economy standards, automakers are required to maintain a minimum average fuel efficiency of 27.5 miles per gallon for the cars they sell and 20.7 miles per gallon for light trucks.[11] These averages are derived by using a formula with the number of cars of each model and their average fuel efficiencies.

    a. To test compliance, the government agency test-runs cars, randomly selected from models and cars (what sampling method is this?), and gets $\bar{x} = 26.9$, $s = 4.8$, and $n = 100$. Do the test.

    b. For trucks, $\bar{x} = 18.8$, $s = 6.1$, and $n = 100$. Do the test.
       Use $\alpha = 0.05$ in both cases.

7–44. Several U.S. airlines carry passengers from the United States to countries in the Pacific region, and the competition in these flight routes is keen. One of the leverage factors for United Airlines in Pacific routes is that, whereas most other airlines fly to Pacific destinations two or three times weekly, United offers daily flights to Tokyo, Hong Kong, and Osaka. Before instituting daily flights, the airline needed to get an idea as to the proportion of frequent fliers in these routes who consider daily service an important aspect of business flights to the Pacific. From previous information, the management of United estimated that 60% of the frequent business travelers to the three destinations believed that daily service was an important aspect of airline service. Following changes in the airline industry, marked by reduced fares and other factors, the airline management wanted to check whether the proportion of frequent business travelers who believe that daily service is an important feature was still about 60%. A random sample of 250 frequent business fliers revealed that 130 thought daily service was important. Compute the *p*-value for this test (is this a one-tailed or a two-tailed test?), and state your conclusion.

7–45. An advertisement for the Toyota Supra model lists the following performance specifications: standing start, 0–50 miles per hour in an average of 5.27 seconds; braking, 60 miles per hour to 0 in 3.15 seconds on the average. An independent testing service hired by a competing manufacturer of high-performance automobiles wants

[10]"In-Flight Plug-In," *Business Week,* June 16, 1997, p. 8.

[11]From V. Reitman and N. Christian, "Chrysler to Equip Minivans to Operate on Ethanol," *The Wall Street Journal* (Europe), June 11, 1997, p. 6. Reprinted from The Wall Street Journal © 1997 The Dow Jones Company.

to prove that Toyota's claims are exaggerated. A random sample of 100 trial runs gives the following results: standing start, 0–50 miles per hour in an average of $\bar{x} = 5.8$ seconds and $s = 1.9$ seconds; braking, 60 miles per hour to 0 in an average of $\bar{x} = 3.21$ seconds and $s = 0.6$ second. Carry out the two hypothesis tests, state the p-value of each test, and state your conclusions.

7–46. Borg-Warner manufactures hydroelectric miniturbines that generate low-cost, clean electric power from the energy in small rivers and streams. One of the models was known to produce an average of 25.2 kilowatts of electricity. Recently the model's design was improved, and the company wanted to test whether the model's average electric output had changed. There was no reason to suspect, a priori, a change in either direction. A random sample of 115 trial runs produced an average of 26.1 kilowatts and a standard deviation of 3.2 kilowatts. Carry out a statistical hypothesis test, give the p-value, and state your conclusion. Do you believe that the improved model has a different average output?

7–47. Recent near misses in the air, as well as several fatal accidents, have brought air traffic controllers under close scrutiny. As a result of a high-level inquiry into the accuracy of speed and distance determinations through radar sightings of airplanes, a statistical test was proposed to check the air traffic controllers' claim that a commercial jet's position can be determined, on the average, to within 110 feet in the usual range around airports in the United States. The proposed test was given as H_0: $\mu \leq 110$ versus the alternative H_1: $\mu > 110$. The test was to be carried out at the 0.05 level of significance using a random sample of 80 airplane sightings. The statistician designing the test wants to determine the power of this test if the actual average distance at detection is 120 feet. An estimate of the standard deviation is 30 feet. Compute the power at $\mu_1 = 120$ feet.

7–48. McDonald's Corporation has been steadily moving into more countries around the world. Recent reports show that McDonald's has been interested in opening franchises in Poland. As part of its efforts to sell fast foods in Poland, McDonald's Corporation has been evaluating the potential of using Polish-grown potatoes not only in Poland, but also for distribution in other European countries. The feasibility of such an option depends on the demand for French-fried potatoes in all McDonald's European franchises. Company analysts believe that if the average weekly demand for fries per franchise per week is more than 500 package units, it may be feasible to use Polish-grown potatoes. Thus, the analysts want to test the null hypothesis H_0: $\mu \leq 500$ versus the alternative hypothesis H_1: $\mu > 500$. The company has data on 100 weekly sales randomly obtained from franchises throughout the continent. The test is to be carried out at the 0.05 level of significance, and an estimate of the population variance is 2,500 (units squared). What is the power of the test if the true mean is 520 units per franchise per week?

7–49. The Polaroid Spectra camera has an electronic device that makes complex focusing and exposure decisions in 50 thousandths of a second. Before each device is installed in a camera, it is tested by quality-control inspectors. The device is linked to a simulator that runs a random sample of 80 situations and measures the sample average reaction time of the device. The statistical test is H_0: $\mu \leq 50$ (thousandths of a second) versus the alternative H_1: $\mu > 50$ (thousandths of a second). If the null hypothesis is not rejected, the device is considered to have good quality and is installed in a camera; otherwise it is replaced. The test is carried out at the 0.01 level of significance, and the population standard deviation is $\sigma = 20$ (thousandths of a second). For quality control considerations, inspectors need to have a high power for this test, that is, a high probability of rejecting a faulty device, when the average speed of the device is 60 thousandths of a second. Find the power at this level of μ. Also compute the power at other levels, and sketch the power curve.

7–50. A large manufacturing firm believes that its market share is 45%. From time to time, a statistical hypothesis test is carried out to check whether the assertion is true. The test consists of gathering a random sample of 500 products sold nationally and finding what percentage of the sample constitutes brands made by the firm. Whenever the test is carried out, there is no suspicion as to the direction of a possible change in market share, that is, increase or decrease; the company wants to detect any change at all. The tests are carried out at the $\alpha = 0.01$ level of significance. What is the probability of being able to statistically determine a true change in the market share of magnitude 5% in either direction? (That is, find the power at $p = 0.50$ or $p = 0.40$. *Hint:* Use the methods of this section in the case of sampling for proportions. You will have to derive the formulas needed for computing the power.)

ADDITIONAL PROBLEMS

7–51. The engine of the Volvo model S70 T-5 is stated to provide 246 horsepower. To test this claim, believing it is too high, a competitor runs the engine $n = 60$ times, randomly chosen, and gets a sample mean of 239 horsepower and standard deviation of 20 horsepower. Conduct the test, using $\alpha = 0.01$.

7–52. How can we increase the power of a test without increasing the sample size?

7–53. The sales of home appliances have increased greatly in recent years. A study sponsored by the home appliance industry is aimed at testing the hypothesis that 6% of all consumers purchased new appliances during the June–September period. Industry analysts have no prior suspicion as to whether the proportion of consumers who bought new appliances during the period in question is higher or lower than stated and merely want to check the claim. A random sample of 2,000 consumers nationwide shows that 142 bought new appliances during the period. Is this a one-tailed or a two-tailed test? What is the p-value? State your conclusion.

7–54. A recent marketing and promotion campaign by Charles of the Ritz more than doubled the sales of the suntan lotion Bain de Soleil, which has become the nation's number 2 suntan product. At the end of the promotional campaign, the company wanted to test the hypothesis that the market share of its product was 0.35 versus the alternative hypothesis that the market share was higher than 0.35. The company polled a random sample of bathers on beaches from Maine to California and Hawaii, and found that out of the sample of 3,850 users of suntan lotions, 1,367 were users of Bain de Soleil. Do you reject the null hypothesis? What is the p-value? Explain your conclusion.

7–55. Efforts are under way to make the U.S. automobile industry more efficient and competitive so that it will be able to survive intense competition from foreign automakers. An industry analyst is quoted as saying, "GM is sized for 60% of the market, and they only have 43%." General Motors needs to know its actual market share because such knowledge would help the company make better decisions about trimming down or expanding so that it could become more efficient. A company executive, pushing for expansion rather than for cutting down, is interested in proving that the analyst's claim that GM's share of the market is 43% is false and that, in fact, GM's true market share is higher. The executive hires a market research firm to study the problem and carry out the hypothesis test she proposed. The market research agency looks at a random sample of 5,500 cars throughout the country and finds that 2,521 are GM cars. What should be the executive's conclusion? How should she present her results to GM's vice president for operations?

7–56.   A graduate student comes out of college with an average debt of $25,000.[12] A sample of 200 graduate students showed that they had an average debt of $30,000 with a standard deviation of $10,000. State the null and alternative hypotheses and carry out the test at the 5% level of significance.

7–57.   Before a beach is declared safe for swimming, a test of the bacteria count in the water is conducted with the null and alternative hypotheses formulated as

H_0: Bacteria count is less than or equal to the specified upper limit for safety

H_1: Bacteria count is more than the specified upper limit for safety

   a. What are type I and type II errors in this case?
   b. Which error is more costly?
   c. In the absence of any further information, which standard value will you recommend for α?

7–58.   Other things remaining the same, which of the following will result in an increase in the power of a hypothesis test?
   a. Increase in the sample size.
   b. Increase in α.
   c. Increase in the population standard deviation.

7–59.   The null and alternative hypotheses of a t test for the mean are

$$H_0\colon \mu \geq 1,000$$
$$H_1\colon \mu < 1,000$$

Other things remaining the same, which of the following will result in an increase in the p-value?
   a. Increase in the sample size.
   b. Increase in the sample mean.
   c. Increase in the sample standard deviation.
   d. Increase in α.

7–60.   The null and alternative hypotheses of a test for population proportion are

$$H_0\colon p \leq 0.25$$
$$H_1\colon p > 0.25$$

Other things remaining the same, which of the following will result in an increase in the p-value?
   a. Increase in sample size.
   b. Increase in sample proportion.
   c. Increase in α.

7–61.   While designing a hypothesis test for population proportion, the cost of a type I error is found to be substantially greater than originally thought. It is possible,

---

[12]From "Now the Real Application," *Money*, January 2001.

as a response, to change the sample size and/or α. Should they be increased or decreased? Explain.

7–62. The p-value obtained in a hypothesis test for population mean is 8%. Select the most precise statement about what it implies. Explain why the other statements are not precise, or are false.

        a. If H_0 is rejected based on the evidence that has been obtained, the probability of type I error would be 8%.

        b. We can be 92% confident that H_0 is false.

        c. There is at most 8% chance of obtaining evidence that is even more unfavorable to H_0 when H_0 is actually true.

        d. If $\alpha = 1\%$, H_0 will not be rejected and there will be 8% chance of type II error.

        e. If $\alpha = 5\%$, H_0 will not be rejected and no error will be committed.

        f. If $\alpha = 10\%$, H_0 will be rejected and there will be 8% chance of type I error.

7–63. Why is it useful to know the power of a test?

7–64. Explain the difference between the p-value and the significance level α.

7–65. Corporate women are still struggling to break into senior management ranks, according to a study of senior corporate executives by Korn/Ferry International, New York recruiter. Of 1,362 top executives surveyed by the firm, only 2%, or 29, were women. Assuming that the sample reported is a random sample, use the results to test the null hypothesis that the percentage of women in top management is 5% or more, versus the alternative hypothesis that the true percentage is less than 5%. If the test is to be carried out at $\alpha = 0.05$, what will be the power of the test if the true percentage of female top executives is 4%?

7–66. Mellon Bank Corporation recently hired a team to analyze computer cathode-ray tube (CRT) systems. This proved to the management that the right choice of a CRT system would increase productivity. Suppose that the average production level at the bank, measured on a scale of 0 to 100, was known to be 78. The team tested a random sample of 24 employees using a proposed new CRT and found that the average productivity level in the sample was 83 and the sample standard deviation was 12. Using $\alpha = 0.05$, conduct the appropriate hypothesis test and state your conclusion. What is the approximate p-value?

7–67. At Armco's steel plant in Middletown, Ohio, statistical quality-control methods have been used very successfully in controlling slab width on continuous casting units. The company claims that a large reduction in the steel slab width variance resulted from the use of these methods. Suppose that the variance of steel slab widths is expected to be 156 (squared units). A test is carried out to determine whether the variance is above the required level, with the intention to take corrective action if it is concluded that the variance is greater than 156. A random sample of 25 slabs gives a sample variance of 175. Using $\alpha = 0.05$, should corrective action be taken?

7–68. According to the mortgage banking firm Lomas & Nettleton, 95% of all households in the second half of last year lived in rental accommodations. The company believes that lower interest rates for mortgages during the following period reduced the percentage of households living in rental units. The company therefore wants to test H_0: $p \geq 0.95$ versus the alternative H_1: $p < 0.95$ for the proportion during the new period. A random sample of 1,500 households shows that 1,380 are rental units. Carry out the test, and state your conclusion. Use an α of your choice.

7–69. A recent study was aimed at determining whether people with increased workers' compensation stayed off the job longer than people without the increased benefits. Suppose that the average time off per employee per year is known to be 3.1 days. A random sample of 21 employees with increased benefits yielded the following number of days spent off the job in one year: 5, 17, 1, 0, 2, 3, 1, 1, 5, 2, 7, 5, 0, 3, 3, 4, 22, 2, 8, 0, 1. Conduct the appropriate test, and state your conclusions.

7–70. Environmental changes have recently been shown to improve firms' competitive advantages. The approach is called the multiple-scenario approach. A study was designed to find the percentage of the *Fortune* top 1,000 firms that use the multiple-scenario approach. The null hypothesis was that 30% or fewer of the firms use the approach. A random sample of 166 firms in the *Fortune* top 1,000 was chosen, and 59 of the firms replied that they used the multiple-scenario approach. Conduct the hypothesis test at $\alpha = 0.05$. What is the p-value? (Do you need to use the finite-population correction factor?)

7–71. Junk bonds are high-risk securities issued by entities with questionable ability to repay their debts. These risky securities pay, on average, high returns to compensate investors for the risk they bear. Investment banks recently came up with a new junk bond: the debt of Third World banks. The investment bankers are trying to sell this debt to investors, promising very high yields. One firm claims that the average return on the risky debt of Peruvian banks is as high as 40% per year. A potential investor wants to check this claim against the alternative that average return on an investment of this kind is less than 40% per year. The potential investor gets a random sample of 14 securities and finds that the average return in the sample is 28% with a standard deviation of 12%. Can the potential investor take action against the investment firm? Give a rough estimate of the p-value.

7–72. Executives at Gammon & Ninowski Media Investments, a top television station brokerage, believe that the current average price for an independent television station in the United States is $125 million. An analyst at the firm wants to check whether the executives' claim is true. The analyst has no prior suspicion that the claim is incorrect in any particular direction and collects a random sample of 25 independent TV stations around the country. The results are (in millions of dollars) 233, 128, 305, 57, 89, 45, 33, 190, 21, 322, 97, 103, 132, 200, 50, 48, 312, 252, 82, 212, 165, 134, 178, 212, 199. Test the hypothesis that the average station price nationwide is $125 million versus the alternative that it is not $125 million. Use a significance level of your choice.

7–73. Microsoft Corporation makes software packages for use in microcomputers. The company believes that if at least 25% of present owners of microcomputers of certain types would be interested in a particular new software package, then the company will make a profit if it markets the new package. A company analyst therefore wants to test the null hypothesis that the proportion of owners of microcomputers of the given kinds who will be interested in the new package is at most 0.25, versus the alternative that the proportion is greater than 0.25. A random sample of 300 microcomputer owners shows that 94 are interested in the new Microsoft package. Should the company market its new product? Report the p-value.

7–74. A recent National Science Foundation (NSF) survey indicates that more than 20% of the staff in U.S. research and development laboratories are foreign-born. Results of the study have been used for pushing legislation aimed at limiting the number of foreign workers in the United States. An organization of foreign-born scientists wants to prove that the NSF survey results do not reflect the true proportion of foreign workers in U.S. laboratories. The organization collects a random sample of 5,000 laboratory workers in all major laboratories in the country and finds that 876 are foreign. Can these results be used to prove that the NSF study overestimated the proportion of foreigners in U.S. laboratories?

7–75. The average number of weeks that banner ads run at a website is estimated to be 5.5. You want to check the accuracy of this estimate. A sample of 50 ads reveals a sample average of 5.1 weeks with a sample standard deviation of 2.3 weeks. State the null and alternative hypotheses and carry out the test at the 5% level of significance.[13]

7–76. It is claimed that there was at least a 56% decrease in online ads targeting children between May and August 2000. A sample of 100 ads selected during this period showed a 45% decrease in online ads targeting children. State the null and alternative hypotheses and carry out the test at the 1% level of significance

7–77. It is claimed that out of all the air-travel bookings in major airlines, at least 58% are done online. A sample of 70 airlines revealed that 52% of bookings for last year were done online. State the null and alternative hypotheses and carry out the test at the 5% level of significance.

7–78. According to an article published in *The Wall Street Journal,* in the United States, unionized employees average 6.63 hours of overtime on top of their regular 40-hour week. To check the accuracy of the statement sample evidence was gathered. A sample of 100 employees revealed their average overtime was 5 hours per week. If the sample standard deviation is 2.1 hours, conduct the hypothesis test at the 1% level of significance.

7–79. According to an article in *Business Week,* the World Health Organization estimates that one out of every three workers may be toiling away in a workplace that is making them sick. You suspect that this is an overestimate. A sample of 200 workers showed that 59 of them suffered from sick-building syndrome. State the null and alternative hypotheses and carry out the test at the 5% level of significance.

7–80. It is reported that one out of every four flights is delayed.[14] To test the accuracy of the report, a sample of 300 flights showed 30% of the flights were delayed. State the null and alternative hypotheses and carry out the test at the 5% level of significance.

7–81. According to *Financial Times,* the AFL-CIO estimates that 60% of adult women in the United States work.[15] To test the accuracy of the estimate, a sample of 200 women was taken and it showed that 65% of them were working. State the null and alternative hypotheses and carry out the test at the 5% level of significance.

7–82. The best places in the United States to be a job seeker are state capitals and university towns, which are claimed to have jobless rates below the national average of 4.2%. A sample of 50 university towns and state capitals showed average jobless rate of 4.4% with a standard deviation of 0.8%. State the null and alternative hypotheses and carry out the test at the 1% level of significance.

---

[13]Problems 7–75 to 7–77 based on information from *Smart Business,* January 2001.

[14]From "Your Guide to Travel," *Money,* January 2001.

[15]From "More Women Working in the United States," *Financial Times,* January 2001.

7–5 Pretest Decisions

Sampling costs money, and so do errors. In the previous chapter we saw how to minimize the total cost of sampling and estimation errors. In this chapter, we do the same for hypothesis testing. Unfortunately, however, finding the cost of errors in hypothesis testing is not as straightforward as in estimation. The reason is that the probabilities of type I and type II errors depend on the actual value of the parameter being tested. Not only do we not know the actual value, but we also do not usually know its distribution. It is therefore difficult, or even impossible, to estimate the expected cost of errors. As a result, people follow a simplified policy of fixing a standard value for α (1%, 5%, or 10%) and a certain minimum sample size for evidence gathering. With the advent of spreadsheets, it is possible to look at the situation more closely and, if needed, change policies.

To look at the situation more closely, we can use the following templates that compute various parameters of the problem and plot helpful charts:

1. Sample size template.
2. β versus α for various sample sizes.
3. The power curve.
4. The operating characteristic curve.

We will see these four templates in the context of testing population means. Similar templates are also available for testing population proportions.

Testing Population Means

Figure 7–14 shows the template that can be used for determining sample sizes when α has been fixed and a limit on the probability of type II error at a predetermined actual value of the population mean has also been fixed. Let us see the use of the template through an example.

The tensile strength of parts made of an alloy is claimed to be at least 1,000 kg/cm². The population standard deviation is known from past experience to be 10 kg/cm². It is desired to test the claim at an α of 5% with the probability of type II error, β, restricted to 8% when the actual strength is only 995 kg/cm². The engineers are not sure about their decision to limit β as described and want to do a sensitivity analysis of the sample size on actual μ ranging from 994 to 997 kg/cm² and limits on β ranging from 5% to 10%. Prepare a plot of the sensitivity.

EXAMPLE 7–9

We use the template shown in Figure 7–14. The null and alternative hypotheses in this case are

$$H_0: \mu \geq 1{,}000 \text{ kg/cm}^2$$
$$H_1: \mu < 1{,}000 \text{ kg/cm}^2$$

Solution

To enter the null hypothesis, choose ">=" in the drop down box, and enter 1000 in cell C4. Enter σ of 10 in cell C5 and α of 5% in cell C6. Enter 995 in cell C9 and the limit of 8% in cell C10. The result 38 appears in cell C12. Since this is greater than 30, the assumption of $n \geq 30$ is satisfied and all calculations are valid.

FIGURE 7–14   The Template for Computing and Plotting Required Sample Size
[Testing Population Mean.xls; Sheet: Sample Size]

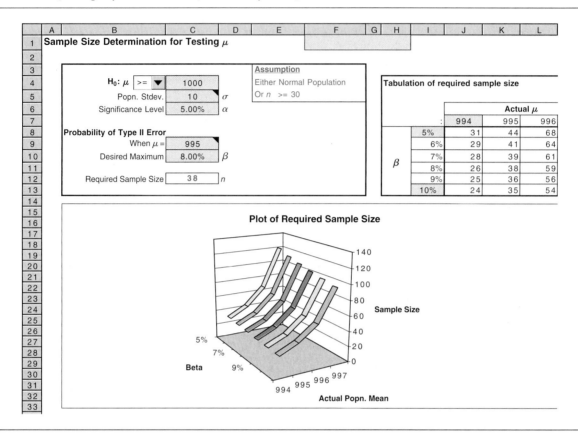

To do the sensitivity analysis, enter 5% in cell I8, 10% in cell I13, 994 in cell J7, and 997 in cell M7. The required tabulation and the chart appear, and they may be printed and reported to the engineers.

Manual Calculation of Required Sample Size

The equation for calculating the required sample size is

$$n = \left\lceil \left(\frac{(|z_0| + |z_1|)\sigma}{\mu_0 - \mu_1} \right)^2 \right\rceil$$

where μ_0 = hypothesized value of μ in H_0
   μ_1 = the value of μ at which type II error is to be monitored
   $z_0 = z_\alpha$ or $z_{\alpha/2}$ depending on whether the test is one-tailed or two-tailed
   $z_1 = z_\beta$ where β is the limit on type II error probability when $\mu = \mu_1$

The symbol $\lceil\ \rceil$ stands for rounding up to the next integer. For example, $\lceil 35.2 \rceil = 36$. Note that the formula calls for the absolute values of z_0 and z_1, so enter positive values regardless of right-tailed or left-tailed test. If the template is not available, this equation can be used to calculate the required n manually.

FIGURE 7–15   The Template for Plotting β versus α for Various *n*
[Testing Population Mean.xls; Sheet: Beta vs. Alpha]

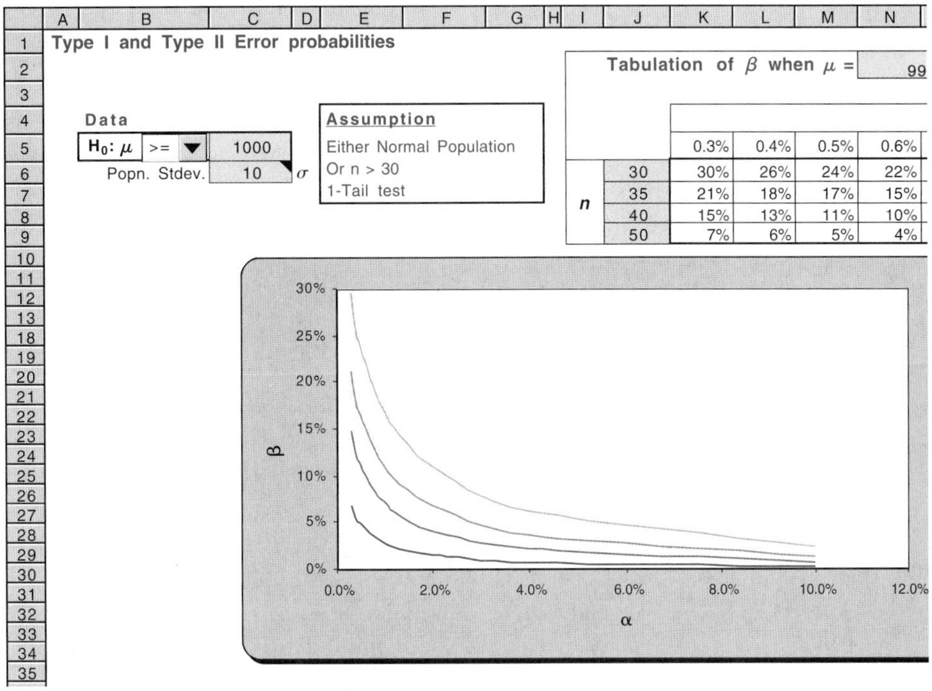

The manual calculation of required sample size for Example 7–9 is

$$n = \left\lceil \left(\frac{(1.645 + 1.4)10}{1{,}000 - 995} \right)^2 \right\rceil = \lceil 37.1 \rceil = 38$$

Figure 7–15 shows the template that can be used to plot β versus α for four different values of *n*. We shall see the use of this template through an example.

EXAMPLE 7–10

The tensile strength of parts made of an alloy is claimed to be at least 1,000 kg/cm². The population standard deviation is known from past experience to be 10 kg/cm². The engineers at a company want to test this claim. To decide *n*, α, and the limit on β, they would like to look at a plot of β when actual μ = 994 kg/cm² versus α for *n* = 30, 35, 40, and 50. Further, they believe that type II errors are more costly and therefore would like β to be not more than half the value of α. Can you make a suggestion for the selection of α and *n*?

Solution

Use the template shown in Figure 7–15. Enter the null hypothesis H₀: μ ≥ 1000 in the range B5:C5. Enter the σ value of 10 in cell C6. Enter the actual μ = 994 in the range N2:O2. Enter the *n* values 30, 35, 40, and 50 in the range J6:J9. The desired plot of β versus α is created.

Looking at the plot, for the standard α value of 5%, a sample size of 40 yields a β of approximately 2.5%. Thus the combination α = 5% and *n* = 40 is a good choice.

FIGURE 7–16   The Template for Plotting the Power Curve
[Testing Population Mean; Sheet: Power]

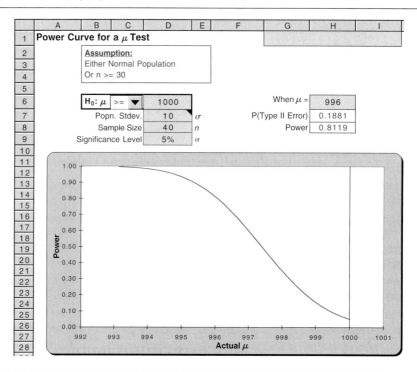

Figure 7–16 shows the template that can be used to plot the **power curve** of a hypothesis test once α and n have been determined. This curve is useful in determining the power of the test for various actual μ values. Since α and n are usually selected without knowing the actual μ, this plot can be used to check if they have been selected well with respect to power. In Example 7–10, if the engineers wanted a power curve of the test, the template shown in Figure 7–16 can be used to produce it. The data and the chart in the figure correspond to Example 7–10. A vertical line appears at the hypothesized value of the population mean, which in this case is 1,000.

The **operating characteristic curve** (OC curve) of a hypothesis test shows how the probability of not rejecting (accepting) the null hypothesis varies with the actual μ. The advantage of an OC curve is that it shows both type I and type II error instances. See Figure 7–17, which shows an OC curve for the case H_0: $\mu \geq 75$; $\sigma = 10$; $n = 40$; $\alpha = 10\%$. A vertical line appears at 75, which corresponds to the hypothesized value of the population mean. Areas corresponding to errors in the test decisions are shaded. The dark area at the top right represents type I error instances, because in that area $\mu > 75$, which makes H_0 true, but H_0 is rejected. The shaded area below represents instances of type I error, because $\mu < 75$, which makes H_0 false, but H_0 is accepted. By looking at both type I and type II error instances on a single chart, we can design a test more effectively.

Figure 7–18 shows the template that can be used to plot OC curves. The template will not shade the areas corresponding to the errors. But that is all right, because we would like to superpose two OC curves on a single chart corresponding to two sample sizes, n_1 and n_2 entered in cells H7 and H8. We shall see the use of the template through an example.

FIGURE 7–17   An Operating Characteristic Curve for the Case H_0: $\mu \geq 75$; $\sigma = 10$; $n = 40$;
$\alpha = 10\%$

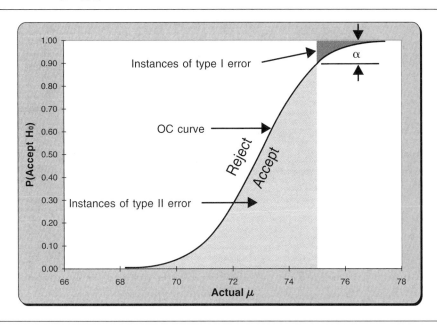

FIGURE 7–18   The Template for Plotting the Operating Characteristic Curve
[Testing Population Mean.xls; Sheet: OC Curve]

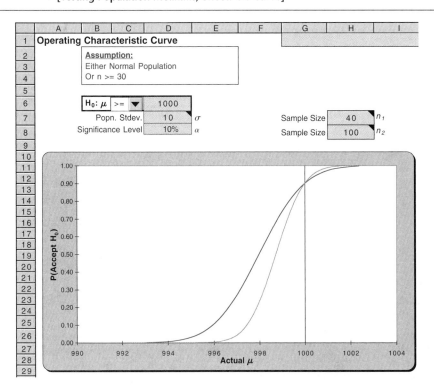

FIGURE 7–19   The Template for Finding the Required Sample Size
[Testing Population Proportion.xls; Sheet: Sample Size]

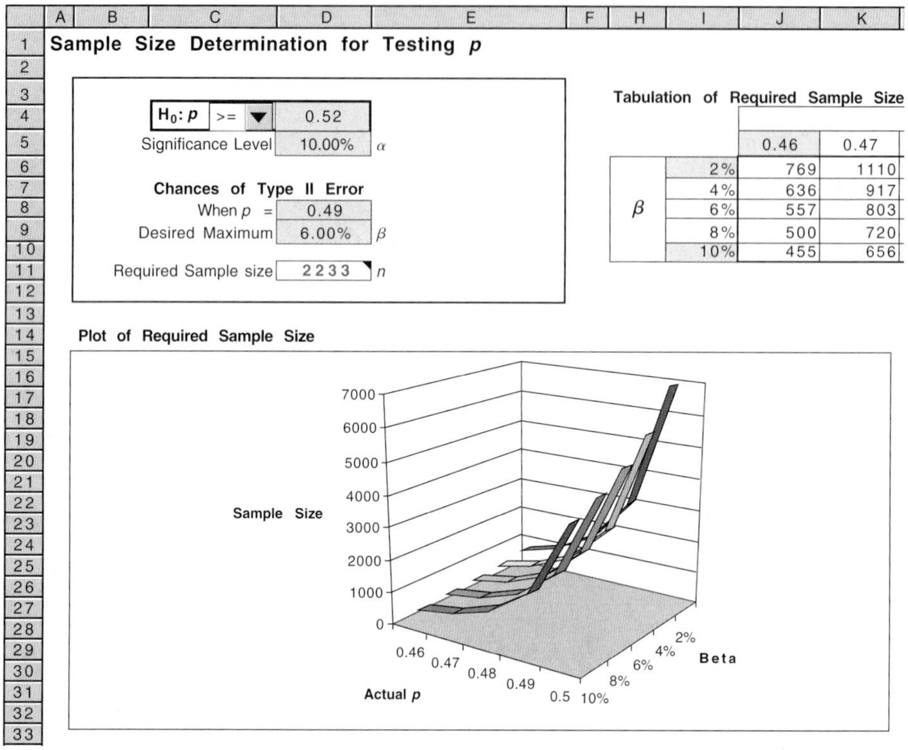

EXAMPLE 7–11   Consider the problem in Example 7–10. The engineers want to see the complete picture of type I and type II error instances. In particular, when $\alpha = 10\%$, they want to know the effect of increasing the sample size from 40 to 100 on type I and type II error possibilities. Construct the OC curves for $n_1 = 40$ and $n_2 = 100$ and comment on the effects.

Solution   Open the template shown in Figure 7–18. Enter the null hypothesis in the range B6:D6, and the σ value of 10 in cell D7. Enter the α value of 10% in cell D8. Enter 40 and 100 in cells H7 and H8. The needed OC curves appear in the chart.

Looking at the OC curves, we see that increasing the sample size from 40 to 100 does not affect the instances of type I error much but substantially reduces type II error instances. For example, the chart reveals that when actual $\mu = 998$ the probability of type II error, β, is reduced by more than 50% and when actual $\mu = 995$ β is almost zero. If these gains outweigh the cost of additional sampling, then it is better to go for a sample size of 100.

Testing Population Proportions

Figure 7–19 shows the template that can be used to calculate the required sample size while testing population means.

It is claimed that at least 52% of a city's population oppose the construction of a high- **EXAMPLE 7–12**
way near the city. It is desired to test the claim at $\alpha = 10\%$. The probability of type II
error when the actual proportion is 49% is to be limited to 6%.

1. How many randomly selected residents of the city should be polled to test
   the claim?
2. Tabulate the required sample size for limits on β varying from 2% to 10%
   and actual proportion varying from 46% to 50%.
3. If the budget allows only a sample size of 2000 and therefore that is the
   number polled, what is the probability of type II error when the actual
   proportion is 49%?

Open the template shown in Figure 7–19. Enter the null hypothesis, H_0: $p \geq 52\%$ *Solution*
in the range C4:D4. Enter α in cell D5, and type II error information in cells D8
and D9.

1. The required sample size of 2233 appears in cell D11.
2. Enter the β values 2% in cell I6 and 10% in cell I10. Enter 0.46 in J5 and
   0.50 in cell N5. The needed tabulation appears in the range I5:N10.
3. In the tabulation of required sample size, in the column corresponding to
   $p = 0.49$, the value 2004 appears in cell M9, which corresponds to a β value
   of 8%. Thus the probability of type II error is about 8%.

Manual Calculation of Sample Size

If the template is not available, the required sample size for testing population proportions can be calculated using the equation

$$n = \left\lceil \left(\frac{(|z_0| \sqrt{p_0(1 - p_0)} + |z_1| \sqrt{p_1(1 - p_1)})}{p_0 - p_1} \right)^2 \right\rceil$$

where p_0 = hypothesized value of μ in H_0
$\quad\quad p_1$ = the value of p at which type II error is to be monitored
$\quad\quad z_0 = z_\alpha$ or $z_{\alpha/2}$ depending on whether the test is one-tailed or two-tailed
$\quad\quad z_1 = z_\beta$ where β is the limit on type II error probability when $p = p_1$.

For the case in Example 7–12, the calculation will be

$$n = \left\lceil \left(\frac{(1.28 \sqrt{0.53(1 - 0.53)} + 1.555 \sqrt{0.49(1 - 0.49)})}{0.53 - 0.49} \right)^2 \right\rceil = \lceil 2{,}230.5 \rceil = 2{,}231$$

The difference of 2 in the manual and template results is due to the approximation of
z_0 and z_1 in manual calculation.

The power curve and the OC curves can be produced for hypothesis tests regarding population proportions using the templates shown in Figures 7–20 and 7–21.
Let us see the use of the charts through an example.

FIGURE 7–20   **The Template for Drawing a Power Curve**
             [Testing Population Proportion.xls; Sheet: Power]

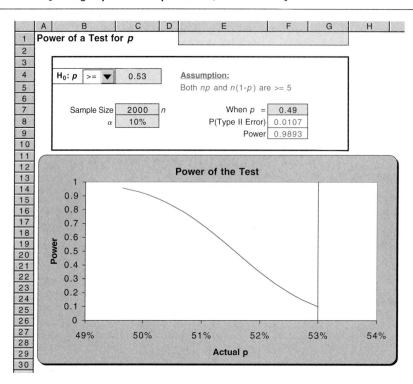

EXAMPLE 7–13   The hypothesis test in Example 7–12 is conducted with sample size 2,000 and $\alpha = 10\%$. Draw the power curve and the OC curve of the test.

Solution   For the power curve, open the template shown in Figure 7–20. Enter the null hypothesis, sample size, and α in their respective places. The power curve appears below the data. For the power at a specific point use the cell F7. Entering 0.49 in cell F7 shows that the power when $p = 0.49$ is 0.9893.

For the OC curve open the template shown in Figure 7–21. Enter the null hypothesis and α in their respective places. Enter the sample size 2000 in cell C7 and leave cell D7 blank. The OC curve appears below the data.

PROBLEMS

7–83.   Consider the null hypothesis $\mu \geq 56$. The population standard deviation is guessed to be 2.16. Type II error probabilities are to be calculated at $\mu = 55$.

   a. Draw a β versus α chart with sample sizes 30, 40, 50, and 60.

   b. The test is conducted with a random sample of size 50 with $\alpha = 5\%$. Draw the power curve. What is the power when $\mu = 55.5$?

FIGURE 7–21   The Template for Drawing OC Curves
[Testing Population Proportion.xls; Sheet: OC Curve]

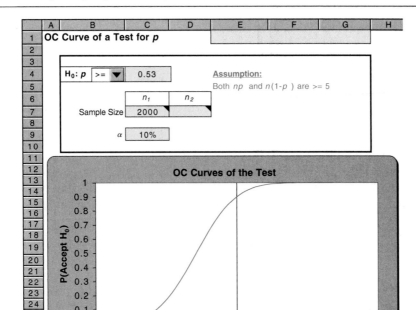

c. Draw an OC curve with $n = 50$ and 60; $\alpha = 5\%$. Is there a lot to gain by going from $n = 50$ to $n = 60$?

7–84. The null hypothesis $p = 0.25$ is tested with $n = 1,000$ and $\alpha = 5\%$.

a. Draw the power curve. What is the power when $p = 0.22$?

b. Draw the OC curve for $n = 1,000$ and 1,200. Is there a lot to gain by going from $n = 1,000$ to $n = 1,200$?

7–85. The null hypothesis $\mu \leq 30$ is to be tested. The population standard deviation is guessed to be 0.52. Type II error probabilities are to be calculated at $\mu = 30.3$.

a. Draw a β versus α chart with sample sizes 30, 40, 50, and 60.

b. The test is conducted with a random sample of size 30 with an α of 5%. Draw the power curve. What is the power when $\mu = 30.2$?

c. Draw an OC curve with $n = 30$ and 60; $\alpha = 5\%$. If type II error is to be almost zero when $\mu = 30.3$, is it better to go for $n = 60$?

7–86. If you look at the power curve or the OC curve of a two-tailed test, you see that there is no *region* that represents instances of type I error, whereas there are large regions that represent instances of type II error. Does this mean that there is no chance of type I error? Think carefully, and explain the chances of type I error and the role of α in a two-tailed test.

7–6   Summary and Review of Terms

In this chapter, we introduced the important ideas of statistical hypothesis testing. We discussed the philosophy behind hypothesis tests, starting with the concepts of **null hypothesis** and **alternative hypothesis.** Depending on the type of null hypothesis, the rejection occurred either on one or both tails of the **test statistic.** Correspondingly, the test became either a **one-tailed test** or a **two-tailed test.** In any test, we saw that there will be chances for **type I** and **type II errors.** We saw how the *p*-value is used in an effort to systematically contain the chances of both types of error. When the *p*-value is less than the **level of significance** α, the null hypothesis is rejected. The probability of not committing a type I error is known as the **confidence level,** and the probability of not committing a type II error is known as the **power** of the test. We also saw how increasing the sample size decreases the chances of both types of errors.

In connection with pretest decisions we saw the compromise between the costs of type I and type II errors. These cost considerations help us in deciding the **optimal sample size** and a suitable level of significance α. In the next chapter we extend these ideas of hypothesis testing to *differences* between two population parameters.

CASE 8   Tiresome Tires I

When a tire is constructed of more than one ply, the interply shear strength is an important property to check. The specification for a particular type of tire calls for a strength of 2,800 pounds per square inch (psi). The tire manufacturer tests the tires using the null hypothesis

$$H_0: \mu \geq 2{,}800 \text{ psi}$$

where μ is the mean strength of a large batch of tires. From past experience, it is known that the population standard deviation is 20 psi.

Testing the shear strength requires a costly destructive test and therefore the sample size needs to be kept at a minimum. A type I error will result in the rejection of a large number of good tires and is therefore costly. A type II error of passing a faulty batch of tires can result in fatal accidents on the roads, and therefore is extremely costly. (For purposes of this case, the probability of type II error, β, is always calculated at $\mu = 2{,}790$ psi.) It is believed that β should be at most 1%. Currently, the company conducts the test with a sample size of 40 and an α of 5%.

1. To help the manufacturer get a clear picture of type I and type II error probabilities, draw a β versus α chart for sample sizes of 30, 40, 60, and 80. If β is to be at most 1% with $\alpha = 5\%$, which sample size among these four values is suitable?

2. Calculate the exact sample size required for $\alpha = 5\%$ and $\beta = 1\%$. Construct a sensitivity analysis table for the required sample size for μ ranging from 2,788 to 2,794 psi and β ranging from 1% to 5%.

3. For the current practice of $n = 40$ and $\alpha = 5\%$ plot the power curve of the test. Can this chart be used to convince the manufacturer about the high probability of passing batches that have a strength of less than 2,800 psi?

4. To present the manufacturer with a comparison of a sample size of 80 versus 40, plot the OC curve for those two sample sizes. Keep an α of 5%.

5. The manufacturer is hesitant to increase the sample size beyond 40 due to the concomitant increase in testing costs and, more important, due to the increased time required for the tests. The production process needs to wait

until the tests are completed, and that means loss of production time. A suggestion is made by the production manager to increase α to 10% as a means of reducing β. Give an account of the benefits and the drawbacks of that move. Provide supporting numerical results wherever possible.

8 The Comparison of Two Populations

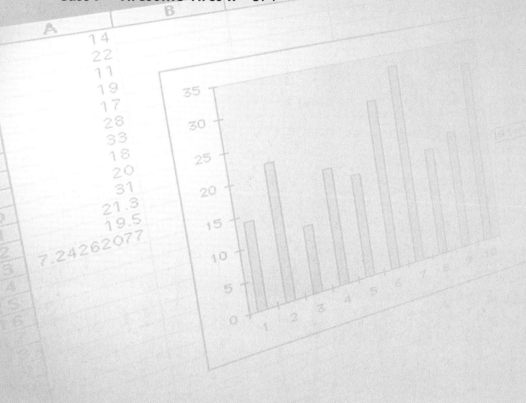

8–1 Using Statistics

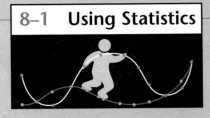

The findings of the study had such potentially far-reaching implications that the Reuters agency, with advance knowledge of the findings, reported the information to its subscribing newspapers more than 24 hours before the article describing the study appeared in the January 28, 1988, issue of the *New England Journal of Medicine*.[1] The large-scale study was begun in 1982. Its subjects were a group of 22,071 male physicians 40 to 84 years of age living in the United States. Each physician was randomly assigned to take one of two kinds of pills: aspirin or beta carotene (as a placebo). Thus, 11,037 physicians were randomly assigned to take aspirin, and 11,034 physicians were assigned to take the placebo. No physician knew which of the two kinds of pills he was taking. Each physician was to take the assigned pill every other day until the scheduled conclusion of the study in 1990. However, at a special meeting on December 18, 1987, the Data Monitoring Board of the Physicians' Health Study took the unusual step of recommending the early termination of the randomized aspirin experiment. At this point, the statistical results of the study had already exceeded everyone's expectations. It was shown, with a p-value of less than 0.00001, that a population (of male physicians, at least) taking a single aspirin pill once every 2 days would have a significantly lower proportion of heart attacks than a similar population not taking aspirin. The study confirmed what doctors have suspected for years—that the wonder drug aspirin can actually help prevent heart attacks.

The comparison of two populations with respect to some population parameter—the population mean, the population proportion, or the population variance—is the topic of this chapter. Testing hypotheses about population parameters in the single-population case, as was done in Chapter 7, is an important statistical undertaking. However, the true usefulness of statistics manifests itself in allowing us to make *comparisons*. Almost daily we compare products, services, investment opportunities, management styles, and so on. In this chapter, we will learn how to conduct such comparisons in an objective and meaningful way. We will learn how to find statistically significant differences between two populations. If you understood the methodology of hypothesis testing presented in the last chapter and the idea of a confidence interval from Chapter 6, you will find the extension to two populations straightforward and easy to understand. We will learn how to conduct a test for the existence of a difference between the means of two populations. In the next section, we will see how such a comparison may be made in the special case where the observations may be paired in some way. Later we will learn how to conduct a test for the equality of the means of two populations, using independent random samples. Then we will see how to compare two population proportions. Finally, we will encounter a test for the equality of the variances of two populations. In addition to statistical hypothesis tests, we will learn how to construct confidence intervals for the difference between two population parameters.

---

[1]The editors of the *New England Journal of Medicine* reacted by barring the London-based Reuters from all new information from the *Journal* for a period of six months.

8–2   Paired-Observation Comparisons

In this section, we describe a method for conducting a hypothesis test and constructing a confidence interval when our observations come from two populations and are *paired* in some way. What is the advantage of pairing observations? Suppose that a taste test of two flavors is carried out. It seems intuitively plausible that if we let every person in our sample rate each one of the two flavors (with random choice of which flavor is tasted first), the resulting *paired* responses will convey more information about the taste difference than if we had used two different sets of people, each group rating only one flavor. Statistically, when we use the same people for rating the two products, we tend to remove much of the *extraneous variation* in taste ratings—the variation in people, experimental conditions, and other extraneous factors—and concentrate on the difference between the two flavors. When possible, it is often advisable to pair the observations, as this makes the experiment more precise. We will demonstrate the paired-observation test with an example.

EXAMPLE 8–1 Home Shopping Network, Inc., pioneered the idea of merchandising directly to customers through cable television. By watching what amounts to 24 hours of commercials, viewers can call a number to buy products. Before expanding their services, network managers wanted to test whether this method of direct marketing increased sales on the average. A random sample of 16 viewers was selected for an experiment. All viewers in the sample had recorded the amount of money they spent shopping during the holiday season of the previous year. The next year, these people were given access to the cable network and were asked to keep a record of their total purchases during the holiday season. The paired observations for each shopper are given in Table 8–1. Faced with these data, Home Shopping Network managers want to test

TABLE 8–1   Total Purchases of 16 Viewers with and without Home Shopping

Shopper	Current Year's Shopping ($)	Previous Year's Shopping ($)	Difference ($)
1	405	334	71
2	125	150	−25
3	540	520	20
4	100	95	5
5	200	212	−12
6	30	30	0
7	1,200	1,055	145
8	265	300	−35
9	90	85	5
10	206	129	77
11	18	40	−22
12	489	440	49
13	590	610	−20
14	310	208	102
15	995	880	115
16	75	25	50

the null hypothesis that their service does not increase shopping volume, versus the alternative hypothesis that it does. The following solution of this problem introduces the *paired-observation t test*.

Solution

The test involves two populations: the population of shoppers who have access to the Home Shopping Network and the population of shoppers who do not. We want to test the null hypothesis that the mean shopping expenditure in both populations is equal versus the alternative hypothesis that the mean for the home shoppers is greater. Using the same people for the test and pairing their observations in a before-and-after way makes the test more precise than it would be without pairing. The pairing removes the influence of factors other than home shopping. The shoppers are the same people; thus, we can concentrate on the effect of the new shopping opportunity, leaving out of the analysis other factors that may affect shopping volume. Of course, we must consider the fact that the first observations were taken a year before. Let us assume, however, that relative inflation between the two years has been accounted for and that people in the sample have not had significant changes in income or other variables since the previous year that might affect their buying behavior.

Under these circumstances, it is easy to see that the variable in which we are interested is the difference between the present year's per-person shopping expenditure and that of the previous year. The population parameter about which we want to draw an inference is the mean difference between the two populations. We denote this parameter by μ_D, the mean difference. This parameter is defined as $\mu_D = \mu_1 - \mu_2$, where μ_1 is the average holiday season shopping expenditure of people who use home shopping and μ_2 is the average holiday season shopping expenditure of people who do not. Our null and alternative hypotheses are, then,

$$H_0: \mu_D \le 0 \tag{8–1}$$

$$H_1: \mu_D > 0$$

Looking at the null and alternative hypotheses and the data in the last column of Table 8–1, we note that the test is a simple t test with $n - 1$ degrees of freedom, where our variable is the *difference* between the two observations for each shopper. In a sense, our two-population comparison test has been reduced to a hypothesis test about one parameter—the difference between the means of two populations. The test, as given by equation 8–1, is a right-tailed test, but it need not be. In general, the paired-observation t test can be done as one-tailed or two-tailed. In addition, the hypothesized difference need not be zero. We can state any other value as the difference in the null hypothesis (although zero is most commonly used). The only assumption we make when we use this test is that *the population of differences is normally distributed.* Recall that this assumption was used whenever we carried out a test or constructed a confidence interval using the t distribution. Also note that, for large samples, the standard normal distribution may be used instead. This is also true for a normal population if you happen to know the population standard deviation of the differences σ_D. The test statistic (assuming σ_D is not known and is estimated by s_D, the sample standard deviation of the differences) is given in equation 8–2.

The test statistic for the paired-observation t test is

$$t = \frac{\overline{D} - \mu_{D_0}}{s_D/\sqrt{n}}$$ (8–2)

where $\overline{D}$ is the sample average difference between each pair of observations, s_D is the sample standard deviation of these differences, and the sample size n is the number of pairs of observations (here, the number of people in the experiment). The symbol μ_{D_0} is the population mean difference under the null hypothesis. When the null hypothesis is true and the population mean difference is μ_{D_0}, the statistic has a t distribution with $n - 1$ degrees of freedom.

Let us now conduct the hypothesis test. From the differences reported in Table 8–1, we find that their mean is $\overline{D} = \$32.81$ and their standard deviation is $s_D = \$55.75$. Since the sample size is small, $n = 16$, we use the t distribution with $n - 1 = 15$ degrees of freedom. The null hypothesis value of the population mean is $\mu_{D0} = 0$. The value of our test statistic is obtained as

$$t = \frac{32.81 - 0}{55.75/\sqrt{16}} = 2.354$$

This computed value of the test statistic is greater than 1.753, which is the critical point for a right-tailed test at $\alpha = 0.05$ using a t distribution with 15 degrees of freedom (see Appendix C, Table 3). The test statistic value is less than 2.602, which is the critical point for a one-tailed test using $\alpha = 0.01$, but greater than 2.131, which is the critical point for a right-tailed area of 0.025. We may conclude that the p-value is between 0.025 and 0.01. This is shown in Figure 8–1. Home Shopping Network managers may

FIGURE 8–1   Carrying Out the Test of Example 8–1

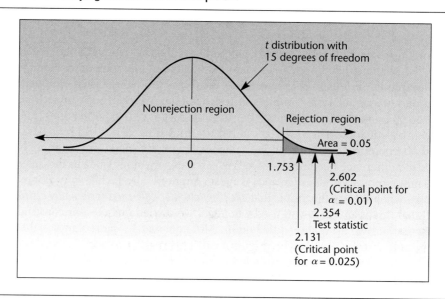

conclude that the test gave significant evidence for increased shopping volume by network viewers.

The Template

Figure 8–2 shows the template that can be used to test paired differences in population means when the sample data are known. The data are entered in columns B and C. The data and the results seen in the figure correspond to Example 8–1. The hypothesized value of the difference is entered in cell F12, and this value is automatically copied into cells F13 and F14 below. The desired α is entered in cell H11. For the present case, the null hypothesis is $\mu_1 - \mu_2 \leq 0$. The corresponding p-value of 0.0163 appears in cell G14. As seen in cell H14, the null hypothesis is to be rejected at an α of 5%.

If a confidence interval is desired, then the confidence level must be entered in cell J12. The α corresponding to the confidence level in cell J12 need not be the same as the α for the hypothesis test entered in cell H11. If a confidence interval is not desired, then cell J12 may be left blank to avoid creating a distraction.

EXAMPLE 8–2

It has recently been asserted that returns on stocks may change once a story about a company appears in *The Wall Street Journal* column "Heard on the Street." An investment portfolio analyst wants to check the statistical significance of this claim. The analyst collects a random sample of 50 stocks that were recommended as winners by the editor of "Heard on the Street." The analyst proceeds to conduct a two-tailed test of whether the annualized return on stocks recommended in the column differs between the month before the recommendation and the month after the recommendation. The analyst decides to conduct a two-tailed rather than a one-tailed test because she wants to allow for the possibility that stocks may be recommended in the column after their price has appreciated (and thus returns may actually decrease in the following month), as well as allowing for an increased return. For each stock in the sample of 50, the analyst computes the return before and after the *event* (the appearance of the story in the column) and the difference between the two return figures. Then the

FIGURE 8–2 **The Template for Testing Paired Differences [Testing Paired Difference.xls; Sheet: Sample Data]**

	A	B	C	D	E	F	G	H	I	J	K
1	**Paired Difference Test**										
2		**Data**									
3		Current	Previous	**Evidence**							
4		Sample1	Sample2		Size	16	n	Assumption			
5	1	405	334		Average Difference	32.8125	μ_D	Populations Normal			
6	2	125	150		Stdev. of Difference	55.7533	s_D				
7	3	540	520					Note: Difference has been defined			
8	4	100	95		Test Statistic	2.3541	t	Sample1 - Sample2			
9	5	200	212		df	15					
10	6	30	30	**Hypothesis Testing**				**At an α of**		**Confidence Interva**	
11	7	1200	1055		**Null Hypothesis**		**p-value**	5%		**(1−α)**	**Cor**
12	8	265	300		$H_0: \mu_1 - \mu_2 = 0$		0.0326	Reject		95%	32.8
13	9	90	85		$H_0: \mu_1 - \mu_2 >= 0$		0.9837				
14	10	206	129		$H_0: \mu_1 - \mu_2 <= 0$		0.0163	Reject			
15	11	18	40								

sample average difference of returns is computed, as well as the sample standard deviation of return differences. The results are $\overline{D} = 0.1\%$ and $s_D = 0.05\%$. What should the analyst conclude?

Solution   The null and alternative hypotheses are H_0: $\mu_D = 0$ *and* H_1: $\mu_D \approx 0$. We now use the test statistic given in equation 8–2, noting that the distribution may be well approximated by the normal distribution because the sample size $n = 50$ is large. We have

$$t = \frac{\overline{D} - \mu_{D_0}}{s_D/\sqrt{n}} = \frac{0.1 - 0}{0.05/7.07} = 14.14$$

The value of the test statistic falls very far in the right-hand rejection region, and the *p*-value, therefore, is very small. The analyst should conclude that there is strong evidence that the average returns on stocks increase (because the rejection occurred in the right-hand rejection region) for stocks recommended in "Heard on the Street," as asserted by financial experts. Incidentally, the assumption of normality of the population of return differences—while not crucial in this case because the sample size is large—is supported by research findings on stock returns.

Confidence Intervals

In addition to tests of hypotheses, it is possible to construct confidence intervals for the average population difference μ_D. Analogous to the case of a single-population parameter, we define a $(1 - \alpha)$ 100% confidence interval for the parameter μ_D as follows.

A $(1 - \alpha)$ 100% confidence interval for the mean difference μ_D is

$$\overline{D} \pm t_{\alpha/2}\frac{s_D}{\sqrt{n}} \tag{8–3}$$

where $t_{\alpha/2}$ is the value of the t distribution with $n - 1$ degrees of freedom that cuts off an area of $\alpha/2$ to its right. When the sample size n is large, we may use $z_{\alpha/2}$ instead.

In Example 8–2, we may construct a 95% confidence interval for the average difference in annualized return on a stock before and after its being recommended in "Heard on the Street." The confidence interval is

$$\overline{D} \pm t_{\alpha/2}\frac{s_D}{\sqrt{n}} = 0.1 \pm 1.96\frac{0.05}{7.07} = [0.086\%, 0.114\%]$$

Based on the data, the analyst may be 95% confident that the average difference in annualized return rate on a stock, measured the month before and the month following a positive recommendation in the column, is anywhere from 0.086% to 0.114%.

The Template

Figure 8–3 shows the template that can be used to test paired differences, when sample statistics rather than sample data are known. The data and results in this figure correspond to Example 8–2.

FIGURE 8–3   **The Template for Testing Paired Differences**
[Testing Paired Difference.xls; Sheet: Sample Statistics]

	A	B	C	D	E	F	G	H	I	J	K
1	**Paired Difference Test**										
2	Evidence										
3		Size	50	n		**Assumption**					
4		Average Difference	0.1	μ_D		Populations Normal					
5		Stdev. of Difference	0.05	s_D							
6						**Note**: Difference has been defined as					
7		Test Statistic	14.1421	t		Sample1 - Sample2					
8		df	49								
9	**Hypothesis Testing**					**At an α of**		Confidence Intervals for the Difference			
10		**Null Hypothesis**		p-value	5%	$(1-\alpha)$		Confidence Interval			
11		$H_0: \mu_1 - \mu_2 = 0$		0.0000	Reject	95%		0.1 $\pm$ 0.01421			
12		$H_0: \mu_1 - \mu_2 \geq 0$		1.0000							
13		$H_0: \mu_1 - \mu_2 \leq 0$		0.0000	Reject						
14											
15											

In this section, we compared population means for paired data. The following sections compare means of two populations where samples are drawn randomly and *independently* of each other from the two populations. When pairing can be done, our results tend to be more precise because the *experimental units* (e.g., the people, each trying two different products) are different from each other, but each acts as an independent measuring device for the two products. This pairing of similar items is called *blocking*, and we will discuss it in detail in Chapter 9.

PROBLEMS

8–1. A market research study is undertaken to test which of two popular electric shavers, a model made by Norelco or a model made by Remington, is preferred by consumers. A random sample of 25 men who regularly use an electric shaver, but not one of the two models to be tested, is chosen. Each man is then asked to shave one morning with the Norelco and the next morning with the Remington, or vice versa. The order, which model is used on which day, is randomly chosen for each man. After every shave, each man is asked to complete a questionnaire rating his satisfaction with the shaver. From the questionnaire, a total satisfaction score on a scale of 0 to 100 is computed. Then, for each man, the difference between the satisfaction score for Norelco and that for Remington is computed. The score differences (Norelco score − Remington score) are 15, −8, 32, 57, 20, 10, −18, −12, 60, 72, 38, −5, 16, 22, 34, 41, 12, −38, 16, −40, 75, 11, 2, 55, 10. Which model, if any, is statistically preferred over the other? How confident are you of your finding? Explain.

8–2. The performance ratings of two sports cars, the Mazda RX7 and the Nissan 300ZX, are to be compared. A random sample of 40 drivers is selected to drive the two models. Each driver tries one car of each model, and the 40 cars of each model are chosen randomly. The time of each test drive is recorded for each driver and model. The difference in time (Mazda time − Nissan time) is computed, and from these differences a sample mean and a sample standard deviation are obtained. The results are $\overline{D} = 5.0$ seconds and $s_D = 2.3$ seconds. Based on these data, which model has higher performance? Explain. Also give a 95% confidence interval for the average time difference, in seconds, for the two models over the course driven.

8–3. Recent studies indicate that in order to be globally competitive, firms must form global strategic partnerships. An investment banker wants to test whether the return on investment for international ventures is different from return on investment for similar domestic ventures. A sample of 12 firms that recently entered into ventures with foreign companies is available. For each firm, the returns on investment for the international venture (I) and a similar domestic venture (D) are given:

D (%):	10	12	14	12	12	17	9	15	8.5	11	7	15
I (%):	11	14	15	11	12.5	16	10	13	10.5	17	9	19

Assuming that these firms represent a random sample from the population of all firms involved in global strategic partnerships, can the investment banker conclude that there are differences between average returns on domestic ventures and average returns on international ventures? Explain.

8–4. A study is undertaken to determine how consumers react to energy conservation efforts. A random group of 60 families is chosen. Their consumption of electricity is monitored in a period before and a period after the families are offered certain discounts to reduce their energy consumption. Both periods are the same length. The difference in electric consumption between the period before and the period after the offer is recorded for each family. Then the average difference in consumption and the standard deviation of the difference are computed. The results are $\overline{D} = 0.2$ kilowatt and $s_D = 1.0$ kilowatt. At $\alpha = 0.01$, is there evidence to conclude that conservation efforts reduce consumption?

8–5. A nationwide retailer wants to test whether new product shelf facings are effective in increasing sales volume. New shelf facings for the soft drink Country Time are tested at a random sample of 15 stores throughout the country. Data on total sales of Country Time for each store, for the week before and the week after the new facings are installed, are given below:

Store:	1	2	3	4	5	6	7	8	9	10	11	12	13	14	15
Before:	57	61	12	38	12	69	5	39	88	9	92	26	14	70	22
After:	60	54	20	35	21	70	1	65	79	10	90	32	19	77	29

Using the 0.05 level of significance, do you believe that the new shelf facings increase sales of Country Time?

8–6. Most international investors managing Asian portfolios invest in both Japan and Hong Kong. Favorable conditions are believed to have brought more of these investments to Hong Kong. A test is carried out during the month of October to determine whether investors have shifted the proportions of their portfolios in favor of Hong Kong investments versus the Japanese investments. A random sample of 25 international investors is collected. These investors voluntarily reveal the proportion of their Hong Kong investments before and after October 15. The difference in proportion (after − before) is computed for each of these investors, and from these differences an average and a standard deviation are obtained. The average difference is +4%, and the standard deviation of differences is 2%. Do you believe that the average international investor has shifted investments to Hong Kong during the period in question? Explain.

8–7. In problem 8–4, suppose that the *population* standard deviation is 1.0 and that the true average reduction in consumption for the entire population in the area is μ_D = 0.1. For a sample size of 60 and $\alpha = 0.01$, what is the power of the test?

8–8. Consider the information in the following table.

Network	Program	Program Rating (Scale: 0 to 100)	
		Men	Women
CBS	60 Minutes	99	96
ABC	ABC Monday Night Football	93	25
NBC	Cheers	88	97
ABC	America's Funniest Home Videos	90	35
ABC	America's Funniest People	81	33
NBC	Unsolved Mysteries	61	10
NBC	Matlock	54	50
CBS	Murder, She Wrote	60	48
ABC	Roseanne	73	73
NBC	The Heat of the Night	44	33
ABC	Davis Rules	30	11
CBS	Murphy Brown	25	58
CBS	Rescue 911	38	18
NBC	L. A. Law	52	12
ABC	ABC Sunday Night Movies	32	61
NBC	Empty Nest	16	96
CBS	Designing Women	8	94
NBC	The Cosby Show	18	80
NBC	A Different World	9	20
NBC	NBC Sunday Night Movies	10	6

Source: Adapted from *Variety*, February 25, 1991, p. 63.

Assume that the television programs were randomly selected from the population of all prime-time TV programs. Also assume that ratings are normally distributed. Conduct a statistical test to determine whether there is a significant difference between average men's and women's ratings of prime-time television programs.

8–3 A Test for the Difference between Two Population Means Using Independent Random Samples

The paired-difference test we saw in the last section is more powerful than the tests we are going to see in this section. It is more powerful because with the same data and the same α, the chances of type II error will be less in a paired-difference test than in other tests. The reason is that pairing gets at the difference between two populations more directly. Therefore, if it is possible to pair the samples and conduct a paired-difference test, then that is what we must do. But there are many situations where the samples cannot be paired, so we cannot take a paired difference. For example, suppose two different machines are producing the same type of parts and we are interested in the difference between the average time taken by each machine to produce one part. To pair two observations we have to make the same part using each of the two machines. But it is impossible to produce the same part once by one machine and once again by the other machine. What we can do is time the machines as randomly and independently selected parts are produced on each machine. We can then compare the average time taken by each machine and test hypotheses about the difference between them.

When independent random samples are taken, the sample sizes need not be the same for both populations. We shall denote the sample sizes by n_1 and n_2. The two population means are denoted by μ_1 and μ_2 and the two population standard deviations are denoted by σ_1 and σ_2. The sample means are denoted by $\bar{X}_1$ and $\bar{X}_2$. We shall use $(\mu_1 - \mu_2)_0$ to denote the claimed difference between the two population means.

The null hypothesis can be any one of the three usual forms:

$$H_0: \mu_1 - \mu_2 = (\mu_1 - \mu_2)_0 \text{ leading to a 2-tailed test}$$
$$H_0: \mu_1 - \mu_2 \geq (\mu_1 - \mu_2)_0 \text{ leading to a left-tailed test}$$
$$H_0: \mu_1 - \mu_2 \leq (\mu_1 - \mu_2)_0 \text{ leading to a right-tailed test}$$

The test statistic can be either Z or t.

Which statistic is applicable to specific cases? This section enumerates the criteria used in selecting the correct statistic and gives the equations for the test statistics. Explanations about why the test statistic is applicable follow the listed cases.

Cases in Which the Test Statistic Is Z

1. The sample sizes n_1 and n_2 are both at least 30 and the population standard deviations σ_1 and σ_2 are known.
2. Both populations are normally distributed and the population standard deviations σ_1 and σ_2 are known.

The formula for the test statistic Z is

$$Z = \frac{(\bar{X}_1 - \bar{X}_2) - (\mu_1 - \mu_2)_0}{\sqrt{\sigma_1^2/n_1 + \sigma_2^2/n_2}} \tag{8-4}$$

where $(\mu_1 - \mu_2)_0$ is the hypothesized value for the difference in the two population means.

In the preceding cases, $\bar{X}_1$ and $\bar{X}_2$ each follows a normal distribution and therefore $(\bar{X}_1 - \bar{X}_2)$ also follows a normal distribution. Because the two samples are independent, we have

$$\text{Var}(\bar{X}_1 - \bar{X}_2) = \text{Var}(\bar{X}_1) + \text{Var}(\bar{X}_2) = \sigma_1^2/n_1 + \sigma_2^2/n_2.$$

Therefore, if the null hypothesis is true, then the quantity

$$\frac{(\bar{X}_1 - \bar{X}_2) - (\mu_1 - \mu_2)_0}{\sqrt{\sigma_1^2/n_1 + \sigma_2^2/n_2}}$$

must follow a Z distribution.

The templates to use for cases where Z is the test statistic are shown in Figures 8–4 and 8–5.

Cases in Which the Test Statistic Is t

Both populations are normally distributed; population standard deviations σ_1 and σ_2 are unknown, but the sample standard deviations S_1 and S_2 are known. The equations for the test statistic t depends on two subcases:

Subcase 1: σ_1 and σ_2 are believed to be equal (although unknown). In this subcase, we calculate t using the formula

$$t = \frac{(\bar{X}_1 - \bar{X}_2) - (\mu_1 - \mu_2)_0}{\sqrt{S_p^2(1/n_1 + 1/n_2)}} \qquad (8\text{--}5)$$

where S_p^2 is the pooled variance of the two samples, which serves as the estimate of the common population variance given by the formula

$$S_p^2 = \frac{(n_1 - 1)S_1^2 + (n_2 - 1)S_2^2}{n_1 + n_2 - 2} \qquad (8\text{--}6)$$

The degrees of freedom for t are $(n_1 + n_2 - 2)$.

Subcase 2: σ_1 and σ_2 are believed to be unequal (although unknown). In this subcase, we calculate t using the formula

$$t = \frac{(\bar{X}_1 - \bar{X}_2) - (\mu_1 - \mu_2)_0}{\sqrt{S_1^2/n_1 + S_2^2/n_2}} \qquad (8\text{--}7)$$

The degrees of freedom for this t are given by

$$df = \left\lfloor \frac{(s_1^2/n_1 + s_2^2/n_2)^2}{(s_1^2/n_1)^2/(n_1 - 1) + (s_2^2/n_2)^2/(n_2 - 1)} \right\rfloor \qquad (8\text{--}8)$$

Subcase 1 is the easier of the two. In this case, let $\sigma_1 = \sigma_2 = \sigma$. Because the two populations are normally distributed, $\bar{X}_1$ and $\bar{X}_2$ each follows a normal distribution and thus $(\bar{X}_1 - \bar{X}_2)$ also follows a normal distribution. Because the two samples are independent, we have

$$\text{Var}(\bar{X}_1 - \bar{X}_2) = \text{Var}(\bar{X}_1) + \text{Var}(\bar{X}_2) = \sigma^2/n_1 + \sigma^2/n_2 = \sigma^2(1/n_1 + 1/n_2)$$

We estimate σ^2 by

$$S_p^2 = \frac{(n_1 - 1)S_1^2 + (n_2 - 1)S_2^2}{(n_1 + n_2 - 2)}$$

which is a weighted average of the two sample variances. As a result, if the null hypothesis is true, then the quantity

$$\frac{(\bar{X}_1 - \bar{X}_2) - (\mu_1 - \mu_2)_0}{S_p\sqrt{1/n_1 + 1/n_2}}$$

must follow a t distribution with $(n_1 + n_2 - 2)$ degrees of freedom.

Subcase 2 does not neatly fall into a t distribution as it combines two sample means from two populations with two different unknown variances. It can be shown that when the null hypothesis is true, the quantity

$$\frac{(\bar{X}_1 - \bar{X}_2) - (\mu_1 - \mu_2)_0}{\sqrt{S_1^2/n_1 + S_2^2/n_2}}$$

will *approximately* follow a t distribution with degrees of freedom given by the complex equation 8–8. The symbol $\lfloor \ \rfloor$ used in this equation means rounding down to the nearest integer. For example, $\lfloor 15.8 \rfloor = 15$. We round the value down to comply with the principle of giving the benefit of doubt to the null hypothesis.

Because approximation is involved in this case, it is better to use subcase 1 whenever possible, to avoid approximation. But, then, subcase 1 requires the strong

FIGURE 8–4　The Template for Testing the Difference in Population Means
[Testing Difference in Means.xls; Sheet: Sample Data]

	A	B	C	D E	F	G	H	I	J	K	L	M
1	**Testing the Difference in Two Population Means**											
2		Data										
3		Current	Previous	Evidence							Assumptions	
4		Sample1	Sample2			Sample1	Sample2					
5	1	405	334		Size	16	16	n			Either 1. Populations	
6	2	125	150		Mean	352.375	319.563	x-bar			Or 2. Large sample	
7	3	540	520			**Popn. 1**	**Popn. 2**				σ_1, σ_2 known	
8	4	100	95	Popn. Std. Devn.		152	128	σ				
9	5	200	212									
10	6	30	30	**Hypothesis Testing**								
11	7	1200	1055									
12	8	265	300	Test Statistic	0.6605	z						
13	9	90	85					At an α of				
14	10	206	129	Null Hypothesis		p-value	5%		Confidence Interval			
15	11	18	40	$H_0: \mu_1 - \mu_2 = 0$		0.5089			$1 - \alpha$	Confidence		
16	12	489	440	$H_0: \mu_1 - \mu_2 \geq 0$		0.7455			95%	32.8125 ±		
17	13	590	610	$H_0: \mu_1 - \mu_2 \leq 0$		0.2545						
18	14	310	208									
19	15	995	880									
20	16	75	25									

assumption that the two population variances are equal. To guard against overuse of subcase 1 we check the assumption using an F test that will be described later in this chapter. In any case, if we use subcase 1, we should understand fully why we believe that the two variances are equal. In general, if the sources or the causes of variance in the two populations are the same, then it is reasonable to expect the two variances to be equal.

The templates that can be used for cases where t is the test statistic are shown in Figures 8–7 and 8–8 on page 346.

Cases Not Covered by Z or t

1. At least one population is not normally distributed and the sample size from that population is less than 30.
2. At least one population is not normally distributed and the standard deviation of that population is unknown.
3. For at least one population, neither the population standard deviation nor the sample standard deviation is known. (This case is rare.)

In the preceding cases, we are unable to find a test statistic that would follow a known distribution. It may be possible to apply the nonparametric method, the Mann-Whitney U test, described in Chapter 14.

The Templates

Figure 8–4 shows the template that can be used to test differences in population means when sample data are known. The data are entered in columns B and C. If a confidence interval is desired, enter the confidence level in cell K16.

Figure 8–5 shows the template that can be used to test differences in population means when sample statistics rather than sample data are known. The data in the figure correspond to Example 8–3.

FIGURE 8–5   **The Template for Testing Difference in Means**
[Testing Difference in Means.xls; Sheet: Sample Stats]

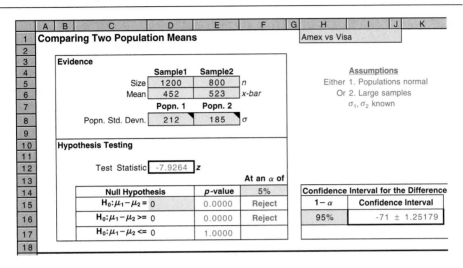

Until a few years ago, the market for consumer credit was considered to be segmented. Higher-income, higher-spending people tended to be American Express
cardholders, and lower-income, lower-spending people were usually Visa cardholders. In the last few years, Visa has intensified its efforts to break into the higher-income segments of the market by using magazine and television advertising to create
a high-class image. Recently, a consulting firm was hired by Visa to determine
whether average monthly charges on the American Express Gold Card are approximately equal to the average monthly charges on Preferred Visa. A random sample of
1,200 Preferred Visa cardholders was selected, and it was found that the sample average monthly charge was $\bar{x}_1 = \$452$. An independent random sample of 800 Gold
Card members revealed a sample mean $\bar{x}_2 = \$523$. Assume $\sigma_1 = \$212$ and $\sigma_2 =
\$185$. (Holders of both the Gold Card and Preferred Visa were excluded from the
study.) Is there evidence to conclude that the average monthly charge in the entire
population of American Express Gold Card members is different from the average
monthly charge in the entire population of Preferred Visa cardholders?

EXAMPLE 8–3

Since we have no prior suspicion that either of the two populations may have a higher
mean, the test is two-tailed. The null and alternative hypotheses are

$$H_0: \mu_1 - \mu_2 = 0$$
$$H_1: \mu_1 - \mu_2 \neq 0$$

The value of our test statistic (equation 8–4) is

$$z = \frac{452 - 523 - 0}{\sqrt{212^2/1,200 + 185^2/800}} = -7.926$$

The computed value of the Z statistic falls in the left-hand rejection region for any
commonly used α, and the p-value is very small. We conclude that there is a statistically significant difference in average monthly charges between Gold Card and

Solution

FIGURE 8–6   Carrying Out the Test of Example 8–3

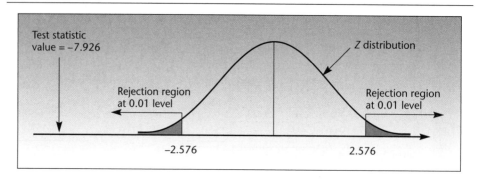

Preferred Visa cardholders. Note that this does not imply any *practical significance*. That is, while a difference in average spending in the two populations may exist, we cannot necessarily conclude that this difference is large. The test is shown in Figure 8–6.

EXAMPLE 8–4   Suppose that the makers of Duracell batteries want to demonstrate that their size AA battery lasts an average of at least 45 minutes longer than Duracell's main competitor, the Energizer. Two independent random samples of 100 batteries of each kind are selected, and the batteries are run continuously until they are no longer operational. The sample average life for Duracell is found to be $\bar{x}_1 = 308$ minutes. The result for the Energizer batteries is $\bar{x}_2 = 254$ minutes. Assume $\sigma_1 = 84$ minutes and $\sigma_2 = 67$ minutes. Is there evidence to substantiate Duracell's claim that its batteries last, on average, at least 45 minutes longer than Energizer batteries of the same size?

Solution   Our null and alternative hypotheses are

$$H_0: \mu_1 - \mu_2 \leq 45$$
$$H_1: \mu_1 - \mu_2 > 45$$

The makers of Duracell hope to demonstrate their claim by rejecting the null hypothesis. Recall that failing to reject a null hypothesis is not a strong conclusion. This is why—in order to demonstrate that Duracell batteries last an average of at least 45 minutes longer—the claim to be demonstrated is stated as the *alternative* hypothesis.

The value of the test statistic in this case is computed as follows:

$$z = \frac{308 - 254 - 45}{\sqrt{84^2/100 + 67^2/100}} = 0.838$$

This value falls in the nonrejection region of our right-tailed test at any conventional level of significance α. The p-value is equal to 0.2011. We must conclude that there is insufficient evidence to support Duracell's claim.

Confidence Intervals

Recall from Chapter 7 that there is a strong connection between hypothesis tests and confidence intervals. In the case of the difference between two population means, we have the following:

A large-sample $(1 - \alpha)$ 100% confidence interval for the difference between two population means $\mu_1 - \mu_2$, using independent random samples, is

$$\bar{x}_1 - \bar{x}_2 \pm z_{\alpha/2} \sqrt{\frac{\sigma_1^2}{n_1} + \frac{\sigma_2^2}{n_2}} \qquad (8\text{--}9)$$

Equation 8–9 should be intuitively clear. The bounds on the difference between the two population means are equal to the difference between the two sample means, plus or minus the z coefficient for $(1 - \alpha)$ 100% confidence times the standard deviation of the difference between the two sample means (which is the expression with the square root sign).

In the context of example 8–3, a 95% confidence interval for the difference between the average monthly charge on the American Express Gold Card and the average monthly charge on the Preferred Visa Card is, by equation 8–9,

$$523 - 452 \pm 1.96 \sqrt{\frac{212^2}{1{,}200} + \frac{185^2}{800}} = [53.44, 88.56]$$

The consulting firm may report to Visa that it is 95% confident that the average American Express Gold Card monthly bill is anywhere from $53.44 to $88.56 higher than the average Preferred Visa bill.

With one-tailed tests, the analogous interval is a one-sided confidence interval. We will not give examples of such intervals in this chapter. In general, we construct confidence intervals for population parameters when we have no *particular* values of the parameters we want to test and are interested in estimation only.

PROBLEMS

8–9. LINC is a software tool developed by Burroughs Corporation. The program automatically writes some of the coding that programmers have to do manually. LINC supposedly saves programming time and allows programmers to operate more efficiently. In a test of the software package, 45 programmers (group 1) were asked to write a program without LINC and then run the program until it performed with no bugs. The times from start to finish for this group were recorded. Group 2 consisted of 32 programmers who were asked to prepare the same program with the aid of LINC. Before getting the data, it was decided to run the test as a one-tailed test to prove that the package reduces the average programming time. The results were $\bar{x}_1 = 26$ minutes, $\bar{x}_2 = 21$ minutes, $s_1 = 8$ minutes, and $s_2 = 6$ minutes. Conduct the test, and state your conclusion. Is LINC effective in reducing average programming time?

8–10. The photography department of a glamour magazine needs to choose a camera. Of the two models the department is considering, one is made by Nikon and one by Minolta. The department contracts with an agency to determine if one of the two models gets a higher average performance rating by professional photographers, or whether the average performance ratings of these two cameras are not statistically different. The agency asks 60 different professional photographers to rate one of the cameras (30 photographers rate each model). The ratings are on a scale of 1 to 10. The average sample rating for Nikon is 8.5, and the sample standard deviation is 2.1. For the Minolta sample, the average sample rating is 7.8, and the standard deviation is 1.8. Is there a difference between the average population ratings of the two cameras? If so, which one is rated higher?

8–11.   Marcus Robert Real Estate Company wants to test whether the average sale price of residential properties in a certain size range in Bel Air, California, is approximately equal to the average sale price of residential properties of the same size range in Marin County, California. The company gathers data on a random sample of 32 properties in Bel Air and finds $\bar{x} = \$345,650$ and $s = \$48,500$. A random sample of 35 properties in Marin County gives $\bar{x} = \$289,440$ and $s = \$87,090$. Is the average sale price of all properties in both locations approximately equal or not? Explain.

8–12.   "A recent study by *What Mortgage?* a British personal finance magazine, found that of 72 lenders, the 25 offering the best value were mutuals. Their rates were, on average, 1% lower than those of nonmutuals."[2] Here $n_1 = 25$ and $n_2 = 72 - 25 = 47$; assume that $s_1 = s_2 = 2\%$. Test for a difference in means.

8–13.   Many companies that cater to teenagers have learned that young people respond to commercials that provide dance-beat music, adventure, and a fast pace rather than words. In one test, a group of 128 teenagers were shown commercials featuring rock music, and their purchasing frequency of the advertised products over the following month was recorded as a single score for each person in the group. Then a group of 212 teenagers was shown commercials for the same products, but with the music replaced by verbal persuasion. The purchase frequency scores of this group were computed as well. The results for the music group were $\bar{x} = 23.5$ and $s = 12.2$; and the results for the verbal group were $\bar{x} = 18.0$ and $s = 10.5$. Assume that the two groups were randomly selected from the entire teenage consumer population. Using the $\alpha = 0.01$ level of significance, test the null hypothesis that both methods of advertising are equally effective versus the alternative hypothesis that they are not equally effective. If you conclude that one method is better, state which one it is, and explain how you reached your conclusion.

8–14.   New corporate strategies take years to develop.[3] Two methods for facilitating the development of new strategies by executive strategy meetings are to be compared. One method is to hold a two-day retreat in a posh hotel; the other is to hold a series of informal luncheon meetings on company premises. The following are the results of two independent random samples of firms following one of these two methods. The data are the number of months, for each company, that elapsed from the time an idea was first suggested until the time it was implemented.

Hotel	On-Site
17	6
11	12
14	13
25	16
9	4
18	8
36	14
19	18
22	10
24	5
16	7
31	12
23	10

[2]"Blood on the High Streets," *The Economist,* June 7, 1997, p. 91.

[3]From G. Hamel, "Strategies That Make Shareholders Rich," *Fortune,* June 23, 1997, p. 25.

Test for a difference between means, using $\alpha = 0.05$.

8–15. A fashion industry analyst wants to prove that models featuring Liz Claiborne clothing earn on average more than models featuring clothes designed by Calvin Klein. For a given period of time, a random sample of 32 Liz Claiborne models reveals average earnings of $4,238.00 and a standard deviation of $1,002.50. For the same period, an independent random sample of 37 Calvin Klein models has mean earnings of $3,888.72 and a sample standard deviation of $876.05.

   a. Is this a one-tailed or a two-tailed test? Explain.

   b. Carry out the hypothesis test at the 0.05 level of significance.

   c. State your conclusion.

   d. What is the *p*-value? Explain its relevance.

   e. Redo the problem, assuming the results are based on a random sample of 10 Liz Claiborne models and 11 Calvin Klein models.

8–16. An article on the growing trade gap in high-technology ideas between the United States and Japan claims that more royalties and licensing fees are paid to the United States by Japanese users of U.S. ideas than the other way round. A random sample of 100 U.S. patents used in Japan reveals an average royalty of $1,838.69 and a standard deviation of $461. A random sample of 80 Japanese patents used in the United States has an average royalty payment of $1,050.22 and a standard deviation of $560. Is there evidence to conclude that the Japanese pay, on the average, more to U.S. patent holders than the other way round?

8–17. A brokerage firm is said to provide both brokerage services and "research" if, in addition to buying and selling securities for its clients, the firm furnishes clients with advice about the value of securities, information on economic factors and trends, and portfolio strategy. The Securities and Exchange Commission (SEC) has been studying brokerage commissions charged by both "research" and "nonresearch" brokerage houses. A random sample of 255 transactions at nonresearch firms is collected as well as a random sample of 300 transactions at research firms. These samples reveal that the difference between the average sample percentage of commission at research firms and the average percentage of commission in the nonresearch sample is 2.54%. The standard deviation of the research firms' sample is 0.85%, and that of the nonresearch firms is 0.64%. Give a 95% confidence interval for the difference in the average percentage of commissions in research versus nonresearch brokerage houses.

The Templates

Figure 8–7 shows the template that can be used to conduct *t* tests for difference in population means, when sample data are known. The top panel can be used if there is reason to believe that the two population variances are equal; the bottom panel should be used in all other cases. As an additional aid to deciding which panel to use, the null hypothesis $H_0: \sigma_1^2 - \sigma_2^2 = 0$ is tested at top right. The *p*-value of the test appears in cell M7. If this value is at least, say, 20%, then there is no problem in using the top panel. If the *p*-value is less than 10%, then it is not wise to use the top panel. In such circumstances, a warning message—"Warning: Equal variance assumption is questionable"—will appear in cell K10.

   If a confidence interval for the difference in the means is desired, enter the confidence level in cell L15 or L24.

   Figure 8–8 shows the template that can be used to conduct *t* tests for difference in population means when sample statistics rather than sample data are known. The top

FIGURE 8–7   The Template for the *t* Test for Difference in Means
[Testing Difference in Means.xls; Sheet: t-Test from Data]

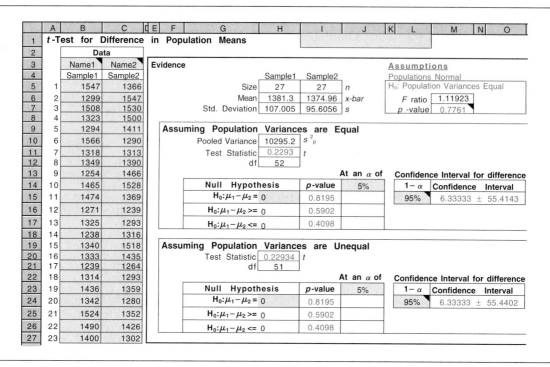

FIGURE 8–8   The Template for the *t* Test for Difference in Means
[Testing Difference in Means.xls; Sheet: Sample Stats]

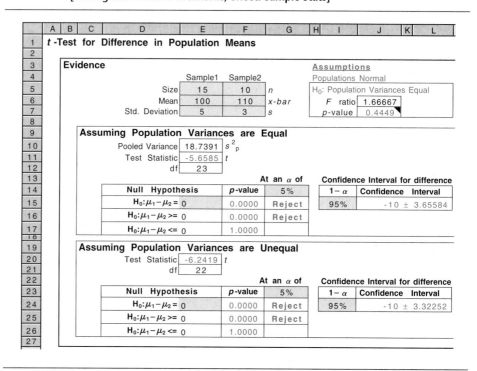

panel can be used if there is reason to believe that the two population variances are equal; the bottom panel should be used otherwise. As an additional aid to deciding which panel to use, the null hypothesis that the population variances are equal is tested at top right. The p-value of the test appears in cell J7. If this value is at least, say, 20%, then there is no problem in using the top panel. If it is less than 10%, then it is not wise to use the top panel. In such circumstances, a warning message—"Warning: Equal variance assumption is questionable"—will appear in cell H10.

If a confidence interval for the difference in the means is desired, enter the confidence level in cell I15 or I24.

EXAMPLE 8–5

Changes in the price of oil have long been known to affect the economy of the United States. An economist wants to check whether the price of a barrel of crude oil affects the Consumer Price Index (CPI), a measure of price levels and inflation. The economist collects two sets of data: one set comprises 14 monthly observations on increases in the CPI, in percentage per month, when the price of crude oil is $27.50 per barrel; the other set consists of 9 monthly observations on percentage increase in the CPI when the price of crude oil is $20.00 per barrel. The economist assumes that her data are a random set of observations from a population of monthly CPI percentage increases when oil sells for $27.50 per barrel, and an independent set of random observations from a population of monthly CPI percentage increases when oil sells for $20.00 per barrel. She also assumes that the two populations of CPI percentage increases are normally distributed and that the variances of the two populations are equal. Considering the nature of the economic variables in question, these are reasonable assumptions. If we call the population of monthly CPI percentage increases when oil sells for $27.50 population 1, and that of oil at $20.00 per barrel population 2, then the economist's data are as follows:[4] $\bar{x}_1 = 0.317\%$, $s_1 = 0.12\%$, $n_1 = 14$; $\bar{x}_2 = 0.210\%$, $s_2 = 0.11\%$, $n_2 = 9$. Our economist is faced with the question: Do these data provide evidence to conclude that average percentage increase in the CPI differs when oil sells at these two different prices?

Solution

Although the economist may have a suspicion about the possible direction of change in the CPI as oil prices decrease, she decides to approach the situation with an open mind and let the data speak for themselves. That is, she wants to carry out the two-tailed test. Her test is $H_0: \mu_1 - \mu_2 = 0$ versus $H_1: \mu_1 - \mu_2 \neq 0$. Using equation 8–7, the economist computes the value of the test statistic, which has a t distribution with $n_1 + n_2 - 2 = 21$ degrees of freedom:

$$t = \frac{0.317 - 0.210 - 0}{\sqrt{\frac{(13)(0.12)^2 + (8)(0.11)^2}{21}\left(\frac{1}{14} + \frac{1}{9}\right)}} = 2.15$$

The computed value of the test statistic $t = 2.15$ falls in the right-hand rejection region at $\alpha = 0.05$, but not very far from the critical point 2.080. The p-value is therefore just less than 0.05. The economist may thus conclude that, based on her data and the validity of the assumptions made, there is some evidence that the average monthly increase in the CPI is greater when oil sells for $27.50 per barrel than it is when oil sells for $20.00 per barrel.

---

[4]The data are based on estimates reported by Data Resources, Inc.

EXAMPLE 8–6   The manufacturers of compact disk players want to test whether a small price reduction is enough to increase sales of their product. Randomly chosen data on 15 weekly
sales totals at outlets in a given area before the price reduction show a sample mean
of $6,598 and a sample standard deviation of $844. A random sample of 12 weekly
sales totals after the small price reduction gives a sample mean of $6,870 and a sam-
ple standard deviation of $669. Is there evidence that the small price reduction is
enough to increase sales of compact disk players?

Solution   This is a one-tailed test, except that we will reverse the notation 1 and 2 so we can
conduct a right-tailed test to determine whether reducing the price increases sales (if
sales increase, then μ_2 will be greater than μ_1, which is what we want the alternative
hypothesis to be). We have H_0: $\mu_1 - \mu_2 \geq 0$ and H_1: $\mu_1 - \mu_2 < 0$. We assume an
equal variance of the populations of sales at the two price levels. Our test statistic has
a t distribution with $n_1 + n_2 - 2 = 15 + 12 - 2 = 25$ degrees of freedom. The computed value of the statistic, by equation 8–7, is

$$t = \frac{(6{,}870 - 6{,}598) - 0}{\sqrt{\dfrac{(14)(844)^2 + (11)(669)^2}{25}\left(\dfrac{1}{15} + \dfrac{1}{12}\right)}} = 0.91$$

This value of the statistic falls inside the nonrejection region for any usual level of
significance.

Confidence Intervals

As usual, it is possible to construct confidence intervals for the parameter in question—here, the difference between the two population means. The confidence interval for this parameter is based on the t distribution with $n_1 + n_2 - 2$ degrees of
freedom (or z when df is large).

A $(1 - \alpha)$ 100% confidence interval for $(\mu_1 - \mu_2)$, assuming equal population variance, is

$$\bar{x}_1 - \bar{x}_2 \pm t_{\alpha/2}\sqrt{s_p^2\left(\frac{1}{n_1} + \frac{1}{n_2}\right)} \qquad (8\text{--}10)$$

The confidence interval in equation 8–10 has the usual form: Estimate $\pm$ Distribution
coefficient $\times$ Standard deviation of estimator.

    In Example 8–6, forgetting that the test was carried out as a one-tailed test, we
compute a 95% confidence interval for the difference between the two means. Since
the test resulted in nonrejection of the null hypothesis (and would have also resulted
so had it been carried out as two-tailed), our confidence interval should contain the
null hypothesis difference between the two population means: zero. This is due to the
connection between hypothesis tests and confidence intervals. Let us see if this really
happens. The 95% confidence interval for $\mu_1 - \mu_2$ is

$$\bar{x}_1 - \bar{x}_2 \pm t_{0.025} \sqrt{s_p^2\left(\frac{1}{n_1} + \frac{1}{n_2}\right)} = (6{,}870 - 6{,}598) \pm 2.06\sqrt{(595{,}835)(0.15)}$$
$$= [-343.85, 887.85]$$

We see that the confidence interval indeed contains the null-hypothesized difference of zero, as expected from the fact that a two-tailed test would have resulted in nonrejection of the null hypothesis.

PROBLEMS

In each of the following problems assume that the two populations of interest are normally distributed with equal variance. Assume independent random sampling from the two populations.

8–18. The recent boom in sales of travel books has led to the marketing of other travel-related guides, such as video travel guides and audio walking-tour tapes. Waldenbooks has been studying the market for these travel guides. In one market test, a random sample of 25 potential travelers was asked to rate audiotapes of a certain destination, and another random sample of 20 potential travelers was asked to rate videotapes of the same destination. Both ratings were on a scale of 0 to 100 and measured the potential travelers' satisfaction with the travel guide they tested and the degree of possible purchase intent (with 100 the highest). The mean score for the audio group was 87, and their standard deviation was 12. The mean score for the video group was 64, and their standard deviation was 23. Do these data present evidence that one form of travel guide is better than the other? Advise Waldenbooks on a possible marketing decision to be made.

8–19. Business schools at certain prestigious universities offer nondegree management training programs for high-level executives. These programs supposedly develop executives' leadership abilities and help them advance to higher management positions within 2 years after program completion. A management consulting firm wants to test the effectiveness of these programs and sets out to conduct a one-tailed test, where the alternative hypothesis is that graduates of the programs under study do receive, on average, salaries more than $4,000 per year higher than salaries of comparable executives without the special university training. To test the hypotheses, the firm traces a random sample of 28 top executives who earn, at the time the sample is selected, about the same salaries. Out of this group, 13 executives—randomly selected from the group of 28 executives—are enrolled in one of the university programs under study. Two years later, average salaries for the two groups and standard deviations of salaries are computed. The results are: $\bar{x} = 48$ and $s = 6$ for the nonprogram executives, and $\bar{x} = 55$ and $s = 8$ for the program executives. All numbers are in thousands of dollars per year. Conduct the test at $\alpha = 0.05$, and evaluate the effectiveness of the programs in terms of increased average salary levels.

8–20. Advertisements generally push a product or at least mention a company's name, but Nike's ads often do neither. Instead, they come across as demonstrations of a passion for sport. These ads are credited with promoting Nike's tremendous corporate success (the company's stock rose an average of 47% per year over the past 10 years).[5] If one ad promoting a product or company and another professing a passion for sports are each tested on an independent random sample, and viewers' scores are

[5]From R. Leiber, "Just Redo It," *Fortune,* June 23, 1997, p. 24.

as follows (on a scale of 0 to 100, with 100 high), test for a difference in population rating means.

Product/company: 75, 71, 65, 61, 40, 38, 90, 76, 73, 77, 80, 91, 86, 51
Passion for sport: 91, 93, 99, 79, 88, 96, 97, 89, 98, 99, 87, 65

Assume equal population variances.

8–21. According to *The Economist*, the French cafe, once "the most solid of the institutions" of France, is disappearing. One reason is a reduction in alcohol consumption from 18 liters a year per person in 1960 to 11 liters per person a year in 1995.[6] Assuming this is based on two independent random samples of 100 people each, with equal standard deviation of 6 liters per year, test for significance. Is there proof of a decline in alcohol use in France?

8–22. Ikarus, the Hungarian bus maker, has lost its important Commonwealth of Independent States market and is reported on the verge of collapse. The company is now trying a new engine in its buses and has gathered the following random samples of miles-per-gallon figures for the old engine versus the new:

Old engine: 8, 9, 7.5, 8.5, 6, 9, 9, 10, 7, 8.5, 6, 10, 9, 8, 9, 5, 9.5, 10, 8
New engine: 10, 9, 9, 6, 9, 11, 11, 8, 9, 6.5, 7, 9, 10, 8, 9, 10, 9, 12, 11.5, 10, 7, 10, 8.5

Is there evidence that the new engine is more economical than the old one?

8–23. *Air Transport World* recently named the Dutch airline KLM "Airline of the Year." One measure of the airline's excellent management is its research effort in developing new routes and improving service on existing routes. The airline wanted to test the profitability of a certain transatlantic flight route and offered daily flights from Europe to the United States over a period of 6 weeks on the new proposed route. Then, over a period of 9 weeks, daily flights were offered from Europe to an alternative airport in the United States. Weekly profitability data for the two samples were collected, under the assumption that these may be viewed as independent random samples of weekly profits from the two populations (one population is flights to the proposed airport, and the other population is flights to an alternative airport). Data are as follows. For the proposed route, $\bar{x} = \$96{,}540$ per week and $s = \$12{,}522$. For the alternative route, $\bar{x} = \$85{,}991$ and $s = \$19{,}548$. Test the hypothesis that the proposed route is more profitable than the alternative route. Use a significance level of your choice.

8–24. T. Peters and R. Waterman stated in their book *In Search of Excellence* that the giant advertising firm of Ogilvy & Mather was more concerned with customer satisfaction than with company profits.[7] Suppose a test is carried out to determine whether new management decisions at OgilvyOne Worldwide increase average customer satisfaction. To test this claim, a random sample of company clients is polled before an important management decision, and customer satisfaction is measured for this sample on a scale of 1 to 10. Then, some time after the management decision has been made and a new company policy implemented, another sample of clients is polled, and their satisfaction scores are computed (on the same scale). Letting the subscript b denote before the new decision and the subscript a denote after the new decision, the results of the surveys are $n_b = 18$, $\bar{x}_b = 7.4$, $s_b = 1.3$; $n_a = 23$, $\bar{x}_a = 8.2$, $s_a = 2.4$. Using

---

[6]"Mais Où Sont Les Cafés D'autan?" *The Economist,* June 10, 1995, p. 50.

[7]T. Peters and R. Waterman, *In Search of Excellence* (New York: Harper & Row, 1982).

these data, do you believe that customer satisfaction is increased, on average, after the new management decision and the resulting new company policy?

8–25. Mark Pollard, financial consultant for Merrill Lynch, Pierce, Fenner & Smith, Inc., is quoted in national advertisements for Merrill Lynch as saying: "I've made more money for clients by saying no than by saying yes." Suppose that Pollard allowed you access to his files so that you could conduct a statistical test of the correctness of his statement. Suppose further that you gathered a random sample of 25 clients to whom Pollard said yes when presented with their investment proposals, and you found that the clients' average gain on investments was 12% and the standard deviation was 2.5%. Suppose you gathered another sample of 25 clients to whom Pollard said no when asked about possible investments; the clients were then offered other investments, which they consequently made. For this sample, you found that the average return was 13.5% and the standard deviation was 1%. Test Pollard's claim at $\alpha = 0.05$. What assumptions are you making in this problem?

8–26. An article reports the results of an analysis of stock market returns before and after antitrust trials that resulted in the breakup of AT&T. The study concentrated on two periods: the pre-antitrust period of 1966 to 1973, denoted period 1, and the antitrust trial period of 1974 to 1981, called period 2. An equation similar to equation 8–7 was used to test for the existence of a difference in mean stock return during the two periods. Conduct a two-tailed test of equality of mean stock return in the population of all stocks before and during the antitrust trials using the following data: $n_1 = 21$, $\bar{x}_1 = 0.105$, $s_1 = 0.09$; $n_2 = 28$, $\bar{x}_2 = 0.1331$, $s_2 = 0.122$. Use $\alpha = 0.05$.

8–27. A banker with Continental Illinois National Bank and Trust Company wants to test which method of raising cash for companies—borrowing from public sources or borrowing from private sources—results in higher average amounts raised by a company. The banker collects a random sample of 12 firms that borrowed only from public sources and finds that the average amount borrowed by a company per source is $12,500 and the standard deviation is $3,400. Another sample of 18 firms that borrowed only from private sources gives a sample average per source of $21,000 and a sample standard deviation of $5,000. Do you believe that either private or public sources lend, on the average, more than the other? Explain.

8–28. In problem 8–25, construct a 95% confidence interval for the difference between the average return to investors following a no recommendation and the average return to investors following a yes recommendation. Interpret your results.

8–4 A Large-Sample Test for the Difference between Two Population Proportions

When sample sizes are large enough that the distributions of the sample proportions $\hat{P}_1$ and $\hat{P}_2$ are both approximated well by a normal distribution, the difference between the two sample proportions is also approximately normally distributed, and this gives rise to a test for equality of two population proportions based on the standard normal distribution. It is also possible to construct confidence intervals for the difference between the two population proportions. Assuming the sample sizes are large and assuming independent random sampling from the two populations, the following are possible hypotheses (we consider situations similar to the ones discussed in the previous two sections; other tests are also possible).

$$\text{Situation I:} \quad \text{H}_0: p_1 - p_2 = 0$$
$$\text{H}_1: p_1 - p_2 \neq 0$$
$$\text{Situation II:} \quad \text{H}_0: p_1 - p_2 \leq 0$$
$$\text{H}_1: p_1 - p_2 > 0$$
$$\text{Situation III:} \quad \text{H}_0: p_1 - p_2 \leq D$$
$$\text{H}_1: p_1 - p_2 > D$$

Here D is some number other than 0.

In the case of tests about the difference between two population proportions, there are two test statistics. One statistic is appropriate when the null hypothesis is that the difference between the two population proportions is equal to (or greater than or equal to, or less than or equal to) zero. This is the case, for example, in situations I and II. The other test statistic is appropriate when the null hypothesis difference is some number D different from zero. This is the case, for example, in situation III (or in a two-tailed test, situation I, with D replacing 0).

The test statistic for the difference between two population proportions where the null hypothesis difference is zero is

$$z = \frac{\hat{p}_1 - \hat{p}_2 - 0}{\sqrt{\hat{p}(1 - \hat{p})(1/n_1 + 1/n_2)}} \tag{8–11}$$

where $\hat{p}_1 = x_1/n_1$ is the sample proportion in sample 1 and $\hat{p}_2 = x_2/n_2$ is the sample proportion in sample 2. The symbol $\hat{p}$ stands for the *combined sample proportion in both samples,* considered as a single sample. That is,

$$\hat{p} = \frac{x_1 + x_2}{n_1 + n_2} \tag{8–12}$$

Note that 0 in the numerator of equation 8–11 is the null hypothesis difference between the two population proportions; we retain it only for conceptual reasons—to maintain the form of our test statistic: (Estimate − Hypothesized value of the parameter)/(Standard deviation of the estimator). When we carry out computations using equation 8–11, we will, of course, ignore the subtraction of zero. Under the null hypothesis that the difference between the two population proportions is zero, both sample proportions $\hat{p}_1$ and $\hat{p}_2$ are estimates of the same quantity, and therefore—assuming, as always, that the null hypothesis is true—we pool the two estimates when computing the estimated standard deviation of the difference between the two sample proportions: the denominator of equation 8–11.

When the null hypothesis is that the difference between the two population proportions is a number other than zero, we cannot assume that $\hat{p}_1$ and $\hat{p}_2$ are estimates of the same population proportion (because the null hypothesis difference between the two population proportions is $D \neq 0$); in such cases we cannot pool the two estimates when computing the estimated standard deviation of the difference between the two sample proportions. In such cases, we use the following test statistic.

The test statistic for the difference between two population proportions when the null hypothesis difference between the two proportions is some number D, other than zero, is

$$z = \frac{\hat{p}_1 - \hat{p}_2 - D}{\sqrt{\hat{p}_1(1 - \hat{p}_1)/n_1 + \hat{p}_2(1 - \hat{p}_2)/n_2}} \qquad (8\text{--}13)$$

We will now demonstrate the use of the test statistics presented in this section with the following examples.

EXAMPLE 8–7

A recent article describes how finance incentives by the major automakers are reducing banks' share of the market for automobile loans. The article reports that in 1980, banks wrote about 53% of all car loans, and in 1995, the banks' share was only 43%. Suppose that these data are based on a random sample of 100 car loans in 1980, where 53 of the loans were found to be bank loans; and the 1995 data are also based on a random sample of 100 loans, 43 of which were found to be bank loans. Carry out a two-tailed test of the equality of banks' share of the car loan market in 1980 and in 1995.

Solution

Our hypotheses are those described as situation I, a two-tailed test of the equality of two population proportions. We have H_0: $p_1 - p_2 = 0$ and H_1: $p_1 - p_2 \neq 0$. Since the null hypothesis difference between the two population proportions is zero, we can use the test statistic of equation 8–11. First we calculate $\hat{p}$, the combined sample proportion, using equation 8–12:

$$\hat{p} = \frac{x_1 + x_2}{n_1 + n_2} = \frac{53 + 43}{100 + 100} = 0.48$$

We also have $1 - \hat{p} = 0.52$.

We now compute the value of the test statistic, equation 8–11:

$$z = \frac{\hat{p}_1 - \hat{p}_2}{\sqrt{\hat{p}(1 - \hat{p})(1/n_1 + 1/n_2)}} = \frac{0.53 - 0.43}{\sqrt{(0.48)(0.52)(0.01 + 0.01)}} = 1.415$$

This value of the test statistic falls in the nonrejection region even if we use $\alpha = 0.10$. In fact, the p-value, found using the standard normal table, is equal to 0.157. We conclude that the data present insufficient evidence that the share of banks in the car loan market has changed from 1980 to 1995. The test is shown in Figure 8–9.

EXAMPLE 8–8

From time to time, BankAmerica Corporation comes out with its Free and Easy Travelers Cheques Sweepstakes, designed to increase the amounts of BankAmerica traveler's checks sold. Since the amount bought per customer determines the customer's chances of winning a prize, a manager hypothesizes that, during sweepstakes time, the proportion of BankAmerica traveler's check buyers who buy more than $2,500 worth of checks will be at least 10% higher than the proportion of traveler's check buyers who buy more than $2,500 worth of checks when there are no sweepstakes. A random sample of 300 traveler's check buyers, taken when the sweepstakes

FIGURE 8–9   Carrying Out the Test of Example 8–7

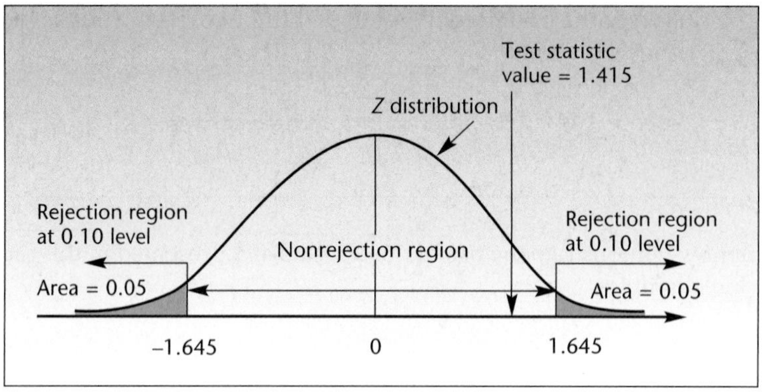

are on, reveals that 120 of these people bought checks for more than $2,500. A random sample of 700 traveler's check buyers, taken when no sweepstakes prizes are offered, reveals that 140 of these people bought checks for more than $2,500. Conduct the hypothesis test.

Solution   The manager wants to prove that the population proportion of traveler's check buyers who buy at least $2,500 in checks when sweepstakes prizes are offered is at least 10% higher than the proportion of such buyers when no sweepstakes are on. Therefore, this should be the manager's alternative hypothesis. We have H_0: $p_1 - p_2 \leq 0.10$ and H_1: $p_1 - p_2 > 0.10$. The appropriate test statistic is the statistic given in equation 8–13:

$$z = \frac{\hat{p}_1 - \hat{p}_2 - D}{\sqrt{\hat{p}_1(1 - \hat{p}_1)/n_1 + \hat{p}_2(1 - \hat{p}_2)/n_2}}$$

$$= \frac{120/300 - 140/700 - 0.10}{\sqrt{[(120/300)(180/300)]/300 + [(140/700)(560/700)]/700}}$$

$$= \frac{(0.4 - 0.2) - 0.1}{\sqrt{(0.4)(0.6)/300 + (0.2)(0.8)/700}} = 3.118$$

This value of the test statistic falls in the rejection region for $\alpha = 0.001$ (corresponding to the critical point 3.09 from the normal table). The p-value is therefore less than 0.001, and the null hypothesis is rejected. The manager is probably right. Figure 8–10 shows the result of the test.

Confidence Intervals

When constructing confidence intervals for the difference between two population proportions, we do not use the pooled estimate because we do not assume that the two proportions are equal. The estimated standard deviation of the difference between the two sample proportions, to be used in the confidence interval, is the denominator in equation 8–13.

FIGURE 8–10   Carrying Out the Test of Example 8–8

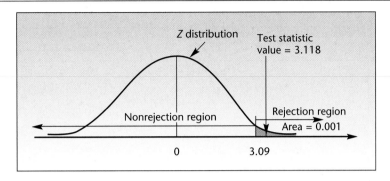

A large-sample $(1 - \alpha)$ 100% confidence interval for the difference between two population proportions is

$$\hat{p}_1 - \hat{p}_2 \pm z_{\alpha/2}\sqrt{\frac{\hat{p}_1(1 - \hat{p}_1)}{n_1} + \frac{\hat{p}_2(1 - \hat{p}_2)}{n_2}} \qquad (8\text{--}14)$$

In the context of Example 8–8, let us now construct a 95% confidence interval for
the difference between the proportion of BankAmerica traveler's check buyers who
buy more than \$2,500 worth of checks during sweepstakes and the proportion of buy-
ers of checks greater than this amount when no sweepstakes prizes are offered. Using
equation 8–14, we get

$$0.4 - 0.2 \pm 1.96\sqrt{\frac{(0.4)(0.6)}{300} + \frac{(0.2)(0.8)}{700}} = 0.2 \pm 1.96(0.032)$$
$$= [0.137, 0.263]$$

The manager may be 95% confident that the difference between the two proportions
of interest is anywhere from 0.137 to 0.263.

The Template

Figure 8–11 shows the template that can be used to test differences in population proportions. The middle panel is used when the hypothesized difference is zero. The
bottom panel is used when the hypothesized difference is nonzero. The data in the
figure correspond to Example 8–8, where H_0: $p_1 - p_2 \leq 0.10$. The bottom panel shows
that the p-value is 0.009, and H_0 is to be rejected.

PROBLEMS

8–29.   Airline mergers cause many problems for the airline industry. One variable
often quoted as a measure of an airline's efficiency is the percentage of on-time
departures. Following the merger of Republic Airlines with Northwest Airlines, the
percentage of on-time departures for Northwest planes declined from approximately
85% to about 68%. Suppose that the percentages reported above are based on two
random samples of flights: a sample of 100 flights over a period of two months before

FIGURE 8–11   The Template for Testing Differences in Proportions
[Testing Difference in Proportions.xls]

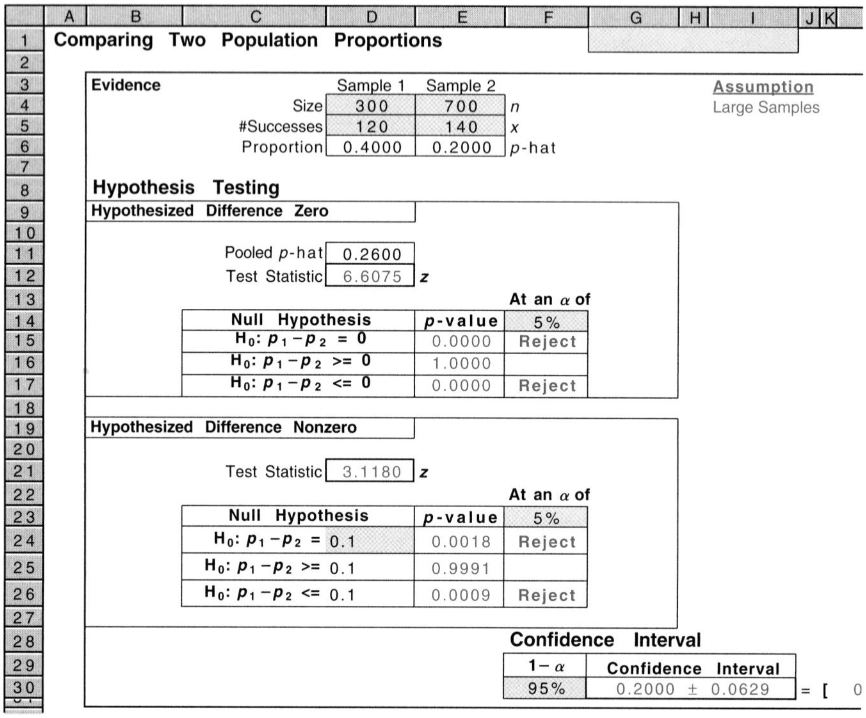

the merger, of which 85 are found to have departed on time; and a sample of 100 flights over a period of two months after the merger, 68 of which are found to have departed on time. Based on these data, do you believe that Northwest's on-time percentage has declined during the period following its merger with Republic?

8–30.   A physicians' group is interested in testing to determine whether more people in small towns choose a physician by word of mouth in comparison with people in large metropolitan areas. A random sample of 1,000 people in small towns reveals that 850 have chosen their physicians by word of mouth; a random sample of 2,500 people living in large metropolitan areas reveals that 1,950 have chosen a physician by word of mouth. Conduct a one-tailed test aimed at proving that the percentage of popular recommendation of physicians is larger in small towns than in large metropolitan areas. Use $\alpha = 0.01$.

8–31.   A corporate raider has been successful in 11 of 31 takeover attempts. Another corporate raider has been successful in 19 of 50 takeover bids. Assuming that the success rate of each raider at each trial is independent of all other attempts, and that the information presented can be regarded as based on two independent random samples of the two raiders' overall performance, can you say whether one of the raiders is more successful than the other? Explain.

8–32.   A random sample of 2,060 consumers shows that 13% prefer California wines. Over the next three months, an advertising campaign is undertaken to show that California wines receive awards and win taste tests. The organizers of the campaign want to prove that the three-month campaign raised the proportion of people who prefer California wines by at least 5%. At the end of the campaign, a random sample of 5,000 consumers shows that 19% of them now prefer California wines. Conduct the test at $\alpha = 0.05$.

8–33. In problem 8–32, give a 95% confidence interval for the increase in the population proportion of consumers preferring California wines following the campaign.

8–34. Federal Reserve Board regulations permit banks to offer to their clients commercial paper. A random sample of 650 customers of Chase Manhattan Bank reveals that 48 own commercial paper as part of their investment portfolios with the bank. A random sample of customers of Chemical Bank reveals that out of 480 customers, only 20 own commercial paper as part of their investments with the bank. Can you conclude that Chase Manhattan has a greater share of the new market for commercial paper? Explain.

8–35. Airbus Industrie, the European maker of the A320 medium-range jet capable of carrying 150 passengers, is currently trying to expand its market worldwide. At one point, Airbus managers wanted to test whether their potential market in the United States, measured by the proportion of airline industry executives who would prefer the A320, is greater than the company's potential market for the A320 in Europe (measured by the same indicator). A random sample of 120 top executives of U.S. airlines looking for new aircraft were given a demonstration of the plane, and 34 indicated that they would prefer the model to other new planes on the market. A random sample of 200 European airline executives were also given a demonstration of the plane, and 41 indicated that they would be interested in the A320. Test the hypothesis that more U.S. airline executives prefer the A320 than their European counterparts.

8–36. Data from the Bureau of Labor Statistics indicate that in one recent year the unemployment rate in Cleveland was 7.5% and the unemployment rate in Chicago was 7.2%. Suppose that both figures are based on random samples of 1,000 people in each city. Test the null hypothesis that the unemployment rates in both cities are equal versus the alternative hypothesis that they are not equal. What is the *p*-value? State your conclusion.

8–37. Of 1,000 firms, 40% chose the word *aggressive* to describe their current business policies; 28% used the word *cautious*.[8] Test for equality of the proportions, assuming two independent samples.

8–38. According to "Holland's Schiphol Sees Room to Grow Out in North Sea," by M. Dubois,[9] 44% of passengers use Schiphol for connecting flights, while for Heathrow the number is 29%. Assume that two independent samples of 1,000 passengers each produced those statistics. Test for significance of difference in proportions.

8–39. Several companies have been developing electronic guidance systems for cars. Motorola and Germany's Blaupunkt are two firms in the forefront of such research. Out of 120 trials of the Motorola model, 101 were successful; and out of 200 tests of the Blaupunkt model, 110 were successful. Is there evidence to conclude that the Motorola electronic guidance system is superior to that of the German competitor?

---

[8] From W. Woods, "Rising Wages," *Fortune,* July 7, 1997, p. 13.

[9] *The Wall Street Journal,* June 16, 1997, p. 1.

8–5   The *F* Distribution and a Test for Equality of Two Population Variances

In this section, we encounter the last of the major probability distributions useful in statistics, the *F distribution*. The *F* distribution is named after the English statistician Sir Ronald A. Fisher.

> The **F distribution** is the distribution of the ratio of two chi-square random variables that are independent of each other, each of which is divided by its own degrees of freedom.

If we let χ_1^2 be a chi-square random variable with k_1 degrees of freedom, and χ_2^2 another chi-square random variable independent of χ_1^2 and having k_2 degrees of freedom, the ratio in equation 8–15 has the *F* distribution with k_1 and k_2 degrees of freedom.

An *F* random variable with k_1 and k_2 degrees of freedom is

$$F_{(k1,\, k2)} = \frac{\chi_1^2/k_1}{\chi_2^2/k_2} \tag{8–15}$$

The *F* distribution thus has two kinds of degrees of freedom: k_1 is called the *degrees of freedom of the numerator* and is always listed as the first item in the parentheses; k_2 is called the *degrees of freedom of the denominator* and is always listed second inside the parentheses. The degrees of freedom of the numerator, k_1, are "inherited" from the chi-square random variable in the numerator; similarly, k_2 is "inherited" from the other, independent chi-square random variable in the denominator of equation 8–15.

Since there are so many possible degrees of freedom for the *F* random variable, tables of values of this variable for given probabilities are even more concise than the chi-square tables. Table 5 in Appendix C gives the critical points for *F* distributions with different degrees of freedom of the numerator and the denominator corresponding to right-tailed areas of 0.10, 0.05, 0.025, and 0.01. The second part of Table 5 gives critical points for $\alpha = 0.05$ and $\alpha = 0.01$ for a wider ranger of *F* random variables. For example, use Table 5 to verify that the point 3.01 cuts off an area of 0.05 to its right for an *F* random variable with 7 degrees of freedom for the numerator and 11 degrees of freedom for the denominator. This is demonstrated in Figure 8–12. Figure 8–13 shows various *F* distributions with different degrees of freedom. The *F* distributions are asymmetric (a quality inherited from their chi-square parents), and their shape resembles that of the chi-square distributions. Note that $F_{(7,\, 11)} \neq F_{(11,\, 7)}$. It is important to keep track of which degrees of freedom are for the numerator and which are for the denominator.

Table 8–2 is a reproduction of a part of Table 5, showing values of *F* distributions with different degrees of freedom cutting off a right-tailed area of 0.05.

The *F* distribution is useful in testing the equality of two population variances. Recall that in Chapter 7 we defined a chi-square random variable as

$$\chi^2 = \frac{(n-1)S^2}{\sigma^2} \tag{8–16}$$

FIGURE 8–12   An *F* Distribution with 7 and 11 Degrees of Freedom

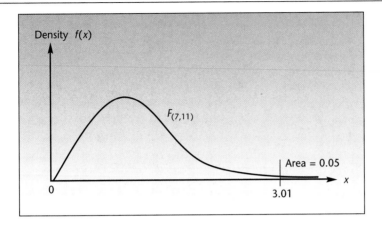

FIGURE 8–13   Several *F* Distributions

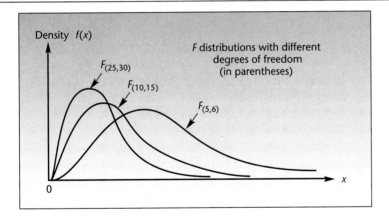

where S^2 is the sample variance from a *normally distributed population*. This was the definition in the single-sample case, where $n - 1$ was the appropriate number of degrees of freedom. Now suppose that we have two *independent* random samples from two *normally distributed populations*. The two samples will give rise to two sample variances, the random variables and S_1^2 and S_2^2 with $n_1 - 1$ and $n_2 - 1$ degrees of freedom, respectively. The ratio of these two random variables is the random variable

$$\frac{S_1^2}{S_2^2} = \frac{\chi_1^2 \sigma_1^2/(n_1 - 1)}{\chi_2^2 \sigma_2^2/(n_2 - 1)} \tag{8–17}$$

When the two population variances σ_1^2 and σ_2^2 are *equal,* the two terms σ_1^2 and σ_2^2 cancel, and equation 8–17 is equal to equation 8–15, which is the ratio of two independent chi-square random variables, each divided by its own degrees of freedom $(k_1$ is $n_1 - 1$, and k_2 is $n_2 - 1)$. This, therefore, is an *F* random variable with $n_1 - 1$ and $n_2 - 1$ degrees of freedom.

TABLE 8–2   Critical Points Cutting Off a Right-Tailed Area of 0.05 for Selected F Distributions

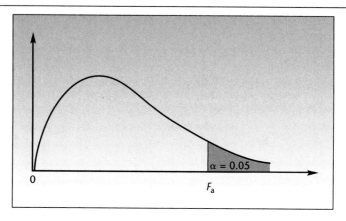

k_2	k_1	\multicolumn{9}{c}{Degrees of Freedom of the Numerator}
		1	**2**	**3**	**4**	**5**	**6**	**7**	**8**	**9**
Degrees	1	161.4	199.5	215.7	224.6	230.2	234.0	236.8	238.9	240.5
of	2	18.51	19.00	19.16	19.25	19.30	19.33	19.35	19.37	19.38
freedom	3	10.13	9.55	9.28	9.12	9.01	8.94	8.89	8.85	8.81
of the	4	7.71	6.94	6.59	6.39	6.26	6.16	6.09	6. 04	6.00
denomi-	5	6.61	5.79	5.41	5.19	5.05	4.95	4.88	4.82	4.77
nator	6	5.99	5.14	4.76	4.53	4.39	4.28	4.21	4.15	4.10
	7	5.59	4.74	4.35	4.12	3.97	3.87	3.79	3.73	3.68
	8	5.32	4.46	4.07	3.84	3.69	3.58	3.50	3.44	3.39
	9	5.12	4.26	3.86	3.63	3.48	3.37	3.29	3.23	3.18
	10	4.96	4.10	3.71	3.48	3.33	3.22	3.14	3.07	3.02
	11	4.84	3.98	3.59	3.36	3.20	3.09	3.01	2.95	2.90
	12	4.75	3.89	3.49	3.26	3.11	3.00	2.91	2.85	2.80
	13	4.67	3.81	3.41	3.18	3.03	2.92	2.83	2.77	2.71
	14	4.60	3.74	3.34	3.11	2.96	2.85	2.76	2.70	2.65
	15	4.54	3.68	3.29	3.06	2.90	2.79	2.71	2.64	2.59

The test statistic for the equality of the variances of two normally distributed populations is

$$F_{(n_1-1,\ n_2-2)} = \frac{S_1^2}{S_2^2} \qquad (8\text{–}18)$$

Now that we have encountered the important F distribution, we are ready to define the test for the equality of two population variances. Incidentally, the F distribution has many more uses than just testing for equality of two population variances. In chapters that follow, we will find this distribution extremely useful in a variety of involved statistical contexts.

A Statistical Test for Equality of Two Population Variances

We assume independent random sampling from the two populations in question. We also assume that the two populations are normally distributed. Let the two populations be labeled 1 and 2. The possible hypotheses to be tested are the following:

$$\text{A two-tailed test:} \quad H_0: \sigma_1^2 = \sigma_2^2$$
$$H_1: \sigma_1^2 \neq \sigma_2^2$$
$$\text{A one-tailed test:} \quad H_0: \sigma_1^2 \leq \sigma_2^2$$
$$H_1: \sigma_1^2 > \sigma_2^2$$

We will consider the one-tailed test first, because it is easier to handle. Suppose that we want to test whether σ_1^2 is greater than σ_2^2. We collect the two independent random samples from populations 1 and 2, and we compute the statistic in equation 8–18. We must be sure to put s_1^2 in the numerator, because in a one-tailed test, rejection may occur only on the right. If s_1^2 is actually less than s_2^2, we can immediately not reject the null hypothesis because the statistic value will be less than 1.00 and, hence, certainly within the nonrejection region for any level α.

In a two-tailed test, we may do one of two things:

1. We may use the convention of always placing the *larger* sample variance in the *numerator*. That is, we label the population with the larger sample variance population 1. Then, if the test statistic value is greater than a critical point cutting off an area of, say, 0.05 to its right, we reject the null hypothesis that the two variances are equal at $\alpha = 0.10$ (that is, at *double* the level of significance from the table). This is so because, under the null hypothesis, either of the two sample variances could have been greater than the other, and we are carrying out a two-tailed test on one tail of the distribution. Similarly, if we can get a *p*-value on the one tail of rejection, we need to *double* it to get the actual *p*-value. Alternatively, we can conduct a two-tailed test as described next.

2. We may choose not to relabel the populations such that the greater sample variance is on top. Instead, we find the right-hand critical point for $\alpha = 0.01$ or 0.05 (or another level) from Appendix C, Table 5. We compute the left-hand critical point for the test (not given in the table) as follows:

The left-hand critical point to go along with $F_{(k_1, k_2)}$ is given by

$$\frac{1}{F_{(k_2, k_1)}} \tag{8–19}$$

where $F_{(k_2, k_1)}$ is the right-hand critical point from the table for an F random variable with the *reverse order of degrees of freedom.*

Thus, the left-hand critical point is the reciprocal of the right-hand critical point obtained from the table and using the reverse order of degrees of freedom for numerator and denominator. Again, the level of significance α must be doubled. For example, from Table 8–2, we find that the right-hand critical point for $\alpha = 0.05$ with degrees of freedom for the numerator equal to 6 and degrees of freedom for the

FIGURE 8–14 The Critical Points for a Two-Tailed Test Using $F_{(6, 9)}$ and $\alpha = 0.10$

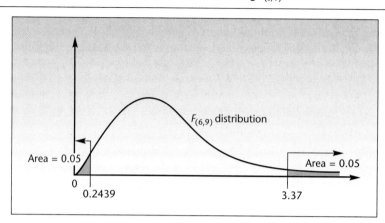

denominator equal to 9 is $C = 3.37$. So, for a two-tailed test at $\alpha = 0.10$ (double the significance level from the table), the critical points are 3.37 and the point obtained using equation 8–19, $1/F_{(9, 6)}$, which, using the table, is found to be $1/4.10 = 0.2439$. This is shown in Figure 8–14.

We will now demonstrate the use of the test for equality of two population variances with examples.

EXAMPLE 8–9   One of the problems that insider trading supposedly causes is unnaturally high stock price volatility. When insiders rush to buy a stock they believe will increase in price, the buying pressure causes the stock price to rise faster than under usual conditions. Then, when insiders dump their holdings to realize quick gains, the stock price dips fast. Price volatility can be measured as the variance of prices.

An economist wants to study the effect of the insider trading scandal and ensuing legislation on the volatility of the price of a certain stock. The economist collects price data for the stock during the period before the event (interception and prosecution of insider traders) and after the event. The economist makes the assumptions that prices are approximately normally distributed and that the two price data sets may be considered independent random samples from the populations of prices before and after the event. As we mentioned earlier, the theory of finance supports the normality assumption. (The assumption of random sampling may be somewhat problematic in this case, but later we will deal with time-dependent observations more effectively.) Suppose that the economist wants to test whether the event has decreased the variance of prices of the stock. The 25 daily stock prices before the event give $s_1^2 = 9.3$ (dollars squared), and the 24 stock prices after the event give $s_2^2 = 3.0$ (dollars squared). Conduct the test at $\alpha = 0.05$.

Solution   Our test is a right-tailed test. We have H_0: $\sigma_1^2 \le \sigma_2^2$ and H_1: $\sigma_1^2 > \sigma_2^2$. We compute the test statistic of equation 8–18:

$$F_{(n_1-1,\ n_2-1)} = F_{(24,\ 23)} = \frac{s_1^2}{s_2^2} = \frac{9.3}{3.0} = 3.1$$

FIGURE 8–15   **Carrying Out the Test of Example 8–9**

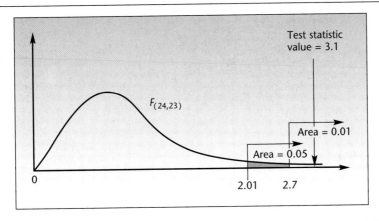

As can be seen from Figure 8–15, this value of the test statistic falls in the rejection region for $\alpha = 0.05$ and for $\alpha = 0.01$. The critical point for $\alpha = 0.05$, from Table 5, is equal to 2.01 (see 24 degrees of freedom for the numerator and 23 degrees of freedom for the denominator). Referring to the F table for $\alpha = 0.01$ with 24 degrees of freedom for the numerator and 23 degrees of freedom for the denominator gives a critical point of 2.70. The computed value of the test statistic, 3.1, is greater than both of these values. The p-value is less than 0.01, and the economist may conclude that (subject to the validity of the assumptions) the data present significant evidence that the event in question has reduced the variance of the stock's price.

EXAMPLE 8–10

Use the data of Example 8–5—$n_1 = 14$, $s_1 = 0.12$; $n_2 = 9$, $s_2 = 0.11$—to test the assumption of equal population variances.

Solution

The test statistic is the same as in the previous example, given by equation 8–18:

$$F_{(13, 8)} = \frac{s_1^2}{s_2^2} = \frac{0.12^2}{0.11^2} = 1.19$$

Here we placed the larger variance in the numerator because it was already labeled 1 (we did not purposely label the larger variance as 1). We can carry this out as a one-tailed test, even though it is really two-tailed, remembering that we must double the level of significance. Choosing $\alpha = 0.05$ from the table makes this a test at true level of significance equal to $2(0.05) = 0.10$. The critical point, using 12 and 8 degrees of freedom for numerator and denominator, respectively, is 3.28. (This is the closest value, since our table does not list critical points for 13 and 8 degrees of freedom.) As can be seen, our test statistic falls inside the nonrejection region, and we may conclude that at the 0.10 level of significance, there is no evidence that the two population variances are different from each other.

   Let us now see how this test may be carried out using the alternative method of solution: finding a left-hand critical point to go with the right-hand one. The right-hand critical point remains 3.28 (let us assume that this is the exact value for 13 and 8 degrees of freedom). The left-hand critical point is found by equation 8–19 as $1/F_{(8, 13)}$

FIGURE 8–16   **Carrying Out the Two-Tailed Test of Example 8–10**

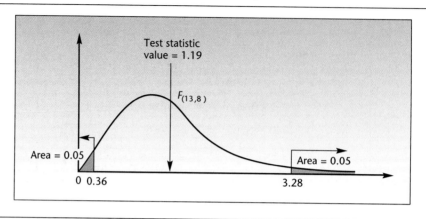

FIGURE 8–17   **The _F_-Distribution Template**
[F.xls]

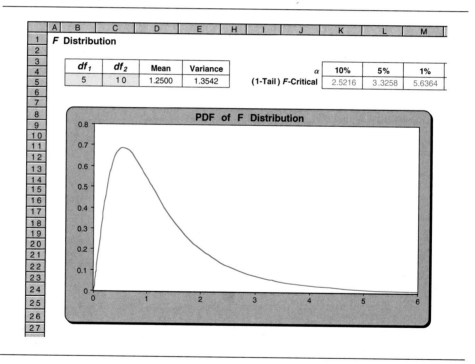

= 1/2.77 = 0.36 (recall that the left-hand critical point is the inverse of the critical point corresponding to reversed-order degrees of freedom). The two tails are shown in Figure 8–16. Again, the value of the test statistic falls inside the nonrejection region for this test at $\alpha = 2(0.05) = 0.10$.

The Templates

Figure 8–17 shows the _F_-distribution template. The numerator and denominator degrees of freedom are entered in cells B4 and C4. For any combination of degrees of

FIGURE 8–18 The Template for Testing Equality of Variances [Testing Equality of Variances.xls; Sheet: Sample Data]

***F*-Test for Equality of Variances**

	Data	
	Name 1	Name 2
	Sample1	Sample2
1	334	405
2	150	125
3	520	540
4	95	100
5	212	200
6	30	30
7	1055	1200
8	300	265
9	85	90
10	129	206
11	40	18
12	440	489

	Sample 1	Sample 2	
Size	16	16	n
Variance	95858.8	118367.7	s^2

Test Statistic	0.809839	F
df1	15	
df2	15	

Null Hypothesis	*p*-value	At an α of 5%
$H_0: \sigma^2_1 - \sigma^2_2 = 0$	0.6882	
$H_0: \sigma^2_1 - \sigma^2_2 >= 0$	0.3441	
$H_0: \sigma^2_1 - \sigma^2_2 <= 0$	0.6559	

FIGURE 8–19 The Template for Testing Equality of Variances [Testing Equality of Variances.xls; Sheet: Sample Statistics]

***F*-Test for Equality of Variances** — Stock Price Volatility

	Sample 1	Sample 2	
Size	25	24	n
Variance	9.3	3	s^2

Test Statistic	3.1	F
df1	24	
df2	23	

Null Hypothesis	*p*-value	At an α of 5%
$H_0: \sigma^2_1 - \sigma^2_2 = 0$	0.0085	Reject
$H_0: \sigma^2_1 - \sigma^2_2 >= 0$	0.9958	
$H_0: \sigma^2_1 - \sigma^2_2 <= 0$	0.0042	Reject

freedom, the template can be used to visualize the distribution, get critical F values, or compute p-values corresponding to a calculated F value.

Figure 8–18 shows the template that can be used to test the *equality* of two population variances from sample data. In other words, the hypothesized difference between the variances can only be zero.

Figure 8–19 shows the template that can be used to test the equality of two population variances when sample statistics, rather than sample data, are known. The data in the figure correspond to Example 8–9.

PROBLEMS

In the following problems, assume that all populations are normally distributed.

8–40. Compaq Computer Corporation has an assembly plant in Houston, where the company's Deskpro computer is built. Engineers at the plant are considering a new production facility and are interested in going on-line with the new facility if and only if they can be fairly sure that the variance of the number of computers assembled per day using the new facility is lower than the production variance of the old system. A random sample of 40 production days using the old production method gives a sample variance of 1,288; and a random sample of 15 production days using the proposed new method gives a sample variance of 1,112. Conduct the appropriate test at $\alpha = 0.05$.

8–41. Test the validity of the equal-variance assumption in problem 8–27.

8–42. Test the validity of the equal-variance assumption for the data presented in problem 8–25.

8–43. Test the validity of the equal-variance assumption for the data presented in problem 8–26.

8–44. The following data are independent random samples of sales of the Nissan Pulsar model made in a joint venture of Nissan and Alfa Romeo. The data represent sales at dealerships before and after the announcement that the Pulsar model will no longer be made in Italy. Sales numbers are monthly.

Before: 329, 234, 423, 328, 400, 399, 326, 452, 541, 680, 456, 220
After: 212, 630, 276, 112, 872, 788, 345, 544, 110, 129, 776

Do you believe that the variance of the number of cars sold per month before the announcement is equal to the variance of the number of cars sold per month after the announcement?

8–45. A large department store wants to test whether the variance of waiting time in two checkout counters is approximately equal. Two independent random samples of 25 waiting times in each of the counters gives $s_1 = 2.5$ minutes and $s_2 = 3.1$ minutes. Carry out the test of equality of variances, using $\alpha = 0.02$.

8–46. An important measure of the risk associated with a stock is the standard deviation, or variance, of the stock's price movements. A financial analyst wants to test the one-tailed hypothesis that stock A has a greater risk (larger variance of price) than stock B. A random sample of 25 daily prices of stock A gives $s_A^2 = 6.52$, and a random sample of 22 daily prices of stock B gives a sample variance of $s_B^2 = 3.47$. Carry out the test at $\alpha = 0.01$.

8–47. Discuss the assumptions made in the solution of the problems in this section.

8–6   Summary and Review of Terms

In this chapter, we extended the ideas of hypothesis tests and confidence intervals to the case of two populations. We discussed the comparisons of two population means, two population proportions, and two population variances. We developed a hypothesis test and confidence interval for the difference between two population means when the population variances were believed to be equal, and in the general case when they are not assumed equal. We introduced an important family of probability distributions: the **F distributions**. We saw that each F distribution has two kinds of degrees of freedom: one associated with the numerator and one associated with the denominator of the expression for F. We saw how the F distribution is used in testing hypotheses about two population variances. In the next chapter, we will make use of the F distribution in tests of the equality of *several* population means: the analysis of variance.

8-48. Zim Container Service ships containers across both the Atlantic and the Pacific oceans. In 1994, the company had to decide whether to include Kingston, Jamaica, as a port of call instead of Savannah, Georgia, where the company's ships had been calling for some time. Zim's general manager of operations decided to use statistics to determine which of the two ports of call—Savannah or Kingston—would have a greater demand for shipped containers, or whether the average number of containers demanded by the two ports might be approximately equal (in which case the decision would be based on the cost of service). The manager had data on the last 17 calls at Savannah. The data, in number of containers unloaded per call, were 9, 6, 7, 7, 8, 7, 9, 4, 7, 6, 6, 5, 7, 5, 9, 8, 6. Then, as a test run, company ships unloaded containers at Kingston during 9 arrivals at that port. The data (number of containers unloaded at each trip) were 10, 5, 8, 8, 9, 10, 4, 9, 11. Use these data to conduct the test, and advise Zim's general manager of operations as to which port should be the regular port of call, or whether the ports will have approximately equal demand for containers.[10] Assume normal populations with equal variances.

8-49. In problem 8-48, construct a 99% confidence interval for the difference between the average number of containers per voyage demanded at Savannah and the average number of containers per voyage demanded at Kingston. Interpret your results.

8-50. "Strategizing for the future" is management lingo for sessions that help managers make better decisions. Managers who deliver the best stock performance get results by bold, rule-breaking strategies, according to an article in *Fortune* entitled "Managing 'R' Us."[11] To test the effectiveness of "strategizing for the future," companies' 5-year average stock performance was considered before, and after, consultant-led "strategizing for the future" sessions were held.

Before (%)	After (%)
10	12
12	16
8	−2
−5	10
−1	11
5	18
−3	−8
16	20
−2	−1
13	21
17	24

Test the effectiveness of the program, using $\alpha = 0.05$.

---

[10]Data provided courtesy of Zim Navigation Company, Ltd.

[11]June 23, 1997, p. 100.

8–51. For problem 8–50, construct a 95% confidence interval for the difference in proportion of program effectiveness.

8–52. A study was undertaken by Montgomery Securities to assess average labor and materials costs incurred by Chrysler and General Motors in building a typical four-door, intermediate-size car. The reported average cost for Chrysler was $9,500, and for GM it was $9,780. Suppose that these data are based on random samples of 25 cars for each company, and suppose that both standard deviations are equal to $1,500. Test the hypothesis that the average GM car of this type is more expensive to build than the average Chrysler car of the same type. Use $\alpha = 0.01$. Assume equal population variances.

8–53. For problem 8-52, give a 99% confidence interval for the difference between the average cost incurred by GM in building the car and the average cost incurred by Chrysler in building the same type of car. Interpret the results.

8–54. Suppose that the data in problem 8–52 are based on two random samples of 100 data points each. Redo the problem, assuming unequal population variances.

8–55. Two movies were screen-tested at two different samples of theaters. *Pocahontas* was viewed at 80 theaters and was considered a success in terms of box office sales in 60 of these theaters. *The Bridges of Madison County* was viewed at a random sample of 100 theaters and was considered a success in 65. Based on these data, do you believe that one of these movies was a greater success than the other? Explain.

8–56. For problem 8–55, give a 95% confidence interval for the difference in proportion of theaters nationwide where one movie will be preferred over the other. Is the point 0 contained in the interval? Discuss.

8–57. Two 12-meter boats, the K boat and the L boat, are tested as possible contenders in the America's Cup races. The following data represent the time, in minutes, to complete a particular track in independent random trials of the two boats:

    K boat: 12.0, 13.1, 11.8, 12.6, 14.0, 11.8, 12.7, 13.5, 12.4, 12.2, 11.6, 12.9
    L boat: 11.8, 12.1, 12.0, 11.6, 11.8, 12.0, 11.9, 12.6, 11.4, 12.0, 12.2, 11.7

Test the null hypothesis that the two boats perform equally well. Is one boat faster, on average, than the other? Assume equal population variances.

8–58. In problem 8–57, assume that the data points are paired as listed and that each pair represents performance of the two boats at a single trial. Conduct the test, using this assumption. What is the advantage of testing using the paired data versus independent samples?

8–59. According to G. Hill in "The Digital Age: How It's Changing Our Lives,"[12] the average number of hours on the Internet per Internet-connected household was 4.2 in 1995 and 6.3 in 1996. Assuming both statistics are based on independent random samples of 1,000 households each with standard deviation = 2 hours, test for change in mean from 1995 to 1996.

8–60. The IIT Technical Institute claims "94% of our graduates get jobs." Assume that the result is based on a random sample of 100 graduates of the program. Suppose that an independent random sample of 125 graduates of a competing technical institute reveals that 92% of these graduates got jobs. Is there evidence to conclude that one institute is more successful than the other in placing its graduates?

---

[12] *The Wall Street Journal,* June 16, 1997, p. 13.

8–61. The power of supercomputers derives from the idea of parallel processing. Engineers at Cray Research are interested in determining whether one of two parallel processing designs produces faster average computing time, or whether the two designs are equally fast. The following are the results, in seconds, of independent random computation times using the two designs.

Design 1	Design 2
2.1, 2.2, 1.9, 2.0, 1.8, 2.4,	2.6, 2.5, 2.0, 2.1, 2.6, 3.0,
2.0, 1.7, 2.3, 2.8, 1.9, 3.0,	2.3, 2.0, 2.4, 2.8, 3.1, 2.7,
2.5, 1.8. 2.2	2.6

Assume that the two populations of computing time are normally distributed and that the two population variances are equal. Is there evidence that one parallel processing design allows for faster average computation than the other?

8–62. Test the validity of the equal-variance assumption in problem 8–61. If you reject the null hypothesis of equal-population variance, redo the test of problem 8–61 using another method.

8–63. The senior vice president for marketing at Westin Hotels believes that the company's recent advertising of the Westin Plaza in New York has increased the average occupancy rate at that hotel by a least 5%. To test the hypothesis, a random sample of daily occupancy rates (in percentages) before the advertising is collected. A similar random sample of daily occupancy rates is collected after the advertising took place. The data are as follows.

Before Advertising (%)	After Advertising (%)
86, 92, 83, 88, 79, 81, 90,	88, 94, 97, 99, 89, 93, 92,
76, 80, 91, 85, 89, 77, 91,	98, 89, 90, 97, 91, 87, 80,
83	88, 96

Assume normally distributed populations of occupancy rates with equal population variances. Test the vice president's hypothesis.

8–64. For problem 8–63, test the validity of the equal-variance assumption.

8–65. Refer to problem 8–48. Test the null hypothesis that the variance of the number of containers demanded at Savannah is less than or equal to the variance of the number of containers demanded at Kingston, versus the alternative hypothesis that the variance of the number of containers demanded at Savannah is greater than that of Kingston.

8–66. Refer to problem 8–57. Do you believe that the variance of performance times for the K boat is about the same as the variance of performance times for the L boat? Explain. What are the implications of your result on the analysis of problem 8–57? If needed, redo the analysis in problem 8–57.

8–67. A company is interested in offering its employees one of two employee benefit packages. A random sample of the company's employees is collected, and each person in the sample is asked to rate each of the two packages on an overall preference scale of 0 to 100. The order of presentation of each of the two plans is randomly selected for each person in the sample. The paired data are:

Program *A:*   45, 67, 63, 59, 77, 69, 45, 39, 52, 58, 70, 46, 60, 65, 59, 80
Program *B:*   56, 70, 60, 45, 85, 79, 50, 46, 50, 60, 82, 40, 65, 55, 81, 68

Do you believe that the employees of this company prefer, on the average, one package over the other? Explain.

8–68. A company that makes electronic devices for use in hospitals needs to decide on one of two possible suppliers for a certain component to be used in the devices. The company gathers a random sample of 200 items made by supplier *A* and finds that 12 items are defective. An independent random sample of 250 items made by supplier *B* reveals that 38 are defective. Is one supplier more reliable than the other? Explain.

8–69. Refer to problem 8–68. Give a 95% confidence interval for the difference in the proportions of defective items made by suppliers *A* and *B*.

8–70. Refer to problem 8–63. Give a 90% confidence interval for the difference in average occupancy rates at the Westin Plaza hotel before and after the advertising.

8–71. Annual profits of the phone companies are as follows: Bell Atlantic, $13.1 billion; Nynex, $13.4 billion.[13] If these numbers are means of annualized observations for a random sample of 24 months each, and the standard deviations are $1.2 billion each, test for equality of annual profits for these firms.

8–72. According to a market research survey of 12,000 households with Web connections,[14] CNET was used by 11.5%. An independent random sample of 12,000 households showed CNN Interactive was accessed by 28.3%. Construct a 95% confidence interval for the difference in percentages.

8–73. The rent for prime office space in San Francisco is $31 per square foot and $33 per square foot in Boston.[15] Test for significance of difference, assuming two random samples (independent, of course) of 100 prime properties each and standard deviations of $2 in both cities.

8–74. According to Labor Department statistics,[16] the average U.S. work week has shortened from 39 hours in the 1950s and early 1960s to 35 hours in the 1990s. Assume the two statistics are based on independent *t* samples of 2,500 workers each, and the standard deviations are both 2 hours.

    a. Test for significance of change.

    b. Give a 95% confidence interval for the difference.

8–75. According to a survey conducted by the Home Improvement Research Institute, 17.8% of 663 homeowners surveyed in June 1999 had made purchases online, whereas 21% of 684 homeowners surveyed in June 2000 had made purchases online.[17] Does this demonstrate that the percentage of homeowners making purchases online increased in 2000? Use $\alpha = 0.05$.

8–76. A survey finds that 62% of lower-income households have Internet access at home as compared to 70% of upper-income households. Assume that the data are based on random samples of size 500 each. Does this demonstrate that lower-income households are less likely to have Internet access than the upper-income households? Use $\alpha = 0.05$.

8–77. The average sale price of existing condominium apartments in 2000 was $128,000 compared to the average of $125,000 in 1999. Assuming that these averages

---

[13]*Business Week,* June 9, 1997, p. 23.

[14]R. Grorer and L. Himelstein, "All the News That's Fit to Browse," *Business Week,* June 16, 1997, p. 74.

[15]G. Koretz, "Less of a Gulf in Office Rents," *Business Week,* June 16, 1997, p. 12.

[16]G. Koretz, "How Many Hours in a Workweek," *Business Week,* June 16, 1997, p. 12.

[17]"Tool Time," *Business Week,* February 19, 2001.

are from a random sample of 400 condos sold in 2000 with a sample standard deviation of $2,500 and a random sample of 350 sold in 1999 with a sample standard deviation of $2,700, test the null hypothesis that the sale price of condos has not increased over the period of time. Use $\alpha = 0.05$.

CASE 9 Tiresome Tires II

A tire manufacturing company invents a new, cheaper method for carrying out one of the steps in the manufacturing process. The company wants to test the new method before adopting it, because the method could alter the interply shear strength of the tires produced.

To test the acceptability of the new method, the company formulates the null and alternative hypotheses as

$$H_0: \mu_1 - \mu_2 \leq 0$$
$$H_1: \mu_1 - \mu_2 > 0$$

where μ_1 is the population mean of the interply shear strength of the tires produced by the old method and μ_2 that of the tires produced by the new method. The evidence is gathered through a destructive test of 40 randomly selected tires from each method. Following are the data gathered:

1. Test the null hypothesis at $\alpha = 0.05$.
2. Later it was found that quite a few tires failed on the road. As a part of the investigation, the above hypothesis test is reviewed. Considering the high cost of type II error, the value of 5% for α is questioned. The response was that the cost of type I error is also high because the new method could save millions of dollars. What value for α would you say is appropriate? Will the null hypothesis be rejected at that α?
3. A review of the tests conducted on the samples reveals that 40 otherwise identical pairs of tires were randomly selected and used. The two tires in each pair underwent the two different methods, and all other steps in the manufacturing process were identically carried out on the two tires. By virtue of this fact, it is argued that a paired difference test is more appropriate. Conduct a paired difference test at $\alpha = 0.05$.
4. There is a move to reduce the variance of the strength by improving the process. Will the reduction in the variance of the process increase or decrease the chances of type I and type II errors?

No.	Sample 1	Sample 2	No.	Sample 1	Sample 2
1	2792	2713	21	2693	2683
2	2755	2741	22	2740	2664
3	2745	2701	23	2731	2757
4	2731	2731	24	2707	2736
5	2799	2747	25	2754	2741
6	2793	2679	26	2690	2767
7	2705	2773	27	2797	2751
8	2729	2676	28	2761	2723
9	2747	2677	29	2760	2763
10	2725	2721	30	2777	2750
11	2715	2742	31	2774	2686
12	2782	2775	32	2713	2727
13	2718	2680	33	2741	2757
14	2719	2786	34	2789	2788
15	2751	2737	35	2723	2676
16	2755	2740	36	2713	2779
17	2685	2760	37	2781	2676
18	2700	2748	38	2706	2690
19	2712	2660	39	2776	2764
20	2778	2789	40	2738	2720

9 Analysis of Variance

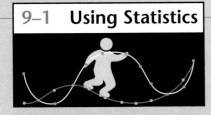

Bee it knowne unto all men by these presents
iune.17.1579.
By the grace of God and in the name of herr
maiesty Queen Elizabeth of England and herr
successors forever. I take possession of this
Kingdome whose King and people freely
resigne their right and title in the whole land unto herr
maiesties keeping. Now named by me an to bee
knowne unto all men as Nova Albion.
G
Francis Drake

In the summer of 1936, a young man named Beryle Shinn chanced upon a brass plate bearing the inscription above while hiking in a hilly area overlooking Point San Quentin and the San Francisco Bay. Mr. Shinn put the plate in the trunk of his car, where it remained for a few years. Later, by chance, the plate came to the attention of Professor Herbert E. Bolton at the Department of History at the University of California, Berkeley.

In a logbook kept during his circumnavigation of the earth, Sir Francis Drake wrote of entering a sheltered area in 1579 on what is now the northern California coast to refit his ship. He also mentioned leaving a brass plate attached to a post to record the event. Thus, upon being presented the Shinn plate, Professor Bolton declared: "One of the world's long-lost historical treasures apparently has been found!"

Although questions about the authenticity of the plate were immediately raised by many scholars—mainly because of the curious forms of many of the letters, and the writing style, which is different from known Elizabethan styles—the plate was nonetheless pronounced genuine and put on permanent display at the Bancroft Library of the University of California.

Contentions that the plate was the work of a modern forger continued to be expressed, and these led the Bancroft Library to order tests of the metallurgic structure of the brass. Finally, in 1976, several tiny holes were drilled in the plate, and a sample of brass particles was sent to the Research Laboratory of Archaeology at Oxford University for analysis. There the sample of brass particles from the plate was statistically compared with two random samples of brass: a sample of brasses made in the 20th century and a sample of English and Continental brasses created between 1540 and 1720. Analysis of the average zinc content of the sample from the discovered plate and that of the two other samples led to the conclusion that the average zinc content of the plate was equal to the average zinc content of modern brasses and very different from the average zinc content of brasses made in the 16th to 18th centuries.[1]

The results of the analysis led Dr. R. Hedges of Oxford University to conclude: "I would regard it as quite unreasonable to continue to believe in the authenticity of the plate." Thus, what was thought to be an ancient artifact was shown to be an ingenious modern forgery. The scientific studies left unanswered the questions of who made the plate and why.

The statistical method of comparing the means of several populations, such as the mean zinc content of the three brasses relevant to the study of the plate, is the

---

[1]Several tests of the plate of brass were conducted. See "The Plate of Brass Reexamined," a report by the Bancroft Library (1979).

analysis of variance. The method is often referred to by its acronym: ANOVA. Analysis of variance is the first of several advanced statistical techniques to be discussed in this book. Along with regression analysis, described in the next two chapters, ANOVA is the most commonly quoted advanced research method in the professional business and economic literature. What is analysis of variance? The name of the technique may seem misleading.

> ANOVA is a statistical method for determining the existence of differences among several population means.

While the aim of ANOVA is to detect differences among several population *means,* the technique requires the analysis of different forms of *variance* associated with the random samples under study—hence the name *analysis of variance.*

The original ideas of analysis of variance were developed by the English statistician Sir Ronald A. Fisher during the first part of this century. (Recall our mention of Fisher in Chapter 8 in reference to the *F* distribution.) Much of the early work in this area dealt with agricultural experiments where crops were given different "treatments," such as being grown using different kinds of fertilizers. The researchers wanted to determine whether all treatments under study were equally effective or whether some treatments were better than others. *Better* referred to those treatments that would produce crops of greater average weight. This question is answerable by the analysis of variance. Since the original work involved different *treatments,* the term remained, and we use it interchangeably with *populations* even when no actual treatment is administered. Thus, for example, if we compare the mean income in four different communities, we may refer to the four populations as four different *treatments.*

In the next section, we will develop the simplest form of analysis of variance—the one-factor, fixed-effects, completely randomized design model. We may ignore this long name for now.

9–2   The Hypothesis Test of Analysis of Variance

The hypothesis test of analysis of variance is as follows:

$$H_0: \mu_1 = \mu_2 = \mu_3 = \cdots = \mu_r \qquad\qquad (9\text{–}1)$$
$$H_1: \text{Not all } \mu_i \ (i = 1, \ldots, r) \text{ are equal}$$

There are *r* populations, or treatments, under study. We draw an independent random sample from each of the *r* populations. The size of the sample from population i $(i = 1, \ldots, r)$ is n_i, and the total sample size is

$$n = n_1 + n_2 + \cdots + n_r$$

From the *r* samples we compute several different quantities, and these lead to a computed value of a test statistic that follows a known *F* distribution when the null hypothesis is true. From the value of the statistic and the critical point for a given level of significance, we are able to make a determination of whether we believe that the *r* population means are equal.

Usually, the number of compared means r is greater than 2. Why greater than 2? If r is equal to 2, then the test in equation 9–1 is just a test for equality of two population means; although we could use ANOVA to conduct such a test, we have seen relatively simple tests of such hypotheses: the two-sample t tests discussed in Chapter 8. In this chapter, we are interested in investigating whether *several* population means may be considered equal. This is a test of a *joint hypothesis* about the equality of several population parameters. But why can we not use the two-sample t tests repeatedly? Suppose we are comparing $r = 5$ treatments. Why can we not conduct all possible pairwise comparisons of means using the two-sample t test? There are 10 such possible comparisons (10 choices of five items taken two at a time, found by using a combinatorial formula presented in Chapter 2). It should be possible to make all 10 comparisons. However, if we use, say, $\alpha = 0.05$ for each test, then this means that the probability of committing a type I error in any particular test (deciding that the two population means are not equal when indeed they are equal) is 0.05. If each of the 10 tests has a 0.05 probability of a type I error, what is the probability of a type I error if we state, "Not all the means are equal" (i.e., rejecting H_0 in equation 9–1)? The answer to this question is not known![2]

If we need to compare more than two population means and we want to remain in control of the probability of committing a type I error, we need to conduct a *joint test*. Analysis of variance provides such a joint test of the hypotheses in equation 9–1. The reason for ANOVA's widespread applicability is that there are many situations in which we need to compare more than two populations simultaneously. Even in cases in which we need to compare only two treatments, say, test the relative effectiveness of two different prescription drugs, our actual test may require the use of a third treatment: a control treatment, or a placebo.

We now present the assumptions that must be satisfied so that we can use the analysis-of-variance procedure in testing our hypotheses of equation 9–1.

The required assumptions of ANOVA:

1. We assume *independent random sampling* from each of the r populations.
2. We assume that the r populations under study are *normally distributed,* with means μ_i that may or may not be equal, but with *equal variances* σ^2.

Suppose, for example, that we are comparing three populations and want to determine whether the three population means μ_1, μ_2, and μ_3 are equal. We draw separate random samples from each of the three populations under study, and we assume that the three populations are distributed as shown in Figure 9–1.

These model assumptions are necessary for the test statistic used in analysis of variance to possess an F distribution when the null hypothesis is true. If the populations are not exactly normally distributed, but have distributions that are close to a normal distribution, the method still yields good results. If, however, the distributions are highly skewed or otherwise different from normality, or if the population variances are not approximately equal, then ANOVA should not be used, and instead we must use a nonparametric technique called the Kruskal-Wallis test. This alternative technique is described in Chapter 14.

---

[2]The problem is complicated because we cannot assume independence of the 10 tests, and therefore, we cannot use a probability computation for independent events. The sample statistics used in the 10 tests are not independent since two such possible statistics are $\overline{X}_1 - \overline{X}_2$ and $\overline{X}_2 - \overline{X}_3$. Both statistics contain a common term $\overline{X}_2$ and thus are not independent of each other.

FIGURE 9–1   **Three Normally Distributed Populations with Different Means but with Equal Variance**

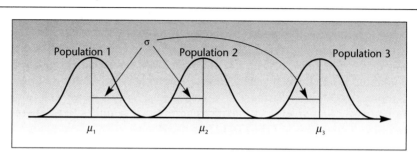

The Test Statistic

As mentioned earlier, when the null hypothesis is true, the test statistic of analysis of variance follows an F distribution. As you recall from Chapter 8, the F distribution has two kinds of degrees of freedom: degrees of freedom for the numerator and degrees of freedom for the denominator.

In the analysis of variance, the numerator degrees of freedom are $r - 1$, and the denominator degrees of freedom are $n - r$. In this section, we will not present the calculations leading to the computed value of the test statistic. Instead, we will assume that the value of the statistic is given. The computations are a topic in themselves and will be presented in the next section. Analysis of variance is an involved technique, and it is difficult and time-consuming to carry out the required computations by hand. Consequently, computers are indispensable in most situations involving analysis of variance, and we will make extensive use of the computer in this chapter. For now, let us assume that a computer is available to us and that it provides us with the value of the test statistic.

$$\text{ANOVA test statistic} = F_{(r - 1, n - r)} \qquad (9\text{–}2)$$

Figure 9–2 shows the F distribution with 3 and 50 degrees of freedom, which would be appropriate for a test of the equality of four population means using a total sample size of 54. Also shown is the critical point for $\alpha = 0.05$, found in Appendix C, Table 5. The critical point is 2.79. For reasons explained in the next section, the test is carried out as a right-tailed test.

We now have the basic elements of a statistical hypothesis test within the context of ANOVA: the null and alternative hypotheses, the required assumptions, and a distribution of the test statistic when the null hypothesis is true. Let us look at an example.

EXAMPLE 9–1   Major roasters and distributors of coffee in the United States have long felt great uncertainty in the price of coffee beans. Over the course of one year, for example, coffee futures prices went from a low of $1.40 per pound up to $2.50 and then down to $2.03. The main reason for such wild fluctuations in price, which strongly affect the performance of coffee distributors, is the constant danger of drought in Brazil. Since

FIGURE 9–2 **Distribution of the ANOVA Test Statistic for $r = 4$ Populations and a Total Sample Size $n = 54$**

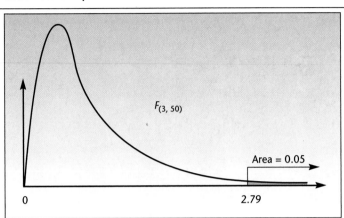

Brazil produces 30% of the world's coffee, the market for coffee beans is very sensitive to the annual rumors of impending drought.

Recently a domestic coffee distributor decided to avert the problem altogether by eliminating Brazilian coffee from all blends the company distributes. Before taking such action, the distributor wanted to minimize the chances of suffering losses in sales volume. Therefore, the distributor hired a market research firm to conduct a statistical test of consumers' taste preferences. The research firm made arrangements with several large restaurants to serve randomly chosen groups of their customers different kinds of after-dinner coffee. Three kinds of coffee were served: a group of 21 randomly chosen customers were served pure Brazilian coffee; another group of 20 randomly chosen customers were served pure Colombian coffee; and a third group of 22 randomly chosen customers were served pure African-grown coffee.

This is the *completely randomized design* part of the name of the ANOVA technique we mentioned at the end of the last section. In completely randomized design, the experimental units (in this case, the people involved in the experiment) are randomly assigned to the three treatments, the treatment being the kind of coffee they are served. Later in this chapter, we will encounter other designs useful in many situations. To prevent a response bias, the people in this experiment were not told the kind of coffee they were being served. The coffee was listed as a "house blend."

Suppose that data for the three groups were consumers' ratings of the coffee on a scale of 0 to 100 and that certain computations were carried out with these data (computations will be discussed in the next section), leading to the following value of the ANOVA test statistic: $F = 2.02$. Is there evidence to conclude that any of the three kinds of coffee leads to an average consumer rating different from that of the other two kinds?

The null and alternative hypotheses here are, by equation 9–1, *Solution*

$$H_0: \mu_1 = \mu_2 = \mu_3$$
$$H_1: \text{Not all three } \mu_i \text{ are equal}$$

FIGURE 9–3   Some of the Possible Relationships among the Relative Magnitudes of the Three Population Means μ_1, μ_2, and μ_3

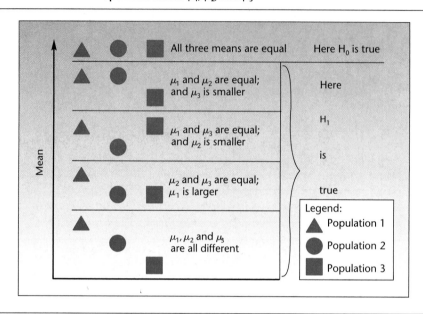

Let us examine the meaning of the null and alternative hypotheses in this example. The null hypothesis states that average consumer responses to each of the three kinds of coffee are equal. The alternative hypothesis says that not all three population means are equal. What are the possibilities covered under the alternative hypothesis? The possible relationships among the relative magnitudes of any three real numbers μ_1, μ_2, and μ_3 are shown in Figure 9–3.

As you can see from Figure 9–3, the alternative hypothesis is composed of several different possibilities—it includes all the cases where *not all* three means are equal. Thus, if we reject the null hypothesis, all we know is that there is statistical evidence to conclude that not all three population means are equal. However, we do not know in what way the means are different. Therefore, once we reject the null hypothesis, we need to conduct further analysis to determine which population means are different from one another. The further analysis following ANOVA will be discussed in a later section.

We have a null hypothesis and an alternative hypothesis. We also assume that the conditions required for ANOVA are met; that is, we assume that the three populations of consumer responses are (approximately) normally distributed with equal population variance. Now we need to conduct the test.

Since there are three populations, or treatments, under study, the degrees of freedom for the numerator are $r - 1 = 3 - 1 = 2$. Since the total sample size is $n = n_1 + n_2 + n_3 = 21 + 20 + 22 = 63$, we find that the degrees of freedom for the denominator are $n - r = 63 - 3 = 60$. Thus, when the null hypothesis is true, our test statistic has an F distribution with 2 and 60 degrees of freedom: $F_{(2, 60)}$. From Appendix C, Table 5, we find that the right-tailed critical point at $\alpha = 0.05$ for an F distribution with 2 and 60 degrees of freedom is 3.15. Since the computed value of the test statistic is equal to 2.02, we may conclude that at the 0.05 level of significance there is insufficient evidence to conclude that the three means are different. The null hypothesis

FIGURE 9–4   Carrying Out the Test of Example 9–1

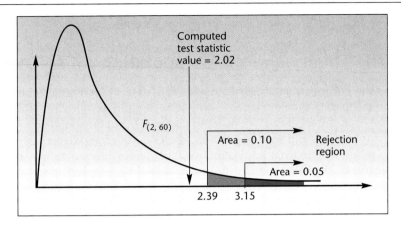

that all three population means are equal cannot be rejected. Since the critical point for $\alpha = 0.10$ is 2.39, we find that the p-value is greater than 0.10.

Based on our data, there is no evidence that consumers tend to prefer the Brazilian coffee to the other two brands. The distributor may substitute one of the other brands for the price-unstable Brazilian coffee. Note that we usually prefer to make conclusions based on the *rejection* of a null hypothesis because nonrejection is often considered a weak conclusion. The results of our test are shown in Figure 9–4.

In this section, we have seen the basic elements of the hypothesis test underlying analysis of variance: the null and alternative hypotheses, the required assumptions, the test statistic, and the decision rule. We have not, however, seen how the test statistic is computed from the data or the reasoning behind its computation. The theory and the computations of ANOVA are explained in the following sections.

PROBLEMS

9–1.   Four populations are compared by analysis of variance. What are the possible relations among the four population means covered under the null and alternative hypotheses?

9–2.   What are the assumptions of ANOVA?

9–3.   Three methods of training managers are to be tested for relative effectiveness. The management training institution proposes to test the effectiveness of the three methods by comparing two methods at a time, using a paired-t test. Explain why this is a poor procedure.

9–4.   In an analysis of variance comparing the output of five plants, data sets of 21 observations per plant are analyzed. The computed F statistic value is 3.6. Do you believe that there are differences in average output among the five plants? What is the approximate p-value? Explain.

9–5.   A real estate development firm wants to test whether there are differences in the average price of a lot of a given size in the center of each of four cities: Philadelphia, New York, Washington, and Baltimore. Random samples of 52 lots in Philadelphia, 38 lots in New York, 43 lots in Washington, and 47 lots in Baltimore lead to a computed test statistic value of 12.53. Do you believe that average lot prices in the four cities are equal? How confident are you of your conclusion? Explain.

9–3   The Theory and the Computations of ANOVA

Recall that the purpose of analysis of variance is to detect differences among several population means based on evidence provided by random samples from these populations. How can this be done? We want to compare r population means. We use r random samples, one from each population. Each random sample has its own mean. The mean of the sample from population i will be denoted by x_i. We may also compute the mean of all data points in the study, regardless of which population they come from. The mean of all the data points (when all data points are considered a single set) is called the *grand mean* and is denoted by $\bar{\bar{x}}$. These means are given by the following equations.

The mean of sample i $(i = 1, \ldots, r)$ is

$$\bar{x}_1 = \frac{\sum\limits_{j=1}^{n_j} x_{ij}}{n_i} \tag{9–3}$$

The **grand mean**, the mean of all the data points, is

$$\bar{\bar{x}} = \frac{\sum\limits_{i=1}^{r} \sum\limits_{j=1}^{n_j} x_{ij}}{n} \tag{9–4}$$

where x_{ij} is the particular data point in position j within the sample from population i. The subscript i denotes the population, or treatment, and runs from 1 to r. The subscript j denotes the data point within the sample from population i; thus, j runs from 1 to n_i.

In Example 9–1, $r = 3$, $n_1 = 21$, $n_2 = 20$, $n_3 = 22$, and $n = n_1 + n_2 + n_3 = 63$. The third data point (person) in the group of 21 people who consumed Brazilian coffee is denoted by x_{13} (that is, $i = 1$ denotes treatment 1 and $j = 3$ denotes the third point in that sample).

We will now define the main principle behind the analysis of variance.

If the r population means are different (i.e., at least two of the population means are *not* equal), then it is likely that the variation of the data points about their respective sample means $\bar{x}_i$ will be *small* when compared with the variation of the r sample means about the grand mean $\bar{\bar{x}}$.

We will demonstrate the ANOVA principle, using three hypothetical populations, which we will call the triangles, the squares, and the circles. Table 9–1 gives the values of the sample points from the three populations. For demonstration purposes, we use very small samples. In real situations, the sample sizes should be much larger. The data given in Table 9–1 are shown in Figure 9–5. The figure also shows the deviations

TABLE 9–1   Data and the Various Sample Means for Triangles, Squares, and Circles

Treatment i	Sample Point j	Value x_{ij}
$i = 1$ Triangle	1	4
Triangle	2	5
Triangle	3	7
Triangle	4	8
Mean of triangles		6
$i = 2$ Square	1	10
Square	2	11
Square	3	12
Square	4	13
Mean of squares		11.5
$i = 3$ Circle	1	1
Circle	2	2
Circle	3	3
Mean of circles		2
Grand mean of all data points		6.909

FIGURE 9–5   Deviations of the Triangles, Squares, and Circles from Their Sample Means and the Deviations of the Sample Means from the Grand Mean

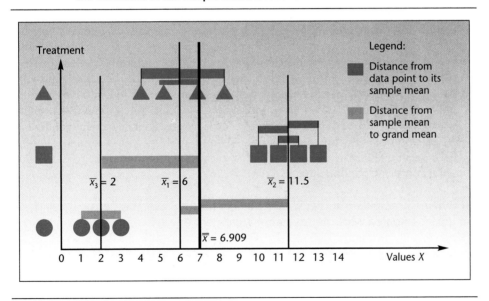

of the data points from their sample means and the deviations of the sample means from the grand mean.

Look carefully at Figure 9–5. Note that the *average* distance (in absolute value) of data points from their respective group means (i.e., the average distance, in absolute value, of a triangle from the mean of the triangles x_1 and similarly for the squares and the circles) is *relatively small* compared with the average distance (in absolute value) of the three sample means from the grand mean. If you are not convinced of this, note that there are only three distances of sample means to the grand mean (in the computation, each distance is weighted by the actual number of points in the group), and that only one of them, the smallest distance—that of $\bar{x}_1$ to $\bar{\bar{x}}$—is of the relative magnitude of the distances between the data points and their respective sample means. The two other distances are much greater; hence, the average distance of the sample means from the grand mean is greater than the average distance of all data points from their respective sample means.

The *average* deviation from a mean is zero. We talk about the average absolute deviation—actually, we will use the average *squared* deviation—to prevent the deviations from canceling. This should remind you of the definition of the sample variance in Chapter 1. Now let us define some terms that will make our discussion simpler.

> We define an **error deviation** as the difference between a data point and its sample mean. Errors are denoted by e, and we have
>
> $$e_{ij} = x_{ij} - \bar{x}_i \qquad (9\text{–}5)$$

Thus, all the distances from the data points to their sample means in Figure 9–5 are errors (some are positive, and others are negative). The reason these distances are called errors is that they are unexplained by the fact that the corresponding data points belong to population i. The errors are assumed to be due to natural variation, or pure randomness, within the sample from treatment i.

On the other hand,

> We define a **treatment deviation** as the deviation of a sample mean from the grand mean. Treatment deviations t_i are given by
>
> $$t_i = \bar{x}_i - \bar{\bar{x}} \qquad (9\text{–}6)$$

The ANOVA principle thus says:

> When the population means are not equal, the "average" error is relatively small compared with the "average" treatment deviation.

Again, if we actually averaged all the deviations, we would get zero. Therefore, when we apply the principle computationally, we will square the error and treatment deviations before averaging them. This way, we will maintain the relative (squared) magnitudes of these quantities. The averaging process is further complicated because we have to average based on degrees of freedom (recall that degrees of freedom were used in the definition of a sample variance). For now, let the term *average* be used in a simplified, intuitive sense.

Since we noted that the average error deviation in our triangle-square-circle example looks small relative to the average treatment deviation, let us see what the populations that brought about our three samples look like. Figure 9–6 shows the three populations, assumed normally distributed with equal variance. (This can be seen from the equal width of the three normal curves. Note also that the three samples seem to

FIGURE 9–6   **Samples of Triangles, Squares, and Circles and Their Respective Populations (the three populations are normal with equal variance but with different means)**

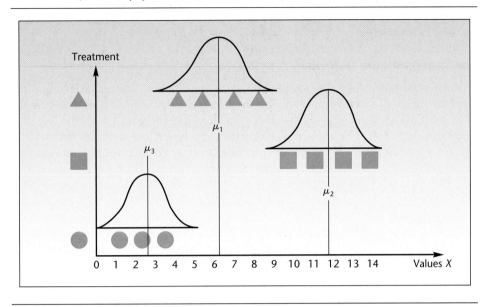

FIGURE 9–7   **Samples of Triangles, Squares, and Circles Where the Average Error Deviation Is Not Smaller than the Average Treatment Deviation**

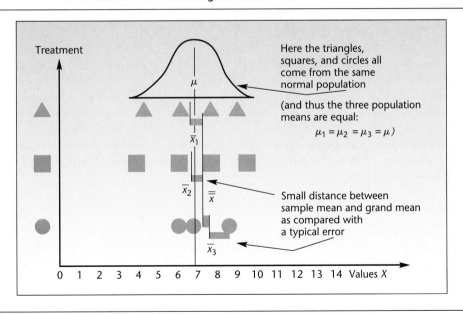

have equal dispersion about their sample means.) The figure also shows that the three population means are not equal.

Figure 9–7, in contrast, shows three samples of triangles, squares, and circles in which the average error deviation is of about the same magnitude (*not* smaller than) as the average treatment deviation. As can be seen from the superimposed normal populations from which the samples have arisen in this case, the three population means μ_1, μ_2, and μ_3 are all equal. Compare the two figures to convince yourself of the ANOVA principle.

The Sum-of-Squares Principle

We have seen how, when the population means are different, the error deviations in the data are small when compared with the treatment deviations. We made general statements about the average error being small when compared with the average treatment deviation. The error deviations measure how close the data *within* each group are to their respective group means. The treatment deviations measure the distances *between* the various groups. It therefore seems intuitively plausible (as seen in Figures 9–5 to 9–7) that when these two kinds of deviations are of about equal magnitude, the population means are about equal. Why? Because when the average error is about equal to the average treatment deviation, the treatment deviation may itself be viewed as just another error. That is, the treatment deviation in this case is due to pure chance rather than to any real differences among the population means. In other words, when the average t is of the same magnitude as the average e, both are estimates of the internal variation within the data and carry no information about a difference between any two groups—about a difference in population means.

We will now make everything quantifiable, using the *sum-of-squares principle*. We start by returning to Figure 9–5, looking at a particular data point, and analyzing distances associated with the data point. We choose the fourth data point from the sample of squares (population 2). This data point is $x_{24} = 13$ (verify this from Table 9–1). We now magnify a section of Figure 9–5, the section surrounding this particular data point. This is shown in Figure 9–8.

We define the **total deviation** of a data point x_{ij} (denoted by Tot_{ij}) as the deviation of the data point from the grand mean:

$$\text{Tot}_{ij} = x_{ij} - \bar{\bar{x}} \qquad (9\text{–}7)$$

Figure 9–8 shows that the total deviation is equal to the treatment deviation plus the error deviation. This is true for *any* point in our data set (even when some of the numbers are negative).

For any data point x_{ij},

$$\text{Tot} = t + e \qquad (9\text{–}8)$$

In words:

Total deviation = Treatment deviation + Error deviation

In the case of our chosen data point x_{24}, we have

$$t_2 + e_{24} = 4.591 + 1.5 = 6.091 = \text{Tot}_{24}$$

Equation 9–8 works for every data point in our data set. Here is how it is derived algebraically:

FIGURE 9–8   **Total Deviation as the Sum of the Treatment Deviation and the Error Deviation for a Particular Data Point**

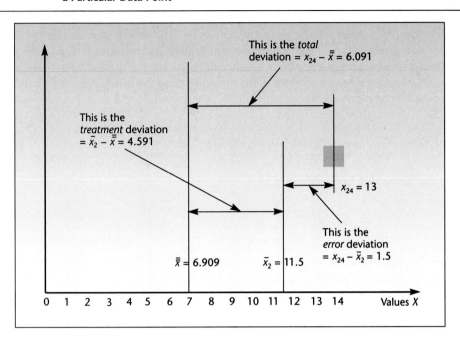

$$t_i + e_{ij} = (\bar{x}_i - \bar{\bar{x}}) + (x_{ij} - \bar{x}_i) = x_{ij} - \bar{\bar{x}} = \text{Tot}_{ij} \tag{9–9}$$

As seen in equation 9–9, the term $\bar{x}_i$ cancels out when the two terms in parentheses are added. This shows that for every data point, the total deviation is equal to the treatment part of the deviation plus the error part. This is also seen in Figure 9–8. The *total* deviation of a data point from the grand mean is thus partitioned into a deviation due to *treatment* and a deviation due to *error*. The deviation due to treatment differences is the *between-treatments* deviation, while the deviation due to error is the *within-treatment* deviation.

We have only considered one point, x_{24}. To determine whether the error deviations are small when compared with the treatment deviations, we need to aggregate the partition over all data points. This is done, as we noted earlier, by averaging the deviations. We take the partition of the deviations in equation 9–9 and we square each of the three terms (otherwise our averaging process would lead to zero).[3] The squaring of the terms in equation 9–9 gives, on one side,

$$t_i^2 + e_{ij}^2 = (\bar{x}_i - \bar{\bar{x}})^2 + (x_{ij} - \bar{x}_i)^2 \tag{9–10}$$

and, on the other side,

$$\text{Tot}_{ij}^2 = (x_{ij} - \bar{\bar{x}})^2 \tag{9–11}$$

---

[3]This can be seen from the data in Table 9–1. Note that the sum of the deviations of the triangles from their mean of 6 is $(4 - 6) + (5 - 6) + (7 - 6) + (8 - 6) = 0$; hence, an average of these deviations, or those of the squares or circles, leads to zero.

Note an interesting thing: The two sides of equation 9–9 are equal, but when all three terms are squared, the two sides (now equations 9–10 and 9–11) are *not* equal. Try this with any of the data points. The surprising thing happens next.

We take the squared deviations of equations 9–10 and 9–11, and we *sum them over all our data points*. Interestingly, the sum of the squared error deviations and the sum of the squared treatment deviations do add up to the sum of the squared total deviations. Mathematically, cross-terms in the equation drop out, allowing this to happen. The result is the sum-of-squares principle.

We have the following:

$$\sum_{i=1}^{r} \sum_{j=1}^{n_i} \text{Tot}_{ij}^2 = \sum_{i=1}^{r} n_i t_i^2 + \sum_{i=1}^{r} \sum_{j=1}^{n_i} e_{ij}^2$$

This can be written in longer form as

$$\sum_{i=1}^{r} \sum_{j=1}^{n_i} (x_{ij} - \bar{\bar{x}})^2 = \sum_{i=1}^{r} n_i(\bar{x}_i - \bar{\bar{x}})^2 + \sum_{i=1}^{r} \sum_{j=1}^{n_i} (x_{ij} - \bar{x}_i)^2$$

The Sum-of-Squares Principle

The sum-of-squares total (SST) is the sum of the two terms: the sum of squares for treatment (SSTR) and the sum of squares for error (SSE).

$$SST = SSTR + SSE \qquad (9\text{--}12)$$

The sum-of-squares principle partitions the sum-of-squares total within the data SST into a part due to treatment effect SSTR and a part due to errors SSE. The squared deviations of the treatment means from the grand mean are *counted for every data point*—hence the term n_i in the first summation on the right side (SSTR) of equation 9–12. The second term on the right-hand side is the sum of the squared errors, that is, the sum of the squared deviations of the data points from their respective sample means.

See Figure 9–8 for the different deviations associated with a single point. Imagine a similar relation among the three kinds of deviations for every one of the data points, as shown in Figure 9–5. Then imagine all these deviations squared and added together—errors to errors, treatments to treatments, and totals to totals. The result is equation 9–12, the sum-of-squares principle.

Sums of squares measure variation within the data. SST is the total amount of variation within the data set. SSTR is that part of the variation within the data that is due to differences among the groups, and SSE is that part of the variation within the data that is due to error—the part that cannot be explained by differences among the groups. Therefore, SSTR is sometimes called the sum of squares *between* (variation among the groups), and SSE is called the sum of squares *within* (within-group variation). SSTR is also called the *explained variation* (because it is the part of the total variation that can be explained by the fact that the data points belong to several different groups). SSE is then called the *unexplained variation*. The partition of the sum of squares in analysis of variance is shown in Figure 9–9.

Breaking down the sum of squares is not enough, however. If we want to determine whether the errors are small compared with the treatment part, we need to find

FIGURE 9–9   Partition of the Sum-of-Squares Total into Treatment and Error Parts

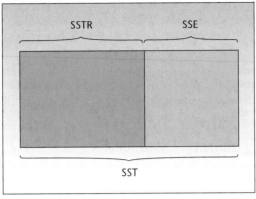

the *average* (squared) error and the *average* (squared) treatment deviation. Averaging, in the context of variances, is achieved by dividing by the appropriate number of degrees of freedom associated with each sum of squares.

The Degrees of Freedom

Recall our definition of degrees of freedom in Chapter 5. The degrees of freedom are the number of data points that are "free to move," that is, the number of elements in the data set minus the number of restrictions. A restriction on a data set is a quantity already computed from the entire data set under consideration; thus, knowledge of this quantity makes one data point fixed and reduces by 1 the effective number of data points that are free to move. This is why, as was shown in Chapter 5, knowledge of the sample mean reduces the degrees of freedom of the sample variance to $n - 1$. What are the degrees of freedom in the context of analysis of variance?

Consider the total sum of squares, SST. In computing this sum of squares, we use the entire data set and information about *one* quantity computed from the data: the grand mean (because, by definition, SST is the sum of the squared deviations of all data points from the grand mean). Since we have a total of n data points and one restriction,

> The number of degrees of freedom associated with SST is $n - 1$.

The sum of squares for treatment SSTR is computed from the deviations of r sample means from the grand mean. The r sample means are considered r independent data points, and the grand mean (which can be considered as having been computed from the r sample means) thus reduces the degrees of freedom by 1.

> The number of degrees of freedom associated with SSTR is $r - 1$.

The sum of squares for error SSE is computed from the deviations of a total of n data points $(n = n_1 + n_2 + \cdots + n_r)$ from r different sample means. Since each of the sample means acts as a restriction on the data set, the degrees of freedom for error are $n - r$. This can be seen another way: There are r groups with n_i data points in group i. Thus, each group, with its own sample mean acting as a restriction, has degrees of freedom equal to $n_i - 1$. The total number of degrees of freedom for error is the sum of the degrees of freedom in the r groups: df $= (n_1 - 1) + (n_2 - 1) + \cdots + (n_r - 1) = n - r$.

> The number of degrees of freedom associated with SSE is $n - r$.

An important principle in analysis of variance is that the degrees of freedom of the three components are *additive* in the same way that the sums of squares are additive.

$$df(total) = df(treatment) + df(error) \qquad (9\text{–}13)$$

This can easily be verified by noting the following: $n - 1 = (r - 1) + (n - r)$—the r drops out. We are now ready to compute the average squared deviation due to treatment and the average squared deviation due to error.

The Mean Squares

In finding the average squared deviations due to treatment and to error, we divide each sum of squares by its degrees of freedom. We call the two resulting averages **mean square treatment (MSTR)** and **mean square error (MSE),** respectively.

$$MSTR = \frac{SSTR}{r - 1} \qquad (9\text{–}14)$$

$$MSE = \frac{SSE}{n - r} \qquad (9\text{–}15)$$

The Expected Values of the Statistics MSTR and MSE under the Null Hypothesis

When the null hypothesis of ANOVA is true, all r population means are equal, and in this case there are *no treatment effects*. In such a case, the average squared deviation due to "treatment" is just another realization of an average squared error. In terms of the expected values of the two mean squares, we have

$$E(MSE) = \sigma^2 \qquad (9\text{–}16)$$

and

$$E(MSTR) = \sigma^2 + \frac{\sum n_i(\mu_i - \mu)^2}{r - 1} \qquad (9\text{–}17)$$

where μ_i is the mean of population i and μ is the combined mean of all r populations.

Equation 9–16 says that *MSE is an unbiased estimator of* σ^2, *the assumed common variance of the* r *populations*. The mean square error in ANOVA is therefore just like the sample variance in the one-population case of earlier chapters.

The mean square treatment, however, comprises two components, as seen from equation 9–17. The first component is σ^2, as in the case of MSE. The second component is a measure of the differences among the *r* population means μ_i. If the null hypothesis is true, all *r* population means are equal—they are all equal to μ. In such a case, the second term in equation 9–17 is equal to *zero*. When this happens, the expected value of MSTR and the expected value of MSE are both equal to σ^2.

> When the null hypothesis of ANOVA is true and all *r* population means are equal, MSTR and MSE are two independent, unbiased estimators of the common population variance σ^2.

If, on the other hand, the null hypothesis is not true and differences do exist among the *r* population means, then *MSTR will tend to be larger than MSE*. This happens because, when not all population means are equal, the second term in equation 9–17 is a positive number.

The **F** *Statistic*

The preceding discussion suggests that the ratio of MSTR to MSE is a good indicator of whether the *r* population means are equal. If the *r* population means are equal, then MSTR/MSE would tend to be close to 1.00. Remember that both MSTR and MSE are sample statistics derived from our data. As such, MSTR and MSE will have some randomness associated with them, and they are not likely to exactly equal their expected values. Thus, when the null hypothesis is true, MSTR/MSE will vary around the value 1.00. When not all the *r* population means are equal, the ratio MSTR/MSE will tend to be greater than 1.00 because the expected value of MSTR, from equation 9–17, will be larger than the expected value of MSE. How large is "large enough" for us to reject the null hypothesis?

This is where statistical inference comes in. We want to determine whether the difference between our observed value of MSTR/MSE and the number 1.00 is due just to chance variation, or whether MSTR/MSE is *significantly* greater than 1.00—implying that not all the population means are equal. We will make the determination with the aid of the *F* distribution.

> Under the assumptions of ANOVA, the ratio MSTR/MSE possesses an *F* distribution with $r - 1$ degrees of freedom for the numerator and $n - r$ degrees of freedom for the denominator when the null hypothesis is true.

In Chapter 8, we saw how the *F* distribution is used in determining differences between two population variances—noting that if the two variances are equal, then the ratio of the two independent, unbiased estimators of the assumed common variance follows an *F* distribution. There, too, the appropriate degrees of freedom for the numerator and the denominator of *F* came from the degrees of freedom of the sample variance in the numerator and the sample variance in the denominator of the ratio. In ANOVA, the numerator is MSTR and has $r - 1$ degrees of freedom; the denominator is MSE and has $n - r$ degrees of freedom. We thus have the following:

The test statistic in analysis of variance is

$$F_{(r-1,\, n-r)} = \frac{\text{MSTR}}{\text{MSE}} \tag{9–18}$$

In this section, we have seen the theoretical rationale for the F statistic we used in Section 9–2. We also saw the computations required for arriving at the value of the test statistic. In the next section, we will encounter a convenient tool for keeping track of computations and reporting our results: the ANOVA table.

PROBLEMS

9–6. Define *treatment* and *error*.

9–7. Explain why trying to compute a simple average of all error deviations and of all treatment deviations will not lead to any results.

9–8. Explain how the total deviation is partitioned into the treatment deviation and the error deviation.

9–9. Explain the sum-of-squares principle.

9–10. Where do errors come from, and what do you think are their sources?

9–11. If, in an analysis of variance, you find that MSTR is greater than MSE, why can you not immediately reject the null hypothesis without determining the F ratio and its distribution? Explain.

9–12. What is the main principle behind analysis of variance?

9–13. Explain how information about the variance components in a data set can lead to conclusions about population means.

9–14. Explain the meaning of the terms *within, between, unexplained,* and *explained* and the context in which these terms arise.

9–15. By the sum-of-squares principle, SSE and SSTR are additive, and their sum is SST. Does such a relation exist between MSE and MSTR? Explain.

9–16. Does the quantity MSTR/MSE follow an F distribution when the null hypothesis of ANOVA is false? Explain.

9–17. (A mathematically demanding problem) Prove the sum-of-squares principle, equation 9–12.

9–4   The ANOVA Table and Examples

Table 9–2 shows the data for our triangles, squares, and circles. In addition, the table shows the deviations from the group means, and their squares. From these quantities, we find the sum of squares and mean squares.

As we see in the last row of the table, the sum of all the deviations of the data points from their group means is zero, as expected. The sum of the *squared* deviations from the sample means (which, from equation 9–12, is SSE) is equal to 17.00:

$$\text{SSE} = \sum_{i=1}^{r} \sum_{j=1}^{n_i} (x_{ij} - \bar{x}_i)^2 = 17.00$$

Now we want to compute the sum of squares for treatment. Recall from Table 9–1 that $\bar{\bar{x}} = 6.909$. Again using the definitions in equation 9–12, we have

TABLE 9–2 Computations for Triangles, Squares, and Circles

Treatment i	j	Value x_{ij}	$x_{ij} - \bar{x}_i$	$(x_{ij} - \bar{x}_i)^2$
Triangle	1	4	$4 - 6 = -2$	$(-2)^2 = 4$
Triangle	2	5	$5 - 6 = -1$	$(-1)^2 = 1$
Triangle	3	7	$7 - 6 = 1$	$(1)^2 = 1$
Triangle	4	8	$8 - 6 = 2$	$(2)^2 = 4$
Square	1	10	$10 - 11.5 = -1.5$	$(-1.5)^2 = 2.25$
Square	2	11	$11 - 11.5 = -0.5$	$(-0.5)^2 = 0.25$
Square	3	12	$12 - 11.5 = 0.5$	$(0.5)^2 = 0.25$
Square	4	13	$13 - 11.5 = 1.5$	$(1.5)^2 = 2.25$
Circle	1	1	$1 - 2 = -1$	$(-1)^2 = 1$
Circle	2	2	$2 - 2 = 0$	$(0)^2 = 0$
Circle	3	3	$3 - 2 = 1$	$(1)^2 = 1$
			Sum $= 0$	Sum $= 17$

$$\text{SSTR} = \sum_{i=1}^{r} n_i(\bar{x}_i - \bar{\bar{x}})^2 = 4(6 - 6.909)^2 + 4(11.5 - 6.909)^2 + 3(2 - 6.909)^2$$

$$= 159.9$$

We now compute the mean squares. From equations 9–14 and 9–15, respectively, we get

$$\text{MSTR} = \frac{\text{SSTR}}{r - 1} = \frac{159.9}{2} = 79.95$$

$$\text{MSE} = \frac{\text{SSE}}{n - r} = \frac{17}{8} = 2.125$$

Using equation 9–18, we get the computed value of the F statistic:

$$F_{(2, 8)} = \frac{\text{MSTR}}{\text{MSE}} = \frac{79.95}{2.125} = 37.62$$

We are finally in a position to conduct the ANOVA hypothesis test to determine whether the means of the three populations are equal. From Appendix C, Table 5, we find that the critical point at $\alpha = 0.01$ (for a right-tailed test) for the F distribution with 2 degrees of freedom for the numerator and 8 degrees of freedom for the denominator is 8.65. We can therefore reject the null hypothesis. Since 37.62 is much greater than 8.65, the p-value is much smaller than 0.01. This is shown in Figure 9–10.

As usual, we must exercise caution in the interpretation of results based on such small samples. As we noted earlier, in real situations we use large data sets, and the computations are usually done by computer. In the rest of our examples, we will assume that sums of squares and other quantities are produced by a computer.[4]

---

[4]If you must carry out ANOVA computations by hand, there are equivalent computational formulas for the sums of squares that may be easier to apply than equation 9–12. These are

$$\text{SST} = \Sigma_i \Sigma_j (x_{ij})^2 - (\Sigma_i \Sigma_j x_{ij})^2/n$$

$$\text{SSTR} = \Sigma_i [(\Sigma_j x_{ij})^2/n_i] - (\Sigma_i \Sigma_j x_{ij})^2/n$$

and we obtain SSE by subtraction: SSE = SST − SSTR.

FIGURE 9–10   Rejecting the Null Hypothesis in the Triangles, Squares, and Circles Example

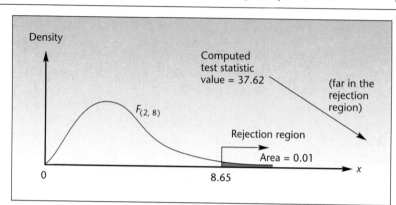

TABLE 9–3   ANOVA Table

Source of Variation	Sum of Squares	Degrees of Freedom	Mean Square	F Ratio
Treatment	SSTR = 159.9	$r - 1 = 2$	$MSTR = \dfrac{SSTR}{r-1}$ $= 79.95$	$F = \dfrac{MSTR}{MSE}$ $= 37.62$
Error	SSE = 17.0	$n - r = 8$	$MSE = \dfrac{SSE}{n-r}$ $= 2.125$	
Total	SST = 176.9	$n - 1 = 10$		

An essential tool for reporting the results of an analysis of variance is the ANOVA table. An ANOVA table lists the sources of variation: treatment, error, and total. (In the two-factor ANOVA, which we will see in later sections, there will be more sources of variation.) The ANOVA table lists the sums of squares, the degrees of freedom, the mean squares, and the F ratio. The table format simplifies the analysis and the interpretation of the results. The structure of the ANOVA table is based on the fact that both the sums of squares and the degrees of freedom are additive. We will now present an ANOVA table for the triangles, squares, and circles example. Table 9–3 shows the results computed above.

Note that the entries in the second and third columns, sum of squares and degrees of freedom, are both additive. The entries in the fourth column, mean square, are obtained by dividing the appropriate sums of squares by their degrees of freedom. We do not define a mean square total, which is why there is no entry in that particular position in the table. The last entry in the table is the main objective of our analysis: the F ratio, which is computed as the ratio of the two entries in the previous column. There are no other entries in the last column. Example 9–2 demonstrates the use of the ANOVA table.

TABLE 9–4 Club Med Survey Results

Resort i	Mean Response $\bar{x}_i$
1. Guadeloupe	89
2. Martinique	75
3. Eleuthera	73
4. Paradise Island	91
5. St. Lucia	85
SST = 112,564	SSE = 98,356

TABLE 9–5 Preliminary ANOVA Table for Club Med Example

Source of Variation	Sum of Squares	Degrees of Freedom	Mean Square	F Ratio
Treatment	SSTR =	$r - 1 = 4$	MSTR =	F =
Error	SSE = 98,356	$n - r = 195$	MSE =	
Total	SST = 112,564	$n - 1 = 199$		

Club Med has more than 30 major resorts worldwide, from Tahiti to Switzerland. **EXAMPLE 9–2**
Many of the beach resorts are in the Caribbean, and at one point the club wanted to test whether the resorts on Guadeloupe, Martinique, Eleuthera, Paradise Island, and St. Lucia were all equally well liked by vacationing club members. The analysis was to be based on a survey questionnaire filled out by a random sample of 40 respondents in each of the resorts. From every returned questionnaire, a general satisfaction score, on a scale of 0 to 100, was computed. Analysis of the survey results yielded the statistics given in Table 9–4.

The results were computed from the responses by using a computer program that calculated the sums of squared deviations from the sample means and from the grand mean. Given the values of SST and SSE, construct an ANOVA table and conduct the hypothesis test. (*Note:* The reported sample means in Table 9–4 will be used in the next section.)

Let us first construct an ANOVA table and fill in the information we have: SST = *Solution*
112,564, SSE = 98,356, $n = 200$, and $r = 5$. This has been done in Table 9–5. We now compute SSTR as the difference between SST and SSE and enter it in the appropriate place in the table. We then divide SSTR and SSE by their respective degrees of freedom to give us MSTR and MSE. Finally, we divide MSTR by MSE to give us the F ratio. All these quantities are entered in the ANOVA table. The result is the complete ANOVA table for the study, Table 9–6.

Table 9–6 contains all the pertinent information for this study. We are now ready to conduct the hypothesis test.

$H_0: \mu_1 = \mu_2 = \mu_3 = \mu_4 = \mu_5$ (average vacationer satisfaction for each of the five resorts is equal)

$H_1:$ Not all μ_i $(i = 1, \ldots, 5)$ are equal (on average, vacationer satisfaction is not equal among the five resorts)

TABLE 9–6   ANOVA Table for Club Med Example

Source of Variation	Sum of Squares	Degrees of Freedom	Mean Square	F Ratio
Treatment	SSTR = 14,208	$r - 1 = 4$	MSTR = 3,552	$F = 7.04$
Error	SSE = 98,356	$n - r = 195$	MSE = 504.4	
Total	SST = 112,564	$n - 1 = 199$		

FIGURE 9–11   **Club Med Test**

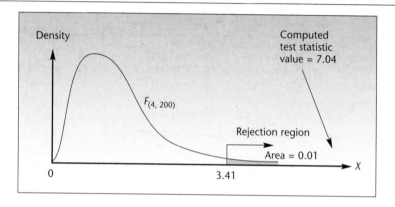

As shown in Table 9–6, the test statistic value is $F_{(4, 195)} = 7.04$. As often happens, the exact number of degrees of freedom we need does not appear in Appendix C, Table 5. We use the nearest entry, which is the critical point for F with 4 degrees of freedom for the numerator and 200 degrees of freedom for the denominator. The critical point for $\alpha = 0.01$ is $C = 3.41$. The test is illustrated in Figure 9–11.

Since the computed test statistic value falls in the rejection region for $\alpha = 0.01$, we reject the null hypothesis and note that the p-value is smaller than 0.01. We may conclude that, based on the survey results and our assumptions, it is likely that the five resorts studied are not equal in terms of average vacationer satisfaction. Which resorts are more satisfying than others? This question will be answered when we return to this example in the next section.

EXAMPLE 9–3   Recent research studied job involvement of salespeople in the four major career stages: exploration, establishment, maintenance, and disengagement. Results of the study included an analysis of variance aimed at determining whether salespeople in each of the four career stages are, on average, equally involved with their jobs. Involvement is measured on a special scale developed by psychologists. The analysis is based on questionnaires returned by a total of 543 respondents, and the reported F value is 8.52. The authors note the result is "significant at $p < .01$." Assuming that MSE = 34.4, construct an ANOVA table for this example. Also verify the authors' claim about the significance of their results.

Solution   In this problem, another exercise in the construction of ANOVA tables, we are doing the opposite of what is usually done: We are going from the final result of an F ratio

TABLE 9-7   ANOVA Table for Job Involvement

Source of Variation	Sum of Squares	Degrees of Freedom	Mean Square	F Ratio
Treatment	SSTR = 879.3	$r - 1 = 3$	MSTR = 293.1	F = 8.52
Error	SSE = 18,541.6	$n - r = 539$	MSE = 34.4	
Total	SST = 19,420.9	$n - 1 = 542$		

to the earlier stages of an analysis of variance. First, multiplying the F ratio by MSE gives us MSTR. Then, from the sample size $n = 543$ and from $r = 4$, we get the number of degrees of freedom for treatment, error, and total. Using our information, we construct the ANOVA table (Table 9–7).

From Appendix C, Table 5, we find that the critical point for a right-tailed test at $\alpha = 0.01$ for an F distribution with 3 and 400 degrees of freedom (the entry for degrees of freedom closest to the needed 3 and 539) is 3.83. Thus, we may conclude that differences do exist among the four career stages with respect to average job involvement. The authors' statement about the p-value is also true: the p-value is much smaller than 0.01.

PROBLEMS

9–18. Gulfstream Aerospace Company produced three different prototypes as candidates for mass production as the company's newest large-cabin business jet, the *Gulfstream IV.* Each of the three prototypes has slightly different features, which may bring about differences in performance. Therefore, as part of the decision-making process concerning which model to produce, company engineers are interested in determining whether the three proposed models have about the same average flight range. Each of the models is assigned a random choice of 10 flight routes and departure times, and the flight range on a full standard fuel tank is measured (the planes carry additional fuel on the test flights, to allow them to land safely at certain destination points). Range data for the three prototypes, in nautical miles (measured to the nearest 10 miles), are as follows.[5]

Prototype A	Prototype B	Prototype C
4,420	4,230	4,110
4,540	4,220	4,090
4,380	4,100	4,070
4,550	4,300	4,160
4,210	4,420	4,230
4,330	4,110	4,120
4,400	4,230	4,000
4,340	4,280	4,200
4,390	4,090	4,150
4,510	4,320	4,220

---

[5]General information about the capabilities of the *Gulfstream IV* is provided courtesy of Gulfstream Aerospace Company.

Do all three prototypes have the same average range? Construct an ANOVA table, and carry out the test. Explain your results.

9–19.   In the theory of finance, a market for any asset or commodity is said to be *efficient* if items of identical quality and other attributes (such as risk, in the case of stocks) are sold at the same price. A Geneva-based oil industry analyst wants to test the hypothesis that the spot market for crude oil is efficient. The analyst chooses the Rotterdam oil market, and he selects Arabian Light as the type of oil to be studied. (Differences in location may cause price differences because of transportation costs, and differences in the type of oil—hence, in the quality of oil—also affect the price. Therefore, both the type and the location must be fixed.) A random sample of eight observations from each of four sources of the spot price of a barrel of oil during February is collected. Data, in U.S. dollars per barrel, are as follows.

U.K.	Mexico	U.A.E.	Oman
$17.80	$18.01	$18.10	$18.05
18.00	17.75	17.92	18.01
17.98	18.00	18.01	17.94
18.20	17.77	17.88	18.23
18.00	18.01	18.30	18.20
17.99	18.01	18.22	18.00
18.10	18.12	18.56	17.84
17.90	18.20	18.10	18.11

Based on these data, what should the analyst conclude about whether the market for crude oil is efficient? Are conclusions valid only for the Rotterdam market? Are conclusions valid for all types of crude oil? What assumptions are necessary for the analysis? Do you believe that all the assumptions are met in this case? What are the limitations, if any, of this study? Discuss.

9–20.   The manager of a store wants to decide what kind of hand-knit sweaters to sell. The manager is considering three kinds of sweaters: Irish, Peruvian, and Shetland. The decision will depend on the results of an analysis of which kind of sweater, if any, lasts the longest before wearing out. The manager has some data collected from various customers who in the past bought different sweaters and reported how many years their sweaters lasted before wearing out. There are 20 observations on Irish sweaters, 18 on Peruvian sweaters, and 21 on Shetland sweaters. The data are assumed to be independent random samples from the three populations of sweaters. The manager hires a statistician, who carries out an ANOVA and finds SSE = 1,240 and SSTR = 740. Construct a complete ANOVA table, and determine whether there is evidence to conclude that the three kinds of sweaters do not have equal average durability.

9–21.   Research has shown[6] that in the new fast-paced world of electronics, the key factor that separates the winners from the losers is actually how *slow* a firm is in making decisions: The most successful firms take longer to arrive at strategic decisions on product development, adopting new technologies, or developing new products. The following values are the number of months to arrive at a decision for firms ranked high, medium, and low in terms of performance:

---

[6]Adapted from I. Sager, "Shattering the Myths of High-Tech Success," *Business Week,* June 26, 1997, p. 73. Reprinted by permission of *Business Week.* All rights reserved.

High	Medium	Low
3.5	3	1
4.8	5.5	2.5
3.0	6	2
6.5	4	1.5
7.5	4	1.5
8	4.5	6
2	6	3.8
6	2	4.5
5.5	9	0.5
6.5	4.5	2
7	5	3.5
9	2.5	1.0
5	7	2
10		
6		

Do an ANOVA. Use $\alpha = 0.05$.

9–22. The Fidelity Overseas mutual fund consists of about equal proportions of Japanese and European stocks. The percentage of the fund invested in any individual country varies according to prevailing rates of return on stocks in different countries. At the end of October 1995, the fund manager was considering the possibility of shifting the proportions invested in French, Dutch, and Italian stocks. This change would be made if it could be statistically substantiated that differences in average annualized rates of return during the period ending in October existed among stocks from the three countries. Random samples of 50 French stocks, 32 Dutch stocks, and 28 Italian stocks were collected, and the annualized rate of return for each stock over the period under study was computed. Then an analysis of variance was carried out, which produced the following results: SSE = 22,399.8 and SST = 32,156.1. Based on these results, should the manager shift the proportions of the fund invested in the three countries? How confident are you of your answer? Construct a complete ANOVA table for this problem.

9–23. A study is undertaken to determine whether differences exist in average consumer quality ratings of the following brands of color television sets: Magnavox, General Electric, Panasonic, Zenith, Sears, Philco, Sylvania, and RCA. For each brand, 100 randomly chosen consumer responses are available, and from these the following quantities are computed: SSTR = 45,210 and SST = 92,340. Construct an ANOVA table for this study, and test the null hypothesis that all eight brands have equal average consumer quality ratings versus the alternative hypothesis that they do not.

9–5   Further Analysis

You have rejected the ANOVA null hypothesis. What next? This is an important question often overlooked in elementary introductions to analysis of variance. After all, what is the meaning of the statement "not all r population means are equal" if we cannot tell *in what way* the population means are not equal? We need to know which of our population means are large, which are small, and the magnitudes of the differences among them. These issues are addressed in this section.

ANOVA can be viewed as a machine or a box: In go the data, and out comes a conclusion—"all r population means are equal" or "not all r population means are

FIGURE 9–12   **The ANOVA Diagram**

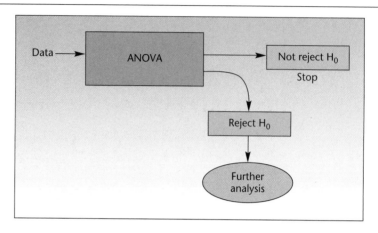

equal." If the ANOVA null hypothesis $H_0: \mu_1 = \mu_2 = \cdots = \mu_r$ is not rejected and we therefore state that there is no strong evidence to conclude that differences exist among the r population means, then there is nothing more to say or do (unless, of course, you believe that differences do exist and you are able to gather more information to prove so). If the ANOVA null hypothesis is rejected, then we have evidence that not all r population means are equal. This calls for *further analysis*—other hypothesis tests and/or the construction of confidence intervals to determine where the differences exist, their directions, and their magnitudes. The schematic diagram of the "ANOVA box" is shown in Figure 9–12.

Several methods have been developed for further analysis following the rejection of the null hypothesis in ANOVA. All the methods make use of the following two properties:

1. The sample means $\overline{X}_i$ are unbiased estimators of the corresponding population means μ_i.
2. The mean square error, MSE, is an unbiased estimator of the common population variance σ^2.

Since MSE can be read directly from the ANOVA table, we have another advantage of using an ANOVA table. This extends the usefulness of the table beyond the primary stages of analysis. The first and simplest post-ANOVA analysis is the estimation of separate population means μ_i. It can be shown that, under the assumptions of analysis of variance, each sample mean $\overline{X}_i$ *has a normal distribution with mean μ_i and standard deviation $\sigma/\sqrt{n_i}$*, where σ is the common standard deviation of the r populations. Since σ is not known, we estimate it by $\sqrt{MSE}$. We get the following relation:

$$\frac{\overline{X}_i - \mu_i}{\sqrt{MSE}/\sqrt{n_i}} \text{ has a } t \text{ distribution with } n - r \text{ degrees of freedom} \qquad (9\text{–}19)$$

This property leads us to the possibility of constructing confidence intervals for individual population means.

> A $(1 - \alpha)$ 100% confidence interval for μ_i, the mean of population i, is
>
> $$\bar{x} \pm t_{\alpha/2} \frac{\sqrt{MSE}}{\sqrt{n_i}} \qquad\qquad (9\text{--}20)$$
>
> where $t_{\alpha/2}$ is the value of the t distribution with $n - r$ degrees of freedom that cuts off a right-tailed area equal to $\alpha/2$.

Confidence intervals given by equation 9–20 are included in the template.

We now demonstrate the use of equation 9–20 with the continuation of Example 9–2, the Club Med example. From Table 9–4, we get the sample means $\bar{x}_i$:

Guadeloupe:	$\bar{x}_1 = 89$
Martinique:	$\bar{x}_2 = 75$
Eleuthera:	$\bar{x}_3 = 73$
Paradise Island:	$\bar{x}_4 = 91$
St. Lucia:	$\bar{x}_5 = 85$

From Table 9–6, the ANOVA table for this example, we get MSE = 504.4 and degrees of freedom for error = $n - r = 195$. We also know that the sample size in each group is $n_i = 40$ for all $i = 1, \ldots, 5$. Since a t distribution with 195 degrees of freedom is, for all practical purposes, a standard normal distribution, we use $z = 1.96$ in constructing 95% confidence intervals for the population mean responses of vacationers on the five islands. We will construct a 95% confidence interval for the mean response on Guadeloupe and will leave the construction of the other four confidence intervals as an exercise. For Guadeloupe, we have the following 95% confidence interval for the population mean μ_1:

$$\bar{x}_1 \pm t_{\alpha/2} \frac{\sqrt{MSE}}{\sqrt{n_1}} = 89 \pm 1.96 \frac{\sqrt{504.4}}{\sqrt{40}} = [82.04, 95.96]$$

The real usefulness of ANOVA, however, does not lie in the construction of individual confidence intervals for population means (these are of limited use because the confidence coefficient does not apply to a *series* of estimates). The power of ANOVA lies in providing us with the ability to make *joint* conclusions about population parameters.

As mentioned earlier, several procedures have been developed for further analysis. The method we will discuss here is the *Tukey method* of pairwise comparisons of the population means. The method is also called the *HSD* (honestly significant differences) *test*. This method allows us to compare every possible pair of means by using a *single level of significance*, say $\alpha = 0.05$ (or a single confidence coefficient, say, $1 - \alpha = 0.95$). The single level of significance applies to the *entire set* of pairwise comparisons.

The Tukey Pairwise-Comparisons Test

We will use the *studentized range distribution*.

> The **studentized range distribution** q is a probability distribution with degrees of freedom r and $n - r$.

Note that the degrees of freedom of q are similar, but not identical, to the degrees of freedom of the F distribution in ANOVA. The F distribution has $r - 1$ and $n - r$

degrees of freedom. The q distribution has degrees of freedom r and $n - r$. Critical points for q with different numbers of degrees of freedom for $\alpha = 0.05$ and for $\alpha = 0.01$ are given in Appendix C, Table 6. Check, for example, that for $\alpha = 0.05$, $r = 3$, and $n - r = 20$, we have the critical point $q_\alpha = 3.58$. The table gives right-hand critical points, which is what we need since our test will be a right-tailed test. We now define the Tukey criterion T.

The Tukey criterion

$$T = q_\alpha \frac{\sqrt{MSE}}{\sqrt{n_i}}$$

(9–21)

Equation 9–21 gives us a critical point, at a given level α, with which we will compare the computed values of test statistics defined later. Now let us define the hypothesis tests. As mentioned, the usefulness of the Tukey test is that it allows us to perform *jointly* all possible pairwise comparisons of the population means using a single, "family" level of significance. What are all the possible pairwise comparisons associated with an ANOVA?

Suppose that we had $r = 3$. We compared the means of three populations, using ANOVA, and concluded that not all the means were equal. Now we would like to be able to compare every *pair* of means to determine where the differences among population means exist. How many pairwise comparisons are there? With three populations, there are

$$\binom{3}{2} = \frac{3!}{2! \, 1!} = 3 \text{ comparisons}$$

These comparisons are

1 with 2

2 with 3

1 with 3

As a general rule, the number of possible pairwise comparisons of r means is

$$\binom{r}{2} = \frac{r!}{2!(r-2)!}$$

(9–22)

You do not really need equation 9–22 for cases where it is relatively easy to list all the possible pairs. In the case of Example 9–2, there are, by equation 9–22, $5!/(2!3!) = (5)(4)(3)(2)/(2)(3)(2) = 10$ possible pairwise comparisons. Let us list all the comparisons:

Guadeloupe (1)–Martinique (2)

Guadeloupe (1)–Eleuthera (3)

Guadeloupe (1)–Paradise Island (4)

Guadeloupe (1)–St. Lucia (5)

Martinique (2)–Eleuthera (3)

Martinique (2)–Paradise Island (4)

Martinique (2)–St. Lucia (5)

Eleuthera (3)–Paradise Island (4)

Eleuthera (3)–St. Lucia (5)

Paradise Island (4)–St. Lucia (5)

These pairings are apparent if you look at Table 9–4 and see that we need to compare the first island, Guadeloupe, with all four islands below it. Then we need to compare the second island, Martinique, with all three islands below it (we already have the comparison of Martinique with Guadeloupe). We do the same with Eleuthera and finally with Paradise Island, which has only St. Lucia listed below it; therefore, this is the last comparison. (In the preceding list, we wrote the number of each population in parentheses after the population name.)

The parameter μ_1 denotes the population mean of all vacationer responses for Guadeloupe. The parameters μ_2 to μ_5 have similar meanings. To compare the population mean vacationer responses for every pair of island resorts, we use the following *set of hypothesis tests:*

I. H_0: $\mu_1 = \mu_2$ VI. H_0: $\mu_2 = \mu_4$

H_1: $\mu_1 \neq \mu_2$ H_1: $\mu_2 \neq \mu_4$

II. H_0: $\mu_1 = \mu_3$ VII. H_0: $\mu_2 = \mu_5$

H_1: $\mu_1 \neq \mu_3$ H_1: $\mu_2 \neq \mu_5$

III. H_0: $\mu_1 = \mu_4$ VIII. H_0: $\mu_3 = \mu_4$

H_1: $\mu_1 \neq \mu_4$ H_1: $\mu_3 \neq \mu_4$

IV. H_0: $\mu_1 = \mu_5$ IX. H_0: $\mu_3 = \mu_5$

H_1: $\mu_1 \neq \mu_5$ H_1: $\mu_3 \neq \mu_5$

V. H_0: $\mu_2 = \mu_3$ X. H_0: $\mu_4 = \mu_5$

H_1: $\mu_2 \neq \mu_3$ H_1: $\mu_4 \neq \mu_5$

The Tukey method allows us to carry out simultaneously all 10 hypothesis tests at a single given level of significance, say, $\alpha = 0.05$. Thus, if we use the Tukey procedure for reaching conclusions as to which population means are equal and which are not, we know that the probability of reaching at least one erroneous conclusion, stating that two means are not equal when indeed they are equal, is at most 0.05.

The **test statistic** for each test is the *absolute difference of the appropriate sample means.*

Thus, the test statistic for the first test (I) is

$$|\bar{x}_1 - \bar{x}_2| = |89 - 75| = 14$$

Conducting the Tests

We conduct the tests as follows. We compute each of the test statistics and compare them with the value of T that corresponds to the desired level of significance α. *We reject a particular null hypothesis if the absolute difference between the corresponding pair of sample means exceeds the value of T.*

Using $\alpha = 0.05$, we now conduct the Tukey test for Example 9–2. All absolute differences of sample means corresponding to the pairwise tests I through X are computed and compared with the value of T. For $\alpha = 0.05$, $r = 5$, and $n - r = 195$ (we use ∞, the last row in the table), we get, from Appendix C, Table 6, $q = 3.86$. We also know that MSE $= 504.4$ and $n_i = 40$ for all i. (Later we will see what to do when not all r samples are of equal size.) Therefore, from equation 9–21,

$$T = q_\alpha \sqrt{\frac{\text{MSE}}{n_i}} = 3.86 \sqrt{\frac{504.4}{40}} = 13.7$$

We now compute all 10 pairwise absolute differences of sample means and compare them with $T = 13.7$ to determine which differences are statistically significant at $\alpha = 0.05$ (these are marked with an asterisk).

$$|\bar{x}_1 - \bar{x}_2| = |89 - 75| = 14 > 13.7^*$$
$$|\bar{x}_1 - \bar{x}_3| = |89 - 73| = 16 > 13.7^*$$
$$|\bar{x}_1 - \bar{x}_4| = |89 - 91| = 2 < 13.7$$
$$|\bar{x}_1 - \bar{x}_5| = |89 - 85| = 4 < 13.7$$
$$|\bar{x}_2 - \bar{x}_3| = |75 - 73| = 2 < 13.7$$
$$|\bar{x}_2 - \bar{x}_4| = |75 - 91| = 16 > 13.7^*$$
$$|\bar{x}_2 - \bar{x}_5| = |75 - 85| = 10 < 13.7^*$$
$$|\bar{x}_3 - \bar{x}_4| = |73 - 91| = 18 > 13.7^*$$
$$|\bar{x}_3 - \bar{x}_5| = |73 - 85| = 12 < 13.7$$
$$|\bar{x}_4 - \bar{x}_5| = |91 - 85| = 6 < 13.7$$

From these comparisons we determine that our data provide statistical evidence to conclude that μ_1 is different from μ_2; μ_1 is different from μ_3; μ_2 is different from μ_4; and μ_3 is different from μ_4. *There are no other statistically significant differences at $\alpha = 0.05$.*

For the purpose of interpretation, it will help to draw a diagram of the significant differences that we found. This has been done in Figure 9–13. Looking at the figure, you may be puzzled by the fact that we believe, for example, that μ_1 is different from μ_2, yet we believe that μ_1 is no different from μ_5 and that μ_5 is no different from μ_2. You may say: If A is equal to B, and B is equal to C, then mathematically we must have A equal to C (the transitivity of equality). But remember that we are doing statistics, not discussing mathematical equality. In statistics, not rejecting the null hypothesis that two parameters are equal does not mean that they are necessarily equal. The nonrejection just means that we have no statistical evidence to conclude that they are different. Thus, in our present example, we conclude that there is statistical evidence to support the claim that, on average, vacationers give higher ratings to Guadeloupe (1) than they give to Martinique (2) or Eleuthera (3); as well as the claim that Paradise Island (4) is, on average, rated higher than Martinique or Eleuthera. There is no statistical evidence for any other claim of differences in average ratings among the five island resorts. Note also that we do not have to hypothesize any of the assertions of tests I through X *before* doing the analysis. The Tukey method allows us to make all the above conclusions at a single level of significance, $\alpha = 0.05$.

The Case of Unequal Sample Sizes, and Alternative Procedures

What can we do if the sample sizes are not equal in all groups? We use the *smallest sample size* of all the n_i in computing the criterion T of equation 9–21. The Tukey procedure is the best follow-up to ANOVA when the sample sizes are all equal. The case

FIGURE 9–13   Differences among the Population Means in Example 9–2
Suggested by the Tukey Procedure

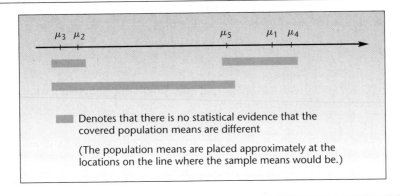

FIGURE 9–14   The Template for Single-Factor ANOVA
[Anova.xls; Sheet: 1-Way]

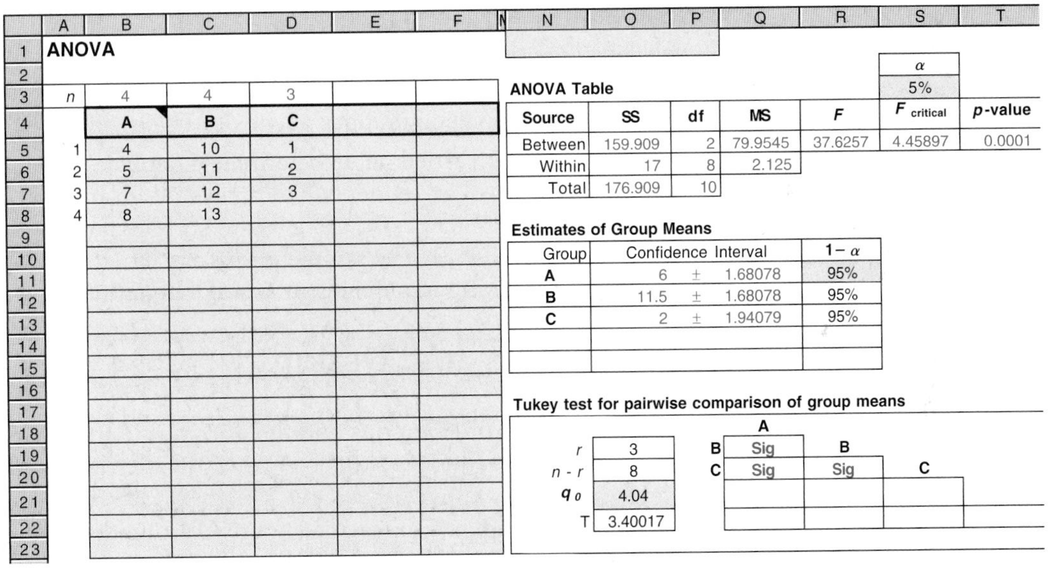

of equal sample sizes is called the *balanced design*. For very unbalanced designs (i.e., when sample sizes are very different), there are other methods of further analysis to be used following ANOVA. Two of the better-known methods are the Bonferroni method and the Scheffé method. We will not discuss these methods.

The Template

Figure 9–14 shows the template that can be used to carry out single-factor ANOVA computations. The ANOVA table appears at the top right. Below it appears a table of confidence intervals for each group mean. The α used for confidence intervals need not be the same as the one used in cell S3 for the F test in the ANOVA table. Below the confidence intervals appears a Tukey comparison table. Enter the q_0 corresponding to r, $n - r$, and desired α in cell O21 before reading off the results from this table. The message "Sig" appears in a cell if the difference between the two corresponding groups is significant at the α used for q_0 in cell O21.

PROBLEMS

9–24. Give 95% confidence intervals for the remaining four population mean responses to the Club Med resorts (the one for Guadeloupe having been given in the text).

9–25. Use the data of Table 9–1 and the Tukey procedure to determine where differences exist among the triangle, circle, and square population means. Use $\alpha = 0.01$.

9–26. For problem 9–18, find which, if any, of the three prototype planes has an average range different from the others. Use $\alpha = 0.05$.

9–27. For problem 9–19, use the Tukey method to determine which oil types, if any, have an average price different from the others. Use $\alpha = 0.05$.

9–28. For problem 9–20, suppose that the appropriate sample means are 6.4, 2.5, and 4.9 (in years and in order of type listed in problem 9–20). Find where differences, if any, exist among the three population means. Use $\alpha = 0.05$.

9–29. For problem 9–23, suppose the sample means, in the order listed in that problem, are 77, 78, 82, 94, 88, 89, 90, and 87. Use the Tukey procedure to determine where differences in average population ratings may exist. Use $\alpha = 0.01$.

9–30. An analysis of variance is carried out to determine differences among average annualized returns on four types of investments. The analysis leads to the rejection of the null hypothesis that no differences exist, using $\alpha = 0.05$. The mean square error is 49.5; the sample sizes are all $n_i = 31$. The sample means are 18, 11, 15, and 14. Find the significant differences that exist among the four types of investment.

9–6   Models, Factors, and Designs

A **statistical model** is a set of equations and assumptions that capture the essential characteristics of a real-world situation.

The model discussed in this chapter is the one-factor ANOVA model. In this model, the populations are assumed to be represented by the following equation.

The one-factor ANOVA model is

$$X_{ij} = \mu_i + \epsilon_{ij} = \mu + \alpha_i + \epsilon_{ij} \tag{9–23}$$

where ϵ_{ij} is the error associated with the jth member of the ith population. The errors are assumed to be normally distributed with mean zero and variance σ^2.

The ANOVA model assumes that the r populations are normally distributed with means μ_i, which may be different, and with equal variance σ^2. The right-hand side of equation 9–23 breaks the mean of population i into a common component μ and a unique component due to the particular population (or treatment) i. This component is written as α_i. When we sample, the sample means $\overline{X}_i$ are unbiased estimators of the respective population means μ_i. The grand mean $\overline{\overline{x}}$ is an unbiased estimator of the common component of the means μ. The treatment deviations a_i are estimators of the differences among population means α_i. The data errors e_{ij} are estimates of the population errors ϵ_{ij}.

Much more will be said about statistical models in the next chapter, dealing with regression analysis. The one-factor ANOVA null hypothesis $H_0: \mu_1 = \mu_2 = \cdots = \mu_r$ may be written in an equivalent form, using equation 9–23, as $H_0: \alpha_i = 0$ for all i. (This is so because if $\mu_i = \mu$ for all i, then the "extra" components α_i are all zero.) This form of the hypothesis will be extended in the two-factor ANOVA model, also called the two-way ANOVA model, discussed in the following section.

We may want to check that the assumptions of the ANOVA model are indeed met. To check that the errors are approximately normally distributed, we may draw a histogram of the observed errors e_{ij}, which are called *residuals*. If serious deviations from the normal-distribution assumption exist, the histogram will not resemble a normal curve. Plotting the residuals for each of the r samples under study will reveal whether the population variances are indeed (at least approximately) equal. If the *spread* of the data sets around their group means is not approximately equal for all r groups, then the population variances may not be equal. When model assumptions are violated, a nonparametric alternative to ANOVA must be used. An alternative method of analysis uses the Kruskal-Wallis test, discussed in Chapter 14. Residual analysis will be discussed in detail in the next chapter.

One-Factor versus Multifactor Models

In each of the examples and problems you have seen so far, we were interested in determining whether differences existed among several populations, or treatments. These treatments may be considered as *levels* of a single *factor*.

A **factor** is a set of populations or treatments of a single kind.

Examples of factors are vacationer ratings of a *set of resorts,* the range of different *types of airplanes,* and the durability of different *kinds of sweaters.*

Sometimes, however, we may be interested in studying more than one factor. For example, an accounting researcher may be interested in testing whether there are differences in average error percentage rate among the Big Eight accounting firms, *and* among different geographical locations, such as the Eastern Seaboard, the South, the Midwest, and the West. In such an analysis, there are *two factors:* the different firms (factor A, with eight levels) and the geographical location (factor B, with four levels).

Another example is that of an advertising firm interested in studying how the public is influenced by color, shape, and size in an advertisement. The firm could carry out an ANOVA to test whether there are differences in average responses to three different colors, as well as to four different shapes of an ad, and to three different ad sizes. This would be a three-factor ANOVA. There are important statistical reasons for jointly studying the effects of several factors in a multifactor ANOVA. These will be explained in the next section, on two-factor ANOVA.

Fixed-Effects versus Random-Effects Models

Recall Example 9–2, where we wanted to determine whether differences existed among the five particular island resorts of Guadeloupe, Martinique, Eleuthera,

Paradise Island, and St. Lucia. Once we reject or do not reject the null hypothesis, *the inference is valid only for the five islands studied*. This is a *fixed-effects model*.

> A **fixed-effects model** is a model in which the levels of the factor under study (the treatments) are *fixed* in advance. Inference is valid only for the levels under study.

Consider another possible context for the analysis. Suppose that Club Med had no particular interest in the five resorts listed, but instead wanted to determine whether differences existed among *any* of its more than 30 resorts. In such a case, we may consider all Club Med resorts as a *population of resorts,* and we may draw a random sample of five (or any other number) of the resorts and carry out an ANOVA to determine differences among population means. The ANOVA would be carried out in exactly the same way. However, since the resorts themselves were randomly selected for analysis from the population of all Club Med resorts, the inference would be valid for *all* Club Med resorts. This is called the *random-effects model*.

> The **random-effects model** is an ANOVA model in which the levels of the factor under study are *randomly chosen* from an entire population of levels (treatments). Inference is valid for the entire population of levels.

The idea should make sense to you if you recall the principle of inference using random sampling, discussed in Chapter 5, and the story of the *Literary Digest*. To make inferences that are valid for an entire population, we must randomly sample from the entire population. Here this principle is applied to a population of treatments.

Experimental Design

Analysis of variance often involves the ideas of **experimental design.** If we want to study the effects of different treatments, we are sometimes in a position to design the experiment by which we plan to study these effects. Designing the experiment involves the choice of elements from a population or populations and the assignment of elements to different treatments. The model we have been using involves a *completely randomized design*.

> A **completely randomized design** is a design in which elements are assigned to treatments *completely at random*. Thus, every element chosen for the study has an equal chance of being assigned to any treatment.

There are other types of design. Some designs, called **blocking designs,** are very useful in reducing experimental errors, that is, reducing variation due to factors other than the ones under study. In the *randomized complete block design,* for example, experimental units are assigned to treatments in blocks of similar elements, with randomized treatment order within each block. In the Club Med situation of Example 9–2, a randomized complete block design could involve sending each vacationer in the sample to all five resorts, the order of the resorts chosen randomly; each vacationer is then asked to rate all the resorts. A design such as this one, with *experimental units* (here, people) given all the treatments, is called a *repeated-measures design*. More will be said about blocking designs later.

9–31. For problem 9–18, suppose that four more prototype planes are built after the study is completed. Could the inference from the ANOVA involving the first three prototypes be extended to the new planes? Explain.

9–32. What is a blocking design?

9–33. For problem 9–18, can you think of a blocking design that would reduce experimental errors?

9–34. How can we determine whether there are violations of the ANOVA model assumptions? What should we do if such violations exist?

9–35. Explain why the factor levels must be randomly chosen in the random-effects model to allow inference about an entire collection of treatments.

9–36. For problem 9–19, based on the given data, can you tell whether the world oil market is efficient?

9–7 Two-Way Analysis of Variance

In addition to being interested in possible differences in the general appeal of its five Caribbean resorts (Example 9–2), suppose that Club Med is also interested in the respective appeal of four vacation attributes: friendship, sports, culture, and excitement.[7] Club Med would like to have answers to the following two questions:

1. Are there differences in average vacationer satisfaction with the five Caribbean resorts?
2. Are there differences in average vacationer satisfaction in terms of the four vacation attributes?

In cases such as this one, where there is interest in *two* factors—resort and vacation attribute—we can answer the two questions *jointly*. In addition, we can answer a *third, very important* question, which may not be apparent to us:

3. Are there any *interactions* between some resorts and some attributes?

The three questions are statistically answerable by conducting a two-factor, or two-way, ANOVA. Why a two-way ANOVA? Why not conduct each of the two ANOVAs separately?

There are several reasons for conducting a two-way ANOVA. One reason is *efficiency*. When we conduct a two-way ANOVA, we may use a smaller total sample size for the analysis than would be required if we were to conduct each of the two tests separately. Basically, we use the same data resources to answer the two main questions. In the case of Club Med, the club may run a friendship program at each of the five resorts for one week; then the next week (with different vacationers) it may run a sports program in each of the five resorts; and so on. All vacationer responses could then be used for evaluating *both* the satisfaction from the resorts and the satisfaction from the attributes, rather than conducting two separate surveys, requiring twice the

[7]Information on the attributes and the resorts was provided through the courtesy of Club Med.

FIGURE 9–15 Two-Way ANOVA Data Layout

Factor B: Attribute	Factor A: Resort: Guadeloupe	Martinique	Eleuthera	Paradise Island	St. Lucia
Friendship	= =	= =	= =	= =	= =
Sports	= =	= =	= =	= =	= =
Culture	= =	= =	= =	= =	= =
Excitement	= =	= =	= =	= =	= =

effort and number of respondents. A more important reason for conducting a two-way ANOVA is that there really are *three* questions to be answered.

Let us call the first factor of interest (here, resorts) factor A and the second factor (here, attributes) factor B. The effects of each factor alone are the factor's *main effects.* The combined effects of the two factors, beyond what we may expect from the consideration of each factor separately, are the *interaction* between the two factors.

Two factors are said to **interact** if the difference between levels (treatments) of one factor depends on the level of the other factor. Factors that do not interact are called *additive.*

An interaction is thus an *extra effect* that appears as a result of a particular combination of a treatment from one factor with a treatment from another factor. An interaction between two factors exists when, for at least one combination of treatments—say Eleuthera and sports—the effect of the combination is not additive: there is some special "chemistry" between the two treatments. Suppose that Eleuthera is rated lowest of all resorts and that sports is rated lowest of all attributes. We then expect the Eleuthera–sports combination to be rated, on average, lowest of all combinations. If this does not happen, the two levels are said to interact.

The three questions answerable by two-way ANOVA:

1. Are there any factor A main effects?
2. Are there any factor B main effects?
3. Are there any interaction effects of factors A and B?

Let n_{ij} be the sample size in the "cell" corresponding to level i of factor A and level j of factor B. Assume there is a uniform sample size for each factor A–factor B combination, say, $n_{ij} = 4$. The layout of the data of a two-way ANOVA, using the Club Med example, is shown in Figure 9–15. Figure 9–16 shows the effects of an

FIGURE 9–16   **Graphical Display of Interaction Effects**

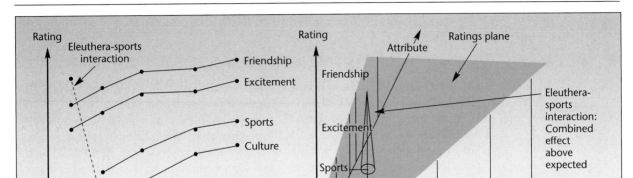

interaction. We arrange the levels of each factor in increasing order of sample mean responses. The general two-variable trend of increasing average response is the response plane shown in Figure 9–16. An exception to the plane is the Eleuthera–sports interaction, which leads to a higher-than-expected average response for this combination of levels.

The Two-Way ANOVA Model

There are a levels of factor A $(a = 5$ resorts in the Club Med example) and b levels of factor B $(b = 4$ attributes in the same example). Thus, there are $a \times b$ combinations of levels, or cells, as shown in Figure 9–15. Each one is considered a treatment. We must assume equal sample sizes in all the cells. If we do not have equal sample sizes, we must use an alternative to the method of this chapter and solve the ANOVA problem by using multiple regression analysis (Chapter 11). Since we assume an equal sample size in each cell, we will simplify our notation and will call the sample size in each cell n, omitting the subscripts i, j. We will denote the total sample size (formerly called n) by the symbol N. In the two-way ANOVA model, the assumptions of normal populations and equal variance for each two-factor combination treatment are still maintained.

> The **two-way ANOVA model** is
>
> $$x_{ijk} = \mu + \alpha_i + \beta_j + (\alpha\beta)_{ij} + \epsilon_{ijk} \qquad (9\text{--}24)$$
>
> where μ is the overall mean; α_i is the effect of level i $(i = 1, \dots, a)$ of factor A; β_j is the effect of level j $(j = 1, \dots, b)$ of factor B; $(\alpha\beta)_{ij}$ is the interaction effect of levels i and j; and ϵ_{ijk} is the error associated with the kth data point

> from level i of factor A and level j of factor B. As before, we assume that the error ϵ_{ijk} is normally distributed[8] with mean zero and variance σ^2 for all i, j, and k.

Our data, assumed to be random samples from populations modeled by equation 9–24, give us estimates of the model parameters. These estimates—as well as the different measures of variation, as in the one-way ANOVA case—are used in testing hypotheses. Since, in two-way ANOVA, three questions are to be answered rather than just one, there are three hypothesis tests relevant to any two-way ANOVA. The hypothesis tests that answer questions 1 to 3 are presented next.

The Hypothesis Tests in Two-Way ANOVA

Factor A main-effects test:

$$H_0: \alpha_i = 0 \text{ for all } i = 1, \ldots, a$$
$$H_1: \text{Not all } \alpha_i \text{ are } 0$$

This hypothesis test is designed to determine whether there are any factor A main effects. That is, the null hypothesis is true if and only if there are no differences in means due to the different treatments (populations) of factor A.

Factor B main-effects test:

$$H_0: \beta_j = 0 \text{ for all } j = 1, \ldots, b$$
$$H_1: \text{Not all } \beta_j \text{ are } 0$$

This test will detect evidence of any factor B main effects. The null hypothesis is true if and only if there are no differences in means due to the different treatments (populations) of factor B.

Test for AB interactions:

$$H_0: (\alpha\beta)_{ij} = 0 \text{ for all } i = 1, \ldots, a \text{ and } j = 1, \ldots, b$$
$$H_1: \text{Not all } (\alpha\beta)_{ij} \text{ are } 0$$

This is a test for the existence of interactions between levels of the two factors. The null hypothesis is true if and only if there are no two-way interactions between levels of factor A and levels of factor B, that is, if the factor effects are additive.

In carrying out a two-way ANOVA, we should test the third hypothesis first. We do so because it is important to first determine whether interactions exist. If interactions do exist, our interpretation of the ANOVA results will be different from the case where no interactions exist (i.e., in the case where the effects of the two factors are additive).

---

[8]Since the terms α_i, β_j, and $(\alpha\beta)_{ij}$ are deviations from the overall mean μ, in the fixed-effects model the sums of all these deviations are all zero: $\Sigma\alpha_i = 0$, $\Sigma\beta_j = 0$, and $\Sigma(\alpha\beta)_{ij} = 0$.

Sums of Squares, Degrees of Freedom, and Mean Squares

We define the data, the various means, and the deviations from the means as follows.

x_{ijk} is the kth data point from level i of factor A and level j of factor B.
$\bar{\bar{x}}$ is the grand mean.
$\bar{x}_{ij}$ is the mean of cell ij.
$\bar{x}_i$ is the mean of all data points in level i of factor A.
$\bar{x}_j$ is the mean of all data points in level j of factor B.

Using these definitions, we have

$$\sum_{1}^{a}\sum_{1}^{b}\sum_{1}^{n}(x_{ijk} - \bar{\bar{x}})^2 = \sum\sum\sum(\bar{x}_{ij} - \bar{\bar{x}})^2 + \sum\sum\sum(x_{ijk} - \bar{x}_{ij})^2 \qquad (9\text{--}25)$$

$$\text{SST} = \text{SSTR} + \text{SSE}$$

(This can be further partitioned.)

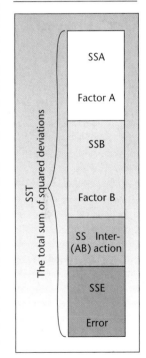

FIGURE 9–17
Partition of the Sum of Squares in Two-Way ANOVA

Equation 9–25 is the usual decomposition of the sum of squares, where each cell (a combination of a level of factor A and a level of factor B) is considered a separate treatment. Deviations of the data points from the cell means are squared and summed. Equation 9–25 is the same as equation 9–12 for the partition of the total sum of squares into sum-of-squares treatment and sum-of-squares error in one-way ANOVA. The only difference between the two equations is that here the summations extend over three subscripts: one subscript for levels of each of the two factors and one subscript for the data point number. The interesting thing is that SSTR can be further partitioned into a component due to factor A, a component due to factor B, and a component due to interactions of the two factors. The partition of the total sum of squares into its components is given in equation 9–26.

Do not worry about the mathematics of the summations. Two-way ANOVA is prohibitively tedious for hand computation, and we will always use a computer. The important thing to understand is that the total sum of squares is partitioned into a part due to factor A, a part due to factor B, a part due to interactions of the two factors, and a part due to error. This is shown in Figure 9–17.

What are the degrees of freedom? Since there are a levels of factor A, the degrees of freedom for factor A are $a - 1$. Similarly, there are $b - 1$ degrees of freedom for factor B, and there are $(a - 1)(b - 1)$ degrees of freedom for AB interactions. The degrees of freedom for error are $ab(n - 1)$. The total degrees of freedom are $abn - 1$. But we knew that [because $(a - 1) + (b - 1) + (a - 1)(b - 1) + ab(n - 1) = a + b - 2 + ab - a - b + 1 + abn - ab = abn - 1$]! Note that since we assume an equal sample size n in each cell and since there are ab cells, we have $N = abn$, and the total number of degrees of freedom is $N - 1 = abn - 1$.

$$\text{SST} = \text{SSTR} + \text{SSE}$$

$$\sum\sum\sum(x - \bar{\bar{x}})^2 = \sum\sum\sum(\bar{x} - \bar{\bar{x}})^2 + \sum\sum\sum(x - \bar{x})^2$$

$$\underbrace{\sum\sum\sum(\bar{x}_i - \bar{\bar{x}})^2}_{\text{SSA}} + \underbrace{\sum\sum\sum(\bar{x}_j - \bar{\bar{x}})^2}_{\text{SSB}} + \underbrace{\sum(\bar{x}_{ij} - \bar{x}_i - \bar{x}_j + \bar{\bar{x}})^2}_{\text{SS(AB)}}$$

Thus,

$$SST = SSA + SSB + SS(AB) + SSE \qquad (9\text{--}26)$$

where SSA = sum of squares due to factor A, SSB = sum of squares due to factor B, and SS(AB) = sum of squares due to the interactions of factors A and B.

Let us now construct an ANOVA table. The table includes the sums of squares, the degrees of freedom, and the mean squares. The mean squares are obtained by dividing each sum of squares by its degrees of freedom. The final products of the table are three F ratios. We define the F ratios as follows.

The F Ratios and the Two-Way ANOVA Table

The F ratio for each one of the hypothesis tests is the ratio of the appropriate mean square to the MSE. That is, for the test of factor A main effects, we use $F = $ MSA/MSE; for the test of factor B main effects, we use $F = $ MSB/MSE; and for the test of interactions of the two factors, we use $F = $ MS(AB)/MSE. We now construct the ANOVA table for two-way analysis, Table 9–8.

The degrees of freedom associated with each F ratio are the degrees of freedom of the respective numerator and denominator (the denominator is the same for all three tests). For the testing of factor A main effects, our test statistic is the first F ratio in the ANOVA table. When the null hypothesis is true (there are no factor A main effects), the ratio $F = $ MSA/MSE follows an F distribution with $a - 1$ degrees of freedom for the numerator and $ab(n - 1)$ degrees of freedom for the denominator. We denote this distribution by $F_{[a-1, \, ab(n-1)]}$. Similarly, for the test of factor B main effects, when the null hypothesis is true, the distribution of the test statistic is $F_{[b-1, \, ab(n-1)]}$. The test for the existence of AB interactions uses the distribution $F_{[(a-1)(b-1), \, ab(n-1)]}$.

We will demonstrate the use of the ANOVA table in two-way analysis, and the three tests, with a new example.

EXAMPLE 9–4   There are claims that the Japanese have now joined the English and people in the United States in paying top dollar for paintings at art auctions. Suppose that an art dealer is interested in testing two hypotheses. The first is that paintings sell for the same price, on average, in London, New York, and Tokyo. The second hypothesis is that works of Picasso, Chagall, and Dali sell for the same average price. The dealer is also aware of a third question. This is the question of a possible interaction between the location (and thus the buyers: people from the United States, English, Japanese) and the artist. Data on auction prices of 10 works of art by each of the three painters at each of the three cities are collected, and a two-way ANOVA is run on a computer. The results include the following: The sums of squares associated with the location (factor A) is 1,824. The sum of squares associated with the artist (factor B) is 2,230. The sum of squares for interactions is 804. The sum of squares for error is 8,262. Construct the ANOVA table, carry out the hypothesis tests, and state your conclusions.

Solution   We enter the sums of squares into the table. Since there are three levels in each of the two factors, and the sample size in each cell is 10, the degrees of freedom are $a - 1 = $

TABLE 9–8   ANOVA Table for Two-Way Analysis

Source of Variation	Sum of Squares	Degrees of Freedom	Mean Square	F Ratio
Factor A	SSA	$a - 1$	$MSA = \dfrac{SSA}{a - 1}$	$F = \dfrac{MSA}{MSE}$
Factor B	SSB	$b - 1$	$MSB = \dfrac{SSB}{b - 1}$	$F = \dfrac{MSB}{MSE}$
Interaction	SS(AB)	$(a - 1)(b - 1)$	$MS(AB) = \dfrac{SS(AB)}{(a - 1)(b - 1)}$	$F = \dfrac{MS(AB)}{MSE}$
Error	SSE	$ab(n - 1)$	$MSE = \dfrac{SSE}{ab(n - 1)}$	
Total	SST	$abn - 1$		

TABLE 9–9   ANOVA Table for Example 9–4

Source of Variation	Sum of Squares	Degrees of Freedom	Mean Square	F Ratio
Location	1,824	2	912	8.94
Artist	2,230	2	1,115	10.93
Interaction	804	4	201	1.97
Error	8,262	81	102	
Total	13,120	89		

2, $b - 1 = 2$, $(a - 1)(b - 1) = 4$, and $ab(n - 1) = 81$. Also, $abn - 1 = 89$, which checks as the sum of all other degrees of freedom. These values are entered in the table as well. The mean squares are computed, and so are the appropriate F ratios. Check to see how each result in the ANOVA table, Table 9–9, is obtained.

Let us now conduct the three hypothesis tests relevant to this problem. We will state the hypothesis tests in words. The factor A test is

H_0: There is no difference in the average price of paintings of the kind studied across the three locations

H_1: There are differences in average price across locations

The test statistic is an F random variable with 2 and 81 degrees of freedom (see Table 9–9). The computed value of the test statistic is 8.94. From Appendix C, Table 5, we find that the critical point for $\alpha = 0.01$ is close to 4.88. Thus, the null hypothesis is rejected, and we know that the p-value is much smaller than 0.01. Computer printouts of ANOVA results often list p-values in the ANOVA table, in a column after the F ratios. Often, the computer output will show $p = 0.0000$. This means that the p-value is smaller than 0.0001. The results of the hypothesis test are shown in Figure 9–18.

Now we perform the hypothesis test for factor B:

H_0: There are no differences in the average price of paintings by the three artists studied

H_1: There are differences in the average price of paintings by the three artists

FIGURE 9–18   **Example 9–4: Location Hypothesis Test**

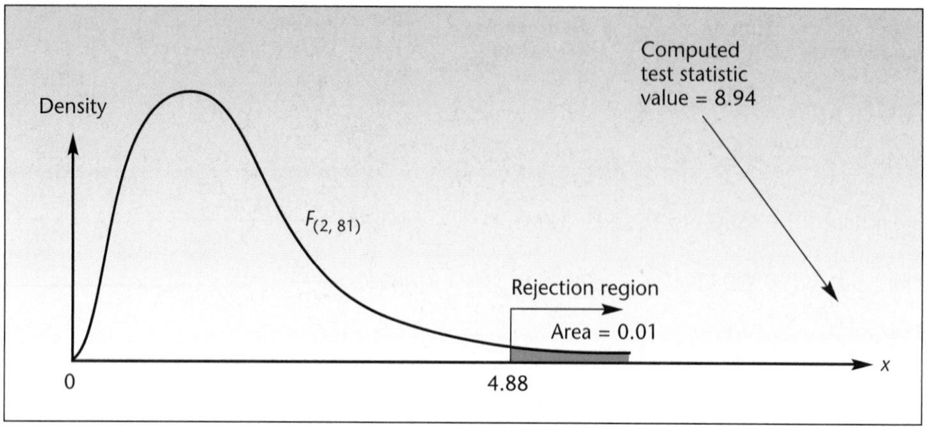

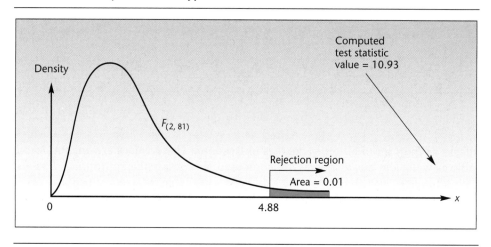

FIGURE 9–19   **Example 9–4: Artist Hypothesis Test**

Here again, the test statistic is an F random variable with 2 and 81 degrees of freedom, and the computed value of the statistic is 10.93. The null hypothesis is rejected, and the p-value is much smaller than 0.01. The test is shown in Figure 9–19.

The hypothesis test for interactions is

H_0: There are no interactions of the locations and the artists under study

H_1: There is at least one interaction of a location and an artist

The test statistic is an F random variable with 4 and 81 degrees of freedom. At a level of significance $\alpha = 0.05$, the critical point (see Appendix C, Table 5) is approximately equal to 2.48, and our computed value of the statistic is 1.97, leading us not to reject the null hypothesis of no interaction at levels of significance greater than 0.05. This is shown in Figure 9–20.

FIGURE 9–20   Example 9–4: Test for Interaction

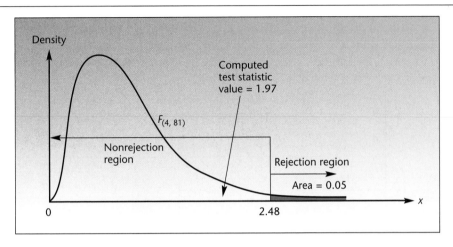

As mentioned earlier, we look at the test for interactions first. Since the null hypothesis of no interactions was not rejected, we have no statistical evidence of interactions of the two factors. This means, for example, that if a work by Picasso sells at a higher average price than works by the other two artists, then his paintings will fetch—on average—higher prices in all three cities. It also means that if paintings sell for a higher average price in London than in the other two cities, then this holds true—again, on average—for all three artists. Now we may interpret the results of the two main-effects tests.

We may conclude that there is statistical evidence that paintings (by these artists) do not fetch the same average price across the three cities. We may similarly conclude that paintings by the three artists under study do not sell, on average, for the same price. Where do the differences exist? This can be determined by a method for further analysis, such as the Tukey method.

In cases where we *do* find evidence of an interaction effect, our results have a different interpretation. In such cases, we must qualify any statement about differences among levels of one factor (say, factor A) as follows: *There exist differences among levels of factor A, averaged over all levels of factor B.*

We demonstrate this with a brief example. An article in *Accounting Review* reports the results of a two-way ANOVA on the factors "accounting" and "materiality." The exact nature of the study need not concern us here, as it is very technical. The results of the study include the following:

Source	df	Mean Square	F	Probability
Materiality	2	1.3499	4.5	0.0155
Accounting–materiality interaction	4	0.8581	2.9	0.0298

From these partial results, we see that the *p*-values ("probability") are each less than 0.05. Therefore, at the 0.05 level of significance for each of the two tests (separately), we find that there is an interaction effect, and we find a main effect for materiality. We may now conclude that, at the 0.05 level of significance, there are differences among the levels of materiality, *averaged over all levels* of accounting.

FIGURE 9–21   **The Template for Two-Way ANOVA**
              [Anova.xls; Sheet: 2-Way]

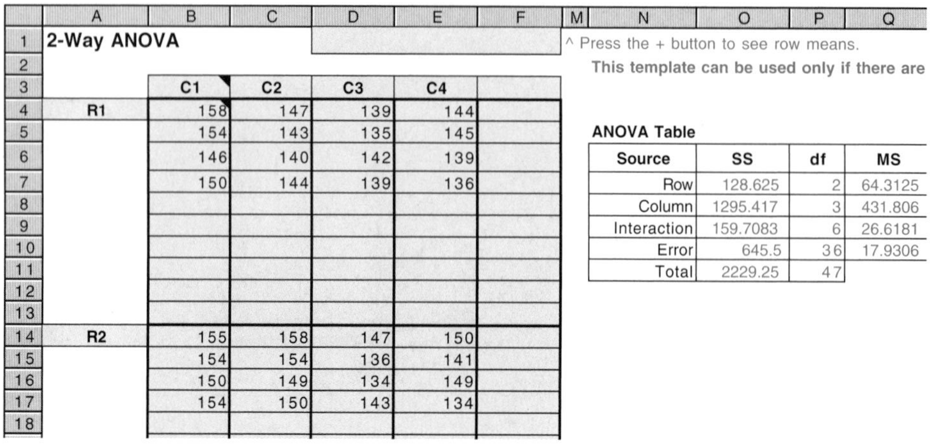

	A	B	C	D	E	F	M	N	O	P	Q
1	2-Way ANOVA							^ Press the + button to see row means.			
2								This template can be used only if there are			
3		C1	C2	C3	C4						
4	R1	158	147	139	144						
5		154	143	135	145			ANOVA Table			
6		146	140	142	139			Source	SS	df	MS
7		150	144	139	136			Row	128.625	2	64.3125
8								Column	1295.417	3	431.806
9								Interaction	159.7083	6	26.6181
10								Error	645.5	36	17.9306
11								Total	2229.25	47	
12											
13											
14	R2	155	158	147	150						
15		154	154	136	141						
16		150	149	134	149						
17		154	150	143	134						
18											

The Template

Figure 9–21 shows the template that can be used for computing two-way ANOVA. *This template can be used only if the number of replications in each cell is equal.* Up to 5 levels of row factor, 5 levels of column factor, and 10 replications in each cell can be entered in this template. Be sure that the data are entered properly in the cells.

To see the row means, unprotect the sheet and click on the "+" button above column M. Scroll down to see the column means and the cell means.

The Overall Significance Level

Remember our discussion of the Tukey analysis and its importance in allowing us to conduct a family of tests at a single level of significance. In two-way ANOVA, as we have seen, there is a family of *three tests,* each carried out at a given level of significance. Here the question arises: What is the level of significance of the *set* of three tests? A bound on the probability of making at least one type I error in the three tests is given by *Kimball's inequality.* If the hypothesis test for factor A main effects is carried out at α_1, the hypothesis test for factor B main effects is carried out at α_2, and the hypothesis test for interactions is carried out at α_3, then the level of significance α of the three tests together is bounded from above as follows.

Kimball's inequality:

$$\alpha \leq 1 - (1 - \alpha_1)(1 - \alpha_2)(1 - \alpha_3) \qquad (9\text{–}27)$$

In Example 9–4 we conducted the first two tests—the tests for main effects—at the 0.01 level of significance. We conducted the test for interactions at the 0.05 level. Using equation 9–27, we find that the level of significance of the family of three tests is *at most* $1 - (1 - 0.01)(1 - 0.01)(1 - 0.05) = 0.0689$.

The Tukey Method for Two-Way Analysis

Equation 9–21, the Tukey statistic for pairwise comparisons, is easily extended to two-way ANOVA. We are interested in comparing the levels of a factor once the ANOVA has led us to believe that differences do exist for that factor. The only difference in the Tukey formula is the number of degrees of freedom. In making pairwise comparisons of the levels of factor A, the test statistics are the pairwise differences between the sample means for all levels of factor A, regardless of factor B. For example, the pairwise comparisons of all the mean prices at the three locations in Example 9–4 will be done as follows. We compute the absolute differences of all the pairs of sample means:

$$|\bar{x}_{\text{London}} - \bar{x}_{\text{NY}}|$$
$$|\bar{x}_{\text{Tokyo}} - \bar{x}_{\text{London}}|$$
$$|\bar{x}_{\text{NY}} - \bar{x}_{\text{Tokyo}}|$$

Now we compare these differences with the Tukey criterion:

Tukey criterion for factor A is

$$T = q_\alpha \sqrt{\frac{\text{MSE}}{bn}} \qquad (9\text{–}28)$$

where the degrees of freedom of the q distribution are now a and $ab(n - 1)$. Note also that MSE is divided by bn.

In Example 9–4 both a and b are 3. The sample size in each cell is $n = 10$. At $\alpha = 0.05$, the Tukey criterion is equal to $(3.4)(\sqrt{102}/\sqrt{30}) = 6.27$.[9] Suppose that the sample mean in New York is 19.6 (hundred thousand dollars), in Tokyo it is 21.4, and in London it is 15.1. Comparing all absolute differences of the sample means leads us to the conclusion that the average prices in London and Tokyo are significantly different; but the average prices in Tokyo and New York are not different, and neither are the average prices in New York and London. The overall significance level of these joint conclusions is $\alpha = 0.05$.

Extension of ANOVA to Three Factors

It is possible to carry out a three-way ANOVA. In such cases, we assume that in addition to a levels of factor A and b levels of factor B, there are c levels of factor C. There are three possible pairwise interactions of factors and one possible triple interaction of factors. These are denoted AB, BC, AC, and ABC. Table 9–10 is the ANOVA table for three-way analysis.

It is beyond the scope of this book to give examples of three-way ANOVA. However, the extension of two-way analysis to this method is straightforward, and if you should need to carry out such an analysis, Table 9–10 will provide you with all the information you need. Three-factor interactions ABC imply that at least some of the two-factor interactions AB, BC, and AC are dependent on the level of the third factor.

---

[9] If the interaction effect is ignored because it was not significant, then MSE $= (8,262 + 804)/(81 + 4) = 107$; $T = 6.41$ with df $= 2, 85$.

TABLE 9–10   Three-Way ANOVA Table

Source of Variation	Sum of Squares	Degrees of Freedom	Mean Square	F Ratio
Factor A	SSA	$a - 1$	$MSA = \dfrac{SSA}{a - 1}$	$F = \dfrac{MSA}{MSE}$
Factor B	SSB	$b - 1$	$MSB = \dfrac{SSB}{b - 1}$	$F = \dfrac{MSB}{MSE}$
Factor C	SSC	$c - 1$	$MSC = \dfrac{SSC}{c - 1}$	$F = \dfrac{MSC}{MSE}$
AB	SS(AB)	$(a - 1)(b - 1)$	$MS(AB) = \dfrac{SS(AB)}{(a - 1)(b - 1)}$	$F = \dfrac{MS(AB)}{MSE}$
BC	SS(BC)	$(b - 1)(c - 1)$	$MS(BC) = \dfrac{SS(BC)}{(b - 1)(c - 1)}$	$F = \dfrac{MS(BC)}{MSE}$
AC	SS(AC)	$(a - 1)(c - 1)$	$MS(AC) = \dfrac{SS(AC)}{(a - 1)(c - 1)}$	$F = \dfrac{MS(AC)}{MSE}$
ABC	SS(ABC)	$(a - 1)(b - 1)(c - 1)$	MS(ABC)	$F = \dfrac{MS(ABC)}{MSE}$
Error	SSE	$abc(n - 1)$	MSE	
Total	SST	$abcn - 1$		

FIGURE 9–22   Data Layout in a Two-Way ANOVA with $n = 1$

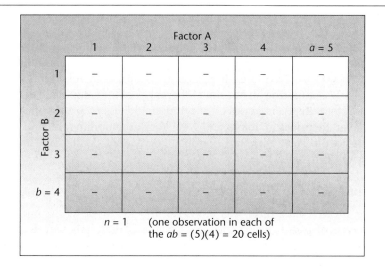

Two-Way ANOVA with One Observation per Cell

The case of one data point in every cell presents a problem in two-way ANOVA. Can you guess why? (*Hint:* Degrees of freedom for error are the answer.) Look at Figure 9–22, which shows the layout of the data in a two-way ANOVA with five levels of factor A and four levels of factor B. Note that the sample size in each of the 20 cells is $n = 1$.

TABLE 9–11 ANOVA Table for Two-Way Analysis with One Observation per Cell, Assuming No Interactions

Source of Variation	Sum of Squares	Degrees of Freedom	Mean Square	F Ratio
Factor A	SSA	$a - 1$	$MSA = \dfrac{SSA}{a - 1}$	$F = \dfrac{MSA}{MS(AB)}$
Factor B	SSB	$b - 1$	$MSB = \dfrac{SSB}{b - 1}$	$F = \dfrac{MSB}{MS(AB)}$
"Error"	SS(AB)	$(a - 1)(b - 1)$	$MS(AB) = \dfrac{SS(AB)}{(a - 1)(b - 1)}$	
Total	SST	$ab - 1$		

As you may have guessed, there are no degrees of freedom for error! With one observation per cell, $n = 1$; the degrees of freedom for error are $ab(n - 1) = ab(1 - 1) = 0$. This can be seen from Table 9–11. What can we do? If we *believe* that there are no interactions (this assumption cannot be statistically tested when $n = 1$), then our sum of squares SS(AB) is due to error and contains no other information. In such a case, we can use SS(AB) and its associated degrees of freedom $(a - 1)(b - 1)$ in place of SSE and its degrees of freedom. We can thus conduct the tests for the main effects by dividing MSA by MS(AB) when testing for factor A main effects. The resulting F statistic has $a - 1$ and $(a - 1)(b - 1)$ degrees of freedom. Similarly, when testing for factor B main effects, we divide MSB by MS(AB) and obtain an F statistic with $b - 1$ and $(a - 1)(b - 1)$ degrees of freedom.

Remember that this analysis assumes that there are no interactions between the two factors. Remember also that in statistics it is always desirable to have as many data as possible. Therefore, the two-way ANOVA with one observation per cell is, in itself, of limited use. The idea of two factors and one observation per cell is useful, however, as it brings us closer to the idea of blocking, presented in the next section.

The two-way ANOVA model with one observation per cell is

$$x_{ij} = \mu + \alpha_i + \beta_j + \epsilon_{ij} \qquad (9\text{–}29)$$

where μ is the overall mean, α_i is the effect of level i of factor A, β_j is the effect of level j of factor B, and ϵ_{ij} is the error associated with x_{ij}. We assume the errors are normally distributed with zero mean and variance σ^2.

PROBLEMS

9–37. Discuss the context in which Example 9–4 can be analyzed by using a random-effects model.

9–38. What are the reasons for conducting a two-way analysis rather than two separate one-way ANOVAs? Explain.

9–39. What are the limitations of two-way ANOVA? What problems may be encountered?

9–40. (This is a hard problem.) Suppose that a limited data set is available. Explain why it is not desirable to increase the number of factors under study (say, four-way ANOVA, five-way ANOVA, and so on). Give two reasons for this—one of the reasons should be a statistical one.

9–41. An article[10] about process tracing and brand extension reports the results of a two-way ANOVA with the two factors "brand extension typicality" and "brand breadth." The article reports an F statistic for interaction of the two factors: $F_{(4, 139)} = 4.27$. Is there evidence that the two factors interact with each other, as believed by the authors of the article?

9–42. The following table reports salaries, in thousands of dollars per year, for executives in three job types and three locations. Conduct a two-way ANOVA on these data.

	Job		
Location	Type I	Type II	Type III
East	54, 61, 59, 56, 70, 62, 63, 57, 68	48, 50, 49, 60, 54, 52, 49, 55, 53	71, 76, 65, 70, 68, 62, 73, 60, 79
Central	52, 50, 58, 59, 62, 57, 58, 64, 61	44, 49, 54, 53, 51, 60, 55, 47, 50	61, 64, 69, 58, 57, 63, 65, 63, 50
West	63, 67, 68, 72, 68, 75, 62, 65, 70	65, 58, 62, 70, 57, 61, 68, 65, 73	82, 75, 79, 77, 80, 69, 84, 83, 76

9–43. The Neilsen Company, which issues television popularity rating reports, is interested in testing for differences in average viewer satisfaction with morning news, evening news, and late news. The company is also interested in determining whether differences exist in average viewer satisfaction with the three main networks: CBS, ABC, and NBC. Nine groups of 50 randomly chosen viewers are assigned to each combination cell CBS–morning, CBS–evening, . . . , NBC–late. The viewers' satisfaction ratings are recorded. The results are analyzed via two-factor ANOVA, one factor being network and the other factor being news time. Complete the following ANOVA table for this study, and give a full interpretation of the results.

Source of Variation	Sum of Squares	Degrees of Freedom	Mean Square	F Ratio
Network	145			
News time	160			
Interaction	240			
Error	6,200			
Total				

9–44. An article reports the results of an analysis of salespersons' performance level as a function of two factors: task difficulty and effort. Included in the article is the following ANOVA table:

---

[10]D. M. Boush and B. Loken, "A Process Tracing Study of Brand Extension Evaluation," *Journal of Marketing Research,* February 1991, p. 21.

Variable	df	F Value	p
Task difficulty	1	0.39	0.5357
Effort	1	53.27	<0.0001
Interaction	1	1.95	0.1649

a. How many levels of task difficulty were studied?

b. How many levels of effort were studied?

c. Are there any significant task difficulty main effects?

d. Are there any significant effort main effects?

e. Are there any significant interactions of the two factors? Explain.

9–45. A study evaluated the results of a two-way ANOVA on the effects of the two factors—exercise price of an option and the time of expiration of an option—on implied interest rates (the measured variable). Included in the article is the following ANOVA table.

Source of Variation	Degrees of Freedom	Sum of Squares	Mean Square	F Ratio
Exercise prices	2	2.866	1.433	0.420
Time of expiration	1	16.518	16.518	4.845
Interaction	2	1.315	0.658	0.193
Explained	5	20.699	4.140	1.214
Residuals (error)	144	490.964	3.409	

a. What is meant by *Explained* in the table, and what is the origin of the information listed under that source?

b. How many levels of exercise price were used?

c. How many levels of time of expiration were used?

d. How large was the total sample size?

e. Assuming an equal number of data points in each cell, how large was the sample in each cell?

f. Are there any exercise price main effects?

g. Are there any time-of-expiration main effects?

h. Are there any interactions of the two factors?

i. Interpret the findings of this study.

j. Give approximate *p*-values for the tests.

k. In this particular study, what other, equivalent distribution may be used for testing for time-of-expiration main effects? (*Hint:* df.) Why?

9–46. An article in the *Accounting Review* reports analysis of variance results given in the following table.[11] Interpret the results.

---

[11]J. Kennedy, "Debiasing the Curse of Knowledge in Audit Judgment," *The Accounting Review* 70, no. 2 (April 1995), pp. 249–73.

Analysis of Variance—*F* Statistics

Source of Variance	df	Dependent Variable[a]					
		Estimates		Likelihood Judgments			
		X	*Y*	*XH*	*XL*	*YH*	*YL*
Accountability[b]	1	0.47	1.28	0.00	0.00	3.59*	0.65
Outcome[c]	2	21.97***	8.21***	6.64***	5.50***	7.23***	1.31
Group	1	2.72*	23.98***	1.96	0.32	1.94	28.68***
Accountability × Outcome	2	0.30	0.31	0.20	0.11	0.23	2.41*
Accountability × Group	1	0.00	0.03	1.11	0.11	0.06	0.514
Outcome × Group	2	0.23	1.18	0.14	0.49	1.21	0.60
Account. × Outcome × Group	2	0.08	0.40	0.36	0.95	1.23	0.14
Error	396						
Total	407						

Planned Comparisons	df	*t* Statistics					
High versus no outcome	1	2.08**	2.63***	1.54*	3.23***	3.24***	−1.40
Low versus no outcome	1	4.51***	1.44*	2.15**	.90	.18	1.40*

[a]X and Y refer to the estimates for the 13th quarter's sales for graphs X and Y, respectively, XH, XL, YH, and YL refer to likelihood judgments that 13th-quarter sales were at least as high as the high outcome (XH and YH) or as low as the low outcome (XL and YL) indicated by the asterisk on graphs X and Y. Predictions regarding outcome knowledge for X, XH, Y, and YH are high > none > low, while predictions for XL and YL are low > none > high.
[b]Accountable subjects were told before they saw the graphs that they might have to justify their responses.
[c]High and low outcome conditions saw asterisks higher or lower than the 12th quarter's sales.
***, **, and * indicate statistically significant at $p < 0.01$, 0.05, and 0.10, respectively.

9–8   Blocking Designs

In this section, we discuss alternatives to the completely randomized design. We seek special designs for data analysis that will help us reduce the effects of extraneous factors (factors not under study) on the measured variable. That is, we seek to reduce the errors. These designs allow for *restricted randomization* by grouping the experimental units (people, items in our data) into homogeneous groups called **blocks** and then randomizing the treatments within each block.

The first, and most important, blocking design we will discuss is the *randomized complete block design*.

Randomized Complete Block Design

Recall the first part of the Club Med example, Example 9–2, where we were interested only in determining possible differences in average ratings among the five resorts (no attributes factor). Suppose that Club Med can get information about its vacationers' age, sex, marital status, socioeconomic level, etc., and then can randomly assign vacationers to the different resorts. The club could form groups of five vacationers each such that the vacationers within each group are similar to one another in age, sex, and marital status, etc. Each group of five vacationers is a *block*. Once the blocks are formed, one member from each block is *randomly assigned* to one of the five resorts (Guadeloupe, Martinique, Eleuthera, Paradise Island, or St. Lucia). Thus, the vacationers sent to each resort will comprise a mixture of ages, of males and females, of married and single people, of different socioeconomic levels, etc. The vacationers within each block, however, will be more or less homogeneous.

The vacationers' ratings of the resorts are then analyzed using an ANOVA that utilizes the blocking structure. Since the members of each block are similar to one another (and different from members of other blocks), we expect them to react to similar conditions in similar ways. This brings about a *reduction in the experimental errors.* Why? If we cannot block, it is possible, for example, that the sample of people we get for Eleuthera will happen to be wealthier (or predominantly married, predominantly male, or whatever) and will tend to react less favorably to a resort of this kind than a more balanced sample would react. In such a case, we will have greater experimental error. If, on the other hand, we can *block* and send one member of each homogeneous group of people to each of the resorts and then compare the responses of the block as a whole, we will be more likely to find real differences among the resorts than differences among the people. Thus, the errors (differences among people and not among the resorts) are reduced by the blocking design. When all members of every block are randomly assigned to all treatments, such as in this example, our design is called the **randomized complete block design.**

The model for randomized complete block design is

$$x_{ij} = \mu + \alpha_i + \beta_j + \epsilon_{ij} \qquad (9\text{--}30)$$

where μ is the overall mean, α_i is the effect of level i of factor A, β_j is the effect of block j on factor B, and ϵ_{ij} is the error associated with x_{ij}. We assume the errors are normally distributed with zero mean and variance σ^2.

Figure 9–23 shows the formation of blocks in the case of Club Med. We assume that the club is able—for the purpose of a specific study—to randomly assign vacationers to resorts.

The analysis of the results in a randomized complete block design, with a single factor, is very similar to the analysis of two-factor ANOVA with one observation per cell (see Table 9–11). Here, one "factor" is the blocks, and the other is the factor of interest (in our example, resorts). The ANOVA table for a randomized complete block design is illustrated in Table 9–12. Compare this table with Table 9–11. There are n blocks of r elements each. We assume that there are no interactions between blocks and treatments; thus, the degrees of freedom for error are $(n - 1)(r - 1)$. The F ratio reported in the table is for use in testing for treatment effects. It is possible to test for block effects with a similar F ratio, although usually such a test is of no interest.

As an example, suppose that Club Med did indeed use a blocking design with $n = 10$ blocks. Suppose the results are SSTR = 3,200, SSBL = 2,800, and SSE = 1,250. Let us conduct the following test:

H$_0$: The average ratings of the five resorts are equal

H$_1$: Not all the average ratings of the five resorts are equal

We enter the information into the ANOVA table and compute the remaining entries we need and the F statistic value, which has an F distribution with $r - 1$ and $(n - 1)(r - 1)$ degrees of freedom when H$_0$ is true. We have as a result Table 9–13.

We see that the value of the F statistic with 4 and 36 degrees of freedom is 23.04. This value exceeds, by far, the critical point of the F distribution with 4 and 36 degrees of freedom at $\alpha = 0.01$, which is 3.89. The p-value is, therefore, much smaller

FIGURE 9–23   **Blocking in the Club Med Example**

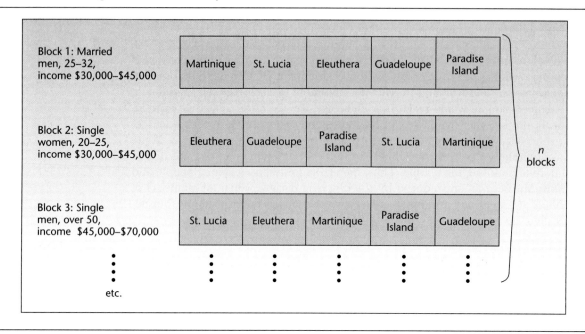

TABLE 9–12   **ANOVA Table for Randomized Complete Block Design**

Source of Variation	Sum of Squares	Degrees of Freedom	Mean Square	F Ratio
Blocks	SSBL	$n-1$	MSBL	
Treatments	SSTR	$r-1$	MSTR	$F = \dfrac{\text{MSTR}}{\text{MSE}}$
Error	SSE	$(n-1)(r-1)$	MSE	
Total		$nr-1$		

TABLE 9–13   **Club Med Blocking Design ANOVA Table**

Source of Variation	Sum of Squares	Degrees of Freedom	Mean Square	F Ratio
Blocks	2,800	9	311.11	
Resorts	3,200	4	800.00	23.04
Error	1,250	36	34.72	
Total	7,250			

than 0.01. We thus reject the null hypothesis and conclude that there is evidence that not all resorts are rated equally, on average. By blocking the respondents into homogeneous groups, Club Med was able to reduce the experimental errors.

TABLE 9–14 The ANOVA Table for Example 9–5

Source of Variation	Sum of Squares	Degrees of Freedom	Mean Square	F Ratio
Blocks	2,750	39	70.51	
Treatments	2,640	2	1,320.00	12.93
Error	7,960	78	102.05	
Total	13,350	119		

FIGURE 9–24 Data Layout for Example 9–5

	Randomized Viewing Order		
First sampled person	Actress B	Actress C	Actress A
Second sampled person	Actress C	Actress B	Actress A
Third sampled person	Actress A	Actress C	Actress B
Fourth sampled person	Actress B	Actress A	Actress C
etc.			

You can probably find many examples where blocking can be useful. For example, recall the situation of problem 9–18. Three prototype airplanes were tested on different flight routes to determine whether differences existed in the average range of the planes. A design that would clearly reduce the experimental errors is a blocking design where all planes are flown over the same routes, at the same time, under the same weather conditions, etc. That is, fly all three planes using each of the sample route conditions. A block in this case is a route condition, and the three treatments are the three planes.

A special case of the randomized complete block design is the **repeated-measures design.** In this design, each experimental unit (person or item) is assigned to *all* treatments in a randomly selected order. Suppose that a taste test is to be conducted, where four different flavors are to be rated by consumers. In a repeated-measures design, each person in the random sample of consumers is assigned to taste all four flavors, in a randomly determined order, independent of all other consumers. A block in this design is one consumer. We demonstrate the repeated-measures design with the following example.

EXAMPLE 9–5

Weintraub Entertainment is a new movie company backed by financial support from Coca-Cola Company. For one of the company's first movies, the director wanted to find the best actress for the leading role. "Best" naturally means the actress who would get the highest average viewer rating. The director was considering three candidates for the role and had each candidate act in a particular test scene. A random group of 40 viewers was selected, and each member of the group watched the same scene enacted by each of the three actresses. The order of actresses was randomly and independently chosen for each viewer. Ratings were on a scale of 0 to 100. The results were analyzed using a block design ANOVA, where each viewer constituted a block of treatments. The results of the analysis are given in Table 9–14. Figure 9–24 shows

FIGURE 9–25   **The Template for Randomized Block Design ANOVA**
[Anova.xls; Sheet: RBD]

	A	B	C	D	E	F	G	N	S	T	U	V	W
1	Randomized Complete Block Design												
2													
3	Mean	10.7143	13	15.4286						ANOVA Table			
4		A	B	C				Mean		Source	SS	df	MS
5	1	12	16	15				14.3333		Treatment	40.9524	6	6.8254
6	2	11	10	14				11.6667		Block	77.8095	2	38.9048
7	3	10	11	17				12.6667		Error	26.1905	12	2.18254
8	4	12	15	19				15.3333		Total	144.952	20	
9	5	11	14	17				14					
10	6	10	12	13				11.6667					
11	7	9	13	13				11.6667					

the layout of the data in this example. Analyze the results. Are all three actresses equally rated, on average?

Solution   The test statistic has an *F* distribution with 2 and 78 degrees of freedom when the following null hypothesis is true.

H_0: There are no differences among average population ratings of the three actresses

Check the appropriate critical point for $\alpha = 0.01$ in Appendix C, Table 5, to see that this null hypothesis is rejected in favor of the alternative that differences do exist and that not all three actresses are equally highly rated, on average. Since the null hypothesis is rejected, there is place for further analysis to determine which actress rates best. Such analysis can be done using the Tukey method or another method of further analysis.

In cases where a repeated-measures design is used on *rankings* of several treatments—here, if we had asked each viewer to rank the three actresses as 1, 2, or 3, rather than rate them on a 0-to-100 scale—there is another method of analysis. This method is the *Friedman test,* discussed in Chapter 14.

The Template

Figure 9–25 shows the template that can be used for computing ANOVA in the case of a randomized block design. The group means appear at the top of each column and the block means appear in each row to the right of the data.

PROBLEMS

9–47.   Explain the advantages of blocking designs.

9–48.   An article on emerging markets reports the findings on returns on equity, bonds, and preferred stock.[12] Data are available on a random sample of firms that

---

[12]P. Misra, "Fund Watch," *Money,* May 1995, p. 61.

issue all three types of instruments. How would you use blocking in this study, aimed at finding which instrument gives the highest average return?

9–49. Suggest a blocking design for the situation in problem 9–19. Explain.

9–50. Suggest a blocking design for the situation in problem 9–21. Explain.

9–51. Is it feasible to design a study utilizing blocks for the situation in problem 9–20? Explain.

9–52. Suggest a blocking design for the situation in problem 9–23.

9–53. How would you design a block ANOVA for the two-way analysis of the situation described in problem 9–42? Which ANOVA method is appropriate for the analysis?

9–54. What important assumption about the relation between blocks and treatments is necessary for carrying out a block design ANOVA?

9–55. Public concern has recently focused on the fact that although people in the United States often try to lose weight, statistics show that the general population has gained weight, on average, during the last 10 years. A researcher hired by a weight-loss organization is interested in determining whether three kinds of artificial sweetener currently on the market are approximately equally effective in reducing weight. As part of a study, a random sample of 300 people is chosen. Each person is given one of the three sweeteners to use for a week, and the number of pounds lost is recorded. To reduce experimental errors, the people in the sample are divided into 100 groups of three persons each. The three people in every group all weigh about the same at the beginning of the test week and are of the same sex and approximately the same age. The results are SSBL = 2,312, SSTR = 3,233, and SSE = 12,386. Are all three sweeteners equally effective in reducing weight? How confident are you of your conclusion? Discuss the merits of blocking in this case as compared with the completely randomized design.

9–56. IBM Corporation has been retraining many of its employees to assume marketing positions. As part of this effort, the company wanted to test four possible methods of training marketing personnel to determine if at least one of the methods was better than the others. Four groups of 70 employees each were assigned to the four training methods. The employees were pretested for marketing ability and put into groups of four, each group constituting a block with approximately equal prior ability. Then the four employees in each group were randomly assigned to the four training methods and retested after completion of the three-week training session. The differences between their initial scores and final scores were computed. The results were analyzed using a block design ANOVA. The results of the analysis include: SSTR = 9,875, SSBL = 1,445, and SST = 22,364. Are all four training methods equally effective? Explain.

9–9   Summary and Review of Terms

In this chapter, we discussed a method of making statistical comparisons of more than two population means. The method is **analysis of variance,** often referred to as **ANOVA.** We defined **treatments** as the populations under study. A set of treatments is a **factor.** We defined **one-factor ANOVA,** also called one-way ANOVA, as the test for equality of means of treatments belonging to one factor. We defined a **two-factor**

ANOVA, also called two-way ANOVA, as a set of three hypothesis tests: (1) a test for main effects for one of the two factors, (2) a test for main effects for the second factor, and (3) a test for the **interaction** of the two factors. We defined the **fixed-effects** model and the **random-effects** model. We discussed one method of further analysis to follow an ANOVA once the ANOVA leads to rejection of the null hypothesis of equal treatment means. The method is the **Tukey HSD procedure.** We also mentioned two other alternative methods of further analysis. We discussed experimental design in the ANOVA context. Among these designs, we mentioned **blocking** as a method of reducing experimental errors in ANOVA by grouping similar items. We also discussed the **repeated-measures design.**

ADDITIONAL PROBLEMS

9–57. An enterprising art historian recently started a new business: the production of walking-tour audiotapes for use by those visiting major cities. She originally produced tapes for eight cities: Paris, Rome, London, Florence, Jerusalem, Washington, New York, and New Haven. A test was carried out to determine whether all eight tapes (featuring different aspects of different cities) were equally appealing to potential users. A random sample of 160 prospective tourists was selected, 20 per city. Each person evaluated the tape he or she was given on a scale of 0 to 100. The results were analyzed using one-way ANOVA and included SSTR = 7,102 and SSE = 10,511. Are all eight tapes equally appealing, on average? What can you say about the p-value?

9–58. NAMELAB is a San Francisco–based company that uses linguistic analysis and computers to invent catchy names for new products. The company is credited with the invention of Acura, Compaq, Sentra, and other names of successful products. Naturally, statistical analysis plays an important role in choosing the final name for a product. In choosing a name for a compact disk player, NAMELAB is considering four names and uses analysis of variance for determining whether all four names are equally liked, on average, by the public. The results include $n_1 = 32$, $n_2 = 30$, $n_3 = 28$, $n_4 = 41$, SSTR = 4,537, and MSE = 412. Are all four names approximately equally liked, on average? What is the approximate p-value?

9–59. As software for microcomputers becomes more and more sophisticated, the element of time becomes more crucial. Consequently, manufacturers of software packages need to work on reducing the time required for running application programs. Speed of execution also depends on the computer used. A two-way ANOVA is suggested for testing whether differences exist among three software packages, and among four microcomputers made by NEC, Toshiba, Kaypro, and Apple, with respect to the average time for performing a certain analysis. The results include SS(software) = 77,645, SS(computer) = 54,521, SS(interaction) = 88,699, and SSE = 434,557. The analysis used a sample of 60 runs of each software package–computer combination. Complete an ANOVA table for this analysis, carry out the tests, and state your conclusions.

9–60. The following table summarizes results of ANOVAs of three separate "dependent variables." For each variable, a four-factor ANOVA is carried out. The table reports p-values. Interpret all the results in the table.

	Dependent Variable		
Effect Source	**RDIFF**	**Bias**	**BE**
METH: factor scoring approach	0.008	0.009	0.457
UNIQ: factor uniqueness	0.474	0.510	0.012
SE: assumed obliqueness relative to true obliqueness	0.106	0.086	0.086
P: number of observed variances	0.106	0.091	0.102
METH × UNIQ	0.002	0.006	0.078
METH × SE	0.010	0.302	0.110
METH × P	0.150	0.112	0.125
UNIQ × SE	0.206	0.087	0.210
UNIQ × P	0.001	0.176	0.008
SE × P	0.210	0.011	0.002

Source: From J. Lastovicka and K. Thamodaran, "Common Factor Score Estimates," *Journal of Marketing Research,* February 1991, p. 109. Reprinted by permission of the American Marketing Association.

9–61. Among the young affluent U.S. professionals, there is a growing new demand for exotic pets. The most popular pets are the Shiba Inu dog breed, Rottweilers, Persian cats, and Maine coons. Prices for these pets vary and depend on supply and demand. A breeder of exotic pets wants to know whether these four pets fetch the same average prices, whether prices for these exotic pets are higher in some geographic areas than in others, and whether there are any interactions—one or more of the four pets being more favored in one location than in others. Prices for 10 of each of these pets at four randomly chosen locations around the country are recorded and analyzed. The results are SS(pet) = 22,245, SS(location) = 34,551, SS(interaction) = 31,778, and SSE = 554,398. Are there any pet main effects? Are there any location main effects? Are there any pet–location interactions? Explain your findings.

9–62. Analysis of variance has long been used in providing evidence of the effectiveness of pharmaceutical drugs. Such evidence is required before the Food and Drug Administration (FDA) will allow a drug to be marketed. In a recent test of the effectiveness of a new sleeping pill, three groups of 25 patients each were given the following treatments. The first group was given the drug, the second group was given a placebo, and the third group was given no treatment at all. The number of minutes it took each person to fall asleep was recorded. The results are as follows.

Drug group:	12, 17, 34, 11, 5, 42, 18, 27, 2, 37, 50, 32, 12, 27, 21, 10, 4, 33, 63, 22, 41,19, 28, 29, 8
Placebo group:	44, 32, 28, 30, 22, 12, 3, 12, 42, 13, 27, 54, 56, 32, 37, 28, 22, 22, 24, 9, 20, 4, 13, 42, 67
No-treatment group:	32, 33, 21, 12, 15, 14, 55, 67, 72, 1, 44, 60, 36, 38, 49, 66, 89, 63, 23, 6, 9, 56, 28, 39, 59

Use a computer (or hand calculations) to determine whether the drug is effective. What about the placebo? Give differences in average effectiveness, if any exist.

9–63. A more efficient experiment than the one described in problem 9–62 was carried out to determine whether a sleeping pill was effective. Each person in a random sample of 30 people was given the three treatments: drug, placebo, nothing. The order in which these treatments were administered was randomly chosen for each person in the sample.

    a. Explain why this experiment is more efficient than the one described for the same investigation in problem 9–62. What is the name of the experimental design used here? Are there any limitations to the present method of analysis?

    b. The results of the analysis include SSTR = 44,572, SSBL = 38,890, and SSE = 112,672. Carry out the analysis, and state your conclusions. Use $\alpha = 0.05$.

9–64. Three new enhanced-definition television models are compared.[13] The distances (in miles) over which a clear signal is received in random trials for each of the models are given below.

General Instrument:	111, 121, 134, 119, 125, 120, 122, 138, 115, 123, 130, 124, 132, 127, 130
Philips:	120, 121, 122, 123, 120, 132, 119, 116, 125, 123, 116, 118, 120, 131, 115
Zenith:	109, 100, 110, 102, 118, 117, 105, 104, 100, 108, 128, 117, 101, 102, 110

Carry out a complete analysis of variance, and report your results in the form of a memorandum. State your hypotheses and your conclusions. Do you believe there are differences among the three models? If so, where do they lie?

9–65. A professor of food chemistry at the University of Wisconsin has recently developed a new system for keeping frozen foods crisp and fresh: coating them with watertight, edible film. The Pillsbury Company wants to test whether the new product is tasty. The company collects a random sample of consumers who are given the following three treatments, in a randomly chosen order for each consumer: regular frozen pizza, frozen pizza packaged in a plastic bag, and the new edible-coating frozen pizza (all reheated, of course). Fifty people take part in the study, and the results include SSTR = 128,899, SSBL = 538,217, and SSE = 42,223,987. (These are ANOVA results for taste scores on a 0–1000 scale.) Based on these results, are all three frozen pizzas perceived as equally tasty?

9–66. Give the statistical reason for the fact that a one-way ANOVA with only two treatments is equivalent to a two-sample t test discussed in Chapter 8.

9–67. Following is a computer output of an analysis of variance based on randomly chosen rents in four cities. Do you believe that the average rent is equal in the four cities studied? Explain.

```
ANALYSIS OF VARIANCE
SOURCE     DF          SS          MS          F
FACTOR      3       37402       12467       1.76
ERROR      44      311303        7075
TOTAL      47      348706
```

[13]A. Kupfer, "The U.S. Wins One in High-Tech TV," *Fortune,* April 8, 1991, p. 63.

9–68. Write several paragraphs explaining the meaning of the following excerpt.[14]

Toy awareness level. A two-way ANOVA was conducted with two levels of language and two levels of income as independent factors and toy awareness level as the dependent measure. As hypothesized, English-speaking children were significantly more aware of the toys than were French-speaking children ($F = 400.47$, $p < 0.0001$). The latter were able to recognize an average of only 8.73 toys correctly, not significantly more than the 1-in-3 chance level (that is, 6.67 toys). In comparison, English-speaking children were able to identify an average of 15.44 toys. A significant income effect is found ($F = 7.08$, $p < 0.01$). Upper-middle-income children recognize more toys (mean = 11.94) than low-income children (mean = 9.94). Contrary to the hypothesis, no significant language by income interaction is found ($F < 1$). The likely reason is a ceiling effect, as all English-speaking children had very high scores. (Of the few mistakes made by English-speaking children, most consisted of boys misidentifying a few girls' toys, and vice versa.)

9–69. One of the oldest and most respected survey research firms is the Gallup Organization. This organization makes a number of special reports available at its corporate website **www.gallup.com.** Select and read one of the special reports available at this site. Based on the information in this report, design a 3×3 ANOVA on a response of interest to you, such as buying behavior. The design should include two factors that you think influence the response, such as location, age, income, or education. Each factor should have three levels for testing in the model.

www.exercise

9–70. Interpret the following computer output.

```
ANALYSIS OF VARIANCE ON SALES
SOURCE     DF        SS        MS         F        P
STORE       2   1017.33    508.67    156.78    0.000
ERROR      15     48.67      3.24
TOTAL      17   1066.00
                                INDIVIDUAL 95 PCT CI'S FOR MEAN
                                BASED ON POOLED STDEV
LEVEL       N      MEAN     STDEV    -+----------+----------+----------+----
    1       6    53.667     1.862               (-*--)
    2       6    67.000     1.673                                      (--*-)
    3       6    49.333     1.862        (-*-)
                                    -+----------+----------+----------+----
POOLED STDEV = 1.801               48.0       54.0       60.0       66.0
```

CASE 10   Uniform Uniforms

A textile manufacturer has a large order for a cloth meant for making uniforms. The cloth is dyed using four different dyeing lines, which produce approximately equal amounts of cloth each day.

Usually not more than one line is used for one product, because no matter how well the process is controlled, there will always be perceptible differences in the shade of the dye from one line to another.

But because the volume of the order is large, four lines are being used. It is important to maintain the shade as uniform as possible by minimizing the variance of the brightness of the shade on all cloth produced. Lately, the customer has been complaining about too much variance in the brightness. It was decided to conduct an ANOVA test of the brightness of the cloth from the four lines. Random samples were taken from each line and were measured for brightness. The measurement is on a 0 to 100 scale. The sample data are

	Line 1	Line 2	Line 3	Line 4
1	66.55	66.16	68.36	72.32
2	71.91	65.94	66.81	66.69
3	67.61	68.62	66.50	72.36
4	66.13	63.86	65.22	70.88
5	71.31	69.38	65.06	69.76
6	68.99	64.55	65.42	71.05
7	71.83	66.82	66.50	68.78
8	68.99	65.56	64.82	74.40
9	69.81	63.66	68.31	66.73
10	72.49	64.71	68.17	73.58
11	69.99	67.32	65.50	66.72
12	73.44	71.39	70.39	70.37
13	70.39	63.78		75.72
14	68.42	70.25		74.65
15	71.66			
16	65.14			

1. Conduct the test at the 5% significance level, and report your conclusion.
2. Which pairs of lines have significant differences in their average brightness?
3. Stopping a line to adjust its average brightness is costly. If only one line can be stopped and adjusted, which one should it be? To what average brightness value should it be adjusted to minimize the variance in all the cloth produced?
4. If two lines can be stopped and adjusted, which ones should be? To what average brightness value should they be adjusted to minimize the total variance in all the cloth produced?

10 Simple Linear Regression and Correlation

In 1855, a 33-year-old Englishman settled down to a life of leisure in London after several years of travel throughout Europe and Africa. The boredom brought about by a comfortable life induced him to write, and his first book was, naturally, *The Art of Travel.* As his intellectual curiosity grew, he shifted his interests to science and many years later published a paper on heredity, "Natural Inheritance" (1889). He reported his discovery that sizes of seeds of sweet pea plants appeared to "revert," or "regress," to the mean size in successive generations. He also reported results of a study of the relationship between heights of fathers and the heights of their sons. A straight line was fit to the data pairs: height of father versus height of son. Here, too, he found a "regression to mediocrity": The heights of the sons represented a movement away from their fathers, toward the average height. The man was Sir Francis Galton, a cousin of Charles Darwin. We credit him with the idea of statistical regression.

While most applications of regression analysis may have little to do with the "regression to the mean" discovered by Galton, the term **regression** remains. It now refers to the statistical technique of modeling the relationship between variables. In this chapter on **simple linear regression,** we model the relationship between two variables: a **dependent variable,** denoted by Y, and an **independent variable,** denoted by X. The model we use is a *straight-line relationship* between X and Y. When we model the relationship between the dependent variable Y and a set of several independent variables, or when the assumed relationship between Y and X is curved and requires the use of more terms in the model, we use a technique called *multiple regression.* This technique will be discussed in the next chapter.

Figure 10–1 is a general example of simple linear regression: fitting a straight line to describe the relationship between two variables X and Y. The points on the graph are randomly chosen observations of the two variables X and Y, and the straight line describes the general *movement* in the data—an increase in Y corresponding to an increase in X. An inverse straight-line relationship is also possible, consisting of a general decrease in Y as X increases (in such cases, the slope of the line is negative).

Regression analysis is one of the most important and widely used statistical techniques and has many applications in business and economics. A firm may be interested in estimating the relationship between advertising and sales (one of the most important topics of research in the field of marketing). Over a short range of values—when advertising is not yet overdone, giving diminishing returns—the relationship between advertising and sales may be well approximated by a straight line. The X variable in Figure 10–1 could denote advertising expenditure, and the Y variable could stand for the resulting sales for the same period. The data points in this case would be pairs of observations of the form $x_1 = \$75,570$, $y_1 = 134,679$ units; $x_2 = \$83,090$, $y_2 = 151,664$ units; etc. That is, the first month the firm spent \$75,570 on advertising, and sales for the month were 134,679 units; the second month the company spent \$83,090 on advertising, with resulting sales of 151,664 units for that month; and so on for the entire set of available data.

The data pairs, values of X paired with corresponding values of Y, are the points shown in a sketch of the data (such as Figure 10–1). A sketch of data on two variables is called a **scatter plot.** In addition to the scatter plot, Figure 10–1 shows the straight line believed to best show how the general trend of increasing sales corresponds, in this example, to increasing advertising expenditures. This chapter will teach you how to find the best line to fit a data set and how to use the line once you have found it.

FIGURE 10–1   Simple Linear Regression

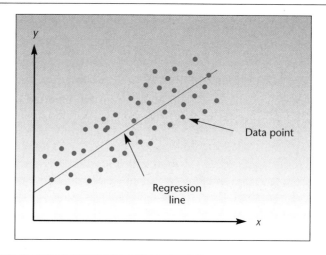

Although, in reality, our sample may consist of all available information on the two variables under study, we always assume that our data set constitutes a random sample of observations from a population of possible pairs of values of X and Y. Incidentally, in our hypothetical advertising sales example, we assume that there is no carryover effect of advertising from month to month; every month's sales depend only on that month's level of advertising. Other common examples of the use of simple linear regression in business and economics are the modeling of the relationship between job performance (the dependent variable Y) and extent of training (the independent variable X); the relationship between returns on a stock (Y) and the riskiness of the stock (X); and the relationship between company profits (Y) and the state of the economy (X).

Model Building

Like the analysis of variance, both simple linear regression and multiple regression are *statistical models*. Recall that a statistical model is a set of mathematical formulas and assumptions that describe a real-world situation. We would like our model to explain as much as possible about the process underlying our data. However, due to the uncertainty inherent in all real-world situations, our model will probably not explain everything, and we will always have some remaining errors. The errors are due to unknown outside factors that affect the process generating our data.

A good statistical model is *parsimonious,* which means that it uses as few mathematical terms as possible to describe the real situation. The model captures the systematic behavior of the data, leaving out the factors that are nonsystematic and cannot be foreseen or predicted—the errors. The idea of a good statistical model is illustrated in Figure 10–2. The errors, denoted by ϵ, constitute the random component in the model. In a sense, the statistical model breaks down the data into a nonrandom, systematic component, which can be described by a formula, and a purely random component.

How do we deal with the errors? This is where probability theory comes in. Since our model, we hope, captures everything systematic in the data, the remaining random errors are probably due to a large number of minor factors that we cannot trace.

FIGURE 10–2
A Statistical Model

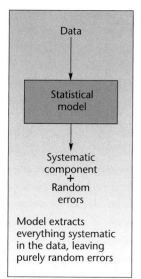

Model extracts everything systematic in the data, leaving purely random errors

FIGURE 10–3   **Steps in Building a Statistical Model**

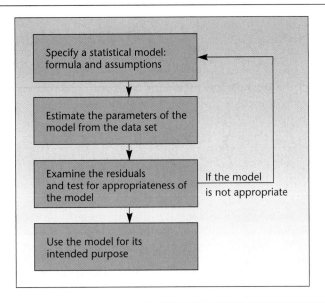

We assume that the random errors ε are *normally distributed*. If we have a properly constructed model, the resulting observed errors will have an average of zero (although few, if any, will actually equal zero), and they should also be *independent* of one another. We note that the assumption of a normal distribution of the errors is not absolutely necessary in the regression model. The assumption is made so that we can carry out statistical hypothesis tests using the F and t distributions. The only necessary assumption is that the errors ε have mean zero and a constant variance σ^2 and that they be uncorrelated with one another. In the next section, we describe the simple linear regression model. We now present a general model-building methodology.

First, we propose a particular model to describe a given situation. For example, we may propose a simple linear regression model for describing the relationship between two variables. Then we estimate the model parameters from the random sample of data we have. The next step is to consider the observed errors resulting from the fit of the model to the data. These observed errors, called **residuals,** represent the information in the data not explained by the model. For example, in the ANOVA model discussed in Chapter 9, the within-group variation (leading to SSE and MSE) is due to the residuals. If the residuals are found to contain some nonrandom, *systematic* component, we reevaluate our proposed model and, if possible, adjust it to incorporate the systematic component found in the residuals; or we may have to discard the model and try another. When we believe that model residuals contain nothing more than pure randomness, we use the model for its intended purpose: *prediction* of a variable, *control* of a variable, or the *explanation* of the relationships among variables.

In the advertising sales example, once the regression model has been estimated and found to be appropriate, the firm may be able to use the model for predicting sales for a given level of advertising within the range of values studied. Using the model, the firm may be able to control its sales by setting the level of advertising expenditure. The model may help explain the effect of advertising on sales within the range of values studied. Figure 10–3 shows the usual steps of building a statistical model.

10–2   The Simple Linear Regression Model

Recall from algebra that the equation of a straight line is $Y = A + BX$, where A is the Y intercept and B is the slope of the line. In simple linear regression, we model the relationship between two variables X and Y as a straight line. Therefore, our model must contain two parameters: an intercept parameter and a slope parameter. The usual notation for the **population intercept** is β_0, and the notation for the **population slope** is β_1. If we include the error term ϵ, the population regression model is given in equation 10–1.

The population simple linear regression model is

$$Y = \beta_0 + \beta_1 X + \epsilon \qquad (10\text{–}1)$$

where Y is the dependent variable, the variable we wish to explain or predict; X is the independent variable, also called the *predictor* variable; and ϵ is the error term, the only random component in the model and thus the only source of randomness in Y.

The model parameters are as follows:

β_0 is the Y intercept of the straight line given by $Y = \beta_0 + \beta_1 X$ (the line does not contain the error term).

β_1 is the slope of the line $Y = \beta_0 + \beta_1 X$.

FIGURE 10–4
Simple Linear Regression Model

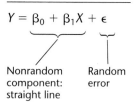

Nonrandom component: straight line / Random error

The simple linear regression model of equation 10–1 is composed of two components: a nonrandom component, which is the line itself, and a purely random component—the error term ϵ. This is shown in Figure 10–4. The nonrandom part of the model, the straight line, is the equation for the *mean of Y, given X*. We denote the conditional mean of Y, given X, by $E(Y \mid X)$. Thus, if the model is correct, the *average* value of Y for a given value of X falls right *on* the regression line. The equation for the mean of Y, given X, is given as equation 10–2.

The conditional mean of Y is

$$E(Y \mid X) = \beta_0 + \beta_1 X \qquad (10\text{–}2)$$

Comparing equations 10–1 and 10–2, we see that our model says that each value of Y comprises the average Y for the given value of X (this is the straight line), plus a random error. We will sometimes use the simplified notation $E(Y)$ for the line, remembering that this is the *conditional* mean of Y for a given value of X. As X increases, the average population value of Y also increases, assuming a positive slope of the line (or decreases, if the slope is negative). The *actual* population value of Y is equal to the average Y conditional on X, plus a random error ϵ. We thus have, for a given value of X,

$$Y = \text{Average } Y \text{ for given } X + \text{Error}$$

Figure 10–5 shows the population regression model.

FIGURE 10–5   Population Regression Line

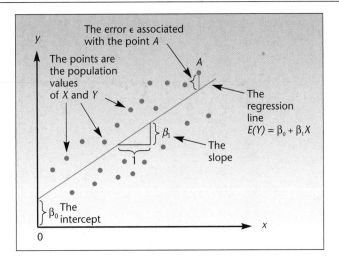

We now state the assumptions of the simple linear regression model.

Model assumptions:

1. The relationship between X and Y is a straight-line relationship.
2. The values of the independent variable X are assumed fixed (not random); the only randomness in the values of Y comes from the error term ϵ.
3. The errors ϵ are normally distributed with mean 0 and a constant variance σ^2. The errors are uncorrelated (not related) with one another in successive observations.[1] In symbols:

$$\epsilon \sim N(0, \sigma^2) \tag{10-3}$$

**FIGURE 10–6
Distributional
Assumptions of the
Linear Regression Model**

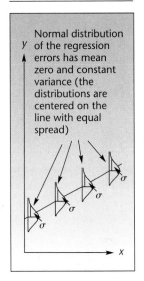

Normal distribution of the regression errors has mean zero and constant variance (the distributions are centered on the line with equal spread)

Figure 10–6 shows the distributional assumptions of the errors of the simple linear regression model. The population regression errors are normally distributed about the population regression line, with mean zero and equal variance. (The errors are equally spread about the regression line; the error variance does not increase or decrease as X increases.)

The simple linear regression model applies only if the true relationship between the two variables X and Y is a straight-line relationship. If the relationship is curved (*curvilinear*), then we need to use the more involved methods of the next chapter. In

---

[1]The idea of statistical *correlation* will be discussed in detail in Section 10–5. In the case of the regression errors, we assume that successive errors $\epsilon_1, \epsilon_2, \epsilon_3, \ldots$ are uncorrelated: they are not related with one another; there is no trend, no joint movement in successive errors. Incidentally, the assumption of zero correlation together with the assumption of a normal distribution of the errors implies the assumption that the errors are independent of one another. Independence implies noncorrelation, but noncorrelation does not imply independence, except in the case of a normal distribution (this is a technical point).

FIGURE 10–7   **Some Possible Relationships between** X **and** Y

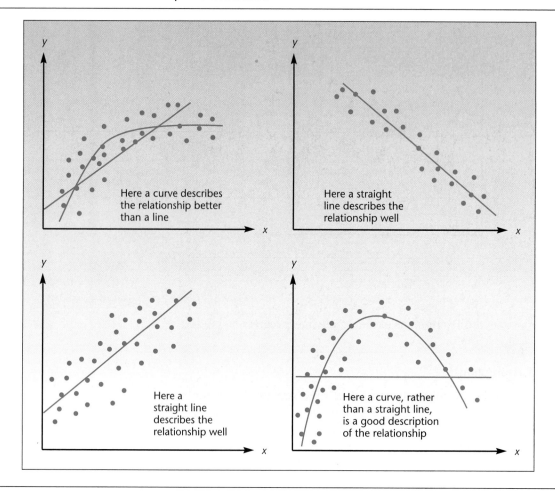

Figure 10–7, we show various relationships between two variables. Some are straight-line relationships that can be modeled by simple linear regression, and others are not.

So far, we have described the population model, that is, the assumed true relationship between the two variables X and Y. Our interest is focused on this unknown population relationship, and we want to *estimate* it, using sample information. We obtain a random sample of observations on the two variables, and we estimate the regression model parameters β_0 and β_1 from this sample. This is done by the *method of least squares*, which is discussed in the next section.

PROBLEMS

10–1.   What is a statistical model?

10–2.   What are the steps of statistical model building?

10–3.   What are the assumptions of the simple linear regression model?

10–4.   Define the parameters of the simple linear regression model.

10–5.   What is the conditional mean of Y, given X?

10–6.   What are the uses of a regression model?

10–7.   What are the purpose and meaning of the error term in regression?

10–8.   Give examples of business situations where you believe a straight-line relationship exists between two variables. What would be the uses of a regression model in each of these situations?

10–3   Estimation: The Method of Least Squares

We want to find good estimates of the regression parameters β_0 and β_1. Remember the properties of good estimators, discussed in Chapter 5. Unbiasedness and efficiency are among these properties. A method that will give us good estimates of the regression coefficients is the **method of least squares.** The method of least squares gives us the *best linear unbiased estimators* (BLUE) of the regression parameters β_0 and β_1. These estimators both are unbiased and have the lowest variance of all possible unbiased estimators of the regression parameters. These properties of the least-squares estimators are specified by a well-known theorem, the *Gauss-Markov theorem*. We denote the least-squares estimators by b_0 and b_1.

The least-squares estimators are

$$b_0 \xrightarrow{\text{estimates}} \beta_0$$

$$b_1 \xrightarrow{\text{estimates}} \beta_1$$

The estimated regression equation is

$$Y = b_0 + b_1 X + e \tag{10–4}$$

where b_0 estimates β_0, b_1 estimates β_1, and e stands for the observed errors—the residuals from fitting the line $b_0 + b_1 X$ to the data set of n points.

In terms of the data, equation 10–4 can be written with the subscript i to signify each particular data point:

$$y_i = b_0 + b_1 x_i + e_i \tag{10–5}$$

where $i = 1, 2, \ldots, n$. Then e_1 is the first residual, the distance from the first data point to the fitted regression line; e_2 is the distance from the second data point to the line; and so on to e_n, the nth error. The errors e_i are viewed as estimates of the true population errors ϵ_i. The equation of the regression line itself is as follows:

The regression line is

$$\hat{Y} = b_0 + b_1 X \tag{10–6}$$

where $\hat{Y}$ (pronounced "Y hat") is the Y value *lying on the fitted regression line* for a given X.

FIGURE 10–8   A Data Set of *X* and *Y* Pairs, and Different Proposed Straight Lines to Describe the Data

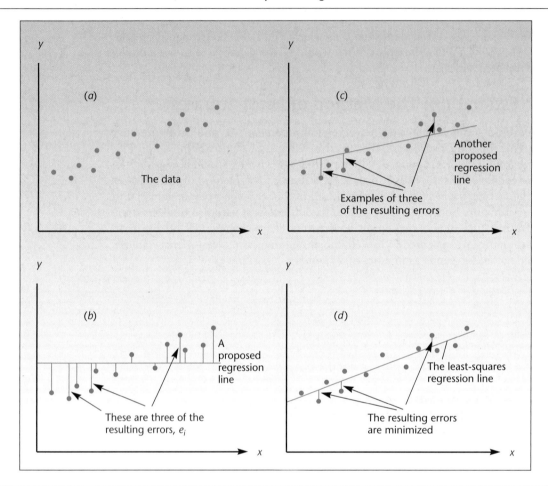

Thus, $\hat{y}_1$ is the fitted value corresponding to x_1, that is, the value of y_1 without the error e_1, and so on for all $i = 1, 2, \ldots, n$. The fitted value Y is also called the *predicted value of $\hat{Y}$* because if we do not know the actual value of Y, it is the value we would predict for a given value of X, using the estimated regression line.

Having defined the estimated regression equation, the errors, and the fitted values of Y, we will now demonstrate the principle of least squares, which gives us the BLUE regression parameters. Consider the data set shown in Figure 10–8(a). In parts (b), (c), and (d) of the figure, we show different lines passing through the data set and the resulting errors e_i.

As can be seen from Figure 10–8, the regression line proposed in part (b) results in very large errors. The errors corresponding to the line of part (c) are smaller than the ones of part (b), but the errors resulting from using the line proposed in part (d) are by far the smallest. The line in part (d) seems to move with the data and *minimize* the resulting errors. This should convince you that the line that best describes the trend in the data is the line that lies "inside" the set of points; since some of the points lie above the fitted line and others below the line, some errors will be positive and others will be negative. If we want to minimize all the errors (both positive and negative ones), we

FIGURE 10–9   Regression Errors Leading to SSE

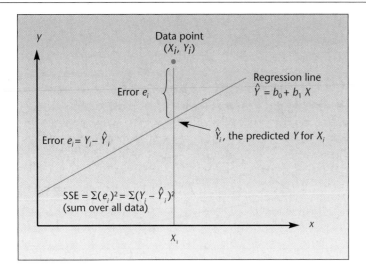

should minimize the *sum of the squared errors* (SSE, as in ANOVA). Thus, we want to find the *least-squares* line—the line that minimizes SSE. We note that least squares is not the only method of fitting lines to data. There are other methods, such as minimizing the sum of the absolute errors. The method of least squares, however, is the most commonly used method to estimate a regression relationship. Figure 10–9 shows how the errors lead to the calculation of SSE.

We define the sum of squares for error in regression as

$$SSE = \sum_{i=1}^{n} e_i^2 = \sum_{i=1}^{n} (y_i - \hat{y}_i)^2 \tag{10–7}$$

Figure 10–10 shows different values of SSE corresponding to values of b_0 and b_1. The least-squares line is the particular line specified by values of b_0 and b_1 that minimize SSE, as shown in the figure.

Calculus is used in finding the expressions for b_0 and b_1 that minimize SSE. These expressions are called the *normal equations* and are given as equations 10–8.[2] This system of two equations with two unknowns is solved to give us the values of b_0 and b_1 that minimize SSE. The results are the least-squares estimators b_0 and b_1 of the simple linear regression parameters β_0 and β_1.

The **normal equations** are

$$\sum_{i=1}^{n} y_i = nb_0 + b_1 \sum_{i=1}^{n} x_i \tag{10–8}$$

$$\sum_{i=1}^{n} x_i y_i = b_0 \sum_{i=1}^{n} x_i + b_1 \sum_{i=1}^{n} x_i^2$$

---

[2]We leave it as an exercise to the reader with background in calculus to derive the normal equations by taking the partial derivatives of SSE with respect to b_0 and b_1 and setting them to zero.

FIGURE 10–10   The Particular Values b_0 and b_1 That Minimize SSE

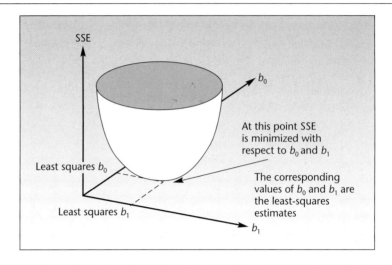

Before we present the solutions to the normal equations, we define the sums of squares SS_X and SS_Y and the sum of the cross-products SS_{XY}. These will be very useful in defining the least-squares estimates of the regression parameters, as well as in other regression formulas we will see later. The definitions are given in equations 10–9.

Definitions of sums of squares and cross-products useful in regression analysis:

$$SS_X = \sum (x - \bar{x})^2 = \sum x^2 - \frac{(\sum x)^2}{n}$$

$$SS_Y = \sum (y - \bar{y})^2 = \sum y^2 - \frac{(\sum y)^2}{n} \qquad (10\text{–}9)$$

$$SS_{XY} = \sum (x - \bar{x})(y - \bar{y}) = \sum xy - \frac{(\sum x)(\sum y)}{n}$$

The first definition in each case is the conceptual one using squared distances from the mean; the second part is a computational definition. Summations are over all data.

We now give the solutions of the normal equations, the least-squares estimators b_0 and b_1.

Least-squares regression estimators include the slope

$$b_1 = \frac{SS_{XY}}{SS_X}$$

and the intercept   (10-10)

$$b_0 = \bar{y} - b_1\bar{x}$$

The formula for the estimate of the intercept makes use of the fact that the *least-squares line always passes through the point* $(\bar{x}, \bar{y})$, the intersection of the mean of X and the mean of Y.

It is important to remember that the obtained estimates b_0 and b_1 of the regression relationship are just realizations of *estimators* of the true regression parameters β_0 and β_1. As always, our estimators have standard deviations (and variances, which, by the Gauss-Markov theorem, are as small as possible). The estimates can be used, along with the assumption of normality, in the construction of confidence intervals for, and the conducting of hypothesis tests about, the true regression parameters β_0 and β_1. This will be done in the next section.

We demonstrate the process of estimating the parameters of a simple linear regression model in Example 10-1.

American Express Company has long believed that its cardholders tend to travel more extensively than others—both on business and for pleasure. As part of a comprehensive research effort undertaken by a New York market research firm on behalf of American Express, a study was conducted to determine the relationship between travel and charges on the American Express card. The research firm selected a random sample of 25 cardholders from the American Express computer file and recorded their total charges over a specified period. For the selected cardholders, information was also obtained, through a mailed questionnaire, on the total number of miles traveled by each cardholder during the same period. The data for this study are given in Table 10-1. Figure 10-11 is a scatter plot of the data.

EXAMPLE 10-1

As can be seen from the figure, it seems likely that a straight line will describe the trend of increase in dollar amount charged with increase in number of miles traveled. The least-squares line that fits these data is shown in Figure 10-12.

Solution

We will now show how the least-squares regression line in Figure 10-12 is obtained. Table 10-2 shows the necessary computations. From equations 10-9, using sums at the bottom of Table 10-2, we get

$$SS_X = \sum x^2 - \frac{(\sum x)^2}{n} = 293{,}426{,}946 - \frac{79{,}448^2}{25} = 40{,}947{,}557.84$$

and

$$SS_{XY} = \sum xy - \frac{(\sum x)(\sum y)}{n} = 390{,}185{,}014 - \frac{(79{,}448)(106{,}605)}{25} = 51{,}402{,}852.4$$

Using equations 10-10 for the least-squares estimates of the slope and intercept parameters, we get

$$b_1 = \frac{SS_{XY}}{SS_X} = \frac{51{,}402{,}852.40}{40{,}947{,}557.84} = 1.255333776$$

and

TABLE 10–1   American Express Study Data

Miles	Dollars
1,211	1,802
1,345	2,405
1,422	2,005
1,687	2,511
1,849	2,332
2,026	2,305
2,133	3,016
2,253	3,385
2,400	3,090
2,468	3,694
2,699	3,371
2,806	3,998
3,082	3,555
3,209	4,692
3,466	4,244
3,643	5,298
3,852	4,801
4,033	5,147
4,267	5,738
4,498	6,420
4,533	6,059
4,804	6,426
5,090	6,321
5,233	7,026
5,439	6,964

FIGURE 10–11   Data for the American Express Study

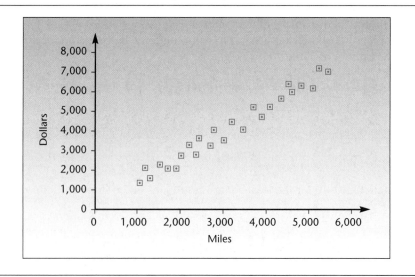

FIGURE 10–12   Least-Squares Line for the American Express Study

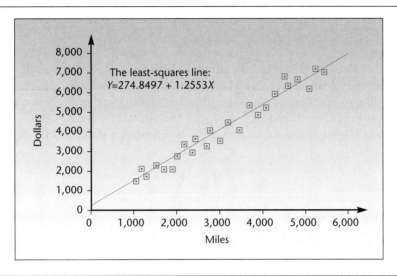

TABLE 10–2   The Computations Required for the American Express Study

Miles X	Dollars Y	X^2	Y^2	XY
1,211	1,802	1,466,521	3,247,204	2,182,222
1,345	2,405	1,809,025	5,784,025	3,234,725
1,422	2,005	2,022,084	4,020,025	2,851,110
1,687	2,511	2,845,969	6,305,121	4,236,057
1,849	2,332	3,418,801	5,438,224	4,311,868
2,026	2,305	4,104,676	5,313,025	4,669,930
2,133	3,016	4,549,689	9,096,256	6,433,128
2,253	3,385	5,076,009	11,458,225	7,626,405
2,400	3,090	5,760,000	9,548,100	7,416,000
2,468	3,694	6,091,024	13,645,636	9,116,792
2,699	3,371	7,284,601	11,363,641	9,098,329
2,806	3,998	7,873,636	15,984,004	11,218,388
3,082	3,555	9,498,724	12,638,025	10,956,510
3,209	4,692	10,297,681	22,014,864	15,056,628
3,466	4,244	12,013,156	18,011,536	14,709,704
3,643	5,298	13,271,449	28,068,804	19,300,614
3,852	4,801	14,837,904	23,049,601	18,493,452
4,033	5,147	16,265,089	26,491,609	20,757,851
4,267	5,738	18,207,289	32,924,644	24,484,046
4,498	6,420	20,232,004	41,216,400	28,877,160
4,533	6,059	20,548,089	36,711,481	27,465,447
4,804	6,426	23,078,416	41,293,476	30,870,504
5,090	6,321	25,908,100	39,955,041	32,173,890
5,233	7,026	27,384,289	49,364,676	36,767,058
5,439	6,964	29,582,721	48,497,296	37,877,196
79,448	106,605	293,426,946	521,440,939	390,185,014

$$b_0 = \bar{y} - b_1\bar{x} = \frac{106{,}605}{25} - 1.2553337776\left(\frac{79{,}448}{25}\right) = 274.8496866$$

It is important to carry out as many significant digits as you can in these computations. Here we carried out the computations by hand, for demonstration purposes. Usually, all computations are done by computer or by calculator. There are many hand calculators with a built-in routine for simple linear regression. From now on, we will present only the computed results, the least-squares estimates. The estimated least-squares relationship for Example 10–1 is reporting estimates to the second significant decimal:

$$Y = 274.85 + 1.26X + e \qquad (10\text{–}11)$$

The equation of the line itself, that is, the predicted value of Y for a given X, is

$$\hat{Y} = 274.85 + 1.26X \qquad (10\text{–}12)$$

The Template

Figure 10–13 shows the template that can be used to carry out a simple regression. The X and Y data are entered in columns B and C. The scatter plot at the bottom shows the regression equation and the regression line. Several additional statistics regarding the regression appear in the remaining parts of the template; these are explained in later sections. The error values appear in column D.

Below the scatter plot, there is a panel for residual analysis. Here you will find the Durbin-Watson statistic, the residual plot, and the normal probability plot. The Durbin-Watson statistic will be explained in the next chapter, and the normal probability plot will be explained later in this chapter. The residual plot shows that there is no relationship between X and the residuals. Figure 10–14 shows the panel.

PROBLEMS

10–9. Explain the advantages of the least-squares procedure for fitting lines to data. Explain how the procedure works.

10–10. (A conceptually advanced problem) Can you think of a possible limitation of the least-squares procedure?

10–11. The following data are from charts in an article by M. Sturani entitled "Italian Business Fears ECU Delay."[3]

	Italy's Deficit/GDP Ratio (%)	Lira/Mark e-Rate
1992	9.5	925
1993	9.4	970
1994	7.5	1050
1995	6.5	1100
1996	4.0	1020
1997	6.0	980

Is there a linear relationship? Are there any outliers? What are the limitations of the study?

---

[3] *The Wall Street Journal Europe,* June 16, 1997, p. 4.

FIGURE 10–13 The Simple Regression Template
[Simple Regression.xls; Sheet: Regression]

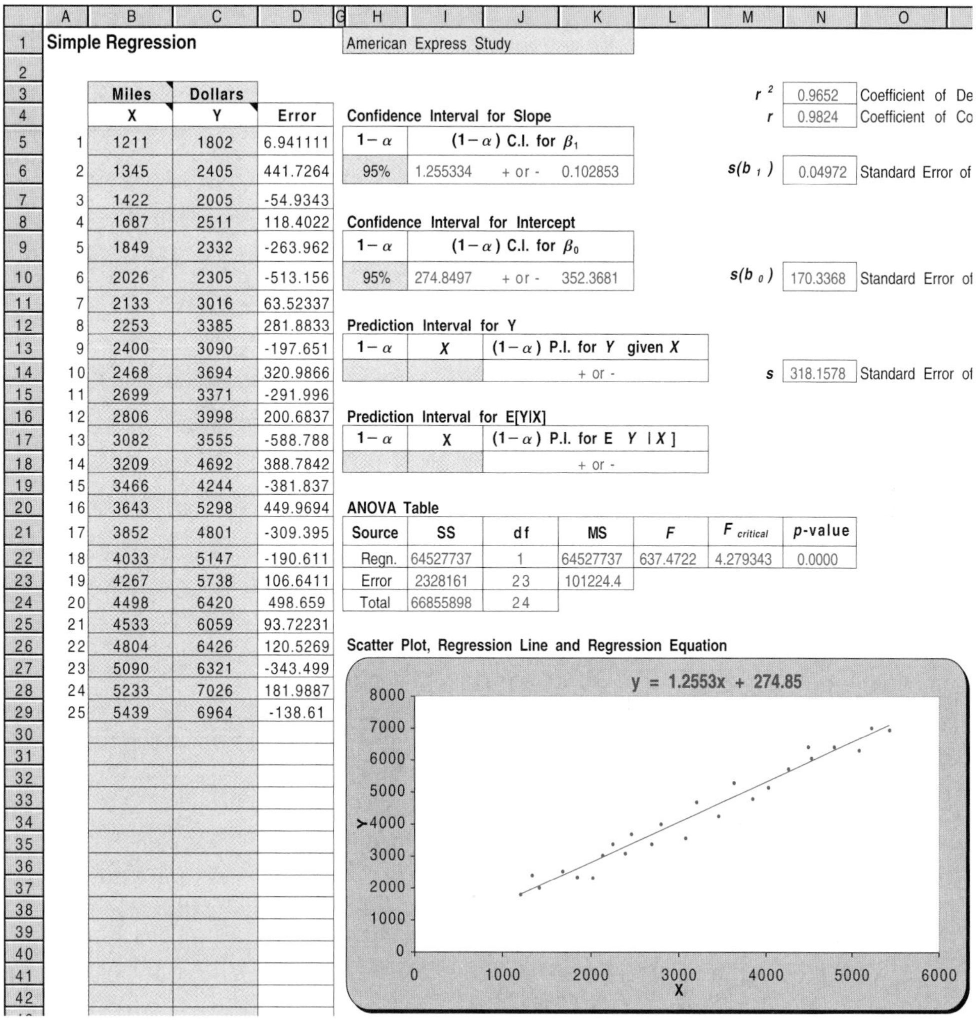

	A	B	C	D	G	H	I	J	K	L	M	N	O
1	Simple Regression					American Express Study							
2													
3		Miles	Dollars								r^2	0.9652	Coefficient of De
4		X	Y	Error		Confidence Interval for Slope					r	0.9824	Coefficient of Co
5	1	1211	1802	6.941111		$1-\alpha$		$(1-\alpha)$ C.I. for β_1					
6	2	1345	2405	441.7264		95%	1.255334	+ or -	0.102853		$s(b_1)$	0.04972	Standard Error of
7	3	1422	2005	-54.9343									
8	4	1687	2511	118.4022		Confidence Interval for Intercept							
9	5	1849	2332	-263.962		$1-\alpha$		$(1-\alpha)$ C.I. for β_0					
10	6	2026	2305	-513.156		95%	274.8497	+ or -	352.3681		$s(b_0)$	170.3368	Standard Error of
11	7	2133	3016	63.52337									
12	8	2253	3385	281.8833		Prediction Interval for Y							
13	9	2400	3090	-197.651		$1-\alpha$	X	$(1-\alpha)$ P.I. for Y given X					
14	10	2468	3694	320.9866				+ or -			s	318.1578	Standard Error of
15	11	2699	3371	-291.996									
16	12	2806	3998	200.6837		Prediction Interval for E[Y|X]							
17	13	3082	3555	-588.788		$1-\alpha$	X	$(1-\alpha)$ P.I. for E Y |X]					
18	14	3209	4692	388.7842				+ or -					
19	15	3466	4244	-381.837									
20	16	3643	5298	449.9694		ANOVA Table							
21	17	3852	4801	-309.395		Source	SS	df	MS	F	$F_{critical}$	p-value	
22	18	4033	5147	-190.611		Regn.	64527737	1	64527737	637.4722	4.279343	0.0000	
23	19	4267	5738	106.6411		Error	2328161	23	101224.4				
24	20	4498	6420	498.659		Total	66855898	24					
25	21	4533	6059	93.72231									
26	22	4804	6426	120.5269		Scatter Plot, Regression Line and Regression Equation							
27	23	5090	6321	-343.499									
28	24	5233	7026	181.9887									
29	25	5439	6964	-138.61									

Scatter plot with regression equation $y = 1.2553x + 274.85$, X axis from 0 to 6000, Y axis from 0 to 8000.

10-12. A financial analyst at Goldman Sachs ran a regression analysis of monthly returns on a certain investment (Y) versus returns for the same month on the Standard & Poor's index (X). The regression results included $SS_X = 765.98$ and $SS_{XY} = 934.49$. Give the least-squares estimate of the regression slope parameter.

10–13. Recently, research efforts have focused on the problem of predicting a manufacturer's market share by using information on the quality of its product. Suppose that the following data are available on market share, in percentage (Y), and product quality, on a scale of 0 to 100, determined by an objective evaluation procedure (X):

X:	27	39	73	66	33	43	47	55	60	68	70	75	82
Y:	2	3	10	9	4	6	5	8	7	9	10	13	12

Estimate the simple linear regression relationship between market share and product quality rating.

FIGURE 10–14   **Residual Analysis in the Template**
                [Simple Regression.xls; Sheet: Regression]

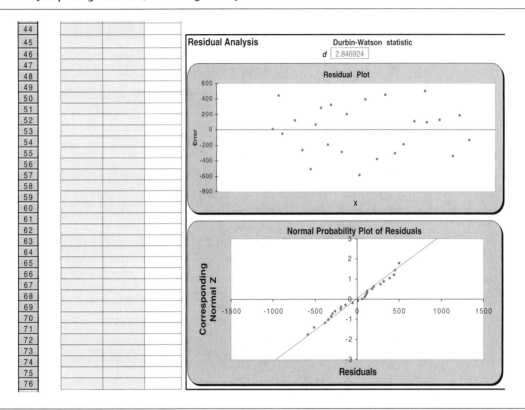

10–14.   A pharmaceutical manufacturer wants to determine the concentration of a key component of cough medicine that may be used without the drug's causing adverse side effects. As part of the analysis, a random sample of 45 patients is administered doses of varying concentration (X), and the severity of side effects (Y) is measured. The results include $\bar{x} = 88.9$, $\bar{y} = 165.3$, $SS_X = 2,133.9$, $SS_{XY} = 4,502.53$, $SS_Y = 12,500$. Find the least-squares estimates of the regression parameters.

10–15.   The following data[4] compare Standard & Poor's 500 index and the U.S. dollar versus the deutsche mark for December 1995 through June 1997. Is there a linear relationship between the two variables? Can you say that one *causes* the other?

Month	Standard & Poor's 500 Index	Dollar versus Deutsche Mark (Bank of England Index January 1992 = 100)
December	610	110
January	620	111
February	660	109
March	640	109
April	640	108

[4]From charts in *The Wall Street Journal*, June 16, 1997, p. 15. Reprinted from *The Wall Street Journal* © 1997 The Dow Jones Company.

Month	Standard & Poor's 500 Index	Dollar versus Deutsche Mark (Bank of England Index January 1992 = 100)
May	670	107
June	665	107
July	640	107
August	670	108
September	690	107
October	725	108
November	745	107
December	740	105
January	760	104
February	785	104
March	810	104
April	760	104
May	840	103
June	900	102

10–16. The following data—price/earnings ratios and growth percent per year—are from "Portfolio Talk" by L. Armour:[5]

P/E:	10, 12, 15, 17, 12, 22, 18
Percentage growth:	15, 20, 35, 25, 30, 20, 40

Run a regression of P/E on the percentage of growth, and test whether the line has a population intercept of 0. Test the rule P/E ratio = % annual growth as a fair-price rule.

10–17. For the data given below,[6] regress one variable on the other. Is there an implication of causality, or are both variables affected by a third?

Year	Off-Line Debit Card (Issued by MasterCard, Visa)	On-Line Debit Card (Issued by Bank, Requiring PIN Code to Use)
	Annual Transactions (millions)	
1991	156	211
1992	204	280
1993	279	386
1994	472	551
1995	822	684
1996	1,213	905

10–18. (A problem requiring knowledge of calculus) Derive the normal equations (10–8) by taking the partial derivatives of SSE with respect to b_0 and b_1 and setting them to zero. [*Hint:* Set $SSE = \Sigma e^2 = \Sigma(y - \hat{y})^2 = \Sigma(y - b_0 - b_1 x)^2$, and take the derivatives of the last expression on the right.]

---

[5] L. Armour, "Portfolio Talk," *Fortune,* July 7, 1997, p. 104. Fortune © 1997 Time Inc. All rights reserved.

[6] From Christine Dugas, "Managing Your Money," *USA Today,* June 2, 1997, p. 4B. © 1997 USA Today. Reprinted with permission.

FIGURE 10-15   **Two Examples of Regression Lines Showing the Error Variance**

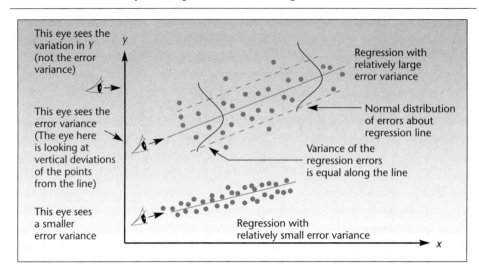

10–4   Error Variance and the Standard Errors of Regression Estimators

Recall that σ^2 is the variance of the population regression errors ϵ and that this variance is assumed to be constant for all values of X in the range under study. The error variance is an important parameter in the context of regression analysis because it is a measure of the spread of the population elements about the regression line. Generally, the smaller the error variance, the more closely the population elements follow the regression line. The error variance is the variance of the dependent variable Y as "seen" by an eye looking in the direction of the regression line (the error variance is not the variance of Y). These properties are demonstrated in Figure 10–15.

The figure shows two regression lines. The top regression line in the figure has a larger error variance than the bottom regression line. The error variance for each regression is the variation in the data points as seen by the eye located at the base of the line, looking *in the direction of the regression line.* The variance of Y, on the other hand, is the variation in the Y values regardless of the regression line. That is, the variance of Y for each of the two data sets in the figure is the variation in the data as seen by an eye looking in a direction parallel to the X axis. Note also that the spread of the data is constant along the regression lines. This is in accordance with our assumption of equal error variance for all X.

Since σ^2 is usually unknown, we need to estimate it from our data. An unbiased estimator of σ^2, denoted by S^2, is the *mean square error* (*MSE*) of the regression. As you will soon see, sums of squares and mean squares in the context of regression analysis are very similar to those of ANOVA, presented in the preceding chapter. The degrees of freedom for error in the context of simple linear regression are $n - 2$ because we have n data points, from which two parameters, β_0 and β_1, are estimated (thus, two restrictions are imposed on the n points, leaving df $= n - 2$). The sum of squares for error (SSE) in regression analysis is defined as the sum of squared deviations of the data values Y from the fitted values $\hat{Y}$. The sum of squares for error may also be defined in terms of a computational formula using SS_X, SS_Y, and SS_{XY} as defined in equations 10–9. We state these relationships in equations 10–13.

$$df(error) = n - 2$$

$$SSE = \sum (Y - \hat{Y})^2 \qquad (10\text{–}13)$$

$$= SS_Y - \frac{(SS_{XY})^2}{SS_X}$$

$$= SS_Y - b_1 SS_{XY}$$

An unbiased estimator of σ^2, denoted by S^2, is

$$MSE = \frac{SSE}{n - 2}$$

In Example 10–1, the sum of squares for error is

$$SSE = SS_Y - b_1 SS_{XY} = 66{,}855{,}898 - (1.255333776)(51{,}402{,}852.4)$$
$$= 2{,}328{,}161.2$$

and

$$MSE = \frac{SSE}{n - 2} = \frac{2{,}328{,}161.2}{23} = 101{,}224.4$$

An estimate of the standard deviation of the regression errors σ is s, which is the square root of MSE. (The estimator S is not unbiased because the square root of an unbiased estimator, such as S^2, is not itself unbiased. The bias, however, is small, and the point is a technical one.) The estimate $s = \sqrt{MSE}$ of the standard deviation of the regression errors is sometimes referred to as *standard error of estimate*. In Example 10–1 we have

$$s = \sqrt{MSE} = \sqrt{101{,}224.4} = 318.1578225$$

The computation of SSE and MSE for Example 10–1 is demonstrated in Figure 10–16.

The standard deviation of the regression errors σ and its estimate s play an important role in the process of estimation of the values of the regression parameters β_0 and β_1. This is so because σ is part of the expressions for the standard errors of both parameter estimators. The standard errors are defined next; they give us an idea of the accuracy of the least-squares estimates b_0 and b_1. *The standard error of b_1 is especially important because it is used in a test for the existence of a linear relationship between X and Y.* This will be seen in Section 10–6.

The standard error of b_0 is

$$s(b_0) = \frac{s\sqrt{\sum x^2}}{\sqrt{n SS_X}} \qquad (10\text{–}14)$$

where $s = \sqrt{MSE}$.

FIGURE 10–16   Computing SSE and MSE in the American Express Study

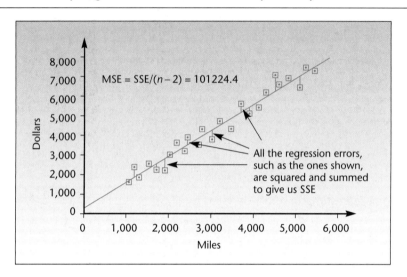

The standard error of b_1 is very important, for the reason just mentioned. The true standard deviation of b_1 is $\sigma/\sqrt{SS_X}$, but since σ is not known, we use the estimated standard deviation of the errors, s.

The standard error of b_1 is

$$s(b_1) = \frac{s}{\sqrt{SS_X}} \qquad (10\text{--}15)$$

Formulas such as equation 10–15 are nice to know, but you should not worry too much about having to use them. Regression analysis is usually done by computer, and the computer output will include the standard errors of the regression estimates. We will now show how the regression parameter estimates and their standard errors can be used in the construction of confidence intervals for the true regression parameters β_0 and β_1. In Section 10–6, as mentioned, we will use the standard error of b_1 for conducting the very important hypothesis test about the existence of a linear relationship between X and Y.

Confidence Intervals for the Regression Parameters

Confidence intervals for the true regression parameters β_0 and β_1 are easy to compute.

A $(1 - \alpha)$ 100% confidence interval for β_0 is

$$b_0 \pm t_{(\alpha/2,\, n-2)}s(b_0) \qquad\qquad (10\text{--}16)$$

where $s(b_0)$ is as given in equation 10–14.

A $(1-\alpha)$ 100% confidence interval for β_1 is

$$b_1 \pm t_{(\alpha/2,\, n-2)}s(b_1) \qquad\qquad (10\text{--}17)$$

where $s(b_1)$ is as given in equation 10–15.

Let us construct 95% confidence intervals for β_0 and β_1 in the American Express example. Using equations 10–14 to 10–17, we get

$$s(b_0) = \frac{s\sqrt{\Sigma x^2}}{\sqrt{n\,SS_X}} = 318.16\frac{\sqrt{293{,}426{,}946}}{\sqrt{(25)(40{,}947{,}557.84)}} = 170.338 \qquad (10\text{--}18a)$$

where the various quantities were computed earlier, including Σx^2, which is found at the bottom of Table 10–2.

A 95% confidence interval for β_0 is

$$b_0 \pm t_{(\alpha/2,\, n-2)}s(b_0) = 274.85 \pm 2.069(170.338) = [-77.58,\, 627.28] \quad (10\text{--}18b)$$

where the value 2.069 is obtained from Appendix C, Table 3, for $1-\alpha = 0.95$ and 23 degrees of freedom. We may be 95% confident that the true regression intercept is anywhere from -77.58 to 627.28. Again using equations 10–14 to 10–17, we get

$$s(b_1) = \frac{s}{\sqrt{SS_X}} = \frac{318.16}{\sqrt{40{,}947{,}557.84}} = 0.04972 \qquad (10\text{--}19a)$$

A 95% confidence interval for β_1 is

$$b_1 \pm t_{(\alpha/2,\, n-2)}s(b_1) = 1.25533 \pm 2.069(0.04972) \qquad (10\text{--}19b)$$
$$= [1.15246,\, 1.35820]$$

From the confidence interval given in equation 10–19b, we may be 95% confident that the *true* slope of the (*population*) regression line is anywhere from 1.15246 to 1.3582. This range of values is far from zero, and so we may be quite confident that the true regression slope is not zero. This conclusion is very important, as we will see in the following sections. Figure 10–17 demonstrates the meaning of the confidence interval given in equation 10–19b.

In the next chapter, we will discuss *joint* confidence intervals for both regression parameters β_0 and β_1, an advanced topic of secondary importance. (Since the two estimates are related, a joint interval will give us greater accuracy and a more meaningful, single confidence coefficient $1-\alpha$. This topic is somewhat similar to the Tukey analysis of Chapter 9.) Again, we want to deemphasize the importance of inference about β_0, even though information about the standard error of the estimator of this parameter is reported in computer regression output. It is the inference about β_1 that is of interest to us. Inference about β_1 has implications for the existence of a linear relationship between X and Y; inference about β_0 has no such implications. In

FIGURE 10–17 Interpretation of the Slope Estimation for Example 10–1

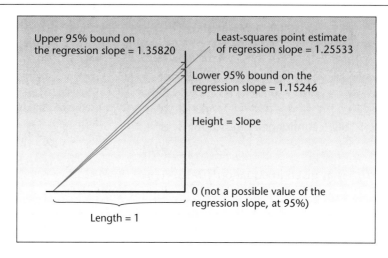

addition, there may be a temptation to use the results of the inference about β_0 to "force" this parameter to equal zero or another number. Such temptation should be resisted for reasons that will be explained in a later section; therefore, we deemphasize inference about β_0.

EXAMPLE 10–2 The data below are international sales versus U.S. sales for the McDonald's chain.

Number of McDonald's Sold at Year End (in billions)

Year	U.S. Sales	International Sales
1987	7.6	2.3
1988	7.9	2.6
1989	8.3	2.9
1990	8.6	3.2
1991	8.8	3.7
1992	9.0	4.1
1993	9.4	4.8
1994	10.2	5.7
1995	11.4	7.0
1996	12.1	8.9

Source: *The Financial Times*, July 16, 1997, p. B1. Reprinted by permission of the Financial Times Limited.

Use the template to regress McDonald's international sales, then answer the following questions:

1. What is the regression equation?
2. What is the 95% confidence interval for the slope?
3. What is the standard error of estimate?

Solution

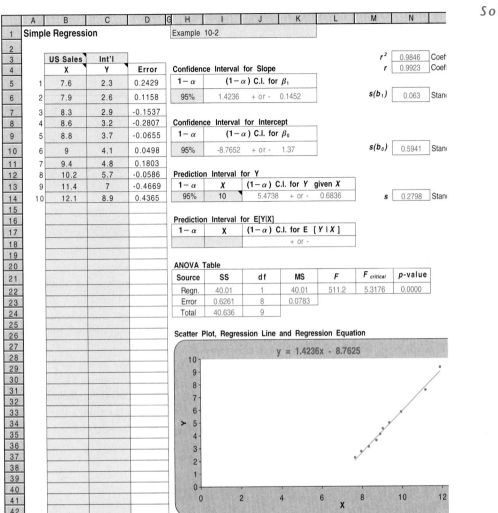

	A	B	C	D	G	H	I	J	K	L	M	N	
1	Simple Regression					Example 10-2							

r^2 | 0.9846 | Coef
r | 0.9923 | Coef

US Sales X / Int'l Y / Error:

	X	Y	Error
1	7.6	2.3	0.2429
2	7.9	2.6	0.1158
3	8.3	2.9	-0.1537
4	8.6	3.2	-0.2807
5	8.8	3.7	-0.0655
6	9	4.1	0.0498
7	9.4	4.8	0.1803
8	10.2	5.7	-0.0586
9	11.4	7	-0.4669
10	12.1	8.9	0.4365

Confidence Interval for Slope

$1-\alpha$	$(1-\alpha)$ C.I. for β_1	
95%	1.4236	+ or - 0.1452

$s(b_1)$ | 0.063 | Stand

Confidence Interval for Intercept

$1-\alpha$	$(1-\alpha)$ C.I. for β_0	
95%	-8.7652	+ or - 1.37

$s(b_0)$ | 0.5941 | Stand

Prediction Interval for Y

$1-\alpha$	X	$(1-\alpha)$ C.I. for Y given X	
95%	10	5.4738	+ or - 0.6836

s | 0.2798 | Stand

Prediction Interval for E[Y|X]

$1-\alpha$	X	$(1-\alpha)$ C.I. for E [Y|X]
		+ or -

ANOVA Table

Source	SS	df	MS	F	$F_{critical}$	p-value
Regn.	40.01	1	40.01	511.2	5.3176	0.0000
Error	0.6261	8	0.0783			
Total	40.636	9				

Scatter Plot, Regression Line and Regression Equation

y = 1.4236x - 8.7625

1. From the template, the regression equation is $\hat{Y} = 1.4326X - 8.7625$.
2. The 95% confidence interval for the slope is 1.4236 ± 0.1452.
3. The standard error of estimate is 0.2798.

PROBLEMS

10–19. Give a 99% confidence interval for the slope parameter in Example 10–1. Is zero a credible value for the true regression slope?

10–20. Give an unbiased estimate for the error variance in the situation of problem 10–11. In this problem and others, you may either use a computer or do the computations by hand.

10–21. Find the standard errors of the regression parameter estimates for problem 10–11.

10–22. Give 95% confidence intervals for the regression slope and the regression intercept parameters for the situation of problem 10–11.

10–23. For the situation of problem 10–13, find the standard errors of the estimates of the regression parameters; give an estimate of the variance of the regression errors. Also give a 95% confidence interval for the true regression slope. Is zero a plausible value for the true regression slope at the 95% level of confidence?

10–24. Repeat problem 10–23 for the situation in problem 10–17. Comment on your results.

10–25. In addition to its role in the formulas of the standard errors of the regression estimates, what is the significance of s^2?

10–5 Correlation

We now digress from regression analysis to discuss an important related concept: statistical *correlation*. Recall that one of the assumptions of the regression model is that the independent variable X is fixed rather than random and that the only randomness in the values of Y comes from the error term ϵ. Let us now relax this assumption and *assume that both* X *and* Y *are random variables*. In this new context, the study of the relationship between two variables is called *correlation analysis*.

In correlation analysis, we adopt a symmetric approach: We make no distinction between an independent variable and a dependent one. The correlation between two variables is a measure of the linear relationship between them. The correlation gives an indication of how well the two variables move together in a straight-line fashion. The correlation between X and Y is the same as the correlation between Y and X. We now define correlation more formally.

> The **correlation** between two random variables X and Y is a measure of the *degree of linear association* between the two variables.

Two variables are highly correlated if they move well together. Correlation is indicated by the **correlation coefficient.**

> The population correlation coefficient is denoted by ρ. The coefficient ρ can take on any value from -1, through 0, to 1.

The possible values of ρ and their interpretations are given below.

1. When ρ is equal to zero, there is no correlation. That is, there is no linear relationship between the two random variables.
2. When $\rho = 1$, there is a perfect, positive, linear relationship between the two variables. That is, whenever one of the variables, X or Y, increases, the other variable also increases; and whenever one of the variables decreases, the other one must also decrease.
3. When $\rho = -1$, there is a perfect negative linear relationship between X and Y. When X or Y increases, the other variable decreases; and when one decreases, the other one must increase.

4. When the value of ρ is between 0 and 1 in absolute value, it reflects the relative strength of the linear relationship between the two variables. For example, a correlation of 0.90 implies a relatively strong positive relationship between the two variables. A correlation of -0.70 implies a weaker, negative (as indicated by the minus sign), linear relationship. A correlation $\rho = 0.30$ implies a relatively weak (positive) linear relationship between X and Y.

A few sets of data on two variables, and their corresponding population correlation coefficients, are shown in Figure 10–18.

How do we arrive at the concept of correlation? Consider the pair of random variables X and Y. In correlation analysis, *we will assume that both* X *and* Y *are normally distributed random variables with means* μ_X *and* μ_Y *and standard deviations* σ_X *and* σ_Y, *respectively.* We define the *covariance* of X and Y as follows:

The **covariance** of two random variables X and Y is

$$Cov(X, Y) = E[(X - \mu_X)(Y - \mu_Y)] \qquad (10\text{--}20)$$

where μ_X is the (population) mean of X and μ_Y is the (population) mean of Y.

The covariance of X and Y is thus the expected value of the product of the deviation of X from its mean and the deviation of Y from its mean. The covariance is positive when the two random variables move together in the same direction, it is negative when the two random variables move in opposite directions, and it is zero when the two variables are not linearly related. Other than this, the covariance does not convey much. Its magnitude cannot be interpreted as an indication of the *degree* of linear association between the two variables, because the covariance's magnitude depends on the magnitudes of the standard deviations of X and Y. But if we divide the covariance by these standard deviations, we get a measure that is constrained to the range of values -1 to 1 and conveys information about the relative strength of the linear relationship between the two variables. This measure is the population correlation coefficient ρ.

The **population correlation coefficient** is

$$\rho = \frac{Cov(X, Y)}{\sigma_X \sigma_Y} \qquad (10\text{--}21)$$

Figure 10–18 gives an idea of what data from populations with different values of ρ may look like.

Like all population parameters, the value of ρ is not known to us, and we need to estimate it from our random sample of (X, Y) observation pairs. It turns out that a sample estimator of $Cov(X, Y)$ is $SS_{XY}/(n - 1)$; an estimator of σ_X is $\sqrt{SS_X/(n - 1)}$; and an estimator of σ_Y is $\sqrt{SS_Y/(n - 1)}$. Substituting these estimators for their population counterparts in equation 10–21, and noting that the term $n - 1$ cancels, we get

FIGURE 10–18   Several Possible Correlations between Two Variables

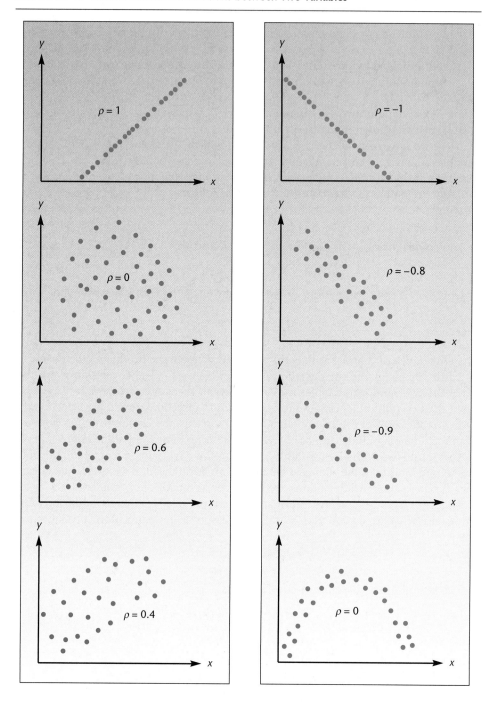

the *sample correlation coefficient,* denoted by r. This estimate of ρ, also referred to as the *Pearson product-moment correlation coefficient,* is given in equation 10–22.

The **sample correlation coefficient** is

$$r = \frac{SS_{XY}}{\sqrt{SS_X SS_Y}}$$ (10–22)

In regression analysis, the square of the sample correlation coefficient, or r^2, has a special meaning and importance. This will be seen in Section 10–7.

We often use the sample correlation coefficient for descriptive purposes as a point estimator of the population correlation coefficient ρ. When r is large and positive (closer to $+1$), we say that the two variables are highly correlated in a positive way; when r is large and negative (toward -1), we say that the two variables are highly correlated in an inverse direction, and so on. That is, we view r as if it were the parameter ρ, which r estimates. It is possible, however, to use r as an estimator in testing hypotheses about the true correlation coefficient ρ. When such hypotheses are tested, the assumption of normal distributions of the two variables is required.

The most common test is a test of whether two random variables X and Y are correlated. The hypothesis test is

---

$$H_0: \rho = 0$$
$$H_1: \rho \neq 0$$ (10–23)

The test statistic for this particular test is

$$t_{(n-2)} = \frac{r}{\sqrt{(1 - r^2)/(n - 2)}}$$ (10–24)

This test statistic may also be used for carrying out a one-tailed test for the existence of a positive only, or a negative only, correlation between X and Y. These would be one-tailed tests instead of the two-tailed test of equation 10–23, and the only difference is that the critical points for t would be the appropriate one-tailed values for a given α. The test statistic, however, is good *only* for tests where the null hypothesis assumes a zero correlation. When the true correlation between the two variables is anything but zero, the t distribution in equation 10–24 does not apply; in such cases the distribution is more complicated.[7] The test in equation 10–23 is the most common

---

[7]In cases where we want to test $H_0: \rho = a$ versus $H_1: \rho \neq a$, where a is some number other than zero, we may do so by using the Fisher transformation: $z' = (1/2) \log [(1 + r)/(1 - r)]$, where z' is approximately normally distributed with mean $\mu' = (1/2) \log [(1 + \rho)/(1 - \rho)]$ and standard deviation $\sigma' = 1/\sqrt{n - 3}$. (Here *log* is taken to mean *natural logarithm.*) Such tests are less common, and a more complete description may be found in advanced texts. As an exercise, the interested reader may try this test on some data. [You need to transform z' to an approximate standard normal $z = (z' - \mu')/\sigma'$; use the null-hypothesis value of ρ in the formula for μ'.]

hypothesis test about the population correlation coefficient because it is a test for the existence of a linear relationship between two variables. We demonstrate this test with the following example.

EXAMPLE 10–3   An article in the *Journal of Marketing Research* reports the results of a study to determine whether there is a linear relationship between the time spent in negotiating a sale and the resulting profits.[8] A random sample of 27 market transactions was collected, and the time taken to conclude the sale as well as the resulting profit were recorded for each transaction. The sample correlation coefficient was computed: $r = 0.424$. Is there a linear relationship between the length of negotiations and transaction profits?

Solution   We want to conduct the hypothesis test $H_0: \rho = 0$ versus $H_1: \rho \neq 0$. Using the test statistic in equation 10–24, we get

$$t_{(25)} = \frac{r}{\sqrt{(1 - r^2)/(n - 2)}} = \frac{0.424}{\sqrt{(1 - 0.424^2)/25}} = 2.34$$

From Appendix C, Table 3, we find that the critical points for a t distribution with 25 degrees of freedom and $\alpha = 0.05$ are ± 2.060. Therefore, we reject the null hypothesis of no correlation in favor of the alternative that the two variables are linearly related. Since the critical points for $\alpha = 0.01$ are ± 2.787, and $2.787 > 2.34$, we are unable to reject the null hypothesis of no correlation between the two variables if we want to use the 0.01 level of significance. If we wanted to test (before looking at our data) only for the existence of a positive correlation between the two variables, our test would have been $H_0: \rho \leq 0$ versus $H_1: \rho > 0$ and we would have used only the right tail of the t distribution. At $\alpha = 0.05$, the critical point of t with 25 degrees of freedom is 1.708, and at $\alpha = 0.01$ it is 2.485. The null hypothesis would, again, be rejected at the 0.05 level but not at the 0.01 level of significance.

In regression analysis, the test for the existence of a linear relationship between X and Y is a test of whether the regression slope β_1 is equal to zero. The regression slope parameter is related to the correlation coefficient (as an exercise, compare the equations of the estimates r and b_1); when two random variables are uncorrelated, the population regression slope is zero.

We end this section with a word of caution. First, the existence of a correlation between two variables does not necessarily mean that one of the variables *causes* the other one. The determination of **causality** is a difficult question that cannot be directly answered in the context of correlation analysis or regression analysis. Also, the statistical determination that two variables are correlated may not always mean that they are correlated in any direct, meaningful way. For example, if we study any two population-related variables and find that both variables increase "together," this may merely be a reflection of the general increase in population rather than any direct correlation between the two variables. We should look for outside variables that may affect both variables under study.

---

[8]L. McAlister, M. Bazerman, and P. Fader, "Power and Goal Setting in Channel Negotiations," *Journal of Marketing Research*, August 1986.

10–26. What is the main difference between correlation analysis and regression analysis?

10–27. Compute the sample correlation coefficient for the data of problem 10–11.

10–28. Compute the sample correlation coefficient for the data of problem 10–13.

10–29. Using the data in problem 10–16, conduct the hypothesis test for the existence of a linear correlation between the two variables. Use $\alpha = 0.01$.

10–30. Is it possible that a sample correlation of 0.51 between two variables will not indicate that the two variables are really correlated, while a sample correlation of 0.04 between another pair of variables will be statistically significant? Explain.

10–31. The following data are indexed prices of gold and copper over a 10-year period. Assume that the indexed values constitute a random sample from the population of possible values. Test for the existence of a linear correlation between the indexed prices of the two metals.

> Gold: 76, 62, 70, 59, 52, 53, 53, 56, 57, 56
> Copper: 80, 68, 73, 63, 65, 68, 65, 63, 65, 66

Also, state one limitation of the data set.

10–32. Follow daily stock price quotations in *The Wall Street Journal* for a pair of stocks of your choice, and compute the sample correlation coefficient. Also, test for the existence of a nonzero linear correlation in the "population" of prices of the two stocks. For your sample, use as many daily prices as you can.

10–33. Again using *The Wall Street Journal* as a source of data, determine whether there is a linear correlation between morning and afternoon price quotations in London for an ounce of gold (for the same day). Any ideas?

10–34. A study was conducted to determine whether a correlation exists between consumers' perceptions of a television commercial (measured on a special scale) and their interest in purchasing the product (measured on a scale). The results are $n = 65$, and $r = 0.37$. Is there statistical evidence of a linear correlation between the two variables?

10–35. (Optional, advanced problem) Using the Fisher transformation (described in footnote 7), carry out a two-tailed test of the hypothesis that the population correlation coefficient for the situation of problem 10–34 is $\rho = 0.22$. Use $\alpha = 0.05$.

10–6   Hypothesis Tests about the Regression Relationship

When there is no linear relationship between X and Y, the population regression slope β_1 is equal to zero. Why? The population regression slope is equal to zero in either of two situations:

1. When Y is *constant* for all values of X. For example, $Y = 457.33$ for all X. This is shown in Figure 10–19(a). If Y is constant for all values of X, the slope of Y with respect to X, parameter β_1, is identically zero; there is no linear relationship between the two variables.

FIGURE 10–19   Two Possibilities Where the Population Regression Slope Is Zero

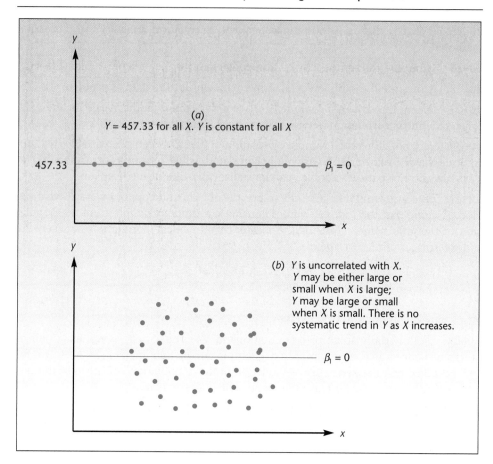

2. When the two variables are *uncorrelated*. When the correlation between X and Y is zero, as X increases Y may increase, or it may decrease, or it may remain constant. There is no *systematic* increase or decrease in the values of Y as X increases. This case is shown in Figure 10–19(b). As can be seen in the figure, data from this process are not "moving" in any pattern; thus, there is no direction for the line to follow. Since there is no direction, the slope of the line is, again, zero.

Also, remember that the relationship may be curved, with no linear correlation, as was seen in the last part of Figure 10–18. In such cases, the slope may also be zero.

In all cases other than these, there is at least *some* linear relationship between the two variables X and Y; the slope of the line in all such cases would be either positive or negative, but not zero. Therefore, *the most important statistical test in simple linear regression is the test of whether the slope parameter* β_1 *is equal to zero*. If we conclude in any particular case that the true regression slope is equal to zero, this means that there is no linear relationship between the two variables: Either the dependent variable is constant, or—more commonly—the two variables are not linearly related. We thus have the following test for determining the existence of a linear relationship between two variables X and Y:

A hypothesis test for the existence of a linear relationship between X and Y is

$$H_0: \beta_1 = 0$$
$$H_1: \beta_1 \neq 0$$

(10–25)

This test is, of course, a two-tailed test. Either the true regression slope is equal to zero, or it is not. If it is equal to zero, there is no linear relationship between the two variables; if the slope is not equal to zero, then it is either positive or negative (the two tails of rejection), in which case there is a linear relationship between the two variables. The test statistic for determining the rejection or nonrejection of the null hypothesis is given in equation 10–26. Given the assumption of normality of the regression errors, the test statistic possesses the t distribution with $n - 2$ degrees of freedom.

The test statistic for the existence of a linear relationship between X and Y is

$$t_{(n-2)} = \frac{b_1}{s(b_1)}$$

(10–26)

where b_1 is the least-squares estimate of the regression slope and $s(b_1)$ is the standard error of b_1. When the null hypothesis is true, the statistic has a t distribution with $n - 2$ degrees of freedom.

This test statistic is a special version of a general test statistic

$$t_{(n-2)} = \frac{b_1 - \beta_{10}}{s(b_1)}$$

(10–27)

where β_{10} is the value of β_1 under the null hypothesis. This statistic follows the format (Estimate − Hypothesized parameter value)/(Standard error of estimator). Since, in the test of equation 10–25, the hypothesized value of β_1 is zero, we have the simplified version of the test statistic, equation 10–26. One advantage of the simple form of our test statistic is that it allows us to conduct the test very quickly. Computer output for regression analysis usually contains a table similar to Table 10–3.

The estimate associated with X (or whatever name the user may have given to the independent variable in the computer program) is b_1. The standard error associated with X is $s(b_1)$. To conduct the test, all you need to do is to divide b_1 by $s(b_1)$. In the example of Table 10–3, $4.88/0.1 = 48.8$. The answer is reported in the table as the t ratio. The t ratio can now be compared with critical points of the t distribution with $n - 2$ degrees of freedom. Suppose that the sample size used was 100. Then the critical points for $\alpha = 0.05$, from the spreadsheet, are ± 1.98, and since $48.8 > 1.98$, we conclude that there is evidence of a linear relationship between X and Y in this hypothetical example. (Actually, the p-value is very small. Some computer programs will also

TABLE 10–3   An Example of a Part of the Computer Output for Regression

Variable	Estimate	Standard Error	t Ratio
Constant	5.22	0.5	10.44
X	4.88	0.1	48.80

FIGURE 10–20   Test for a Linear Relationship for Example 10–1

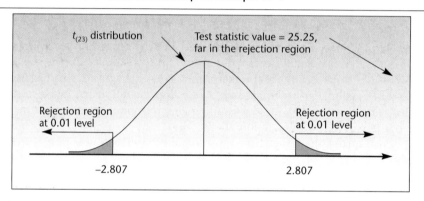

report the p-value in an extra column on the right.) What about the first row in the table? The test suggested here is a test of whether the intercept β_0 (this is the constant) is equal to zero. The test statistic is the same as equation 10–26, but with subscripts 0 instead of 1. As we mentioned earlier, this test, although suggested by the output of computer routines, is usually not meaningful and should generally be avoided.

We now conduct the hypothesis test for the existence of a linear relationship between miles traveled and amount charged on the American Express card in Example 10–1. Our hypotheses are H_0: $\beta_1 = 0$ and H_1: $\beta_1 \neq 0$. Recall that for the American Express study, $b_1 = 1.25533$ and $s(b_1) = 0.04972$ (from equations 10–11 and 10–19a). We now compute the test statistic, using equation 10–26:

$$t = \frac{b_1}{s(b_1)} = \frac{1.25533}{0.04972} = 25.25$$

From the magnitude of the computed value of the statistic, we know that there is statistical evidence of a linear relationship between the variables, because 25.25 is certainly greater than any critical point of a t distribution with 23 degrees of freedom. We show the test in Figure 10–20. The critical points of t with 23 degrees of freedom and $\alpha = 0.01$ are obtained from Appendix C, Table 3. We conclude that there is evidence of a linear relationship between the two variables "miles traveled" and "dollars charged" in Example 10–1.

Other Tests[9]

Although the test of whether the slope parameter is equal to zero is a very important test, because it is a test for the existence of a linear relationship between the two vari-

---

[9]This subsection may be skipped without loss of continuity.

ables, there are other possible tests in the context of regression. These tests serve secondary purposes. In financial analysis, for example, it is often important to determine from past performance data of a particular stock whether the stock generally moves with the market as a whole. If the stock does move with the stock market as a whole, the slope parameter of the regression of the stock's returns (Y) versus returns on the market as a whole (X) would be equal to 1.00. That is, $\beta_1 = 1$. We demonstrate this test with Example 10–4.

EXAMPLE 10–4

The *Market Sensitivity Report,* issued by Merrill Lynch, Inc., lists estimated beta coefficients of common stocks as well as their standard errors. *Beta* is the term used in the finance literature for the estimate b_1 of the regression of returns on a stock versus returns on the stock market as a whole. Returns on the stock market as a whole are taken by Merrill Lynch as returns on the Standard & Poor's 500 index. The November 1995 issue of the report lists the following findings for common stock of Time, Inc.: beta = 1.24, standard error of beta = 0.21, $n = 60$. Is there statistical evidence to reject the claim that the Time stock moves, in general, with the market as a whole?

Solution

We want to carry out the special-purpose test $H_0: \beta_1 = 1$ versus $H_1: \beta_1 \neq 1$. We use the general test statistic of equation 10–27:

$$t_{(n-2)} = \frac{b_1 - \beta_{10}}{s(b_1)} = \frac{1.24 - 1}{0.21} = 1.14$$

Since $n - 2 = 58$, we use the standard normal distribution. The test statistic value is in the nonrejection region for any usual level α, and we conclude that there is no statistical evidence against the claim that Time moves with the market as a whole.

PROBLEMS

10–36.   A regression analysis of fuel efficiency (X) versus sales (Y) of different types of corporate aircraft includes the following results: $b_1 = 2.435$, $s(b_1) = 1.567$, and $n = 12$. Do you believe there is a linear relationship between sales of corporate aircraft and the aircraft's fuel efficiency?

10–37.   An article[10] in the *Financial Analysts Journal* reports the results of a regression analysis of returns on stocks (Y) versus the ratio of book to market value (X). The resulting prediction equation is

$$\text{Return} = 1.21 + 3.1 \text{ BMV}$$
$$(2.89)$$

where the number in parentheses is the standard error of the slope estimate. The sample size used is $n = 18$. Is there evidence of a linear relationship between returns and book to market value?

10–38.   In the situation of problem 10–11, test for the existence of a linear relationship between the two variables.

---

[10]R. Harris and F. Marston, "Value versus Growth Stocks: Book-to-Market, Growth, and Beta," *Financial Analysts Journal,* September–October 1994, pp. 18–24.

10–39.   In the situation of problem 10–13, test for the existence of a linear relationship between the two variables.

10–40.   In the situation of problem 10–16, test for the existence of a linear relationship between the two variables.

10–41.   For Example 10–4, test for the existence of a linear relationship between returns on the stock and returns on the market as a whole.

10–42.   An advertising research firm conducts a study to determine the relationship between the length of a commercial and the resulting product recall scores of people viewing the commercial. Results of the analysis include $b_1 = 3.467$, $s(b_1) = 0.775$, and $n = 23$. Is there evidence of a linear relationship between commercial length and viewer recall score?

10–43.   An article in *Financial Analysts Journal* discusses results of a regression analysis of average price per share P on the independent variable X/k, where X/k is the contemporaneous earnings per share divided by firm-specific discount rate. The regression was run using a random sample of 213 firms listed in the *Value Line Investment Survey*. The reported results are

$$P = 16.67 + 0.68X/k(12.03)$$

where the number in parentheses is the standard error. Is there a linear relationship between the two variables?

10–44.   A management recruiter wants to estimate a linear regression relationship between an executive's experience and the salary the executive may expect to earn after placement with an employer. From data on 28 executives, which are assumed to be a random sample from the population of executives that the recruiter places, the following regression results are obtained: $b_1 = 5.49$, and $s(b_1) = 1.21$. Is there a linear relationship between the experience and the salary of executives placed by the recruiter?

10–7   How Good Is the Regression?

Once we have determined that a linear relationship exists between the two variables, the question is: How strong is the relationship? If the relationship is a strong one, prediction of the dependent variable can be relatively accurate, and other conclusions drawn from the analysis may be given a high degree of confidence.

We have already seen one measure of the regression fit: the mean square error. The MSE is an estimate of the variance of the true regression errors and is a measure of the variation of the data about the regression line. The MSE, however, depends on the nature of the data, and what may be a large error variation in one situation may not be considered large in another. What we need, therefore, is a *relative* measure of the degree of variation of the data about the regression line. Such a measure allows us to compare the fits of different models.

The relative measure we are looking for is a measure that compares the variation of Y about the regression line with the variation of Y without a regression line. This should remind you of analysis of variance, and we will soon see the relation of ANOVA to regression analysis. It turns out that the relative measure of regression fit we are looking for is the square of the estimated correlation coefficient r. It is called the *coefficient of determination*.

FIGURE 10–21   The Three Deviations Associated with a Data Point

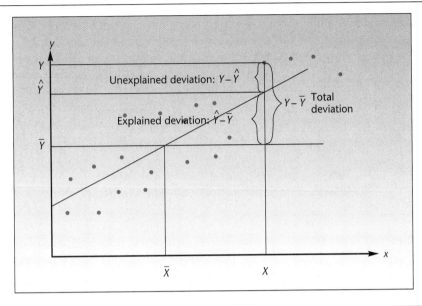

The **coefficient of determination** r^2 is a descriptive measure of the strength of the regression relationship, a measure of how well the regression line fits the data.

The coefficient of determination r^2 is an estimator of the corresponding population parameter ρ^2, which is the square of the population coefficient of correlation between two variables X and Y. Usually, however, we use r^2 as a descriptive statistic—a relative measure of how well the regression line fits the data. Ordinarily, we do not use r^2 for inference about ρ^2.

We will now see how the coefficient of determination is obtained directly from a decomposition of the variation in Y into a component due to error and a component due to the regression. Figure 10–21 shows the least-squares line that was fit to a data set. One of the data points (x, y) is highlighted. For this data point, the figure shows three kinds of deviations: the deviation of y from its mean $y - \bar{y}$, the deviation of y from its predicted value using the regression $y - \hat{y}$, and the deviation of the regression-predicted value of y from the mean of y, which is $\hat{y} - \bar{y}$. Note that the least-squares line passes through the point $(\bar{x}, \bar{y})$.

We will now follow exactly the same mathematical derivation we used in Chapter 9 when we derived the ANOVA relationships. There we looked at the deviation of a data point from its respective group mean—the error; here the error is the deviation of a data point from its regression-predicted value. In ANOVA, we also looked at the to-tal deviation, the deviation of a data point from the grand mean; here we have the de-viation of the data point from the mean of Y. Finally, in ANOVA we also considered the treatment deviation, the deviation of the group mean from the grand mean; here we have the *regression deviation*—the deviation of the predicted value from the mean of Y.

The error is also called the *unexplained deviation* because it is a deviation that cannot be explained by the regression relationship; the regression deviation is also called the *explained deviation* because it is that part of the deviation of a data point from the mean that can be explained by the regression relationship between X and Y. We *explain* why the Y value of a particular data point is above the mean of Y by the fact that its X component happens to be above the mean of X and by the fact that X and Y are linearly (and positively) related. As can be seen from Figure 10–21, and by simple arithmetic, we have

$$y - \bar{y} \;=\; y - \hat{y} \;+\; \hat{y} - \bar{y} \qquad (10\text{–}28)$$

$$\underset{\text{deviation}}{\text{Total}} = \underset{\text{deviation (error)}}{\text{Unexplained}} + \underset{\text{deviation (regression)}}{\text{Explained}}$$

As in the analysis of variance, we square all three deviations for each one of our data points, and we sum over all n points. Here, again, cross terms drop out, and we are left with the following important relationship for the sums of squares:[11]

$$\sum_{i=1}^{n} (y_i - \bar{y})^2 = \sum_{i=1}^{n} (y_i - \hat{y}_i)^2 \;+\; \sum_{i=1}^{n} (\hat{y}_i - \bar{y})^2$$

$$\text{SST} \;=\; \text{SSE} \;+\; \text{SSR} \qquad (10\text{–}29)$$

$$\underset{\text{of squares)}}{\text{(Total sum}} = \underset{\text{squares for error)}}{\text{(Sum of}} + \underset{\text{squares for regression)}}{\text{(Sum of}}$$

The term SSR is also called the *explained variation;* it is the part of the variation in Y that is explained by the relationship of Y with the explanatory variable X. Similarly, SSE is the *unexplained variation,* due to error; the sum of the two is the *total variation* in Y.

We define the coefficient of determination as the sum of squares due to the regression divided by the total sum of squares. Since by equation 10–29 SSE and SSR add to SST, the coefficient of determination is equal to 1 minus SSE/SST. We have

$$r^2 = \frac{\text{SSR}}{\text{SST}} = 1 - \frac{\text{SSE}}{\text{SST}} \qquad (10\text{–}30)$$

The coefficient of determination can be interpreted as *the proportion of the variation in* Y *that is explained by the regression relationship of* Y *with* X.

Recall that the correlation coefficient r can be between -1 and 1. Its square, r^2, can therefore be anywhere from 0 to 1. This is in accordance with the interpretation of r^2 as the *percentage of the variation in* Y *explained by the regression*. The coefficient is a measure of how closely the regression line fits the data; it is a measure of how much the variation in the values of Y is reduced once we regress Y on variable X. When

---

[11]The proof of the relation is left as an exercise for the mathematically interested reader.

$r^2 = 1$, we know that 100% of the variation in Y is explained by X. This means that the data all lie right on the regression line, and there are no resulting errors (because, from equation 10–30, SSE must be equal to zero). Since r^2 cannot be negative, we do not know whether the line slopes upward or downward (the direction can be found from b_1 or r), but we know that the line gives a *perfect fit* to the data. Such cases do not occur in business or economics. In fact, when there are no errors, no natural variation, there is no need for statistics.

At the other extreme is the case where the regression line explains nothing. Here the errors account for everything, and SSR is zero. In this case, we see from equation 10–30 that $r^2 = 0$. In such cases, there is no linear relationship between X and Y, and the true regression slope is probably zero (we say *probably* because r^2 is only an estimator, given to chance variation; it could possibly be estimating a nonzero ρ^2). Between the two cases $r^2 = 0$ and $r^2 = 1$ are values of r^2 that give an indication of the *relative fit* of the regression model to the data. *The higher r^2 is, the better the fit and the higher our confidence in the regression.* Be wary, however, of situations where the reported r^2 is exceptionally high, such as 0.99 or 0.999. In such cases, something may be wrong. We will see an example of this in the next chapter. Incidentally, in the context of multiple regression, discussed in the next chapter, we will use the notation R^2 for the coefficient of determination to indicate that the relationship is based on several explanatory X variables.

How high should the coefficient of determination be before we can conclude that a regression model fits the data well enough to use the regression with confidence? There is no clear-cut answer to this question. The answer depends on the intended use of the regression model. If we intend to use the regression for *prediction,* the higher the r^2, the more accurate will be our predictions.

An r^2 value of 0.9 or more is very good, a value greater than 0.8 is good, and a value of 0.6 or more may be satisfactory in some applications, although we must be aware of the fact that, in such cases, errors in prediction may be relatively high. When the r^2 value is 0.5 or less, the regression explains only 50% or less of the variation in the data; therefore, predictions may be poor. If we are interested only in understanding the relationship between the variables, lower values of r^2 may be acceptable, as long as we realize that the model does not explain much.

Figure 10–22 shows several regressions and their corresponding r^2 values. If you think of the total sum of squared deviations as being in a box, then r^2 is the proportion of the box that is filled with the explained sum of squares, the remaining part being the squared errors. This is shown for each regression in the figure.

It is easy to compute r^2 if we express SSR, SSE, and SST in terms of the computational sums of squares and cross products (equations 10–9):

$$\text{SST} = \text{SS}_Y \qquad \text{SSR} = b_1\text{SS}_{XY} \qquad \text{SSE} = \text{SS}_Y - b_1\text{SS}_{XY} \qquad (10\text{–}31)$$

We will now use equation 10–31 in computing the coefficient of determination for Example 10–1. For this example, we have

$$\text{SST} = \text{SS}_Y = 66{,}855{,}898$$
$$\text{SSR} = b_1\text{SS}_{XY} = (1.255333776)(51{,}402{,}852.4) = 64{,}527{,}736.8$$

and

$$\text{SSE} = \text{SST} - \text{SSR} = 2{,}328{,}161.2$$

(These were computed when we found the MSE for this example.) We now compute r^2 as

FIGURE 10–22   Value of the Coefficient of Determination in Different Regressions

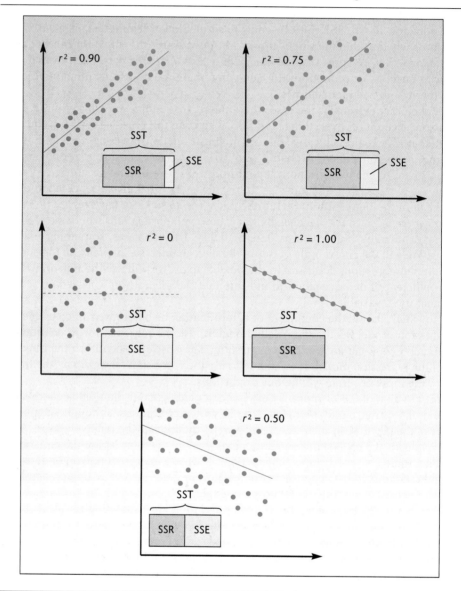

$$r^2 = \frac{\text{SSR}}{\text{SST}} = \frac{64,527,736.8}{66,855,898} = 0.96518$$

The r^2 in this example is very high. The interpretation is that over 96.5% of the variation in charges on the American Express card can be explained by the relationship between charges on the card and extent of travel (miles). Again we note that while the computational formulas are easy to use, r^2 is always reported in a prominent place in regression computer output.

In the next section, we will see how the sums of squares, along with the corresponding degrees of freedom, lead to mean squares—and to an analysis of variance in the context of regression. In closing this section, we note that in Chapter 11, we will introduce an adjusted coefficient of determination that accounts for degrees of freedom.

10–45. In the regression analysis of problem 10-37, the reported coefficient of determination is $r^2 = 0.067$. Does this surprise you? Explain.

10–46. Results of a study reported in *Financial Analysts Journal* include a simple linear regression analysis of firms' pension funding (Y) versus profitability (X). The regression coefficient of determination is reported to be $r^2 = 0.02$. (The sample size used is 515.)

   a. Would you use the regression model to predict a firm's pension funding?

   b. Does the model explain much of the variation in firms' pension funding on the basis of profitability?

   c. Do you believe these regression results are worth reporting? Explain.

10–47. What percentage of the variation in the lira/mark exchange rate is explained by the regression in problem 10–11?

10–48. What is r^2 in the regression of problem 10–13? Interpret its meaning.

10–49. What is r^2 in the regression of problem 10–16?

10–50. What is r^2 for the regression in problem 10–17? Explain its meaning.

10–51. Mita, the manufacturer of copiers, has been spending increasing amounts of money on radio and television advertising in recent years. An analyst employed by Mita wanted to estimate a simple linear regression of the company's annual copier sales versus advertising dollars. The regression results included SSE = 12,745 and SSR = 87,691. What is the coefficient of determination for this regression? Interpret its meaning. Do you believe this regression model would prove a useful tool for predicting sales based on advertising expenditure? Explain.

10–52. An article in *Journal of Finance* reports the results of simple linear regressions of New York Stock Exchange share price versus after-hours price movements for various firms. Following are the regression r^2 values for various firms:[12] American Express 0.068, Du Pont 0.091, Kodak 0.049, IBM 0.188. Comment on the explanatory powers of the linear regression models for these firms.

10–53. Find r^2 for the regression in problem 10–15.

10–54. (A mathematically demanding problem) Starting with equation 10–28, derive equation 10–29.

10–55. Using equation 10–31 for SSR, show that $SSR = (SS_{XY})^2/SS_X$.

10–8   Analysis-of-Variance Table and an *F* Test of the Regression Model

We know from our discussion of the *t* test for the existence of a linear relationship that the degrees of freedom for *error* in simple linear regression are $n - 2$. For the *regression*, we have 1 degree of freedom because there is one independent *X* variable in the

[12]D. Neumark, P. A. Tinsley, and S. Tosini, "After-Hours Stock Prices and Post-Crash Hangovers," *Journal of Finance,* March 1991, p. 165.

TABLE 10–4   ANOVA Table for Regression

Source of Variation	Sum of Squares	Degrees of Freedom	Mean Square	F Ratio
Regression	SSR	1	$MSR = \dfrac{SSR}{1}$	$F_{(1, n-2)} = \dfrac{MSR}{MSE}$
Error	SSE	$n - 2$	$MSE = \dfrac{SSE}{n - 2}$	
Total	SST	$n - 1$		

TABLE 10–5   ANOVA Table for American Express Example

Source of Variation	Sum of Squares	Degrees of Freedom	Mean Square	F Ratio	p
Regression	64,527,736.8	1	64,527,736.8	637.47	0.000
Error	2,328,161.2	23	101,224.4		
Total	66,855,898.0	24			

regression. The *total* degrees of freedom are $n - 1$ because here we only consider the mean of Y, to which 1 degree of freedom is lost. These are similar to the degrees of freedom for ANOVA in the last chapter. Mean squares are obtained, as usual, by dividing the sums of squares by their corresponding degrees of freedom. This gives us the mean square regression (MSR) and mean square error (MSE), which we encountered earlier. Further dividing MSR by MSE gives us an F ratio with degrees of freedom 1 and $n - 2$. All these can be put in an ANOVA table for regression. This has been done in Table 10–4.

In regression, there are three sources of variation (see Figure 10–21): *regression—*the explained variation; *error—*the unexplained variation; and their sum, the *total* variation. We know how to obtain the sums of squares and the degrees of freedom, and from them the mean squares. Dividing the mean square regression by the mean square error should give us another measure of the accuracy of our regression because MSR is the average squared explained deviation and MSE is the average squared error (where averaging is done using the appropriate degrees of freedom). The ratio of the two has an F distribution with 1 and $n - 2$ degrees of freedom *when there is no regression relationship between* X *and* Y. This suggests an F test for the existence of a linear relationship between X and Y. *In simple linear regression, this test is equivalent to the* t *test.* In multiple regression, as we will see in the next chapter, the F test serves a general role, and separate t tests are used to evaluate the significance of different variables. In simple linear regression, we may conduct either an F test or a t test; the results of the two tests will be the same. The hypothesis test is as given in equation 10–25; the test is carried on the right tail of the F distribution with 1 and $n - 2$ degrees of freedom. We illustrate the analysis with data from Example 10–1. The ANOVA results are given in Table 10–5.

To carry out the test for the existence of a linear relationship between miles traveled and dollars charged on the card, we compare the computed F ratio of 637.47 with a critical point of the F distribution with 1 degree of freedom for the numerator and

23 degrees of freedom for the denominator. Using $\alpha = 0.01$, the critical point from Appendix C, Table 5, is found to be 7.88. Clearly, the computed value is far in the rejection region, and the p-value is very small. We conclude, again, that there is evidence of a linear relationship between the two variables.

Recall from Chapter 8 that an F distribution with 1 degree of freedom for the numerator and k degrees of freedom for the denominator is the *square* of a t distribution with k degrees of freedom. In Example 10–1 our computed F statistic value is 637.47, which is the square of our obtained t statistic 25.25 (to within rounding error). The same relationship holds for the critical points: for $\alpha = 0.01$, we have a critical point for $F_{(1, 23)}$ equal to 7.88, and the (right-hand) critical point of a two-tailed test at $\alpha = 0.01$ for t with 23 degrees of freedom is $2.807 = \sqrt{7.88}$.

PROBLEMS

10–56. Conduct the F test for the existence of a linear relationship between the two variables in problem 10–11.

10–57. Carry out an F test for a linear relationship in problem 10–13. Compare your results with those of the t test.

10–58. Repeat problem 10–57 for the data of problem 10–17.

10–59. Conduct an F test for the existence of a linear relationship in the case of problem 10–15.

10–60. For problem 10–51, assume the sample size used was $n = 104$, and conduct an F test for the existence of a linear relationship between the two variables.

10–61. In a simple linear regression analysis, it is found that $b_1 = 2.556$ and $s(b_1) = 4.122$. The sample size is $n = 22$. Conduct an F test for the existence of a linear relationship between the two variables.

10–62. (A mathematically demanding problem) Using the definition of the t statistic in terms of sums of squares, prove (in the context of simple linear regression) that $t^2 = F$.

10–9 Residual Analysis and Checking for Model Inadequacies

Recall our discussion of statistical models in Section 10–1. We said that a good statistical model accounts for the systematic movement in the process, leaving out a series of uncorrelated, purely random errors ϵ, which are assumed to be normally distributed with mean zero and a constant variance σ^2. In Figure 10–3, we saw a general methodology for statistical model building, consisting of model identification, estimation, tests of validity, and, finally, use of the model. We are now at the third stage of the analysis of a simple linear regression model: examining the residuals and testing the validity of the model.

Analysis of the residuals could reveal whether the assumption of normally distributed errors holds. In addition, the analysis could reveal whether the variance of the errors is indeed constant, that is, whether the spread of the data around the regression line is uniform. The analysis could also indicate whether there are any missing variables that should have been included in our model (leading to a multiple regression equation). The analysis may reveal whether the order of data collection

FIGURE 10–23   A Residual Plot Indicating Heteroscedasticity

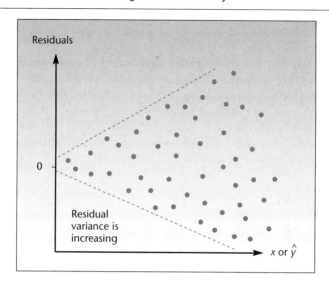

(e.g., time of observation) has any effect on the data and whether the order should have been incorporated as a variable in the model. Finally, analysis of the residuals may determine whether the assumption that the errors are uncorrelated is satisfied. A test of this assumption, the Durbin-Watson test, entails more than a mere examination of the model residuals, and discussion of this test is postponed until the next chapter. We now describe some graphical methods for the examination of the model residuals that may lead to discovery of model inadequacies.

A Check for the Equality of Variance of the Errors

A graph of the regression errors, the residuals, versus the independent variable X, or versus the predicted values $\hat{Y}$, will reveal whether the variance of the errors is constant. The variance of the residuals is indicated by the width of the scatter plot of the residuals as X increases. If the width of the scatter plot of the residuals either increases or decreases as X increases, then the assumption of constant variance is not met. This problem is called **heteroscedasticity.** When heteroscedasticity exists, we cannot use the ordinary least-squares method for estimating the regression and should use a more complex method, called *generalized least squares.* Figure 10–23 shows how a plot of the residuals versus X or $\hat{Y}$ looks in the case of heteroscedasticity. Figure 10–24 shows a residual plot in a good regression, with no heteroscedasticity.

Testing for Missing Variables

Figure 10–24 also shows how the residuals should look when plotted against time (or the order in which data are collected). There should be no trend in the residuals when plotted versus time. A linear trend in the residuals plotted versus time is shown in Figure 10–25.

If the residuals exhibit a pattern when plotted versus time, then time should be incorporated as an explanatory variable in the model in addition to X. The same is true for any other variable against which we may plot the residuals: If there is any trend in

FIGURE 10–24   A Residual Plot Indicating No Heteroscedasticity

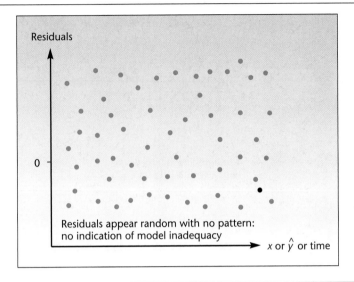

FIGURE 10–25   A Residual Plot Indicating a Trend with Time

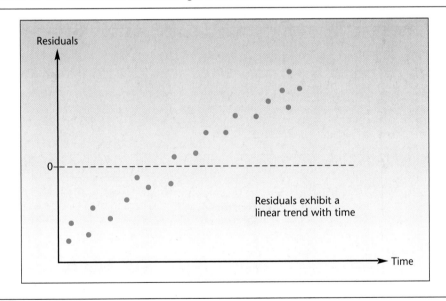

the plot, the variable should be included in our model along with X. Incorporating additional variables leads to a multiple regression model.

Detecting a Curvilinear Relationship between Y and X

If the relationship between X and Y is curved, "forcing" a straight line to fit the data will result in a poor fit. This is shown in Figure 10–26. In this case, the residuals are at first large and negative, then decrease, become positive, and again become negative.

FIGURE 10–26   **Results of Forcing a Straight Line to Fit a Curved Data Set**

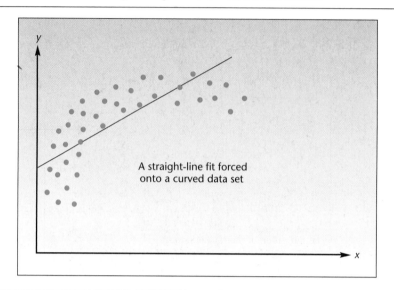

FIGURE 10–27   **Resulting Pattern of the Residuals When a Straight Line Is Forced to Fit a Curved Data Set**

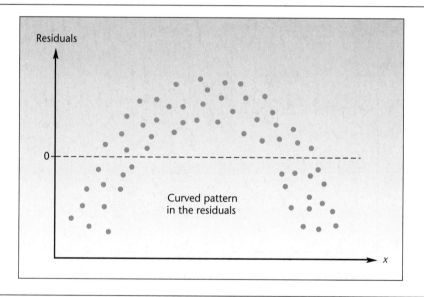

The residuals are not random and independent; they show curvature. This pattern appears in a plot of the residuals versus X, shown in Figure 10–27.

The situation can be corrected by adding the variable X^2 to the model. This also entails the techniques of multiple regression analysis. We note that, in cases where we have repeated Y observations at some levels of X, there is a statistical test for model lack of fit such as that shown in Figure 10–26. The test entails decomposing the sum of squares for error into a component due to lack of fit and a component due to pure error. This gives rise to an F test for lack of fit. This test is described in advanced texts.

FIGURE 10–28   A Histogram of the Residuals

	Interval	Freq.				Data
	<=-600	0			1	6.941111
	(-600, -500]	2			2	441.7264
	(-500, -400]	0			3	-54.93432
	(-400, -300]	3			4	118.4022
	(-300, -200]	2			5	-263.9618
	(-200, -100]	3			6	-513.1559
	(-100, 0]	1			7	63.52337
	(0, 100]	3			8	281.8833
	(100, 200]	4			9	-197.6507
	(200, 300]	2			10	320.9866
	(300, 400]	2			11	-291.9955
	(400, 500]	3			12	200.6837
	(500, 600]	0			13	-588.7884
	>600	0			14	388.7842
					15	-381.8366
					16	449.9694
					17	-309.3954
					18	-190.6108
					19	106.6411
					20	498.659
					21	93.72231
					22	120.5269
	Total	25			23	-343.4986
					24	181.9887
					25	-138.6101

Start [-600]   Interval Width [100]   End [600]

We point out, however, that examination of the residuals is an excellent tool for detecting such model deficiencies, and this simple technique does not require the special data format needed for the formal test.

The Normal Probability Plot

One of the assumptions in the regression model is that the errors are normally distributed. This assumption is necessary for calculating prediction intervals and for hypothesis tests about the regression. One of several ways to test for the normality of the residuals is to plot a histogram of the residuals and visually observe whether the shape of the histogram is close to the shape of a normal distribution. To do this we can use the histogram template, Histogram.xls, from Chapter 1, shown in Figure 10–28.

Let us plot the histogram for the residuals in the American Express study (Example 10–1) seen in Figure 10–13. First we copy the residuals from column D and paste them (using the Paste Special command and choosing "values" only to be pasted) into the data area (column Y) of the histogram template. We then enter suitable Start, Interval Width, and End values, which in this case could be −600, 100, and 600. The resulting histogram, shown in Figure 10–28, looks more like a uniform distribution than like a normal distribution. But this is only a visual test rather than a formal hypothesis test, and therefore we do not get a p-value for this test. In Chapter 14, we will see a formal χ^2 test for normality, which yields a p-value and thus can be used to possibly reject the null hypothesis that the residuals are normally distributed. Coming back to the histogram, to the extent the shape of the histogram deviates from the normal distribution, the prediction intervals and t or F tests about the regression are questionable.

FIGURE 10–29   Patterns of Nonnormal Distributions on the Normal Probability Plot

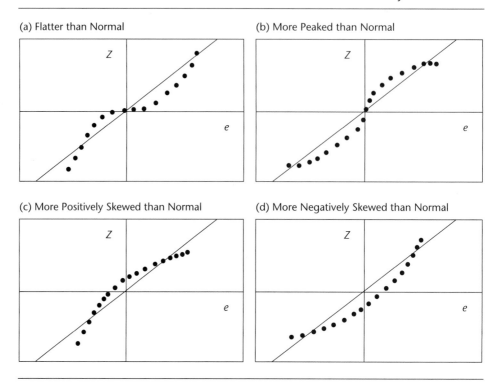

(a) Flatter than Normal

(b) More Peaked than Normal

(c) More Positively Skewed than Normal

(d) More Negatively Skewed than Normal

Checking the normality of residuals using a histogram may work, but a wrong choice of Start, Interval Width, and End values can distort the shape of the distribution to some extent. A slightly better method to use is a **normal probability plot** of the residuals. The simple regression template creates this plot automatically (see Figure 10–14). In this plot, the residual values are on the horizontal axis and the corresponding z values from the normal distribution are on the vertical axis. If the residuals are normal, then they should align themselves along the straight line that appears on the plot. To the extent the points deviate from this straight line, the residuals deviate from a normal distribution. Note that this also is only a visual test and does not provide a p-value. In Figure 10–14, the points do deviate from the straight line, causing some concern and confirming what we saw in the histogram.

The normal probability plot is constructed as follows. For each value e of the residual, its quantile (cumulative probability) is calculated using the equation

$$q = \frac{l + 1 + m/2}{n + 1}$$

where l is the number of residuals less than e, m is the number of residuals equal to e, and n is the total number of observations. Then the z value corresponding to the quantile q, denoted by z_q, is calculated. A point with this z_q on the vertical axis and e on the horizontal axis is plotted. This process is repeated, with one point plotted for each observation. The diagonal straight line is drawn by connecting 3 standard deviations on either side of zero both on vertical and horizontal axes.

It is useful to recognize different nonnormal cases on a normal probability plot. Figure 10–29 shows four different patterns of lines along which the points will align.

FIGURE 10–30   **Distribution of the Residuals Is Flatter and More Positively Skewed Than Normal**

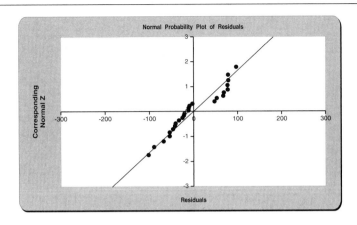

Figure 10–30 shows a case where the residuals are clearly nonnormal. From the pattern of the points we can infer that the distribution of the residuals is flatter and more positively skewed than the normal distribution.

10–63.   For each of the following plots of regression residuals versus X, state whether there is any indication of model inadequacy; if so, identify the inadequacy.

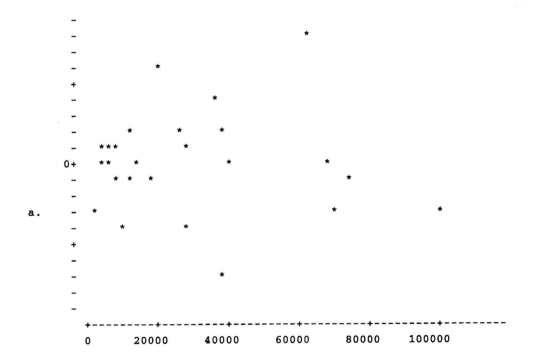

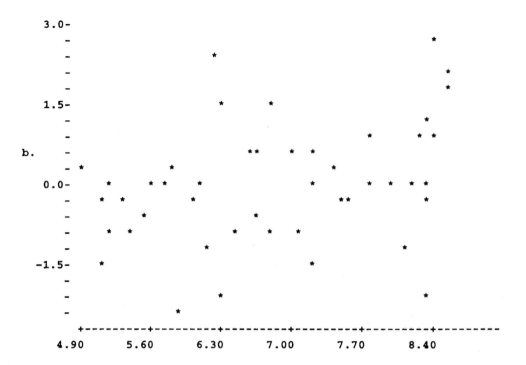

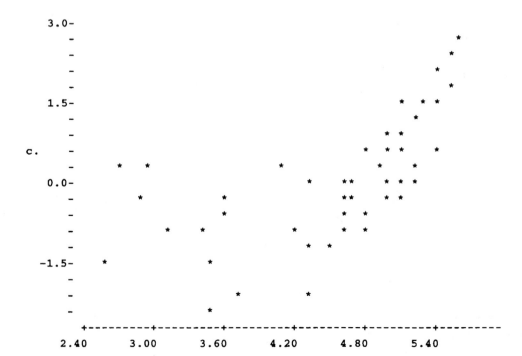

10–64. In the following plots of the residuals versus time of observation, state whether there is evidence of model inadequacy. How would you correct any inadequacy?

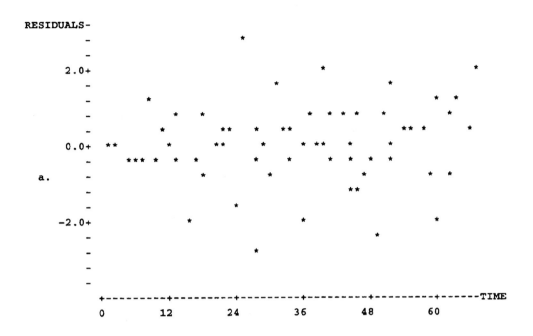

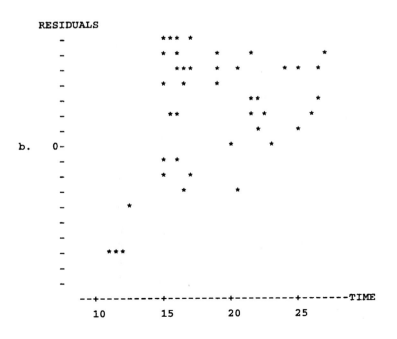

10-65. Is there any indication of model inadequacy in the following plots of residuals on a normal probability scale?

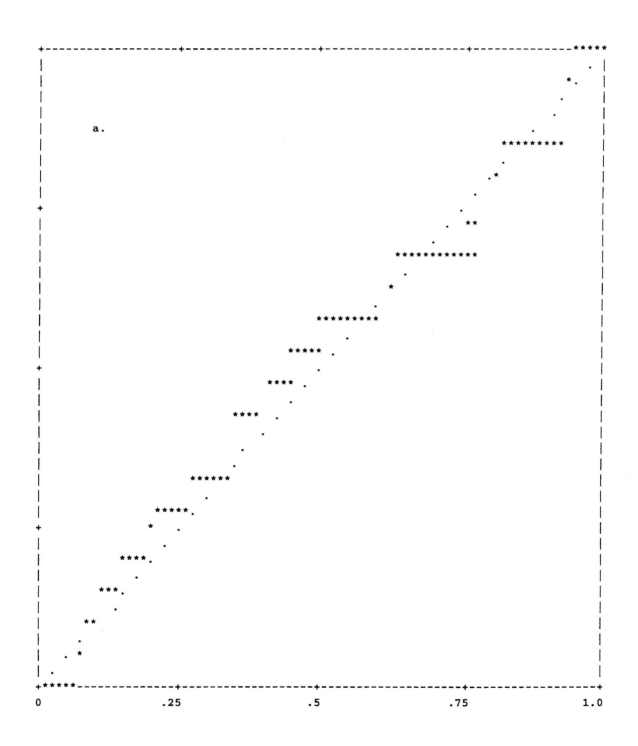

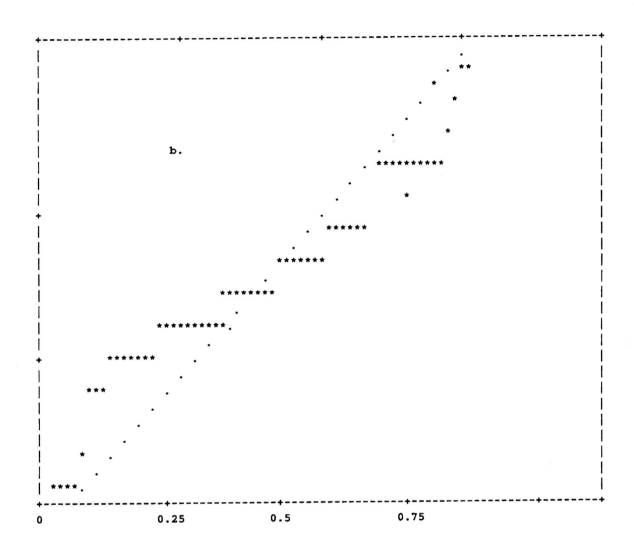

10–66. Produce residual plots for the regression of problem 10–11. Is there any apparent model inadequacy?

10–67. Repeat problem 10–66 for the regression of problem 10–13.

10–68. Repeat problem 10–66 for the regression of problem 10–16.

10–10 Use of the Regression Model for Prediction

As mentioned in the first section of this chapter, there are several possible uses of a regression model. One is to understand the relationship between the two variables. As with correlation analysis, understanding a relationship between two variables in regression does not imply that one variable causes the other. Causality is a much more complicated issue and cannot be determined by a simple regression analysis.

A more frequent use of a regression analysis is *prediction*: providing estimates of values of the dependent variable by using the prediction equation $\hat{Y} = b_0 + b_1 X$. It is important that prediction be done in the region of the data used in the estimation

FIGURE 10–31   The Danger of Extrapolation

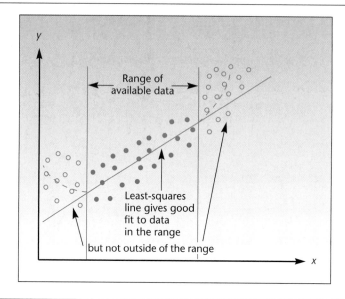

process. *You should be aware that using a regression for extrapolating outside the estimation range is risky, as the estimated relationship may not be appropriate outside this range.* This is demonstrated in Figure 10–31.

Point Predictions

It is very easy to produce point predictions using the estimated regression equation. All we need to do is to substitute the value of X for which we want to predict Y into the prediction equation. In Example 10–1 suppose that American Express wants to predict charges on the card for a member who traveled 4,000 miles during a period equal to the one studied (note that $x = 4,000$ is in the range of X values used in the estimation). We use the prediction equation, equation 10–12, but with greater accuracy for b_1:

$$\hat{y} = 274.85 + 1.2553x = 274.85 + 1.2553(4,000) = 5,296.05 \text{ (dollars)}$$

The process of prediction in this example is demonstrated in Figure 10–32.

Prediction Intervals

Point predictions are not perfect and are subject to error. The error is due to the uncertainty in estimation as well as the natural variation of points about the regression line. A $(1 - \alpha)$ 100% prediction interval for Y is given in equation 10–32.

A $(1 - \alpha)$ 100% prediction interval for Y is

$$\hat{y} \pm t_{\alpha/2}\, s \sqrt{1 + \frac{1}{n} + \frac{(x - \bar{x})^2}{SS_X}} \tag{10-32}$$

FIGURE 10–32   **Prediction in American Express Study**

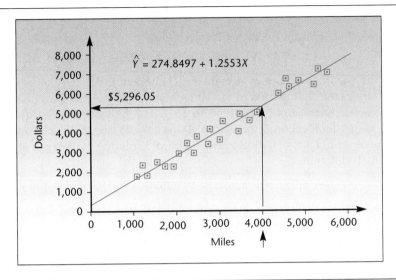

FIGURE 10–33   **Prediction Band and Its Width**

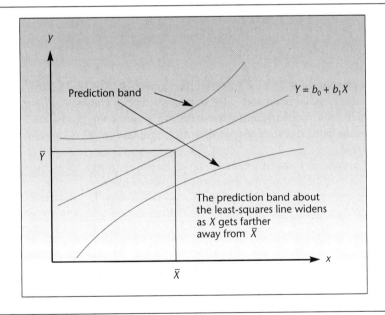

As can be seen from the formula, the width of the interval depends on the distance of our value x (for which we wish to predict Y) from the mean $\bar{x}$. This is shown in Figure 10–33.

We will now use equation 10–32 to compute a 95% prediction interval for the amount charged on the American Express card by a member who traveled 4,000 miles. We know that in this example $\bar{x} = \Sigma x/n = 79,448/25 = 3,177.92$. We also know that $SS_X = 40{,}947{,}557.84$ and $s = 318.16$. From Appendix C, Table 3, we get the critical point for t with 23 degrees of freedom: 2.069. Applying equation 10–32, we get

$$5{,}296.05 \pm (2.069)(318.16)\sqrt{1 + 1/25 + (4{,}000 - 3{,}177.92)^2/40{,}947{,}557.84}$$
$$= 5{,}296.05 \pm 676.62 = [4{,}619.43, 5{,}972.67]$$

Based on the validity of the study, we are 95% confident that a cardholder who traveled 4,000 miles during a period of the given length will have charges on her or his card totaling anywhere from $4,619.43 to $5,972.67. What about the *average* total charge of all cardholders who traveled 4,000 miles? This is $E(Y \mid x = 4{,}000)$. The point estimate of $E(Y \mid x = 4{,}000)$ is also equal to $\hat{Y}$, but the confidence interval for this quantity is different.

A Confidence Interval for the Average Y, Given a Particular Value of X

We may compute a confidence interval for $E(Y \mid X)$, the expected value of Y for a given X. Here the variation is smaller because we are dealing with the average Y for a given X, rather than a particular Y. Thus, the confidence interval is narrower than a prediction interval of the same confidence level. The confidence interval for $E(Y \mid X)$ is given in equation 10–33:

A $(1 - \alpha)$ 100% confidence interval for $E(Y \mid X)$ is

$$\hat{y} \pm t_{\alpha/2}\, s \sqrt{\frac{1}{n} + \frac{(x - \bar{x})^2}{SS_X}} \qquad (10\text{--}33)$$

The confidence band for $E(Y \mid X)$ around the regression line looks like Figure 10–33 except that the band is narrower. The standard error of the estimator of the conditional mean $E(Y \mid X)$ is smaller than the standard error of the predicted Y. Therefore, the 1 is missing from the square root quantity in equation 10–33 as compared with equation 10–32.

For the American Express example, let us now compute a 95% confidence interval for $E(Y \mid x = 4{,}000)$. Applying equation 10–33, we have

$$5{,}296.05 \pm (2.069)(318.16)\sqrt{1/25 + (4{,}000 - 3{,}177.92)^2/40{,}947{,}557.84}$$
$$= 5{,}296.05 \pm 156.48 = [5{,}139.57, 5{,}452.53]$$

Being a confidence interval for a conditional mean, the interval is much narrower than the prediction interval, which has the same confidence level for covering *any given* observation at the level of X.

PROBLEMS

10–69. For the American Express example, give a 95% prediction interval for the amount charged by a member who traveled 5,000 miles. Compare the result with the one for $x = 4{,}000$ miles.

10–70.   Give a 95% confidence interval for $E(Y|\,x = 5{,}000)$ in the American Express example. Compare with your answer to that of problem 10–69.

10–71.   For problem 10–69, give a 99% prediction interval.

10–72.   For problem 10–11, give a point prediction and a 99% prediction interval for the exchange rate when the deficit/GDP ratio is 5%.

10–73.   For problem 10–72, give a 99% confidence interval for the average exchange rate when the deficit/GDP ratio is 5%.

10–74.   For problem 10–16, give a 95% confidence interval for the P/E ratio when the percent growth is 32%

10–75.   For problem 10–16, give a 95% prediction interval for the P/E ratio when percentage growth is 22%.

10–76.   For problem 10–15 predict the dollars/deutsche mark rate when the Standard & Poor's 500 index is 700.

10–11   The Solver Method for Regression[13]

The Solver macro available in Excel can also be used to conduct a simple linear regression. The advantage of using this method is that additional constraints can be imposed on the slope and the intercept. For instance, if we want the intercept to be a particular value, or if we want to force the regression line to go through a desired point, we can do that by imposing appropriate constraints. As an example, suppose we are regressing the weight of a certain amount of a chemical against its volume (in order to find the average density). We know that when the volume is zero, the weight should be zero. This means the intercept for the regression line must be zero. We can impose this as a constraint, if we use the Solver method, and be assured that the intercept will be zero. The slope obtained with the constraint can be quite different from the slope obtained without the constraint.

As another example, consider a common type of regression carried out in the area of finance. The risk of a stock (or any capital asset) is measured by regressing its returns against the market return (which is the average return from all the assets in the market) during the same period. The Capital Asset Pricing Model (CAPM) stipulates that when the market return equals the risk-free interest rate (such as the interest rate of short-term Treasury bills), the stock will also return the same amount. In other words, if the market return = risk-free interest rate = 7%, then the stock's return, according to the CAPM, will also be 7%. This means that according to the CAPM, the regression line must pass through the point $(7, 7)$. This can be imposed as a constraint in the Solver method of regression.

Note that forcing a regression line through the origin, $(0, 0)$, is the same as forcing the intercept to equal zero, and forcing the line through the point $(0, 5)$ is the same as forcing the intercept to equal 5.

The criterion for the line of best fit by the Solver method is still the same as before—minimize the sum of squared errors (SSE).

---

[13]This section is optional.

A limitation of this method is that we cannot find confidence intervals for the regression coefficients or prediction intervals for Y. All we get is the constrained line of best fit and point predictions based on that line. Also, we cannot conduct hypothesis tests about the regression, because we are deviating from the model assumptions given in equation 10–3. In particular, the errors may not be normally distributed.

We shall see the use of this method through an example.

EXAMPLE 10–5   A certain fuel produced by a chemical company varies in its composition and therefore in its density. The average density of the fuel is to be determined for engineering purposes. Rather than take the average of all the densities observed, it was decided to estimate the density through regression, thus minimizing SSE. Different amounts of the fuel were sampled at different times and the weights (in grams) and volumes (in cubic centimeters) were accurately measured. The results are in Figure 10–34.

1. Regress the weight against the volume and find the regression equation. Predict the weight when the volume is 7 cm³.
2. Force the regression line through the origin and find the regression line that minimizes SSE. What is the new regression equation? What is the density implied by this regression equation? Predict the weight when the volume is 7 cm³.

Solution   1. Without any constraint, the regression equation is $\hat{Y} = 1.352X - 0.847$ (obtained from the template for regular regression). For a volume of 7 cm³, the predicted volume is 8.62 grams.

2. Use the template shown in Figure 10–34. Enter the data in columns B and C. Enter zero in cell J5. Choose the Solver command under the Tools menu. In the dialog box that appears, press the Add button. The dialog box shown in Figure 10–35 appears. In this Add Constraint dialog box, enter the constraint K5 = 0. Note the use of the drop down box in the middle to choose the relationship. Press the OK button. The Solver dialog box appears again, as seen in Figure 10–36. Click the Solve button. When the problem is solved the dialog box shown in Figure 10–37 appears. Make sure Keep Solver Solution is selected. Click the OK button.

The result is seen in Figure 10–34. The intercept is zero, as expected, and the slope is 1.21969. Thus the new regression equation is $\hat{Y} = 1.21969X$. The average density implied by the equation is 1.21969 g/cm³.

To get the predicted value for $X = 7$, enter 7 in cell J5. The predicted value of 8.53785 g appears in cell K5.

Several other types of constraints can be imposed. For example, we can impose the condition that the slope must be less than or equal to 10. We can even impose a condition that the slope must be numerically less than the intercept, although why we would want such a constraint is not clear. In any case, all constraints are entered using the dialog box seen in Figure 10–35. There are some syntax rules for entering constraints in this dialog box. For example, the entry in the right-hand-side box of Figure 10–35 (0 in the figure) cannot be a formula, whereas the entry in the left-hand-side box (K5) can be a formula such as 3*K5 + 6. Such details can be obtained from the help screens.

FIGURE 10–34 The Template for Using the Solver for Regression
[Simple Regression.xls; Sheet: Solver]

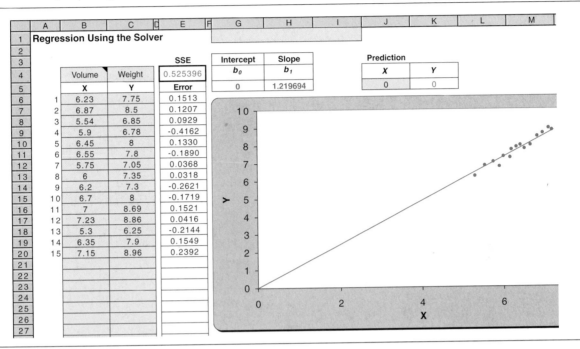

FIGURE 10–35 The Add Constraint Dialog Box

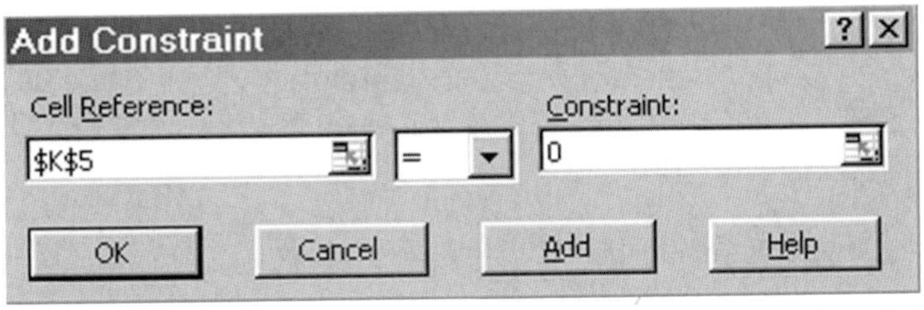

FIGURE 10–36 The Solver Dialog Box

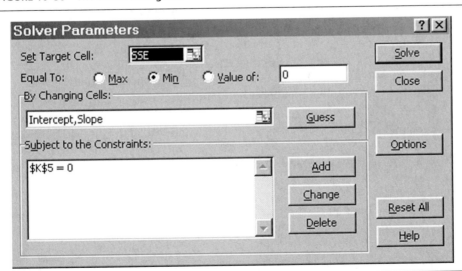

FIGURE 10–37   The Solver Results Dialog Box

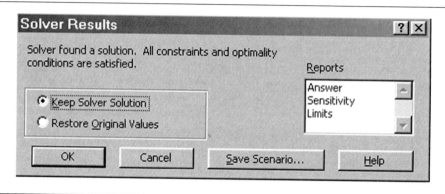

PROBLEMS

10–77.   Consider the following sample data of X and Y:

X	Y
8	22.30
6	16.71
9	25.21
6	15.84
1	2.75
8	21.22
2	5.27
1	2.32
10	27.39
7	19.35

a. Regress Y against X without any constraints. What is the regression equation? Predict Y when $X = 10$.

b. Force the regression line through the origin. What is the new regression equation? Predict Y when $X = 10$ with this equation.

c. Force the regression line to go through the point $(5, 13)$. What is the new regression line? What is the new regression equation? Predict Y when $X = 10$ with this equation.

d. Regress Y against X with the constraint that the slope must be less than or equal to 2. What is the new regression equation? Predict Y when $X = 10$ with this equation.

10–78.   Why would it be silly to force a regression line through two distinct points at the same time?

10–12   Summary and Review of Terms

In this chapter, we introduced simple linear regression, a technique for estimating the straight-line relationship between two variables. We defined the **dependent variable** as the variable we wish to predict by, or understand the relation with, the **independent variable** (also called the **explanatory** or **predictor** variable). We described the **least-squares** estimation procedure as the procedure that produces the **best linear unbiased estimators (BLUE)** of the regression coefficients, the **slope** and **intercept** parameters. We learned how to conduct two statistical tests for the existence of a linear relationship between the two variables: the t test and the F test. We noted that in the case of simple linear regression, the two tests are equivalent. We saw how to evaluate the fit of the regression model by considering the **coefficient of determination** r^2. We learned how to check the validity of the assumptions of the simple linear regression model by examining the **residuals**. Finally, we saw how the regression model can be used for prediction. In addition, we discussed a linear **correlation** model. We saw that the correlation model is appropriate when the two variables are viewed in a symmetric role: both being normally distributed random variables, rather than one of them (X) being considered nonrandom, as in regression analysis.

ADDITIONAL PROBLEMS

10–79. The following data are adapted from an article in *The Economist* entitled "The Dash for the Off Switch."[14] Regress Fox on the sum of the three networks.

	Share of U.S. Primetime Audience	
Year	ABC + CBS + NBC	Fox
1990	62	8
1991	61	11
1992	57	10
1993	54	11
1994	51	12
1995	49	12
1996	49	13
1997	47	14

10–80. Consider the following data from *The Financial Times*:[15]

	Percentage of Penetration of Population for Telecommunications Companies	
Country	Percentage of Fixed Lines	Percentage of Cellular
France	57	4.3
Ireland	36	7.4
Spain	39	7.7
Portugal	37	6.7

---

[14]June 7, 1997, pp. 63–64. © 1997 The Economist Newspaper Groups, Inc. Reprinted with permission. Further reproduction prohibited. www.economist.com.

[15]*The Financial Times,* June 11, 1997, p. 11. Reprinted by permission of The Financial Times Limited.

Country	Percentage of Penetration of Population for Telecommunications Companies	
	Percentage of Fixed Lines	Percentage of Cellular
Italy	53	11.2
Sweden/Finland/Norway/Denmark	64	28.2
Netherlands	55	6.6
Greece	49	4.9

Is there evidence for a commonly made claim that in Europe, where fixed lines are harder to obtain, cellular phones are more common? Use simple linear regression to answer this question.

10–81. The Commerce Department looks at global U.S. corporate profits per unit of U.S. real (adjusted for inflation) GDP (gross domestic product) as an indicator of economic growth. The following quarterly data are from *Business Week* (June 16, 1997, p. 15). Run a regression of the variable against time.

Year-Quarter	Profit (%) per Unit Real GDP for Nonfinancial Corporations
95-2	11.3
95-3	11.0
95-4	12.1
96-1	11.9
96-2	12.6
96-3	12.6
96-4	12.2
97-1	12.6

10–82. The following data[16] are operating income X and monthly stock close Y for Clorox, Inc. Graph the data. Then regress log Y on X.

X ($ millions): 240, 250, 260, 270, 280, 300, 310, 320, 330, 340, 350, 360, 370, 400, 410, 420, 430, 450

Y ($s): 45, 42, 44, 46, 47, 50, 48, 60, 61, 59, 67, 75, 74, 85, 95, 110, 125, 130

Predict Y for $X = 305$.

10–83. Consider this table of "whisper stocks" from *Business Week*.[17] Regress takeover value on price. For the universe in question, what is *Business Week*'s formula?

Stock	Price*	Takeover Value
Aames Financial	13⅝	25
Alumax	38⅝	52
Bay Networks	24⅜	35
Bowater	48¾	65

[16]From J. Hamilton, "Brighter Days at Clorox," *Business Week*, June 16, 1997, pp. 66–67. Reprinted by permission of *Business Week*. All rights reserved.

[17]G. Marcial, "Everybody's Talking Takeover," June 16, 1997, p. 58. Reprinted by permission of *Business Week*. All rights reserved.

Stock	Price*	Takeover Value
Digital Equipment	34⅞	50
Genovese Drug Stores	15¾	25
Georgia Gulf	28½	40
Glenayre Technologies	14⅝	22
Group Long Distance	4⅝	11
Inland Steel	25	40
Integon	12½	20
John Alden Financial	20⅞	28
Lehman Brothers	39¼	50
Little Switzerland	5¾	12
Mellon Bank	43¾	80

*As of June 2 DATA: BW SURVEY, BLOOMBERG FINANCIAL MARKETS

10–84. A simple regression produces the regression equation $\hat{Y} = 5X + 7$.

  a. If we add 2 to all the X values in the data (and keep the Y values the same as the original), what will the new regression equation be?

  b. If we add 2 to all the Y values in the data (and keep the X values the same as the original), what will the new regression equation be?

  c. If we multiply all the X values in the data by 2 (and keep the Y values the same as the original), what will the new regression equation be?

  d. If we multiply all the Y values in the data by 2 (and keep the X values the same as the original), what will the new regression equation be?

10–85. In a simple regression the regression equation is $\hat{Y} = 5X + 7$. Now if we interchange the X and Y data (that is, what was originally X is now Y and vice versa) and repeat the regression, we would expect the slope of the new regression line to be exactly equal to $1/5 = 0.2$. But the slope will only be approximately equal to 0.2 and almost never exactly equal to 0.2. Why?

10–86. Regress Y against X with the following data from a random sample of 15 observations:

X	Y
12	100
4	60
10	96
15	102
6	68
4	70
13	102
11	92
10	95
18	125
20	134
22	133
8	87
20	122
11	101

  a. What is the regression equation?

b. What is the 90% confidence interval for the slope?

c. Test the null hypothesis "*X* does not affect *Y*" at an α of 1%.

d. Test the null hypothesis "the slope is zero" at an α of 1%.

e. Make a point prediction of *Y* when *X* = 10.

f. Assume that the value of *X* is controllable. What should be the value of *X* if the desired value for *Y* is 100?

g. Construct a residual plot. Are the residuals random?

h. Construct a normal probability plot. Are the residuals normally distributed?

CASE 11   Return on Capital in Health Care

The debt-to-capital ratio of a company signifies the amount of financial risk a company is taking. If the ratio is high, the risk is high. The reason? When the profit margin falls, debt will worsen the situation by making the return on capital fall even more adversely.

The data on debt-to-capital ratio and return on capital for 24 health care companies are given in the accompanying table. Assume that this is a random sample of health care companies. Regress return on capital against debt-to-capital ratio, then answer the following questions:

1. What is the regression equation? Give a 95% confidence interval for the slope.

2. Test the null hypothesis "debt-to-capital ratio does not affect return on capital of health care companies" at an α of 5%. Report the *p*-value.

3. Test the null hypothesis "the slope of the regression line is zero" at an α of 5%. Report the *p*-value.

4. Construct a residual plot. Are the residuals random?

5. Construct a normal probability plot. Are the residuals normally distributed?

6. Give a 95% prediction interval for the *expected* return on capital for a health care company with zero debt.

7. Which two companies in the data set produce the two largest (absolute) errors?

8. Remove the two companies found in part 7 and repeat the regression. What is the new regression equation? Give a 95% confidence interval for the slope.

	Debt/Capital (%)	Return on Capital (%)
Abbott Laboratories	11.4	31.2
Allergan	24	25.5
AmeriSource Health	59.4	17.5
Amgen	5.3	37.3
Becton, Dickinson	27.8	15.5
Bristol-Myers Squibb	12.1	44.6
Cardinal Health	30.1	14.4
Guidant	27.4	34.6
Health Management	32.7	14.2
Johnson & Johnson	11.4	24.8
Eli Lilly	31.2	39.8
McKesson HBOC	22.2	15.5
Medtronic	0.3	26.2
Merck	14.7	33.3
Mid Atlantic Medical	0	18.8
Pall	25.5	16
Paterson Dental	0	20.4
Pfizer	7	23.5
Schering-Plough	0.1	43.2
SmithKline Beecham	10.3	22.3
Stryker	55.6	13.8
UnitedHealth Group	9.9	16.4
Universal Health	38.6	9.8
WellPoint Health	21.3	20.9

Source: From "400 Best Big Companies," *Forbes Magazine,* January 8, 2001. Reprinted by permission of Forbes Magazine © 2001 Forbes Inc.

CASE 12   Risk and Return

According to the Capital Asset Pricing Model (CAPM), the risk associated with a capital asset is proportional to the slope β_1 (or simply β) obtained by regressing the asset's past returns with the corresponding returns of the average portfolio called the *market portfolio*. (The return of the market portfolio represents the return earned by the average investor. It is a weighted average of the returns from all the assets in the market.) The larger the slope β of an asset, the larger is the risk associated with that asset. A β of 1.00 represents average risk.

The returns from an electronics firm's stock and the corresponding returns for the market portfolio for the past 15 years are given below.

Market Return (%)	Stock's Return (%)
16.02	21.05
12.17	17.25
11.48	13.1
17.62	18.23
20.01	21.52
14	13.26
13.22	15.84
17.79	22.18
15.46	16.26
8.09	5.64
11	10.55
18.52	17.86
14.05	12.75
8.79	9.13
11.6	13.87

1. Carry out the regression and find the β for the stock. What is the regression equation?

2. Does the value of the slope indicate that the stock has above-average risk? (For the purposes of this case assume that the risk is average if the slope is in the range 1 ± 0.1, below average if it is less than 0.9, and above average if it is more than 1.1.)

3. Give a 95% confidence interval for this β. Can we say the risk is above average with 95% confidence?

4. If the market portfolio return for the current year is 10%, what is the stock's return predicted by the regression equation? Give a 95% confidence interval for this prediction.

5. Construct a residual plot. Do the residuals appear random?

6. Conduct the Durbin-Watson test. What do you conclude?

7. Construct a normal probability plot. Do the residuals appear to be normally distributed?

8. (Optional) The *risk-free rate of return* is the rate associated with an investment that has no risk at all, such as lending money to the government. Assume that for the current year the risk-free rate is 6%. According to the CAPM, when the return from the market portfolio is equal to the risk-free rate, the return from every asset must also be equal to the risk-free rate. In other words, if the market portfolio return is 6%, then the stock's return should also be 6%. It implies that the regression line must pass through the point (6, 6). Repeat the regression forcing this constraint. Comment on the risk based on the new regression equation.

11 Multiple Regression

In regression analysis, often the variable of interest depends on more than just one other variable. There may be several independent variables that contain information about the variable we are trying to predict or understand. In such cases, it may be worthwhile to formulate a model that allows us to consider the relation of our variable of interest with a set of independent variables. In the American Express example of Chapter 10, the company may be able to predict more accurately the amount charged on the American Express card from knowledge of cardholders' incomes, in addition to miles traveled. When several independent variables are included in a regression equation, our model is called a **multiple regression** model.

How many independent variables should we include in our regression equation? It seems logical that if our equation incorporates information about as many variables as possible, we will have maximum prediction power of the variable of interest. There are, however, some serious limitations to this assertion.

In the summer of 1983, a student of economics had an idea about predicting the nation's economic future. The student decided to collect data on as many economic variables as possible and to formulate a multiple regression equation linking these variables with the gross national product (GNP). The student thought that if the number of variables used were large enough, he would be able to predict the GNP with great accuracy. The student collected 12 years' worth of quarterly data, 48 observations in total, and information on 47 economic variables: national income, prime lending rate, unemployment rate, etc., corresponding to the 48 quarters for which GNP values were known. He formulated the regression relation $Y = \beta_0 + \beta_1 X_1 + \beta_2 X_2 + \beta_3 X_3 + \cdots + \beta_{47} X_{47} + \epsilon$.

In the next section, we will see that this equation is a generalization to k variables (in this case, $k = 47$) of the population regression equation with one variable, introduced in Chapter 10. The student went on to carry out an estimation of the model parameters β_0 to β_{47}, using a computer, and to compute the multiple coefficient of determination, denoted R^2. (The coefficient R^2 is an extension to multiple regression of the coefficient of determination r^2 in simple linear regression. This coefficient is a measure of how well the regression equation fits the data.) To his great delight, the student noted that the value of R^2 for his regression was 0.9999. The regression equation apparently had a perfect fit with the data! The student was sure, therefore, that he would be able to predict the GNP from knowledge of the 47 variables with great accuracy. As it turned out, not only were model predictions very poor–the forecasts produced from the model were worse than pure guesses of the GNP[1]–but also the model was deemed erroneous, misleading, and in violation of statistical methodology. Why?

Recall that in Chapter 10, we said that a good statistical model fits the data well but is also parsimonious; that is, it has as few parameters as possible. At the time, you may have wondered why a model should be parsimonious; you may have reasoned that the more parameters in your model, the better the model. Examine Figure 11–1, which demonstrates an important mathematical fact. Given any two points, it is possible to find a one-dimensional surface, a straight line, that will pass through the two points and fit the two points perfectly. Once a third point is obtained, it may not lie

---

[1]There are statistical methods for evaluating how good a prediction is, once the actual values become known. These methods consist of looking at the difference between forecast and actuality. Methods of evaluating forecast accuracy will be discussed in the next chapter.

FIGURE 11–1   The Dimensionality of Surfaces Fitting Two, Three, and Four Points

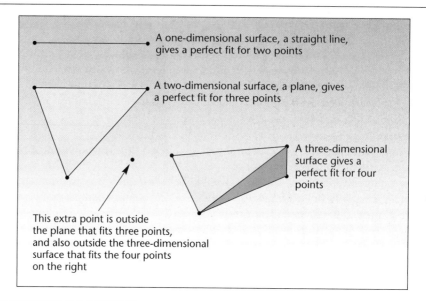

A one-dimensional surface, a straight line, gives a perfect fit for two points

A two-dimensional surface, a plane, gives a perfect fit for three points

A three-dimensional surface gives a perfect fit for four points

This extra point is outside the plane that fits three points, and also outside the three-dimensional surface that fits the four points on the right

on the straight line connecting the original two points. Thus, the line—though providing a perfect fit for two points—may be a poor *predictor* of future observations. With three points, however, we can always find a two-dimensional surface, a plane, that will pass through the three points and thus provide a perfect fit. When a fourth point is added, the point may not lie on the original plane. With four points, we can find a three-dimensional surface (a surface of more than two dimensions is called a *hyperplane*) that will provide a perfect fit. Add a fifth point, and a three-dimensional surface is not enough, but a four-dimensional surface will provide a perfect fit. Thus, given n points, we can find an $(n - 1)$-dimensional surface that will provide a perfect fit for the data. When we do this, however, we are not doing statistics. We are overfitting our data. This procedure leaves no degrees of freedom for error.

Remember that in simple linear regression, there are $n - 2$ degrees of freedom for error. This is so because there are two parameters to be estimated from the data—an intercept and a slope. In a multiple regression model with k variables, there are $k + 1$ parameters to be estimated from the data set—one slope parameter for each of the k variables, and an intercept. Hence, the degrees of freedom for error in a multiple regression model with k independent variables are $n - (k + 1)$. Our student, fitting a regression model with $k = 47$ variables to a data set of $n = 48$ points, is left with $n - (k + 1) = 48 - (47 + 1) = 0$ degrees of freedom for error! This means that there is no allowance for error whatsoever. The model tracks the data, adding a dimension to the regression surface for each data point. The model does not allow for any chance variation. Since real-world data always have some chance variation, future observations do not follow such an overfitted model. Recall that a statistical model should capture as much as possible of the *systematic* movement within the data set, leaving out pure errors—variation due to chance. An overfitted model forces the regression surface to go through every point. This is why the R^2 coefficient is high: The error part of the regression is forced to zero. (The R^2 in the student's model was actually 1.00; the computer-reported value 0.9999 is due to rounding in the computations.)

FIGURE 11–2   Overfitting a Data Set with a 15-Degree Polynomial

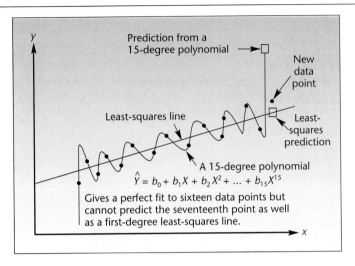

The idea of overfitting can be seen in another way. Multiple regression includes a technique called *polynomial regression*. In polynomial regression, we regress a dependent variable Y on powers of the independent variables. Each power of a variable, say, X_1^2, is considered an independent variable in its own right. For example, a fourth-degree polynomial regression in X is modeled as $Y = \beta_0 + \beta_1 X + \beta_2 X^2 + \beta_3 X^3 + \beta_4 X^4 + \epsilon$. This is treated as a four-variable regression equation. Overfitting in this context is shown in Figure 11–2.

As shown in the figure, a 15-degree polynomial provides a perfect fit to a set of 16 data points [here again, df for error $= 16 - (15 + 1) = 0$]. The prediction of a new data point using this model is worse than the prediction provided by a 1-degree model—a least-squares line. The line captures the systematic trend in the data, leaving the rest of the movements within the data to error. In this case, the straight line is a good statistical model, whereas the 15-degree polynomial—analogous to the model proposed by the student—is not. In the next section, we will formally develop the multiple regression model with k independent variables.

11–2   The *k*-Variable Multiple Regression Model

The population regression model of a dependent variable Y on a set of k independent variables $X_1, X_2, \ldots, X_k$ is given by

$$Y = \beta_0 + \beta_1 X_1 + \beta_2 X_2 + \cdots + \beta_k X_k + \epsilon \qquad (11\text{–}1)$$

where β_0 is the Y intercept of the regression surface and each β_i, $i = 1, \ldots, k$, is the slope of the regression surface—sometimes called the **response surface**—with respect to variable X_i.

As with the simple linear regression model, we have some assumptions.

Model assumptions:

1. For each observation, the error term ϵ is normally distributed with mean zero and standard deviation σ and is independent of the error terms associated with all other observations. That is,

$$\epsilon_j \sim N(0, \sigma^2) \qquad \text{for all } j = 1, 2, \ldots, n \qquad (11\text{–}2)$$

independent of other errors.[2]

2. In the context of regression analysis, the variables X_j are considered *fixed quantities*, although in the context of correlational analysis, they are random variables. In any case, X_j *are independent of the error term* ϵ. When we assume that X_j are fixed quantities, we are assuming that we have realizations of k variables X_j and that the only randomness in Y comes from the error term ϵ.

For a case with $k = 2$ variables, the response surface is a plane in three dimensions (the dimensions are Y, X_1, and X_2). The plane is the surface of average response $E(Y)$ for any combination of the two variables X_1 and X_2. The response surface is given by the equation for $E(Y)$, which is the expected value of equation 11–1 with two independent variables. Taking the expected value of Y gives the value 0 to the error term ϵ. The equations for Y and $E(Y)$ in the case of regression with two independent variables are

$$Y = \beta_0 + \beta_1 X_1 + \beta_2 X_2 + \epsilon \qquad (11\text{–}3)$$
$$E(Y) = \beta_0 + \beta_1 X_1 + \beta_2 X_2 \qquad (11\text{–}4)$$

These are equations analogous to the case of simple linear regression. Here, instead of a regression line, we have a regression plane. Some values of Y (i.e., combinations of the X_i variables times their coefficients β_i, and the errors ϵ) are shown in Figure 11–3. The figure shows the response surface, the plane corresponding to equation 11–4.

We estimate the regression parameters of equation 11–3 by the method of least squares. This is an extension of the procedure used in simple linear regression. In the case of two independent variables where the population model is equation 11–3, we need to estimate an equation of a plane that will minimize the sum of the squared errors $(Y - \hat{Y})^2$ over the entire data set of n points. The method is extendable to any k independent variables. In the case of $k = 2$, there are three equations, and their solutions are the least-squares estimates b_0, b_1, and b_2. These are estimates of the Y intercept, the slope of the plane with respect to X_1, and the slope of the plane with respect to X_2. The normal equations for $k = 2$ follow.

When the various sums Σy, Σx_1, and the other sums and products are entered into these equations, it is possible to solve the three equations for the three unknowns b_0, b_1, and b_2. These computations are always done by computer. We will, however, demonstrate the solution of equations 11–5 with a simple example.

---

[2]The multiple regression model is valid under less restrictive assumptions than these. The assumptions of normality of the errors allows us to perform t tests and F tests of model validity. Also, all we need is that the errors be *uncorrelated* with one another. However, normal distribution + noncorrelation = independence.

FIGURE 11–3 **A Two-Dimensional Response Surface** $E(Y) = \beta_0 + \beta_1 X_1 + \beta_2 X_2$ **and Some Points**

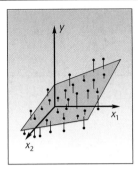

The normal equations for the case of two independent variables:

$$\Sigma\, y = nb_0 + b_1 \,\Sigma\, x_1 + b_2 \,\Sigma\, x_2$$

$$\Sigma\, x_1 y = b_0 \,\Sigma\, x_1 + b_1 \,\Sigma\, x_1^2 + b_2 \,\Sigma\, x_1 x_2 \qquad (11\text{–}5)$$

$$\Sigma\, x_2 y = b_0 \,\Sigma\, x_2 + b_1 \,\Sigma\, x_1 x_2 + b_2 \,\Sigma\, x_2^2$$

EXAMPLE 11–1

Alka-Seltzer recently embarked on an in-store promotional campaign, with displays of its antacid featured prominently in supermarkets. The company also ran its usual radio and television commercials. Over a period of 10 weeks, the company kept track of its expenditure on radio and television advertising, variable X_1, as well as its spending on in-store displays, variable X_2. The resulting sales for each week in the area studied were recorded as the dependent variable Y. The company analyst conducting the study hypothesized a linear regression model of the form

$$Y = \beta_0 + \beta_1 X_1 + \beta_2 X_2 + \epsilon$$

linking sales volume with the two independent variables, advertising and in-store promotions. The analyst wanted to use the available data, considered a random sample of 10 weekly observations, to estimate the parameters of the regression relationship.

Solution

Table 11–1 gives the data for this study in terms of Y, X_1, and X_2, all in thousands of dollars. The table also gives additional columns of products and squares of data values needed for the solution of the normal equations. These columns are $X_1 X_2$, X_1^2, X_2^2, $X_1 Y$, and $X_2 Y$. The sums of these columns are then substituted into equations 11–5, which are solved for the estimates b_0, b_1, and b_2 of the regression parameters.

From Table 11–1, the sums needed for the solution of the normal equations are $\Sigma y = 743$, $\Sigma x_1 = 123$, $\Sigma x_2 = 65$, $\Sigma x_1 y = 9{,}382$, $\Sigma x_2 y = 5{,}040$, $\Sigma x_1 x_2 = 869$, $\Sigma x_1^2 = 1{,}615$, and $\Sigma x_2^2 = 509$. When these sums are substituted into equations 11–5, we get the resulting normal equations:

$$743 = 10 b_0 + 123 b_1 + 65 b_2$$

$$9{,}382 = 123 b_0 + 1{,}615 b_1 + 869 b_2$$

$$5{,}040 = 65 b_0 + 869 b_1 + 509 b_2$$

TABLE 11–1   Various Quantities Needed for the Solution of the Normal Equations for Example 11–1 (numbers are in thousands of dollars)

Y	X_1	X_2	X_1X_2	x_1^2	x_2^2	X_1Y	X_2Y
72	12	5	60	144	25	864	360
76	11	8	88	121	64	836	608
78	15	6	90	225	36	1,170	468
70	10	5	50	100	25	700	350
68	11	3	33	121	9	748	204
80	16	9	144	256	81	1,280	720
82	14	12	168	196	144	1,148	984
65	8	4	32	64	16	520	260
62	8	3	24	64	9	496	186
90	18	10	180	324	100	1,620	900
743	123	65	869	1,615	509	9,382	5,040

Solution of this system of equations by substitution, or by any other method of solution, gives

$$b_0 = 47.164942 \qquad b_1 = 1.5990404 \qquad b_2 = 1.1487479$$

These are the *least-squares estimates* of the true regression parameters β_0, β_1, and β_2. Recall that the normal equations (equations 11–5) are originally obtained by calculus methods. (They are the results of differentiating the sum of squared errors with respect to the regression coefficients and setting the results to zero.)

Figure 11–4 shows the results page of the template on which the same problem has been solved. The template is described later.

The meaning of the estimates b_0, b_1, and b_2 as the Y intercept, the slope with respect to X_1, and the slope with respect to X_2, respectively, of the estimated regression surface is illustrated in Figure 11–5.

The general multiple regression model, equation 11–1, has one Y intercept parameter and k slope parameters. Each slope parameter β_i, $i = 1, \ldots, k$, represents the amount of increase (or decrease, in case it is negative) in $E(Y)$ for an increase of 1 unit in variable X_i when all other variables are kept constant. The regression coefficients β_i are therefore sometimes referred to as *net regression coefficients* because they represent the net change in $E(Y)$ for a change of 1 unit in the variable they represent, all

FIGURE 11–4   The Results from the Template
[Multiple Regression.xls; Sheet: Results]

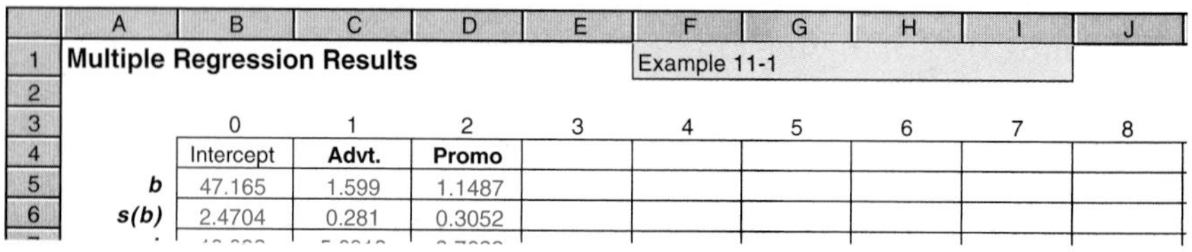

	A	B	C	D	E	F	G	H	I	J
1	Multiple Regression Results					Example 11-1				
2										
3		0	1	2	3	4	5	6	7	8
4		Intercept	Advt.	Promo						
5	b	47.165	1.599	1.1487						
6	s(b)	2.4704	0.281	0.3052						

else remaining constant.[3] The X_i variables should be independent of one another, as we want each coefficient β_i to reflect change in $E(Y)$ for a unit change in X_i, *with all other independent variables left constant.* This is often difficult to achieve in multiple regression analysis since the explanatory variables are often interrelated in some way.

The Estimated Regression Relationship

> The **estimated regression relationship** is
>
> $$\hat{Y} = b_0 + b_1X_1 + b_2X_2 + \cdots + b_kX_k \qquad (11\text{--}6)$$
>
> where $\hat{Y}$ is the predicted value of Y, the value lying *on* the estimated regression surface. The terms b_i, $i = 0, \ldots, k$, are the least-squares estimates of the population regression parameters β_i.

The least-squares estimators giving us the b_i are BLUEs (best linear unbiased estimators).

It is also possible to write the estimated regression relationship in a way that shows how each value of Y is expressed as a linear combination of the values of X_i plus an error term. This is given in equation 11–7.

> $$y_j = b_0 + b_1x_{1j} + b_2x_{2j} + \cdots + b_kx_{kj} + e_j \qquad j = 1, \ldots, n \quad (11\text{--}7)$$

In Example 11–1 the estimated regression relationship of sales volume Y on advertising X_1 and in-store promotions X_2 is given by

$$\hat{Y} = 47.164942 + 1.5990404X_1 + 1.1487479X_2$$

FIGURE 11–5   The Least-Squares Regression Surface for Example 11–1

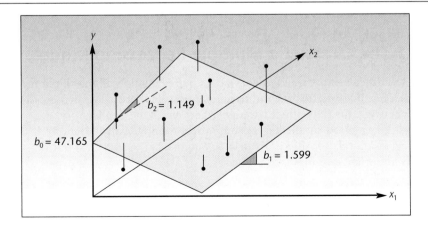

---

[3]For the reader with knowledge of calculus, we note that the coefficient β_i is the partial derivative of $E(Y)$ with respect to X_i: $\beta_i = \partial E(Y)/\partial X_i$.

PROBLEMS

11–1. What are the assumptions underlying the multiple regression model? What is the purpose of the assumption of normality of the errors?

11–2. In a regression analysis of job performance Y versus the explanatory variables age X_1 and experience X_2, the estimated coefficient b_2 is equal to 1.34. Explain the meaning of this estimate in terms of the impact of experience on performance.

11–3. In terms of model assumptions, what is the difference between a multiple regression model with k independent variables and a correlation analysis involving these variables?

11–4. What is a response surface? For a regression model with seven independent variables, what is the dimensionality of the response surface?

11–5. Again, for a multiple regression model with $k = 7$ independent variables, how many normal equations are there leading to the values of the estimates of the regression parameters?

11–6. What are the BLUEs of the regression parameters?

11–7. For a multiple regression model with two independent variables, results of the analysis include $\Sigma y = 852$, $\Sigma x_1 = 155$, $\Sigma x_2 = 88$, $\Sigma x_1 y = 11{,}423$, $\Sigma x_2 y = 8{,}320$, $\Sigma x_1 x_2 = 1{,}055$, $\Sigma x_1^2 = 2{,}125$, and $\Sigma x_2^2 = 768$, $n = 100$. Solve the normal equations for this regression model, and give the estimates of the parameters.

11–8. A Realtor is interested in assessing the impact of size (in square feet) and distance from the center of town (in miles) on the value of homes (in thousands of dollars) in a certain area. Nine randomly chosen houses are selected; data are as follows.

$$Y \text{ (value):}\quad 345, 238, 452, 422, 328, 375, 660, 466, 290$$
$$X_1 \text{ (size):}\quad 1{,}650, 1{,}870, 2{,}230, 1{,}740, 1{,}900, 2{,}000, 3{,}200, 1{,}860, 1{,}230$$
$$X_2 \text{ (distance):}\quad 3.5, 0.5, 1.5, 4.5, 1.8, 0.1, 3.4, 3.0, 1.0$$

Compute the estimated regression coefficients, and explain their meaning.

11–9. The estimated regression coefficients in Example 11–1 are $b_0 = 47.165$, $b_1 = 1.599$, and $b_2 = 1.149$ (rounded to three decimal places). Explain the meaning of each of the three numbers in terms of the situation presented in the example.

11–3   The *F* Test of a Multiple Regression Model

The first statistical test we need to conduct in our evaluation of a multiple regression model is a test that will answer the basic question: Is there a linear regression relationship between the dependent variable Y and *any* of the explanatory, independent variables X_i suggested by the regression equation under consideration? If the proposed regression relationship is given in equation 11–1, a statistical test that can answer this important question is as follows.

A statistical hypothesis test for the existence of a linear relationship between Y and any of the X_i is

$$H_0: \quad \beta_1 = \beta_2 = \beta_3 = \cdots = \beta_k = 0 \qquad (11\text{–}8)$$
$$H_1: \quad \text{Not all the } \beta_i \ (i = 1, \ldots, k) \text{ are zero}$$

If the null hypothesis is true, there is no linear relationship between Y and any of the independent variables in the proposed regression equation. In such a case, there is nothing more to do. There is no regression. If, on the other hand, we reject the null hypothesis, there is statistical evidence to conclude that there is a regression relationship between Y and at least one of the independent variables proposed in the regression model.

To carry out the important test in equation 11–8, we will perform an analysis of variance. The ANOVA is the same as the one given in Chapter 10 for simple linear regression, except that here we have k independent variables instead of just 1. Therefore, the F test of the analysis of variance is not equivalent to the t test for the significance of the slope parameter, as was the case in Chapter 10. Since in multiple regression there are k slope parameters, we have k different t tests to follow the ANOVA.

Figure 11–6 is an extension of Figure 10–21 to the case of $k = 2$ independent variables—to a regression plane instead of a regression line. The figure shows a particular data point y, the predicted point $\hat{y}$ which lies on the estimated regression surface, and the mean of the dependent variable $\bar{y}$. The figure shows the three deviations associated with the data point: the error deviation $y - \hat{y}$, the regression deviation $\hat{y} - \bar{y}$, and the total deviation $y - \bar{y}$. As seen from the figure, the three deviations satisfy the relation: Total deviation = Regression deviation + Error deviation. As in the case of simple linear regression, when we square the deviations and sum them over all n data points, we get the following relation for the sums of squares. The sums of squares are denoted by SST for the total sum of squares, SSR for the regression sum of squares, and SSE for the error sum of squares.

$$SST = SSR + SSE \tag{11–9}$$

This is the same as equation 10–29. The difference lies in the degrees of freedom. In simple linear regression, the degrees of freedom for error were $n - 2$ because two parameters, an intercept and a slope, were estimated from a data set of n points. In multiple regression, we estimate k slope parameters and an intercept from a data set of

FIGURE 11–6   Decomposition of the Total Deviation in Multiple Regression Analysis

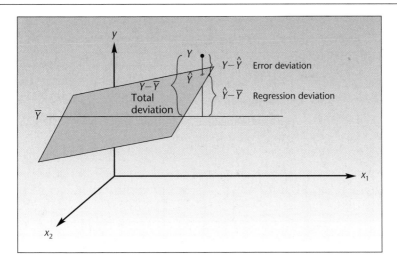

n points. Therefore, the degrees of freedom for error are $n - (k + 1)$. The degrees of freedom for the regression are k, and the total degrees of freedom are $n - 1$. Again, the degrees of freedom are additive. Table 11–2 is the ANOVA table for a multiple regression model with k independent variables.

For Example 11–1, we present the ANOVA table computed by using the template. The results are shown in Table 11–3. Since the p-value is small, we reject the null hypothesis that both slope parameters β_1 and β_2 are zero (equation 11–8), in favor of the alternative that the slope parameters are not both zero. We conclude that there is evidence of a linear regression relationship between sales and at least one of the two variables, advertising or in-store promotions (or both). The F test is shown in Figure 11–7.

TABLE 11–2 ANOVA Table for Multiple Regression

Source of Variation	Sum of Squares	Degrees of Freedom	Mean Square	F Ratio
Regression	SSR	k	$MSR = \dfrac{SSR}{k}$	$F = \dfrac{MSR}{MSE}$
Error	SSE	$n - (k + 1)$	$MSE = \dfrac{SSE}{n - (k + 1)}$	
Total	SST	$n - 1$		

TABLE 11–3 ANOVA Table Produced by the Template
[Multiple Regression.xls; Sheet: Results]

	ANOVA Table									
11										
12	Source	SS	df	MS	F	F_{Critical}	p-value			
13	Regn.	630.538	2	315.27	86.335	4.7374	0.0000		s	1.9109
14	Error	25.5619	7	3.6517						
15	Total	656.1	9			R^2	0.9610		Adjusted R^2	0.9499
16										

FIGURE 11–7 Regression F Test for Example 11–1

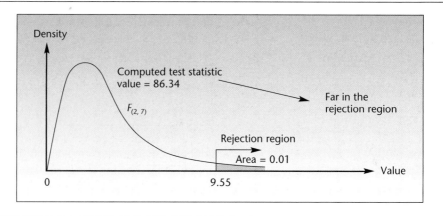

Note that since there are two independent variables in Example 11–1, we do not yet know whether there is a regression relationship between sales and both advertising and in-store promotions, or whether the relationship exists between sales and one of the two variables only—and if so, which one. All we know is that our data present statistical evidence to conclude that a relationship exists between sales and at least one of the two independent variables. This is, of course, true for all cases where there are two or more independent variables. The F test only tells us that there is evidence of a relationship between the dependent variable and at least one of the independent variables in the full regression equation under consideration. Once we conclude that a relationship exists, we need to conduct separate tests to determine which of the slope parameters β_i, where $i = 1, \ldots, k$, are different from zero. There are, therefore, k further tests.

Compare the use of ANOVA tables in multiple regression with the analysis of variance discussed in Chapter 9. Once we rejected the null hypothesis that all r population means are equal, we required further analysis (the Tukey procedure or an alternative technique) to determine where the differences existed. In multiple regression, the further tests necessary for determining which variables are important are t tests. These tests tell us which variables help explain the variation in the values of the dependent variable and which variables have no explanatory power and should be eliminated from the regression model. Before we get to the separate tests of multiple regression parameters, we want to be able to evaluate how good the regression relationship is as a whole.

PROBLEMS

11–10. Explain what is tested by the hypothesis test in equation 11–8. What conclusion should be reached if the null hypothesis is not rejected? What conclusion should be reached if the null hypothesis is rejected?

11–11. In a multiple regression model with 12 independent variables, what are the degrees of freedom for error? Explain.

11–12. An article[4] about banks' profitability lists four independent variables that may affect profitability. A regression analysis with four independent variables is carried out. The data are a random sample of 120 observations. The results of the analysis include SSE = 4,560 and SSR = 562. Is there a regression relationship between the dependent variable and any of the four proposed explanatory variables? Explain.

11–13. Avis is interested in estimating weekly costs of maintenance of its rental cars of a certain size based on these variables: number of miles driven during the week, number of renters during the week, the car's total mileage, and the car's age. A regression analysis is carried out, and the results include $n = 45$ cars (each car selected randomly, during a randomly selected week of operation), SSR = 7,768, and SST = 15,673. Construct a complete ANOVA table for this problem, and test for the existence of a linear regression relationship between weekly maintenance costs and any of the four independent variables considered.

11–14. Nissan Motor Company wanted to find leverage factors for marketing the Maxima model in the United States. The company hired a market research firm in New York City to carry out an analysis of the factors that make people favor the model in question. As part of the analysis, the market research firm selected a random

---

[4]"Regional, but Less Power," *The Economist*, March 30, 1991, p. 74.

sample of 17 people and asked them to fill out a questionnaire about the importance of three automobile characteristics: prestige, comfort, and economy. Each respondent reported the importance he or she gave to each of the three attributes on a 0–100 scale. Each respondent then spent some time becoming acquainted with the car's features and drove it on a test run. Finally, each of the respondents gave an overall appeal score for the model on a 0–100 scale. The appeal score was considered the dependent variable, and the three attribute scores were considered independent variables. A multiple regression analysis was carried out, and the results included the following ANOVA table. Complete the table. Based on the results, is there a regression relationship between the appeal score and at least one of the attribute variables? Explain.

```
Analysis of Variance
   SOURCE        DF        SS        MS
Regression              7474.0
Error
Total                   8146.5
```

11–4   How Good Is the Regression?

The mean square error MSE is an unbiased estimator of the variance of the population errors ϵ, which we denote by σ^2. The mean square error is defined in equation 11–10.

The **mean square error** is

$$\text{MSE} = \frac{\text{SSE}}{n - (k + 1)} = \frac{\sum_{j=1}^{n}(y_j - \hat{y}_j)^2}{n - (k + 1)} \qquad (11\text{–}10)$$

The errors resulting from the fit of a regression surface to our set of n data points are shown in Figure 11–8. The smaller the errors, the better the fit of the regression model. Since the mean square error is the average squared error, where averaging is done by dividing by the degrees of freedom, MSE is a measure of how well the regression fits the data. The square root of MSE is an estimator of the standard deviation of the population regression errors σ. (Note that a square root of an unbiased estimator is not unbiased; therefore, $\sqrt{\text{MSE}}$ is not an unbiased estimator of σ, but is still a good estimator.) The square root of MSE is usually denoted by s and is referred to as the *standard error of estimate*.

The **standard error of estimate** is

$$s = \sqrt{\text{MSE}} \qquad (11\text{–}11)$$

This statistic is usually reported in computer output of multiple regression analysis. The mean square error and its square root are measures of the size of the errors in

FIGURE 11–8   Errors in a Multiple Regression Model (shown for *k* = 2)

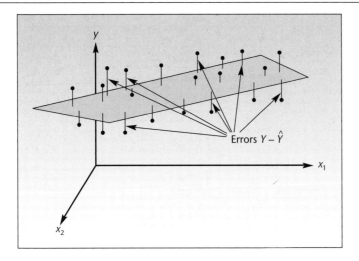

regression and do not give an indication about the *explained* component of the regression fit (see Figure 11–6, showing the breakdown of the total deviation of any data point to the error and regression components). A measure of regression fit that does incorporate the explained as well as the unexplained components is the *multiple coefficient of determination,* denoted by R^2. This measure is an extension to multiple regression of the coefficient of determination in simple linear regression, denoted by r^2.

> The **multiple coefficient of determination R^2** measures the proportion of the variation in the dependent variable that is explained by the combination of the independent variables in the multiple regression model:
>
> $$R^2 = \frac{SSR}{SST} = 1 - \frac{SSE}{SST} \qquad (11–12)$$

Note that R^2 is also equal to SSR/SST because SST = SSR + SSE. We prefer the definition in equation 11–12 for consistency with another measure of how well the regression model fits our data, the *adjusted* multiple coefficient of determination, which will be introduced shortly.

The measures SSE, SSR, and SST are reported in the ANOVA table for multiple regression. Because of the importance of R^2, however, it is reported separately in computer output of multiple regression analysis. The square root of the multiple coefficient of determination, $R = \sqrt{R^2}$, is the **multiple correlation coefficient.** In the context of multiple regression analysis (rather than correlation analysis), the multiple coefficient of determination R^2 is the important measure, not R. The coefficient of determination measures the percentage of variation in Y explained by the X variables; thus, it is an important measure of how well the regression model fits the data. In correlation analysis, where the X_i variables as well as Y are assumed to be random variables, the multiple correlation coefficient R measures the strength of the linear relationship between Y and the k variables X_i.

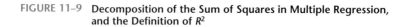

FIGURE 11–9 Decomposition of the Sum of Squares in Multiple Regression, and the Definition of R^2

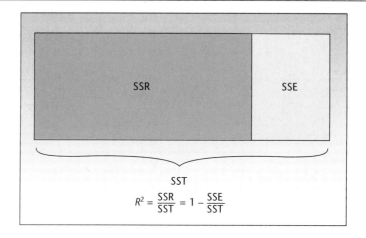

Figure 11–9 shows the breakdown of the total sum of squares (the sum of squared deviations of all n data points from the mean of Y; see Figure 11–6) into the sum of squares due to the regression (the explained variation) and the sum of squares due to error (the unexplained variation). The interpretation of R^2 is the same as that of r^2 in simple linear regression. The difference is that here the regression errors are measured as deviations from a regression surface that has higher dimensionality than a regression line. The multiple coefficient of determination R^2 is a very useful measure of performance of a multiple regression model. It does, however, have some limitations.

Recall the story at the beginning of this chapter about the student who wanted to predict the nation's economic future with a multiple regression model that had many variables. It turns out that, for any given data set of n points, as the number of variables in the regression model increases, so does R^2. You have already seen how this happens: The greater the number of variables in the regression equation, the more the regression surface "chases" the data until it overfits them. Since the fit of the regression model increases as we increase the number of variables, R^2 cannot decrease and approaches 1.00, or 100% explained variation in Y. This can be very deceptive, as the model—while appearing to fit the data very well—would produce poor predictions.

Therefore, a new measure of fit of a multiple regression model must be introduced: the *adjusted* (or corrected) *multiple coefficient of determination.* The adjusted multiple coefficient of determination, denoted $\bar{R}^2$, is the multiple coefficient of determination corrected for degrees of freedom. It accounts, therefore, not only for SSE and SST, but also for their appropriate degrees of freedom. This measure does not always increase as new variables are entered into our regression equation. When $\bar{R}^2$ does increase as a new variable is entered into the regression equation, it may be worthwhile to include the variable in the equation. The adjusted measure is defined as follows:

The **adjusted multiple coefficient of determination** is

$$\bar{R}^2 = 1 - \frac{SSE/[n - (k + 1)]}{SST/(n - 1)} \qquad (11\text{--}13)$$

The adjusted R^2 is the R^2 (defined in equation 11–12) where both SSE and SST are divided by their respective degrees of freedom. Since $SSE/[n - (k + 1)]$ is the MSE, we can say that, in a sense, $\bar{R}^2$ is a mixture of the two measures of the performance of a regression model: MSE and R^2. The denominator on the right-hand side of equation 11–13 would be *mean square total,* were we to define such a measure.

Computer output for multiple regression analysis usually includes the adjusted R^2. If it is not reported, we can get $\bar{R}^2$ from R^2 by a simple formula:

$$\bar{R}^2 = 1 - (1 - R^2)\frac{n - 1}{n - (k + 1)} \qquad (11\text{--}14)$$

It is instructional to prove this relation between R^2 and $\bar{R}^2$, and the proof is left as an exercise. *Note:* Unless the number of variables is relatively large compared to the number of data points (as in the economics student's problem), R^2 and $\bar{R}^2$ are close to each other in value. Thus, in many situations, it suffices to consider the uncorrected measures R^2. We evaluate the fit of a multiple regression model based on this measure. When we are considering whether to include an independent variable in a regression model that already contains other independent variables, the increase in R^2 when the new variable is added must be weighed against the loss of 1 degree of freedom for error resulting from the addition of the variable (a new parameter would be added to the equation). With a relatively small data set and several independent variables in the model, it may not be worthwhile to add a new variable if R^2 increases, say, from 0.85 to 0.86. As mentioned earlier, in such cases, the adjusted measure $\bar{R}^2$ may be a good indicator of whether to include the new variable. We may decide to include the variable if $\bar{R}^2$ increases when the variable is added.

Of several possible multiple regression models with different independent variables, the model that minimizes MSE will also maximize $\bar{R}^2$. This should not surprise you, since MSE is related to the adjusted measure $\bar{R}^2$. The use of the two criteria MSE and $\bar{R}^2$ in selecting variables to be included in a regression model will be discussed in a later section.

We now return to the analysis of Example 11–1. Note that in Table 11–3 $R^2 = 0.961$, which means that 96.1% of the variation in sales volume is explained by the combination of the two independent variables, advertising and in-store promotions. Note also that the adjusted R^2 is 0.95, which is very close to the unadjusted measure. We conclude that the regression model fits the data very well since a high percentage of the variation in Y is explained by X_1, and/or X_2 (we do not yet know which of the two variables, if not both, is important). The standard error of estimate s is an estimate of σ, the standard deviation of the population regression errors. Note that R^2 is also a *statistic,* like s or MSE. It is a sample estimate of the population multiple coefficient of determination ρ^2, a measure of the proportion of the explained variation in Y in the entire population of Y and X_i values.

FIGURE 11–10　Measures of Performance of a Regression Model and the ANOVA Table

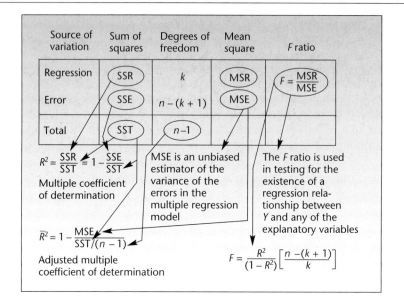

All three measures of the performance of a regression model—MSE (and its square root s), the coefficient of determination R^2, and the adjusted measure $\bar{R}^2$—are obtainable from quantities reported in the ANOVA table. This is shown in Figure 11–10, which demonstrates the relations among the different measures.

PROBLEMS

11–15.　Under what conditions is it important to consider the adjusted multiple coefficient of determination?

11–16.　Explain why the multiple coefficient of determination never decreases as variables are added to the multiple regression model.

11–17.　Would it be useful to consider an adjusted coefficient of determination in a simple linear regression situation? Explain.

11–18.　Prove equation 11–14.

11–19.　Can you judge how well a regression model fits the data by considering the mean square error only? Explain.

11–20.　A multiple regression analysis is carried out with three independent variables and a data set of 44 points. The results include SSE = 6,980 and SSR = 11,778. Construct an ANOVA table, and find MSE, s, R^2, $\bar{R}^2$, and the F ratio. Analyze the results. Is this a good regression model? Why or why not?

11–21.　A portion of the regression output for the Nissan Motor Company study of problem 11–14 follows. Interpret the findings, and show how these results are obtainable from the ANOVA table results presented in problem 11–14. How good is the regression relationship between the overall appeal score for the automobile and the attribute-importance scores? Also, obtain the adjusted R^2 from the multiple coefficient of determination.

```
     S  = 7.192          R² = 91.7%          R² (ADJ)  = 89.8%
```

11–22. An article in *Journal of Marketing Research* reports the results of a regression analysis of brand attitude on the independent variables "ad reactions" and "brand reactions."[5] The reported results are $n = 103$ and $R^2 = 0.67$. Find the adjusted R^2. Also, conduct the F test for the existence of a linear relationship between the dependent variable and at least one of the independent variables. How good is the relationship?

11–23. In the Nissan Motor Company situation in problem 11–21, suppose that a new variable is considered for inclusion in the equation and a new regression relationship is analyzed with the new variable included. Suppose that the resulting multiple coefficient of determination is $R^2 = 91.8\%$. Find the adjusted multiple coefficient of determination. Should the new variable be included in the final regression equation? Give your reasons for including or excluding the variable.

11–24. In the situation of problem 11–22, suppose that a new regression is run with the variable "brand reactions" removed from the equation. The resulting new coefficient of determination is $R^2 = 0.61$. Should the new equation, with the variable omitted, be used for prediction of the dependent variable?

11–25. The following excerpt reports the results of a regression of excess stock returns on firm size and stock price, both variables being ranked on some scale. Explain, critique, and evaluate the reported results.

Estimated Coefficient Value (t Statistic)

INTCPT	SIZRNK	PRCRNK	ADJUSTED-R^2

Ordinary Least-Squares Regression Results

INTCPT	SIZRNK	PRCRNK	ADJUSTED-R^2
0.484	−0.030	−0.017	0.093
(5.71)***	(−2.91)***	(−1.66)*	

*Denotes significance at the 10% level.
**Denotes significance at the 5% level.
***Denotes significance at the 1% level.
Source: M. Bajaj and A. Vijh, "Trading Behavior and Market Reaction," *Journal of Finance* 50, no. 1 (March 1995), p. 265.

11–26. A study of Dutch tourism behavior included a regression analysis using a sample of 713 respondents. The dependent variable, number of miles traveled on vacation, was regressed on the independent variables, family size and family income; and the multiple coefficient of determination was $R^2 = 0.72$. Find the adjusted multiple coefficient of determination $\bar{R}^2$. Is this a good regression model? Explain.

11–27. In a regression analysis with six independent variables and a data set of 250 points, it is found that SSE = 5,445 and SST = 22,679. Construct an ANOVA table, conduct the F test, find R^2 and $\bar{R}^2$, and find MSE.

11–5 Tests of the Significance of Individual Regression Parameters

Until now, we have discussed the multiple regression model in general. We saw how to test for the existence of a regression relationship between Y and at least one of a set of independent X_i variables by using an F test. We also saw how to evaluate the fit of the general regression model by using the multiple coefficient of determination and the adjusted multiple coefficient of determination. We have not yet seen, however, how to evaluate the significance of individual regression parameters β_i. A test for the significance of an individual parameter is important because it tells us whether the

---

[5]K. Keller, "Cue Compatibility and Framing in Advertising," *Journal of Marketing Research,* February 1991, p. 51.

variable in question, X_h, has explanatory power with respect to the dependent variable. Such a test tells us whether the variable in question should be included in the regression equation.

In the last section, we saw that some indication about the benefit from inclusion of a particular variable in the regression equation is gained by comparing the adjusted coefficient of determination of a regression that includes the variable of interest with the value of this measure when the variable is not included. In this section, we will perform individual t tests for the significance of each slope parameter β_i. As we will see, however, we must use caution in interpreting the results of the individual t tests.

In Chapter 10 we saw that the hypothesis test

$$H_0: \beta_1 = 0$$
$$H_1: \beta_1 \neq 0$$

can be carried out using either a t statistic $t = b_1/s(b_1)$ or an F statistic. Both tests were shown to be equivalent because F with 1 degree of freedom for the numerator is a squared t random variable with the same number of degrees of freedom as the denominator of F. In simple linear regression, there is only one slope, β_1, and if that slope is zero, there is no linear regression relationship. In multiple regression, where $k > 1$, the two tests are not equivalent. The F test tells us whether a relationship exists between Y and at least one of the X_i, and the k ensuing t tests tell us which of the X_i variables are important and should be included in the regression equation. From the similarity of this situation with the situation of analysis of variance discussed in Chapter 9, you probably have guessed at least one of the potential problems: The individual t tests are each carried out at a single level of significance α, and we cannot determine the level of significance of the family of all k tests of the regression slopes jointly. The problem is further complicated by the fact that the tests are not independent of each other because the regression estimates come from the same data set.

Recall that hypothesis tests and confidence intervals are related. We may test hypotheses about regression slope parameters (in particular, the hypothesis that a slope parameter is equal to zero), or we may construct confidence intervals for the values of the slope parameters. If a 95% confidence interval for a slope parameter β_h contains the point zero, then the hypothesis test $H_0: \beta_h = 0$ carried out using $\alpha = 0.05$ would lead to nonrejection of the null hypothesis and thus to the conclusion that there is no evidence that the variable X_h has a linear relationship with Y.

We will demonstrate the interdependence of the separate tests of significance of the slope parameters with the use of confidence intervals for these parameters. When $k = 2$, there are two regression slope parameters: β_1 and β_2. (As in simple linear regression, usually there is no interest in testing hypotheses about the intercept parameter.) The sample estimators of the two regression parameters are b_1 and b_2. These estimators (and their standard errors) are correlated with each other (and assumed to be normally distributed). Therefore, the joint confidence region for the pair of parameters (β_1, β_2) is an *ellipse*. If we consider the estimators b_1 and b_2 separately, the joint confidence region will be a rectangle, with each side a separate confidence interval for a single parameter. This is demonstrated in Figure 11–11. A point inside the rectangle formed by the two separate confidence intervals for the parameters, such as point A in the figure, seems like a plausible value for the pair of regression slopes (β_1, β_2) but is not *jointly* plausible for the parameters. Only points inside the ellipse in the figure are jointly plausible for the pair of parameters.

Another problem that may arise in making inferences about individual regression slope coefficients is due to **multicollinearity**—the problem of correlations among the

FIGURE 11–11   Joint Confidence Region and Individual Confidence Intervals
for the Slope Parameters β_1 and β_2

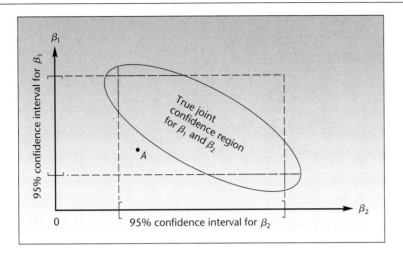

independent variables themselves. In multiple regression, we hope to have a strong correlation between each independent variable and the dependent variable Y. Such correlations give the independent X_i variables predictive power with respect to Y. However, we do not want the independent variables to be correlated with one another. When the independent variables are correlated with one another, we have multicollinearity. When this happens, the independent variables rob one another of explanatory power. Many problems may then arise. One problem is that the standard errors of the individual slope estimators become unusually high, making the slope coefficients seem statistically not significant (not different from zero). For example, if we run a regression of job performance Y versus the variables age X_1 and experience X_2, we may encounter multicollinearity. Since, in general, as age increases so does experience, the two independent variables are not independent of each other; the two variables rob each other of explanatory power with respect to Y. If we run this regression, it is likely that—even though experience affects job performance—the individual test for significance of the slope parameter β_2 would lead to nonrejection of the null hypothesis that this slope parameter is equal to zero. Much will be said later about the problem of multicollinearity. It is important to remember that in the presence of multicollinearity, the significance of any regression parameter depends on the other variables included in the regression equation. Multicollinearity may also cause the signs of some estimated regression parameters to be the opposite of what we expect.

Another problem that may affect the individual tests of significance of model parameters occurs when one of the model assumptions is violated. Recall from Section 11–2 that one of the assumptions of the regression model is that the error terms ϵ_j are uncorrelated with one another. When this condition does not hold, as may happen when our data are time series observations (observations ordered by time: yearly data, monthly data, etc.), we encounter the problem of autocorrelation of the errors. This causes the standard errors of the slope estimators to be unusually small, making some parameters seem more significant than they really are. This problem, too, should be considered, and we will discuss it in detail later.

Forewarned of problems that may arise, we now consider the tests of the individual regression parameters. In a regression model of Y versus k independent variables $X_1, X_2, \ldots, X_k$, there are k tests of significance of the slope parameters $\beta_1, \beta_2, \ldots, \beta_k$:

Hypothesis tests about individual regression slope parameters:

$$(1) \quad H_0: \beta_1 = 0$$
$$ H_1: \beta_1 \neq 0$$
$$(2) \quad H_0: \beta_2 = 0$$
$$ H_1: \beta_2 \neq 0 \cdot \qquad\qquad (11\text{--}15)$$
$$\cdot \qquad\qquad \cdot$$
$$\cdot \qquad\qquad \cdot$$
$$\cdot \qquad\qquad \cdot$$
$$(k) \quad H_0: \beta_k = 0$$
$$ H_1: \beta_k \neq 0$$

These tests are carried out by comparing each test statistic with a critical point of the distribution of the test statistic. The distribution of each test statistic, when the appropriate null hypothesis is true, is the t distribution with $n - (k + 1)$ degrees of freedom. The distribution depends on our assumption that the regression errors are normally distributed. The test statistic for each hypothesis test (i) in equations 11–15 (where $i = 1, 2, \ldots, k$) is the slope estimate b_i, divided by the standard error of the estimator $s(b_i)$. The estimates and the standard errors are reported in the computer output. Each $s(b_i)$ is an estimate of the population standard deviation of the estimator $\sigma(b_i)$, which is unknown to us.[6] The test statistics for the hypothesis tests (1) through (k) in equations 11–15 are as follows:

Test statistics for tests about individual regression slope parameters:

For test i $(i = 1, \ldots, k)$:

$$t_{[n - (k + 1)]} = \frac{b_i - 0}{s(b_i)} \qquad\qquad (11\text{--}16)$$

We write each test statistic as the estimate minus zero (the null-hypothesis value of β_i) to stress the fact that we may test the null hypothesis that β_i is equal to any number, not necessarily zero. Testing for equality to zero is most important because it tells us whether there is evidence that variable X_i has a linear relationship with Y. It tells us whether there is statistical evidence that variable X_i has explanatory power with respect to the dependent variable.

Let us look at a quick example. Suppose that a multiple regression analysis is carried out relating the dependent variable Y to five independent variables X_1, X_2, X_3, X_4,

---

[6]Each $s(b_i)$ is the product of $s = \sqrt{\text{MSE}}$ and a term denoted by c_i, which is a diagonal element in a matrix obtained in the regression computations. You need not worry about matrices. However, the matrix approach to multiple regression is discussed in a section at the end of this chapter for the benefit of students familiar with matrix theory.

and X_5. In addition, suppose that the F test resulted in rejection of the null hypothesis that none of the predictor variables has any explanatory power with respect to Y; suppose also that R^2 of the regression is respectably high. As a result, we believe that the regression equation gives a good fit to the data and potentially may be used for prediction purposes. Our task now is to test the importance of each of the X_i variables separately. Suppose that the sample size used in this regression analysis is $n = 150$. The results of the regression estimation procedure are given in Table 11–4.

From the information in Table 11–4, which variables are important, and which are not? Note that the first variable listed is "Constant." This is the Y intercept. As we noted earlier, testing whether the intercept is zero is less important than testing whether the coefficient parameter of any of the k variables is zero. Still, we may do so by dividing the reported coefficient estimate, 53.12, by its standard error, 5.43. The result is the value of the test statistic that has a t distribution with $n - (k + 1) = 150 - 6 = 144$ degrees of freedom when the null hypothesis that the intercept is zero is true. For manual calculation purposes, we shall approximate this t random variable as a standard normal variable Z. The test statistic value is $z = 53.12/5.43 = 9.78$. This value is greater than 1.96, and we may reject the null hypothesis that β_0 is equal to zero at the $\alpha = 0.05$ level of significance. Actually, the p-value is very small. The regression hyperplane, therefore, most probably does not pass through the origin.

Let us now turn to the tests of significance of the slope parameters of the variables in the regression equation. We start with the test for the significance of variable X_1 as a predictor variable. The hypothesis test is H_0: $\beta_1 = 0$ versus H_1: $\beta_1 \neq 0$. We now compute our test statistic (again, we will use Z for $t_{(144)}$):

$$z = \frac{b_1 - 0}{s(b_1)} = \frac{2.03}{0.22} = 9.227$$

The value of the test statistic, 9.227, lies far in the right-hand rejection region of Z for any conventional level of significance; the p-value is very small. We therefore conclude that there is statistical evidence that the slope of Y with respect to X_1, the population parameter β_1, is not zero. Variable X_1 is shown to have some explanatory power with respect to the dependent variable.

If it is not zero, what is the value of β_1? The parameter, as in the case of all population parameters, is not known to us. An unbiased estimate of the parameter's value is $b_1 = 2.03$. We can also compute a confidence interval for β_1. A 95% confidence interval for β_1 is $b_1 \pm 1.96s(b_1) = 2.03 \pm 1.96(0.22) = [1.599, 2.461]$. Based on our data and the validity of our assumptions, we can be 95% confident that the true slope of Y with respect to X_1 is anywhere from 1.599 to 2.461. Figure 11–12 shows the hypothesis test for the significance of variable X_1.

TABLE 11–4 **Regression Results for Individual Parameters**

Variable	Coefficient Estimate	Standard Error
Constant	53.12	5.43
X_1	2.03	0.22
X_2	5.60	1.30
X_3	10.35	6.88
X_4	3.45	2.70
X_5	−4.25	0.38

FIGURE 11–12 **Testing Whether $\beta_1 = 0$**

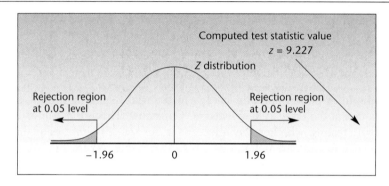

For the other variables X_2 through X_5, we show the hypothesis tests without figures. The tests are carried out in the same way, with the same distribution. We also do not show the computation of confidence intervals for the slope parameters. These are done exactly as shown for β_1. Note that when the hypothesis test for the significance of a slope parameter leads to nonrejection of the null hypothesis that the slope parameter is zero, the point zero will be included in a confidence interval with the same confidence level as the level of significance of the test.

The hypothesis test for β_2 is H_0: $\beta_2 = 0$ versus H_1: $\beta_2 \neq 0$. The test statistic value is $z = 5.60/1.30 = 4.308$. This value, too, is in the right-hand rejection region for usual levels of significance; the p-value is small. We conclude that X_2 is also an important variable in the regression equation.

The hypothesis test for β_3 is H_0: $\beta_3 = 0$ versus H_1: $\beta_3 \neq 0$. Here the test statistic value is $z = 10.35/6.88 = 1.504$. This value lies in the nonrejection region for levels of α even larger than 0.10. The p-value is greater than 0.133, as you can verify from a normal table. We conclude that variable X_3 is probably not important. Remember our cautionary comments that preceded this discussion—there is a possibility that X_3 is actually an important variable. The variable may *appear* to have a slope that is not different from zero because its standard error, $s(b_3) = 6.88$, may be unduly inflated; the variable may be correlated with another explanatory variable (the problem of multicollinearity). A way out of this problem is to drop another variable, one that we suspect to be correlated with X_3, and see if X_3 becomes significant in the new regression model. We will come back to this problem in the section on multicollinearity and in the section on selection of variables to be included in a regression model.

The hypothesis test about β_4 is H_0: $\beta_4 = 0$ versus H_1: $\beta_4 \neq 0$. The value of the test statistic for this test is $z = 3.45/2.70 = 1.278$. Again, we cannot reject the null hypothesis that the slope parameter of X_4 is zero and that the variable has no explanatory power. Note, however, the caution in our discussion of the test of β_3. It is possible, for example, that X_3 and X_4 are collinear and that this is the reason for their respective tests resulting in nonsignificance. It would be wise to drop one of these two variables and check whether the other variable then becomes significant. If it does, the reason for our test result is multicollinearity, and not the absence of explanatory power of the variable in question. Another point worth mentioning is the idea of joint inference, discussed earlier. Although the separate tests of β_3 and β_4 both may lead to

TABLE 11–5   Multiple Regression Results from the Template
[Multiple Regression.xls; Sheet: Results]

	A	B	C	D	E	F	G	H	I	J	K	L	M
1	Multiple Regression Results					Example 11-1							
2													
3		0	1	2	3	4	5	6	7	8	9	10	
4		Intercept	Advt.	Promo									
5	b	47.165	1.599	1.1487									
6	s(b)	2.4704	0.281	0.3052									
7	t	19.092	5.6913	3.7633									
8	p-value	0.0000	0.0007	0.0070									

the nonrejection of the hypothesis that the parameters are zero, it may be that the two parameters are not jointly equal to zero. This would be the situation if, in Figure 11–11, the rectangle contained the point zero while the ellipse—the true joint confidence region for both parameters—did not contain that point. It is important to note that the t tests are *conditional*. The significance or nonsignificance of a variable in the equation is conditional on the fact that the regression equation contains the other variables.

Finally, the test for parameter β_5 is H_0: $\beta_5 = 0$ versus H_1: $\beta_5 \neq 0$. The computed value of the test statistic is $z = -4.25/0.38 = -11.184$. This value falls far in the left-hand rejection region, and we conclude that variable X_5 has explanatory power with respect to the dependent variable and therefore should be included in the regression equation. The slope parameter is negative, which means that, everything else staying constant, the dependent variable Y decreases on average as X_5 increases. We note that these tests can be carried out very quickly by just considering the p-values.

We now return to Example 11–1 and look at the rest of the results from the template. For easy reference the results from Table 11–4 are repeated here in Table 11–5. As seen in the table, the test statistic t is very significant for both advertisement and promotion variables, because the p-value is less than 1% in both cases. We therefore declare that both of these variables affect the sales.

EXAMPLE 11–2

In recent years, many U.S. firms have intensified their efforts to market their products in the Pacific Rim. Among the major economic powers in that area are Japan, Hong Kong, and Singapore. A consortium of U.S. firms that produce raw materials used in Singapore is interested in predicting the level of exports from the United States to Singapore, as well as understanding the relationship between U.S. exports to Singapore and certain variables affecting the economy of that country. Understanding this relationship would allow the consortium members to time their marketing efforts to coincide with favorable conditions in the Singapore economy. Understanding the relationship would also allow the exporters to determine whether there is room for expansion of exports to Singapore. The economist hired to do the analysis obtained from the Monetary Authority of Singapore (MAS) monthly data on five economic variables for the period of January 1989 to August 1995. The variables were U.S. exports to Singapore in billions of Singapore dollars (the dependent variable, Exports), money supply figures in billions of Singapore dollars (variable M1), minimum Singapore bank lending rate in percentages (variable Lend), an index of local prices where the base year is 1974 (variable Price), and the exchange rate of Singapore dollars per U.S. dollar (variable Exchange). The monthly data are given in Table 11–6.

TABLE 11–6　Example 11–2 Data

Row	Exports	M1	Lend	Price	Exchange
1	2.6	5.1	7.8	114	2.16
2	2.6	4.9	8.0	116	2.17
3	2.7	5.1	8.1	117	2.18
4	3.0	5.1	8.1	122	2.20
5	2.9	5.1	8.1	124	2.21
6	3.1	5.2	8.1	128	2.17
7	3.2	5.1	8.3	132	2.14
8	3.7	5.2	8.8	133	2.16
9	3.6	5.3	8.9	133	2.15
10	3.4	5.4	9.1	134	2.16
11	3.7	5.7	9.2	135	2.18
12	3.6	5.7	9.5	136	2.17
13	4.1	5.9	10.3	140	2.15
14	3.5	5.8	10.6	147	2.16
15	4.2	5.7	11.3	150	2.21
16	4.3	5.8	12.1	151	2.24
17	4.2	6.0	12.0	151	2.16
18	4.1	6.0	11.4	151	2.12
19	4.6	6.0	11.1	153	2.11
20	4.4	6.0	11.0	154	2.13
21	4.5	6.1	11.3	154	2.11
22	4.6	6.0	12.6	154	2.09
23	4.6	6.1	13.6	155	2.09
24	4.2	6.7	13.6	155	2.10
25	5.5	6.2	14.3	156	2.08
26	3.7	6.3	14.3	156	2.09
27	4.9	7.0	13.7	159	2.10
28	5.2	7.0	12.7	161	2.11
29	4.9	6.6	12.6	161	2.15
30	4.6	6.4	13.4	161	2.14
31	5.4	6.3	14.3	162	2.16
32	5.0	6.5	13.9	160	2.17
33	4.8	6.6	14.5	159	2.15
34	5.1	6.8	15.0	159	2.10
35	4.4	7.2	13.2	158	2.06
36	5.0	7.6	11.8	155	2.05
37	5.1	7.2	11.2	155	2.06
38	4.8	7.1	10.1	154	2.11
39	5.4	7.0	10.0	154	2.12
40	5.0	7.5	10.2	154	2.13
41	5.2	7.4	11.0	153	2.04
42	4.7	7.4	11.0	152	2.14
43	5.1	7.3	10.7	152	2.15

(Continued)

Row	Exports	M1	Lend	Price	Exchange
44	4.9	7.6	10.2	152	2.16
45	4.9	7.8	10.0	151	2.17
46	5.3	7.8	9.8	152	2.20
47	4.8	8.2	9.3	152	2.21
48	4.9	8.2	9.3	152	2.15
49	5.1	8.3	9.5	152	2.08
50	4.3	8.3	9.2	150	2.08
51	4.9	8.0	9.1	147	2.09
52	5.3	8.2	9.0	147	2.10
53	4.8	8.2	9.0	146	2.09
54	5.3	8.0	8.9	145	2.12
55	5.0	8.1	9.0	145	2.13
56	5.1	8.1	9.0	146	2.14
57	4.8	8.1	9.0	147	2.14
58	4.8	8.1	8.9	147	2.13
59	5.2	8.6	8.9	147	2.13
60	4.9	8.8	9.0	146	2.13
61	5.5	8.4	9.1	147	2.13
62	4.3	8.2	9.0	146	2.13
63	5.2	8.3	9.2	146	2.09
64	4.7	8.3	9.6	146	2.09
65	5.4	8.4	10.0	146	2.10
66	5.2	8.3	10.0	147	2.11
67	5.6	8.2	10.1	146	2.15

TABLE 11–7 Regression Results from the Template for Exports to Singapore [Multiple Regression.xls]

Multiple Regression Results | Exports

	0	1	2	3	4	5	6	7	8	9	10
	Intercept	M1	Lend	Price	Exch.						
b	-4.0155	0.3685	0.0047	0.0365	0.2679						
s(b)	2.7664	0.0638	0.0492	0.0093	1.1754						
t	-1.4515	5.7708	0.0955	3.9149	0.2279						
p-value	0.1517	0.0000	0.9242	0.0002	0.8205						

ANOVA Table

Source	SS	df	MS	F	$F_{Critical}$	p-value		
Regn.	32.946	4	8.2366	73.059	2.5201	0.0000	*s*	0.3358
Error	6.9898	62	0.1127					
Total	39.936	66		R^2	0.8250		Adjusted R^2	0.8137

Solution

Use the template to perform a multiple regression analysis with exports as the dependent variable and the four economic variables M1, Lend, Price, and Exchange as the predictor variables. Table 11–7 shows the results.

Let us analyze the regression results. We start with the ANOVA table and the F test for the existence of linear relationships between the independent variables and

TABLE 11–8   Regression Results for Singapore Exports without M1

| Multiple Regression Results | | | | | Exports | | | | | |

	0	1	2	3	4	5	6	7	8	9	10
	Intercept	Lend	Price	Exch.							
b	-0.2891	-0.2114	0.0781	-2.095							
$s(b)$	3.3085	0.0393	0.0073	1.3551							
t	-0.0874	-5.3804	10.753	-1.546							
p-value	0.9306	0.0000	0.0000	0.1271							

ANOVA Table

Source	SS	df	MS	F	$F_{Critical}$	p-value		
Regn.	29.192	3	9.7306	57.057	2.7505	0.0000	s	0.413
Error	10.744	63	0.1705					
Total	39.936	66		R^2 0.7310		Adjusted R^2 0.7182		

exports from the United States to Singapore. We have $F_{(4, 62)} = 73.059$ with a p-value of "0.000." We conclude that there is strong evidence of a linear regression relationship here. This is further confirmed by noting that the coefficient of determination is high: $R^2 = 0.825$. Thus, the combination of the four economic variables explains 82.5% of the variation in exports to Singapore. The adjusted coefficient of determination $\bar{R}^2$ is a little smaller: 0.8137. Now the question is, Which of the four variables are important as predictors of export volume to Singapore and which are not? Looking at the reported p-values, we see that the Singapore money supply M1 is an important variable; the level of prices in Singapore is also an important variable. The remaining two variables, minimum lending rate and exchange rate, have very large p-values. It is surprising that the lending rate and the exchange rate of Singapore dollars to U.S. dollars seem to have no effect on the volume of Singapore's imports from the United States. Remember, however, that we may have a problem of multicollinearity. This is especially true when we are dealing with economic variables, which tend to be correlated with one another.[7]

When M1 is dropped from the equation and the new regression analysis considers the independent variables Lend, Price, and Exchange, we see that the lending rate, which was not significant in the full regression equation, now becomes significant! This is seen in Table 11–8. Note that R^2 has dropped greatly with the removal of M1. The fact that the lending rate is significant in the new equation is an indication of *multicollinearity;* variables M1 and Lend are correlated with each other. Therefore, Lend is not significant when M1 is in the equation, but in the absence of M1, Lend does have explanatory power.

Note that the exchange rate is still not significant. Since R^2 and the adjusted R^2 both decrease significantly when the money supply M1 is dropped, let us put that variable back into the equation and run U.S. exports to Singapore versus the independent variables M1 and Price only. The results are shown in Table 11–9. In this regression equation, both independent variables are significant. Note that R^2 in this regression is virtually the same as R^2 with all four variables in the equation (see

---

[7]The analysis of economic variables presents special problems. Economists have developed methods that account for the intricate interrelations among economic variables. These methods, based on multiple regression and time series analysis, are usually referred to as *econometric methods.*

TABLE 11–9   Regressing Exports against M1 and Price

Multiple Regression Results					Exports					
	0	1	2	3	4	5	6	7	8	9	10
	Intercept	M1	Price								
b	-3.423	0.3614	0.0037								
s(b)	0.5409	0.0392	0.0041								
t	-6.3288	9.209	9.0461								
p-value	0.0000	0.0000	0.0000								

ANOVA Table

Source	SS	df	MS	F	$F_{Critical}$	p-value		
Regn.	32.94	2	16.47	150.67	3.1404	0.0000	s	0.3306
Error	6.9959	64	0.1093					
Total	39.936	66		R^2	0.8248		Adjusted R^2	0.8193

Table 11–7). However, the adjusted coefficient of determination $\bar{R}^2$ is different. The adjusted R^2 actually *increases* as we drop the variables Lend and Exchange. In the full model with the four variables (Table 11–7), $\bar{R}^2 = 0.8137$, while in the reduced model, with variables M1 and Price only (Table 11–9), $\bar{R}^2 = 0.8193$. This demonstrates the usefulness of the adjusted R^2. When unimportant variables are added to the equation (unimportant in the presence of other variables), $\bar{R}^2$ decreases even if R^2 increases. The best model, in terms of explanatory power gauged against the loss of degrees of freedom, is the reduced model in Table 11–9, which relates exports to Singapore with only the money supply and price level. This is also seen by the fact that the other two variables are not significant once M1 and Price are in the equation. Later, when we discuss stepwise regression–a method of letting the computer choose the best variables to be included in the model–we will see that this automatic procedure also chooses the variables M1 and Price as the best combination for predicting U.S. exports to Singapore.

PROBLEMS

11–28.   A regression analysis is carried out, and a confidence interval for β_1 is computed to be $[1.25, 1.55]$; a confidence interval for β_2 is $[2.01, 2.12]$. Both are 95% confidence intervals. Explain the possibility that the point $(1.26, 2.02)$ may not lie inside a joint confidence region for (β_1, β_2) at a confidence level of 95%.

11–29.   A regression analysis of total bank deposits versus the independent variable area affluence is carried out. The estimated slope coefficient is 1.27, and its standard error is 0.11. A second independent variable, average value of single-family home in the area, is added to the model. The estimated slope coefficient of area affluence in the new regression equation is 1.02, and its standard deviation is 0.87. Explain what happened and why.

11–30.   Give three reasons why caution must be exercised in interpreting the significance of single regression slope parameters.

11–31.   Give 95% confidence intervals for the slope parameters β_2 through β_5, using the information in Table 11–4. Which confidence intervals contain the point $(0, 0)$? Explain the interpretation of such outcomes.

11–32. Suppose that in a regression equation, two slope parameters are significant and two are not. One of the two insignificant variables is dropped, and the other one remains insignificant in the new equation. Give two possible reasons for this finding, and suggest a way of determining the correct reason.

11–33. A computer program for regression analysis produces a joint confidence region for the two slope parameters considered in the regression equation, β_1 and β_2. The elliptical region of confidence level 95% does not contain the point $(0, 0)$. Not knowing the value of the F statistic, or R^2, do you believe there is a linear regression relationship between Y and at least one of the two explanatory variables? Explain.

11–34. In the Nissan Motor Company situation of problems 11–14 and 11–21, the regression results are as follows. Give a complete interpretation of these results.

```
The regression equation is
RATING = 24.1 - 0.166 PRESTIGE + 0.324 COMFORT + 0.514 ECONOMY

Predictor       Coef        Stdev
Constant        24.14       18.22
PRESTIGE       -0.1658      0.1215
COMFORT         0.3236      0.1228
ECONOMY         0.5139      0.1143
```

11–35. Refer to Example 11–2, where exports to Singapore were regressed on several economic variables. Interpret the results of the following analysis, and compare them with the results reported in the text. How does the present model fit with the rest of the analysis? Explain.

```
The regression equation is
EXPORTS = - 3.40 + 0.363 M1 + 0.0021 LEND + 0.0367 PRICE

Predictor       Coef        Stdev      t-ratio      P
CONSTANT       -3.4047      0.6821      -4.99      0.000
M1              0.36339     0.05940      6.12      0.000
LEND            0.00211     0.04753      0.04      0.965
PRICE           0.036666    0.009231     3.97      0.000

s = 0.3332      R-sq = 82.5%      R-sq (adj) = 81.6%
```

11–36. After the model of problem 11–35, the next model was run:

```
The regression equation is
EXPORTS = - 1.09 + 0.552 M1 + 0.171 LEND

Predictor       Coef        Stdev      t-ratio      P
Constant       -1.0859      0.3914      -2.77      0.007
M1              0.55222     0.03950     13.98      0.000
LEND            0.17100     0.02357      7.25      0.000

s = 0.3697      R-sq = 78.1%      R-sq (adj) = 77.4%
```

Analysis of Variance

SOURCE	DF	SS	MS	F	P
Regression	2	31.189	15.594	114.09	0.000
Error	64	8.748	0.137		
Total	66	39.936			

    a. What happened when Price was dropped from the regression equation? Why?

    b. Compare this model with all previous models of exports versus the economic variables, and draw conclusions.

    c. Which model is best overall? Why?

    d. Conduct the *F* test for this particular model.

    e. Compare the reported value of *s* in this model with the reported *s* value in the model of problem 11–35. Why is *s* higher in this model?

    f. For the model in problem 11–35, what is the mean square error?

11–37. A regression analysis of monthly sales versus four independent variables is carried out. One of the variables is known not to have any effect on sales, yet its slope parameter in the regression is significant. In your opinion, what may have caused this to happen?

11–38. The marketing department of a tool manufacturing company forecasts the quarterly demand for the company's products using multiple regression. The independent variables used are Car Sales lagged by 6 years, Money Supply Index of the previous quarter, and Oil Price of previous quarter. The data for the last 14 quarters are given below. Carry out the multiple regression.

	Sales, Million $	Car Sales, Millions	Money Supply Index	Oil Price, $/barrel
1997 Q2	35.16	40.4	2.29	18.83
1997 Q3	32.3	38	2.32	19.75
1997 Q4	32.78	35.6	2.32	18.53
1998 Q1	30.91	35.6	2.33	17.61
1998 Q2	30.5	36.7	2.36	17.95
1998 Q3	28.8	35.9	2.36	15.84
1998 Q4	30.22	36.9	2.36	14.28
1999 Q1	29.52	37.9	2.38	13.02
1999 Q2	30.04	37.2	2.37	15.89
1999 Q3	31.17	39	2.39	16.91
1999 Q4	29.17	39.3	2.40	16.29
2000 Q1	31.07	41.9	2.42	17
2000 Q2	30.33	40	2.43	17.93
2000 Q3	31.42	43.2	2.44	16.98

    a. What is the regression equation? Report R^2 and adjusted R^2.

    b. Predict the sales for the fourth quarter of 2000 using the regression equation. The car sales were $42.7 million, Money Supply Index was 2.46, and the oil price was $17.51/barrel.

    c. It is claimed that oil price does not affect demand for tools. Conduct a *t* test with $\alpha = 5\%$.

    d. Remove the oil price variable from the model and repeat the regression. Report the new R^2 and adjusted R^2.

11–39. Using the following table, run a regression of total return to investors against revenue growth and operating margin growth. Interpret all your findings.

Company	Total Return to Investors 1986–96 annual rate	Revenue Growth 1986–96 annual rate	Operating Margin Growth 1986–96 annual rate
Amgen	67.8%	108.1%	29.0%
Oracle	53.5%	59.7%	17.9%
Worldcom	53.0%	50.7%	23.0%
Microsoft	51.0%	46.8%	34.2%
Conseco	47.3%	55.6%	28.1%
Nike	46.9%	21.3%	14.5%
Champion Enterprises	46.7%	49.6%	4.6%
Intel	43.8%	33.1%	22.1%
Harley-Davidson	43.3%	22.0%	9.1%
Micron Technology	41.8%	65.1%	13.0%
Pacificare Health Systems	41.3%	40.3%	2.1%
Applied Materials	40.6%	42.4%	13.5%
Home Depot	40.2%	36.4%	7.1%
Compaq Computer	36.9%	43.0%	12.8%
Iomega	36.6%	42.4%	2.0%
United Healthcare	36.0%	52.1%	7.4%
Sunamerica	35.0%	60.9%	26.7%

From *Fortune*, June 23, 1997, p. 33. © 1997 Time, Inc. All rights reserved.

11–6   Testing the Validity of the Regression Model

In Chapter 10, we stressed the importance of the three stages of statistical model building: model specification, estimation of parameters, and testing the validity of the model assumptions. We will now discuss the third and very important stage of checking the validity of the model assumptions in multiple regression analysis.

Residual Plots

As with simple linear regression, the analysis of regression residuals is an important tool for determining whether the assumptions of the multiple regression model are met. Residual plots are easy to use, and they convey much information quickly. The saying "A picture is worth a thousand words" is a good description of the technique of examining plots of regression residuals. As with simple linear regression, we may plot the residuals against the predicted values of the dependent variable, against each independent variable, against time (or the order of selection of the data points), and on a probability scale, to check the normality assumption. Since we have already discussed the use of residual plots in Chapter 10, we will demonstrate only some of the residual plots, using Example 11–2. Figure 11–13 is a plot of the residuals produced from the model with the two independent variables M1 and Price (Table 11–9) against variable M1. It appears that the residuals are randomly distributed with no pattern and with equal variance as M1 increases.

FIGURE 11–13   Residuals versus M1

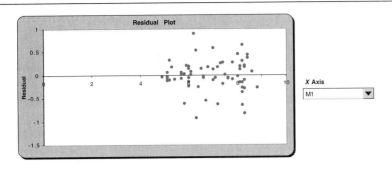

FIGURE 11–14   Residuals versus Price

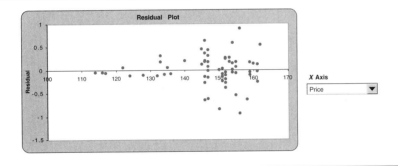

Figure 11–14 is a plot of the regression residuals against the variable Price. Here the picture is quite different. As we examine this figure carefully, we see that the spread of the residuals increases as Price increases. Thus, the variance of the residuals is not constant. We have the situation called *heteroscedasticity*—a violation of the assumption of equal error variance. In such cases, the ordinary least-squares (OLS) estimation method is not efficient, and an alternative method, called *weighted least squares (WLS)*, should be used instead. The WLS procedure is discussed in advanced texts on regression analysis.

Figure 11–15 is a plot of the regression residuals against the variable Time, that is, the order of the observations. (The observations are a time sequence of monthly data.) This variable was not included in the model, and the plot could reveal whether time should have been included as a variable in our regression model. The plot of the residuals against time reveals no pattern in the residuals as time increases. The residuals seem to be more or less randomly distributed about their mean of zero.

Figure 11–16 is a plot of the regression residuals against the predicted export values $\hat{Y}$. We leave it as an exercise to the reader to interpret the information in this plot.

Standardized Residuals

Remember that under the assumptions of the regression model, the population errors ϵ_j are normally distributed with mean zero and standard deviation σ. As a result, the errors divided by their standard deviation should follow the standard normal distribution:

FIGURE 11–15   Residuals versus Time

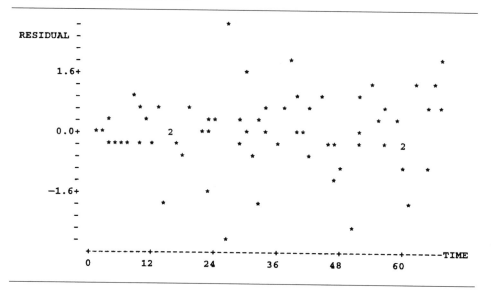

FIGURE 11–16   Residuals versus Predicted *Y* Values

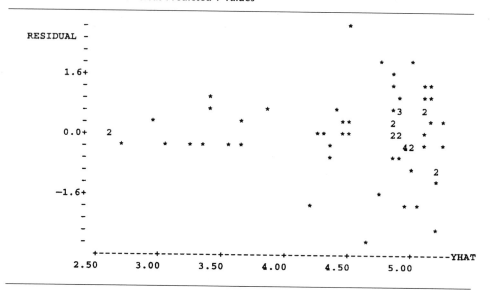

$$\frac{\epsilon_j}{\sigma} \sim N(0, 1) \quad \text{for all } j$$

Therefore, dividing the observed regression errors e_j by their estimated standard deviation s will give us standardized residuals. Examination of a histogram of these residuals may give us an idea as to whether the normal assumption is valid.[8]

---

[8]Actually, the residuals are not independent and do not have equal variance; therefore, we really should divide the residuals e_j by something a little more complicated than s. However, the simpler procedure outlined here and implemented in some computer packages is usually sufficiently accurate.

FIGURE 11–17   The Normal Probability Plot of the Residuals
[Multiple Regression.xls; Sheet: Residuals]

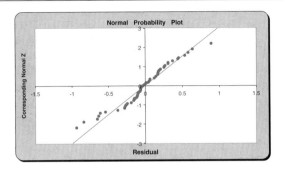

FIGURE 11–18   A Least-Squares Regression Line Estimated with and without the Outlier

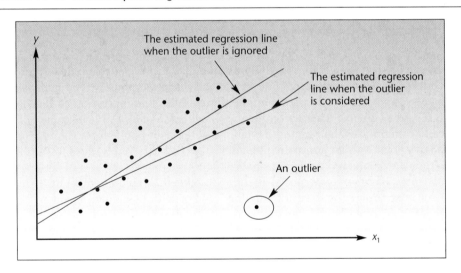

The Normal Probability Plot

Just as we saw in the simple regression template, the multiple regression template also produces a normal probability plot of the residuals. If the residuals are perfectly normally distributed, they will lie along the diagonal straight line in the plot. The more they deviate from the diagonal line, the more they deviate from normal distribution. In Figure 11–17, the deviations do not appear to be significant. Consequently, we assume that the residuals are normally distributed.

Outliers and Influential Observations

An **outlier** is an extreme observation. It is a point that lies away from the rest of the data set. Because of this, outliers may exert greater influence on the least-squares estimates of the regression parameters than do other observations. To see why, consider the data in Figure 11–18. The graph shows the estimated least-squares regression line without the outlier and the line obtained when the outlier is considered.

FIGURE 11–19   Influence of an Observation Far in the X_1 Direction

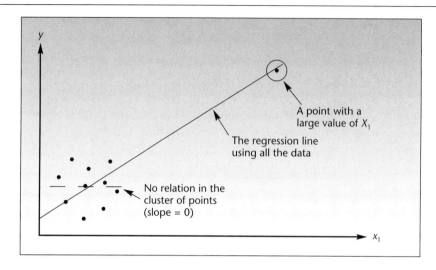

As can be seen from Figure 11–18, the outlier has a strong effect on the estimation of model parameters. (We used a line showing Y versus variable X_1. The same is true for a regression plane or hyperplane: the outlier "tilts" the regression surface away from the other points.) The reason for this effect is the nature of least squares: The procedure minimizes the squared deviations of the data points from the regression surface. A point with an unusually large deviation "attracts" the surface toward itself so as to make its squared deviation smaller.

We must, therefore, pay special attention to outliers. If an outlier can be traced to an error in recording the data or to another type of error, it should, of course, be removed. On the other hand, if an outlier is not due to error, it may have been caused by special circumstances, and the information it provides may be important. For example, an outlier may be an indication of a missing variable in the regression equation. The data shown in Figure 11–18 may be maximum speed for an automobile as a function of engine displacement. The outlier may be an automobile with four cylinders, while all others are six-cylinder cars. Thus, the fact that the point lies away from the rest may be explained. Because of the possible information content in outliers, they should be carefully scrutinized before being discarded. There are some alternative regression methods that do not use a squared-distance approach and are therefore more robust—less sensitive to the influence of outliers.

Sometimes an outlier is actually a point that is distant from the rest because the value of one of its independent variables is larger than the rest of the data. For example, suppose we measure chemical yield Y as a function of temperature X_1. There may be other variables, but we will consider only these two. Suppose that most of our data are obtained at low temperatures within a certain range, but one observation is taken at a high temperature. This outlying point, far in the X_1 direction, exerts strong influence on the estimation of the model parameters. This is shown in Figure 11–19. Without the point at high temperature, the regression line may have slope zero, and no relationship may be detected, as can be seen from the figure. We must also be careful in such cases to guard against estimating a straight-line relation where a curvilinear one may be more appropriate. This could become evident if we had more data

FIGURE 11–20   **Possible Relation in the Region between the Available Cluster of Data and the Far Point**

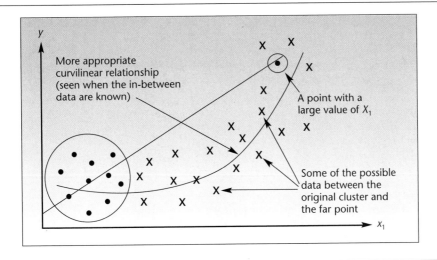

points in the region between the far point and the rest of the data. This is shown in Figure 11–20.

Figure 11–20 serves as a good reminder that regression analysis should not be used for extrapolation. We do not know what happens in the region in which we have no data. This region may be between two regions where we have data, or it may lie beyond the last observation in a given direction. The relationship may be quite different from what we estimate from the data. This is also a reason why it is not a good idea to force the regression surface to go through the origin (that is, carry out a regression with no constant term $\beta_0 = 0$), as is done in some applications. The reasoning in such cases follows the idea expressed in the statement "In this particular case, when there is zero input, there must be zero output," which may very well be true. Forcing the regression to go through the origin, however, may make the estimation procedure biased. This is because in the region where the data points are located—assuming they are not near the origin—the best straight line to describe the data may not have an intercept of zero. This happens when the relationship is not a straight-line relationship. We mentioned this problem in Chapter 10.

A data point far from the other point in some X_i direction is called an *influential observation* if it strongly affects the regression fit. There are statistical techniques of testing whether the regression fit is strongly affected by a given observation. Computer routines such as MINITAB automatically search for outliers and influential observations, reporting them in the regression output so that the user is alerted to the possible effects of these observations. Table 11–10 shows part of the MINITAB output for the analysis of Example 11–2. The table reports "unusual observations": large residuals and influential observations that affect the estimation of the regression relationship.

Lack of Fit and Other Problems

Model lack of fit occurs if, for example, we try to fit a straight line to curved data. There is a statistical method of determining the existence of lack of fit. The method consists of breaking down the sum of squares for error to a sum of squares due to pure

TABLE 11–10   Part of the MINITAB Output for Example 11–2

```
Unusual Observations
 Obs.     M1      EXPORTS       Fit     Stdev.Fit    Residual     St.Resid
   1     5.10     2.6000     2.6420      0.1288       -0.0420      -0.14 X
   2     4.90     2.6000     2.6438      0.1234       -0.0438      -0.14 X
  25     6.20     5.5000     4.5949      0.0676        0.9051       2.80R
  26     6.30     3.7000     4.6311      0.0651       -0.9311      -2.87R
  50     8.30     4.3000     5.1317      0.0648       -0.8317      -2.57R
  67     8.20     5.6000     4.9474      0.0668        0.6526       2.02R

R denotes an obs. with a large st.resid.
X denotes an obs. whose X value gives it large influence.
```

error and a sum of squares due to lack of fit. The method requires that we have observations at equal values of the independent variables or near-neighbor points. This method is described in advanced texts on regression.

A statistical method for determining whether the errors in a regression model are correlated through time (thus violating the regression model assumptions) is the Durbin-Watson test. This test is discussed in a later section of this chapter. Once we determine that our regression model is valid and that there are no serious violations of assumptions, we can use the model for its intended purpose.

PROBLEMS

11–40. Analyze the following plot of the residuals versus $\hat{Y}$.

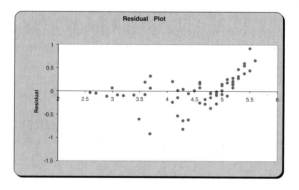

11–41. The normal probability plots of two regression experiments are given below. For each case, give your comments.

a.

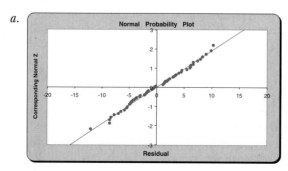

b.
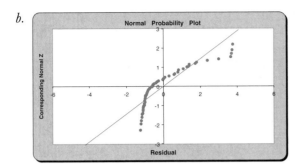

11–42. Explain what an outlier is.

11–43. How can you detect outliers? Discuss two ways of doing so.

11–44. Why should outliers not be discarded and the regression run without them?

11–45. Discuss the possible effects of an outlier on the regression analysis.

11–46. What is an influential observation? Give a few examples.

11–47. What are the limitations of forcing the regression surface to go through the origin?

11–48. Analyze the residual plot of Figure 11–16.

11–7 Using the Multiple Regression Model for Prediction

The use of the multiple regression model for prediction follows the same lines as in the case of simple linear regression, discussed in Chapter 10. We obtain a regression model prediction of a value of the dependent variable Y, based on given values of the independent variables, by substituting the values of the independent variables into the prediction equation. That is, we substitute the values of X_i variables into the equation for $\hat{Y}$. We demonstrate this in Example 11–1.

The predicted value of Y is given by substituting the given values of advertising X_1 and in-store promotions X_2 for which we want to predict sales Y into equation 11–6, using the parameter estimates obtained in Section 11–2. Let us predict sales when advertising is at a level of \$10,000 and in-store promotions are at a level of \$5,000.

$$\hat{Y} = 47.165 + 1.599X_1 + 1.149X_2$$
$$= 47.165 + (1.599)(10) + (1.149)(5) = 68.9 \text{ (thousand dollars)}$$

This prediction is not bad, since the value of Y actually occurring for these values of X_1 and X_2 is known from Table 11–1 to be $Y = 70$ (thousand dollars). Our point estimate of the expected value of Y, denoted $E(Y)$, given these values of X_1 and X_2, is also 68.9 (thousand dollars). Note that our predictions lie *on* the estimated regression surface. The estimated regression surface for Example 11–1 is the plane shown in Figure 11–21.

We may also compute prediction intervals as well as confidence intervals for $E(Y)$, given values of the independent variables. As you recall, while the predicted value and the estimate of the mean value of Y are equal, the prediction interval is wider than a confidence interval for $E(Y)$ using the same confidence level. There is more uncertainty about the predicted value than there is about the average value of Y given the values X_i. The equation for a $(1 - \alpha)$ 100% prediction interval is an

FIGURE 11–21   **Estimated Regression Plane for Example 11–1**

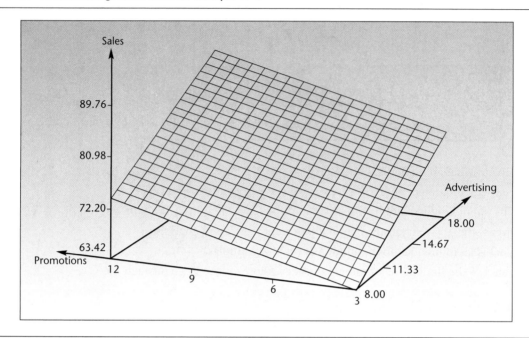

extension of equation 10–32 for simple linear regression. The only difference is that the degrees of freedom of the t distribution are $n - (k + 1)$ rather than just $n - 2$, as is the case for $k = 1$. The standard error, when there are several explanatory variables, is a complicated expression, and we will not give it here; we will denote it by $s(\hat{Y})$. The prediction interval is given in equation 11–17.

A $(1 - \alpha)$ 100% prediction interval for a value of Y given values of X_i is

$$\hat{y} \pm t_{[\alpha/2, \, n - (k + 1)]} \sqrt{s^2(\hat{Y}) + \text{MSE}} \qquad (11\text{–}17)$$

While the expression in the square root is complex, it is computed by most computer packages for regression. The prediction intervals for any values of the independent variables and a given level of confidence are produced as output.

Similarly, the equation for a $(1 - \alpha)$ 100% confidence interval for the conditional mean of Y is an extension of equation 10–33 for the simple linear regression. It is given as equation 11–18. Again, the degrees of freedom are $n - (k + 1)$. The formula for the standard error is complex and will not be given here. We will call the standard error $s[E(\hat{Y})]$. The confidence interval for the conditional mean of Y is computable and may be reported, upon request, in the output of most computer packages that include regression analysis.

A $(1 - \alpha)$ 100% confidence interval for the conditional mean of Y is

$$\hat{y} \pm t_{[\alpha/2, \, n - (k + 1)]} \, s \, [E(\hat{Y})] \qquad (11\text{–}18)$$

**TABLE 11–11 Prediction Using Multiple Regression
[Multiple Regression.xls; Sheet: Results]**

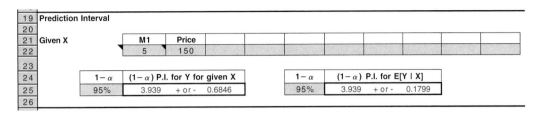

Equations 11–17 and 11–18 are implemented in the template on the Results sheet. To make a prediction, we enter the values of the independent variables in row 22 and the confidence level desired in row 25. Table 11–11 shows the case of Example 11–2 with independent variables M1 and Price. The 95% prediction interval has been computed for the exports when M1 = 5 and Price = 150. A similar interval for the expected value of the exports for the given M1 and Price values has also been computed. The two prediction intervals appear in row 24.

The predictions are not very reliable because of the heteroscedasticity we discovered in the last section, but they are useful as a demonstration of the procedure. Remember that it is never a good idea to try to predict values outside the region of the data used in the estimation of the regression parameters, because the regression relationship may be different outside that range. In this example, all predictions use values of the independent variables within the range of the estimation data.

When using regression models, it is important to know that a regression relationship between the dependent variable and some independent variables does not imply causality. Thus, if we find a linear relationship between Y and X, it does not necessarily mean that X causes Y. Causality is very different to determine and to prove. There is also the issue of spurious correlations between variables—correlations that are not real. Montgomery and Peck give an example of a regression analysis of the number of mentally disturbed people in the United Kingdom versus the number of radio receiver licenses issued in that country.[9] The regression relationship is close to a perfect straight line, with $r^2 = 0.9842$. Can the conclusion be drawn that there is a relationship between the number of radio receiver licenses and the incidence of mental illness? Probably not. Both variables—the number of licenses and the incidence of mental illness—are related to a third variable: population size. The increase in both of these variables reflects the growth of the population in general, and there is probably no *direct* connection between the two variables. We must be very careful in our interpretation of regression results.

The Template

The multiple regression template [Multiple Regression.xls] consists of a total of five sheets. The sheet titled "Data" is used to enter the data (see Figure 11–30 for an example). The sheet titled "Results" contains the regression coefficients, their standard errors, the corresponding t tests, the ANOVA table, and a panel for prediction intervals. The sheet titled "Residuals" contains a plot of the residuals, the Durbin-Watson

---

[9]D. Montgomery and E. Peck, *Introduction to Linear Regression Analysis* (New York: Wiley, 1982).

statistic (described later), and a normal probability plot for testing the normality assumption of the error term. The sheet titled "Correl" displays the correlation coefficient between every pair of variables. The use of the correlation matrix is described later in this chapter. The sheet titled "Partial F" can be used to find partial F, which is also described later in this chapter.

Setting Recalculation to "Manual" on the Template

Since the calculations performed in the multiple regression template are voluminous, a recalculation can take a little longer than in other templates. Therefore, entering data in the Data sheet may be difficult, especially on slower PCs (Pentium II or earlier), because the computer will recalculate every result before taking in the next data entry. If this problem occurs, set the Recalculation feature to manual. This can be done by going into the Tools|Options menu and choosing Manual under the Calculation tab. When this is done, a change made in the data or in any cell *will not cause the spreadsheet to automatically update itself.* Only when recalculation is manually initiated will the spreadsheet update itself. To initiate recalculation, *press the F9 key* on the keyboard. A warning message about pressing the F9 key is displayed at a few places in the template. If the recalculation has not been set to manual, this message can be ignored.

Note also that when the recalculation is set to manual, *none of the open spreadsheets* will update itself. That is, if other spreadsheets were open, they will not update themselves either. The F9 key needs to be pressed on every open spreadsheet to initiate recalculation. This state of manual recalculation will continue until the Excel program is closed and reopened. For this reason, set the recalculation to manual only after careful consideration.

PROBLEMS

11–49.   Explain why it is not a good idea to use the regression equation for predicting values outside the range of the estimation data set.

11–50.   Use equation 11–6 to predict sales in Example 11–1 when the level of advertising is $8,000 and in-store promotions are at a level of $12,000.

11–51.   Using the regression relationship you estimated in problem 11–8, predict the value of a home 1,800 square feet located 2.0 miles from the center of the town.

11–52.   Using the regression equation from problem 11–25, predict excess stock return when SIZRNK = 5 and PRCRNK = 6.

11–53.   Using the information in Table 11–11, what is the standard error of $\hat{Y}$? What is the standard error of $E(\hat{Y})$?

11–54.   Use a computer to produce a prediction interval and a confidence interval for the conditional mean of Y for the prediction in problem 11–50. Use the data in Table 11–1.

11–55.   What is the difference between a predicted value of the dependent variable and the conditional mean of the dependent variable?

11–56.   Why is the prediction interval of 95% wider than the 95% confidence interval for the conditional mean, using the same values of the independent variables?

11–8   Qualitative Independent Variables

The variables we have encountered so far in this chapter have all been *quantitative* variables: variables that can take on values on a scale. Sales volume, advertising expenditure, exports, the money supply, and people's ratings of an automobile are all examples of quantitative variables. In this section, we will discuss the use of *qualitative* variables as explanatory variables in a regression model. Qualitative variables are variables that describe a quality rather than a quantity. This should remind you of analysis of variance in Chapter 9. There we had qualitative variables: the kind of resort in the Club Med example, type of airplane, type of coffee, and so on.

In some cases, it is very useful to include information on one or more qualitative variables in our multiple regression model. For example, a hotel chain may be interested in predicting the number of occupied rooms as a function of the economy of the area in which the hotel is located, as well as advertising level and some other quantitative variables. The hotel may also want to know whether the peak season is in progress—a qualitative variable that may have a lot to do with the level of occupancy at the hotel. A property appraiser may be interested in predicting the value of different residential units on the basis of several quantitative variables, such as age of the unit and area in square feet, as well as the qualitative variable of whether the unit is owned or rented.

Each of these qualitative variables has only two *levels:* peak season versus nonpeak season, rental unit versus nonrental unit. An easy way to quantify such a qualitative variable is by way of a single **indicator variable**, also called a **dummy variable.** An indicator variable is a variable that indicates whether some condition holds. It has the value 1 when the condition holds and the value 0 when the condition does not hold. If you are familiar with computer science, you probably know the indicator variable by another name: *binary variable*, because it takes on only two possible values, 0 and 1.

When included in the model of hotel occupancy, the indicator variable will equal 0 if it is not peak season and 1 if it is (or vice versa; it makes no difference). Similarly, in the property value analysis, the dummy variable will have the value 0 when the unit is rented and the value 1 when the unit is owned, or vice versa. We define the general form of an indicator variable in equation 11–19.

An indicator variable of qualitative level A is

$$X_h = \begin{cases} 1 & \text{if level A is obtained} \\ 0 & \text{if level A is not obtained} \end{cases} \qquad (11\text{–}19)$$

The use of indicator variables in regression analysis is very simple. No special computational routines are required. All we do is code the indicator variable as 1 whenever the quality of interest is obtained for a particular data point and as 0 when it is not obtained. The rest of the variables in the regression equation are left the same. We demonstrate the use of an indicator variable in modeling a qualitative variable with two levels in the following example.

A motion picture industry analyst wants to estimate the gross earnings generated by a movie. The estimate will be based on different variables involved in the film's   **EXAMPLE 11–3**

TABLE 11–12 Data for Example 11–3

Movie	Gross Earnings Y, Million $	Production Cost X_1, Million $	Promotion Cost X_2, Million $	Book X_3
1	28	4.2	1	0
2	35	6.0	3	1
3	50	5.5	6	1
4	20	3.3	1	0
5	75	12.5	11	1
6	60	9.6	8	1
7	15	2.5	0.5	0
8	45	10.8	5	0
9	50	8.4	3	1
10	34	6.6	2	0
11	48	10.7	1	1
12	82	11.0	15	1
13	24	3.5	4	0
14	50	6.9	10	0
15	58	7.8	9	1
16	63	10.1	10	0
17	30	5.0	1	1
18	37	7.5	5	0
19	45	6.4	8	1
20	72	10.0	12	1

production. The independent variables considered are X_1 = production cost of the movie and X_2 = total cost of all promotional activities. A third variable that the analyst wants to consider is the qualitative variable of whether the movie is based on a book published before the release of the movie. This third, qualitative variable is handled by the use of an indicator variable: $X_3 = 0$ if the movie is not based on a book, and $X_3 = 1$ if it is. The analyst obtains information on a random sample of 20 Hollywood movies made within the last 5 years (the inference is to be made only about the population of movies in this particular category). The data are given in Table 11–12. The variable Y is gross earnings, in millions of dollars. The two quantitative independent variables are also in millions of dollars.

Solution The data are entered into the template. The resulting output is presented in Figure 11–22. The coefficient of determination of this regression is very high; the F statistic value is very significant, and we have a good regression relationship. From the individual t ratios and their p-values, we find that all three independent variables are important in the equation.

From the intercept of 7.84, we could (erroneously, of course) deduce that a movie costing nothing to produce or promote, and that is not based on a book, would still gross $7.84 million! The point 0 ($X_1 = 0$, $X_2 = 0$, $X_3 = 0$) is outside the estimation region, and the regression relationship may not hold for that region. In our case, it evidently does not. The intercept is merely a reference point used to move the regression surface upward to where it should be in the estimation region.

FIGURE 11–22   Multiple Regression Results for Example 11–3.
[Multiple Regression.xls; Sheet: Results]

	A	B	C	D	E	F	G	H	I	J	K	L
1	Multiple Regression Results					Movies						
2												
3		0	1	2	3	4	5	6	7	8	9	10
4		Intercept	rod.Cos	Promo	Book							
5	b	7.8362	2.8477	2.2782	7.1661							
6	s(b)	2.3334	0.3923	0.2534	1.818							
7	t	3.3583	7.2582	8.9894	3.9418							
8	p-value	0.0040	0.0000	0.0000	0.0012							
9												
10												
11	ANOVA Table											
12		Source	SS	df	MS	F	$F_{Critical}$	p-value				
13		Regn.	6325.2	3	2108.4	154.89	3.2389	0.0000		s	3.6895	
14		Error	217.8	16	13.612							
15		Total	6543	19			R^2	0.9667		Adjusted R^2	0.9605	
16												

The estimated slope for the cost variable, 2.85, means that—within the estimation region—an increase of $1 million in a movie's production cost (the other variables held constant) increases the movie's gross earnings by an average of $2.85 million. Similarly, the estimated slope coefficient for the promotion variable means that, in the estimation region of the variables, an increase of $1 million in promotional activities (with the other variables constant) increases the movie's gross earnings by an average of $2.28 million.

How do we interpret the estimated coefficient of variable X_3? The estimated coefficient of 7.17 means that having the movie based on a published book ($X_3 = 1$) increases the movie's gross earnings by an average of $7.17 million. Again, the inference is valid only for the region of the data used in the estimation. When $X_3 = 0$, that is, when the movie is not based on a book, the last term in the estimated equation for $\hat{Y}$ drops out—there is no added $7.17 million.

What do we learn from this example about the function of the indicator variable? Note that the predicted value of Y, given the values of the quantitative independent variables, shifts upward (or downward, depending on the sign of the estimated coefficient) by an amount equal to the coefficient of the indicator variable whenever the variable is equal to 1. In this particular case, the surface of the regression—the plane formed by the variables Y, X_1, and X_2—is split into two surfaces: one corresponding to movies based on books and the other corresponding to movies not based on books. The appropriate surface depends on whether $X_3 = 0$ or $X_3 = 1$; the two estimated surfaces are separated by a distance equal to $b_3 = 7.17$. This is demonstrated in Figure 11–23. The regression surface in this example is a plane, so we can draw its image (for a higher-dimensional surface, the same idea holds).

We will now look at the simpler case, with one independent quantitative variable and one indicator variable. Here we assume an estimated regression relationship of the form $\hat{Y} = b_0 + b_1X_1 + b_2X_2$, where X_1 is a quantitative variable and X_2 is an indicator variable. The regression relationship is a straight line, and the indicator variable splits the line into two parallel straight lines, one for each level (0 or 1) of the

FIGURE 11–23   Two Regression Planes of Example 11–3

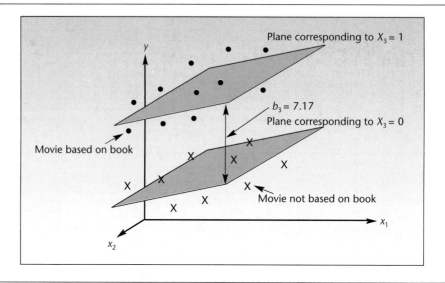

FIGURE 11–24   A Regression with One Quantitative Variable and One Dummy Variable

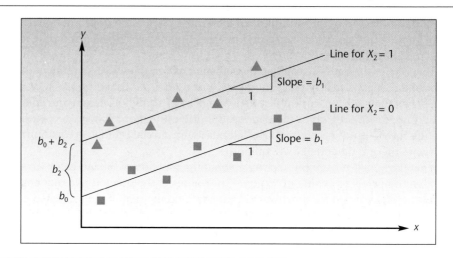

qualitative variable. The points belonging to one level (a level could be Book, as in Example 11–3) are shown as triangles, and the points belonging to the other level are shown as squares. The distance between the two parallel lines (measured as the difference between the two intercepts) is equal to the estimated coefficient of the dummy variable X_2. The situation is demonstrated in Figure 11–24.

We have been dealing with qualitative variables that have only two levels. Therefore, it has sufficed to use an indicator variable with two possible values, 0 and 1. What about situations where we have a qualitative variable with more than two levels? Should we use an "indicator" variable with more than two values? The answer is no. Were we to do this and give our variable values such as 0, 1, 2, 3, . . . to indicate qualitative levels, we would be using a quantitative variable that has several discrete values but no values in between. Also, the assignment of the qualities to the values would be arbitrary. Since there may be no justification for using the values 1, 2, 3, etc., we would be imposing a very special measuring scale on the regression problem—a scale that may not be appropriate. Instead, we will use several indicator variables.

> We account for a qualitative variable with r levels by the use of $r - 1$ indicator (0/1) variables.

We will now demonstrate the use of this rule by changing Example 11–3 somewhat. Suppose that the analyst is interested not in whether a movie is based on a book, but rather in using an explanatory variable that represents the category to which each movie belongs: adventure, drama, or romance. Since this qualitative variable has $r = 3$ levels, the rule tells us that we need to model this variable by using $r - 1 = 2$ indicator variables. Each of the two indicator variables will have one of two possible values, as before: 0 or 1. The setup of the two dummy variables indicating the level of the qualitative variable, movie category, is shown in the following table. For simplicity, let us also assume that the only quantitative variable in the equation is production cost (we leave out the promotion variable). This will allow us to have lines rather than planes. We let $X_1 =$ production cost, as before. We now define the two dummy variables X_2 and X_3.

Category	X_2	X_3
Adventure	0	0
Drama	0	1
Romance	1	0

The definition of the values of X_2 and X_3 for representing the different categories is arbitrary; we could just as well have assigned the values $X_2 = 0$, $X_3 = 0$ to drama or to romance as to adventure. The important thing to remember is that the number of dummy variables is 1 less than the number of categories they represent. Otherwise our model will be overspecified, and problems will occur. In this example, variable X_2 is the indicator variable for romance; when a movie is in the romance category, this variable has the value 1. Similarly, X_3 is the indicator for drama and has the value 1 in cases where a movie is in the drama category. There are only three categories under consideration, so when both X_2 and X_3 are zero, the movie is neither a drama nor a romance; therefore, it must be an adventure movie.

If we use the model

$$Y = \beta_0 + \beta_1 X_1 + \beta_2 X_2 + \beta_3 X_3 + \epsilon \qquad (11\text{–}20)$$

with X_2 and X_3 as defined, we will be estimating three regression lines, one line per category. The line for adventure movies will be $\hat{Y} = b_0 + b_1 X_1$ because here both X_2 and X_3 are zero. The drama line will be $\hat{Y} = b_0 + b_3 + b_1 X_1$ because here $X_3 = 1$ and $X_2 = 0$. In the case of romance movies, our line will be $\hat{Y} = b_0 + b_2 + b_1 X_1$ because in this case $X_2 = 1$ and $X_3 = 0$. Since the estimated coefficients b_i may be negative as well as positive, the different parallel lines may position themselves above or below one another, as determined by the data. Of course, the b_i may be estimates of zero. If we did not reject the null hypothesis H_0: $\beta_3 = 0$, using the usual t test, it would mean that there was no evidence that the adventure and the drama lines were different. That is, it would mean that, on average, adventure movies and drama movies have the same gross earnings as determined by the production costs. If we determine that β_2 is not different from zero, the adventure and romance lines will be the same and the drama line may be different. In case the adventure line is different from drama and romance, these two being the same, we would determine statistically that both β_2 and β_3 are different from zero, but not different from each other.

If we have three regression lines, why bother with indicator variables at all? Why not just run three separate regressions, each for a different movie category? One answer to this question has already been given: The use of indicator variables and their estimated regression coefficients with their standard errors allows us to *test statistically* whether the qualitative variable of interest has any effect on the dependent variable. We are able to test whether we have one distinct line, two lines, three lines, or as many lines as there are levels of the qualitative variable. Another reason is that even if we know that there are, say, three distinct lines, estimating them together via a regression analysis with dummy variables allows us to pool the degrees of freedom for the three regressions, leading to better estimation and a more efficient analysis.

Figure 11–25 shows the three regression lines of our new version of Example 11–3; each line shows the regression relationship between a movie's production cost and the resulting movie's gross earnings in its category. In case there are two independent quantitative variables, say, if we add promotions as a second quantitative variable, we will have three regression *planes* like the two planes shown in Figure 11–23. In Figure 11–25, we show adventure movies as triangles, romance movies as squares, and drama movies as circles. Assuming that adventure movies have the highest average gross earnings, followed by romance and drama, the estimated coefficients b_2 and b_3 have to be negative, as can be seen from the figure.

Can we run a regression on a qualitative variable (by use of dummy variables) only? Yes. You have already seen this model, essentially. Running a regression on a qualitative variable only means modeling some quantitative response by levels of a qualitative factor: it is the *analysis of variance,* discussed in Chapter 9. Doing the

FIGURE 11–25   **The Three Possible Regression Lines, Depending on Movie Category (modified Example 11–3)**

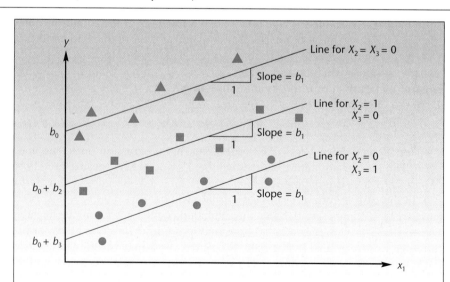

analysis by regression means using a different computational procedure than was done in Chapter 9, but it is still the analysis of variance. Two qualitative variables make the analysis a two-way ANOVA, and interaction terms are cross-products of the appropriate dummy variables, such as X_2X_3. We will say more about cross-products a little later. For now, we note that the regression approach to ANOVA allows us more freedom. Remember that a two-way ANOVA, using the method in Chapter 9, required a balanced design (equal sample size in each cell). If we use the regression approach, we are no longer restricted to the balanced design and may use any sample size.

Let us go back to regressions using quantitative independent variables with some qualitative variables. In some situations, we are not interested in using a regression equation for prediction or for any of the other common uses of regression analysis. Instead, we are intrinsically interested in a qualitative variable used in the regression. Let us be more specific. Recall our original Example 11–3. Suppose we are not interested in predicting a movie's gross earnings based on the production cost, promotions, and whether the movie is based on a book. Suppose instead that we are interested in answering the question: Is there a difference in average gross earnings between movies based on books and movies not based on books?

To answer this question, we use the estimated regression relationship. We use the estimate b_3 and its standard error in testing the null hypothesis H_0: $\beta_3 = 0$ versus the alternative H_1: $\beta_3 \neq 0$. The question is really an ANOVA question. We want to know whether a difference exists in the population means of the two groups of movies based on books and movies not based on books. However, we have some quantitative variables that affect the variable we are measuring (gross earnings). We therefore incorporate information on these variables (production cost and promotions) in a regression model aimed at answering our ANOVA question. When we do this, that is, when we attempt to answer the question of whether differences in population means exist, using a regression equation to account for other sources of variation in our data (the quantitative independent variables), we are conducting an **analysis of covariance.** The independent variables used in the analysis of covariance are called **concomitant variables,** and their purpose in the analysis is not to explain or predict the independent variable, but rather to reduce the errors in the test of significance of the indicator variable or variables.

One of the interesting applications of analysis of covariance is in providing statistical evidence in cases of sex or race discrimination. We demonstrate this particular use in the following example.

EXAMPLE 11–4

A large service company was sued by its female employees in a class action suit alleging sex discrimination in salary levels. The claim was that, on average, a man and a woman of the same education and experience received different salaries: the man's salary was believed to be higher than the woman's salary. The attorney representing the women employees hired a statistician to provide statistical evidence supporting the women's side of the case. The statistician was allowed access to the company's payroll files and obtained a random sample of 100 employees, 40 of whom were women. In addition to salary, the files contained information on education and experience. The statistician then ran a regression analysis of salary Y versus three variables: education level X_1 (on a scale based on the total number of years in school, with an additional value added to the score for each college degree earned, by type), years of experience X_2 (on a scale that combined the number of years of experience directly related to the job assignment with the number of years of similar job experience), and gender X_3 (0 if the employee was a man and 1 if the employee was a woman). The computer output for the regression included the results F ratio = 1,237.56 and $R^2 = 0.67$, as well as the coefficient estimates and standard errors given in Table 11–13. Based on this information, does the attorney for the women employees have a case against the company?

Solution   Let us analyze the regression results. Remember that we are using a regression with a dummy variable to perform an analysis of covariance. There is certainly a regression relationship between salary and at least some of the variables, as evidenced by the very large F value, which is beyond any critical point we can find in a table. The p-value is very small. The coefficient of determination is not extremely high, but then we are using very few variables to explain variation in salary levels. This being the case, 67% explained variation, based on these variables only, is quite respectable. Now we consider the information in Table 11–13.

Dividing the four coefficient estimates by their standard errors, we find that all three variables are important, and the intercept is different from zero. However, we are particularly interested in the hypothesis test:

$$H_0: \beta_3 = 0$$
$$H_1: \beta_3 \neq 0$$

Our test statistic is $t_{(96)} = b_3/s(b_3) = -3,256/212.4 = -15.33$. Since t with 96 degrees of freedom [df $= n - (k + 1) = 100 - 4 = 96$] is virtually a standard normal random variable, we conduct this as a Z test. The computed test statistic value of -15.33 lies very far in the left-hand rejection region. This means that there are two regressions: one for men and one for women. Since we coded X_3 as 0 for a man and 1 for a woman, the women's estimated regression plane lies \$3,256 below the regression plane for men. Since the parameter of the sex variable is significantly different from zero (with an extremely small p-value) and is negative, there is statistical evidence of sex discrimination in this case. The situation here is as seen in Figure 11–23 for the previous example: We have two regression planes, one below the other. The only difference is that in this example, we were not interested in using the regression for prediction, but rather for an ANOVA-type statistical test.

Interactions between Qualitative and Quantitative Variables

Do the different regression lines or higher-dimensional surfaces have to be parallel? The answer is no. Sometimes, there are *interactions* between a qualitative variable and one or more quantitative variables. The idea of an interaction in regression analysis is the same as the idea of interaction between factors in a two-way ANOVA model (as

TABLE 11–13   Regression Results for Example 11–4

Variable	Coefficient Estimate	Standard Error
Constant	8,547	32.6
Education	949	45.1
Experience	1,258	78.5
Sex	−3,256	212.4

well as higher-order ANOVAs). In regression analysis with qualitative variables, the interaction between a qualitative variable and a quantitative variable makes the regression lines or planes at different levels of the dummy variables have *different slopes*. Let us look at the simple case where we have one independent quantitative variable X_1 and one qualitative variable with two levels, modeled by the dummy variable X_2. When an interaction exists between the qualitative and the quantitative variables, the slope of the regression line for $X_2 = 0$ is different from the slope of the regression line for $X_2 = 1$. This is shown in Figure 11–26.

We model the interactions between variables by the cross-product of the variables. The interaction of X_1 with X_2 in this case is modeled by adding the term X_1X_2 to the regression equation. We are thus interested in the model

$$Y = \beta_0 + \beta_1X_1 + \beta_2X_2 + \beta_3X_1X_2 + \epsilon \tag{11–21}$$

We can use the results of the estimation procedure to test for the existence of an interaction. We do so by testing the significance of parameter β_3.

When regression parameters β_1, β_2, and β_3 are all nonzero, we have two distinct lines with different intercepts and different slopes. When β_2 is zero, we have two lines with the same intercept and different slopes (this is unlikely to happen, except when both intercepts are zero). When β_3 is zero, we have two parallel lines, as in the case of equation 11–20. If β_1 is zero, of course, we have no regression—just an ANOVA model; we then assume that β_3 is also zero. Assuming the full model of equation 11–21, representing two distinct lines with different slopes and different intercepts, the intercept and the slope of each line will be as shown in Figure 11–26. By substituting $X_2 = 0$ or $X_2 = 1$ into equation 11–21, verify the definition of each slope and each intercept.

Again, there are advantages to estimating a single model for the different levels of the indicator variable. These are the pooling of degrees of freedom (we assume that the spread of the data about the two or more lines is equal) and an understanding of

FIGURE 11–26   **Effects of an Interaction between a Qualitative Variable and a Quantitative Variable**

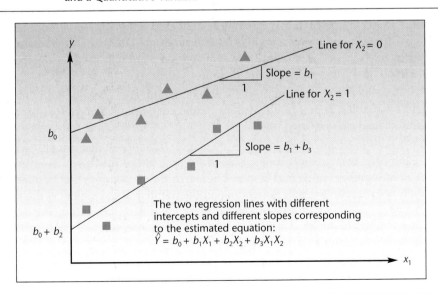

the joint process generating the data. More important, we may use the model to statistically test for the equality of intercepts and slopes. Note that when several indicator variables are used in modeling one or more qualitative variables, there are several possible interaction terms in the model. We will learn more about interactions in general in the next section.

PROBLEMS

11–57. Echlin, Inc., makes parts for automobiles. The company is engaged in strong competition with Japanese, Taiwanese, and Korean manufacturers of the same automobile parts. Recently, the company hired a statistician to study the relationship between monthly sales and the independent variable, number of cars on the road. Data on the explanatory variable are published in national statistical reports. Because of the keen competition with Asian firms, an indicator variable was also used. This variable was given the value 1 during months when restrictions on imports from Asia were in effect and 0 when such restrictions were not in effect. Denoting sales by Y, total number of cars on the road by X_1, and the import restriction dummy variable by X_2, the following regression equation was estimated:

$$\hat{Y} = -567.3 + 0.006X_1 + 26,540X_2$$

The standard error of the intercept estimate was 38.5, that of the coefficient of X_1 was 0.0002, and the standard error of the coefficient of X_2 was 1,534.67. The multiple coefficient of determination was $R^2 = 0.783$. The sample size used was $n = 60$ months (5 years of data). Analyze the results presented. What kind of regression model was used? Comment on the significance of the model parameters and the value of R^2. How many distinct regression lines are there? What likely happens during times of restricted trade with Asia?

11–58. A chemical company has three plants. Due to a shortage of labor and the intense heat involved in production, only one plant is operated at a time, and operations are shifted every few days to another plant. The company wants to estimate the relationship between daily production Y and the independent variable, labor force present X_1. Since it is known that differences may exist among the different plants, it is proposed that a qualitative variable X_2 also be used. This variable is defined as $X_2 = 1$ if plant A is used, 2 if plant B is used, and 3 if plant C is used during the day. Comment on the proposed regression model.

11–59. If we have a regression model with no quantitative variables and only two qualitative variables, represented by some indicator variables and cross-products, what kind of analysis is carried out?

11–60. Recall our Club Med example of Chapter 9. Suppose that not all vacationers at Club Med resorts stay an equal length of time at the resort—different people stay different numbers of days. The club's research director knows that people's ratings of the resorts tend to differ depending on the number of days spent at the resort. Design a new method for studying whether there are differences among the average population ratings of the five Caribbean resorts. What is the name of your method of analysis, and how is the analysis carried out? Explain.

11–61. A financial institution specializing in venture capital is interested in predicting the success of business operations that the institution helps to finance. Success is defined by the institution as return on its investment, as a percentage, after 3 years of operation. The explanatory variables used are Investment (in thousands of dollars), Early investment (in thousands of dollars), and two dummy variables denoting the category of business. The values of these variables are (0, 0) for high-technology industry, (0, 1) for biotechnology companies, and (1, 0) for aerospace firms. Following is part of the computer output for this analysis. Interpret the output, and give a complete analysis of the results of this study based on the provided information.

```
The regression equation is
Return = 6.16 + 0.617 INVEST + 0.151 EARLY + 11.1 DUM1 + 4.15 DUM2
Predictor       Coef         Stdev
 Constant      6.162         1.642
 INVEST        0.6168        0.1581
 EARLY         0.1509        0.1465
 DUM1         11.051         1.355
 DUM2          4.150         1.315

s = 2.148      R-sq = 91.6%      R-sq (adj) = 89.4%


Analysis of Variance

  SOURCE        DF         SS
Regression      4       755.99
Error          15        69.21
Total          19       825.20
```

11–9   Polynomial Regression

Often, the relationship between the dependent variable Y and one or more of the independent X variables is not a straight-line relationship but, rather, has some curvature to it. Several such situations are shown in Figure 11–27 (we show the curved relationship between Y and a *single* explanatory variable X). In each of the situations shown, a straight line provides a poor fit to the data. Instead, polynomials of order higher than 1, that is, functions of higher powers of X, such as X^2 and X^3, provide much better fit to our data. Such polynomials in the X variable or in several X_i variables are still considered linear regression models. Only models where the parameters β_i are not all of the first power are called *nonlinear models*. The multiple linear regression model thus covers situations of fitting data to polynomial functions. The general form of a polynomial regression model in one variable X is given in equation 11–22.

> A one-variable polynomial regression model is
>
> $$Y = \beta_0 + \beta_1 X + \beta_2 X^2 + \beta_3 X^3 + \cdots + \beta_m X^m + \epsilon \qquad (11\text{–}22)$$
>
> where m is the *degree* of the polynomial—the highest power of X appearing in the equation. The degree of the polynomial is the *order* of the model.

Figure 11–28 shows how second- and third-degree polynomial models provide good fits for the data sets in Figure 11–27. A straight line is also shown in each case,

FIGURE 11–27   **Situations Where the Relationship between X and Y Is Curved**

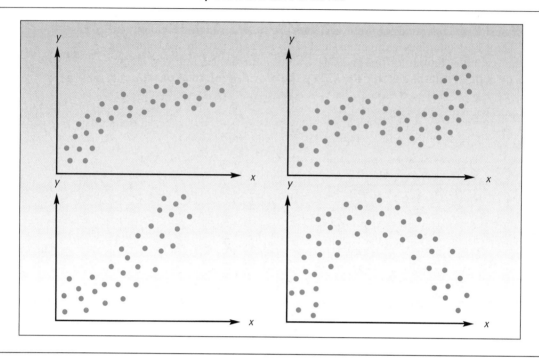

FIGURE 11–28   **The Fits Provided for the Data Sets in Figure 11–27 by Polynomial Models**

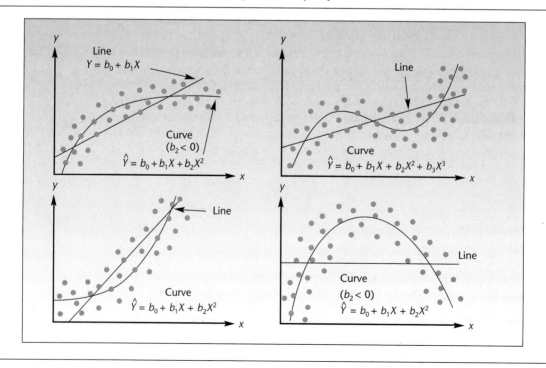

for comparison. Compare the fit provided in each case by a polynomial with the poor fit provided by a straight line. Also compare the use of low-order polynomials in these examples with the polynomial fit shown in Figure 11–2 in the first section of this chapter. There the polynomial of a high order was matching the variation in the data and thus modeling the random variation rather than the general trend, unlike the examples shown in Figure 11–28. Some authors, for example, Weisberg, recommend using polynomials of order no greater than 2 (the third-order example in Figure 11–28 would be an exception) because of the overfitting problem.[10] At any rate, models should never be of order 6 or higher (unless the powers of X have been transformed in a special way). Seber shows that when a polynomial of degree 6 or greater is fit to a data set, a matrix involved in regression computations becomes *ill-conditioned*, which means that very small errors in the data cause relatively large errors in the estimated model parameters.[11] In short, we must be very careful with polynomial regression models and try to obtain the most parsimonious polynomial model that will fit our data. In the next section, we will discuss *transformations* of data that often can change curved data sets into a straight-line form. If we can find such a transformation for a data set, it is always better to use a first-order model on the transformed data set than to use a higher-order polynomial model on the original data. It should be intuitively clear that problems may arise in polynomial regression. The variables X and X^2, for example, are clearly not independent of each other. This may cause the problem of multicollinearity in cases where the data are confined to a narrow range of values.

Having seen what to beware of in using polynomial regression, now we see how these models are used. Since powers of X can be obtained directly from the value of variable X, it is relatively easy to run polynomial models. We enter the data into the computer and add a command that uses X to form a new variable. In a second-order model, we create an X^2 column using spreadsheet commands. Then we run a multiple regression model with two "independent" variables: X and X^2. We demonstrate this with a new example.

EXAMPLE 11–5

It is well known that sales response to advertising usually follows a curve reflecting the diminishing returns to advertising expenditure. As a firm increases its advertising expenditure, sales increase, but the rate of increase drops continually after a certain point. If we consider company sales profits as a function of advertising expenditure, we find that the response function can be very well approximated by a second-order (quadratic) model of the form

$$Y = \beta_0 + \beta_1 X + \beta_2 X^2 + \epsilon$$

A quadratic response function such as this one is shown in Figure 11–29.

It is very important for a firm to identify its own point X_m, shown in the figure. At this point, a maximum benefit is achieved from advertising in terms of the resulting sales profits. Figure 11–29 shows a general form of the sales response to advertising. To find its own maximum point X_m, a firm needs to estimate its response-to-advertising function from its own operation data, obtained by using different levels

---

[10] Sanford Weisberg, *Applied Linear Regression*, 2nd ed. (New York: Wiley, 1985).

[11] G. A. F. Seber, *Linear Regression Analysis* (New York: Wiley, 1977).

FIGURE 11–29   **A Quadratic Response Function of Sales Profits to Advertising Expenditure**

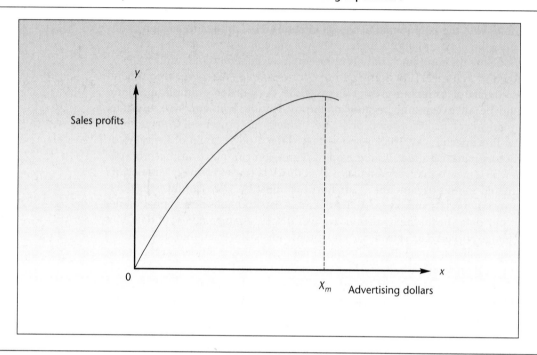

of advertising at different time periods and observing the resulting sales profits. For a particular firm, the data on monthly sales Y and monthly advertising expenditure X, both in hundred thousand dollars, are given in Table 11–14. The table also shows the values of X^2 used in the regression analysis.

Solution   Figure 11–30 shows the data entered in the template. In cell E5, the formula "=D5^2" has been entered. This calculates X^2. The formula has been copied down through cell E25. The regression results from the Results sheet of the template are shown in Table 11–15. The coefficient of determination is $R^2 = 0.9587$, the F ratio is significant, and both Advert and Advsqr are very significant. The minus sign of the squared variable, Advsqr, is logical because a quadratic function with a maximum point has a negative leading coefficient (the coefficient of X^2). We may write the estimated quadratic regression model of Y in terms of X and X^2 as follows:

$$Y = 3.52 + 2.51X - 0.0875X^2 + e \qquad (11\text{–}23)$$

The equation of the estimated regression curve itself is given by dropping the error term e, giving an equation for the predicted values $\hat{Y}$ that lie on the quadratic curve

$$\hat{Y} = 3.52 + 2.51X - 0.0875X^2 \qquad (11\text{–}24)$$

In our particular example, the equation of the curve (equation 11–24) is of importance, as it can be differentiated with respect to X, with the derivative then set to zero and the result solved for the maximizing value X_m shown in Figure 11–29. (If you have not studied calculus, you may ignore the preceding statement.) The result here is $x_m = 14.34$ (hundred thousand dollars). This value maximizes sales profits with respect to advertising (within estimation error of the regression). Thus, the firm should set its advertising level at $1.434 million. The fact that polynomials can always be

TABLE 11–14 **Data for Example 11–5**

Row	Sales	Advert	Advsqr
1	5.0	1.0	1.00
2	6.0	1.8	3.24
3	6.5	1.6	2.56
4	7.0	1.7	2.89
5	7.5	2.0	4.00
6	8.0	2.0	4.00
7	10.0	2.3	5.29
8	10.8	2.8	7.84
9	12.0	3.5	12.25
10	13.0	3.3	10.89
11	15.5	4.8	23.04
12	15.0	5.0	25.00
13	16.0	7.0	49.00
14	17.0	8.1	65.61
15	18.0	8.0	64.00
16	18.0	10.0	100.00
17	18.5	8.0	64.00
18	21.0	12.7	161.29
19	20.0	12.0	144.00
20	22.0	15.0	225.00
21	23.0	14.4	207.36

differentiated gives these models an advantage over alternative models. Remember, however, to keep the order of the model low.

Other Variables and Cross-Product Terms

The polynomial regression model in one variable X, given in equation 11–22, can easily be extended to include more than one independent explanatory variable. The new model, which includes several variables at different powers, is a mixture of the usual multiple regression model in k variables (equation 11–1) and the polynomial regression model (equation 11–22). When several variables are in a regression equation, we may also consider interactions among variables. We have already encountered interactions in the previous section, where we discussed interactions between an indicator variable and a quantitative variable. We saw that an interaction term is just the cross-product of the two variables involved. In this section, we discuss the general concept of interactions between variables, quantitative or not.

The interaction term $X_i X_j$ is a second-order term (the product of two variables is classified the same way as an X^2 term). Similarly, $X_i X_j^2$, for example, is a third-order term. Thus, models that incorporate interaction terms find their natural place within the class of polynomial models. Equation 11–25 is a second-order regression model in two variables X_1 and X_2. This model includes both first and second powers of both variables and an interaction term.

$$Y = \beta_0 + \beta_1 X_1 + \beta_2 X_2 + \beta_3 X_1^2 + \beta_4 X_2^2 + \beta_5 X_1 X_2 + \epsilon \qquad (11\text{--}25)$$

FIGURE 11–30   **Data for the Regression**
[Multiple Regression.xls; Sheet: Data]

	E5	▼	=	=D5^2		
	A	B	C	D	E	F	G
1	**Multiple Regression**					Example 11-5	
2							
3		Y	1	X1	X2	X3	X4
4	Sl.No.	**Sales**		**Advert**	**Advsqr**		
5	1	5	1	1	1		
6	2	6	1	1.8	3.24		
7	3	6.5	1	1.6	2.56		
8	4	7	1	1.7	2.89		
9	5	7.5	1	2	4		
10	6	8	1	2	4		
11	7	10	1	2.3	5.29		
12	8	10.8	1	2.8	7.84		
13	9	12	1	3.5	12.25		
14	10	13	1	3.3	10.89		
15	11	15.5	1	4.8	23.04		
16	12	15	1	5	25		
17	13	16	1	7	49		
18	14	17	1	8.1	65.61		
19	15	18	1	8	64		
20	16	18	1	10	100		
21	17	18.5	1	8	64		
22	18	21	1	12.7	161.29		
23	19	20	1	12	144		
24	20	22	1	15	225		
25	21	23	1	14.4	207.36		

TABLE 11–15   **Results of the Regression**
[Multiple Regression; Sheet: Results]

	0	1	2	3	4	5	6	7	8	9	10
	Intercept	**Advert**	**Advsqr**								
b	3.51505	2.51478	-0.0875								
s(b)	0.73840	0.25796	0.0166								
t	4.7599	9.7487	-5.2751								
p-value	0.0002	0.0000	0.0001								

ANOVA Table

Source	SS	df	MS	F	$F_{Critical}$	p-value		s	1.228
Regn.	630.258	2	315.13	208.99	3.5546	0.0000			
Error	27.142	18	1.5079						
Total	657.4	20							

R^2 0.9587   Adjusted R^2 0.9541

A regression surface of a model like that of equation 11–25 is shown in Figure 11–31. There are, of course, many possible surfaces depending on the values of the coefficients of all terms in the equation. Equation 11–25 may be generalized to more than two explanatory variables, to higher powers of each variable, and to more interaction terms.

When we are considering polynomial regression models in several variables, it is very important not to get carried away by the number of possible terms we can

FIGURE 11–31   **An Example of the Regression Surface of a Second-Order Model in Two Variables**

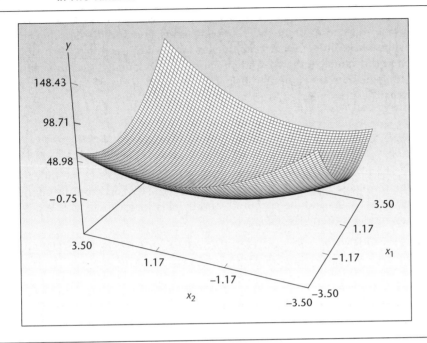

TABLE 11–16   **Example of Regression Output for a Second-Order Model in Two Variables**

Variable	Estimate	Standard Error	t Ratio
X_1	2.34	0.92	2.54
X_2	3.11	1.05	2.96
X_1^2	4.22	1.00	4.22
X_2^2	3.57	2.12	1.68
$X_1 X_2$	2.77	2.30	1.20

include in the model. The number of variables, as well as the powers of these variables and the number of interaction terms, should be kept to a minimum.

How do we choose the terms to include in a model? This question will be answered in Section 11–13, where we discuss methods of variable selection. You already know several criteria for the inclusion of variables, powers of variables, and interaction terms in a model. One thing to consider is the adjusted coefficient of determination. If this measure decreases when a term is included in the model, then the term should be dropped. Also, the significance of any particular term in a model depends on which other variables, powers, or interaction terms are in the model. We must consider the significance of each term by its t statistic, and we must consider what happens to the significance of regression terms once other terms are added to the model or removed from it. For example, let us consider the regression output in Table 11–16.

From the results in the table, it is clear that only X_1, X_2, and X_1^2 are significant. The apparent nonsignificance of X_2^2 and X_1X_2 may be due to multicollinearity. At any rate, a regression without these last two variables should be carried out. We must also look at R^2 and the adjusted R^2 of the different regressions, and find the most parsimonious model with statistically significant parameters that explain as much as possible of the variation in the values of the dependent variable. Incidentally, the surface in Figure 11–31 was generated by computer, using all the coefficient estimates given in Table 11–16 (regardless of their significance) and an intercept of zero.

PROBLEMS

11–62. The following results pertain to a regression analysis of the difference between the mortgage rate and the Treasury bill rate (SPREAD) on the shape of the yield curve (SHAPE) and the corporate bond yields spread (RISK).[12] What kind of regression model is used? Explain.

$$\text{SPREAD}_t = b_0 + b_1\text{STD}_t + b_2\text{SHAPE}_t + b_3\text{RISK}_t + b_4\text{STD}_t^2$$
$$- b_5\text{SHAPE}_t^2 - b_6\text{STD}_t * \text{SHAPE}_t$$

11–63. Use the data in Table 11–6 to run a polynomial regression model of exports to Singapore versus M1 and M1 squared, as well as Price and Price squared, and an interaction term. Also try to add a squared exchange rate variable into the model. Find the best, most parsimonious regression model for the data.

11–64. Use the data of Example 11–3, presented in Table 11–12, to try to fit a polynomial regression model of movie gross earnings on production cost and production cost squared. Also try promotion and promotion squared. What is the best, most parsimonious model?

11–65. Give at least two reasons why a polynomial regression model in one or more independent variables should be kept as parsimonious as possible.

11–66. A regression model of sales Y versus advertising X_1, advertising squared X_1^2, competitors' advertising X_2, competitors' advertising squared X_2^2, and the interaction of X_1 and X_2 is run. The results are as follows.

Variable	Parameter Estimate	Standard Error
X_1	5.324	2.478
X_2	3.229	1.006
X_1^2	4.544	3.080
X_2^2	1.347	0.188
X_1X_2	2.692	1.517
$R^2 = 0.657$	Adjusted $R^2 = 0.611$	$n = 197$

Interpret the regression results. Which regression equation should be tried next? Explain.

---

[12]"The Thrift Crisis, Mortgage-Credit Intermediation, and Housing Activity," by Michael G. Bradley, Stuart A. Gabriel, and Mark E. Wohar. *Journal of Money, Credit, and Banking* Volume 27, Number 2, p. 484. Copyright 1995 by the Ohio State University Press. All rights reserved. Reprinted by permission.

11-67. What regression model would you try for the following data? Give your reasons why.

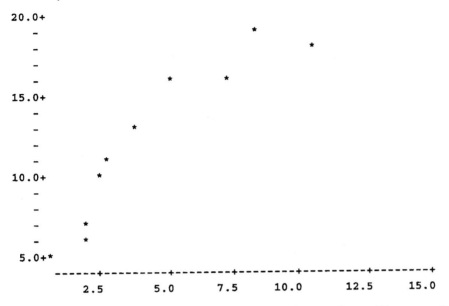

11-68. The regression model $Y = \beta_0 + \beta_1 X + \beta_2 X^2 + \beta_3 X^3 + \beta_4 X^4 + \epsilon$ was fit to the following data set. Can you suggest a better model? If so, which?

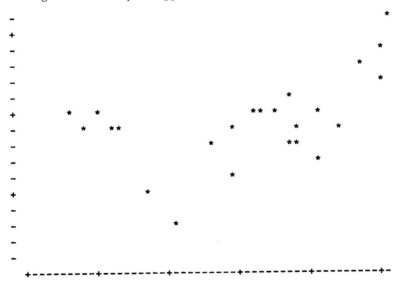

11-10   Nonlinear Models and Transformations

Sometimes the relationship between Y and one or more of the independent X_i variables is nonlinear. Remember that powers of the X_i variables in the regression model still keep the model linear, but that powers of the coefficients β_i make the model nonlinear. We may have prior knowledge about the process generating the data that indicates that a nonlinear model is appropriate; or we may observe that the data follow one of the general nonlinear curves shown in the figures in this section.

In many cases, a nonlinear model may be changed to a linear model by use of an appropriate **transformation**. Models that can be transformed to linear models are called **intrinsically linear** models. These models are the subject of this section. The "hard-core" nonlinear models, those that cannot be transformed into linear models, are difficult to analyze and therefore are outside the scope of this book.

The first model we will encounter is the *multiplicative model,* given by equation 11–26.

The multiplicative model is

$$Y = \beta_0 X_1^{\beta_1} X_2^{\beta_2} X_3^{\beta_3} \epsilon \qquad (11\text{--}26)$$

This is a multiplicative model in the three variables X_1, X_2, and X_3. The generalization to k variables is clear. The β_i are unknown parameters, and ϵ is a multiplicative random error.

The multiplicative model of equation 11–26 can be transformed to a linear regression model by the use of a **logarithmic transformation**. A logarithmic transformation is the most common transformation of data in statistical analysis. We will use natural logarithms—logs to base e—although any log transformation would do (we may use logs to any base, as long as we are consistent throughout the equation). Taking natural logs (sometimes denoted by ln) of both sides of equation 11–26 gives us the following linear model:

$$\log Y = \log \beta_0 + \beta_1 \log X_1 + \beta_2 \log X_2 + \beta_3 \log X_3 + \log \epsilon \qquad (11\text{--}27)$$

Equation 11–27 is now in the form of equation 11–1: It is a linear regression equation of $\log Y$ in terms of $\log X_1$, $\log X_2$, and $\log X_3$ as independent variables. The error term in the linearized model is $\log \epsilon$. To conform with the assumptions of the multiple regression model and to allow us to perform tests of significance of model parameters, we must assume that the linearized errors $\log \epsilon$ are normally distributed with mean 0 and equal variance σ^2 for successive observations and that these errors are independent of each other.

When we consider only one independent variable, the model of equation 11–26 is a power curve in X, of the form

$$Y = \beta_0 X^{\beta_1} \epsilon \qquad (11\text{--}28)$$

Depending on the values of parameters β_0 and β_1, equation 11–28 gives rise to a wide-range family of power curves. Several members of this family of curves, showing the relationship between X and Y, leaving out the errors ϵ, are shown in Figure 11–32. When more than one independent variable is used, as in equation 11–26, the graph of the relationship between the X_i variables and Y is a multidimensional extension of Figure 11–32.

As you can see from the figure, many possible data relationships may be well modeled by a power curve in one variable or its extension to several independent variables. The resemblance of the curves in Figure 11–32 to at least two curves shown in Figure 11–28 is also evident. As you look at Figure 11–32, we repeat our suggestion from the last section that, when possible, a transformed model with few parameters is better than a polynomial model with more parameters.

When dealing with a multiplicative, or power, model, we take logs of both sides of the equation and run a linear regression model on the logs of the variables. Again,

FIGURE 11–32 A Family of Power Curves of the Form $Y = \beta_0 X^{\beta_1}$

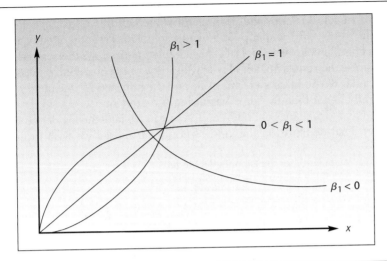

it is important to understand that the errors must be multiplicative. This makes sense in situations where the magnitude of an error is proportional to the magnitude of the response variable. We assume that the logs of the errors are normally distributed and satisfy all the assumptions of the linear regression model. Models where the error term is additive rather than multiplicative, such as $Y = \beta_0 X_1^{\beta_1} X_2^{\beta_2} X_3^{\beta_3} + \epsilon$, are *not* intrinsically linear because there is no expression for logs of *sums*.

When using a model such as equation 11–26 or equation 11–28, we enter the data into the computer, form new variables by having the computer take logs of the Y and X_i variables, and run a regression on the transformed variables. In addition to making sure that the model assumptions seem to hold, we must remember that computer-generated predictions and confidence intervals will be in terms of the transformed variables unless the computer algorithm is designed to convert information back to the original variables. The conversion back to the original variables is done by taking antilogs.

In many situations, it is possible to determine the need for a log transformation by inspecting a scatter plot of the data. We demonstrate the analysis, using the data of Example 11–5. We will assume that a model of the form of equation 11–28 fits the relationship between sales profits Y and advertising dollars X. We assume a power curve with multiplicative errors. Thus, we assume that the relationship between X and Y is given by

$$Y = \beta_0 X^{\beta_1} \epsilon$$

Taking logs of both sides of the equation, we get the linearized model

$$\log Y = \log \beta_0 + \beta_1 \log X + \log \epsilon \qquad (11\text{–}29)$$

Our choice of a power curve and a transformation using logarithms is prompted by the fact that our data in this example exhibit curvature that may resemble a member of the family of curves in Figure 11–32, and by the fact that a quadratic regression model, which is similar to a power curve, was found to fit the data well. (Data for this example are shown in Figure 11–30.)

To solve this problem using the template we use the "=LN()" function available in Excel. This function calculates the natural logarithm of the number in parentheses.

Unprotect the Data sheet and enter the Sales and Advert data in some unused columns, say, columns N and O, as in Figure 11–33. Then enter the formula "=LN(N5)" in cell B5. Copy that formula down to cell B25. Next, in cell D5, enter the formula "=LN(O5)" and copy it down to cell D25. Press the F9 key and the regression is complete. Protect the Data sheet. Table 11–17 shows the regression results.

Comparing the results in Table 11–17 with those of the quadratic regression, given in Table 11–15, we find that, in terms of R^2 and the adjusted R^2, the quadratic regression is slightly better than the log Y versus log X regression.

Do we have to take logs of both X and Y, or can we take the log of one of the variables only? That depends on the kind of nonlinear model we wish to linearize. It

FIGURE 11–33   Data Entry for the Exponential Model
[Multiple Regression.xls; Sheet: Data]

	A	B	C	D	E	F	G		N	O
1	Multiple Regression					Exponential Model				
2										
3		Y	1	X1	X2	X3	X4			
4	Sl.No.	Log Sales		Log Advt					Sales	Advert
5	1	1.60944	1	0					5	1
6	2	1.79176	1	0.5878					6	1.8
7	3	1.8718	1	0.47					6.5	1.6
8	4	1.94591	1	0.5306					7	1.7
9	5	2.0149	1	0.6931					7.5	2
10	6	2.07944	1	0.6931					8	2
11	7	2.30259	1	0.8329					10	2.3
12	8	2.37955	1	1.0296					10.8	2.8
13	9	2.48491	1	1.2528					12	3.5
14	10	2.56495	1	1.1939					13	3.3
15	11	2.74084	1	1.5686					15.5	4.8
16	12	2.70805	1	1.6094					15	5
17	13	2.77259	1	1.9459					16	7
18	14	2.83321	1	2.0919					17	8.1
19	15	2.89037	1	2.0794					18	8
20	16	2.89037	1	2.3026					18	10
21	17	2.91777	1	2.0794					18.5	8
22	18	3.04452	1	2.5416					21	12.7
23	19	2.99573	1	2.4849					20	12
24	20	3.09104	1	2.7081					22	15
25	21	3.13549	1	2.6672					23	14.4
26										

TABLE 11–17   Regression Results.
[Multiple Regression.xls; Sheet: Results]

	0	1	2	3	4	5	6	7	8	9	10
	Intercept	Log Advt									
b	1.70082	0.55314									
s(b)	0.05123	0.03011									
t	33.2006	18.3727									
p-value	0.0000	0.0000									

ANOVA Table

Source	SS	df	MS	F	$F_{Critical}$	p-value		
Regn.	4.27217	1	4.2722	337.56	4.3808	0.0000	s	0.1125
Error	0.24047	19	0.0127					
Total	4.51263	20		R^2 0.9467			Adjusted R^2	0.9439

turns out that there is indeed a nonlinear model that may be linearized by taking the log of one of the variables. Equation 11–30 is a nonlinear regression model of Y versus the independent variable X that may be linearized by taking logs of both sides of the equation. To use the resulting linear model, given in equation 11–31, we run a regression of log Y versus X (not log X).

The **exponential model** is

$$Y = \beta_0 e^{\beta_1 X} \epsilon \qquad\qquad (11\text{--}30)$$

The linearized model of the exponential relationship, obtained by taking logs of both sides of equation 11–30, is given by

$$\log Y = \log \beta_0 + \beta_1 X + \log \epsilon \qquad\qquad (11\text{--}31)$$

When the relationship between Y and X is of the exponential form, the relationship is mildly curved upward or downward. Thus, taking log of Y only and running a regression of log Y versus X may be useful when our data display mild curvature.

The exponential model of equation 11–30 is extendable to several independent X_i variables. The model is given in equation 11–32.

An exponential model in two independent variables is

$$Y = e^{\beta_0 + \beta_1 X_1 + \beta_2 X_2} \epsilon \qquad\qquad (11\text{--}32)$$

The letter e in equation 11–32, as in equation 11–30, denotes the natural number $e = 2.7182\ldots$, the base of the natural logarithm. Taking the natural logs of both sides of equation 11–32 gives us the following linear regression model:

$$\log Y = \beta_0 + \beta_1 X_1 + \beta_2 X_2 + \log \epsilon \qquad\qquad (11\text{--}33)$$

This relationship is extendable to any number of independent variables. The transformation of log Y, leaving the X_i variables in their natural form, allows us to perform linear regression analysis. The data of Example 11–5, shown in Figure 11–30, do not display a mild curvature. The next model we discuss, however, may be more promising.

Figure 11–34 shows curves corresponding to the logarithmic model given in equation 11–34.

FIGURE 11–34
Curves Corresponding to a Logarithmic Model

The **logarithmic model** is

$$Y = \beta_0 + \beta_1 \log X + \epsilon \qquad\qquad (11\text{--}34)$$

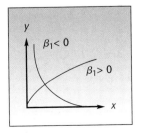

This nonlinear model can be linearized by substituting the variable $X' = \log X$ into the equation. This gives us the linear model in X':

$$Y = \beta_0 + \beta_1 X' + \epsilon \qquad\qquad (11\text{--}35)$$

From Figure 11–32, it seems that the logarithmic model with $\beta_1 > 0$ may fit the data of Example 11–5. We will therefore try to fit this model. The required transformation to obtain the linearized model in equation 11–35 is to take the log of X only, leaving Y as is. We will tell the computer program to run Y versus log X. By doing so, we assume that our data follow the logarithmic model of equation 11–34. The results of the regression analysis of sales profits versus the natural logarithm of advertising expenditure are given in Table 11–18.

As seen from the regression results, the model of equation 11–35 is probably the best model to describe the data of Example 11–5. The coefficient of determination is $R^2 = 0.978$, which is higher than those of both the quadratic model and the power curve model we tried earlier. Figure 11–35 is a plot of the sales variable versus the log of advertising (the regression model of equation 11–35). As can be seen from the figure, we have a straight-line relationship between log advertising and sales. Compare this figure with Figure 11–36, which is the relationship of log sales versus log adver-

TABLE 11–18   Results of the Logarithmic Model
[Multiple Regression.xls; Sheet: Results]

	0	1	2	3	4	5	6	7	8	9	10
	Intercept	Log Advt									
b	3.66825	6.784									
s(b)	0.40159	0.23601									
t	9.13423	28.7443									
p-value	0.0000	0.0000									

ANOVA Table

Source	SS	df	MS	F	$F_{Critical}$	p-value
Regn.	642.622	1	642.62	826.24	4.3808	0.0000
Error	14.7777	19	0.7778			
Total	657.4	20				

s 0.8819

R^2 0.9775   Adjusted R^2 0.9763

FIGURE 11–35   Plot of Sales versus the Natural Log of Advertising Expenditure (Example 11–5)

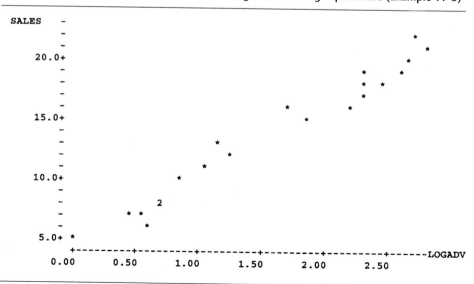

tising, the model of equation 11–31 we tried earlier. In the latter graph, there is some extra curvature, and a straight line does not quite fit the transformed variable. We conclude that the model given by equation 11–34 fits the sales profits–advertising expenditure relationship best. The estimated regression relationship is as given in Table 11–18: Sales = 3.67 + 6.78 Log Advt.

Remember that when we transform our data, the least-squares method minimizes the sum of the squared errors for the *transformed* variables. It is, therefore, very important for us to check for any violations of model assumptions that may occur as a result of the transformations. We must be especially careful with the assumptions about the regression errors and their distribution. This is why residual plots are very important when transformations of variables are used. In our present model for the data of Example 11–5, a plot of the residuals versus the predicted sales values $\hat{Y}$ is given in Figure 11–37. The plot of the residuals does not indicate any violation of assumptions, and we therefore conclude that the model is adequate. We note also that confidence intervals for transformed models do not always correspond to correct intervals for the original model.

FIGURE 11–36   Plot of Log Sales versus Log Advertising Expenditure (Example 11–5)

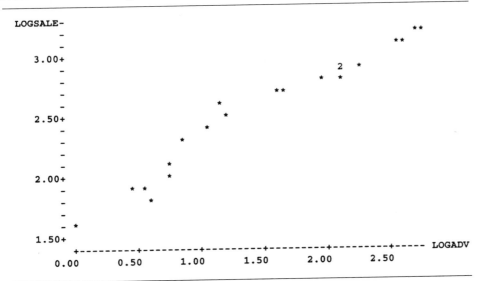

FIGURE 11–37   Residual Plot of the Logarithmic Model; X Axis Is Sales
[Multiple Regresson.xls; Sheet: Residuals]

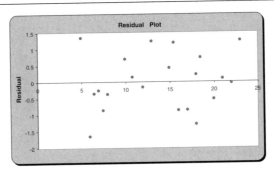

Another nonlinear model that may be linearized by an appropriate transformation is the *reciprocal model*. A reciprocal model in several variables is given in equation 11–36.

The **reciprocal model** is

$$Y = \frac{1}{\beta_0 + \beta_1 X_1 + \beta_2 X_2 + \beta_3 X_3 + \epsilon} \qquad (11\text{–}36)$$

This model becomes a linear model upon taking the reciprocals of both sides of the equation. In practical terms, we run a regression of $1/Y$ versus the X_i variables unchanged. A particular reciprocal model with one independent variable has a complicated form, which will not be explicitly stated here. This model calls for linearization by taking the reciprocals of both X and Y. Two curves corresponding to this particular reciprocal model are shown in Figure 11–38. When our data display the acute curvature of one of the curves in the figure, running a regression of $1/Y$ versus $1/X$ may be fruitful.

Next we will discuss transformations of the dependent variable Y only. These are transformations designed to stabilize the variance of the regression errors.

Variance-Stabilizing Transformations

Remember that one of the assumptions of the regression model is that the regression errors ϵ have equal variance. If the variance of the errors increases or decreases as one or more of the independent variables change, we have the problem of heteroscedasticity. When heteroscedasticity is present, our regression coefficient estimators are not efficient. This violation of the regression assumptions may sometimes be corrected by the use of a transformation. We will consider three major transformations of the dependent variable Y to correct for heteroscedasticity.

Transformations of Y that may help correct the problem of heteroscedasticity:

1. The square root transformation: $Y' = \sqrt{Y}$
   This is the least "severe" transformation. It is useful when the variance of the regression errors is approximately proportional to the mean of Y, conditional on the values of the independent variables X_i.

2. The logarithmic transformation: $Y' = \log Y$ (to any base)
   This is a transformation of a stronger nature and is useful when the variance of the errors is approximately proportional to the square of the conditional mean of Y.

3. The reciprocal transformation: $Y' = 1/Y$
   This is the most severe of the three transformations and is required when the violation of equal variance is serious. This transformation is useful when the variance of the errors is approximately proportional to the conditional mean of Y to the fourth power.

FIGURE 11–38
Two Examples of a Relationship Where a Regression of $1/Y$ versus $1/X$ Is Appropriate

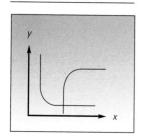

There are transformations other than these, although the preceding transformations are most commonly used. In a given situation, we want to find the transformation that makes the errors have approximately equal variance as evidenced by the residual plots. An alternative to using transformations to stabilize the variance is the use of the weighted least-squares procedure mentioned in our earlier discussion of the heteroscedasticity problem. We note that a test for heteroscedasticity exists. The test is the Goldfeld-Quandt test, discussed in econometrics books.

It is important to note that transformations may also correct problems of nonnormality of the errors. A variance-stabilizing transformation may thus make the distribution of the new errors closer to a normal distribution. In using transformations—whether to stabilize the variance, to make the errors approximate a normal distribution, or to make a nonlinear model linear—it is important to remember that all results should be converted back to the original variables. As a final example of a nonlinear model that can be linearized by using a transformation, we present the *logistic regression model.*

Regression with Dependent Indicator Variable

In Section 11–8, we discussed models with indicator variables as independent X_i variables. In this subsection, we discuss regression analysis where the dependent variable Y is an indicator variable and may obtain only the value 0 or the value 1. This is the case when the response to a set of independent variables is in binary form: success or failure. An example of such a situation is the following.

A bank is interested in predicting whether a given loan applicant would be a good risk, i.e., pay back his or her loan. The bank may have data on past loan applicants, such as applicant's income, years of employment with the same employer, and value of home. All these independent variables may be used in a regression analysis where the dependent variable is binary: $Y = 0$ if the applicant did not repay the loan, and $Y = 1$ if she or he did pay back the loan. When only one explanatory variable X is used, the model is the *logistic function,* given in equation 11–37.

> The **logistic function** is
> $$E(Y \mid X) = \frac{e^{\beta_0 + \beta_1 X}}{1 + e^{\beta_0 + \beta_1 X}} \qquad (11\text{–}37)$$

The expected value of Y given X, that is, $E(Y \mid X)$, has a special meaning: It is the probability that Y will equal 1 (the probability of success), given the value of X. Thus, we write $E(Y \mid X) = p$. The transformation given below linearizes equation 11–37.

> Transformation to linearize the logistic function:
> $$p' = \log\left(\frac{p}{1 - p}\right) \qquad (11\text{–}38)$$

FIGURE 11–39
The Logistic Function

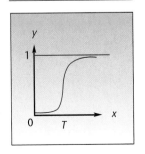

We leave it to the reader to show that the resulting regression equation is linear. In practical terms, the transformed model is difficult to employ because resulting errors are intrinsically heteroscedastic. A better approach is to use the more involved methods of nonlinear regression analysis. We present the example to show that, in many cases, the dependent variable may be an indicator variable as well. Much research is being done today on the logistic regression model, which reflects the model's growing importance. Fitting data to the curve of the logistic function is called *logit analysis.* A graph of the logistic function of equation 11–37 is shown in Figure 11–39. Note the typical elongated S shape of the graph. This function is useful as a "threshold model," where the probability that the dependent variable Y will be equal to 1 (a success in the experiment) increases as X increases. This increase becomes very dramatic as X reaches a certain threshold value (the point T in the figure).

PROBLEMS

11–69. What are the two main reasons for using transformations?

11–70. Explain why a transformed model may be better than a polynomial model. Under what conditions is this true?

11–71. Refer to the residual plot in Figure 11–14. What transformation would you recommend be tried to correct the situation?

11–72. For the Singapore data of Example 11–2, presented in Table 11–6, use several different data transformations of the variables Exports, M1, and Price, and find a better model to describe the data. Comment on the properties of your new model.

11–73. Which transformation would you try for modeling the following data set?

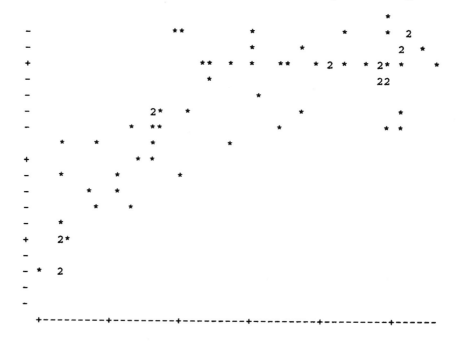

11–74. Which transformation would you recommend for the following data set?

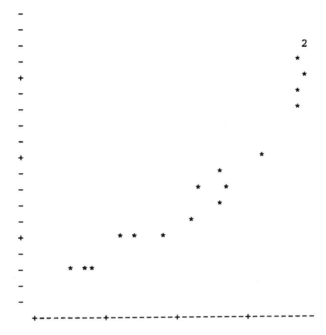

11–75. An analysis of the effect of advertising on consumer price sensitivity is carried out. The log of the quantity purchased ($\ln q$), the dependent variable, is run against the log of an advertising-related variable called RP (the log is variable $\ln RP$). There is an additive error term ϵ in the transformed regression. What assumptions about the model relating q and RP are implied by the transformation?

11–76. The following regression model is run.

$$\log Y = 3.79 + 1.66X_1 + 2.91X_2 + \log e$$

Give the equation of the original, nonlinear model linking the explanatory variables with Y.

11–77. Consider the following nonlinear model.

$$Y = e^{\beta_1 X_1} + e^{\beta_2 X_2} + \epsilon$$

Is this model intrinsically linear? Explain.

11–78. The model used in economics to describe production is

$$Q = \beta_0 C^{\beta_1} K^{\beta_2} L^{\beta_3} \epsilon$$

where the dependent variable Q is the quantity produced, C is the capacity of a production unit, K is the capital invested in the project, and L is labor input, in days. Transform the model to linear regression form.

11–79. Consider the nonlinear model

$$Y = \frac{1}{\beta_0 + \beta_1 X_1 + \beta_2 X_2 + \epsilon}$$

What transformation linearizes this model?

11–80. If the residuals from fitting a linear regression model display mild heteroscedasticity, what data transformation may correct the problem?

11–81. The model in problem 11–78 is transformed to a linear regression model and analyzed with a computer. Do the estimated regression coefficients minimize the sum of the squared deviations of the data from the original curve? Explain.

11–82. An article reports the results of a regression of a firm's restructuring level on the log of total assets in the previous year.[13] The equation is

$$Y = -2.039 + 0.034 \log X$$
$$(0.001) \qquad (0.71)$$

Explain this regression.

11–11 Multicollinearity

The idea of multicollinearity permeates every aspect of multiple regression, and we have encountered this idea in earlier sections of this chapter. The reason multicollinearity (or simply *collinearity*) has such a pervasive effect on multiple regression is that whenever we study the relationship between Y and several X_i variables, we are bound to encounter some relationships among the X_i variables themselves. Ideally, the X_i variables in a regression equation are uncorrelated with one another; each variable contains a unique piece of information about Y—information that is not contained in any of the other X_i. When the ideal occurs in practice, we have no multicollinearity. On the other extreme, we encounter the case of perfect collinearity. Suppose that we run a regression of Y on two explanatory variables X_1 and X_2. Perfect collinearity occurs when one X variable can be expressed precisely in terms of the other X variable for all elements in our data set.

> Variables X_1 and X_2 are perfectly collinear if
>
> $$X_1 = a + bX_2 \qquad (11\text{–}39)$$
>
> for some real numbers a and b.

In the case of equation 11–39, the two variables are on a straight line, and one of them perfectly determines the other. Here there is no new information about Y to be gained by adding X_2 to a regression equation that already contains X_1 (or vice versa).

In practice, most situations fall between the two extremes. Often, there is some degree of collinearity among several of the independent variables in a regression equation. A measure of the collinearity between two X_i variables is the *correlation* between the two. Recall that in regression analysis we assume that the X_i are constants and not random variables. Here we relax this assumption and measure the correlation between the independent variables (this assumes they are random variables in their own right). When two independent X_i variables are found to be highly correlated with each other, we may expect the adverse effects of multicollinearity on the regression estimation procedure.

---

[13]E. Hotchkiss, "Postbankruptcy Performance and Management Turnover," *Journal of Finance* 50, no. 1 (March 1995), p. 17.

In the case of perfect collinearity, the regression algorithm breaks down completely. Even if we were able to get regression coefficient estimates in such a case, their variance would be infinite. When the degree of collinearity is less severe, we may expect the variance of the regression estimators (and the standard errors) to be large. There are other problems that may occur, and we will discuss them shortly. Multicollinearity is a problem of degree. When the correlations among the independent regression variables are minor, the effects of multicollinearity may not be serious. In cases of strong correlations, the problem may affect the regression more adversely, and we may need to take some corrective action. Note that in a multiple regression analysis with several independent variables, *several* of the X_i may be correlated. A set of independent variables that are correlated with one another is called a *multicollinearity set*.

Let us imagine a variable and its information content as a direction in space. Two uncorrelated variables can be viewed as *orthogonal* directions in space—directions that are at 90° to each other. Perfectly correlated variables represent directions that have an angle of 0° or 180° between them, depending on whether the correlation is +1 or −1. Variables that are partly correlated are directions that form an angle greater than 0° but less than 90° (or between 90° and 180° if the correlation is negative). The closer the angle between the directions is to 0° or 180°, the greater the collinearity. This is illustrated in Figure 11–40.

Causes of Multicollinearity

There are several different causes of multicollinearity. A data collection method may produce multicollinearity if, without intention, we tend to gather data with related values on several variables. For example, we may be interested in running a regression

FIGURE 11–40 **Collinearity Viewed as the Relationship between Two Directions in Space**

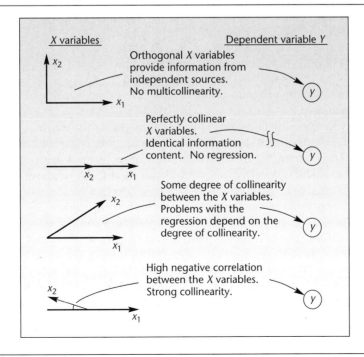

of size of home Y versus family income X_1 and family size X_2. If, unwittingly, we always sample families with high income and large size (rather than also obtaining sample families with low income and large size or high income and small size), then we have multicollinearity. In such cases, improving the sampling method would solve the problem. In other cases, the variables may by nature be related to one another, and sampling adjustments may not work. In such cases, one of the correlated variables should probably be excluded from the model to avoid the collinearity problem.

In industrial processes, sometimes there are physical constraints on the data. For example, if we run a regression of chemical yield Y versus the concentration of two elements X_1 and X_2, and the total amount of material in the process is constant, then as one chemical increases in concentration, we must reduce the concentration of the other. In this case, X_1 and X_2 are (negatively) correlated, and multicollinearity is present.

Yet another source of collinearity is the inclusion of higher powers of the X_i. Including X^2 in a model that contains the variable X may cause collinearity if our data are restricted to a narrow range of values. This was seen in one of the problems in an earlier section.

Whatever the source of the multicollinearity, it is important to be aware of its existence so that we may guard against its adverse effects on the estimation procedure and the ensuing use of the regression equation in prediction, control, or understanding the underlying process. We now present several methods of detecting multicollinearity and a description of its major symptoms.

Detecting the Existence of Multicollinearity

Many statistical computer packages have built-in warnings about severe cases of multicollinearity. When multicollinearity is extreme (i.e., when we have near-perfect correlation between some of the explanatory variables), the program may automatically drop collinear variables so that computations may be possible. In such cases, the MINITAB program, for example, will print the following message:

```
[variable name] is highly correlated with other X variables.
[variable name] has been omitted from the equation.
```

In less serious cases, the program prints the first line of the warning above but does not drop the variable.

In cases where multicollinearity is not serious enough to cause computational problems, it may still disturb the statistical estimation procedure and make our estimators have large variances. In such cases, the computer may not print a message telling us about multicollinearity, but we will still want to know about it. There are two methods available in most statistical packages to help us determine the extent of multicollinearity present in our regression.

The first method is the computation of a **correlation matrix** of the independent regression variables. The correlation matrix is an array of all estimated pairwise correlations between the independent variables X_i. The format of the correlation matrix is shown in Figure 11–41. The correlation matrix allows us to identify those explanatory variables that are highly correlated with one another and thus cause the problem of multicollinearity when they are included together in the regression equation. For example, in the correlation matrix shown in Figure 11–41, we see that the correlation between variable X_1 and variable X_2 is very high (0.92). This means that the two variables represent very much the same direction in space, as was shown in Figure 11–40.

FIGURE 11–41   A Correlation Matrix

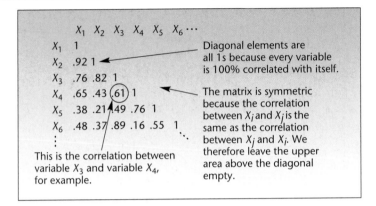

$$
\begin{array}{c}
\quad\quad X_1\ X_2\ X_3\ X_4\ X_5\ X_6\cdots\\
\end{array}
$$

X₁ 1

X₂ .92 1 ← Diagonal elements are all 1s because every variable is 100% correlated with itself.

X₃ .76 .82 1

X₄ .65 .43 (61) 1 ← The matrix is symmetric because the correlation between X_i and X_j is the same as the correlation between X_j and X_i. We therefore leave the upper area above the diagonal empty.

X₅ .38 .21 .49 .76 1

X₆ .48 .37 .89 .16 .55 1

This is the correlation between variable X_3 and variable X_4, for example.

TABLE 11–19   The Correlation Matrix for Example 11–2
[Multiple Regression.xls; Sheet: Correl]

		0 M1	1 Lend	2 Price	3 Exch.	4	5	6	7	8	9	10
1	**M1**	1										
2	**Lend**	-0.112	1									
3	**Price**	0.4471	0.7451	1								
4	**Exch.**	-0.4097	-0.2786	-0.4196	1							
5												
6												
7												
8												
9												
10												
Y	**Exports**	0.7751	0.335	0.7699	-0.4329							

Being highly correlated with each other, the two variables contain much of the same information about Y and therefore cause multicollinearity when both are in the regression equation. A similar statement can be made about X_3 and X_6, which have a 0.89 correlation. Remember that multicollinearity is a matter of extent or degree. It is hard to give a rule of thumb as to how high a correlation may be before multicollinearity has adverse effects on the regression analysis. Correlations as high as the ones just mentioned are certainly large enough to cause multicollinearity problems.

The template has a sheet titled "Correl" in which the correlation matrix is computed and displayed. Table 11–19 shows the correlation matrix among all the variables in Example 11–2.

The highest pairwise correlation exists between Lend and Price. This correlation of 0.745 is the source of the multicollinearity detected in problem 11–36. Recall that the model we chose as best in our solution of Example 11–2 did not include the lending rate. In our solution of Example 11–2, we discussed other collinear variables as well. The multicollinearity may have been caused by the smaller pairwise correlations in Table 11–19, or it may have been caused by more complex correlations in the data than just the pairwise correlations. This brings us to the second statistical method of detecting multicollinearity: variance inflation factors.

FIGURE 11–42   **Relationship between R_h^2 and VIF**

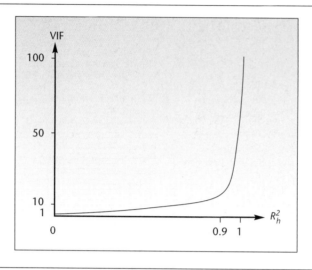

The degree of multicollinearity introduced to the regression by variable X_h, once variables $X_1, \ldots, X_k$ are in the regression equation, is a function of the multiple correlation between X_h and the other variables $X_1, \ldots, X_k$. Thus, suppose we run a multiple regression—not of Y, but of X_h—on all the other X variables. From this multiple regression, we get an R^2 value. This R^2 is a measure of the multicollinearity "exerted" by variable X_h. Recall that a major problem caused by multicollinearity is the inflation of the variance of the regression coefficient estimators. To measure this ill effect of multicollinearity, we use the *variance inflation factor (VIF)* associated with variable X_h.

The **variance inflation factor** associated with X_h is

$$VIF(X_h) = \frac{1}{1 - R_h^2} \qquad (11\text{–}40)$$

where R_h^2 is the R^2 value obtained for the regression of X_h, as dependent variable, on the other X variables in the original equation aimed at predicting Y.

It can be shown that the VIF of variable X_h is equal to the ratio of the variance of the coefficient estimator b_h in the original regression (with Y as dependent variable) and the variance of the estimator b_h in a regression where X_h is orthogonal to the other X variables.[14] The VIF is the inflation factor of the variance of the estimator as compared with what that variance would have been if X_h were not collinear with any of the other X variables in the regression. A graph of the relationship between R_h^2 and the VIF is shown in Figure 11–42.

---

[14]J. Johnston, *Econometric Methods* (New York: McGraw-Hill, 1984).

TABLE 11–20 VIF Values for the Regression from Example 11–2

	A	B	C	D	E	F
1	**Multiple Regression Results**					Exports
2						
3		0	1	2	3	4
4		Intercept	M1	Lend	Price	Exch.
5	**b**	-4.0155	0.36846	0.0047	0.0365	0.2679
6	**s(b)**	2.7664	0.06385	0.04922	0.00933	1.17544
7	**t**	-1.45151	5.7708	0.09553	3.91491	0.22791
8	**p-value**	0.1517	0.0000	0.9242	0.0002	0.8205
9						
10	**VIF**		3.2072	5.3539	6.2887	1.3857

As can be seen from the figure, when the R^2 value of X_h versus the other X variables increases from 0.9 to 1, the VIF rises very dramatically. In fact, for $R_h^2 = 1.00$, the VIF is infinite. The graph, however, should not deceive you. Even for values of R_h^2 less than 0.9, the VIF is still large. A VIF of 6, for example, means that the variance of the regression coefficient estimator b_h is 6 times what it should be (when no collinearity exists). Most computer packages will report, on request, the VIFs for all the independent variables in a regression model.

Table 11–20 shows the template output for the regression from Example 11–2, which contains the VIF values in row 10. We note that the VIF for variables Lend and Price are greater than 5 and thus indicate that some degree of multicollinearity exists with respect to these two variables. Some action, as described in the next subsection, is required to take care of this multicollinearity.

What symptoms and effects of multicollinearity would we find without looking at a variable correlation matrix or the VIFs? There are several noticeable effects of multicollinearity. The major ones are presented in the following list.

The effects of multicollinearity:

1. The variances (and standard errors) of regression coefficient estimators are inflated.
2. The magnitudes of the regression coefficient estimates may be different from what we expect.
3. The signs of the regression coefficient estimates may be the opposite of what we expect.
4. Adding or removing variables produces large changes in the coefficient estimates or their signs.
5. Removing a data point causes large changes in the coefficient estimates or their signs.
6. In some cases, the F ratio is significant, but none of the t ratios is.

When any of or all these effects are present, multicollinearity is likely to be present. How bad is the problem? What are the adverse consequences of multicollinearity? The problem is not always as bad as it may seem. Actually, if we wish to use the regression model for prediction purposes, multicollinearity may not be a serious problem.

From the effects of multicollinearity just listed (some of them were mentioned in earlier sections), we know that the regression coefficient estimates are not reliable when multicollinearity is present. The most serious effect is the variance inflation, which makes some variables seem not significant. Then there is the problem of the magnitudes of the estimates, which may not be accurate, and the problem of the signs of the estimates. We see that in the presence of multicollinearity, it may not be possible for us to assess the impact of a particular variable on the dependent variable Y because we do not have a reliable estimate of the variable's coefficient. If we are interested in prediction only and do not care about understanding the net effect of each independent variable on Y, the regression model may be adequate even in the presence of multicollinearity. Even though individual regression parameters may be poorly estimated when collinearity exists, the combination of all regression coefficients in the regression may, in some cases, be estimated with sufficient accuracy that satisfactory predictions are possible. In such cases, however, we must be very careful to predict values of Y only within the range of the X variables where the multicollinearity is the same as in the region of estimation. If we try to predict in regions of the X variables where the multicollinearity is not present or is different from that present in the estimation region, large errors may result. We will now explore some of the solutions commonly used to remedy the problem of multicollinearity.

Solutions to the Multicollinearity Problem

1. One of the best solutions to the problem of multicollinearity is to *drop collinear variables from the regression equation.* Suppose that we have a regression of Y on X_1, X_2, X_3, and X_4 and we find that X_1 is highly correlated with X_4. In this case, much of the information about Y in X_1 is also contained in X_4. If we dropped one of the two variables from the regression model, we would solve the multicollinearity problem and lose little information about Y. By comparing the R^2 and the adjusted R^2 of different regressions with and without one of the variables, we can decide which of the two independent variables to drop from the regression. We want to maintain a high R^2 and therefore should drop a variable if R^2 is not reduced much when the variable is removed from the equation. When the adjusted R^2 increases when a variable is deleted, we certainly want to drop the variable. For example, suppose that the R^2 of the regression with all four independent variables is 0.94, the R^2 when X_1 is removed is 0.87, and the R^2 of the regression of X_1, X_2, and X_3 on Y (X_4 removed) is 0.92. In this case, we clearly want to drop X_4 and not X_1. The variable selection methods to be discussed in Section 11–13 will help us determine which variables to include in a regression model.

We note a limitation of this remedy to multicollinearity. In some areas, such as economics, theoretical considerations may require that certain variables be in the equation. In such cases, the bias resulting from deletion of a collinear variable must be weighed against the increase in the variance of the coefficient estimators when the variable is included in the model. The method of weighing the consequences and choosing the best model is presented in advanced books.

2. When the multicollinearity is caused by sampling schemes that, by their nature, tend to favor elements with similar values of some of the independent variables, a change in the sampling plan to include elements outside the multicollinearity range may reduce the extent of this problem.

3. Another method that sometimes helps to reduce the extent of the multicollinearity, or even eliminate it, is to change the form of some of the variables. This can be done in several ways. The best way is to form new combinations of the X variables that are uncorrelated with one another and then run the regression on the new combinations instead of on the original variables. Thus the information content in the original variables is maintained, but the multicollinearity is removed. There are other ways of changing the form of the variables, such as centering the data—a technique of

subtracting the means from the variables and running a regression on the resulting new variables.

4. The problem of multicollinearity may be remedied by using an alternative to the least-squares procedure called *ridge regression*. The coefficient estimators produced by ridge regression are biased, but in some cases, it may be worthwhile to tolerate some bias in the regression estimators in exchange for a reduction in the high variance of the estimators that results from multicollinearity.

In summary, the problem of multicollinearity is an important one. We need to be aware of the problem when it exists and to try to solve it when we can. Removing collinear variables from the equation, when possible, is the simplest method of solving the multicollinearity problem.

PROBLEMS

11–83. For the data of Example 11–3 presented in Table 11–12, find the sample correlations between every pair of variables (the correlation matrix), and determine whether you believe that multicollinearity exists in the regression.

11–84. For the data of Example 11–3, find the variance inflation factors, and comment on their relative magnitudes.

11–85. Find the correlation between X_1 and X_2 for the data of Example 11–1 presented in Table 11–1. Is multicollinearity a problem here? Also find the variance inflation factors, and comment on their magnitudes.

11–86. Regress Y against X_1, X_2 and X_3 with the following sample data:

Y	X_1	X_2	X_3
13.79	76.45	44.47	8.00
21.23	24.37	37.45	7.56
66.49	98.46	95.04	19.00
35.97	49.21	2.17	0.44
37.88	76.12	36.75	7.50
72.70	82.93	42.83	8.74
81.73	23.04	82.17	16.51
58.91	80.98	7.84	1.59
30.47	47.45	88.58	17.86
8.51	65.09	25.59	5.12
39.96	44.82	74.93	15.05
67.85	85.17	55.70	11.16
10.77	27.71	30.60	6.23
72.30	62.32	12.97	2.58

a. What is the regression equation?

b. Change the first observation of X_3 from 8.00 to 9.00. Repeat the regression. What is the new regression equation?

c. Compare the old and the new regression equations. Does the comparison prove that there is multicollinearity in the data? What is your suggestion for getting rid of the multicollinearity?

d. Looking at the results of the original regression only, could you have figured out that there is a multicollinearity problem? How?

11–87. How does multicollinearity manifest itself in a regression situation?

11–88. Explain what is meant by perfect collinearity. What happens when perfect collinearity is present?

11–89. Is it true that the regression equation can never be used adequately for prediction purposes if multicollinearity exists? Explain.

11–90. In a regression of Y on the two explanatory variables X_1 and X_2, the F ratio was found not to be significant. Neither t ratios were found to be significant, and R^2 was found to be 0.12. Do you believe that multicollinearity is a problem here? Explain.

11–91. In a regression of Y on X_1, X_2, and X_3, the F ratio is very significant, and R^2 is 0.88, but none of the t ratios are significant. Then X_1 is dropped from the equation, and a new regression is run of Y on X_2 and X_3 only. The R^2 remains approximately the same, and F is still very significant, but the two t ratios are still not significant. What do you think is happening here?

11–92. A regression is run of Y versus X_1, X_2, X_3, and X_4. The R^2 is high, and F is significant, but only the t ratio corresponding to X_1 is significant. What do you propose to do next? Why?

11–93. In a regression analysis with several X variables, the sign of the coefficient estimate of one of the variables is the opposite of what you believe it should be. How would you test to determine whether multicollinearity is the cause?

11–12   Residual Autocorrelation and the Durbin-Watson Test

Remember that one of the assumptions of the regression model is that the errors ϵ are independent from observation to observation. This means that successive errors are not correlated with one another at any lag; that is, the error at position i is not correlated with the error at position $i - 1$, $i - 2$, $i - 3$, etc. The idea of correlation of the values of a variable (in this case we consider the errors as a variable) with values of the same variable lagged one, two, three, or more time periods back is called *autocorrelation*.

> An **autocorrelation** is a correlation of the values of a variable with values of the same variable lagged one or more time periods back.

Here we demonstrate autocorrelation in the case of regression errors. Suppose that we have 10 observed regression errors $e_{10} = 1$, $e_9 = 0$, $e_8 = -1$, $e_7 = 2$, $e_6 = 3$, $e_5 = -2$, $e_4 = 1$, $e_3 = 1.5$, $e_2 = 1$, and $e_1 = -2.5$. We arrange the errors in descending order of occurrence i. Then we form the lag 1 errors, the regression errors lagged one period back in time. The first error is now $e_{10 - 1} = e_9 = 0$, the second error is now $e_{9 - 1} = e_8 = -1$, and so on. We demonstrate the formation of variable $e_{i - 1}$ from variable e_i (that is, the formation of the lag 1 errors from the original errors), as well as the variables $e_{i - 2}$, $e_{i - 3}$, etc., in Table 11–21.

We now define the autocorrelations. The error autocorrelation of lag 1 is the correlation between the *population* errors ϵ_i and $\epsilon_{i - 1}$. We denote this correlation by ρ_1. This autocorrelation is estimated by the *sample* error autocorrelation of lag 1, denoted r_1, which is the computed correlation between variables e_i and $e_{i - 1}$. Similarly ρ_2 is the lag 2 error autocorrelation. This autocorrelation is estimated by r_2, computed from

TABLE 11–21 Formation of the Lagged Errors

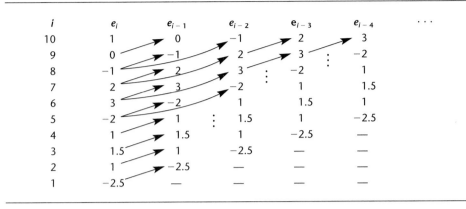

i	e_i	e_{i-1}	e_{i-2}	e_{i-3}	e_{i-4}	$\cdots$
10	1	0	−1	2	3	
9	0	−1	2	3	−2	
8	−1	2	3	−2	1	
7	2	3	−2	1	1.5	
6	3	−2	1	1.5	1	
5	−2	1	1.5	1	−2.5	
4	1	1.5	1	−2.5	—	
3	1.5	1	−2.5	—	—	
2	1	−2.5	—	—	—	
1	−2.5	—	—	—	—	

the data for e_i and e_{i-2} in the table. Note that lagging the data makes us lose data points; one data point is lost for each lag. When computing the estimated error autocorrelations r_j, we use as many points as we have for e_{i-j} and shorten e_i appropriately. We will not do any of these computations.

The assumption that the regression errors are uncorrelated means that they are uncorrelated at *any* lag. That is, we assume $\rho_1 = \rho_2 = \rho_3 = \rho_4 = \cdots = 0$. A statistical test was developed in 1951 by Durbin and Watson for the purpose of detecting when the assumption is violated. The test, called the *Durbin-Watson test*, checks for evidence of the existence of a first-order autocorrelation.

The **Durbin-Watson test** is

$$H_0: \rho_1 = 0 \qquad\qquad (11\text{–}41)$$
$$H_1: \rho_1 \neq 0$$

In testing for the existence of a first-order error autocorrelation, we use the Durbin-Watson test statistic. Critical points for this test statistic are given in Appendix C, Table 7. Part of the table is reproduced here as Table 11–22. The formula of the Durbin-Watson test statistic is equation 11–42.

The Durbin-Watson test statistic is

$$d = \frac{\sum_{i=2}^{n}(e_i - e_{i-1})^2}{\sum_{i=1}^{n} e_i^2} \qquad\qquad (11\text{–}42)$$

Note that the test statistic[15] d is not the sample autocorrelation r_1. The statistic d has a known, tabulated distribution. Also note that the summation in the numerator extends from 2 to n rather than from 1 to n, as in the denominator. An inspection of the

---

[15]Actually, d is approximately equal to $2(1 - r_1)$.

TABLE 11–22 **Critical Points of the Durbin-Watson Statistic d at $\alpha = 0.05$ (n = sample size, k = number of independent variables in the regression) (partial table)**

	$k = 1$		$k = 2$		$k = 3$		$k = 4$		$k = 5$	
n	d_L	d_U	d_L	d_U	d_L	d_U	d_L	d_U	d_L	d_U
15	1.08	1.36	0.95	1.54	0.82	1.75	0.69	1.97	0.56	2.21
16	1.10	1.37	0.98	1.54	0.86	1.73	0.74	1.93	0.62	2.15
17	1.13	1.38	1.02	1.54	0.90	1.71	0.78	1.90	0.67	2.10
18	1.16	1.39	1.05	1.53	0.93	1.69	0.82	1.87	0.71	2.06
.	.	.	.	.	.	.	.	.	.	.
.	.	.	.	.	.	.	.	.	.	.
.	.	.	.	.	.	.	.	.	.	.
65	1.57	1.63	1.54	1.66	1.50	1.70	1.47	1.73	1.44	1.77
70	1.58	1.64	1.55	1.67	1.52	1.70	1.49	1.74	1.46	1.77
75	1.60	1.65	1.57	1.68	1.54	1.71	1.51	1.74	1.49	1.77
80	1.61	1.66	1.59	1.69	1.56	1.72	1.53	1.74	1.51	1.77
85	1.62	1.67	1.60	1.70	1.57	1.72	1.55	1.75	1.52	1.77
90	1.63	1.68	1.61	1.70	1.59	1.73	1.57	1.75	1.54	1.78
95	1.64	1.69	1.62	1.71	1.60	1.73	1.58	1.75	1.56	1.78
100	1.65	1.69	1.63	1.72	1.61	1.74	1.59	1.76	1.57	1.78

first two columns in Table 11–21, corresponding to e_i and e_{i-1}, and our comment on the "lost" data points (here, one point) reveal the reason for this.

Using a given level α from the table (0.05 or 0.01), we may conduct either a test for $\rho_1 < 0$ or a test for $\rho_1 > 0$. The test has two critical points for testing for a positive autocorrelation (the one-tailed half of H_1 in equation 11–41). When the test statistic d falls to the left of the lower critical point d_L, we conclude that there is evidence of a positive error autocorrelation of order 1. When d falls between d_L and the upper critical point d_U, the test is inconclusive. When d falls above d_U, we conclude that there is no evidence of a positive first-order autocorrelation (conclusions are at the appropriate level α). Similarly, when testing for negative autocorrelation, if d is greater than $4 - d_L$, we conclude that there is evidence of negative first-order error autocorrelation. When d is between $4 - d_U$ and $4 - d_L$, the test is inconclusive; and when d is below $4 - d_U$, there is no evidence of negative first-order autocorrelation of the errors. When we test the two-tailed hypothesis in equation 11–41, the actual level of significance α is double what is shown in the table. In cases where we have no prior suspicion of one type of autocorrelation (positive or negative), we carry out the two-tailed test and double the α. The critical points for the two-tailed test are shown in Figure 11–43.

For example, suppose we run a regression using $n = 18$ data points and $k = 3$ independent variables, and the computed value of the Durbin-Watson statistic is $d = 3.1$. Suppose that we want to conduct the two-tailed test. From Table 11–22 (or Appendix C, Table 7), we find that at $\alpha = 0.10$ (twice the level of the table) we have $d_L = 0.93$ and $d_U = 1.69$. We compute $4 - d_L = 3.07$ and $4 - d_U = 2.31$. Since the computed value $d = 3.1$ is greater than $4 - d_L$, we conclude that there is evidence of a negative first-order autocorrelation in the errors. As another example, suppose $n = 80$, $k = 2$, and $d = 1.6$. In this case, the statistic value falls between d_L and d_U, and the test is inconclusive.

FIGURE 11–43   **Critical Regions of the Durbin-Watson Test**

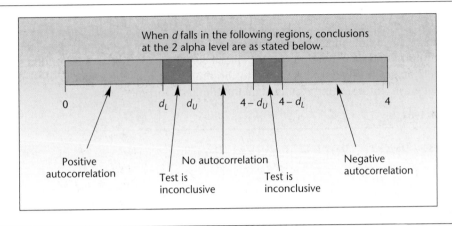

The Durbin-Watson statistic helps us test for first-order autocorrelation in the errors. In most cases, when autocorrelation exists, there is a first-order autocorrelation. There are cases, however, when second- or higher-order autocorrelation exists without there being a first-order autocorrelation. In such cases, the test does not help us. Fortunately, such cases are not common.

In the template, the Durbin-Watson statistic appears in cell H4 of the Residuals sheet. For the Exports problem of Example 11–2, the template computes the statistic to be 2.58. Recall that for this example, $n = 67$ and $k = 4$ (in this version of the equation). At $\alpha = 0.10$, for a two-tailed test, we have $d_U = 1.73$, $d_L = 1.47$, $4 - d_L = 2.53$, and $4 - d_U = 2.27$. We conclude that there is evidence that our regression errors are negatively correlated at lag 1. This, of course, sheds doubt on the regression results; an alternative to least-squares estimation should be used. One alternative procedure that is useful in cases where the ordinary least-squares routine produces autocorrelated errors is a procedure called *generalized least squares* (*GLS*). This method is described in advanced books.

PROBLEMS

11–94.   What is the purpose of the Durbin-Watson test?

11–95.   Discuss the meaning of autocorrelation. What is a third-order autocorrelation?

11–96.   What is a first-order autocorrelation? If a fifth-order autocorrelation exists, is it necessarily true that a first-order autocorrelation exists as well? Explain.

11–97.   State three limitations of the Durbin-Watson test.

11–98.   Find the value of the Durbin-Watson statistic for the data of Example 11–5, and conduct the Durbin-Watson test. State your conclusion.

11–99.   Find the value of the Durbin-Watson statistic for the model of Example 11–3, and conduct the Durbin-Watson test. Is the assumption of no first-order error autocorrelation satisfied? Explain.

11–100.   Do problem 11–99 for the data of Example 11–1.

11–101. State the conditions under which a one-sided Durbin-Watson test is appropriate (i.e., a test for positive autocorrelation only, or a test for a negative autocorrelation only).

11–102. The following table, reprinted by permission from *Journal of Finance,* reports the results of several regressions of percentage change in the Tokyo and London stock markets on several variables relating to the time following the October 1987 stock market crash. Interpret the results of all the implied Durbin-Watson tests.

Trading Days Following October 20, 1987

Regressors	Intercept	Days 1–30	Days 30+	Days 1–10	Days 11–20	Days 20+	N	R^2	DW
Tokyo Index									
(1)	0.000	1.058	0.868	—	—	—	104	0.93	1.17
	(0.001)	(0.045)	(0.032)						
(2)	0.000	—	—	1.065	1.033	0.882	104	0.92	1.13
	(0.001)			(0.057)	(0.135)	(0.031)			
London Index									
(3)	0.000	1.001	0.805	—	—	—	129	0.89	2.45
	(0.001)	(0.044)	(0.035)						
(4)	0.000	—	—	0.994	1.038	821	129	0.89	2.40
	(0.001)			(0.056)	(0.103)	(0.034)			

Source: D. Neumark, P. Tinsley, and S. Tosini, "After-Hours Stock Prices and Post-Crash Hangovers," *Journal of Finance,* March 1991, p. 167.
© 1991 American Finance Association. Reprinted with permission.

11–13   Partial *F* Tests and Variable Selection Methods

Our method of deciding which variables to include in a given multiple regression model has been trial and error. We started by asserting that several variables may have an effect on our variable of interest *Y*, and we tried to run a multiple linear regression model of *Y* versus these variables. The "independent" variables have included dummy variables, powers of a variable, transformed variables, and a combination of all the above. Then we scrutinized the regression model and tested the significance of any individual variable (while being cautious about multicollinearity). We also tested the predictive power of the regression equation as a whole. If we found that an independent variable seemed insignificant due to a low *t* ratio, we dropped the variable and reran the regression without it, observing what happened to the remaining independent variables. By a process of adding and deleting variables, powers, or transformations, we hoped to end up with the best model: the most parsimonious model with the highest relative predictive power.

In this section, we present a statistical test, based on the *F* distribution and, in simple cases, the *t* distribution, for evaluating the relative significance of parts of a regression model. The test is sometimes called a *partial F test* because it is an *F* test (or a *t* test, in simple cases) of a part of our regression model.

Suppose that a regression model of *Y* versus *k* independent variables is postulated, and the analysis is carried out (the *k* variables may include dummy variables, powers, etc.). Suppose that the equation of the regression model is as given in equation 11–1:

$$Y = \beta_0 + \beta_1 X_1 + \beta_2 X_2 + \beta_3 X_3 + \cdots + \beta_k X_k + \epsilon$$

We will call this model the *full model.* It is the full model in the sense that it includes the maximal set of independent variables X_i that we consider as predictors of *Y.* Now suppose that we want to test the relative significance of a subset of r of the k independent variables in the full model. (By relative significance we mean the significance of the r variables given that the remaining $k - r$ variables are in the model.) We will do this by comparing the *reduced model*, consisting of Y and the $k - r$ independent variables that remain once the r variables have been removed, with the full model, equation 11–1. The statistical comparison of the reduced model with the full model is done by the **partial F test.**

We will present the partial F test, using a more specific example. Suppose that we are considering the following two models.

Full model:

$$Y = \beta_0 + \beta_1 X_1 + \beta_2 X_2 + \beta_3 X_3 + \beta_4 X_4 + \epsilon \qquad (11\text{–}43)$$

Reduced model:

$$Y = \beta_0 + \beta_1 X_1 + \beta_2 X_2 + \epsilon \qquad (11\text{–}44)$$

By comparing the two models, we are asking the question: Given that variables X_1 and X_2 are already in the regression model, would we be gaining anything by adding X_3 and X_4 to the model? Will the reduced model be improved in terms of its predictive power by the addition of the two variables X_3 and X_4?

The statistical way of posing and answering this question is, of course, by way of a test of a hypothesis. The null hypothesis that the two variables X_3 and X_4 have no additional value once X_1 and X_2 are in the regression model is the hypothesis that both β_3 and β_4 are zero (given that X_1 and X_2 are in the model). The alternative hypothesis is that the two slope coefficients are not both zero. The hypothesis test is stated in equation 11–45.

Partial F test:

$$H_0: \beta_3 = \beta_4 = 0 \text{ (given that } X_1 \text{ and } X_2 \text{ are in the model)} \qquad (11\text{–}45)$$
$$H_1: \beta_3 \text{ and } \beta_4 \text{ are not both zero}$$

The test statistic for this hypothesis test is the partial F statistic.

The partial F statistic is

$$F_{[r,\, n - (k + 1)]} = \frac{(SSE_R - SSE_F)/r}{MSE_F} \qquad (11\text{–}46)$$

where SSE_R is the sum of squares for error of the reduced model; SSE_F is the sum of squares for error of the full model; MSE_F is the mean square error of the full model: $MSE_F = SSE_F/[n - (k + 1)]$; k is the number of independent variables in the full model ($k = 4$ in the present example); and r is the number of variables dropped from the full model in creating the reduced model (in the present example, $r = 2$).

The difference $\text{SSE}_R - \text{SSE}_F$ is called the *extra sum of squares* associated with the reduced model. Since this additional sum of squares for error is due to r variables, it has r degrees of freedom. (Like the sums of squares, degrees of freedom are additive. Thus, the extra sum of squares for error has degrees of freedom $[n - (k + 1)] - [n - (k - r + 1)] = r$.)

Suppose that the sum of squares for error of the full model, equation 11–43, is 37,653 and that the sum of squares for error of the reduced model, equation 11–44, is 42,900. Suppose also that the regression analysis is based on a data set of $n = 45$ points. Is there a statistical justification for including X_3 and X_4 in a model already containing X_1 and X_2?

To answer this question, we conduct the hypothesis test, equation 11–45. To do so, we compute the F statistic of equation 11–46:

$$F_{(2, 40)} = \frac{(\text{SSE}_R - \text{SSE}_F)/2}{\text{SSE}_F/40} = \frac{(42,900 - 37,653)/2}{37,653/40} = 2.79$$

This value of the statistic falls in the nonrejection region for $\alpha = 0.05$, and so we do not reject the null hypothesis and conclude that the decrease in the sum of squares for error when we go from the reduced model to the full model, adding X_3 and X_4 to the model that already has X_1 and X_2, is not statistically significant. It is not worthwhile to add the two variables.

Figure 11–44 shows the Partial F sheet of the Multiple Regression template. When we enter the value of r in cell D4, the partial F value appears in cell C9 and the corresponding p-value appears in cell C10. We can also see the SSE values for the full and reduced models in cells C6 and C7.

Figure 11–44 shows the partial F calculation for the exports problem of Example 11–2. Recall that the four independent variables in the problem are M1, Price, Lend, and Exchange. The p-value of 0.0010 for the partial F indicates that we should reject the null hypothesis H_0: the slopes for Lend and Exchange are zero (when M1 and Price are in the model).

In this example, we conducted a partial F test for the conditional significance of a set of $r = 2$ independent variables. This test can be carried out for the significance of any number of independent variables, powers of variables, or transformed variables, considered *jointly* as a set of variables to be added to a model. Frequently, however, we are interested in considering the relative merit of a single variable at a time. We may be interested in sequentially testing the conditional significance of a single independent variable, once other variables are already in the model (when no other variables are in the model, the F test is just a test of the significance of a single-variable regression). The F statistic for this test is still given by equation 11–46, but since the

FIGURE 11–44   **Partial *F* from the Template**
                 [Multiple Regression.xls; Sheet: Partial F]

	A	B	C	D	E	F
1	**Partial *F* Calculations**		Exports			
2						
3		#Independent variables in full model	4	*k*		
4		#Independent variables **dropped** from the model	2	*r*		
5						
6		SSE$_F$	6.989784			
7		SSE$_R$	8.747573			
8						
9		**Partial *F***	7.79587			
10		p-value	0.0010			

degrees of freedom are 1 and $n - (k + 1)$, this statistic is equal to the square of a t statistic with $n - (k + 1)$ degrees of freedom. Thus, the partial F test for the significance of a *single* variable may be carried out as a t test.

It may have occurred to you that a computer may be programmed to sequentially test the significance of each variable as it is added to a potential regression model, starting with one variable and building up until a whole set of variables has been tested and the best subset of variables chosen for the final regression model. We may also start with a full model, consisting of the entire set of potential variables, and delete variables from the model, one by one, whenever these variables are found not to be significant. Indeed, computers have been programmed to carry out both kinds of sequential single-variable tests and even a combination of the two methods. We will now discuss these three methods of variable selection called, respectively, *forward selection, backward elimination,* and their combination, *stepwise regression.* We will also discuss a fourth method, called *all possible regressions.*

Variable Selection Methods

1. **All possible regressions:** This method consists of running all possible regressions when k independent variables are considered and choosing the best model. If we assume that every one of the models we consider has an intercept term, then there are 2^k possible models. This is so because each of the k variables may be either included in the model or not included, which means that there are two possibilities for each variable—2^k possibilities for a model consisting of k potential variables. When four potential variables are considered, such as in Example 11–2, there are $2^4 = 16$ possible models: four models with a single variable, six models with a pair of variables, four models with three variables, one model with all four variables, and one model with no variables (an intercept term only). As you can see, the number of possible regression models increases very quickly as the number of variables considered increases.

   The different models are evaluated according to some criterion of model performance. There are several possible criteria: We may choose to select the model with the highest adjusted R^2 or the model with the lowest MSE (an equivalent condition). We may also choose to find the model with the highest R^2 for a given number of variables and then assess the increase in R^2 as we go to the best model with one more variable, to see if the increase in R^2 is worth the addition of a parameter to the model. There are other criteria, such as Mallows' C_p statistic, described in advanced books. The SAS System software has a routine called RSQUARE that runs all possible regressions and identifies the model with the highest R^2 for each number of variables included in the model. The all-possible-regressions procedure is thorough but tedious to carry out. The next three methods we describe are all stepwise procedures for building the best model. While the procedure called stepwise regression is indeed stepwise, the other two methods, forward selection and backward elimination, are also stepwise methods. These procedures are usually listed in computer manuals as variations of the stepwise method.

2. **Forward selection:** Forward selection starts with a model with no variables. The method then considers all k models with one independent variable and chooses the model with the highest significant F statistic, assuming that at least one such model has an F statistic with a p-value smaller than some

predetermined value (this may be set by the user; otherwise a default value is used). Then the procedure looks at the variables remaining outside the model, considers all partial F statistics (i.e., keeping the added variables in the model, the statistic is equation 11–46), and adds the variable with the highest F value to the equation, again assuming that at least one variable is found to meet the required level of significance. The procedure is then continued until no variable left outside the model has a partial F statistic that satisfies the level of significance required to enter the model.

3. **Backward elimination:** This procedure works in a manner opposite to forward selection. We start with a model containing all k variables. Then the partial F statistic, equation 11–46, is computed for each variable, treated as if it were the last variable to enter the regression (i.e., we evaluate each variable in terms of its contribution to a model that already contains all other variables). When the significance level of a variable's partial F statistic is found not to meet a preset standard (i.e., when the p-value is above the preset p-value), the variable is removed from the equation. All statistics are then computed for the new, reduced model, and the remaining variables are screened to see if they meet the significance standard. When a variable is found to have a higher p-value than required, the variable is dropped from the equation. The process continues until all variables left in the equation are significant in terms of their partial F statistic.

4. **Stepwise regression:** This is probably the most commonly used, wholly computerized method of variable selection. The procedure is an interesting mixture of the backward elimination and the forward selection methods. In forward selection, once a variable enters the equation, it remains there. This method does not allow for a reevaluation of a variable's significance once it is in the model. Recall that multicollinearity may cause a variable to become redundant in a model once other variables with much of the same information are included. This is a weakness of the forward selection technique. Similarly, in the backward elimination method, once a variable is out of the model, it stays out. Since it is possible that a variable that was not significant due to multicollinearity and was dropped may have predictive power once other variables are removed from the model, there are limitations to backward elimination as well.

Stepwise regression is a combination of forward selection and backward elimination that reevaluates the significance of every variable at every stage. This minimizes the chance of leaving out important variables or keeping unimportant ones. The procedure works as follows. The algorithm starts, as with the forward selection method, by finding the most significant single-variable regression model. Then the variables left out of the model are checked via a partial F test, and the most significant variable, assuming it meets the entry significance requirement, is added to the model. At this point, the procedure diverges from the forward selection scheme, and the logic of backward elimination is applied. The original variable in the model is reevaluated to see if it meets preset significance standards for staying in the model once the new variable has been added. If not, the variable is dropped. Then variables still outside the model are screened for the entry requirement, and the most significant one, if found, is added. All variables in the model are then checked again for staying significance once the new variable has been added. The procedure continues until there are no variables outside that should be added to the model and no variables inside the model that should be out.

The minimum significance requirements to enter the model and to stay in the model are often called P_{IN} and P_{OUT}, respectively. These are significance levels of the partial F statistic. For example, suppose that P_{IN} is 0.05 and P_{OUT} is also 0.05. This means that a variable will enter the equation if the p-value associated with its partial F statistic is less than 0.05, and it will stay in the model as long as the p-value of its partial F statistic is less than 0.05 after the addition of other variables. The two significance levels P_{IN} and P_{OUT} do not have to be equal, but we must be careful when setting them (or leave their values as programmed) because if P_{IN} is less strict than P_{OUT} (that is, $P_{IN} > P_{OUT}$), then we may end up with a circular routine where a variable enters the model, then leaves it, then reenters, etc., in an infinite loop. We demonstrate the stepwise regression procedure as a flowchart in Figure 11–45. Note that since we test the significance of one variable at a time, our partial F test may be carried out as a t test. This is done in some computer packages.

It is important to note that computerized variable selection algorithms may not find the best model. When a model is found, it may not be a unique best model; there may be several possibilities. The best model based on one evaluation criterion may not be best based on other criteria. Also, since there is order dependence in the selection process, we may not always arrive at the same "best" model. We must remember that computers do only what we tell them to do; and so if we have not considered some good variables to include in the model, including cross-products of variables, powers, and transformations, our model may not be as good as it could be. We must always use judgment in model selection and not rely blindly on the computer to find the best model. The computer should be used as an aid.

FIGURE 11–45 **The Stepwise Regression Algorithm**

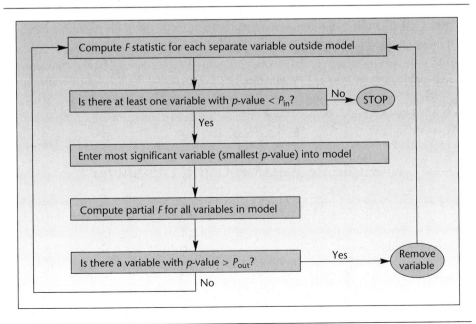

TABLE 11–23   Stepwise Regression Using MINITAB for Example 11–2

```
MTB > STEPWISE regression of 'EXPORTS', PREDICTORS 'M1', 'LEND',
 'PRICE' , ' EXCHANGE'
  STEPWISE REGRESSION OF EXPORTS ON 4 PREDICTORS, WITH N = 67
  STEP        1         2
CONSTANT    0.9348    -3.4230
M1          0.520      0.361
T-RATIO      9.89       9.21

PRICE                  0.0370
T-RATIO                  9.05

S           0.495      0.331

R-SQ       60.08      82.48
```

Table 11–23 shows output from a MINITAB stepwise regression for the Singapore exports example, Example 11–2. Note that the procedure chose the same "best" model (out of the 16 possible regression models) as we did in our analysis. The table also shows the needed commands for the stepwise regression analysis. Note that MINITAB uses t tests rather than the equivalent F tests.

PROBLEMS

11–103. Use equation 11–46 and the information in Tables 11–7 and 11–9 to conduct a partial F test for the significance of the lending rate and the exchange rate in the model of Example 11–2.

11–104. Use a stepwise regression program to find the best set of variables for Example 11–1.

11–105. Redo problem 11–104, using the data of Example 11–3.

11–106. Discuss the relative merits and limitations of the four variable selection methods described in this section.

11–107. In the stepwise regression method, why do we need to test the significance of a variable that is already determined to be included in the model, assuming $P_{IN} = P_{OUT}$?

11–108. Discuss the commonly used criteria for determining the best model.

11–109. Is there always a single "best" model in a given situation? Explain.

11–14   Multiple Regression Using the Solver[16]

Just as we did in simple regression, we can conduct a multiple regression using the Solver command in Excel. The advantage of the method is that we can impose all kinds of constraints on the regression coefficients. The disadvantage is that our assumptions about the errors being normally distributed will not be valid. Hence, hypothesis tests about the regression coefficients and calculation of prediction intervals are not possible. We can make point predictions, though.

Figure 11–46 shows the template that can be used to carry out a multiple regression using the Solver. After entering the data, we may start with zeroes for all the

---

[16]This section is optional.

FIGURE 11–46   The Template for Multiple Regression by Solver
[Mult Regn by Solver.xls]

	A	B	C	D	E	F	G	H	I	J	K	L	M	N	O
1	**Using the Solver**					Exports									
2	Unprotect the sheet before using the Solver.														
3															
4			Regression Coefficients												
5			b_0	b_1	b_2	b_3	b_4	b_5	b_6	b_7	b_8	b_9	b_{10}		
6			-4.024	0.3686	0.0048	0.0365	0.2714	0	0	0	0	0	0		*SSE*
7															6.98979
8		Y	1	X1	X2	X3	X4	X5	X6	X7	X8	X9	X10		Error
9	Sl.No.	Exports		M1	Lend	Price	Exch.								-0.04107
10	1	2.6	1	5.1	7.8	114	2.16								-0.04404
11	2	2.6	1	4.9	8	116	2.17								-0.05744
12	3	2.7	1	5.1	8.1	117	2.18								0.05459
13	4	3	1	5.1	8.1	122	2.2								-0.12114
14	5	2.9	1	5.1	8.1	124	2.21								

FIGURE 11–47   Solver Dialog Box Containing Two Constraints

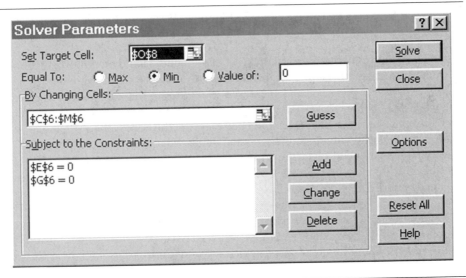

regression coefficients in row 6. Enter zeroes even in unused columns. (Strictly, this row of cells should have been shaded in green. For the sake of clarity, they have not been shaded.) Then the Solver is invoked by selecting the Solver command under the Tools menu. If a constraint needs to be entered, the Add button in the Solver dialog box should be used to enter the constraint. Click the Solve button to start the Solver. When the problem is solved, select Keep Solver Solution and press the OK button.

The results seen in Figure 11–46 are for the exports to Singapore problem (Example 11–2) using all four independent variables. One way to drop an independent variable from the model is to force the regression coefficient (slope) of that variable to equal zero. Figure 11–47 presents the Solver dialog box showing the constraints needed to drop the variables Lend and Exchange from the model. Figure 11–48 shows the results of this constrained regression. Note that the regression coefficients for Lend and Exchange show up as zeroes.

FIGURE 11–48   Results of Constrained Regression

	A	B	C	D	E	F	G	H	I	J	K	L	M	N	O
1	**Using the Solver**					Exports									
2	Unprotect the sheet before using the Solver.														
3															
4			**Regression Coefficients**												
5			b_0	b_1	b_2	b_3	b_4	b_5	b_6	b_7	b_8	b_9	b_{10}		
6			-3.423	0.361	0	0.037	0	0	0	0	0	0	0		
7															**SSE**
8		Y	1	X1	X2	X3	X4	X5	X6	X7	X8	X9	X10		6.9959
9	Sl.No.	Exports		M1	Lend	Price	Exch.								**Error**
10	1	2.6	1	5.1	7.8	114	2.16								-0.042
11	2	2.6	1	4.9	8	116	2.17								-0.0438
12	3	2.7	1	5.1	8.1	117	2.18								-0.0531
13	4	3	1	5.1	8.1	122	2.2								0.0617

As we saw in simple regression, many different types of constraint can be imposed. The possibilities are more extensive in the case of multiple regression since there are more regression coefficients. For instance, we can enter a constraint such as

$$2b_1 + 5b_2 - 6b_3 \leq 10$$

Such constraints may be needed in econometric regression problems.

A Comment on R^2

The idea of constrained regression makes it easy to understand an important concept in multiple regression. Note that the SSE in the unconstrained regression (Figure 11–46) is 6.9898, whereas it has increased to 6.9959 in the constrained regression (Figure 11–48). When a constraint is imposed, it cannot decrease SSE. Why? Whatever values we have for the regression coefficients in the constrained version are certainly feasible in the unconstrained version. Thus, whatever SSE is achieved in the constrained version can be achieved in the unconstrained version just as well. Thus the SSE in the constrained version will be more than, or at best equal to, the SSE in the unconstrained version. Therefore, *a constraint cannot decrease SSE*. Dropping an independent variable from the model is the same as constraining its slope to zero. Thus dropping a variable cannot decrease SSE. Note that $R^2 = 1 - \text{SSE}/\text{SST}$. We can therefore say that dropping a variable cannot increase R^2. Conversely, introducing a new variable cannot increase SSE, which is to say, *introducing a new variable cannot decrease R^2*.

In an effort to increase R^2, an experimenter may be tempted to include more and more independent variables in the model, reasoning that this cannot decrease R^2 but will very likely increase R^2. This is an important reason why we have to look carefully at the included variables in any model. In addition, we should also look at adjusted R^2 and the difference between R^2 and adjusted R^2. A large difference means that some questionable variables have been included.

11–15   The Matrix Approach to Multiple Regression Analysis[17]

Computers carry out regression calculations by working with arrays of numbers. An array of numbers is called a *matrix* (the plural is *matrices*). The use of matrices vastly

---

[17]This section is optional and requires familiarity with linear algebra.

simplifies the operations of the computer because rules of matrix algebra make it possible to program the computer to perform operations on matrices that are similar to operations on individual numbers. The use of matrices and their underlying theory also allows the statistician to define properties of regression statistics in clear and concise terms.

If you are familiar with matrix algebra, the following brief description of the matrix approach to regression analysis will enhance your understanding of regression theory. You will also see definitions of various statistics, such as standard errors of estimators, that are not easily defined without the use of matrices and therefore were not given earlier. If you have not yet studied matrix algebra but will do so sometime in the future, you may always come back to this section and gain a deeper understanding of regression analysis.

Suppose we have a data set of n observations on the dependent variable Y, and k independent variables $X_1, X_2, X_3, \ldots, X_k$. We may write the multiple regression equation, equation 11–1, in a format that includes all our data points. We do so in terms of arrays of numbers as follows:

$$
\begin{bmatrix} y_1 \\ y_2 \\ y_3 \\ \cdot \\ \cdot \\ \cdot \\ y_n \end{bmatrix}
=
\begin{bmatrix}
1 & x_{11} & x_{12} & x_{13} & \cdots & x_{1k} \\
1 & x_{21} & x_{22} & x_{23} & \cdots & x_{2k} \\
1 & x_{31} & x_{32} & x_{33} & \cdots & x_{3k} \\
\cdot & \cdot & \cdot & \cdot & \cdots & \cdot \\
\cdot & \cdot & \cdot & \cdot & \cdots & \cdot \\
\cdot & \cdot & \cdot & \cdot & \cdots & \cdot \\
1 & x_{n1} & x_{n2} & x_{n3} & \cdots & x_{nk}
\end{bmatrix}
\cdot
\begin{bmatrix} \beta_0 \\ \beta_1 \\ \beta_2 \\ \cdot \\ \cdot \\ \cdot \\ \beta_k \end{bmatrix}
+
\begin{bmatrix} \epsilon_1 \\ \epsilon_2 \\ \epsilon_3 \\ \cdot \\ \cdot \\ \cdot \\ \epsilon_n \end{bmatrix}
\qquad (11\text{–}47)
$$

Compare equation 11–47 with equation 11–1, and note that here, for each data point, the value of Y is written as the sum of the products of the unknown β parameters and the point's X_i values, plus the error term associated with the data point. The operations are matrix operations.

As an example of the application of equation 11–47, we write the data of Example 11–1, presented in Table 11–1, in this format. These data are shown in equation 11–48.

$$
\begin{bmatrix} 72 \\ 76 \\ 78 \\ 70 \\ 68 \\ 80 \\ 82 \\ 65 \\ 62 \\ 90 \end{bmatrix}
=
\begin{bmatrix}
1 & 12 & 5 \\
1 & 11 & 8 \\
1 & 15 & 6 \\
1 & 10 & 5 \\
1 & 11 & 3 \\
1 & 16 & 9 \\
1 & 14 & 12 \\
1 & 8 & 4 \\
1 & 8 & 3 \\
1 & 18 & 10
\end{bmatrix}
\cdot
\begin{bmatrix} \beta_0 \\ \beta_1 \\ \beta_2 \end{bmatrix}
+
\begin{bmatrix} \epsilon_1 \\ \epsilon_2 \\ \epsilon_3 \\ \epsilon_4 \\ \epsilon_5 \\ \epsilon_6 \\ \epsilon_7 \\ \epsilon_8 \\ \epsilon_9 \\ \epsilon_{10} \end{bmatrix}
\qquad (11\text{–}48)
$$

As an exercise, multiply the matrices as shown, and add the error vector. You will get a set of n equations, one for each data point. Note that the $\mathbf{X}$ matrix contains a column of 1s. This is necessary for picking up the intercept term β_0, which has no X value multiplying it.

We now name the matrices in equation 11–47. We use boldface letters to denote matrices and vectors. The $n \times 1$ matrix (n-dimensional vector) of Y values is called $\mathbf{Y}$; the $n \times (k + 1)$ matrix of X values is called $\mathbf{X}$; the $(k + 1)$-dimensional vector of β coefficients is called $\boldsymbol{\beta}$; and the n-dimensional vector of errors is called $\boldsymbol{\epsilon}$. We now may write the regression equation, equation 11–47, in terms of matrices as follows:

$$\mathbf{Y} = \mathbf{X}\boldsymbol{\beta} + \boldsymbol{\epsilon} \tag{11–49}$$

The estimated regression model, with the b estimates replacing the unknown β's, and the observed errors e replacing the unobserved ϵ's, can be written in matrix notation as

$$\mathbf{Y} = \mathbf{X}\mathbf{b} + \boldsymbol{\epsilon} \tag{11–50}$$

As you can see, matrix notation greatly simplifies our regression formulas. By using the definitions of matrix multiplication and addition, the relations between possibly large arrays of numbers can be written with very few symbols. Adding the concept of a transposed matrix, $\mathbf{A}' =$ transpose of matrix $\mathbf{A}$, we can present more results.

The *normal equations* (for a regression with any number of variables and data points) are written in matrix format as follows:

$$\mathbf{X}'\mathbf{X}\mathbf{b} = \mathbf{X}'\mathbf{Y} \tag{11–51}$$

Adding the concept of an inverse of a matrix, $\mathbf{A}^{-1} =$ inverse of matrix $\mathbf{A}$, we can solve the normal equations by premultiplying each side of equation 11–51 by the matrix $(\mathbf{X}'\mathbf{X})^{-1}$:

$$(\mathbf{X}'\mathbf{X})^{-1}(\mathbf{X}'\mathbf{X})\mathbf{b} = (\mathbf{X}'\mathbf{X})^{-1}\mathbf{X}'\mathbf{Y}$$
$$\mathbf{I}\mathbf{b} = (\mathbf{X}'\mathbf{X})^{-1}\mathbf{X}'\mathbf{Y}$$

where $\mathbf{I}$ is the *identity matrix,* resulting from the multiplication of a matrix by its inverse. Since $\mathbf{I}\mathbf{b} = \mathbf{b}$, we arrive at the solution of the normal equations in terms of the least-squares estimate vector $\mathbf{b}$:

$$\mathbf{b} = (\mathbf{X}'\mathbf{X})^{-1}\mathbf{X}'\mathbf{Y} \tag{11–52}$$

Thus, for any number of variables and data points, we obtain the vector of parameter estimates, the transpose of which is $(b_0, b_1, b_2, \ldots, b_k)$, as the result of the matrix operation in equation 11–52. We now present more results.

The predicted values of Y, denoted $\hat{Y}$, corresponding to the observed data (these are called the *fitted values* of Y) are given by

$$\hat{\mathbf{Y}} = \mathbf{X}\mathbf{b} \tag{11–53}$$

Since $\mathbf{b} = (\mathbf{X}'\mathbf{X})^{-1}\mathbf{X}'\mathbf{Y}$, we can write the fitted values in terms of a transformation of the observed values:

$$\hat{\mathbf{Y}} = \mathbf{X}(\mathbf{X}'\mathbf{X})^{-1}\mathbf{X}'\mathbf{Y} \tag{11–54}$$

The matrix $\mathbf{H} = \mathbf{X}(\mathbf{X}'\mathbf{X})^{-1}\mathbf{X}'$ is called the *hat matrix.* Using it, we may write

$$\hat{\mathbf{Y}} = \mathbf{H}\mathbf{Y}$$

The hat matrix, as well as other matrices, such as $\mathbf{X}$ and $(\mathbf{X}'\mathbf{X})^{-1}$, can be obtained from many regression computer programs.

The variance-covariance matrix of the coefficient estimates vector $\mathbf{b}$ is given by

$$\mathbf{V}(\mathbf{b}) = \sigma^2(\mathbf{X}'\mathbf{X})^{-1} \tag{11–55}$$

Since σ^2 is unknown and estimated by MSE, we have an estimated variance-covariance matrix of **b** given by

$$\mathbf{s}^2(\mathbf{b}) = \text{MSE}(\mathbf{X'X})^{-1} \tag{11-56}$$

If we define the matrix **C** as $\mathbf{C} = (\mathbf{X'X})^{-1}$, the standard error of each b_i is given, using the diagonal of **C**, C_{ii}, as

$$s(b_i) = \sqrt{\text{MSE } C_{ii}} \tag{11-57}$$

Another formula that has not been given until now is the definition of the elliptical joint confidence region for p regression parameters (shown in Figure 11–11). The region is defined by the equation

$$\frac{(\boldsymbol{\beta} - \mathbf{b})'\mathbf{X'X}(\boldsymbol{\beta} - \mathbf{b})}{p\,\text{MSE}} \leq F_{(\alpha;\, p,\, n-p)} \tag{11-58}$$

A computer may generate values using equation 11–58 that will give us a joint confidence region for several regression parameters.

What about Multicollinearity?

Earlier we said that in cases of severe multicollinearity, regression calculations cannot be carried out. Stated in terms of matrices, there is a very simple explanation for this.

When perfect multicollinearity exists, the columns of the **X** matrix are *linearly dependent*. Perfect multicollinearity means that at least two of the X_i variables are dependent on each other. From equation 11–47, you can see that this means that the columns of the **X** matrix are dependent. This dependency carries through to the matrix **X'X**, and this matrix, therefore, *cannot be inverted* (its determinant is zero). Thus, we cannot compute $(\mathbf{X'X})^{-1}$ and cannot obtain the vector of parameter estimates **b**. In cases where multicollinearity exists but is not perfect, the matrix is invertible but has much instability in it (the matrix is ill-conditioned). It is the matrix $(\mathbf{X'X})^{-1}$ that causes the variance inflation in cases of multicollinearity.

The purpose of this section is just to give you an idea about regression results that are obtained via the matrix approach. Other results are given in advanced books.

11–16   Summary and Review of Terms

In this chapter, we extended the simple linear regression method of Chapter 10 to include several independent variables. We saw how the F test and the t test are adapted to the extension: The F test is aimed at determining the existence of a linear relationship between Y and any of the explanatory variables, and the separate t tests are each aimed at checking the significance of a single variable. We saw how the geometry of least-squares estimation is extended to planes and to higher-dimensional surfaces as more independent variables are included in a model. We extended the coefficient of determination to multiple regression situations, as well as the correlation coefficient. We discussed the problem of **multicollinearity** and its effects on estimation and prediction. We extended our discussion of the use of residual plots and mentioned the problem of **outliers** and the problem of **autocorrelation** of the errors and its detection. We discussed **qualitative variables** and their modeling using **indicator (dummy) variables**. We also talked about higher-order models: **polynomials** and **cross-product terms**. We emphasized the need for parsimony. We showed the relationship between regression and ANOVA and between regression and **analysis of**

covariance. We also talked about **nonlinear** models and about **transformations.** Finally, we discussed methods for selecting variables to find the "best" multiple regression model: **forward selection, backward elimination, stepwise regression,** and **all possible regressions.**

ADDITIONAL PROBLEMS

11–110.   See the figure "Sheltering from the Trade Winds?" It is from an article in *The Economist* entitled "Second Thoughts about Globalisation."[18] Run a regression of exports and imports as a percentage of GDP on spending on social protection as a percentage of GDP, using estimated values from the graph. Evaluate your model.

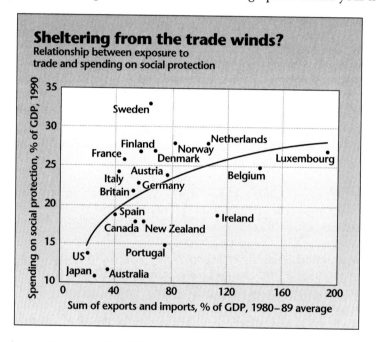

11–111.   An article reports the following results of a regression analysis aimed at predicting performance. The sample size used in section A was $n = 86$, and the sample size used in section B was $n = 74$. Interpret all the findings in the table.

Variable Entered	Step	R^2	ΔR^2	T
Section A				
Feedback condition	1	0.39	0.39	12.01**
Power distance				1.96*
Collectivism				2.17**
Trust in supervisor	2	0.52	0.13	3.45**
Importance of praise				2.11**
Importance of criticism				1.95*
Feedback × power distance	3	0.54	0.02	−1.99*
Feedback × collectivism				−2.01*

Variable Entered	Step	R^2	ΔR^2	T
Section B				
Importance of praise	1	0.41	0.41	3.42**
Importance of criticism				2.35**
Trust in supervisor				6.71**
Feedback condition	2	0.52	0.11	3.14**
Power distance				0.84
Collectivism				1.07
Feedback × power distance	3	0.54	0.02	−1.99*
Feedback × collectivism				−2.01*

$*p < 0.05.$ $**p < 0.01.$

11–112. A regression analysis of share price was conducted on standardized dividends per share (D) and standardized retained earnings (R). The results are given in the following equation. The author also reports that the R^2 is 0.56. What can you say about this regression?

$$P = 22.59 + 5.26D + 5.38R$$
$$(20.87) \quad (21.37)$$

11–113. An article in *Fortune* expounds upon a problem that has become serious in recent years: near collisions of airplanes and airline safety. Data on the total number of flights and the total number of near collisions for the years 1988 through 1994 are given in the article and are reproduced here:

Year	Flights (millions)	Near Collisions (per 100,000 flight hours)
1988	4.95	2.0
1989	4.65	1.5
1990	4.55	1.0
1991	4.90	0.9
1992	5.56	0.7
1993	5.75	0.6
1994	6.25	0.9

Source: Adapted from "Airline Safety," *Fortune*, June 26, 1995, p. 20.
© 1995 Time, Inc. All rights reserved.

Find the best regression model to fit the relationship between the total number of flights and the total number of near collisions in the sky. What are the implications of your model?

11–114. The following data are the asking price and other variables for condominiums in Brookline, Massachusetts, in early 1990.[19] Try to construct a prediction equation for the asking price based on any of or all the other reported variables.

---

[19]Thanks to Lisa Glucksman of Prudential Edna Kranz Real Estate for the data on condominiums in Brookline in 1990.

Price ($)	Number of Rooms	Number of Bedrooms	Number of Baths	Age	Assessed Value ($)	Area (square feet)
145,000	4	1	1	69	116,500	790
144,900	4	2	1	70	127,200	915
145,900	3	1	1	78	127,600	721
146,500	4	1	1	75	121,700	800
146,900	4	2	1	40	94,800	718
147,900	4	1	1	12	169,700	915
148,000	3	1	1	20	151,800	870
148,900	3	1	1	20	147,800	875
149,000	4	2	1	70	140,500	1,078
149,000	4	2	1	60	120,400	705
149,900	4	2	1	65	160,800	834
149,900	3	1	1	20	135,900	725
149,900	4	2	1	65	125,400	900
152,900	5	2	1	37	134,500	792
153,000	3	1	1	100	132,100	820
154,000	3	1	1	18	140,800	782
158,000	5	2	1	89	158,000	955
158,000	4	2	1	69	127,600	920
159,000	4	2	1	60	152,800	1,050
159,000	5	2	2	49	157,000	1,092
179,900	5	2	2	90	165,800	1,180
179,900	6	3	1	89	158,300	1,328
179,500	5	2	1	60	148,100	1,175
179,000	6	3	1	87	158,500	1,253
175,000	4	2	1	80	156,900	650

www.exercise

11–115. By definition, the U.S. trade deficit is the sum of the trade deficits it has with all its trading partners. Consider a model of the trade deficit based on regions such as Asia, Africa, Europe, and so on. Whether there is collinearity, meaning the deficits move in the same direction, among these trading regions depends on the similarities of the goods that the United States trades in each region. You can investigate the collinearity of these regional deficits from data available from the U.S. Census Bureau, **www.census.gov/.** At this site, locate the trade data in the International Trade Reports. *Hint:* Start at the A–Z area; locate foreign trade by clicking on "F."

Read the highlights of the current report, and examine current-year country-by-commodity detailed data for a selection of countries in each of the regions of Asia, Africa, and Europe. Based on the country-by-commodity detailed information, would you expect the deficits in these regions to be correlated? How would you design a statistical test for collinearity among these regions?

CASE 13   Return on Capital for Four Different Sectors

The table below presents financial data of some companies drawn from four different industry sectors. The data include return on capital, sales, operating margin, and debt-to-capital ratio all pertaining to the same latest 12 months for which data were available for that company. The period may be different for different companies, but we shall ignore that fact.

Using suitable indicator variables to represent the sector of each company, regress the return on capital against all other variables, including the indicator variables.

1. The sectors are to be ranked in descending order of return on capital. Based on the regression results, what will that ranking be?

2. It is claimed that the sector that a company belongs to does not affect its return on capital.

Conduct a partial F test to see if all the indicator variables can be dropped from the regression model.

3. For each of the four sectors, give a 95% prediction interval for the *expected* return on capital for a company with the following annual data: sales of $2 billion, operating margin of 35%, and a debt-to-capital ratio of 50%.

	Return on Capital (%)	Sales ($ millions)	Operating Margin (%)	Debt/ Capital (%)
BANKING				
Bank of New York	17.2	7,178	38.1	28.5
Bank United	11.9	1,437	26.7	24.3
Comerica	17.1	3,948	38.9	65.6
Compass Bancshares	15.4	1,672	27	26.4
Fifth Third Bancorp	16.6	4,123	34.8	46.4
First Tennessee National	15.1	2,317	21.3	20.1
Firstar	13.7	6,804	36.6	17.7
Golden State Bancorp	15.9	4,418	21.5	65.8
Golden West Financial	14.6	3,592	23.8	17
GreenPoint Financial	11.3	1,570	36	14.1
Hibernia	14.7	1,414	26	0
M&T Bank	13.4	1,910	30.2	21.4
Marshall & Ilsley	14.7	2,594	24.4	19.2
Northern Trust	15.3	3,379	28.4	35.7
Old Kent Financial	16.6	1,991	26	21.9
PNC Financial Services	15	7,548	32	29.5
SouthTrust	12.9	3,802	24	26.1
Synovus Financial	19.7	1,858	27.3	5.1
UnionBanCal	16.5	3,085	31.4	14.6
Washington Mutual	13.8	15,197	24.7	39.6
Wells Fargo	11.9	24,532	38.9	50.7
Zions Bancorp	7.7	1,845	23.5	19.3
COMPUTERS				
Agilent Technologies	22.4	10,773	14	0
Altera	32.4	1,246	41.7	0
American Power Conversion	21.2	1,459	22.2	0
Analog Devices	36.8	2,578	35.3	34
Applied Materials	42.2	9,564	32.5	7.4
Atmel	16.4	1,827	30.8	28.1
Cisco Systems	15.5	21,529	27.3	0
Dell Computer	38.8	30,016	9.6	7.8
EMC	24.9	8,127	31	0.2
Gateway	26.6	9,461	9.8	0.1
Intel	28.5	33,236	46.3	1.5
Jabil Circuit	25	3,558	8.4	1.9
KLA-Tencor	21.8	1,760	26.7	0
Micron Technology	26.5	7,336	44.8	11.8

	Return on Capital (%)	Sales ($ millions)	Operating Margin (%)	Debt/ Capital (%)
Palm	10.1	1,282	7.8	0
Sanmina	14.1	3,912	13.9	39
SCI Systems	12.5	8,707	5.8	35.9
Solectron	14.6	14,138	7	46.6
Sun Microsystems	30.5	17,621	19.6	14.7
Tech Data	13	19,890	2	22.6
Tektronix	41.3	1,118	12.3	13.2
Teradyne	40.4	2,804	27	0.54
Texas Instruments	25.5	11,406	29.9	8.4
Xilinx	35.8	1,373	36.8	0
CONSTRUCTION				
Carlisle Companies	15.7	1,752	13.9	34.3
Granite Construction	14.1	1,368	9.8	13.7
DR Horton	12.3	3,654	9.3	58
Kaufman & Broad Home	12.1	3,910	9.2	58.4
Lennar	14.7	3,955	10.7	59.7
Martin Marietta Materials	10.3	1,354	26.4	39.3
Masco	14.3	7,155	18.4	38.3
MDC Holdings	21.4	1,674	12.3	28.4
Mueller Industries	15	1,227	15.9	14.2
NVR	40.8	2,195	11.9	31.5
Pulte Homes	11.5	4,052	8.9	37.6
Standard Pacific	13.7	1,198	10.7	52.9
Stanley Works	16.9	2,773	14	18.9
Toll Brothers	11	1,642	14.7	53.5
URS	8.7	2,176	9.8	62.9
Vulcan Materials	11.8	2,467	23.5	27.1
Del Webb	8.2	2,048	10.3	64.8
ENERGY				
Allegheny Energy	7.8	3,524	26.4	47.9
Apache	12.5	2,006	79.8	32.3
BJ Services	9.8	1,555	19.1	10.6
BP Amoco	19.4	131,450	15.4	17.9
Chevron	16.6	43,957	23	16
Cinergy	7.7	7,130	16.7	42.3
Conoco	17.5	30	14	36.7
Consol Energy	20.4	2,036	17.1	55.9
Duke Energy	7.8	40,104	10.4	37.7
Dynegy	18.4	24,074	3.7	39.4
Enron	8.1	71,011	3.2	40.6
Exelon	8.6	5,620	33.9	56.8
ExxonMobil	14.9	196,956	14.7	7.9
FPL Group	8.6	6,744	33.1	32.8
Halliburton	11.9	12,424	8	18.2
Kerr-McGee	17.2	3,760	54.7	45
KeySpan	8.9	4,123	23.1	39.9

	Return on Capital (%)	Sales ($ millions)	Operating Margin (%)	Debt/ Capital (%)
MDU Resources	8.7	1,621	17.7	40.5
Montana Power	10.5	1,055	23.5	24
Murphy Oil	17.5	3,172	20.5	22.1
Noble Affiliates	13.5	1,197	42.4	36
OGE Energy	7.9	2,894	18.7	48.6
Phillips Petroleum	14.9	19,414	21.6	47
PPL	10.1	5,301	26.4	54
Progress Energy	8.3	3,661	40.8	38.7
Reliant Energy	8.3	23,576	11.9	37.8
Royal Dutch Petroleum	17.9	129,147	19.8	5.7
Scana	7.2	2,839	42.5	47.4
Smith International	7	2,539	9.3	22.8
Sunoco	13.4	11,791	6.4	34.3
TECO Energy	9	2,189	31.2	40.4
Tosco	16.7	21,287	5.9	41.5
Valero Energy	14.5	13,188	4.7	35.8

Source: *Forbes*, January 8, 2001.

12 Time Series, Forecasting, and Index Numbers

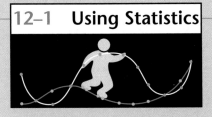

The desire to forecast the future is as old as the human race—older if you allow that animals also form anticipations of what the future may bring. A predator may try to predict where the prey will run, and there are other such examples. In ancient times, people relied on prophets, soothsayers, and crystal balls. Today we have computers and with them an impressive, ever-expanding array of quantitative capabilities. Are these powerful quantitative tools useful in predicting the future?

It is important to understand that many economic and business-related variables (as well as others) cannot be forecast well. The future values of these variables cannot be predicted from their history. This is especially true of the prices of assets traded in efficient markets, such as stock. The *efficient markets hypothesis* states that the future price of a stock cannot be predicted from its past. This hypothesis is widely held by academics. Stock market analysts who do not believe this hypothesis are known as *chartists,* people who chart stock price movements in an effort to extrapolate them into the future. The rationale behind the efficient markets hypothesis is that the demand for a given stock at any time reflects people's expectations of the future value of the stock. If the future value of the stock is expected to increase, people exert buying pressure, and the price increases immediately in fulfillment of this expectation. Mathematically, the theory states that movements of a stock are a *random walk:* The stock moves up or down like a drunk who takes each step randomly and moves with no apparent pattern.

> A **time series** is a set of measurements of a variable that are ordered through time.

A time series variable is often denoted by Z_t to distinguish it from the X and Y variables used in regression analysis and other areas. The particular value of the variable at time t is denoted by z_t. We denote the *error* occurring at time t by a_t. The error a_t in a time series model serves the same purpose as the regression error ϵ. The random error at time t, denoted a_t, is assumed to have mean zero and constant variance σ^2, and successive errors are assumed to be uncorrelated with one another.

Using our new time series notation, we model a random walk by equation 12–1.

A random walk is

$$Z_t - Z_{t-1} = a_t \qquad\qquad (12\text{–}1)$$

or equivalently

$$Z_t = Z_{t-1} + a_t$$

Equation 12–1 says that the difference between the value of variable Z at time t and its value at time $t - 1$ is a random error. The time series Z_t is a random walk if each step, as we go from time $t - 1$ to t, is just a random error a_t. An example of a random walk is shown in Figure 12–1. As can be seen from the figure, there is no persistent pattern or trend in the movement of variable Z_t through time. The difference between the stock price today and the stock price tomorrow is a purely random "error." Since the error a_t

FIGURE 12–1   A Random Walk

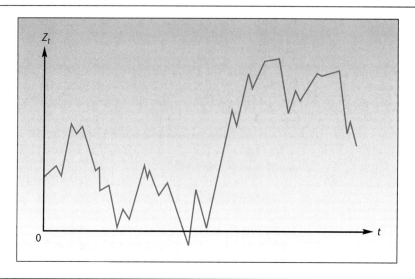

cannot be forecast, because it is purely random, we cannot forecast Z_t. Our best forecast within this framework assumes that a_t will equal its mean of zero, and, therefore, that the price tomorrow will be equal to the price today. This is called the naive approach to forecasting.

Many variables, however, such as sales and other business variables, can be forecast, and the use of statistics plays an important role in forecasting these variables. While there are many different methods of forecasting—subjective, naive, and others—this is a book on statistics, and we will therefore concentrate our discussion on the statistical methods of forecasting.

On January 15, 1884, a physicist, Professor J. H. Poynting of Trinity College, Cambridge, delivered a paper at a meeting of the Royal Statistical Society in London. The paper was entitled "A Comparison of the Fluctuations in the Price of Wheat and in the Cotton and Silk Imports into Great Britain." The paper contained an analysis of these variables and of sunspots. On page 43 of the paper, Poynting described his statistical method:

> Suppose that while a full band is playing, we could draw a curve to represent the pressure of the air at a given moment along a line drawn in the direction in which the sound is traveling. Such a curve would appear to fluctuate irregularly. But we know that it is really made up by the superposition of the perfectly regular waves of pressure corresponding to the separate notes sent out by all the various instruments, and a well-trained ear stationed in the line which can pick out the notes practically analyses this irregular pressure curve, that is, breaks it up into its simple harmonics. What is true of this pressure curve is true of all curves however irregularly they fluctuate, and the ear performs for the pressure curve what the mathematician seeks to perform for any other curve by Fourier's theorem.[1]

One of the aims of time series analysis is to forecast future values of time series variables. Incidentally, in this book we make a distinction between the terms *forecast-*

---

[1]J. H. Poynting, "A Comparison of the Fluctuations in the Price of Wheat and in the Cotton and Silk Imports into Great Britain," *Statistical Journal* XLVII (March 1884), pp. 34–48.

FIGURE 12–2   **A Time Series as a Superposition of Two Regular Wave Functions and a Random Error Component (not shown)**

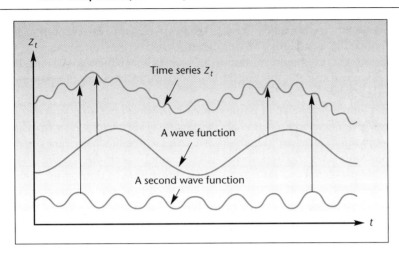

ing and *prediction.* We use the word *prediction* in the context of regression analysis; the word *forecasting* is reserved for time series analysis, where *forecasting is an extrapolation of the series values beyond the region of the estimation data.* Remember the warning we gave about extrapolation in Chapter 10; we will return to that problem later in this chapter.

Now let us return to Professor Poynting. His statements suggest that a time series may look very irregular, but we should be able to break it down into its harmonics. These are regular wavelike functions (sines and cosines) of different frequencies, and they are superimposed on one another. Poynting suggests that we break down a time series into its different wave components. Then we can forecast the time series by forecasting where each wave component is moving. In his paper, Poynting does just that with the different time series: English cotton imports, Italian silk imports, Bengal silk imports, Brutia silk imports, English bank rate, sunspots, and other time series occurring between 1780 and 1880. The decomposition of a time series into wave components is demonstrated in Figure 12–2.

Much of time series analysis makes use of the idea of breaking down a time series into functions like the two shown in Figure 12–2, plus a random-error term. The regular (wavelike) functions in the figure have a cycle, and they can be forecast. This reduces the forecasting errors to the magnitude of the nonforecastable residual error once the regularity has been accounted for. Another regularity in a time series may be an increasing or decreasing *trend,* that is, a general movement of increase or decrease. The trend, too, may be accounted for, thus helping us get better forecasts. Remember the idea of a statistical model presented in Chapter 10. Here we see an application of that important idea in the context of time series modeling. A good time series model is one that accounts for as much as possible of the *regular movement* in the time series, leaving out only a random error, which cannot be forecast. By comparison, the random walk series, given in equation 12–1, has *no regular components,* only random errors, as the series moves through time. Since there is no regularity, the series cannot be forecast.

12–2   Trend Analysis

Sometimes a time series displays a steady tendency of increase or decrease through time. Such a tendency is called a **trend.** When we plot the observations against time, we may notice that a straight line can describe the increase or decrease in the series as time goes on. This should remind us of simple linear regression, and, indeed, in such cases we will use the method of least squares to estimate the parameters of a straight-line model.

At this point, we make an important remark. *When one is dealing with time series data, the errors of the regression model may not be independent of one another: Time series observations tend to be sequentially correlated.* Therefore, we cannot give much credence to regression results. Our estimation and hypothesis tests may not be accurate. We must be aware of such possible problems and must realize that fitting lines to time series data is less an accurate statistical method than a simple *descriptive* method that may work in some cases. We will now demonstrate the procedure of trend analysis with an example.

EXAMPLE 12–1   An economist is researching banking activity and wants to find a model that would help her forecast total net loans by commercial banks. The economist gets the data presented in Table 12–1. A plot of the data is shown in Figure 12–3.

Solution   As can be seen from the figure, the observations may be described by a straight line. A simple linear regression equation is fit to the data by least squares. A straight-line model to account for a trend is of the form

$$Z_t = \beta_0 + \beta_1 t + a_t \tag{12–2}$$

where t is time and a_t is the error term. The coefficients β_0 and β_1 are the regression intercept and slope, respectively. The regression can be carried out as we saw in Chapter 10. To simplify the calculation, the first year in the data (1994) is coded $t = 1$, and next $t = 2$, and so on. We shall see the regression results in a template.

Figure 12–4 shows the template that can be used for trend forecast. The data are entered in columns B and D. The coded t values appear in column C. As seen in the range M7:M8, the slope and the intercept of the regression line are, respectively, 109.19 and 696.89. In other words, the regression equation is

$$Z_t = 696.89 + 109.19t$$

TABLE 12–1   Annual Total Net Loans by Commercial Banks

Year	Loans ($ billions)
1994	833
1995	936
1996	1,006
1997	1,120
1998	1,212
1999	1,301
2000	1,490
2001	1,608

FIGURE 12–3   Annual Total Net Loans by Commercial Banks

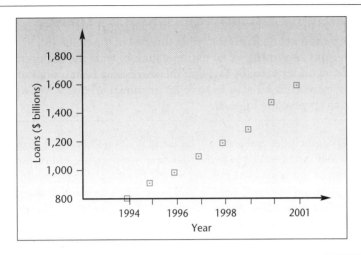

**FIGURE 12–4   The Template for Trend Analysis
[Trend Forecast.xls]**

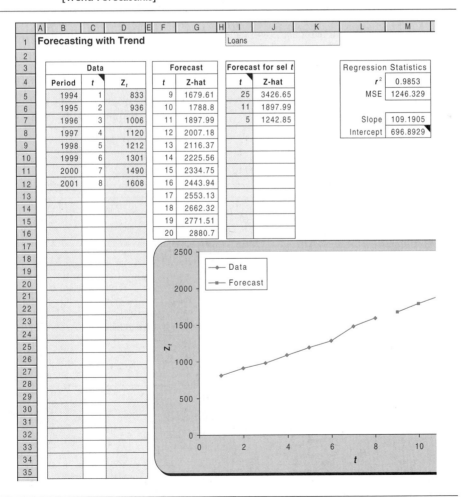

By substituting 9 for t, we get the forecast for year 2002 as 1,679.61. In the template, this appears in cell G5. Indeed, the template contains forecasts for $t = 9$ through 20, which correspond to the years 2002 to 2013. In the range I5:I16, we can enter any desired values for t and get the corresponding forecast in the range J5:J16.

Remember that forecasting is an extrapolation outside the region of the estimation data. This, in addition to the fact that the regression assumptions are not met in trend analysis, causes our forecast to have an unknown accuracy. We will, therefore, not construct any prediction interval.

Trend analysis includes cases where the trend is not necessarily a straight line. It is possible to model *curved* trends as well, and here we may use either polynomials or transformations, as we have seen in Chapter 11. In fact, a careful examination of the data in Figure 12–3 and of the fitted line in Figure 12–4 reveals that the data are actually curved upward somewhat. We will, therefore, fit an exponential model $Z = \beta_0 e^{\beta_1 t} a_t$, where β_0 and β_1 are constants and e is the number 2.71828 . . . , the base of the natural logarithm. We assume a multiplicative error a_t. We run a regression of the natural log of Z on variable t. The transformed regression, in terms of the original exponential equation, is shown in Figure 12–5. The coefficient of determination of this model is very close to 1.00. The figure also shows the forecast for 2002, obtained from the equation by substituting $t = 9$, as we did when we tried fitting the straight line.

A polynomial regression with t and t^2 leads to a fit very similar to the one shown in Figure 12–5, and the forecast is very close to the one obtained by the exponential equation. We do not elaborate on the details of the analysis here because much was explained about regression models in Chapters 10 and 11. Remember that trend analysis does not enjoy the theoretical strengths that regression analysis does in non-time-series contexts; therefore, your forecasts are of questionable accuracy. In the case of Example 12–1, we conclude that an exponential or quadratic fit is probably

FIGURE 12–5 Fitting an Exponential Model to the Data of Example 12–1

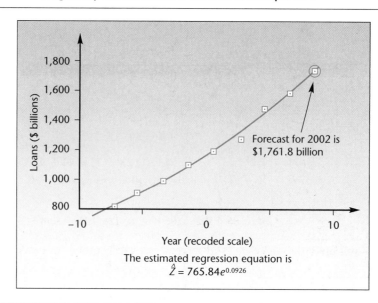

The estimated regression equation is
$$\hat{Z} = 765.84 e^{0.0926}$$

better than a straight line, but there is no way to objectively evaluate our forecast. The main advantage of trend analysis is that when the model is appropriate and the data exhibit a clear trend, we may carry out a simple analysis.

The following data are a price index for industrial metals for the years 1993 to 2001. A company that uses the metals in this index wants to construct a time series model to forecast future values of the index.

EXAMPLE 12–2

	A	B
1	Time Series	
2	A Price Index	
3		
4	Year	Price
5	1993	122.55
6	1994	140.64
7	1995	164.93
8	1996	167.24
9	1997	211.28
10	1998	242.17
11	1999	247.08
12	2000	277.72
13	2001	353.40

Figure 12–6 shows the results on the template. As seen in the template, the regression equation is $Z_t = 82.96 + 26.23t$. The forecast for $t = 10$, or year 2002, is 345.27.

Solution

PROBLEMS

12–1. What are the advantages and disadvantages of trend analysis? When would you use this method of forecasting?

12–2. The following data[2] are average 1990–1997 (based on the first quarter) market shares of the Japanese cars Honda, Acura, Nissan, Infiniti, Toyota, and Lexus in the car market:

25, 27.5, 27, 25.5, 26.4, 27.3, 28.2, 30.1

Do trend analysis to predict next year's value.

12–3. The following data are a local newspaper's readership figures, in thousands:

Year:	1986	1987	1988	1989	1990	1991	1992	1993	1994	1995	1996	1997
Readers:	53	65	74	85	92	105	120	128	144	158	179	195

Do a trend regression on the data, and forecast the total number of readers for 1998 and for 1999.

---

[2]From L. Schiff, "The Outlook for the Big Three Is Getting Worse," *Fortune*, June 23, 1997, p. 14.

FIGURE 12–6   **Price Index Forecast**
                [Trend Forecast.xls]

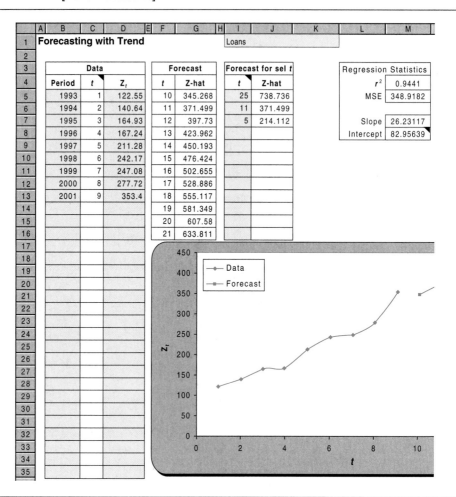

12–4. The following data on the average price of ground beef in U.S. cities is from the Bureau of Labor Statistics website. Forecast the prices for the 12 months of year 2001. Go to the website (**http://stats.bls.gov/cpihome.htm**), access public data query, and compare your forecast with the actual data for that year. Tabulate the errors.

Price of Ground Beef ($/pound)

	Jan	Feb	Mar	Apr	May	Jun	Jul	Aug	Sep	Oct	Nov	Dec
1999	1.382	1.431	1.404	1.429	1.444	1.448	1.435	1.442	1.484	1.496	1.518	1.528
2000	1.483	1.516	1.532	1.584	1.57	1.556	1.574	1.611	1.579	1.582	1.621	1.625

12–5. Would trend analysis, by itself, be a useful forecasting tool for monthly sales of swimming suits? Explain.

12–6. A firm's profits are known to vary with a business cycle of several years. Would trend analysis, by itself, be a good forecasting tool for the firm's profits? Why?

12–3 Seasonality and Cyclical Behavior

Monthly times series observations very often display seasonal variation. The seasonal variation follows a complete cycle throughout a whole year, with the same general pattern repeating itself year after year. The obvious examples of such variation are sales of seasonal items, for example, suntan oil. We expect that sales of suntan oil will be very high during the summer months. We expect sales to taper off during the on-set of fall and to decline drastically in winter—with another peak during the winter holiday season, when many people travel to sunny places on vacation—and then increase again as spring progresses into summer. The pattern repeats itself the following year.

Seasonal variation, which is very obvious in a case such as suntan oil, actually exists in many time series, even those that may not appear at first to have a seasonal characteristic. Electricity consumption, gasoline consumption, credit card spending, corporate profits, and sales of most discretionary items display distinct seasonal variation. Seasonality is not confined to monthly observations. Monthly time series observations display a 12-month period: a 1-year cycle. If our observations of a seasonal variable are quarterly, these observations will have a four-quarter period. Weekly observations of a seasonal time series will display a 52-week period. The term *seasonality,* or *seasonal variation,* frequently refers to a 12-month cycle.

The second wave function in Figure 12–2 describes the seasonality in the time series, and the first wave function—with a longer period—reflects *cyclical variation.* In addition to a linear or curvilinear trend and seasonality, a time series may exhibit cyclical variation (where the period is not 1 year). In the context of business and economics, cyclical behavior is often referred to as the *business cycle.* The business cycle is marked by troughs and peaks of business activity in a cycle that lasts several years. The cycle is often of irregular, unpredictable pattern, and the period may be anything from 2 to 15 years and may change within the same time series. We repeat the distinction between the terms *seasonal variation* and *cyclical variation:*

> When a cyclical pattern in our data has a period of 1 year, we usually call the pattern **seasonal variation.** When a cyclical pattern has a period other than 1 year, we refer to it as **cyclical variation.**

We now give an example of a time series with a linear trend and with seasonal variation and no cyclical variation. Figure 12–7 shows sales data for suntan oil. Note that the data display both a trend (increasing sales as one compares succeeding years) and a seasonal variation.

Figure 12–8 shows a time series of annual corporate gross earnings for a given company. Since the data are annual, there is no seasonal variation. As seen in the figure, the data exhibit both a trend and a cyclical pattern. The business cycle here has a period of approximately 4 years (the period does change during the time span under study). Figure 12–9 shows monthly total numbers of airline passengers traveling between two cities. See what components you can visually detect in the plot of the time series.

How do we incorporate seasonal behavior in a time series model? There are several different approaches to the problem. Having studied regression analysis, and

FIGURE 12–7   **Monthly Sales of Suntan Oil**

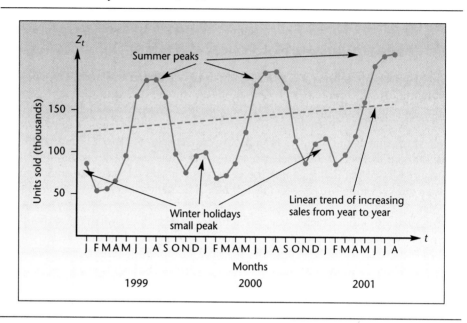

FIGURE 12–8   **Annual Corporate Gross Earnings**

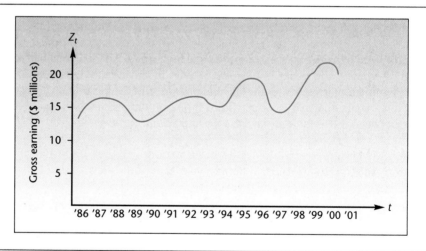

having used it (somewhat informally) in trend analysis, we now extend regression analysis to account for seasonal behavior. If you think about it for a while and recall the methods described in Chapter 11, you probably realize that one of the tools of multiple regression—the dummy variable—is applicable here. We can formulate a regression model for the trend, whether linear or curvilinear, and add 11 dummy variables to the model to account for seasonality if our data are monthly. (Why 11? Reread the appropriate section of Chapter 11 if you do not know.) If data are quar-

FIGURE 12–9   Monthly Total Numbers of Airline Passengers Traveling between Two Cities

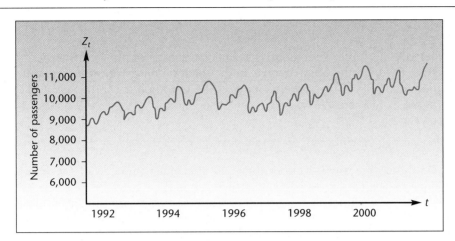

terly, we use three dummy variables to denote the particular quarter. You have probably spotted a limitation to this analysis, in addition to the fact that the assumptions of the regression model are not met in the context of time series. The new limitation is lack of parsimony. If you have 2 years' worth of monthly data and you use the dummy variable technique along with linear trend, then you have a regression analysis of 24 observations using a model with 12 variables. If, on the other hand, your data are quarterly and you have many years of data, then the problem of the proliferation of variables does not arise. Since the regression assumptions are not met anyway, we will not worry about this problem.

Using the dummy variable regression approach to seasonal time series assumes that the effect of the seasonal component of the series is additive. The seasonality is added to the trend and random error, as well as to the cycle (nonseasonal periodicity)—if one exists. We are thus assuming a model of the following form.

An additive model is

$$Z_t = T_t + S_t + C_t + I_t \tag{12–3}$$

where T is the trend component of the series, S is the seasonal component, C is the cyclical component, and I is the irregular component

---

(The irregular component is the error a_t; we use I_t because it is the usual notation in decomposition models.) Equation 12–3 states the philosophy inherent in the use of dummy variable regression to account for seasonality: The time series is viewed as comprising four components that are added to each other to give the observed values of the series.

The particular regression model, assuming our data are quarterly, is given by the following equation.

A regression model with dummy variables for seasonality is

$$Z_t = \beta_0 + \beta_1 t + \beta_2 Q_1 + \beta_3 Q_2 + \beta_4 Q_3 + a_t \qquad (12\text{--}4)$$

where $Q_1 = 1$ if the observation is in the first quarter of the year and 0 otherwise; $Q_2 = 1$ if the observation is in the second quarter of the year and 0 otherwise; $Q_3 = 1$ if the observation is in the third quarter of the year and 0 otherwise; and all three Q_i are 0 if the observation is in the fourth quarter of the year

Since the procedure is a straightforward application of the dummy variable regression technique of Chapter 11, we will not give an example.

A second way of modeling seasonality assumes a *multiplicative* model for the components of the time series. This is more commonly used than the additive model, equation 12–3, and is found to describe appropriately time series in a wide range of applications. The overall model is of the following form.

A multiplicative model is

$$Z_t = (T_t)(S_t)(C_t)(I_t) \qquad (12\text{--}5)$$

Here the observed time series values are viewed as the *product* of the four components, when all exist. If there is no cyclicity, for example, then $C_t = 1$. When equation 12–5 is the assumed overall model for the time series, we deal with the seasonality by using a method called *ratio to moving average*. Once we account for the seasonality, we may also model the cyclical variation and the trend. We describe the procedure in the next section.

PROBLEMS

12–7. Explain the difference between the terms *seasonal variation* and *cyclical variation*.

12–8. What particular problem would you encounter in fitting a dummy variable regression to 70 weekly observations of a seasonal time series?

12–9. In your opinion, what could be the reasons why the seasonal component is not constant? Give examples where you believe the seasonality may change.

12–10. Cyclical variation may also refer to cycles less than 1 year. Give examples of time series where you believe there may be a week-long cycle.

12–4   The Ratio-to-Moving-Average Method

A **moving average** of a time series is an average of a fixed number of observations (say, five observations) that moves as we progress down the series.[3]

---

[3]The term *moving average* has another meaning within the Box-Jenkins methodology (an advanced forecasting technique not discussed in this book).

TABLE 12–2 **Demonstration of a Five-Observation Moving Average**

Time t:	1	2	3	4	5	6	7	8	9	10	11	12	13	14
Series values, Z_t:	15	12	11	18	21	16	14	17	20	18	21	16	14	19
Corresponding series of five-observation moving average:				15.4	15.6	16	17.2	17.6	17	18	18.4	17.8	17.6	

FIGURE 12–10 **Computing the Five-Observation Moving Averages for the data in Table 12–2**

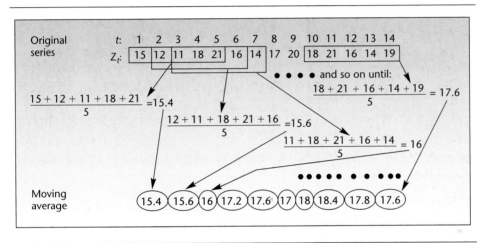

A moving average based on five observations is demonstrated in Table 12–2. Figure 12–10 shows how the moving average in Table 12–2 is obtained and how this average *moves* as the series progresses. Note that the first moving average is obtained from the first five observations, so we must wait until $t = 5$ to produce the first moving average. Therefore, there are fewer observations in the moving-average series than there are in the original series, Z_t. *A moving average smooths the data of their variations.* The original data of Table 12–2 along with the smoothed moving-average series are displayed in Figure 12–11.

The idea may have already occurred to you that if we have a seasonal time series and we compute a moving-average series for the data, then we will smooth out the seasonality. This is indeed the case. Assume a multiplicative time series model of the form given in equation 12–5:

$$Z = TSCI$$

(here we drop the subscript t). If we smooth out the series by using a 12-month moving average when data are monthly, or four-quarter moving average when data are quarterly, then the resulting smoothed series will contain trend and cycle but not seasonality or the irregular component; the last two will have been smoothed out by the moving average. If we then divide each observation by the corresponding value of

FIGURE 12–11   Original Series and Smoothed Moving-Average Series

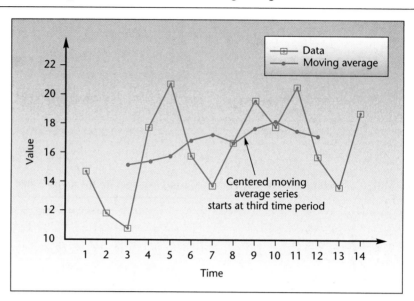

the moving-average (MA) series, we will have isolated the seasonal and irregular components. Notationally,

$$\frac{Z_t}{\text{MA}} = \frac{TSCI}{TC} = SI \tag{12–6}$$

This is the **ratio to moving average.** If we average each seasonal value with all values of $Z_t/$MA for the same season (i.e., for quarterly data, we average all values corresponding to the first quarter, all values of the second quarter, and so on), then we cancel out *most* of the irregular component I_t and isolate the seasonal component of the series. There are two more steps to be followed in the general procedure just described: (1) we compute the seasonal components as percentages by multiplying $Z_t/$MA by 100; and (2) we *center* the dates of the moving averages by averaging them. In the case of quarterly data, we average every two consecutive moving averages and center them midway between quarters. Centering is required because the number of terms in the moving average is even (4 quarters or 12 months).

Summary of the ratio-to-moving-average procedure for quarterly data (a similar procedure is carried out when data are monthly):

1. Compute a four-quarter moving-average series.
2. Center the moving averages by averaging every consecutive pair.
3. For each data point, divide the original series value by the corresponding moving average. Then multiply by 100.
4. For each quarter, average all data points corresponding to the quarter. The averaging can be done in one of several ways: find the simple average; find a modified average, which is the average after dropping the highest and lowest points; or find the median. Once we average the ra-

TABLE 12–3 Data and Four-Quarter Moving Averages for Example 12–3

Quarter	Sales (billions Btu)	Four-Quarter Moving Average	Centered Moving Average	Ratio to Moving Average (%)
1998 W	170			
Sp	148			
Su	141	152.25	(141/151.125)100 → 151.125 → 93.3	93.3
F	150	150	148.625	100.9
1999 W	161	147.25	146.125	110.2
Sp	137	145	146	93.8
Su	132	147	146.5	90.1
F	158	146	147	107.5
2000 W	157	148	147.5	106.4
Sp	145	147	144	100.7
Su	128	141	141.375	90.5
F	134	141.75	141	95.0
2001 W	160	140.25	140.5	113.9
Sp	139	140.75	142 → 97.9	97.9
Su	130	143.25	(139/142)100	
F	144			

tio-to-moving-average figures for each quarter, we will have four *quarterly indexes*. Finally, we adjust the indexes so that their mean will be 100. This is done by multiplying each by 400 and dividing by their sum.

We demonstrate the procedure with Example 12–3.

EXAMPLE 12–3

The distribution manager of Northern Natural Gas Company needs to analyze the time series of quarterly sales of natural gas in a Midwestern region served by the company. Quarterly data for 1998 through 2001 are given in Table 12–3. The table also shows the four-quarter moving averages, the centered moving averages, and the ratio to moving average (multiplied by 100 to give percentages), Figure 12–12 shows both the original series and the centered four-quarter moving-average series. Note how the seasonal variation is smoothed out.

The ratio-to-moving-average column in Table 12–3 gives us the contribution of the seasonal component and the irregular component within the multiplicative model, as seen from equation 12–6. We now come to step 4 of the procedure—averaging each seasonal term so as to average out the irregular effects and isolate the purely seasonal component as much as possible. We will use the simple average in obtaining the four seasonal indexes. This is done in Table 12–4, with the ratio-to-moving-average figures from Table 12–3.

Due to rounding, the indexes do not add to exactly 400, but their sum is very close to 400. The seasonal indexes quantify the seasonal effects in the time series of natural gas sales. We will see shortly how these indexes and other quantities are used in forecasting future values of the time series.

The ratio-to-moving-average procedure, which gives us the seasonal indexes, may also be used for **deseasonalizing** the data. Deseasonalizing a time series is a procedure

FIGURE 12–12 **Northern Natural Gas Sales: Original Series and Moving Average**

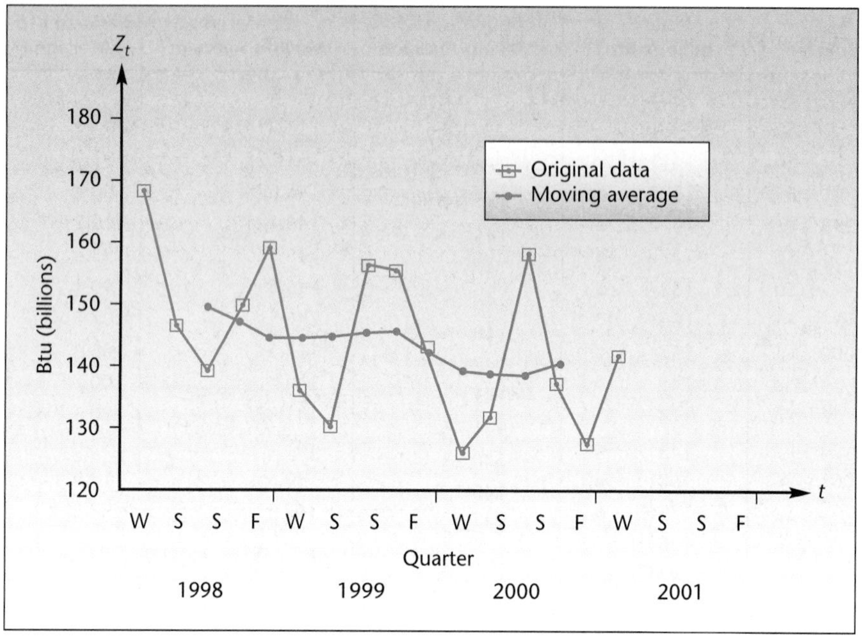

TABLE 12–4 **Obtaining the Seasonal Indexes for Example 12–3**

	Quarter			
	Winter	**Spring**	**Summer**	**Fall**
1998			93.3	100.9
1999	110.2	93.8	90.1	107.5
2000	106.4	100.7	90.5	95.0
2001	113.9	97.9		
Sum	330.5	292.4	273.9	303.4
Average	110.17	97.47	91.3	101.13
	Sum of averages = 400.07			
	Seasonal index = (Average)(400)/(400.07):			
	110.15	97.45	91.28	101.11

that is often used to display the general movement of a series without regard to the seasonal effects. Many government economic statistics are reported in the form of deseasonalized time series. To deseasonalize the data, we divide every data point by its appropriate seasonal index. If we assume a multiplicative time series model (equation 12–5), then dividing by the seasonal index gives us a series containing the other components only:

$$\frac{Z}{S} = \frac{TSCI}{S} = CTI \qquad (12\text{–}7)$$

TABLE 12–5 Deseasonalizing the Series for Example 12–3

Quarter		Sales Z (billions Btu)	Seasonal Indexes S	Deseasonalized Series $(Z/S)(100)$
1998	Winter	170	110.15	154.33
	Spring	148	97.45	151.87
	Summer	141	91.28	154.47
	Fall	150	101.11	148.35
1999	Winter	161	110.15	146.16
	Spring	137	97.45	140.58
	Summer	132	91.28	144.61
	Fall	158	101.11	156.27
2000	Winter	157	110.15	142.53
	Summer	145	97.45	148.79
	Spring	128	91.28	140.23
	Fall	134	101.11	132.53
2001	Winter	160	110.15	145.26
	Summer	139	97.45	142.64
	Spring	130	91.28	142.42
	Fall	144	101.11	142.42

FIGURE 12–13 Original and Deseasonalized Series for the Northern Natural Gas Example

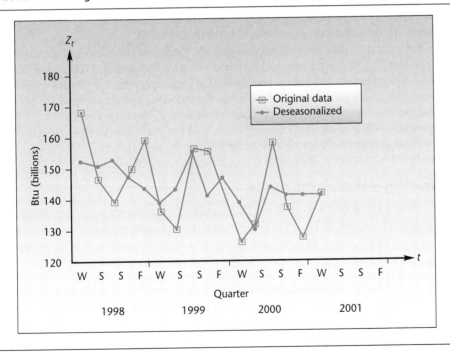

Table 12–5 shows how the series of Example 12–3 is deseasonalized. The deseasonalized natural gas time series, along with the original time series, is shown in Figure 12–13. Note that we have to multiply our results Z/S by 100 to cancel out the fact that our seasonal indexes were originally multiplied by 100 by convention.

The deseasonalized series in Figure 12–13 does have some variation in it. Comparing this series with the moving-average series in Figure 12–12, containing only the trend and cyclical components *TC,* we conclude that the relatively high residual variation in the deseasonalized series is due to the irregular component *I* (because the deseasonalized series is *TCI* and the moving-average series is *TC*). The large irregular component is likely due to variation in the weather throughout the period under study.

The Template

The template that can be used for Trend + Season Forecasting using the ratio-to-moving-average method is shown in Figure 12–14. In this figure, the sheet meant for quarterly data is shown. In the same workbook, Trend+Season Forecasting.xls, there is another sheet meant for monthly data, which is shown in Figure 12–15.

The Cyclical Component of the Series

Since the moving-average series is *TC,* we could isolate the cyclical component of the series by dividing the moving-average series by the trend *T.* We must, therefore, first estimate the trend. By visually inspecting the moving-average series in Figure 12–12, we notice a slightly decreasing linear trend and what looks like two cycles. We should therefore try to fit a straight line to the data. The line, fit by simple linear regression of moving average against *t,* is shown in Figure 12–16. The estimated trend line is

$$\hat{Z} = 152.26 - 0.837t \qquad (12\text{--}8)$$

From Figure 12–16, it seems that the cyclical component is relatively small in comparison with the other components of this particular series.

If we want to isolate the cyclical component, we divide the moving-average series (the product *TC*) by the corresponding trend value for the period. Multiplying the answer by 100 gives us a kind of cyclical index for each data point. There are problems, however, in dealing with the cycle. Unlike the seasonal component, which is fairly regular (with a 1-year cycle), the cyclical component of the time series may not have a dependable cycle at all. Both the amplitude and the cycle (peak-to-peak distance) of the cyclical component may be erratic, and it may be very difficult, if not impossible, to predict. This makes it difficult to forecast future series observations.

Forecasting a Multiplicative Series

Forecasting a series of the form *Z = TSCI* entails trying to forecast the three "regular" components *C, T,* and *S.* We try to forecast each component separately and then multiply them to get a series forecast:

> The forecast of a multiplicative series is
> $$\hat{Z} = TSC \qquad (12\text{--}9)$$

As we noted, it is very difficult to obtain reliable forecasts of the cyclical component. The trend is forecast simply by substituting the appropriate value of *t* in the least-squares line, as was done in Section 12–2. Then we multiply the value by the seasonal index (expressed as a decimal—divided by 100) to give us *TS.* Finally, we follow the cyclical component and try to guess what it may be at the point we need to

FIGURE 12–14 **The Template for the Ratio-to-Moving-Average Method of Trend + Season Forecasting**
[Trend + Season Forecast.xls; Sheet: Quarterly]

Forecasting with Trend and Seasonality

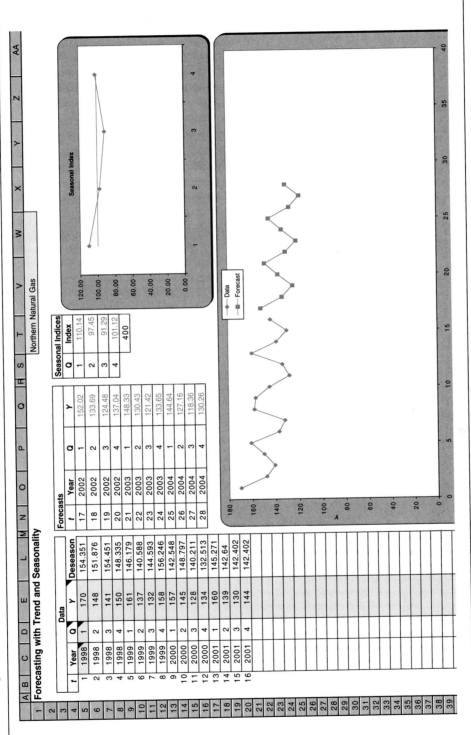

Data

t	Year	Q	Y	Deseason
1	1998	1	170	154.351
2	1998	2	148	151.876
3	1998	3	141	154.451
4	1998	4	150	148.335
5	1999	1	161	146.179
6	1999	2	137	140.588
7	1999	3	132	144.593
8	1999	4	158	156.246
9	2000	1	157	142.548
10	2000	2	145	148.797
11	2000	3	128	140.211
12	2000	4	134	132.513
13	2001	1	160	145.271
14	2001	2	139	142.64
15	2001	3	130	142.402
16	2001	4	144	142.402

Forecasts

t	Year	Q	Y
17	2002	1	152.02
18	2002	2	133.69
19	2002	3	124.48
20	2002	4	137.04
21	2003	1	148.33
22	2003	2	130.43
23	2003	3	121.42
24	2003	4	133.65
25	2004	1	144.64
26	2004	2	127.16
27	2004	3	118.36
28	2004	4	130.26

Seasonal Indices

Q	Index
1	110.14
2	97.45
3	91.29
4	101.12
	400

Northern Natural Gas

FIGURE 12-15 The Template for Trend + Season Forecasting with Monthly Data
[Trend+Season Forecast.xls; Sheet: Monthly]

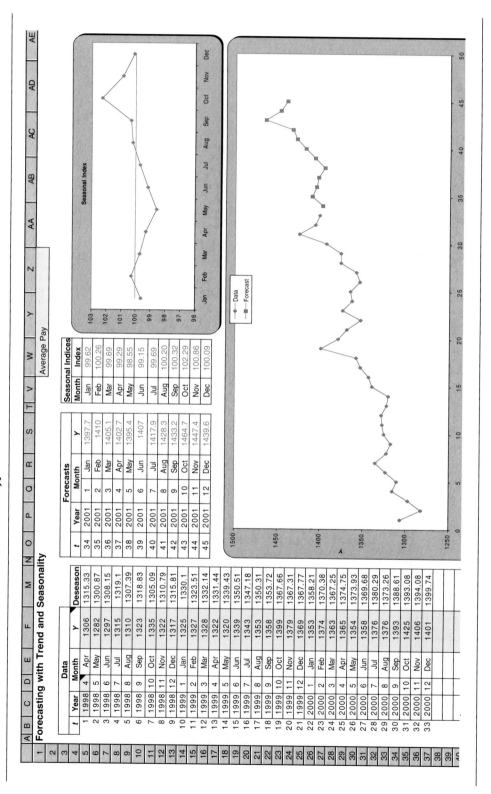

Forecasting with Trend and Seasonality

t	Year	Month	y	Deseason
1	1998	4 Apr	1306	1315.33
2	1998	5 May	1282	1300.87
3	1998	6 Jun	1297	1308.15
4	1998	7 Jul	1315	1319.1
5	1998	8 Aug	1310	1307.39
6	1998	9 Sep	1323	1318.83
7	1998	10 Oct	1335	1305.09
8	1998	11 Nov	1322	1310.79
9	1998	12 Dec	1317	1315.81
10	1999	1 Jan	1325	1330.1
11	1999	2 Feb	1327	1323.51
12	1999	3 Mar	1328	1332.14
13	1999	4 Apr	1322	1331.44
14	1999	5 May	1320	1339.43
15	1999	6 Jun	1339	1350.51
16	1999	7 Jul	1343	1347.18
17	1999	8 Aug	1353	1350.31
18	1999	9 Sep	1358	1353.72
19	1999	10 Oct	1399	1367.66
20	1999	11 Nov	1379	1367.31
21	1999	12 Dec	1369	1367.77
22	2000	1 Jan	1353	1358.21
23	2000	2 Feb	1374	1370.38
24	2000	3 Mar	1363	1367.25
25	2000	4 Apr	1365	1374.75
26	2000	5 May	1354	1373.93
27	2000	6 Jun	1358	1369.68
28	2000	7 Jul	1376	1380.29
29	2000	8 Aug	1376	1373.26
30	2000	9 Sep	1393	1388.61
31	2000	10 Oct	1425	1393.08
32	2000	11 Nov	1406	1394.08
33	2000	12 Dec	1401	1399.74

Forecasts

t	Year	Month	y
34	2001	1 Jan	1397.7
35	2001	2 Feb	1410
36	2001	3 Mar	1405.1
37	2001	4 Apr	1402.7
38	2001	5 May	1395.4
39	2001	6 Jun	1407
40	2001	7 Jul	1417.9
41	2001	8 Aug	1428.3
42	2001	9 Sep	1433.2
43	2001	10 Oct	1464.7
44	2001	11 Nov	1447.4
45	2001	12 Dec	1439.6

Seasonal Indices

Month	Index
Jan	99.62
Feb	100.26
Mar	99.69
Apr	99.29
May	98.55
Jun	99.15
Jul	99.69
Aug	100.20
Sep	100.32
Oct	102.29
Nov	100.86
Dec	100.09

Average Pay

Seasonal Index

FIGURE 12–16   **Trend Line and Moving Average for the Northern Natural Gas Example**

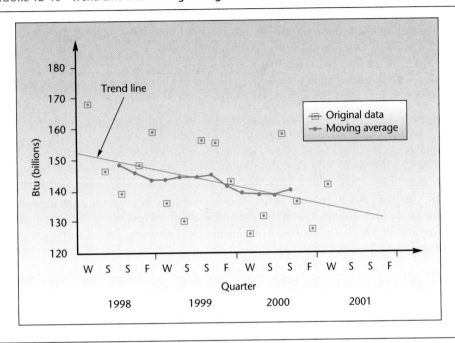

forecast; then we multiply by this component to get *TSC*. In our example, since the cyclical component seems small, we may avoid the nebulous task of guessing the future value of the cyclical component. We will therefore forecast using only *S* and *T*.

Let us forecast natural gas sales for winter 2002. Equation 12–8 for the trend was estimated with each quarter sequentially numbered from 1 to 16. Winter 2002 is $t =$ 17. Substituting this value into equation 12–8, we get

$$\hat{z} = 152.26 - 0.837(17) = 138.03 \text{ (billion Btu)}$$

The next stage is to multiply this result by the seasonal index (divided by 100). Since the point is a winter quarter, we use the winter index. From the bottom of Table 12–4 (or the second column of Table 12–5), we get the seasonal index for winter: 110.15. Ignoring the (virtually unforecastable) cyclical component by letting it equal 1, we find, using the forecast equation, equation 12–9:

$$\hat{z} = TSC = (1)(138.03)(1.1015) = 152.02 \text{ (billion Btu)}$$

This is our forecast of sales for winter 2002. See Figure 12–14 for further forecasts.

PROBLEMS

12–11.   The following data represent average factory worker weekly pay, in dollars, for the months between June 1994 and May 1997:

375, 370, 374, 378, 376, 380, 384, 380, 378, 380, 382, 383, 382, 381, 385, 387, 390, 392, 403, 398, 393, 389, 394, 392, 393, 391, 392, 396, 395, 400, 409, 404, 404, 405, 399, 402

Decompose this time series into trend and seasonal components, as well as a cyclical component (if it exists) and the irregular component. Describe each component. Develop the seasonal (monthly) indexes. Use the ratio-to-moving-average method. Also, deseasonalize the series. Forecast average weekly pay for factory workers for June 1997.

12–12.   The following data are monthly figures of U.S. steel production, in millions of tons, from July 1994 through April 1997:

> 7.4, 6.8, 6.4, 6.6, 6.5, 6.0, 7.0, 6.7, 8.2, 7.8, 7.7, 7.3, 7.0, 7.1, 6.9, 7.3, 7.0, 6.7, 7.6, 7.2, 7.9, 7.7, 7.6, 6.7, 6.3, 5.7, 5.6, 6.1, 5.8, 5.9, 6.2, 6.0, 7.3, 7.4

Decompose the series into its components, using the methods of this section, and forecast steel production for May 1997.

12–13.   A company importing Italian cameos needs to forecast the exchange rate of Italian lire to U.S. dollars. Quarterly data from the third quarter of 1991 through the first quarter of 1997 are obtained from *International Financial Statistics* and are as follows (in Italian lire per U.S. dollar):[4]

> 1,654, 1,695, 1,697, 1,659, 1,708, 1,772, 1,807, 1,757, 1,794, 1,761, 1,747, 1,670, 1,625, 1,624, 1,630, 1,608, 1,594, 1,607, 1,630, 1,629, 1,612, 1,608, 1,604

Try to forecast the exchange rate for the second quarter of 1997.

12–14.   The following are monthly electrical consumption data, in megawatts, for a town in Arizona from January 1994 through December 1997:

> 21, 23, 20, 24, 33, 41, 45, 50, 42, 28, 25, 26, 25, 27, 24, 30, 48, 53, 58, 62, 55, 32, 30, 32, 29, 32, 28, 35, 37, 50, 58, 66, 60, 45, 39, 40, 40, 43, 39, 45, 61, 68, 70, 72, 65, 54, 40, 39

Decompose the series to its components, and forecast consumption for January 1998.

12–15.   The following are quarterly data of spot prices for a barrel of oil (in U.S. dollars) on the Gulf Coast of the United States, from 1993 through 1997:

> 21.6, 23.0, 22.9, 24.0, 23.1, 21.7, 21.1, 20.6, 21.2, 19.8, 20.6, 22.0, 21.1, 19.6, 18.4, 21.1, 20.0, 19.0, 18.8, 19.9

Decompose the series and forecast the price of a barrel of oil for the first quarter of 1998.

12–5   Exponential Smoothing Methods

One method that is often useful in forecasting time series is *exponential smoothing*. There are exponential smoothing methods of varying complexity, but we will discuss only the simplest model, called *simple exponential smoothing*. Simple exponential smoothing is a useful method for forecasting time series that have no pronounced trend or seasonality. The concept is an extension of the idea of a moving average, introduced in the last section. Look at Figures 12–11 and 12–12, and notice how the moving average *smooths* the original series of its sharp variations. The idea of exponential smoothing is to smooth the original series the way the moving average does and to use the smoothed series in forecasting future values of the variable of interest. In exponential

---

[4]From *International Financial Statistics*, March 1997.

smoothing, however, we want to allow the more recent values of the series to have greater influence on the forecasts of future values than the more distant observations.

> **Exponential smoothing** is a forecasting method in which the forecast is based on a *weighted average* of current and past series values. The largest weight is given to the present observation, less weight to the immediately preceding observation, even less weight to the observation before that, and so on. *The weights decline geometrically as we go back in time.*

We define a **weighting factor w** as a selected number between 0 and 1

$$0 < w < 1 \qquad (12\text{–}10)$$

Once we select w—for example, $w = 0.4$—we define the forecast equation. The forecast equation is

$$\hat{Z}_{t+1} = w(Z_t) + w(1 - w)(Z_{t-1}) + w(1 - w)^2(Z_{t-2}) \qquad (12\text{–}11)$$
$$+ \; w(1 - w)^3(Z_{t-3}) + \cdots$$

where $\hat{Z}_{t+1}$ is the *forecast* value of the variable Z at time $t + 1$ from knowledge of the *actual* series values Z_t, Z_{t-1}, Z_{t-2}, and so on back in time to the first known value of the time series Z_1.

The series of weights used in producing the forecast $\hat{Z}_{t+1}$ is w, $w(1 - w)$, $w(1 - w)^2, \ldots$ These weights decline toward 0 in an *exponential* fashion; thus, as we go back in the series, each value has a smaller weight in terms of its effect on the forecast. If $w = 0.4$, then the rest of the weights are $w(1 - w) = 0.24$, $w(1 - w)^2 = 0.144$, $w(1 - w)^3 = 0.0864$, $w(1 - w)^4 = 0.0518$, $w(1 - w)^5 = 0.0311$, $w(1 - w)^6 = 0.0187$, and so on. The exponential decline of the weights toward 0 is evident. This is shown in Figure 12–17.

FIGURE 12–17 **Exponentially Declining Weights**

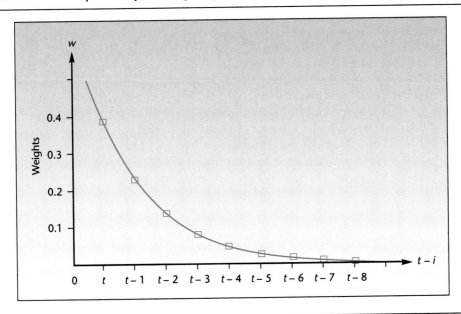

Before we show how the exponential smoothing model is used, we will rewrite the model in a recursive form that uses both previous observations and previous forecasts. Let us look at the forecast of the series value at time $t + 1$, denoted $\hat{Z}_{t+1}$. It can be shown that the exponential smoothing model of equation 12–11 is equivalent to the following model.

The exponential smoothing model is

$$\hat{Z}_{t+1} = w(Z_t) + (1 - w)(\hat{Z}_t) \tag{12–12}$$

where Z_t is the actual, known series value at time t and $\hat{Z}_t$ is the forecast value for time t.

The recursive equation, equation 12–12, can be restated in words as

$$\text{Next forecast} = w(\text{Present actual value}) + (1 - w)(\text{Present forecast})$$

The forecast value for time period $t + 1$ is thus seen as a weighted average of the actual value of the series at time t and the forecast value of the series at time t (the forecast having been made at time $t - 1$). There is yet a third way of writing the formula for the simple exponential smoothing model.

An equivalent form of the exponential smoothing model is

$$\hat{Z}_{t+1} = Z_t + (1 - w)(\hat{Z}_t - Z_t) \tag{12–13}$$

The proofs of the equivalence of equations 12–11, 12–12, and 12–13 are left as exercises at the end of this section. The importance of equation 12–13 is that it describes the forecast of the value of the variable at time $t + 1$ as the actual value of the variable at the previous time period t plus a fraction of the previous *forecast error*. The forecast error is the difference between the forecast $\hat{Z}_t$ and the actual series value Z_t. We will formally define the forecast error soon.

　　The recursive equation (equation 12–12) allows us to compute the forecast value of the series for each time period in a sequential manner. This is done by substituting values for t ($t = 1, 2, 3, 4, \ldots$) and using equation 12–12 for each t to produce the forecast at the next period $t + 1$. Then the forecast and actual values at the last known time period, $\hat{Z}_t$ and Z_t, are used in producing a forecast of the series into the future. The recursive computation is done by applying equation 12–12 as follows:

$$\begin{aligned}
\hat{Z}_2 &= w(Z_1) + (1 - w)(\hat{Z}_1) \\
\hat{Z}_3 &= w(Z_2) + (1 - w)(\hat{Z}_2) \\
\hat{Z}_4 &= w(Z_3) + (1 - w)(\hat{Z}_3) \\
\hat{Z}_5 &= w(Z_4) + (1 - w)(\hat{Z}_4)
\end{aligned} \tag{12–14}$$

.

.

.

The problem is how to determine the first forecast $\hat{Z}_1$. Customarily, we use $\hat{Z}_1 = Z_1$. Since the effect of the first forecast in a series of values diminishes as the series progresses toward the future, the choice of the first forecast is of little importance (it is an *initial value* of the series of forecasts, and its influence diminishes exponentially).

The choice of w, which is up to the person carrying out the analysis, is very important, however. *The larger the value of* w, *the faster the forecast series responds to change in the original series.* Conversely, the smaller the value of w, the less sensitive is the forecast to changes in the variable Z_t. If we want our forecasts not to respond quickly to changes in the variable, we set w to be a relatively small number. Conversely, if we want the forecast to quickly follow abrupt changes in variable Z_t, we set w to be relatively large (closer to 1.00 than to 0). We demonstrate this, as well as the computation of the exponentially smoothed series and the forecasts, in Example 12–4.

EXAMPLE 12-4

A sales analyst is interested in forecasting weekly firm sales in thousands of units. The analyst collects 15 weekly observations in 1995 and recursively computes the exponentially smoothed series of forecasts, using $w = 0.4$ and the exponentially smoothed forecast series for $w = 0.8$. The original data and both exponentially smoothed series are given in Table 12–6.

Solution

The original series and the two exponentially smoothed forecast series, corresponding to $w = 0.4$ and $w = 0.8$, are shown in Figure 12–18. The figure also shows the forecasts of the unknown value of the series at the 16th week produced by the two exponential smoothing procedures ($w = 0.4$ and $w = 0.8$). As was noted earlier, the smoothing coefficient w is set at the discretion of the person carrying out the analysis. Since w has a strong effect on the magnitude of the forecast values, the forecast accuracy depends on guessing a "correct" value for the smoothing coefficient. We have presented a simple

TABLE 12–6 **Exponential Smoothing Sales Forecasts Using $w = 0.4$ and $w = 0.8$**

Day	Z_t Original Series	$\hat{Z}_t$ Forecast Using $w = 0.4$	$\hat{Z}_t$ Forecast Using $w = 0.8$
1	925	925	925
2	940	$0.4(925) + 0.6(925) = 925$	925
3	924	$0.4(940) + 0.6(925) = 931$	937
4	925	928.2	926.6
5	912	926.9	925.3
6	908	920.9	914.7
7	910	915.7	909.3
8	912	913.4	909.9
9	915	912.8	911.6
10	924	913.7	914.3
11	943	917.8	922.1
12	962	927.9	938.8
13	960	941.5	957.4
14	958	948.9	959.5
15	955	952.5	958.3
16 (Forecasts)		953.5	955.7

FIGURE 12–18   The Sales Data: Original Series and Two Exponentially Smoothed Series

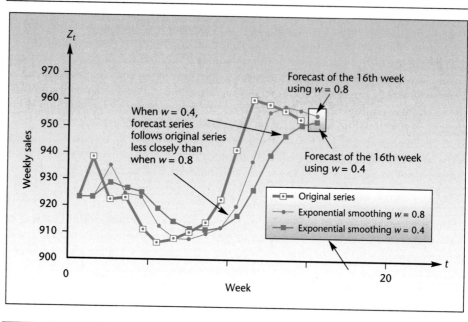

exponential smoothing method. When the data exhibit a trend or a seasonal variation, or both, more complicated exponential smoothing methods apply.

The Template

The template that can be used for exponential smoothing is shown in Figure 12–19. An additional feature available on the template is the use of the Solver to find the optimal w. We saw the results of using $w = 0.4$ and $w = 0.8$ in Example 12–4. Suppose we want to find the optimal w that minimizes MSE. To do this, unprotect the sheet and invoke the Solver. Click the Solve button, and when the Solver is done, choose Keep Solver Solution. The value of w found in the template is the optimal w that minimizes MSE. These instructions are also available in the comment at cell C4.

The Solver works better with MSE than MAPE (mean absolute percent error), because MSE is a "smooth" function of w whereas MAPE is not. With MAPE, the Solver may be stuck at a local minimum and miss the global minimum.

PROBLEMS

12–16. The following are monthly Dow Jones Commodity Price Index values from June 1996 through June 1997:[5]

142, 137, 143, 142, 149, 143, 151, 150, 151, 146, 144, 145, 147

---

[5]From *The Wall Street Journal,* June 19, 1998.

FIGURE 12–19   **The Template for Exponential Smoothing**
[Exponential Smoothing.xls]

t	Z_t	Forecast	|Error|	%Error	Error²
	w 0.4		**MAE** 11.3034	**MAPE** 1.20%	**MSE** 216.931
1	925	925			
2	940	925			
3	924	931	7	0.76%	49
4	925	928.2	3.2	0.35%	10.24
5	912	926.92	14.92	1.64%	222.61
6	908	920.95	12.952	1.43%	167.75
7	910	915.77	5.7712	0.63%	33.307
8	912	913.46	1.4627	0.16%	2.1395
9	915	912.88	2.1224	0.23%	4.5044
10	924	913.73	10.273	1.11%	105.54
11	943	917.84	25.164	2.67%	633.23
12	962	927.9	34.098	3.54%	1162.7
13	960	941.54	18.459	1.92%	340.74
14	958	948.92	9.0754	0.95%	82.364
15	955	952.55	2.4453	0.26%	5.9793
16		953.53			

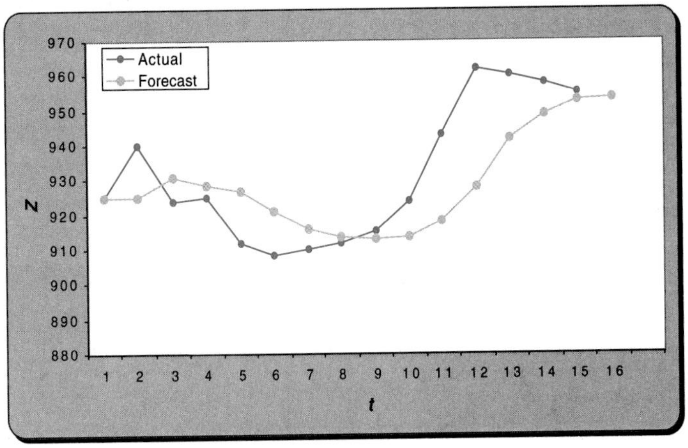

Construct an exponential smoothing model for these data, using $w = 0.6$, and forecast the Commodity Index value for July 1997. Experiment with other values of w. See if you can improve forecasts of the data points that you have.

12–17. The following are weekly sales data, in thousands of units, for microcomputer disks:

57, 58, 60, 54, 56, 53, 55, 59, 62, 57, 50, 48, 52, 55, 58, 61

Use $w = 0.3$ and $w = 0.8$ to produce an exponential smoothing model for these data. Which value of w produces better forecasts? Explain.

12–18. Construct an exponential smoothing forecasting model, using $w = 0.7$, for new orders reported by manufacturers. Monthly data (in billions of dollars) to April 1997 are

195, 193, 190, 185, 180, 190, 185, 186, 184, 185, 198, 199, 200, 201, 199, 187, 186, 191, 195, 200, 200, 190, 186, 196, 198, 200, 200

12–19. The following data are average monthly prices, in U.S. dollars, for a barrel of Urals crude oil from May 1996 through April 1997:

16.4, 17.1, 16.9, 17.3, 17.5, 17.2, 17.3, 17.1, 16.9, 17.0, 17.1, 17.2

Construct an exponential smoothing model for these data, and use it to forecast the price of a barrel of Urals crude oil for May 1997.

12–20. Use *The Wall Street Journal* or another source to gather information on the daily price of gold. Collect a series of prices, and construct an exponential smoothing model. Choose the weighting factor w that seems to fit the data best. Forecast the next day's price of gold, and compare the forecast with the actual price once it is known.

12–21. Prove that equation 12–11 is equivalent to equation 12–12.

12–22. Prove the equivalence of equations 12–12 and 12–13.

12–6 Index Numbers

It was dubbed the "Crash of '87." Measured as a percentage, the decline was worse than the one that occurred during the same month in 1929 and ushered in the Great Depression. Within a few hours on Monday, October 19, 1987, the Dow Jones Industrial Average plunged 508.32 points, a drop of 22.6%—the greatest percentage drop ever recorded in one day.

What is the Dow Jones Industrial Average, and why is it useful? The Dow Jones average is an example of an **index.** It is one of several quantitative measures of price movements of stocks through time. Another commonly used index is the New York Stock Exchange (NYSE) Index, and there are others. The Dow Jones captures in one number (e.g., the 508.32 points just mentioned) the movements of 30 industrial stocks considered by some to be representative of the entire market. Other indexes are based on a wider proportion of the market than just 30 big firms.

Indexes are useful in many other areas of business and economics. Another commonly quoted index is the **consumer price index (CPI),** which measures price fluctuations. The CPI is a single number representing the general level of prices that affect consumers.

> An **index number** is a number that measures the relative change in a set of measurements over time.

When the measurements are of a *single variable,* for example, the price of a certain commodity, the index is called a *simple index number.* A simple index number is the ratio of two values of a variable, expressed as a percentage. First, a *base period* is chosen. The value of the index at any time period is equal to the ratio of the current value of the variable divided by the base-period value, times 100.

EXAMPLE 12–5 The following data are annual cost figures for residential natural gas for the years 1984 to 1997 (in dollars per thousand cubic feet):

121, 121, 133, 146, 162, 164, 172, 187, 197, 224, 255, 247, 238, 222

If we want to describe the relative change in price of residential natural gas, we construct a simple index of these prices. Suppose that we are interested in comparing prices of residential natural gas of any time period to the price in 1984 (the first year in our series). In this case, 1984 is our base year, and the index for that year is defined as 100. The index for any year is defined by equation 12–15.

$$\text{Index number for period } i = 100\left(\frac{\text{Value in period } i}{\text{Value in base period}}\right) \quad (12\text{–}15)$$

Thus, the index number for 1986 (using the third data point in the series) is computed as

$$\text{Index number for 1986} = 100\left(\frac{\text{Price in 1986}}{\text{Price in 1984}}\right)$$

TABLE 12–7 **Price Index for Residential Natural Gas, Base Year 1984**

Year	Price	Index
1984	121	100
1985	121	100
1986	133	109.9
1987	146	120.7
1988	162	133.9
1989	164	135.5
1990	172	142.1
1991	187	154.5
1992	197	162.8
1993	224	185.1
1994	255	210.7
1995	247	204.1
1996	238	196.7
1997	222	183.5

$$= 100\left(\frac{133}{121}\right) = 109.9$$

This means that the price of residential natural gas increased by 9.9% from 1984 to 1986. Incidentally, the index for 1985 is also 100 since the price did not change from 1984 to 1985. Let us now compute the index for 1987:

$$\text{Index number for } 1987 = 100\left(\frac{146}{121}\right) = 120.66$$

Thus, compared with the price in 1984, the price in 1987 was 20.66% higher. It is very important to understand that changes in the index from year to year *may not be interpreted as percentages* except when one of the two years is the base year. The fact that the index for 1987 is 120.66 and for 1986 is 109.9 does not imply that the price in 1987 was $20.66 - 9.9 = 10.76\%$ higher than in 1986. Comparisons in terms of percentages may be made only with the base year. We can only say that the price in 1986 was 9.9% higher than 1984, and the price in 1987 was 20.66% higher than in 1984. Table 12–7 shows the year, the price, and the price index for residential natural gas from 1984 to 1997, inclusive.

From the table, we see, for example, that the price in 1994 was more than 210% of what it was in 1984 and that by 1997 the price declined to only 183.5% of what it was in 1984. Figure 12–20 shows both the raw price and the index with base year 1984. (The units of the two plots are different, and no comparison between them is suggested.)

As time goes on, the relevance of any base period in the past decreases in terms of comparison with values in the present. Therefore, it is sometimes useful to change the base period and move it closer to the present. Many indexed economic variables, for example, use the base year 1967. As we move into more recent years, the base year for these variables is changed to 1980 or later. There is a simple way of changing the base period of an index. All we need to do is to change the index number of the

FIGURE 12–20   Price and Index (Base Year 1984) of Residential Natural Gas

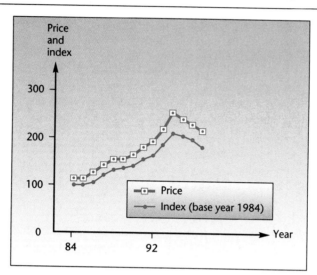

new base period so that it will equal 100 and to change all other numbers using the *same operation*. Thus, we divide all numbers in the index by the index value of the proposed new base period and multiply them by 100. This is shown in equation 12–16.

Changing the base period of an index:

$$\text{New index value} = \frac{\text{Old index value}}{\text{Index value of new base}} \times 100 \qquad (12\text{–}16)$$

Suppose that we want to change the base period of the residential natural gas index (Table 12–7) from 1984 to 1991. We want the index for 1991 to equal 100, so we divide all index values in the table by the current value for 1991, which is 154.5, and we multiply these values by 100. For 1992, the new index value is $(162.8/154.5)100 = 105.4$. The new index, using 1991 as base, is shown in Table 12–8.

Figure 12–21 shows the two indexes of the price of residential natural gas using the two different base years. Note that the changes in the index numbers that use 1984 as the base year are more pronounced. This is so because 1991, when used as the base year, is close to the middle of the series, and percentage changes with respect to that year are smaller.

An important use of index numbers is as *deflators*. This allows us to compare prices or quantities through time in a meaningful way. Using information on the relative price of natural gas at different years, as measured by the price index for this commodity, we can better assess the effect of changes in consumption by consumers. The most important use of index numbers as deflators, however, is in the case of *composite index numbers*. In particular, the consumer price index is an overall measure of relative changes in prices of many goods and thus reflects changes in the value of the

TABLE 12–8   **Residential Natural Gas Price Index**

Year	Index Using 1984 Base	Index Using 1991 Base
1984	100	64.7
1985	100	64.7
1986	109.9	71.1
1987	120.7	78.1
1988	133.9	86.7
1989	135.5	87.7
1990	142.1	92.0
1991	154.5	100
1992	162.8	105.4
1993	185.1	119.8
1994	210.7	136.4
1995	204.1	132.1
1996	196.7	127.3
1997	183.5	118.7

Note: All entries in the rightmost column are obtained from the entries in the middle column by multiplication by 100/154.5.

FIGURE 12–21   **Comparison of the Two Price Indexes for Residential Natural Gas**

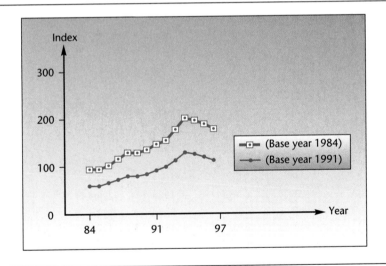

dollar. We all know that a dollar today is not worth the same as a dollar 20 years ago. Using the consumer price index, or another composite index, allows us to compare prices through time in "constant" dollars.

The Consumer Price Index

The CPI is probably the best-known price index. It is published by the U.S. Bureau of Labor Statistics and is based on the prices of several hundred items. The base year is 1967. For obtaining the base-year quantities used as weights, the Bureau of Labor Statistics interviewed thousands of families to determine their consumption patterns.

Since the CPI reflects the general price level in the country, it is used, among other purposes, in converting nominal amounts of money to what are called *real* amounts of money: amounts that can be compared through time without requiring us to consider changes in the value of money due to inflation. This use of the CPI is what we referred to earlier as using an index as a *deflator*. By simply dividing X dollars in year i by the CPI value for year i and multiplying by 100, we convert our X *nominal* (year i) dollars to *constant* (base-year) dollars. This allows us to compare amounts of money across time periods. Let us look at an example.

EXAMPLE 12–6   Table 12–9 gives the CPI values for the years 1950 to 1996. The base year is 1967. This is commonly denoted by [1967 = 100]. The data in Table 12–9 are from the U.S. Bureau of Labor Statistics publication *Monthly Labor Review* (March 1996).

We see, for example, that the general level of prices in the United States in 1994 was almost 4½ times what it was in 1967 (the base year). Thus, a dollar in 1994 could buy, on average, only what $1/4.44 = \$0.225$, or 22.5 cents, could buy in 1967. By dividing any amount of money in a given year by the CPI value for that year and multiplying by 100, we convert the amount to constant (1967) dollars. The term *constant* means dollars of a constant point in time—the base year.

TABLE 12–9 **The Consumer Price Index [1967 = 100]**

Year	CPI	Year	CPI
1950	72.1	1974	147.7
1951	77.8	1975	161.2
1952	79.5	1976	170.5
1953	80.1	1977	181.5
1954	80.5	1978	195.4
1955	80.2	1979	217.4
1956	81.4	1980	246.8
1957	84.3	1981	272.4
1958	86.6	1982	289.1
1959	87.3	1983	298.4
1960	88.7	1984	311.1
1961	89.6	1985	322.2
1962	90.6	1986	328.4
1963	91.7	1987	340.4
1964	92.9	1988	354.3
1965	94.5	1989	371.3
1966	97.2	1990	391.4
1967	100.0	1991	408.0
1968	104.2	1992	420.3
1969	109.8	1993	432.7
1970	116.3	1994	444.0
1971	121.3	1995	456.0
1972	125.3	1996	471.0
1973	133.1		

Let us illustrate the use of the CPI as a price deflator. Suppose that during the years 1980 to 1985, an analyst was making the following annual salaries:

1980	$29,500	1983	$35,000
1981	31,000	1984	36,700
1982	33,600	1985	38,000

Looking at the raw numbers, we may get the impression that this analyst has done rather well. His or her salary has increased from $29,500 to $38,000 in just 5 years. Actually the analyst's salary has not even kept up with inflation! That is, in *real* terms of *actual buying power,* this analyst's 1985 salary is smaller than what it was in 1980. To see why this is true, we use the CPI.

Solution

If we divide the 1980 salary of $29,500 by the CPI value for that year and multiply by 100, we will get the equivalent salary in 1967 dollars: $(29,500/246.8)(100) = \$11,953$. We now take the 1985 salary of $38,000 and divide it by the CPI value for 1985 and multiply by 100. This gives us $(38,000/322.2)(100) = \$11,794$—a *decrease* of $159 (1967)!

If you perform a similar calculation for the salaries of all other years, you will find that none of them have kept up with inflation. If we transform all salaries to 1967 dollars (or for that matter, to dollars of any single year), the figures can be compared with one another. Often, time series data such as these are converted to constant dollars of a single time period and then are analyzed by using methods of time series analysis such as the ones presented earlier in this chapter. To convert to dollars of another year (not the base year), you need to divide the salary by the CPI for the current year and multiply by the CPI value for the constant year in which you are interested. For example, let us convert the 1985 salary to 1980 (rather than 1967) dollars. We do this as follows: $(38,000/322.2)(246.8) = \$29,107$. Thus, in terms of 1980 dollars, the analyst was making only $29,107 in 1985, whereas in 1980 he or she was making $29,500 (1980)! Figure 12–22 shows the CPI for the years 1950 to 1997.

FIGURE 12–22 **The Consumer Price Index, 1950–1987 (Base Year = 1967)**

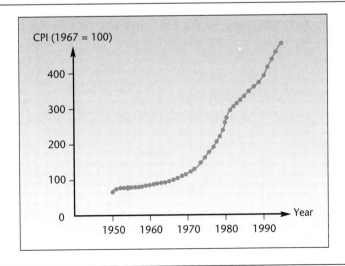

FIGURE 12–23   **The Template for Index Calculations**
[Index.xls]

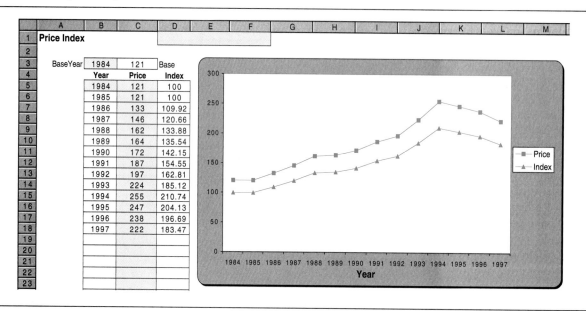

The Template

The template that can be used for index calculations is shown in Figure 12–23. The data entered in the template are from Table 12–7.

PROBLEMS

12–23.   In 1987, the base year was changed to [1982 = 100]. Change the base for all the figures in Table 12–9.

12–24.   The percentage change in the price index of food and beverages in the United States from July 2000 to July 2001 was +3.1%. The change in the same index from June 2001 to July 2001 was +0.3%. The index was 174.0 in July 2001, with base year 1982. Calculate what the index was in July 2000 and June 2001, with base year 1982.

12–25.   What is a simple price index?

12–26.   What are the uses of index numbers?

12–27.   The following is Nigeria's Industrial Output Index for the years 1984 to 1997:

Year	Index of Output
1984	175
1985	190
1986	132
1987	96
1988	100
1989	78

—Continued

Year	Index of Output
1990	131
1991	135
1992	154
1993	163
1994	178
1995	170
1996	145
1997	133

a. What is the base year used here?

b. Change the base year to 1993.

c. What happened to Nigeria's industrial output from 1996 to 1997?

d. Describe the trends in industrial output throughout the years 1984 to 1997.

12–28. The following data are June 1994 to June 1995 commodity prices as described by the Dow Jones Commodity Index:[6] 142, 137, 143, 142, 145, 151, 147, 144, 149, 154, 148, 153, 154. Form a new index, using January 1995 as the base month.

12–7   Summary and Review of Terms

In this chapter, we discussed forecasting methods. We saw how simple **Cycle × Trend × Seasonality × Irregular components models** are created and used. We then talked about **exponential smoothing models.** We also discussed **index numbers.**

ADDITIONAL PROBLEMS

12–29. The following data[7] are monthly existing-home sales, in millions of dwelling units, for January 1996 through June 1997. Construct a forecasting model for these data, and use it in forecasting sales for July 1997.

4.4, 4.2, 3.8, 4.1, 4.1, 4.0, 4.0, 3.9, 3.9, 3.8, 3.7, 3.7, 3.8, 3.9, 3.8, 3.7, 3.5, 3.4

12–30. Discuss and compare all the forecasting methods presented in this chapter. What are the relative strengths and weaknesses of each method? Under what conditions would you use any of the methods?

12–31. Discuss the main principle of the exponential smoothing method of forecasting. What effect does the smoothing constant *w* have on the forecasts?

12–32. The following data[8] are the numbers of cents (U.S.) that one German mark buys from May 25 to June 4:

59.3, 58.6, 59.0, 58.9, 58.6, 57.3, 57.9, 57.8

Predict the cents per mark for the next period.

---

[6]"Commodities," *The Wall Street Journal,* June 19, 1995, p. C14.

[7]"Existing-Home Sales," *The Wall Street Journal,* June 27, 1998.

[8]From Corey Story, "Backlash," *Business Week,* June 16, 1997, p. 20.

12–33. The following data, from C. Hill, "The Digital Age: How It's Changing Our Lives,"[9] are the percentages of households with a personal computer from 1979 to 1997. Do a trend line analysis (or other method) to forecast next year's percentage.

   2, 3, 5, 18, 19, 16, 15, 15, 16, 17, 19, 20, 19, 23, 30, 35, 40, 47, 52

12–34. Open the Trend+Season Forecasting template shown in Figure 12–15, used for monthly data. We will do a sensitivity analysis on this template.

   a. Change the value for April 1998 (in cell F5) from 1306 to 2000. What effect did this have on the seasonal index for April? What effect should we expect on the seasonal index? Explain why the result in the template did not agree with the expectation. (If necessary, unprotect the sheet and open up columns G through L to look at the calculations.)

   b. (Return April 1998 value to 1306) Change the value for April 1999 (in cell F17) from 1322 to 2000. What effect did this have on the seasonal index for April?

   c. (Return April 1999 value to 1322) Change the value for December 2000 (in cell F37) from 1401 to 2000. What effect did this have on the seasonal index for December? What effect should we expect on the seasonal index? Explain why the result in the template did not agree with the expectation.

   d. Look at the answers to parts *a, b,* and *c.* Conduct similar analysis on other selected data points. Make a formal statement about the sensitivity of the seasonal indexes to data points near the beginning and near the end compared to data points near the middle.

12–35. The following data[10] are annual percentage changes in GDP for Indonesia from 1991 to 1997:

   6.3, 6.6, 7.3, 7.4, 7.8, 6.9, 7.8

Do trend-line forecasting.

www.exercise

12–36. Use your library to research the Standard & Poor's 500 index. Write a short report explaining this index, its construction, and its use.

12–37. The CPI is the most pervasively used index number in existence. You can keep current with it at the Bureau of Labor Statistics site **http://stats.bls.gov/**. Beginning from the Data area, locate the CPI index-All urban consumers, all items data from the Most Requested Series Information.

   If the majority of new college graduates in 1978 could expect to land a job earning $20,000 per year, what must the starting salary be for the majority of new college graduates in 2002 in order for them to be at the same standard of living as their counterparts in 1978? Is today's typical college graduate better or worse off than her or his 1978 counterpart?

---

[9] *The Wall Street Journal,* June 16, 1997, p. 13.

[10] M. Shari, "Scharto's Capitalism," *Business Week,* June 16, 1997, pp. 26–29.

CASE 14 — Auto Parts Sales Forecast

The quarterly sales of a large manufacturer of spare parts for automobiles are tabulated below. Since the sales are in millions of dollars, forecast errors can be costly. The company wants to forecast the sales as accurately as possible.

Quarter	Sales	M2 Index	Non-Farm-Activity Index	Oil Price
98 Q1	$35,452,300	2.356464	34.2	19.15
98 Q2	$41,469,361	2.357643	34.27	16.46
98 Q3	$40,981,634	2.364126	34.3	18.83
98 Q4	$42,777,164	2.379493	34.33	19.75
99 Q1	$43,491,652	2.373544	34.4	18.53
99 Q2	$57,669,446	2.387192	34.33	17.61
99 Q3	$59,476,149	2.403903	34.37	17.95
99 Q4	$76,908,559	2.42073	34.43	15.84
00 Q1	$63,103,070	2.431623	34.37	14.28
00 Q2	$84,457,560	2.441958	34.5	13.02
00 Q3	$67,990,330	2.447452	34.5	15.89
00 Q4	$68,542,620	2.445616	34.53	16.91
01 Q1	$73,457,391	2.45601	34.6	16.29
01 Q2	$89,124,339	2.48364	34.7	17
01 Q3	$85,891,854	2.532692	34.67	18.2
01 Q4	$69,574,971	2.564984	34.73	17

1. Carry out a Trend+Season forecast with the sales data, and forecast the sales for the four quarters of 2002.

The director of marketing research of the company believes that the sales can be predicted better using a multiple regression of sales against three selected econometric variables that the director believes have significant impact on the sales. These variables are M2 Index, Non-Farm-Activity Index, and Oil Price. The values of these variables for the corresponding periods are available in the data tabulated to the left.

2. Conduct a multiple regression of sales against the three econometric variables, following the procedure learned in Chapter 11. What is the regression equation?

3. Make a prediction for the four quarters of 2002 based on the regression equation using projected values of the following econometric variables:

Quarter	M2 Index	Non-Farm-Activity Index	Oil Price
02 Q1	2.597688	34.7	17.1
02 Q2	2.630159	34.4	17.3
02 Q3	2.663036	34.5	18
02 Q4	2.696324	34.5	18.2

4. Which method do you think is better? Why?

13 Quality Control and Improvement

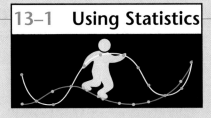

Not long after the Norman Conquest of England, the Royal Mint was established in London. The Mint has been in constant operation from its founding to this very day, producing gold and silver coins for the Crown (and in later periods, coins from cheaper metals). Sometime during the reign of Henry II (1154–1189), a mysterious ceremony called the "Trial of the Pyx" was initiated.

The word *pyx* is Old English for "box," and the ceremony was an actual trial by jury of the contents of a box. The ancient trial had religious overtones, and the jurors were all members of the Worshipful Company of Goldsmiths. The box was thrice locked and held under guard in a special room, the Chapel of the Pyx, in Westminster Abbey. It was ceremoniously opened at the trial, which was held once every three or four years.

What did the Pyx box contain, and what was the trial? Every day, a single coin of gold (or silver, depending on what was being minted) was randomly selected by the minters and sent to Westminster Abbey to be put in the Pyx. In three or four years, the Pyx contained a large number of coins. For a given type of coin, say a gold sovereign, the box also contained a royal standard, which was the exact desired weight of a sovereign. At the trial, the contents of the box were carefully inspected and counted, and later some coins were assayed. The total weight of all gold sovereigns was recorded. Then the weight of the royal standard was multiplied by the number of sovereigns in the box and compared with the actual total weight of the sovereigns. A given tolerance was allowed in the total weight, and the trial was declared a success if the total weight was within the tolerance levels established above and below the computed standard.

The trial was designed so that the King or Queen could maintain control of the use of the gold and silver ingots furnished to the Mint for coinage. If, for example, coins were too heavy, then the monarch's gold was being wasted. A shrewd merchant could then melt down such coins and sell them back to the Mint at a profit. This actually happened often enough that such coins were given the name *come again guineas* as they would return to the Mint in melted-down form, much to the minters' embarrassment. On the other hand, if coins contained too little gold, then the currency was being debased and would lose its value. In addition, somebody at the Mint could then be illegally profiting from the leftover gold.

When the trial was successful, a large banquet would be held in celebration. We may surmise that when the trial was not successful . . . the Tower of London was not too far away. The Trial of the Pyx is practiced (with modifications) to this very day. Interestingly, the famous scientist and mathematician Isaac Newton was at one time (1699 to 1727) Master of the Mint. In fact, one of the trials during Newton's tenure was not successful, but he survived.[1]

The Trial of the Pyx is a classic example, and probably the earliest on record, of a two-tailed statistical test for the population mean. The Crown wants to test the null hypothesis that, on average, the weight of the coins is as specified. The Crown wants to test this hypothesis against the two-tailed alternative that the average coin is either too heavy or too light—both having negative consequences for the Crown. The test

---

[1]Adapted from the article "Eight Centuries of Sampling Inspection: The Trial of the Pyx," by S. Stigler, originally published in the *Journal of the American Statistical Association*, copyright 1977 by the American Statistical Association. All rights reserved.

statistic used is the sum of the weights of n coins, and the critical points are obtained as n times the standard weight, plus or minus the allowed tolerance.[2] The Trial of the Pyx is also a wonderful example of *quality control.* We have a production process, the minting of coins, and we want to ensure that high quality is maintained throughout the operation. We sample from the production process, and we take corrective action whenever we believe that the process is *out of control*—producing items that, on average, lie outside our specified target limits.

13–2   W. Edwards Deming Instructs

We now jump 800 years, to the middle of the 20th century and to the birth of modern quality control theory. In 1950 Japan was trying to recover from the devastation of World War II. Japanese industry was all but destroyed, and its leaders knew that industry must be rebuilt well if the nation was to survive. But how? By an ironic twist of fate, Japanese industrialists decided to hire a U.S. statistician as their consultant. The man they chose was the late W. Edwards Deming, at the time a virtually unknown government statistician. No one in the United States paid much attention to Deming's theories on how statistics could be used to improve industrial quality. The Japanese wanted to listen. They brought Deming to Japan in 1950, and in July of that year he met with the top management of Japan's leading companies. He then gave the first of many series of lectures to Japanese management. The title of the course was "Elementary Principles of the Statistical Control of Quality," and it was attended by 230 Japanese managers of industrial firms, engineers, and scientists.

The Japanese listened closely to Deming's message. In fact, they listened so well that in a few short decades, Japan became one of the most successful industrial nations on earth. Whereas "Made in Japan" once meant low quality, the phrase has now come to denote the highest quality. In 1960, Emperor Hirohito awarded Dr. Deming with the Medal of the Sacred Treasure. The citation with the medal stated that the Japanese people attribute the rebirth of Japanese industry to W. Edwards Deming. In addition, the Deming Award was instituted in Japan to recognize outstanding developments and innovations in the field of quality improvement. On the walls of the main lobby of Toyota's headquarters in Tokyo hang three portraits. One portrait is of the company's founder, another is of the current chairman, and the largest portrait is of Dr. Deming.

Ironically, Dr. Deming's ideas did get recognized in the United States—alas, when he was 80 years old. For years, U.S. manufacturing firms had been feeling the pressure to improve quality, but not much was actually being done while the Japanese were conquering the world markets. In June 1980, Dr. Deming appeared in a network television documentary entitled "If Japan Can, Why Can't We?" Starting the next morning, Dr. Deming's mail quadrupled, and the phone was constantly ringing. Offers came from Ford, General Motors, Xerox, and many others.

While well into his 90s, Dr. Ed Deming was one of the most sought-after consultants to U.S. industry. His appointment book was filled years in advance, and companies were willing to pay very high fees for an hour of his time. He traveled around the country, lecturing on quality and how to achieve it. The first U.S. company to adopt the Deming philosophy and to institute a program of quality improvement at

---

[2]According to Professor Stigler, the tolerance was computed in a manner incongruent with statistical theory, but he feels we may forgive this error as the trial seems to have served its purpose well through the centuries.

all levels of production was Nashua Corporation. The company kindly agreed to provide us with actual data of a production process and its quality improvement. This is presented as Case 15 at the end of this chapter. How did Deming do it? How did he apply statistical quality control schemes so powerful that they could catapult a nation to the forefront of the industrialized world and are now helping U.S. firms improve as well? This chapter should give you an idea.

13–3 Statistics and Quality

In all fairness, Dr. Deming did not invent the idea of using statistics to control and improve quality; that honor goes to a colleague of his. What Deming did was to expand the theory and demonstrate how it could be used very successfully in industry. Since then, Deming's theories have gone beyond statistics and quality control, and they now encompass the entire firm. His tenets to management are the well-known "14 points" he advocated, which deal with the desired corporate approach to costs, prices, profits, labor, and other factors. Deming even liked to expound about antitrust laws and capitalism and to have fun with his audience. At a lecture attended by one author (A.D.A.) around Deming's 90th birthday, Dr. Deming opened by writing on a transparency: "Deming's Second Theorem: 'Nobody gives a hoot about profits.'" He then stopped and addressed the audience, "Ask me what is Deming's First Theorem." He looked expectantly at his listeners and answered, "I haven't thought of it yet!" The philosophical approach to quality and the whole firm, and how it relates to profits and costs, is described in the ever-growing literature on this subject. It is sometimes referred to as *total quality management* (TQM). Dr. Deming's vision of what a firm can do by using the total quality management approach to continual improvement is summarized in his famous 14 points. The Deming approach centers on creating an environment in which the 14 points can be implemented toward the achievement of quality.

Deming's 14 Points

1. Create constancy of purpose for continual improvement of products and service to society, allocating resources to provide for long-range needs rather than only short-term profitability, with a plan to become competitive, to stay in business, and to provide jobs.

2. Adopt the new philosophy. We are in a new economic age, created in Japan. We can no longer live with commonly accepted levels of delays, mistakes, defective materials, and defective workmanship. Transformation of Western management style is necessary to halt the continued decline of industry.

3. Eliminate the need for mass inspection as the way of life to achieve quality by building quality into the product in the first place. Require statistical evidence of built-in quality in both manufacturing and purchasing functions.

4. End the practice of awarding business solely on the basis of price tag. Instead, require meaningful measures of quality along with the price. Reduce the number of suppliers for the same item by eliminating those that do not qualify with statistical and other evidence of quality. The aim is to minimize *total* cost, not merely initial cost, by minimizing variation. This may be achievable by moving toward a single supplier for any one item, on a long-term relationship of loyalty and trust. Purchasing managers have a new job and must learn it.

5. Improve constantly and forever every process for planning, production, and service. Search continually for problems in order to improve every activity in the company, to improve quality and productivity, and thus to constantly decrease costs. Institute innovation and constant improvement of product, service, and process. It is the management's job to work continually on the system (design, incoming materials, maintenance, improvement of machines, supervision, training, and retraining).

6. Institute modern methods of training on the job for all, including management, to make better use of every employee. New skills are required to keep up with changes in materials, methods, product design, machinery, techniques, and service.

7. Adopt and institute leadership aimed at helping people to do a better job. The responsibility of managers and supervisors must be changed from sheer numbers to quality. Improvement of quality will automatically improve productivity. Management must ensure that immediate action is taken on reports of inherited defects, maintenance requirements, poor tools, fuzzy operational definitions, and all conditions detrimental to quality.

8. Encourage effective two-way communication and other means to drive out fear throughout the organization so that everybody may work effectively and more productively for the company.

9. Break down barriers between departments and staff areas. People in different areas, such as research, design, sales, administration, and production, must work in teams to tackle problems that may be encountered with products or service.

10. Eliminate the use of slogans, posters, and exhortations for the workforce, demanding zero defects and new levels of productivity, without providing methods. Such exhortations only create adversarial relationships; the bulk of the causes of low quality and low productivity belong to the system, and thus lie beyond the power of the workforce.

11. Eliminate work standards that prescribe quotas for the workforce and numerical goals for people in management. Substitute aids and helpful leadership in order to achieve continual improvement of quality and productivity.

12. Remove the barriers that rob hourly workers, and people in management, of their right to pride of workmanship. This implies, *inter alia,* abolition of the annual merit rating (appraisal of performance) and of management by objective. Again, the responsibility of managers, supervisors, and foremen must be changed from sheer numbers to quality.

13. Institute a vigorous program of education, and encourage self-improvement for everyone. What an organization needs is not just good people; it needs people who are improving with education. Advances in competitive position will have their roots in knowledge.

14. Clearly define top management's permanent commitment to ever-improving quality and productivity, and their obligation to implement all these principles. Indeed, it is not enough that top managers commit themselves for life to quality and productivity. They must know what it is that they are committed to—that is, what they must do. Create a structure in top management that will push every day on the preceding 13 points, and

take action in order to accomplish the transformation. Support is not enough: Action is required.

Process Capability

Process capability is the best in-control performance that an existing process can achieve without major expenditures. The *capability* of any process is the natural behavior of the particular process after disturbances are eliminated. In an effort to improve quality and productivity in the firm, it is important to first try to establish the capability of the process. An investigation is undertaken to actually achieve a state of statistical control in a process based on current data. This gives a live image of the process. For example, in trying to improve the quality of car production, we first try to find how the process operates in the best way, then make improvements. Control charts are a useful tool in this analysis.

Control Charts

The first modern ideas on how statistics could be used in quality control came in the mid-1920s from a colleague of Deming's, Walter Shewhart of Bell Laboratories. Shewhart invented the **control chart** for industrial processes. A control chart is a graphical display of measurements (usually aggregated in the form of means or other statistics) of an industrial process through time. By carefully scrutinizing the chart, a quality control engineer can identify any potential problems with the production process. The idea is that when a process is in control, the variable being measured—the mean of every four observations, for example—should remain stable through time. The mean should stay somewhere around the middle line (the grand mean for the process) and not wander off "too much." By now you understand what "too much" means in statistics: more than several standard deviations of the process. The required number of standard deviations is chosen so that there will be a small probability of exceeding them when the process is in control. Addition and subtraction of the required number of standard deviations (generally three) give us the **upper control limit** (UCL) and the **lower control limit** (LCL) of the control chart. The UCL and LCL are similar to the "tolerance" limits in the story of the Pyx. When the bounds are breached, the process is deemed **out of control** and must be corrected. A control chart is illustrated in Figure 13–1. We assume throughout that the variable being charted is at least approximately normally distributed.

In addition to looking for the process exceeding the bounds, quality control workers look for patterns and trends in the charted variable. For example, if the mean of four observations at a time keeps increasing or decreasing, or it stays too long above or below the centerline (even if the UCL and LCL are not breached), the process may be out of control.

> A **control chart** is a time plot of a statistic, such as a sample mean, range, standard deviation, or proportion, with a centerline and *upper and lower control limits*. The limits give the desired range of values for the statistic. When the statistic is outside the bounds, or when its time plot reveals certain patterns, the process may be out of control.

Central to the idea of a control chart—and, in general, to the use of statistics in quality control—is the concept of **variance**. If we were to summarize the entire field of statistical quality control (also called *statistical process control,* or SPC) in one word, that word would have to be *variance*. Shewhart, Deming, and others wanted to bring

FIGURE 13–1   A Control Chart

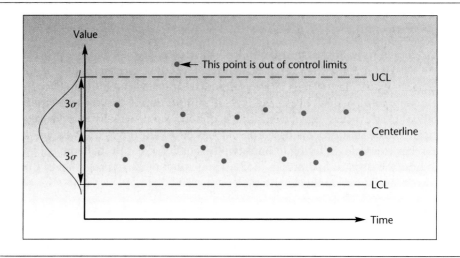

the statistical concept of variance down to the shop floor. If supervisors and production line workers could understand the existence of variance in the production process, then this awareness by itself could be used to help minimize the variance. Furthermore, the variance in the production process could be partitioned into two kinds: the natural, random variation of the process and variation due to assignable causes. Examples of assignable causes are fatigue of workers and breakdown of components. Variation due to assignable causes is especially undesirable because it is due to something's being wrong with the production process, and may result in low quality of the produced items. Looking at the chart helps us detect an assignable cause, by asking what has happened at a particular time on the chart where the process looks unusual.

> A process is considered in **statistical control** when it has no assignable causes, only natural variation.

Figure 13–2 shows how a process could be in control or out of control. Recall the assumption of a normal distribution—this is what is meant by the normal curves shown on the graphs. These curves stand for the hypothetical populations from which our data are assumed to have been randomly drawn.

Actually, any kind of variance is undesirable in a production process. Even the natural variance of a process due to purely random causes rather than to assignable causes can be detrimental. The control chart, however, will detect only assignable causes. As the following story shows, one could do very well by removing all variance.

An American car manufacturer was having problems with transmissions made at one of its domestic plants, and warranty costs were enormous. The identical type of transmission made in Japan was not causing any problems at all. Engineers carefully examined 12 transmissions made at the company's American plant. They found that variations existed among the 12 transmissions, but there were no assignable causes and a control chart revealed nothing unusual. All transmissions were well within specifications. Then they looked at 12 transmissions made at the Japanese plant. The engineer who made the measurements reported

FIGURE 13–2   A Production Process in, and out of, Statistical Control

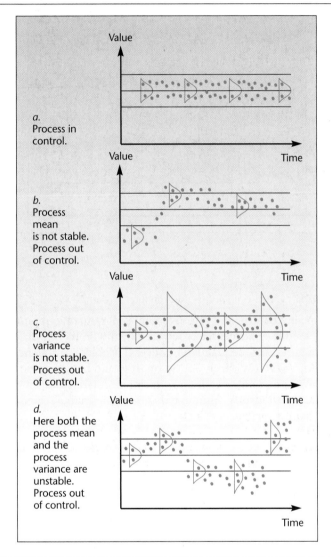

a.
Process in
control.

b.
Process
mean
is not stable.
Process out
of control.

c.
Process
variance
is not stable.
Process out
of control.

d.
Here both the
process mean
and the
process
variance are
unstable.
Process out
of control.

that the measuring equipment was broken: in testing one transmission after the other, the needle did not move at all. A closer investigation revealed that the measuring equipment was perfectly fine: the transmissions simply had *no variation*. They did not just satisfy specifications; for all practical purposes, the 12 transmissions were identical![3]

Such perfection may be difficult to achieve, but the use of control charts can go a long way toward improving quality. Control charts are the main topic of this chapter, and we will discuss them in later sections. We devote the remainder of this section to brief descriptions of other quality control tools.

---

[3]From the article "Ed Deming wants big changes and he wants them fast," by L. Dobyns. Originally appeared in *Smithsonian*, August 1990, pp. 74–83. Reprinted by permission. All rights reserved.

FIGURE 13–3   **Pareto Diagram for Ceramics Example**

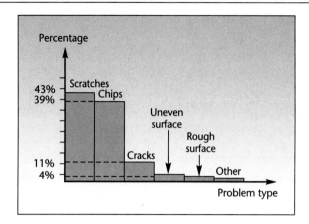

Pareto Diagrams

In instituting a quality control and improvement program, one important question to answer is: What are the exact causes of lowered quality in the production process? A ceramics manufacturer may be plagued by several problems: scratches, chips, cracks, surface roughness, uneven surfaces, and so on. It would be very desirable to find out which of these problems were serious and which not. A good and simple tool for such analysis is the **Pareto diagram.** Although the diagram is named after an Italian economist, its use in quality control is due to J. M. Juran.

> A **Pareto diagram** is a bar chart of the various problems in production and their percentages, which must add to 100%.

A Pareto diagram for the ceramics example above is given in Figure 13–3. As can be seen from the figure, scratches and chips are serious problems, accounting for most of the nonconforming items produced. Cracks occur less frequently, and the other problems are relatively rare. A Pareto diagram thus helps management to identify the most significant problems and concentrate on their solution rather than waste time and resources on unimportant causes.

The Template

Figure 13–4 shows the template that can be used to draw Pareto diagrams. After entering the data in columns B and C, make sure that the frequencies are in descending order. To sort them in descending order, select the whole range of data, B4:C23, choose the Sort command under the Data menu and click the OK button.

The chart has a cumulative relative frequency line plotted on top of the histogram. This line helps to calculate the cumulative percentage of cases that are covered by a set of defects. For example, we see from the cumulative percentage line that approximately 90% of the cases are covered by the first three types: scratches, chips, and cracks. Thus, a quality control manager can hope to remedy 90% of the problems by concentrating on and eliminating the first three types.

FIGURE 13–4 **The Template for Pareto Diagrams**
[Pareto.xls]

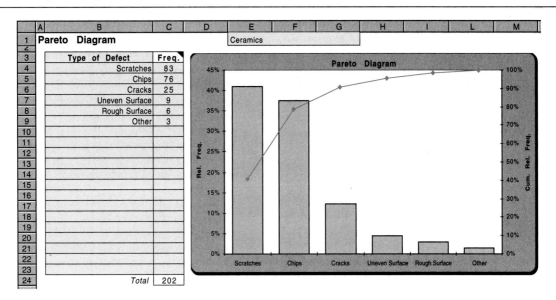

Acceptance Sampling

Finished products are grouped in lots before being shipped to customers. The lots are numbered, and random samples from these lots are inspected for quality. Such checks are made both before lots are shipped out and when lots arrive at their destination. The random samples are measured to find out which and how many items do not meet specifications.

A lot is rejected whenever the sample mean exceeds or falls below some pre-specified limit. For attribute data, the lot is rejected when the number of defective or nonconforming items in the sample exceeds a prespecified limit. Acceptance sampling does not, by itself, improve quality; it simply removes bad lots. To improve quality, it is necessary to control the production process itself, removing any assignable causes and striving to reduce the variation in the process.

Analysis of Variance and Experimental Design

As statistics in general is an important collection of tools to improve quality, so in particular is experimental design. Industrial experiments are performed to find production methods that can bring about high quality. Experiments are designed to identify the factors that affect the variable of interest, for example, the diameter of a rod. We may find that method B produces rods with diameters that conform to specifications more often than those produced by method A or C. Analysis of variance (as well as regression and other techniques) is used in making such a determination. These tools are more "active" in the quest for improved quality than the control charts, which are merely diagnostic and look at a process already in place. However, both types of tool should be used in a comprehensive quality improvement plan.

Taguchi Methods

The Japanese engineer Genichi Taguchi developed new notions about quality engineering. Taguchi's ideas transcend the customary wisdom of tolerance limits, where we implicitly assume that any value for a parameter within the specified range is as good as any other value. Taguchi aims at the ideal *optimal* value for a parameter in question. For example, if we look at a complete manufactured product, such as a car, the car's quality may not be good even if all its components are within desired levels when considered alone. The idea is that the quality of a large system deteriorates as we add the small variations in quality for all its separate components.

To try to solve this problem, Taguchi developed the idea of a total loss to society due to the lowered quality of any given item. That loss to society is to be minimized. That is, we want to *minimize* the variations in product quality, not simply keep them within limits. This is done by introducing a *loss function* associated with the parameter in question (e.g. , rod diameter) and by trying to create production systems that minimize this loss both for components and for finished products.

In the following sections, we describe Shewhart's control charts in detail, since they are the main tool currently used for maintaining and improving quality. Information on the other methods we mentioned in this section can be found in the ever-increasing literature on quality improvement. (For example, see the appropriate references in Appendix A at the end of this book.) Our discussion of control charts will roughly follow the order of their frequency of use in industry today. The charts we will discuss are the $\bar{x}$ chart, the R chart, the s chart, the p chart, the c chart, and the x chart.

PROBLEMS

13–1. Discuss what is meant by *quality control* and *quality improvement.*

13–2. What is the main statistical idea behind current methods of quality control?

13–3. Describe the two forms of variation in production systems and how they affect quality.

13–4. What is a quality control chart, and how is it used?

13–5. What are the components of a quality control chart?

13–6. What are the limitations of quality control charts?

13–7. What is acceptance sampling?

13–8. Describe how one would use experimental design in an effort to improve industrial quality.

13–9. The errors in the inventory records of a company were analyzed for their causes. It was found that there were 164 cases of omissions, 103 cases of wrong quantity entered, 45 cases of wrong part numbers entered, 24 cases of wrong dates entered, and 8 cases of withdrawal versus deposit mixup. Draw a Pareto diagram for the different causes.

> *a.* What percentage of the cases is covered by the top two causes?
>
> *b.* If at least 90% of the cases are to be covered, which causes must be addressed?

13–10. Out of 1,000 automobile engines tested for quality, 62 had cracked blocks, 17 had leaky radiators, 106 had oil leaks, 29 had faulty cylinders, and 10 had ignition problems. Draw a Pareto diagram for these data, and identify the key problems in this particular production process.

13–11. In an effort to improve quality, AT&T has been trying to control pollution problems.[4] Problem causes and their relative seriousness, as a percentage of the total, are as follows: chlorofluorocarbons, 61%; air toxins, 30%; manufacturing wastes, 8%; other, 1%. Draw a Pareto diagram of these causes.

13–12. Discuss the quality implications of the data displayed in the following figure.

A Long, Hard Battle to Squash the Bugs

A study of 80 software projects of NASA's Goddard Space Center shows that reliability of the software used for ground-based tracking of space missions is slowly improving. And the gap between best and worst is narrowing, making software performance more predictable as well as more reliable.

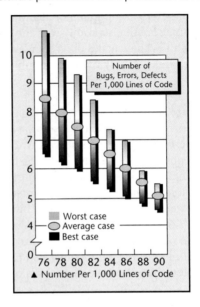

Source: "Turning Software from a Black Art into a Science," *Business Week,* Quality Issue, 1991, pp. 80–81, with data from Univ. of Maryland Software Engineering Lab. Reprinted by permission.

13–4 The $\bar{x}$ Chart

We want to compute the centerline and the upper and lower control limits for a process believed to be in control. Then future observations can be checked against these bounds to make sure the process remains in control. To do this, we first conduct an *initial run.* We determine trial control limits to test for control of past data, and then we remove out-of-control observations and recompute the control limits. We apply these improved control limits to future data. This is the philosophy behind all control charts discussed in this chapter. Although we present the $\bar{x}$ chart first, in an actual quality control program we would first want to test that the process variation is under control. This is done by using the R (range) or the s (standard deviation) chart. Unless the process variability is under statistical control, there is no stable distribution of values with a fixed mean.

An **$\bar{x}$ chart** can help us to detect shifts in the process mean. One reason for a control chart for the process mean (rather than for a single observation) has to do with the central limit theorem. We want to be able to use the known properties of the normal

---

[4]"AT&T's Clean Goals," *Business Week,* Quality Issue, 1991, p. 49.

curve in designing the control limits. By the central limit theorem, the distribution of the sample mean tends toward a normal distribution as the sample size increases. Thus, when we aggregate data from a process, the aggregated statistic, or sample mean, becomes closer to a normal random variable than the original, unaggregated quantity. Typically, a set number of observations will be aggregated and averaged. For example, a set of four measurements of rod diameter will be made every hour of production. The four rods will be chosen randomly from all rods made during that hour. If the distribution of rod diameters is roughly mound-shaped, then the sample means of the groups of four diameters will have a distribution closer to normal.

The mean of the random variable $\overline{X}$ is the population mean μ, and the standard deviation of $\overline{X}$ is $\sigma/\sqrt{4}$, where σ is the population standard deviation. We know all this from the theory in Chapter 5. We also know from the theory that the probability that a normal random variable will exceed 3 of its standard deviations on either side of the mean is 0.0026 (check this by using the normal table). Thus, the interval

$$\mu \pm 3\sigma/\sqrt{n} \tag{13-1}$$

should contain about 99.74% of the sample means. This is, in fact, the logic of the control chart for the process mean. The idea is the same as that of a hypothesis test (conducted in a form similar to a confidence interval). We try to select the bounds so that they will be as close as possible to equation 13–1. We then chart the bounds, with an estimate of μ in the center (the centerline) and the upper and lower bounds (UCL and LCL) as close as possible to the bounds of the interval specified by equation (13–1). Out of 1,000 $\bar{x}$'s, fewer than 3 are expected to be out of bounds. Therefore, with a limited number of $\bar{x}$'s on the control chart, observing even one of them out of bounds is cause to reject the null hypothesis that the process is in control, in favor of the alternative that it is out of control. (One could also compute a p-value here, although it is more complicated since we have several $\bar{x}$'s on the chart, and in general this is not done.)

We note that the assumption of random sampling is important here as well. If somehow the process is such that successively produced items have values that are correlated—thus violating the independence assumption of random sampling—the interpretation of the chart may be misleading. Various new techniques have been devised to solve this problem.

To construct the control chart for the sample mean, we need estimates of the parameters in equation 13–1. The grand mean of the process, that is, the mean of all the sample means (the mean of all the observations of the process), is our estimate of μ. This is our centerline. To estimate σ, we use s, the standard deviation of all the process observations. However, this estimate is good only for large samples, $n > 10$. For smaller sample sizes we use an alternative procedure. When sample sizes are small, we use the *range* of the values in each sample used to compute an $\bar{x}$. Then we average these ranges, giving us a mean range $\overline{R}$. When the mean range $\overline{R}$ is multiplied by a constant, which we call A_2, the result is a good estimate for 3σ. Values of A_2 for all sample sizes up to 25 are found in Appendix C, Table 13, at the end of the book. The table also contains the values for all other constants required for the quality control charts discussed in this chapter.

The box on page 649 shows how we compute the centerline and the upper and lower control limits when constructing a control chart for the process mean.

In addition to a sample mean being outside the bounds given by the UCL and LCL, other occurrences on the chart may lead us to conclude that there is evidence that the process is out of control. Several such sets of rules have been developed, and

TABLE 13–1 **Tests for Assignable Causes**

Test 1:   One point beyond 3σ ($3s$)
Test 2:   Nine points in a row on one side of the centerline
Test 3:   Six points in a row steadily increasing or decreasing
Test 4:   Fourteen points in a row alternating up and down
Test 5:   Two out of three points in a row beyond 2σ ($2s$)
Test 6:   Four out of five points in a row beyond 1σ ($1s$)
Test 7:   Fifteen points in a row within 1σ ($1s$) of the centerline
Test 8:   Eight points in a row on both sides of the centerline, all beyond 1σ ($1s$)

the idea behind them is that they represent occurrences that have a very low probability when the process is indeed in control. The set of rules we use is given in Table 13–1.[5]

Elements of a control chart for the process mean:

Centerline:
$$\bar{\bar{x}} = \frac{\sum_{i=1}^{k} \bar{x}_i}{k}$$

UCL:
$$\bar{\bar{x}} + A_2\bar{R} \qquad \bar{R} = \frac{\sum_{i=1}^{k} R_i}{k}$$

LCL:
$$\bar{\bar{x}} - A_2\bar{R}$$

where k = number of samples, each of size n
$\bar{x}_i$ = sample mean for ith sample
R_i = range of the ith sample

If the sample size in each group is over 10, then

$$UCL = \bar{\bar{x}} + 3\frac{\bar{s}/c_4}{\sqrt{n}} \qquad LCL = \bar{\bar{x}} - 3\frac{\bar{s}/c_4}{\sqrt{n}}$$

where $\bar{s}$ is the average of the standard deviations of all groups and c_4 is a constant found in Appendix C, Table 13.

The Template

The template for drawing X-bar charts is shown in Figure 13–5. Its use is illustrated through Example 13–1.

A pharmaceutical manufacturer needs to control the concentration of the active ingredient in a formula used to restore hair to bald people. The concentration should be around 10%, and a control chart is desired to check the sample means of 30 observations, aggregated in groups of 3. The template containing the data, as well as the

EXAMPLE 13–1

---

[5]This particular set of rules was provided courtesy of Dr. Lloyd S. Nelson of Nashua Corporation, one of the pioneers in the area of quality control. See L. S. Nelson, "The Shewhart Control Chart—Tests for Special Causes," *Journal of Quality Technology* Issue 16 (1984), pp. 237–39. The MINITAB package tests for special causes using Nelson's criteria. © 1984 American Society for Quality. Reprinted by permission.

FIGURE 13–5 **The Template for *X*-bar Chart [Control Charts.xls; Sheet: X-bar R s Charts]**

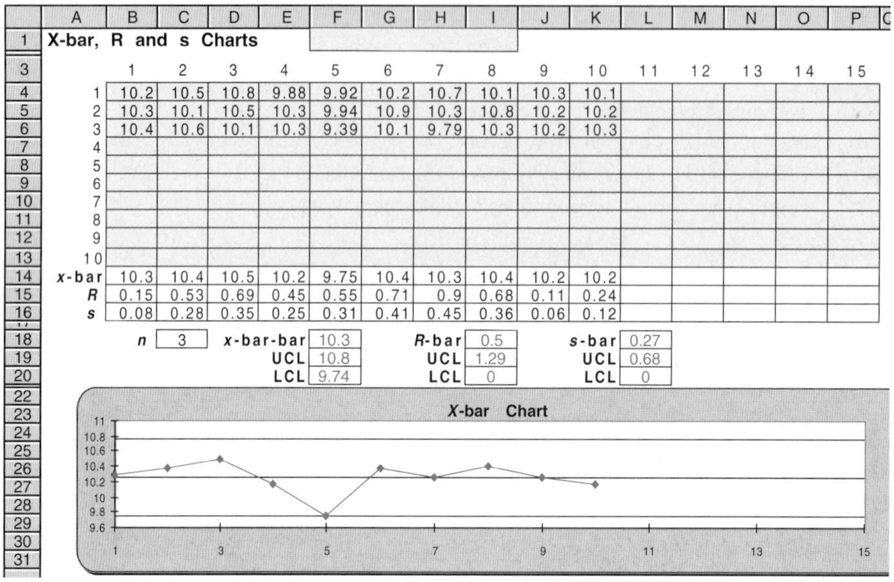

	1	2	3	4	5	6	7	8	9	10	11	12	13	14	15
1	10.2	10.5	10.8	9.88	9.92	10.2	10.7	10.1	10.3	10.1					
2	10.3	10.1	10.5	10.3	9.94	10.9	10.3	10.8	10.2	10.2					
3	10.4	10.6	10.1	10.3	9.39	10.1	9.79	10.3	10.2	10.3					
4															
5															
6															
7															
8															
9															
10															
x-bar	10.3	10.4	10.5	10.2	9.75	10.4	10.3	10.4	10.2	10.2					
R	0.15	0.53	0.69	0.45	0.55	0.71	0.9	0.68	0.11	0.24					
s	0.08	0.28	0.35	0.25	0.31	0.41	0.45	0.36	0.06	0.12					

n 3	*x*-bar-bar	10.3	*R*-bar	0.5	*s*-bar 0.27
	UCL	10.8	UCL	1.29	UCL 0.68
	LCL	9.74	LCL	0	LCL 0

control chart it produced, are given in Figure 13–5. As can be seen from the control chart, there is no evidence here that the process is out of control.

The grand mean is $\bar{\bar{x}} = 10.253$. The ranges of the groups of three observations each are 0.15, 0.53, 0.69, 0.45, 0.55, 0.71, 0.90, 0.68, 0.11, and 0.24. Thus, $\bar{R} = 0.501$. From Table 13 we find for $n = 3$, $A_2 = 1.023$. Thus, UCL $= 10.253 + 1.023(0.501) = 10.766$, and LCL $= 10.253 - 1.023(0.501) = 9.74$. Note that the $\bar{x}$ chart cannot be interpreted unless the R or s chart has been examined and is in control. These two charts are presented in the next section.

PROBLEMS

13–13. What is the logic behind the control chart for the sample mean, and how is the chart constructed?

13–14. Boston-based Legal Seafoods prides itself on having instituted an advanced quality control system that includes the control of both food quality and service quality. The following are successive service times at one of the chain's restaurants on a Saturday night in May (time is stated in minutes from customer entry to appearance of waitperson):

5, 6, 5, 5.5, 7, 4, 12, 4.5, 2, 5, 5.5, 6, 6, 13, 2, 5, 4, 4.5, 6.5, 4, 1,
2, 3, 5.5, 4, 4, 8, 12, 3, 4.5, 6.5, 6, 7, 10, 6, 6.5, 5, 3, 6.5, 7

Aggregate the data into groups of four, and construct a control chart for the process mean. Is the waiting time at the restaurant under control?

13–15. What assumptions are necessary for constructing an $\bar{x}$ chart?

13–16. The manufacturer of jet engines needs to control the maximum power delivered by engines. The following are readings related to power for successive engines produced:

121, 122, 121, 125, 123, 121, 129, 123, 122, 122, 120, 121, 119, 118, 121,
125, 139, 150, 121, 122, 120, 123, 127, 123, 128, 129, 122, 120, 128, 120

Aggregate the data in groups of 3, and create a control chart for the process mean. Use the chart to test the assumption that the production process is under control.

13–17. The following data are tensile strengths, in pounds, for a sample of string for industrial use made at a plant. Construct a control chart for the mean, using groups of 5 observations each. Test for statistical control of the process mean.

5, 6, 4, 6, 5, 7, 7, 7, 6, 5, 3, 5, 5, 5, 6, 5, 5, 6, 7, 7, 7, 7, 6, 7, 5,
5, 5, 6, 7, 7, 7, 7, 7, 5, 5, 6, 4, 6, 6, 6, 7, 6, 6, 6, 6, 6, 7, 5, 7, 6

13–5 The *R* Chart and the *s* Chart

In addition to the process mean, we want to control the process variance. When the variation in the production process is high, it means that produced items will have a wider range of values, and this jeopardizes the product's quality. Recall also that in general we want as small a variance as possible. As noted earlier, it is advisable first to check the process variance and then to check its mean. Two charts are commonly used to achieve this aim. The more common of the two is a control chart for the process range, called the **R chart.** The other is a control chart for the process standard deviation, the **s chart.** A third chart is a chart for the actual variance, called the s^2 chart, but we will not discuss it since it is the least frequently used of the three.

The **R** Chart

Like the $\bar{x}$ chart, the R chart contains a centerline and upper and lower control limits. One would expect the limits to be of the form

$$\bar{R} \pm 3\sigma_{\bar{R}} \qquad (13\text{–}2)$$

But the distribution of R is not normal and hence the limits need not be symmetric. Additionally, the lower limit cannot go below zero and is therefore bounded by zero. With these considerations in mind, the limits are calculated using the formulas in the box below, where the constants D_3 and D_4 are obtained from Table 13 of Appendix C. Notice that the constant D_3 is bounded below at zero for small samples.

The elements of an R chart:	
Centerline:	$\bar{R}$
LCL:	$D_3\bar{R}$
UCL:	$D_4\bar{R}$
where $\bar{R}$ is the sum of group ranges, divided by the number of groups.	

Returning to Example 13–1, we find that $\bar{R} = 0.501$, and from Table 13, $D_3 = 0$ and $D_4 = 2.574$. Thus, the centerline is 0.501, the lower control limit is 0, and the

FIGURE 13–6   **The *R* Chart for Example 13–1**
[Control Charts.xls; Sheet: X-bar R s charts

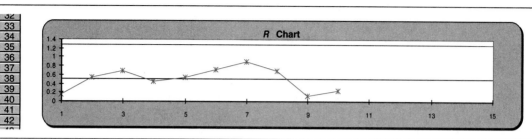

upper control limit is $(0.501)(2.574) = 1.29$. Figure 13–6 gives the control chart for the process range for this example.

The test for control in the case of the process range is just to look for at least one observation outside the bounds. Based on the *R* chart for Example 13–1, we conclude that the process range seems to be in control.

The *s* Chart

The *R* chart is in common use because it is easier (by hand) to compute ranges than to compute standard deviations. Today (as compared with the 1920s, when these charts were invented), computers are usually used to create control charts, and an *s* chart should be at least as good as an *R* chart. We note, however, that the standard deviation suffers from the same nonnormality (skewness) as does the range. Again, symmetric bounds as suggested by equation 13–2, with *s* replacing *R*, are still used. The control chart for the process standard deviation is similar to that for the range. Here we use constants B_3 and B_4, also found in Appendix C, Table 13. The bounds and the centerline are given in the following box.

Elements of the *s* chart:

Centerline:	$\bar{s}$
LCL:	$B_3\bar{s}$
UCL:	$B_4\bar{s}$

where $\bar{s}$ is the sum of group standard deviations, divided by the number of groups.

The *s* chart for Example 13–1 is given in Figure 13–7.

Again, we note that the process standard deviation seems to be in control. Since *s* charts are done by computer, we will not carry out the computations of the standard deviations of all the groups.

EXAMPLE 13–2   The nation's largest retailer wants to make sure that supplier delivery times are under control. Consistent supply times are critical factors in forecasting. The actual number of days it took each of the 30 suppliers to deliver goods last month is grouped into 10 sets in Figure 13–8.

FIGURE 13–7   **The s Chart for Example 13–1**
[Control Charts.xls; Sheet: X-bar R s charts]

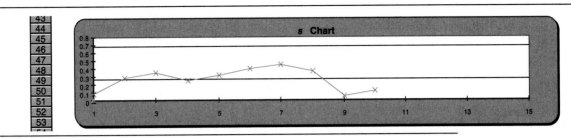

FIGURE 13–8   **The Charts for Example 13–2**
[Control Charts.xls; Sheet: X-bar R s charts]

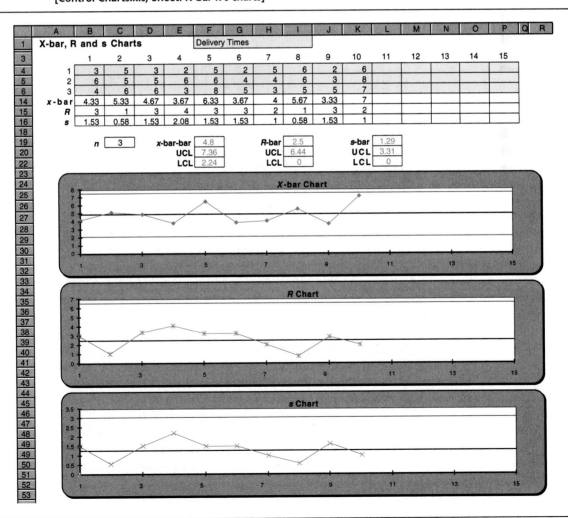

PROBLEMS

13–18. Why do we need a control chart for the process range?

13–19. Compare and contrast the control charts for the process range and the process standard deviation.

13–20. What are the limitations of symmetric LCL and UCL? Under what conditions are symmetric bounds impossible in practice?

13–21. Create R and s charts for Problem 13–14. Is the process in control?

13–22. Create R and s charts for Problem 13–16. Is the process in control?

13–23. Create R and s charts for Problem 13–17. Is the process in control?

13–24. Create X-bar, R, and s charts for the following data on the diameter of pins produced by an automatic lathe. The data summarize 12 samples of size 5 each, obtained at random on 12 different days. Is the process in control? If it is not, remove the sample that is out of control and redraw the charts.

1	2	3	4	5	6	7	8	9	10	11	12
1.300	1.223	1.310	1.221	1.244	1.253	1.269	1.325	1.306	1.255	1.221	1.268
1.207	1.232	1.290	1.218	1.206	1.289	1.318	1.285	1.288	1.260	1.256	1.208
1.287	1.289	1.255	1.200	1.294	1.279	1.270	1.301	1.243	1.296	1.245	1.207
1.237	1.310	1.317	1.273	1.302	1.303	1.224	1.315	1.288	1.270	1.239	1.218
1.258	1.228	1.260	1.219	1.269	1.229	1.224	1.224	1.238	1.307	1.265	1.238

13–25. The capacity of a certain fuel tank is designed to be 12.625 gallons. The actual capacity of tanks produced is controlled using a control chart. The data of 9 random samples of size 5 each collected on 9 different days are tabulated below. Draw X-bar, R, and s charts. Is the process in control? If it is not, remove the sample that is out of control, and redraw the charts.

1	2	3	4	5	6	7	8	9
12.667	12.600	12.599	12.607	12.738	12.557	12.646	12.710	12.529
12.598	12.711	12.583	12.524	12.605	12.745	12.647	12.627	12.725
12.685	12.653	12.515	12.718	12.640	12.626	12.651	12.605	12.306
12.700	12.703	12.653	12.615	12.653	12.694	12.607	12.648	12.551
12.722	12.579	12.599	12.554	12.507	12.574	12.589	12.545	12.600

13–26. The amount of cola filled in 2-liter bottles by an automatic bottling machine is monitored using a control chart. The actual contents of random samples of 4 bottles collected on 8 different days are tabulated below. Draw X-bar, R, and s charts. Is the process in control? If it is not, remove the sample that is out of control, and redraw the charts.

1	2	3	4	5	6	7	8
2.015	2.006	1.999	1.983	2.000	1.999	2.011	1.983
2.012	1.983	1.988	2.008	2.016	1.982	1.983	1.991
2.001	1.996	2.012	1.999	2.016	1.997	1.983	1.989
2.019	2.003	2.015	1.999	1.988	2.005	2.000	1.998
2.018	1.981	2.004	2.005	1.986	2.017	2.006	1.990

13–6 The *p* Chart

The example of a quality control problem used most frequently throughout the book has been that of controlling the proportion of defective items in a production process. This, indeed, is the topic of this section. Here we approach the problem by using a control chart.

The number of defective items in a random sample chosen from a population has a binomial distribution: the number of successes x out of a number of trials n with a constant probability of success p in each trial. The parameter p is the proportion of defective items in the population. If the sample size n is fixed in repeated samplings, then the sample proportion $\hat{P}$ derives its distribution from a binomial distribution. Recall that the binomial distribution is symmetric when $p = 0.5$, and it is skewed for other values of p. By the central limit theorem, as n increases, the distribution of $\hat{P}$ approaches a normal distribution. Thus, a normal approximation to the binomial should work well with large sample sizes; a relatively small sample size would suffice if $p = 0.5$ because of the symmetry of the binomial in this case.

The chart for p is called the **p chart.** Using the normal approximation, we want bounds of the form

$$\hat{p} \pm 3\sigma_{\hat{p}} \qquad (13–3)$$

The idea is, again, that the probability of a sample proportion falling outside the bounds is small when the process is under control. When the process is not under control, the proportion of defective or nonconforming items will tend to exceed the upper bound of the control chart. The lower bound is sometimes zero, which happens when $\hat{p}$ is sufficiently small. Being at the lower bound of zero defectives is, of course, a very good occurrence.

Recall that the sample proportion $\hat{p}$ is given by the number of defectives x, divided by the sample size n. We estimate the population proportion p by the total number of defectives in all the samples of size n we have obtained, divided by the entire sample size (all the items in all our samples). This is denoted by $\bar{p}$, and it serves as the centerline of the chart. Also recall that the standard deviation of this statistic is given by

$$\sqrt{\frac{\bar{p}(1 - \bar{p})}{n}}$$

Thus, the control chart for the proportion of defective items is given in the following box. The process is believed to be out of control when at least one sample proportion falls outside the bounds.

The elements of a control chart for the process proportion:

Centerline: $\bar{p}$

LCL: $\bar{p} - 3\sqrt{\dfrac{\bar{p}(1 - \bar{p})}{n}}$

UCL: $\bar{p} + 3\sqrt{\dfrac{\bar{p}(1 - \bar{p})}{n}}$

where n is the number of items in each sample and $\bar{p}$ is the proportion of defectives in the combined, overall sample.

FIGURE 13–9   **The Template for *p* Charts**
              [Control Charts.xls; Sheet: p Chart]

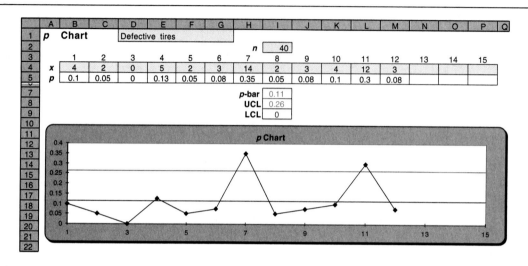

The Template

The template for drawing *p* charts can be seen in Figure 13–9. Example 13–3 demonstrates use of the template.

EXAMPLE 13–3   A tire manufacturer randomly samples 40 tires at the end of each shift to test for tires that are defective. The number of defectives in 12 shifts is as follows: 4, 2, 0, 5, 2, 3, 14, 2, 3, 4, 12, 3. Construct a control chart for this process. Is the production process under control?

Solution   The results are shown in Figure 13–9. Our estimate of *p*, the centerline, is the sum of all the defective tires, divided by 40 × 12. It is $\bar{p} = 0.1125$. Our standard error of $\bar{p}$ is $\sqrt{\bar{p}(1 - \bar{p})/n} = 0.05$; thus, LCL = $0.1125 - 3(0.05) = -0.0375$, which means that the LCL should be 0. Similarly, UCL = $0.1125 + 3(0.05) = 0.2625$.

As we can see from the figure, two sample proportions are outside the UCL. These correspond to the samples with 14 and 12 defective tires, respectively. There is ample evidence that the production process is out of control.

PROBLEMS

13–27.   The manufacturer of steel rods looks at random samples of 20 items from each production shift and notes the number of nonconforming rods in these samples. The results of 10 shifts are 8, 7, 8, 9, 6, 7, 8, 6, 6, 8. Is there evidence that the process is out of control? Explain.

13–28.   A battery manufacturer looks at samples of 30 batteries at the end of every day of production and notes the number of defective batteries. Results are 1, 1, 0, 0, 1, 2, 0, 1, 0, 0, 2, 5, 0, 1. Is the production process under control?

13–29.   BASF Inc. makes 3.5-in two-sided double-density disks for use in micro-computers. A quality control engineer at the plant tests batches of 50 disks at a time and plots the proportions of defective disks on a control chart. The first 10 batches used to create the chart had the following numbers of defective disks: 8, 7, 6, 7, 8, 4, 3, 5, 5, 8. Construct the chart and interpret the results.

13–30.   If the proportion of defective items in a production process is very small, and few items are tested in each batch, what problems do you foresee? Explain.

13–7   The c Chart

It often happens in production activities that we want to control the *number of defects or imperfections per item*. When fabric is woven, for example, it is of interest to keep a record of the number of blemishes per yard and to take corrective action when this number is out of control.

Recall from Chapter 3 that the random variable representing the count of the number of errors occurring in a fixed time or space is often modeled using the *Poisson distribution*. This is the model we use here. For the Poisson distribution, we know that the mean and the variance are both equal to the same parameter. Here we call that parameter c, and our chart for the number of defects per item (or yard, etc.) is the **c chart**. In this chart we plot a random variable, the number of defects per item. We estimate c by $\bar{c}$, which is the average number of defects per item, the total number averaged over all the items we have. The standard deviation of the random variable is thus the square root of c. Now, the Poisson distribution can be approximated by the normal distribution for large sample sizes, and this again suggests the form

$$\bar{c} \pm 3\sqrt{\bar{c}} \qquad\qquad (13\text{–}4)$$

Equation 13–4 leads to the control bounds and centerline given in the box that follows.

Elements of the c chart:

Centerline: $\bar{c}$

LCL: $\bar{c} - 3\sqrt{\bar{c}}$

UCL: $\bar{c} + 3\sqrt{\bar{c}}$

where $\bar{c}$ is the average number of defects or imperfections per item (or area, volume, etc.).

The Template

The template that produces c charts can be seen in Figure 13–10. We shall see the use of the template through Example 13–4.

FIGURE 13–10   The Template for *c* Charts
[Control Charts.xls; Sheet: c Chart]

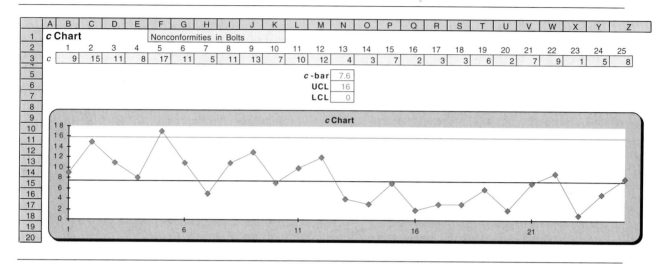

EXAMPLE 13–4   The following data are the numbers of nonconformities in bolts for use in cars made by the Ford Motor Company:[6] 9, 15, 11, 8, 17, 11, 5, 11, 13, 7, 10, 12, 4, 3, 7, 2, 3, 3, 6, 2, 7, 9, 1, 5, 8. Is there evidence that the process is out of control?

Solution   We need to find the mean number of nonconformities per item. This is the sum of the numbers, divided by 25, or 7.56. The standard deviation of the statistic is the square root of this number, or 2.75, and the control limits are obtained as shown in the box. Figure 13–10 gives the template solution.

From the figure we see that one observation is outside the upper control limit, indicating that the production process may be out of control. We also note a general downward trend, which should be investigated (maybe the process is improving).

PROBLEMS

13–31.   The following are the numbers of imperfections per yard of yarn: 5, 3, 4, 8, 2, 3, 1, 2, 5, 9, 2, 2, 2, 3, 4, 2, 1. Is there evidence that the process is out of control?

13–32.   The following are the numbers of blemishes in the coat of paint of new automobiles: 12, 25, 13, 20, 5, 22, 8, 17, 31, 40, 9, 62, 14, 16, 9, 28. Is there evidence that the painting process is out of control?

13–33.   The following are the numbers of imperfections in rolls of wallpaper: 5, 6, 3, 4, 5, 2, 7, 4, 5, 3, 5, 5, 3, 2, 0, 5, 5, 6, 7, 6, 9, 3, 3, 4, 2, 6. Construct a *c* chart for the process, and determine whether there is evidence that the process is out of control.

13–34.   What are the assumptions underlying the use of the *c* chart?

---

[6]From T. P. Ryan, *Statistical Methods for Quality Improvement* (New York: Wiley, 1989), p. 198.

13–8   The x Chart

Sometimes we are interested in controlling the process mean, but our observations come so slowly from the production process that we cannot aggregate them into groups. In such a case, and in other situations as well, we may consider an **x chart.** An x chart is a chart for the raw values of the variable in question.

As you may guess, the chart is effective if the variable in question has a distribution that is close to normal. We want to have the bounds as mean of the process ± 3 standard deviations of the process. The mean is estimated by $\bar{x}$, and the standard deviation is estimated by s/c_4.

The tests for special causes in Table 13–1 can be used in conjunction with an x chart as well. Case 15 at the end of this chapter will give you an opportunity to study an actual x chart using all these tests.

13–9   Summary and Review of Terms

Quality control and improvement is a fast-growing, important area of application of statistics in both production and services. We discussed **Pareto diagrams,** which are relatively simple graphical ways of looking at problems in production. We discussed quality control in general and how it relates to statistical theory and hypothesis testing. We mentioned **process capability** and **Deming's 14 points.** Then we described **control charts,** graphical methods of determining when there is evidence that a process is out of statistical control. The control chart has a **centerline,** an **upper control limit,** and a **lower control limit.** The process is believed to be out of control when one of the limits is breached at least once. The control charts we discussed were the $\bar{x}$ **chart,** for the mean; the **R chart,** for the range; the **s chart,** for the standard deviation; the **p chart,** for the proportion; the **c chart,** for the number of defects per item; and the **x chart,** a chart of individual observations for controlling the process mean.

ADDITIONAL PROBLEMS

13–35.   Discuss and compare the various control charts discussed in this chapter.

13–36.   The number of blemishes in rolls of tape coming out of a production process is as follows: 17, 12, 13, 18, 12, 13, 14, 11, 18, 29, 13, 13, 15, 16. Is there evidence that the production process is out of control?

13–37.   The number of defective items out of random samples of 100 windshield wipers selected at the end of each production shift at a factory is as follows: 4, 4, 5, 4, 4, 6, 6, 3, 3, 3, 3, 2, 2, 4, 5, 3, 4, 6, 4, 12, 2, 2, 0, 1, 1, 1, 2, 3, 1. Is there evidence that the production process is out of control?

13–38.   Weights of pieces of tile (in ounces) are as follows: 2.5, 2.66, 2.8, 2.3, 2.5, 2.33, 2.41, 2.88, 2.54, 2.11, 2.26, 2.3, 2.41, 2.44, 2.17, 2.52, 2.55, 2.38, 2.89, 2.9, 2.11, 2.12, 2.13, 2.16. Create an R chart for these data, using subgroups of size 4. Is the process variation under control?

13–39.   Use the data in problem 13–38 to create an $\bar{x}$ chart to test whether the process mean is under control.

13–40.   Create an s chart for the data in problem 13–38.

CASE 15 — Quality Control and Improvement at Nashua Corporation*

n 1979, Nashua Corporation, with an increasing awareness of the importance of always maintaining and improving quality, invited Dr. W. Edwards Deming for a visit and a consultation. Dr. Deming, then almost 80 years old, was the most sought-after quality guru in the United States.

Following many suggestions by Deming, Nashua hired Dr. Lloyd S. Nelson the following year as director of statistical methods. The idea was to teach everyone at the company about quality and how it can be maintained and improved by using statistics.

Dr. Nelson instituted various courses and workshops lasting 4 to 10 weeks for all the employees. Workers on the shop floor became familiar with statistical process control (SPC) charts and their use in maintaining and improving quality. Nashua uses individual x charts as well as $\bar{x}$, R, and p charts. These are among the most commonly used SPC charts today. Here we will consider the x chart. This chart is used when values come slowly, as in the following example, and it is not practical to take the time to form the subgroups necessary for an $\bar{x}$ or R chart.

Among the many products Nashua makes is thermally responsive paper, which is used in printers and recording instruments. The paper is coated with a chemical mixture that is sensitive to heat, thus producing marks in a printer or instrument when heat is applied by a print head or stylus. The variable of interest is the amount of material coated on the paper (the *weight coat*). Large rolls, some as long as 35,000 feet, are coated, and samples are taken from the ends of the rolls. A template 12×18 inches is used in cutting through four layers of the paper—first from an area that was coated and second from an uncoated area. A gravimetric comparison of the coated and uncoated samples gives four measurements of the weight coat. The average of these is the individual x value for that roll.

Assume that 12 rolls are coated per shift and that each roll is tested as described above. For two shifts, the 24 values of weight coat, in pounds per 3,000 square feet, were

3.46, 3.56, 3.58, 3.49, 3.45, 3.51, 3.54, 3.48, 3.54, 3.49, 3.55, 3.60, 3.62, 3.60, 3.53, 3.60, 3.51, 3.54, 3.60, 3.61, 3.49, 3.60, 3.60, 3.49.

Exhibit 1 shows the individual control chart for this process, using all 24 values to calculate the limits. Is the production process in statistical control? Explain. Discuss any possible actions or solutions.

---

*We are indebted to Dr. Lloyd S. Nelson of Nashua Corporation for providing us with this interesting and instructive case.

EXHIBIT 1 **Standardized *x* Chart**

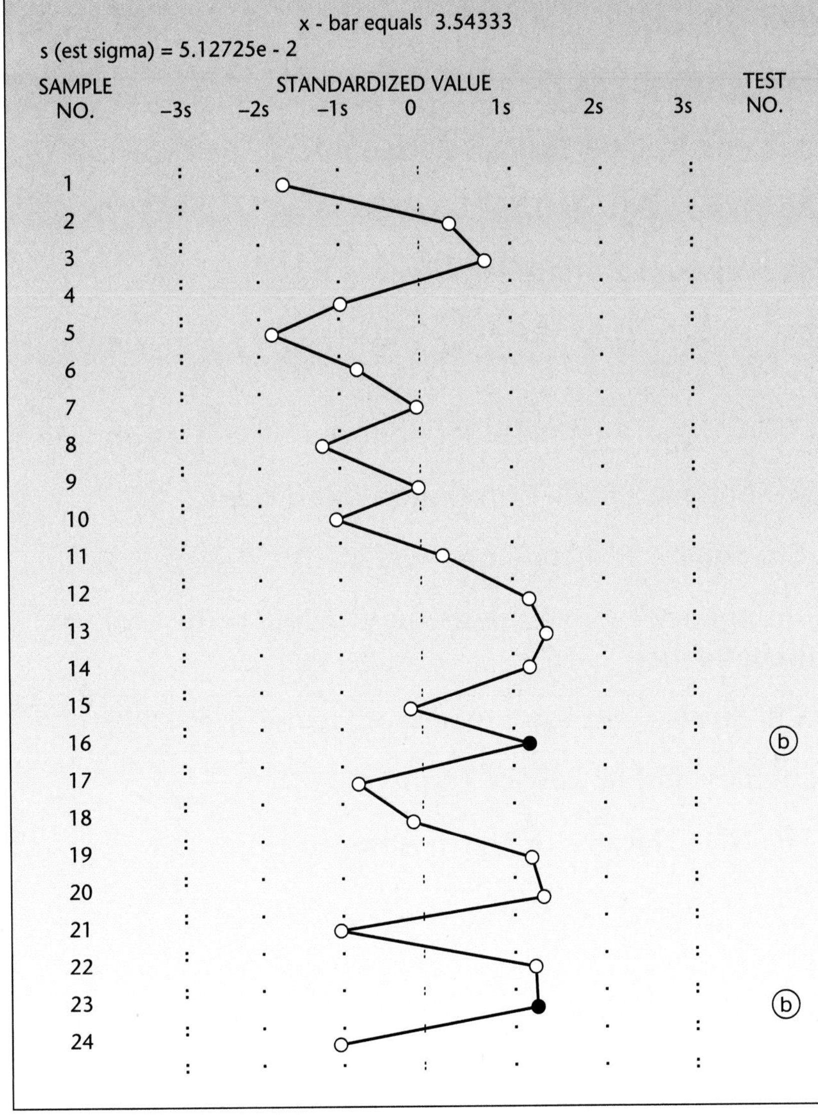

x - bar equals 3.54333
s (est sigma) = 5.12725e - 2
SAMPLE NO.
STANDARDIZED VALUE
−3s −2s −1s 0 1s 2s 3s
TEST NO.
1 2 3 4 5 6 7 8 9 10 11 12 13 14 15 16 17 18 19 20 21 22 23 24
b
b

SAMPLE NO.	WEIGHT COAT
1.	3.46
2.	3.56
3.	3.58
4.	3.49
5.	3.45
6.	3.51
7.	3.54
8.	3.48
9.	3.54
10.	3.49
11.	3.55
12.	3.60
13.	3.62
14.	3.60
15.	3.53
16.	3.60
17.	3.51
18.	3.54
19.	3.60
20.	3.61
21.	3.49
22.	3.60
23.	3.60
24.	3.49

<table>
<tr><td>

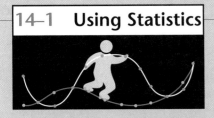

</td><td>

In 1710, Dr. John Arbuthnott, Physician in Ordinary to Her Majesty, published an article in the *Philosophical Transactions of the Royal Society of London.* The article was aimed at proving the existence of God and was titled "An Argument for Divine Providence, taken from the constant

</td></tr>
</table>

Regularity observ'd in the Births of both Sexes." Here is an excerpt from his paper.

> Among innumerable Footsteps of Divine Providence to be found in the Works of Nature, there is a very remarkable one to be observed in the exact Ballance that is maintained, between the Numbers of Men and Women, for by this means it is provided, that the Species may never fail, nor perish, since every Male may have its Female, and of a proportionable Age. This Equality of Males and Females is not the Effect of Chance but Divine Providence, working for a good End, which I thus demonstrate.[1]

The distribution of the number of "successes" (girls, or boys) in a fixed number of independent trials with the same probability of success is binomial. Arbuthnott tried in his article to use the binomial distribution to show that the roughly equal number of boys and girls born could not be by chance, hence an argument for Divine Providence. While Dr. Arbuthnott may not have proved the existence of God, the scientific method presented in his paper is believed to constitute the oldest known statistical test, the *sign test.* The sign test is a *nonparametric test.*

So far, the statistical procedures we have discussed have dealt with particular population parameters and made use of specific assumptions about the probability distributions of sample estimators, or assumptions about the nature of the population. In particular, we have usually assumed a *normal distribution.* The statistical procedures described in this chapter do not require stringent assumptions. Many of the tests do not deal with particular *parameters* of the population and, thus, are nonparametric. Some tests do not require assumptions about the distributions of the populations of interest. Therefore, nonparametric methods are often called *distribution-free methods.*

The chi-square methods we will discuss do make an assumption of a probability distribution: a limiting normal distribution of sampling outcomes. This assumption leads to a chi-square test statistic. These methods, however, are more general, and our analysis often does not deal with population parameters. In this sense, the chi-square analysis of enumerative data is a nonparametric one.

Typically, nonparametric methods require less stringent assumptions than do their parametric counterparts; on the other hand, they also use less information from the data. This makes the nonparametric tests somewhat less powerful than the corresponding parametric tests for the same situations, when the assumptions of the parametric tests are met. When the assumptions of the parametric tests are *not* met, the nonparametric tests are the ones we should use. Remember, for example, that in the analysis of variance, we assume that the populations of interest are normally distributed with equal variance. If these assumptions are met, then the *F* test of ANOVA is a powerful method for determining whether several population means are equal. If these assumptions are seriously violated, we must use an alternative method. Here, nonparametric statistics offers a good solution. There is a nonparametric ANOVA procedure called the Kruskal-Wallis test. This test requires no assumptions about the

---

[1]From J. Arbuthnott, "An Argument for Divine Providence, Taken from the Constant Regularity Observed in the Births of Both Sexes," *Philosophical Transactions* 27 (1710), pp. 186–90.

populations involved and uses less of the information in the data: The procedure uses only the *ranks* of the observations.

Many other hypothesis-testing situations have nonparametric alternatives to be used when the usual assumptions we make are not met. In other situations, nonparametric methods offer *unique* solutions to problems at hand. Because a nonparametric test usually requires fewer assumptions and uses less information in the data, it is often said that *a parametric procedure is an exact solution to an approximate problem,* whereas *a nonparametric procedure is an approximate solution to an exact problem.*

> In short, we define a **nonparametric method** as one that satisfies at least one of the following criteria.
>
> 1. The method deals with *enumerative data* (data that are frequency counts).
> 2. The method *does not deal with specific population parameters* such as μ or σ.
> 3. The method *does not require assumptions about specific population distributions* (in particular, the assumption of normality).

Since nonparametric methods require fewer assumptions than do parametric ones, the methods are useful when the scale of measurement is weaker than required for parametric methods. As we will refer to different measurement scales, you may want to review Section 1–7 at this point.

14–2   The Sign Test

In Chapter 8, we discussed statistical methods of comparing the means of two populations. There we used the *t* test, which required the assumption that the populations were normally distributed with equal variance. In many situations, one or both of these assumptions are not satisfied. In some situations, it may not even be possible to make exact measurements except for determining the relative magnitudes of the observations. In such cases, the **sign test** is a good alternative. The sign test is also useful in testing for a trend in a series of ordinal values and in testing for a correlation, as we will see soon.

As a test for comparing two populations, the sign test is stated in terms of the probability that values of one population are greater than values of a second population that are paired with the first in some way. For example, we may be interested in testing whether consumer responses to one advertisement are about the same as responses to a second advertisement. We would take a random sample of consumers, show them both ads, and ask them to rank the ads on some scale. For each person in our sample, we would then have two responses: one response for each advertisement. The null hypothesis is that the probability that a consumer's response to one ad will be greater than his or her response to the other ad is equal to 0.50. The alternative hypothesis is that the probability is not 0.50. Note that these null and alternative hypotheses are more general than those of the analogous parametric test—the paired-*t* test—which is stated in terms of the means of the two populations. When the two populations under study are symmetric, the test is equivalent to a test of the equality of two means, like the parametric *t* test. As stated, however, the sign test is more general and requires fewer assumptions.

We define p as the probability that X will be greater than Y, where X is the value from population 1 and Y is the value from population 2. Thus,

$$p = P(X > Y) \tag{14-1}$$

The test could be a two-tailed test, or a one-tailed test in either direction. Under the null hypothesis, it is as likely that X will exceed Y as it is likely that Y will exceed X: The probability of either occurrence is 0.50. We leave out the possibility of a tie, that is, the possibility that $X = Y$. When we gather our random sample of observations, we denote every pair (X, Y) where X is greater than Y by a plus sign $(+)$, and we denote every pair where Y is greater than X by a minus sign $(-)$ (hence the name *sign test*). In terms of signs, the null hypothesis is that the probability of a plus sign [that is, $P(X > Y)$] is equal to the probability of a minus sign [that is, $P(X < Y)$], and both are equal to 0.50. These are the possible hypothesis tests:

Possible hypotheses for the sign test:

Two-tailed test

$$H_0: p = 0.50$$
$$H_1: p \neq 0.50 \tag{14-2}$$

Right-tailed test

$$H_0: p \leq 0.50$$
$$H_1: p > 0.50 \tag{14-3}$$

Left-tailed test

$$H_0: p \geq 0.50$$
$$H_1: p < 0.50 \tag{14-4}$$

The test assumes that the pairs of (X, Y) values are independent and that the measurement scale within each pair is at least *ordinal*. After discarding any ties, we are left with the number of plus signs and the number of minus signs. These are used in defining the test statistic.

The test statistic is

$$T = \text{number of plus signs} \tag{14-5}$$

Suppose the null hypothesis is $p \leq 0.5$, which makes it a right-tailed test. Then the larger the T, the more unfavorable it would be to the null hypothesis. The p-value[2] is therefore the binomial probability of observing a value greater than or equal to the observed T. The calculation of a binomial probability requires two parameters, n and p. The value of n used is the sample size minus the tied cases, and the value of p used is 0.5 (which gives the maximum benefit of doubt to the null hypothesis).

---

[2]Unfortunately, we cannot avoid using the p symbol in two different senses. Take care not to confuse the two senses.

If the null hypothesis is $p \geq 0.5$, which makes it a left-tailed test, the p-value is the probability of observing a value less than or equal to the observed T. If the null hypothesis is $p = 0.5$, which makes it a two-tailed test, the p-value is twice the tail area.

Figure 14–1 shows the template that can be used for conducting a sign test. Its use will be illustrated in Example 14–1.

EXAMPLE 14–1

According to a recent survey of 220 chief executive officers (CEOs) of *Fortune 1000* companies, 18.3% of the CEOs in these firms hold MBA degrees. A management consultant wants to test whether there are differences in attitude toward CEOs who hold MBA degrees. In order to control for extraneous factors affecting attitudes toward different CEOs, the consultant designed a study that recorded the attitudes toward the same group of 19 CEOs before and after these people completed an MBA program. The consultant had no prior intention of proving one kind of attitudinal change; she believed it was possible that the attitude toward a CEO could change for the better, change for the worse, or not change at all following the completion of an MBA program. Therefore, the consultant decided to use the following two-tailed test.

H_0: There is no change in attitude toward a CEO following his or her being awarded an MBA degree

versus

H_1: There is a change in attitude toward a CEO following the award of an MBA degree

The consultant defined variable X_i as the attitude toward CEO i before receipt of the MBA degree, as rated by his or her professional associates on a scale of 1 to 5 (5 being highest). Similarly, she defined Y_i as the attitude toward CEO i following receipt of the MBA degree, as rated by his or her professional associates on the same scale.

Solution

In this framework, the null and alternative hypotheses may be stated in terms of the probability that the attitude score *after* (Y) is greater than the attitude *before* (X). The null hypothesis is that the probability that the attitude after receipt of the degree is higher than the attitude before is 0.50 (i.e., the attitude is as likely to improve as it is to become worse, where *worse* means a lower numerical score). The alternative hypothesis is that the probability is not 0.50 (i.e., the attitude is likely to change in one or the other direction). The null and alternative hypotheses can now be stated in the form of equation 14–2:

$$H_0: p = 0.50$$

versus

$$H_1: p \neq 0.50$$

The consultant looked at her data of general attitude scores toward the 17 randomly chosen CEOs both before and after these CEOs received their MBAs. Data are given in Table 14–1. The first thing to note is that there are two ties: for CEOs 2 and 5. We thus remove these two from our data set and reduce the sample size to $n = 15$. We now (arbitrarily) define a plus sign to be any data point where the after-attitude score is greater than the before score. In terms of plus and minus symbols, the data in Table 14–1 are as follows:

TABLE 14–1   Data for Example 14–1

CEO:	1	2	3	4	5	6	7	8	9	10	11	12	13	14	15	16	17
Attitude before:	3	5	2	2	4	2	1	5	4	5	3	2	2	2	1	3	4
Attitude after:	4	5	3	4	4	3	2	4	5	4	4	5	5	3	2	2	5

$$+ + + + + - + - + + + + + - +$$

According to our definition of the test statistic (equation 14–5), we have

$$T = \text{number of pluses} = 12$$

We now carry out the statistical hypothesis test. From Appendix C, Table 1 (page 835), the binomial table, we find for $p = 0.5$ $n = 15$ that the point $C_1 = 3$ corresponds to a "tail" probability of 0.018. That is, $F(3) = 0.018$. The p-value is 0.036. Since the rejection happened in the right-hand rejection region, the consultant may conclude that there is evidence that attitudes toward CEOs who recently received their MBA degrees have become more positive (as defined by the attitude test).

Figure 14–1 shows the same results obtained through the template. The data, entered in column C, should consist of only $+$ and $-$ symbols. All ties should be removed from the data before entry. The p-value for the null hypothesis $p = 0.5$ appears in cell G12; it is 0.0352. It is more accurate than the manually calculated 0.036. As seen in cell H12, the null hypothesis is rejected at an α of 5%.

The sign test can be viewed as a test of the hypothesis that the *median* difference between two populations is zero. As such, the test may be adapted for testing whether the median of a single population is equal to any prespecified number. The null and alternative hypotheses here are

$$H_0: \text{Population median} = a$$
$$H_1: \text{Population median} \neq a$$

where a is some number. One-tailed tests of this hypothesis are also possible, and the extension is straightforward.

To conduct the test, we pair our data observations with the null-hypothesis value of the median and perform the sign test. If the null hypothesis is true, then we expect that about one-half of the signs will be pluses and one-half minuses because, by the definition of the median, one-half of the population values are above it and one-half are below it.

Suppose that we wish to test the null hypothesis that median income in a certain region is $24,000 per family per year. The following random sample of family incomes is available (in thousands of dollars): 22, 30, 28, 22, 34, 19, 42, 18, 16, 26, 30, 25, 29, 20, 17, 33, 32, 24, 15, 31. In terms of $+$ signs, $-$ signs, and ties (t), the data are as follows when paired with the hypothesized median of 24: $- + + - + - + - -$ $+ + + + - - + + t - +$. (The choice of how to define a $+$ versus a $-$ is, again, arbitrary.) Discarding the single tie, we see that the number of plus signs is 11, and the sample size is $n = 19$.

Since 11 is more than the average of 9.5 ($= np = 19 * 0.5$), the tail area is to the right of 11. The tail area is the binomial probability $P(T \geq 11)$ with $n = 19$, $p = 0.5$. From the binomial template, this probability is 0.3238. The p-value is twice the tail

FIGURE 14–1   **The Template for the Sign Test**
[Nonparametric.xls; Sheet: Sign]

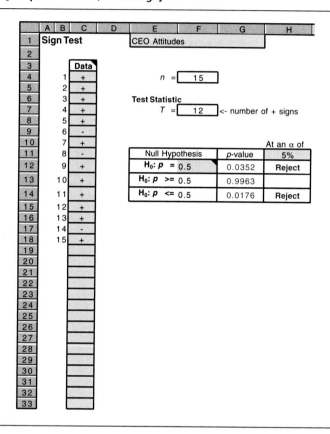

area and therefore equal to $2 * 0.3238 = 0.6476$. Since this p-value is so large, we cannot reject the null hypothesis that the median equals 24.

The same results could have also been obtained using the template.

PROBLEMS

14–1.   An investment analyst wants to test whether differences exist between the returns on two mutual funds. Paired data of annualized rates of return for the two mutual funds during 15 randomly chosen months are as follows:

Month	Fund A Return (%)	Fund B Return (%)
1	12	14
2	11	15
3	14	16
4	10	9
5	12	10
6	8	8
7	16	18
8	13	12
9	12	17

—Continued

Month	Fund A Return (%)	Fund B Return (%)
10	10	13
11	6	10
12	9	12
13	16	15
14	13	19
15	10	14

Conduct the sign test for determining whether returns on the two mutual funds are equal.

14-2. Breakstone Company makes whipped butter and whipped margarine. A company market analyst wanted to test whether people prefer the taste of one of these products over the other. A random sample of consumers was selected, and each one was asked to taste both the butter and the margarine and then to state a preference. The data follow. Is there evidence that one of the two products is preferred over the other? (M denotes margarine and B is for butter.) M B B B M B B M B B B B M M M B M M M B B M B B M B B B B (no pref.) M B B B B (no pref.) M M M B M B B B B B M B

14-3. The median amount of accounts payable to a retailer is believed to be $78.50. Conduct a test to assess whether this assumption is still true after several changes in company operations have taken place. A random sample of 30 accounts is collected. The data follow (in dollars):

34.12, 58.90, 73.25, 33.70, 69.00, 70.53, 12.68, 100.00, 82.55, 23.12, 57.55, 124.20, 89.60, 79.00, 150.13, 30.35, 42.45, 50.00, 90.25, 65.20, 22.28, 165.00, 120.00, 97.25, 78.45, 24.57, 12.11, 5.30, 234.00, 76.65

14-4. Biometrics is a technology that helps identify people by facial and body features and is used by banks to reduce fraud.[3] If in 15 trials the machine correctly identified 10 people, test the hypothesis that the machine's identification rate is 50%.

14-5. The median age of a tourist to Aruba is claimed to be 41 years. A random sample of 18 tourists gives the following ages:

25, 19, 38, 52, 57, 39, 46, 46, 30, 49, 40, 27, 39, 44, 63, 31, 67, 42

Test the hypothesis against a two-tailed alternative using $\alpha = 0.05$.

14-3 The Runs Test—A Test for Randomness

In his well-known book *Introduction to Probability Theory and Its Applications* (New York: John Wiley & Sons, Inc., 1973), William Feller tells of noticing how people occupy bar stools. Let S denote an occupied seat and E an empty seat. Suppose that, entering a bar, you find the following sequence:

S E S E S E S E S E S E S E S E S E   (case 1)

Do you believe that this sequence was formed at random? Is it likely that the 10 seated persons took their seats by a random choice, or did they purposely make sure they sat at a distance of one seat away from their neighbors? Just looking at the perfect regularity of this sequence makes us doubt its randomness.

---

[3]From "Here's Looking at You," *Fortune*, February 16, 1998, p. 104.

Let us now look at another way the people at the bar might have been occupying 10 out of 20 seats:

$$S\ S\ S\ S\ S\ S\ S\ S\ S\ S\ E\ E\ E\ E\ E\ E\ E\ E\ E\ E \qquad \text{(case 2)}$$

Is it likely that this sequence was formed at random? In this case, rather than perfect separation between people, there is a perfect clustering together. This, too, is a form of regularity not likely to have arisen by chance.

Let us now look at yet a third case:

$$S\ E\ E\ S\ S\ E\ E\ S\ E\ S\ S\ E\ S\ E\ E\ S\ S\ S\ E \qquad \text{(case 3)}$$

This last sequence seems more random. It is much more likely that this sequence was formed by chance than the sequences in cases 1 and 2. There does not seem to be any consistent regularity in the series in case 3.

What we feel intuitively about order versus randomness in these cases can indeed be quantified. There is a statistical test that can help us determine whether we believe that a sequence of symbols, items, or numbers resulted from a random process. The statistical test for randomness depends on the concept of a *run*.

> A **run** is a sequence of like elements that are preceded and followed by different elements or no element at all.

Using the symbols S and E, Figure 14–2 demonstrates the definition of a run by showing all runs in a particular sequence of symbols. There are seven runs in the sequence of elements in Figure 14–2.

Applying the definition of runs to cases 1, 2, and 3, we see that in case 1 there are 20 runs in a sequence of 20 elements! This is clearly the largest possible number of runs. The sequence in case 2 has only two runs (the smallest possible number). In the first case, there are too many runs, and in the second case, there are too few runs for randomness to be a probable generator of the process. In case 3, there are 12 runs— neither too few nor too many. This sequence could very well have been generated by a random process. To quantify how many runs are acceptable before we begin to doubt the randomness of the process, we use a probability distribution. This distribution leads to a *statistical test for randomness*.

Let us call the number of elements of one kind (S) n_1 and the number of elements of the second kind (E) n_2. The total sample size is $n = n_1 + n_2$. In all three cases, both n_1 and n_2 are equal to 10. For a given pair (n_1, n_2) and a given number of runs, Appendix C, Table 8 (page 856), gives the probability that the number of runs will be less than or equal to the given number (i.e., left-hand "tail" probabilities).

Based on our example, look at the row in Table 8 corresponding to $(n_1, n_2) = (10, 10)$. We find that the probability that four or fewer runs will occur is 0.001; the probability that five or fewer will occur is 0.004; the probability that six or fewer runs will occur is 0.019; and so on.

FIGURE 14–2 **Examples of Runs**

SSSS	EE	S	EEE	SSSS	E	SSS
↑	↑	↑	↑	↑	↑	↑
run	run	run	run	run	run	run

The logic of the test for randomness is as follows. We know the probabilities of obtaining any number of runs, and if we obtain an extreme number of runs—too many or too few—we will decide that the elements in our sequence were not generated in a random fashion.

A two-tailed hypothesis test for randomness:

H_0: Observations are generated randomly   (14–6)

H_1: Observations are not randomly generated

The test statistic is

$$R = \text{number of runs} \qquad (14\text{–}7)$$

The decision rule is to reject H_0 at level α if $R \leq C_1$ or $R \geq C_2$, where C_1 and C_2 are critical values obtained from Appendix C, Table 8, with total tail probability $P(R \leq C_1) + P(R \geq C_2) = \alpha$.

Let us conduct the hypothesis test for randomness (equation 14–6) for the sequences in cases 1, 2, and 3. Note that the tail probability for 6 or fewer runs is 0.019, and the probability for 16 or more runs is $P(R \geq 16) = 1 - F(15) = 1 - 0.981 = 0.019$. Thus, if we choose $\alpha = 2(0.019) = 0.038$, which is as close to 0.05 as we can get with this discrete distribution, our decision rule will be to reject H_0 for $R \geq 16$ or $R \leq 6$.

In case 1, we have $R = 20$. We reject the null hypothesis. In fact, the p-value obtained by looking in the table is less than 0.001. The same is true in case 2, where $R = 2$. In case 3, we have $R = 12$. We find the p-value as follows: $2[P(R \geq 12)] = 2[(1 - F(11)] = 2(1 - 0.586) = 2(0.414) = 0.828$. The null hypothesis cannot be rejected.

Large-Sample Properties

As you may have guessed, as the sample sizes n_1 and n_2 increase, the distribution of the number of runs approaches a normal distribution.

The mean of the normal distribution of the number of runs is

$$E(R) = \frac{2n_1 n_2}{n_1 + n_2} + 1 \qquad (14\text{–}8)$$

The standard deviation is

$$\sigma_R = \sqrt{\frac{2n_1 n_2 (2n_1 n_2 - n_1 - n_2)}{(n_1 + n_2)^2 (n_1 + n_2 - 1)}} \qquad (14\text{–}9)$$

Therefore, when the sample size is large, we may use a *standard normal test statistic* given by

$$z = \frac{R - E(R)}{\sigma_R} \qquad (14\text{–}10)$$

We demonstrate the large-sample test for randomness with Example 14–2.

EXAMPLE 14–2

One of the most important uses of the test for randomness is its application in residual analysis. Recall that a regression model, or a time series model, is adequate if the errors are random (no regular pattern). A time series model was fitted to sales data of multiple-vitamin pills. After the model was fitted to the data, the following residual series was obtained from the computer. Is there any statistical evidence to conclude that the time series errors are not random and, hence, that the model should be corrected?

-23, 30, 12, -10, -5, -17, -22, 57, 43, -23, 31, 42, 50, 61, -28, -52, 10, 34, 28, 55, 60, 32, 88, -75, -22, -56, -89, -34, -20, -2, -5, 29, 12, 45, 77, 78, 91, 25, 60, -25, 45, 42, 30, -59, -60, -40, -75, -25, -34, -66, -90, 10, -20

(The sequence of residuals continues, and their sum is zero.) Using this part of the sequence, we reason that since the mean residual is zero, we may look at the sign of the residuals and write them as plus or minus signs. Then we may count the number of runs of positive and negative residuals and perform the runs test for randomness.

Solution

We have the following signs:

$- + + - - - - + + - + + + + - - + + + + + + + - - - - - - - - +$
$+ + + + + + + - + + + - - - - - - - - + -$

Letting n_1 be the number of positive residuals and n_2 the number of negative ones, we have $n_1 = 27$ and $n_2 = 26$. We count the number of runs and find that $R = 15$.

We now compute the value of the Z statistic from equation 14–10. We have, for the mean and standard deviation given in equations 14–8 and 14–9, respectively,

$$E(R) = \frac{2(27)(26)}{27 + 26} + 1 = 27.49$$

and

$$\sigma_R = \sqrt{\frac{2(27)(26)[2(27)(26) - 27 - 26]}{(27 + 26)^2 (27 + 26 - 1)}} = 3.6$$

The computed value of the Z test statistic is

$$z = \frac{R - E(R)}{\sigma_R} = \frac{15 - 27.49}{3.6} = -3.47$$

From the Z table we know that the p-value is 0.0006 (this is a two-tailed test). We reject the null hypothesis that the residuals are random and conclude that the time series model needs to be corrected.

The Template

The same results for Example 14–2 could have been obtained using the template shown in Figure 14–3. The data, which should be $+$ or $-$ only, are entered in column B. The p-value appears in cell F18. For the current example the p-value of 0.0005 is more accurate than the manually calculated value of 0.0006.

**FIGURE 14–3   The Template for the Runs Test
[Nonparametric Tests.xls; Sheet: Runs]**

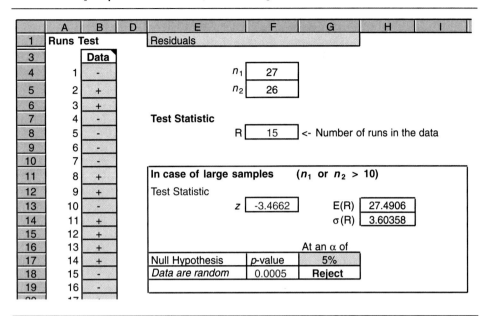

The Wald-Wolfowitz Test

An extension of the runs test for determining whether two populations have the same distribution is the **Wald-Wolfowitz test.**

The null and alternative hypotheses for the Wald-Wolfowitz test are

H_0: The two populations have the same distribution

H_1: The two populations have different distributions

(14–11)

This is one nonparametric analog to the t test for equality of two population means. Since the test is nonparametric, it is stated in terms of the distributions of the two populations rather than their means; however, the test is aimed at determining the difference between the two means. The test is two-tailed, but it is carried out on one tail of the distribution of the number of runs.

The only assumptions required for this test are that the two samples are independently and randomly chosen from the two populations of interest, and that values are on a continuous scale. The test statistic is, as before, R = number of runs.

We arrange the values of the two samples in increasing order in one sequence, regardless of the population from which each is taken. We denote each value by the symbol representing its population, and this gives us a sequence of symbols of two types. We then count the number of runs in the sequence. This gives us the value of R.

Logically, if the two populations have the same distribution, we may expect a higher degree of overlapping of the symbols of the two populations (i.e., a large

FIGURE 14–4   **Overlap versus Clustering of Two Samples**

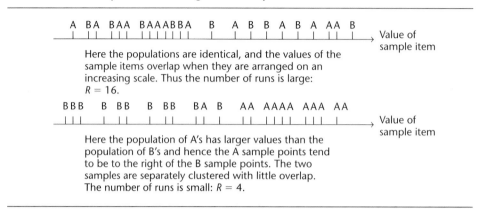

Here the populations are identical, and the values of the sample items overlap when they are arranged on an increasing scale. Thus the number of runs is large: $R = 16$.

Here the population of A's has larger values than the population of B's and hence the A sample points tend to be to the right of the B sample points. The two samples are separately clustered with little overlap. The number of runs is small: $R = 4$.

number of runs). If, on the other hand, the two populations are different, we may expect a clustering of the sample items from each of the groups. If, for example, the values in population 1 tend to be larger than the values in population 2, then we may expect the items from sample 1 to be clustered to the right of the items from sample 2. This produces a small number of runs. We would like to reject the null hypothesis when the number of runs is too small. We illustrate the idea of overlapping versus clustering in Figure 14–4.

We demonstrate the Wald-Wolfowitz test with Example 14–3.

EXAMPLE 14–3

The manager of a record store wants to test whether her two salespeople are equally effective. That is, she wants to test whether the number of sales made by each salesperson is about the same or whether one salesperson is better than the other. The manager gets the following random samples of daily sales made by each salesperson.

Salesperson A: 35, 44, 39, 50, 48, 29, 60, 75, 49, 66
Salesperson B: 17, 23, 13, 24, 33, 21, 18, 16, 32

Solution

We have $n_1 = 10$ and $n_2 = 9$. We arrange the items from the two samples in increasing order and denote them by A or B based on which population they came from. We get

B B B B B B B A B B A A A A A A A A A

The total number of runs is $R = 4$.

From Appendix C, Table 8, we find that the probability of four or fewer runs for sample sizes of 9 and 10 is 0.002. As the p-value is 0.002, we reject the null hypothesis that the two salespeople are equally effective. Since salesperson A had the larger values, we conclude that he or she tends to sell more than salesperson B.

We have assumed here that the sales of the two salespersons cannot be paired as taking place on the same days. Otherwise, a paired test would be more efficient, as it would reduce day-to-day variations.

The Wald-Wolfowitz test is a *weak test*. There are other nonparametric tests, as we will see, that are more powerful than this test in determining differences between two populations. The advantage of the present test is that it is easy to carry out. There is no need to compute any quantity from the data values—all we need to do is to order the data on an increasing scale and to count the number of runs of elements from the two samples.

14-6. Some items produced by a machine are defective. If the machine follows some pattern where defective items are not randomly produced throughout the process, the machine needs to be adjusted. A quality control engineer wants to determine whether the sequence of defective (D) versus good (G) items is random. The data are

G G G G G D D D G G G G G G G D D D D G G G G G G G G G G G D D D D D G G G
G G G G G G G G G D D D G G G G G G G G G G G G G D D D D

Conduct the test for randomness, and state your conclusions.

14-7. A computer is used for generating random numbers. It is necessary to test whether the numbers produced are indeed random. A common method of doing this is to look at runs of odd versus even digits. Conduct the test using the following sequence of numbers produced by the computer.

27658983764454499867521387975637458764534267898763348219109347 3640898763

14-8. In a regression analysis, 12 out of 30 residuals are greater than 1.00 in value, and the rest are not. With A denoting a residual greater than 1 and B a residual less than 1, the residuals are as follows:

B B B B B B B B B A A A A A A A A A A B B B B B B B B B B A A

Do you believe that the regression errors are random? Explain.

14-9. A messenger service employs eight men and nine women. Every day, the assignments of errands are supposed to be done at random. On a certain day, all the best jobs, in order of desirability, were given to the eight men. Is there evidence of sex discrimination? Discuss this also in the context of a continuing, daily operation. What would happen if you tested the randomness hypothesis every day?

14-10. Bids for a government contract are supposed to be opened in a random order. For a given contract, there were 42 bids, 30 of them from domestic firms and 12 from foreign firms. The order in which the sealed bids were opened was as follows (D denotes a domestic firm and F a foreign one):

D D D D D D D D F D D D D D D D F F D D D D D D D D D F D D F D D D D D D
F F F F F F F

Could the foreign firms claim that they have been discriminated against? Explain.

14-11. Two advertisements are to be compared for their appeal. A random sample of eight people was selected, and their responses to ad 1 were recorded. Another random sample, of nine people, was shown ad 2, and their responses were also recorded. The response data are as follows (10 is highest appeal).

Ad 1:   7, 8, 6, 7, 8, 9, 9, 10
Ad 2:   3, 4, 3, 5, 5, 4, 2, 5, 4

Is there a quick statistical proof that one ad is better than the other?

14–12. The following data are salaries of seven randomly chosen owners of furniture-making firms and eight randomly chosen owners of paper product firms. The data are in thousands of dollars per year.

> Furniture: 175, 170, 166, 168, 204, 96, 147
> Paper products: 89, 120, 136, 160, 111, 101, 98, 80

Use the Wald-Wolfowitz test to determine whether there is evidence that average owner salaries in the two business lines are not equal.

14–4   The Mann-Whitney *U* Test

In this section, we present the first of several statistical procedures that are based on *ranks*. In these procedures, we rank the observations from smallest to largest and then use the ranks instead of the actual sample values in our computations. Sometimes, our data are themselves ranks. Methods based on ranks are useful when the data are at least on an ordinal scale of measurement. Surprisingly, when we substitute ranks for actual observations, the loss of information does not weaken the tests very much. In fact, when the assumptions of the corresponding parametric tests are met, the non-parametric tests based on ranks are often about 95% as efficient as the parametric tests. When the assumptions needed for the parametric tests (usually, a normal distribution) are not met, the tests based on ranks are excellent, powerful alternatives.

We demonstrate the ranking procedure with a simple set of numbers: 17, 32, 99, 12, 14, 44, 50. We rank the observations from smallest to largest. This gives us 3, 4, 7, 1, 2, 5, 6. (The reason is that the smallest observation is 12, the next one up is 14, and so on. The largest observation—the seventh—is 99.) This simple ranking procedure is the basis of the test presented in this section, as well as of the tests presented in the next few sections. Tests based on ranks are probably the most widely used nonparametric procedures.

In this section, we present the **Mann-Whitney *U* test**, also called the *Wilcoxon rank sum test*, or just the *rank sum test*. This test is different from the test we discuss in the next section, called the Wilcoxon *signed-rank* test. Try not to get confused by these names. The Mann-Whitney test is an adaptation of a procedure due to Wilcoxon, who also developed the signed-rank test. The most commonly used name for the rank sum test, however, is the Mann-Whitney *U* test.

The Mann-Whitney *U* test is a test of equality of two population distributions. The test is most useful, however, in testing for equality of two population means. As such, the test is an alternative to the two-sample *t* test and is used when the assumption of normal population distributions is not met. The test is only slightly weaker than the *t* test and is more powerful than the Wald-Wolfowitz runs test described in the previous section.

The null and alternative hypotheses for the Mann-Whitney *U* test are

H_0: The distributions of the two populations are identical

H_1: The two population distributions are not identical   (14–12)

Often, the hypothesis test in equation 14–12 is written in terms of equality versus non-equality of two population means or equality versus nonequality of two population medians. As such, we may also have one-tailed versions of the test. We may test whether one population mean is greater than the other. We may state these hypotheses in terms of population medians.

The only assumptions required by the test are that the samples be random samples from the two populations of interest and that they be drawn independently of each other. If we want to state the hypotheses in terms of population means or medians, however, we need to add an assumption, namely, that if a difference exists between the two populations, the difference is in *location* (mean, median).

The Computational Procedure

We combine the two random samples and rank all our observations from smallest to largest. To any ties we assign the *average* rank of the tied observations. Then we sum all the ranks of the observations from one of the populations and denote that population as population 1. The sum of the sample ranks is R_1.

The Mann-Whitney U statistic is

$$U = n_1 n_2 + \frac{n_1(n_1 + 1)}{2} - R_1 \qquad (14\text{–}13)$$

where n_1 is the sample size from population 1 and n_2 is the sample size from population 2

The U statistic is a measure of the difference between the ranks of the two samples. Large values of the statistic, or small ones, provide evidence of a difference between the two populations. If we assume that differences between the two populations are only in location, then large or small values of the statistic provide evidence of a difference in the location (mean, median) of the two populations.

The distribution of the U statistic for small samples is given in Appendix C, Table 9 (pages 858–862). The table assumes that n_1 is the smaller sample size. For large samples, we may, again, use a normal approximation. The convergence to the normal distribution is relatively fast, and when both n_1 and n_2 are greater than 10 or so, the normal approximation is good.

The mean of the distribution of U is

$$E(U) = \frac{n_1 n_2}{2} \qquad (14\text{–}14)$$

The standard deviation of U is

$$\sigma_U = \sqrt{\frac{n_1 n_2 (n_1 + n_2 + 1)}{12}} \qquad (14\text{–}15)$$

The large-sample test statistic is

$$z = \frac{U - E(U)}{\sigma_U} \qquad (14\text{--}16)$$

For large samples, the test is straightforward. In a two-tailed test, we reject the null hypothesis if z is greater than or less than the values that correspond to our chosen level of α (for example, ± 1.96 for $\alpha = 0.05$). Similarly, in a one-tailed test, we reject H_0 if z is greater than (or less than) the appropriate critical point. Note that U is large when R_1 is small, and vice versa. Thus, if we want to prove the alternative hypothesis that the location parameter of population 1 is greater than the location parameter of population 2, we reject on the left tail of the normal distribution.

With small samples, we have a problem because the U table lists only left-hand-side probabilities of the statistic [the table gives $F(U)$ values]. Here we will use the following procedure. For a two-tailed test, we define R_1 as the larger of the two sums of ranks. This will make U small so it can be tested against a left-hand critical point with tail probability $\alpha/2$. For a one-tailed test, if we want to prove that the location parameter of population 1 is greater than that of population 2, we look at the sum of the ranks of sample 1 and do not reject H_0 if this sum is smaller than that for sample 2. Otherwise, we compute the statistic and test on the left side of the distribution. We choose the left-hand critical point corresponding to α. Relabel populations 1 and 2 if you want to prove the other one-tailed possibility.

We demonstrate the Mann-Whitney test with two examples.

EXAMPLE 14–4   Federal aviation officials tested two proposed versions of the Copter-plane, a twin-engine plane with tilting propellers that make takeoffs and landings easy and save time during short flights. The two models, made by Bell Helicopter Textron, Inc., were tested on the New York–Washington route. The officials wanted to know whether the two models were equally fast or whether one was faster than the other. Each model was flown six times, at randomly chosen departure times. The data, in minutes of total flight time for models A and B, are as follows.

Model A: 35, 38, 40, 42, 41, 36
Model B: 29, 27, 30, 33, 39, 37

Solution   First we order the data so that they can be ranked. This has been done in Figure 14–5. We note that the sum of the ranks of the sample points from the population of model A should be higher since the ranks for this model are higher. We will thus define R_1 as the sum of the ranks from this sample because we need a small value of U (which happens when R_1 is large) for comparison with table values. We find the value of R_1 as $R_1 = 5 + 6 + 8 + 10 + 11 + 12 = 52$. This is the sum of the circled ranks in Figure 14–5, the ranks belonging to the sample for model A.

FIGURE 14–5   **Ordering and Ranking the Data for Example 14–4**

Model A:					35	36		38		40	41	42
Model B:	27	29	30	33			37		39			
Rank:	1	2	3	4	(5)	(6)	7	(8)	9	(10)	(11)	(12)

We now compute the test statistic U. From equation 14–13, we find

$$U = n_1 n_2 + \frac{n_1(n_1 + 1)}{2} - R_1 = (6)(6) + \frac{(6)(7)}{2} - 52 = 5$$

Looking at Appendix C, Table 9, we find that the probability that U will attain a value of 5 or less is 0.0206. Since this is a two-tailed test, we want to reject the null hypothesis if the value of the statistic is less than or equal to the (left-hand) critical point corresponding to $\alpha/2$; if we choose $\alpha = 0.05$, then $\alpha/2 = 0.025$. Since 0.0206 is less than 0.025, we reject the null hypothesis at the 0.05 level. The p-value for this test is $2(0.0206) = 0.0412$. (Why?)

Suppose that we had chosen to conduct this as a one-tailed test. If we had originally wanted to test whether model B were slower than model A, then we would not be able to reject the null hypothesis that model B was *not* slower because the sum of the ranks of model B is smaller than the sum of the ranks of model A, and, hence, U would be large and not in the (left-side) rejection region. If, on the other hand, we wanted to test whether model A were slower, the test statistic would have been the same as the one we used, except that we could have rejected with a value of U as high as 7 (from Table 9, the tail probability for $U = 7$ is 0.0465, which is less than $\alpha = 0.05$). Remember that in a one-tailed test, we use the (left-hand) critical point corresponding to α and not to $\alpha/2$. In any case, we reject the null hypothesis and state that there is evidence to conclude that model B is generally faster.

When the sample sizes are large and we use the normal approximation, conducting the test is much easier since we do not have to redefine U so that it is always on the left-hand side of the distribution. We just compute the standardized Z statistic, using equations 14–14 through 14–16, and consult the standard normal table. This is demonstrated in Example 14–5.

EXAMPLE 14–5

A multinational corporation is about to open a subsidiary in Greece. Since the operation will involve a large number of executives who will have to move to that country, the company plans to offer an extensive program of teaching the language to the executives who will operate in Greece. For its previous operation starts in France and Italy, the company used cassettes and books provided by Educational Services Teaching Cassettes, Inc. Recently one of the company directors suggested that the book-and-cassette program offered by Metacom, Inc., sold under the name *The Learning Curve,* might provide a better introduction to the language. The company therefore decided to test the null hypothesis that the two programs were equally effective versus the one-tailed alternative that students who go through The Learning Curve program achieve better proficiency scores in a comprehensive examination following the course. Two groups of 15 executives were randomly selected, and each group studied the language under a different program. The final scores for the two groups, Educational Services (ES) and Learning Curve (LC), are as follows. Is there evidence that The Learning Curve method is more effective?

ES: 65, 57, 74, 43, 39, 88, 62, 69, 70, 72, 59, 60, 80, 83, 50
LC: 85, 87, 92, 98, 90, 88, 75, 72, 60, 93, 88, 89, 96, 73, 62

Solution

We order the scores and rank them. When ties occur, we assign to each tied observation the average rank of the ties.

ES: 39 43 50 57 59 60 62 65 69 70 72   74   80 83   88

LC: 60 62   72 73   75   85 87 88 89 90 92 93 96 98

                                                                   88

The tied observations are 60 (two—one from each group), 62 (two—one from each group), 72 (two—one from each group), and 88 (three—one from ES and two from LC). If we disregarded ties, the two observations of 60 would have received ranks 6 and 7. Since either one of them could have been rank 6 or rank 7, each gets the *average* rank of 6.5 (and the next rank up is 8). The next two observations are also tied (both are 62). They would have received ranks 8 and 9, so each gets the average rank of 8.5, and we continue with rank 10, which goes to the observation 65. The two 72 observations each get the average rank of 13.5 [(13 + 14)/2]. There are three 88 observations; they occupy ranks 22, 23, and 24. Therefore, each of them gets the average rank of 23.

We now list the ranks of all the observations in each of the two groups:

ES: 1   2   3   4   5   6.5   8.5   10   11   12   13.5   16   18   19   23
LC: 6.5 8.5 13.5   15   17   20   21   23   23   25   26   27   28   29   30

Note that 2 of the 23 ranks belong to LC and 1 belongs to ES. We may now compute the test statistic U. To be consistent with the small-sample procedure, let us define LC as population 1. We have

$$R_1 = 6.5 + 8.5 + 13.5 + 15 + 17 + 20 + 21 + 23 + 23 + 25 + 26 + 27 + 28$$
$$+ 29 + 30$$
$$= 312.5$$

Thus, the value of the statistic is

$$U = (15)(15) + \frac{(15)(16)}{2} - 312.5 = 32.5$$

We now compute the value of the standardized Z statistic, equation 14–16. From equation 14–14,

$$E(U) = \frac{(15)(15)}{2} = 112.5$$

and from equation 14–15,

$$\sigma_U = \sqrt{\frac{(15)(15)(31)}{12}} = 24.1$$

We get

$$z = \frac{U - E(U)}{\sigma_U} = \frac{32.5 - 112.5}{24.1} = -3.32$$

We want to reject the null hypothesis if we believe that LC gives higher scores. Our test statistic is defined to give a negative value in such a case. Since the computed value of the statistic is in the rejection region for any usual α value, we reject the null hypothesis and conclude that there is evidence that the LC program is more effective. Our *p*-value is 0.0005.

FIGURE 14–6   The Template for the Mann-Whitney U Test
[Nonparametric Tests.xls; Sheet: Mann-Whitney]

	A	B	C	D	E F	G	H	I	J	K	L
1	Mann-Whitney U Test							The Learning Curve method			
2											
3		Group	Data	Rank		Counts					
4	1	2	65	10	1	15					
5	2	2	57	4	2	15					
6	3	2	74	16							
7	4	2	43	2							
8	5	2	39	1		Test Statistic			Rank Sums		
9	6	2	88	23			U	32.5	1	312.5	
10	7	2	62	8.5					2	152.5	
11	8	2	69	11							
12	9	2	70	12							
13	10	2	72	13.5		In case of large samples (n_1 or $n_2 > 10$)					
14	11	2	59	5		Test Statistic			E[U]	112.5	
15	12	2	60	6.5			z	-3.3182	σ(U)	24.10913	
16	13	2	80	18							
17	14	2	83	19					At an α of		
18	15	2	50	3		Null Hypothesis	p-value	5%			
19	16	1	85	20		$H_0 : \mu_1 = \mu_2$	0.0009	Reject			
20	17	1	87	21		$H_0 : \mu_1 >= \mu_2$	0.9995				
21	18	1	92	27		$H_0 : \mu_1 <= \mu_2$	0.0005	Reject			
22	19	1	98	30							

In Example 14–5 we used the Mann-Whitney test instead of the parametric t test because some people have a facility with language and tend to score high on language tests, whereas others do not and tend to score low. This can create a bimodal distribution (one with two modes) rather than a normal curve, which is required for the t test.

In Example 14–4, we had small samples. When small samples are used, the parametric tests are sensitive to deviations from the normal assumption required for the t distribution. In such cases, it is usually better to use a nonparametric method such as the Mann-Whitney test, unless there is a good indication that the populations in question are approximately normally distributed.

The Template

The template for the Mann-Whitney U test is shown in Figure 14–6. The data are entered using group numbers 1 and 2. The groups need not be in any sorted order. The data seen in the figure correspond to Example 14–5. In this problem, LC was labeled as Group 1 and ES as 2. Note the group labels entered in column B.

For small samples, the message "Look up U tables for p-values" appears in cell N9 and the range I19:I21 goes blank. For large samples, the p-values appear in the range I19:I21. For this problem, we see that the p-value for the null hypothesis $\mu_1 \leq \mu_2$ is 0.0005, which agrees with our manual calculation. Because the p-value is so small, we reject the null hypothesis.

PROBLEMS

14–13.   Gotex is considering two possible bathing suit designs for the new season. One is called Nautical Design, and the other is Geometric Prints. Since the fashion industry is very competitive, Gotex needs to test before marketing the bathing suits. Ten randomly chosen top models are selected for modeling Nautical Design, and 10 other randomly chosen top models are selected to model Geometric Prints bathing suits. The results of the judges' ratings of the 20 bathing suits follow.

ND: 86, 90, 77, 81, 86, 95, 99, 92, 93, 85
GP: 67, 72, 60, 59, 78, 69, 70, 85, 65, 62

Is there evidence to conclude that one design is better than the other? If so, which one is it, and why?

14–14. Following reports by scientists that a large hole in the atmospheric ozone revolves around Antarctica, two U.S. jets of the U-2 type were sent to collect samples of air for comparison with samples collected elsewhere, to see if the ozone layer is indeed depleted. If so, controls on the manufacture of ozone-depleting fluorocarbons would have to be imposed to prevent further dangerous depletion of ozone around the world. Using the following data, test the null hypothesis that the ozone concentration over Antarctica is approximately equal to the concentration found in other areas against the alternative hypothesis that the concentration over Antarctica is lower than that found elsewhere.

Antarctic (%):   6, 10, 12, 23, 11, 31, 25, 11, 8, 7, 34, 14, 16, 18, 40, 31, 22, 5
Other places (%): 17, 32, 41, 29, 16, 30, 19, 51, 65, 22, 40, 47, 29, 30, 16, 70

14–15. Explain when you would use the Mann-Whitney test, when you would use the two-sample *t* test, and when you would use the Wald-Wolfowitz test. Discuss your reasons for choosing each test in the appropriate situation.

14–16. Superconductors, materials that carry electricity without losing energy, are believed to be the key to technology in the 21st century. Currently, two types of ceramics are considered for potential use—one designed at an IBM laboratory in the United States and one designed at the University of Tokyo, Japan. The efficiency of electrical conductivity is measured using a special formula; the higher the measurement, the more efficient the conductor. Using the following data, determine whether there is statistical evidence to conclude that one of the two superconductors is more efficient.

IBM conductor: 143, 121, 120, 101, 107, 142, 118, 130, 128, 107, 108, 126
Tokyo conductor: 102, 119, 121, 113, 126, 116, 117, 129, 104, 109, 110

14–17. Shearson Lehman Brothers, Inc., now encourages its investors to consider real estate limited partnerships. The company offers two limited partnerships—one in a condominium project in Chicago and one in Dallas. Annualized rates of return for the two investments during separate eight-month periods are as follows. Is one type of investment better than the other? Explain.

Chicago (%):   12, 13, 10, 14, 15, 9, 11, 10
Dallas (%):   10, 9, 8, 7, 9, 11, 6, 13

14–18. According to a recent *Business Week* article, the average Taiwanese is equally likely to use a bank or a black-marketeer for changing money. Since rates vary from place to place, it may be of interest to determine whether black-market foreign exchange commissions are approximately equal to those of banks. A random sample of

16 bank commissions and a random sample of 17 black-market commissions are collected. Data, in percentage charged per transaction, are as follows. Conduct a two-tailed test of equality of commission rates.

Black market: 1, 1.5, 1.3, 1.2, 2, 2.1, 1.8, 1.4, 1.5, 2.3, 1.7, 1.2, 1.4, 1.1, 1.9, 2.3, 2.6
Banks: 1.8, 1.7, 1.9, 1.5, 1.7, 1.3, 2.4, 2.8, 1.9, 2.1, 2.2, 2.6, 1.9, 2.5, 2.0, 2.1

14–5   The Wilcoxon Signed-Rank Test

The **Wilcoxon signed-rank test** is useful in comparing two populations for which we have paired observations. This happens when there is a natural way to pair off our data, for example, husband's score and wife's score in a consumer rating study. As such, the test is a good alternative to the paired-observations t test in cases where the differences between paired observations are not believed to be normally distributed. We have already seen a nonparametric test for such a situation—the sign test. Unlike the sign test, the Wilcoxon test accounts for the magnitude of differences between paired values, not only their signs. The test does so by considering the *ranks* of these differences. The test is therefore more efficient than the sign test when the differences may be quantified rather than just given a positive or negative sign. The sign test, on the other hand, is easier to carry out.

The Wilcoxon procedure may also be adapted for testing whether the location parameter of a single population (its median or its mean) is equal to any given value. There are one-tailed and two-tailed versions of each test. We start with the paired-observations test for the equality of two population distributions (or the equality of the location parameters of the two populations).

The Paired-Observations Two-Sample Test

The null hypothesis is that the median difference between the two populations is zero. The alternative hypothesis is that it is not zero.

> The hypothesis test is
>
> H_0: The median difference between populations 1 and 2 is zero   (14–17)
> H_1: The median difference between populations 1 and 2 is not zero

We assume that the distribution of differences between the two populations is symmetric, that the differences are mutually independent, and that the measurement scale is at least interval. By the assumption of symmetry, hypotheses may be stated in terms of means. The alternative hypothesis may also be a directed one: that the mean (or median) of one population is greater than the mean (or median) of the other population.

First, we list the pairs of observations we have on the two variables (the two populations). The data are assumed to be a random sample of paired observations. For each pair, we compute the difference

$$D = x_1 - x_2 \qquad (14\text{–}18)$$

Then we rank the absolute values of the differences D.

In the next step, we form sums of the ranks of the positive and of the negative differences.

The Wilcoxon T statistic is defined as the smaller of the two sums of ranks— the sum of the negative or the positive ones

$$T = \min\left(\sum(+),\ \sum(-) \right) \qquad (14\text{–}19)$$

where $\sum(+)$ is the sum of the ranks of the positive differences and $\sum(-)$ is the sum of the ranks of the negative differences

The **decision rule:** Critical points of the distribution of the test statistic T (when the null hypothesis is true) are given in Appendix C, Table 10 (page 863). We carry out the test on the left tail; that is, we reject the null hypothesis if the computed value of the statistic is less than a critical point from the table, for a given level of significance.

For a one-tailed test, suppose that the alternative hypothesis is that the mean (median) of population 1 is greater than that of population 2; that is,

$$\begin{aligned} &\text{H}_0\text{: } \mu_1 \leq \mu_2 \\ &\text{H}_1\text{: } \mu_1 > \mu_2 \end{aligned} \qquad (14\text{–}20)$$

Here we use the sum of the ranks of negative differences. If the alternative hypothesis is reversed (populations 1 and 2 are switched), then we use the sum of the ranks of the positive differences as the statistic. In either case, the test is carried out on the left "tail" of the distribution. Appendix C, Table 10, gives critical points for both one-tailed and two-tailed tests.

Large-Sample Version of the Test

As in other situations, when the sample size increases, the distribution of the Wilcoxon statistic T approaches the normal probability distribution. In the Wilcoxon test, n is defined as the number of *pairs* of observations from populations 1 and 2. As the number of pairs n gets large (as a rule of thumb, $n > 25$ or so), T may be approximated by a normal random variable as follows.

The mean of T is

$$E(T) = \frac{n(n+1)}{4} \qquad (14\text{–}21)$$

The standard deviation of T is

$$\sigma_T = \sqrt{\frac{n(n+1)(2n+1)}{24}} \qquad (14\text{–}22)$$

The standardized z statistic is

$$z = \frac{T - E(T)}{\sigma_T} \qquad (14\text{–}23)$$

We now demonstrate the Wilcoxon signed-rank test with Example 14–6.

TABLE 14–2   Data and Computations for Example 14–6

Store	Number of Violet Sold X_1	Number of Pink Sold X_2	Difference $D = X_1 - X_2$	Rank of Absolute Difference $|D|$	Rank of Positive D	Rank of Negative D
1	56	40	16	9	9	
2	48	70	−22	12		12
3	100	60	40	15	15	
4	85	70	15	8	8	
5	22	8	14	7	7	
6	44	40	4	2	2	
7	35	45	−10	6		6
8	28	7	21	11	11	
9	52	60	−8	5		5
10	77	70	7	3.5	3.5	
11	89	90	−1	1		1
12	10	10	0			
13	65	85	−20	10		10
14	90	61	29	13	13	
15	70	40	30	14	14	
16	33	26	7	3.5	3.5	
					$\Sigma(+) = 86$	$\Sigma(-) = 34$

EXAMPLE 14–6

The Sunglass Hut of America, Inc., operates kiosks occupying previously unused space in the well-traveled aisles of shopping malls. Sunglass Hut owner Sanford Ziff hopes to expand within a few years to every major shopping mall in the United States. He is using the present $4.5 million business as a test of the marketability of different types of sunglasses. Two types of sunglasses are sold: violet and pink. Ziff wants to know whether there is a difference in the quantities sold of each type. The numbers of sunglasses sold of each kind are paired by store; these data for each of 16 stores during the first month of operation are given in Table 14–2. The table also shows how the differences and their absolute values are computed and ranked, and how the signed ranks are summed, leading to the computed value of T.

Solution

Note that a difference of zero is discarded, and the sample size is reduced by 1. The effective sample size for this experiment is now $n = 15$. Note also that ties are handled as before: We assign the average rank to tied differences. Since the smaller sum is the one associated with the negative ranks, we define T as that sum. We therefore have the following value of the Wilcoxon test statistic:

$$T = \Sigma(-) = 34$$

We now conduct the test of the hypotheses in equation 14–17. We compare the computed value of the statistic $T = 34$ with critical points of T from Appendix C, Table 10. For a two-tailed test, we find that for $\alpha = 0.05$ ($P = 0.05$ in the table) and $n = 15$, the critical point is 25. Since the test is carried out on the "left tail"—that is, we do not reject the null hypothesis if the computed value of T is *less than or equal to* the table value—we do not reject the null hypothesis that the distribution of sales of the violet sunglasses is identical to the distribution of sales of the pink sunglasses.

A Test for the Mean or Median of a Single Population

As stated earlier, the Wilcoxon signed-rank test may be adapted for testing whether the mean (or median) of a single population is equal to any given number. There are three possible tests. The first is a left-tailed test where the alternative hypothesis is that the mean (or median—both are equal if we assume a *symmetric* population distribution) is smaller than some value specified in the null hypothesis. The second is a right-tailed test where the alternative hypothesis is that the mean (or median) is greater than some value. The third is a two-tailed test where the alternative hypothesis is that the mean (or median) is not equal to the value specified in the null hypothesis.

The computational procedure is as follows. Using our n data points $x_1, x_2, \ldots, x_n$, we form pairs: $(x_1, m), (x_2, m), \ldots, (x_n, m)$, where m is the value of the mean (or median) specified in the null hypothesis. Then we perform the usual Wilcoxon signed-rank test on these pairs.

In a right-tailed test, if the negative ranks have a larger sum than the positive ranks, we do not reject the null hypothesis. If the negative ranks have a smaller sum than the positive ones, we conduct the test (on the left tail of the distribution, as usual) and use the critical points in the table corresponding to the one-tailed test. We use the same procedure in the left-tailed test. For a two-tailed test, we use the two-tailed critical points. In any case, we always reject the null hypothesis if the computed value of T is less than or equal to the appropriate critical point from Appendix C, Table 10.

We will now demonstrate the single-sample Wilcoxon test for a mean using the large-sample normal approximation.

EXAMPLE 14–7 The average hourly number of messages transmitted by a private communications satellite is believed to be 149. The satellite's owners have recently been worried about the possibility that demand for this service may be declining. They therefore want to test the null hypothesis that the average number of messages is 149 (or more) versus the alternative hypothesis that the average hourly number of relayed messages is less than 149. A random sample of 25 operation hours is selected. The data (numbers of messages relayed per hour) are

> 151, 144, 123, 178, 105, 112, 140, 167, 177, 185, 129, 160, 110, 170, 198, 165, 109, 118, 155, 102, 164, 180, 139, 166, 182

Is there evidence of declining use of the satellite?

Solution   We form 25 pairs, each pair consisting of a data point and the null-hypothesis mean of 149. Then we subtract the second number from the first number in each pair (i.e., we subtract 149 from every data point). This gives us the differences D.

> 2, −5, −26, 29, −44, −37, −9, 18, 28, 36, −20, 11, −39, 21, 49, 16, −40, −31, 6, −47, 15, 31, −10, 17, 33

The next step is to rank the absolute value of the differences from smallest to largest. We have the following ranks, in the order of the data:

> 1, 2, 13, 15, 23, 20, 4, 10, 14, 19, 11, 6, 21, 12, 25, 8, 22, 16.5, 3, 24, 7, 16.5, 5, 9, 18

Note that the differences 31 and -31 are tied, and since they would occupy positions 16 and 17, each is assigned the average of these two ranks, or 16.5.

The next step is to compute the sum of the ranks of the positive differences and the sum of the ranks of the negative differences. The ranks associated with the positive differences are 1, 15, 10, 14, 19, 6, 12, 25, 8, 3, 7, 16.5, 9, and 18. (Check this.) The sum of these ranks is $\Sigma(+) = 163.5$. When using the normal approximation, we may use either sum of ranks. Since this is a left-tailed test, we want to reject the null hypothesis that the mean is 149 only if there is evidence that the mean is less than 149, that is, when the sum of the positive ranks is too small. We will therefore carry out the test on the left tail of the normal distribution.

Using equations 14–21 to 14–23, we compute the value of the test statistic Z as

$$z = \frac{T - E(T)}{\sigma_T} = \frac{T - n(n + 1)/4}{\sqrt{n(n + 1)(2n + 1)/24}} = \frac{163.5 - (25)(26)/4}{\sqrt{(25)(26)(51)/24}} = 0.027$$

This value of the statistic lies inside the nonrejection region, far from the critical point for any conventional level of significance. (If we had decided to carry out the test at $\alpha = 0.05$, our critical point would have been -1.645.) We do not reject the null hypothesis and conclude that there is no evidence that use of the satellite is declining.

In closing this section, we note that the Wilcoxon signed-rank test assumes that the distribution of the population is symmetric in the case of the single-sample test, and that the distribution of differences between the two populations in the paired, two-sample case is symmetric. This assumption allows us to make inferences about population means or medians. Another assumption inherent in our analysis is that the random variables in question are continuous. The measurement scale of the data is at least ordinal.

The Template

Figure 14–7 shows the use of the template for testing the mean (or median). The data entered correspond to Example 14–7. Note that we enter the claimed value of the mean (median) in every used row of the second column of data. In the problem, the null hypothesis is $\mu_1 \geq \mu_2$. Since the sample size is large, the p-values appear in the range J17:J19. The p-value we need is 0.5107. Since this is too large, we cannot reject the null hypothesis.

Note that the template always uses the sum of the negative ranks to calculate the test statistic Z. Hence its sign differs from the manual calculation. The final results—the p-value and whether or not we reject the null hypothesis—will be the same in both cases.

PROBLEMS

14–19. Explain the purpose of the Wilcoxon signed-rank test. When is this test useful? Why?

14–20. For problem 14–17, suppose that the returns for the Chicago and Dallas investments are paired by month: the first observation for each investment is for the first month (say, January), the second is for the next month, and so on. Conduct the analysis again, using the Wilcoxon signed-rank test. Is there a difference in your conclusion? Explain.

FIGURE 14–7   **Wilcoxon Test for the Mean or Median**
[Nonparametric Tests.xls; Sheet: Wilcoxon]

	A	B	C	G	H	I	J	K	L	M	N
1	**Wilcoxon's Signed Rank Test**						# Message transmitted				
2											
3		# Message	Mean								
4		X_1	X_2							Test	
5	1	151	149							Statistic	
6	2	144	149		$\Sigma(+)$	163.5		Null Hypothesis		T	
7	3	123	149		$\Sigma(-)$	161.5		$H_0: \mu_1 = \mu_2$		161.5	
8	4	178	149					$H_0: \mu_1 >= \mu_2$		163.5	
9	5	105	149		n	25		$H_0: \mu_1 <= \mu_2$		161.5	
10	6	112	149								
11	7	140	149		**For Large Samples ($n >= 25$)**						
12	8	167	149		Test Statistic						
13	9	177	149				z	−0.026907	$E[T]$	162.5	
14	10	185	149						$\sigma(T)$	37.1652	
15	11	129	149					At an α of			
16	12	160	149		Null Hypothesis	p-value		5%			
17	13	110	149		$H_0: \mu_1 = \mu_2$	0.9785					
18	14	170	149		$H_0: \mu_1 >= \mu_2$	0.5107					
19	15	198	149		$H_0: \mu_1 <= \mu_2$	0.4893					
20	16	165	149								

14–21.   An article discusses differences in management style between North American and European corporations. The following are paired management achievement scores for a European subsidiary and a North American subsidiary of the same firm for a random sample of 10 firms. Test the hypothesis of no difference in management style.

(25, 41), (28, 18), (37, 35), (10, 56), (14, 15), (51, 72), (30, 43), (28, 66), (33, 31), (20, 29)

14–22.   The average life of a 100-watt lightbulb is stated on the package to be 750 hours. The quality control director at the plant making the lightbulbs needs to check whether the statement is correct. The director is only concerned about a possible reduction in quality and will stop the production process only if statistical evidence exists to conclude that the average life of a lightbulb is less than 750 hours. A random sample of 20 bulbs is collected and left on until they burn out. The lifetime of each bulb is recorded. The data are (in hours of continuous use) 738, 752, 710, 701, 689, 779, 650, 541, 902, 700, 488, 555, 870, 609, 745, 712, 881, 599, 659, 793. Should the process be stopped and corrected? Explain why or why not.

14–23.   A retailer of tapes and compact disks wants to test whether people can differentiate by the quality of sound only—between the two products. A random sample of consumers who agreed to participate in the test and who have no particular experience with high-quality audio equipment is selected. The same musical performance is played for each person, once on a disk and once on a tape. The listeners do not know which is playing, and the order has been determined randomly. Each person is asked to state which of the two performances he or she prefers. What statistical test is most appropriate here? Why?

14–24.   From experience, a manager knows that the commissions earned by her salespeople are very well approximated by a normal distribution. The manager wants to test whether the average commission is $439 per month. A random sample of 100 observations is available. What statistical test is best in this situation? Why?

14–25. Returns on stock of small firms have been shown to be symmetrically distributed, but the distributions are believed to be "long-tailed"—not well approximated by the normal distribution. If it is desired to test whether the average return on a stock of a small firm is equal to 12% per year, what test would you recommend? Why?

14–26. Sky Pies is a recently opened shop at O'Hare International Airport in Chicago that sells frozen, packaged-to-go pizzas. "You can bring sourdough bread from San Francisco, then why not pizza from Chicago?" says founder Bill Gramas. Bill calculated that his business will turn a profit if he can sell at least 120 pizzas per day. Sales data, in number sold per day for the first 15 days, are as follows (assume these data are a random sample of daily sales): 145, 190, 206, 167, 120, 178, 110, 102, 119, 201, 100, 118, 127, 148, 155. Test the null hypothesis that average sales are less than or equal to 120 per day versus the alternative that the average is greater than 120. What is the main, serious limitation of this analysis?

14–27. Air New Zealand offers two package tours from the United States to New Zealand. One, which includes airfare and five nights' hotel accommodation in Auckland, is advertised at $1,799. The other includes only two nights' accommodation and is advertised at $1,710. For a random sample of 12 days, the airline records the number of bookings for each package. The paired observations are as follows. Do you believe that one package is more popular than the other? Explain.

$1,799 package:	56, 79, 85, 77, 32, 48, 88, 95, 57, 70, 52, 90
$1,710 package:	60, 85, 70, 82, 41, 60, 89, 80, 77, 86, 66, 75

14–28. A stock market analyst wants to test whether there are higher-than-usual returns on stocks following a two-for-one split. A random sample of 10 stocks that recently split is available. For each stock, the analyst records the percentage return during the month preceding the split and the percentage return for the month following the split. The data are

Before split (%):	0.5, −0.2, 0.9, 1.1, −0.7, 1.5, 2.0, 1.3, 1.6, 2.1
After split (%):	1.1, 0.3, 1.2, 1.9, −0.2, 1.4, 1.8, 1.8, 2.4, 2.2

Is there evidence that a stock split causes excess returns for the month following the split? Redo the problem, using the sign test. Compare the results of the sign test with those of the Wilcoxon test.

14–29. Much has been said about airline deregulation and the effects it has had on the airline industry and its performance. Following a deluge of complaints from passengers, the public relations officer of one of the major airlines asked the company's operations manager to look into the problem. The operations manager obtained average takeoff delay figures for a random sample of the company's routes over time periods of equal length before and after the deregulation. The data, in minutes of average delay per route, are as follows.

Before:	3, 2, 4, 5, 1, 0, 1, 5, 6, 3, 10, 4, 11, 7
After:	6, 8, 2, 9, 8, 2, 6, 12, 5, 9, 8, 12, 11, 10

Is there evidence in these data that the airline's delays have increased after deregulation?

14–30. The average score on a vocational training test has been known to be 64. Recently, several changes have been instituted in the program; the effect of these changes on performance on the test is unknown. It is therefore desirable to test the null hypothesis that the average score for all people who will complete the program will be 64 versus the alternative that it will not be 64. The following random sample of scores is available.

87, 91, 65, 31, 8, 53, 99, 44, 42, 60, 77, 73, 42, 50, 79, 90, 54, 39, 77, 60, 33, 41, 42, 85, 71, 50, 63, 58, 89, 5, 66, 99, 57, 12, 47, 72, 80, 84

Conduct the test.

14–6   The Kruskal-Wallis Test—A Nonparametric Alternative to One-Way ANOVA

Remember that the ANOVA procedure discussed in Chapter 9 requires the assumption that the populations being compared are all normally distributed with equal variance. When there is reason to believe that the populations under study are not normally distributed, we cannot use the ANOVA procedure. There is, however, a nonparametric test designed to detect differences among populations that does not require any assumptions about the shape of the population distributions. This test is the **Kruskal-Wallis test**. The test is the nonparametric alternative to the (completely randomized design) one-way analysis of variance. In the next section, we will see a nonparametric alternative to the randomized block design analysis of variance, the *Friedman test*. Both of these tests use ranks.

The Kruskal-Wallis test is an analysis of variance that uses the ranks of the observations rather than the data themselves. This assumes, of course, that the observations are on an interval scale. If our data are in the form of ranks, we use them as they are. The Kruskal-Wallis test is identical to the Mann-Whitney test when only two populations are involved. We thus use the Kruskal-Wallis test for comparing k populations, where k is greater than 2. The null hypothesis is that the k populations under study have the same distribution, and the alternative hypothesis is that at least two of the population distributions are different from each other.

The Kruskal-Wallis hypothesis test is

H_0: All k populations have the same distribution

H_1: Not all k populations have the same distribution

(14–24)

Although the hypothesis test is stated in terms of the distributions of the populations of interest, the test is most sensitive to differences in the locations of the populations. Therefore, the procedure is actually used to test the ANOVA hypothesis of equality of k population means. The only assumptions required for the Kruskal-Wallis test are that the k samples are random and are independently drawn from the respective populations. The random variables under study are continuous, and the measurement scale used is at least ordinal.

We rank all data points in the entire set from smallest to largest, without regard to which sample they come from. Then we sum all the ranks from each separate sample. Let n_1 be the sample size from population 1, n_2 the sample size from population 2, and so on up to n_k, which is the sample size from population k. Define n as the total sample size: $n = n_1 + n_2 + \cdots + n_k$. We define R_1 as the sum of the ranks from sample 1,

R_2 as the sum of the ranks from sample 2, and so on to R_k, the sum of the ranks from sample k. We now define the Kruskal-Wallis test statistic H.

The Kruskal-Wallis test statistic is

$$H = \frac{12}{n(n+1)}\left(\sum_{j=1}^{k}\frac{R_j^2}{n_j}\right) - 3(n+1) \qquad (14\text{-}25)$$

For very small samples $(n_j < 5)$, there are tables for the exact distribution of H under the null hypothesis; these are found in books devoted to nonparametric statistics. Usually, however, we have samples that are greater than 5 for each group (remember the serious limitations of inference based on very small samples). For larger samples, as long as each n_j is at least 5, the distribution of the test statistic H under the null hypothesis is well approximated by the chi-square distribution with $k - 1$ degrees of freedom.

We reject the null hypothesis on the right-hand tail of the chi-square distribution. That is, we reject the null hypothesis if the computed value of H is too large, exceeding a critical point of $\chi^2_{(k-1)}$ for a given level of significance α. We demonstrate the Kruskal-Wallis test with an example.

EXAMPLE 14–8

A company is planning to buy a word processing software package to be used by its office staff. Three available packages, made by different companies, are considered: Multimate, WordPerfect, and Microsoft Word. Demonstration packages of the three alternatives are available, and the company selects a random sample of 18 staff members, 6 members assigned to each package. Every person in the sample learns how to use the particular package to which she or he is assigned. The time it takes every member to learn how to use the word processing package is recorded. The question is: Does it take approximately the same amount of time to learn how to use each package proficiently?

None of the office staff has used any of these packages before, and because of similarity in use, each person is assigned to learn only one package. The staff, however, have varying degrees of experience. In particular, some are very experienced typists, and others are beginners. Therefore, it is believed that the three populations of time it takes to learn how to use a package are not normally distributed. If a conclusion is reached that one package takes longer to learn than the others, then learning time will be a consideration in the purchase decision. Otherwise, the decision will be based only on package capabilities and price. Table 14–3 (page 692) gives the data, in minutes, for every person in the three samples. It also shows the ranks and the sum of the ranks for each group.

Solution

Using the obtained sums of ranks for the three groups, we compute the Kruskal-Wallis statistic H. From equation 14–25 we get

$$H = \frac{12}{n(n+1)}\left(\sum\frac{R_j^2}{n_j}\right) - 3(n+1) = \frac{12}{(18)(19)}\left(\frac{90^2}{6} + \frac{56^2}{6} + \frac{25^2}{6}\right) - 3(19)$$

$$= 12.3625$$

TABLE 14–3   **The Data (in minutes) and Ranks for Example 14–8**

Multimate		WordPerfect		Microsoft Word	
Time	Rank	Time	Rank	Time	Rank
45	14	30	8	22	4
38	10	40	11	19	3
56	16	28	7	15	1
60	17	44	13	31	9
47	15	25	5	27	6
65	18	42	12	17	2
	$R_1 = 90$		$R_2 = 56$		$R_3 = 25$

FIGURE 14–8   **Carrying Out the Test for Example 14–8**

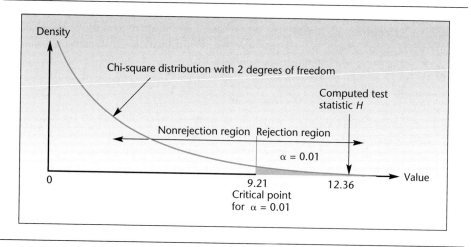

We now perform the test of the hypothesis that the populations of the learning times of the three software packages are identical. We compare the computed value of H with critical points of the chi-square distribution with $k - 1 = 3 - 1 = 2$ degrees of freedom. Using Appendix C, Table 4 (pages 840–841), we find that $H = 12.36$ exceeds the critical point for $\alpha = 0.01$, which is given as 9.21. We therefore reject the null hypothesis and conclude that there is evidence that the time it takes to learn how to use the word processing packages is not the same for all three; at least one package takes longer to learn. Our p-value is smaller than 0.01. The test is demonstrated in Figure 14–8.

We note that even though our example had a balanced design (equal sample sizes in all groups), the Kruskal-Wallis test can also be performed if sample sizes are different. We also note that we had no ties in this example. If ties do exist, we assign them the average rank, as we have done in previous tests based on ranks. It is possible to correct for the effect of ties by using a correction formula, which may be found in advanced books.

FIGURE 14–9 **The Template for the Kruskal-Wallis Test [Nonparametric Tests.xls; Sheet: Kruskal-Wallis]**

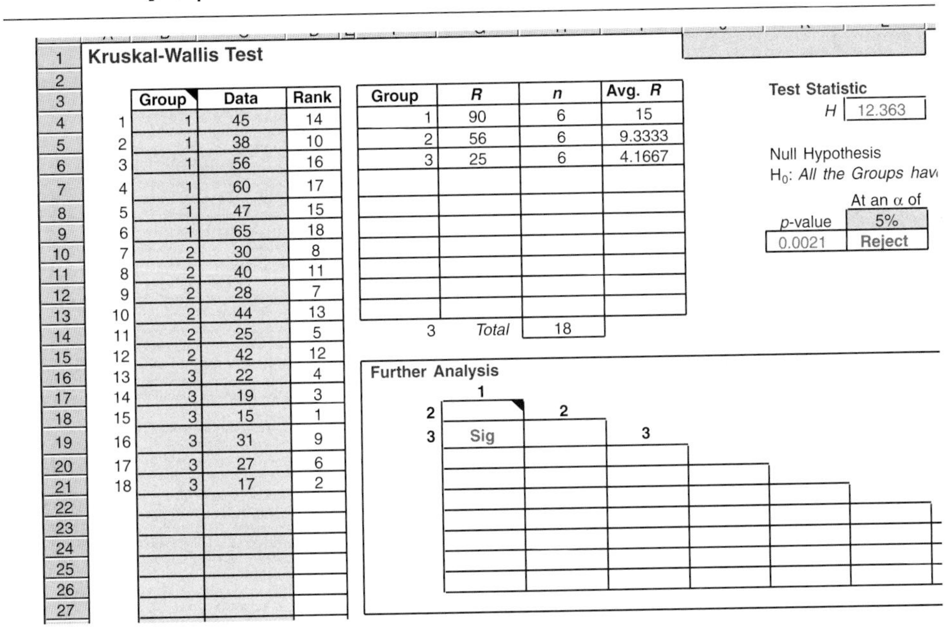

	Group	Data	Rank
1	1	45	14
2	1	38	10
3	1	56	16
4	1	60	17
5	1	47	15
6	1	65	18
7	2	30	8
8	2	40	11
9	2	28	7
10	2	44	13
11	2	25	5
12	2	42	12
13	3	22	4
14	3	19	3
15	3	15	1
16	3	31	9
17	3	27	6
18	3	17	2

Group	R	n	Avg. R
1	90	6	15
2	56	6	9.3333
3	25	6	4.1667
3	Total	18	

	1	2	3
2			
3	Sig		

The Template

The template for the Kruskal-Wallis test is shown in Figure 14–9. Note that the group numbers are to be entered in the first column of data as 1, 2, 3, ... The group numbers need not be in a particular order.

The data seen in the template correspond to Example 14–3. The advantage in using the template is that we get to know the exact p-value. It is 0.0021, seen in cell K10. In addition, the tabulation in the range G18:O26 shows if the difference in the means of every pair of groups is significant. The appearance of "Sig" means the corresponding difference is significant. In the current problem, the difference in the means of groups 1 and 3 is significantly more than zero. This aspect of the problem is discussed a little later, in the subsection "Further Analysis."

EXAMPLE 14–9

Because its delivery times are too slow, a trucking company is on the verge of losing an important customer. A manager wants to explore upgrading the fleet of trucks. There are three new models to choose from, each of which claims significant fuel efficiency improvements. Better gas mileage translates to fewer stops on long trips, cutting delivery times.

The manager is allowed to test-drive the trucks for a few days and randomly picks 15 drivers to do so. Five drivers will test each truck. The mpg results are as follows. Conduct a Kruskal-Wallis rank test for differences in the three population medians.

	A	B
1	Truck	MPG
2	Truck A	17.00

	A	B
3	Truck A	18.20
4	Truck A	18.50
5	Truck C	18.70
6	Truck A	19.40
7	Truck C	19.90
8	Truck C	20.30
9	Truck C	21.10
10	Truck B	22.70
11	Truck A	23.50
12	Truck B	23.80
13	Truck C	23.90
14	Truck B	24.20
15	Truck B	25.10
16	Truck B	26.30

Solution Figure 14–10 shows the template solution to the problem. Since the p-value of 0.014 is less than 5%, we reject the null hypothesis that the medians of the mileage for the three groups are equal at an α of 5%.

Further Analysis

As in the case of the usual ANOVA, once we reject the null hypothesis of no difference among populations, the question arises: Where are the differences? That is, which populations are different from which? Here we use a procedure that is similar to the Tukey method of further analysis following ANOVA. For every pair of populations we wish to compare (populations i and j, for example), we compute the average rank of the sample.

FIGURE 14–10 **The Template Solution to Example 14–9 [Nonparametric Tests.xls; Sheet: Kruskal-Wallis]**

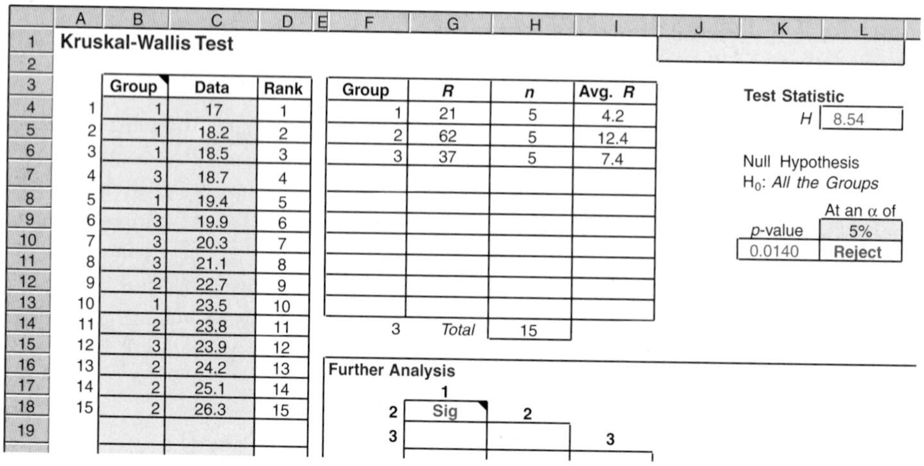

Kruskal-Wallis Test

	Group	Data	Rank
1	1	17	1
2	1	18.2	2
3	1	18.5	3
4	3	18.7	4
5	1	19.4	5
6	3	19.9	6
7	3	20.3	7
8	3	21.1	8
9	2	22.7	9
10	1	23.5	10
11	2	23.8	11
12	3	23.9	12
13	2	24.2	13
14	2	25.1	14
15	2	26.3	15

Group	R	n	Avg. R
1	21	5	4.2
2	62	5	12.4
3	37	5	7.4
3	*Total*	15	

Test Statistic
H 8.54

Null Hypothesis
H_0: *All the Groups*

p-value	At an α of 5%
0.0140	Reject

Further Analysis

	1	2	3
2	Sig		
3			

$$\bar{R}_i = \frac{R_i}{n_i} \quad \text{and} \quad \bar{R}_j = \frac{R_j}{n_j} \tag{14-26}$$

where R_i and R_j are the sums of the ranks from samples i and j, respectively, computed as part of the original Kruskal-Wallis test. We now define the test statistic D as the absolute difference between $\bar{R}_i$ and $\bar{R}_j$.

The test statistic for determining whether there is evidence to reject the null hypothesis that populations i and j are identical is

$$D = |\bar{R}_i - \bar{R}_j| \tag{14-27}$$

We carry out the test by comparing the test statistic D with a quantity that we compute from the critical point of the chi-square distribution at the same level α at which we carried out the Kruskal-Wallis test. The quantity is computed as follows.

The critical point for the paired comparisons is

$$C_{KW} = \sqrt{\chi^2_{\alpha,k-1} \frac{n(n+1)}{12} \left(\frac{1}{n_i} + \frac{1}{n_j}\right)} \tag{14-28}$$

where $\chi^2_{\alpha,\,k-1}$ is the critical point of the chi-square distribution used in the original, overall test.

By comparing the value of the statistic D with C_{KW} for every pair of populations, we can perform all pairwise comparisons *jointly* at the level of significance α at which we performed the overall test. We reject the null hypothesis if and only if $D > C_{KW}$. We demonstrate the procedure by performing all three pairwise comparisons of the populations in Example 14–8.

Since we have a balanced design, $n_i = n_j = 6$ for all three samples, the critical point C_{KW} will be the same for all pairwise comparisons. Using equation 14–28 and 9.21 as the value of chi-square for the overall test at $\alpha = 0.01$, we get

$$C_{KW} = \sqrt{9.21 \frac{(18)(19)}{12} \left(\frac{1}{6} + \frac{1}{6}\right)} = 9.35$$

Comparing populations 1 and 2: From the bottom of Table 14–3, we find that $R_1 = 90$ and $R_2 = 56$. Since the sample sizes are each 6, we find that the average rank for sample 1 is 90/6 = 15, and the average rank for sample 2 is 56/6 = 9.33. Hence, the test statistic for comparing these two populations is the absolute value of the difference between 15 and 9.33, which is 5.67. This value is less than C_{KW}, and we must conclude that there is no evidence, at $\alpha = 0.01$, of a difference between populations 1 and 2.

Comparing populations 1 and 3: Here the absolute value of the difference between the average ranks is $|(90/6) - (25/6)| = 10.83$. Since 10.83 is greater than $C_{KW} = 9.35$, we conclude that there is evidence, at $\alpha = 0.01$, that population 1 is different from population 3.

Comparing populations 2 and 3: Here we have $D = |(56/6) - (25/6)| = 5.17$, which is less than 9.35. Therefore we conclude that there is no evidence, at $\alpha = 0.01$, that populations 2 and 3 are different.

Our interpretation of the data is that at $\alpha = 0.01$, there are significant differences only between the time it takes to learn Multimate and the time it takes to learn Microsoft Word. Since the values for Multimate are larger, we conclude that the study provides evidence that Multimate takes longer to learn.

PROBLEMS

14–31. Electronic Data Interchange (EDI) is an electronic pipeline linking suppliers with their customers. The system cuts the delivery time of goods by as much as 50% compared with other means of communication. The system allows specially formatted documents, such as purchase orders, to be sent from one company's computer to that of another. Currently, the system is used mainly by three industries: grocery, transportation, and pharmaceutical. EDI's management plans great expansion in the next few years and wants to know whether the three industry groups make about equal use of the system, or whether one group uses it more frequently than another group. A random sample of 10 grocery retailers using the system is gathered, as well as a random sample of 8 transportation users and 7 pharmaceutical users. For each company, the number of times EDI was used in the last month is recorded. The data are as follows. Is the frequency of use of EDI's services about the same for all three industry groups?

Groceries:	12, 14, 30, 5, 9, 18, 52, 19, 65, 25
Transportation:	48, 72, 99, 30, 62, 79, 120, 88
Pharmaceuticals:	40, 48, 67, 112, 31, 141, 69

14–32. An analyst in the publishing industry wants to find out whether the cost of a newspaper advertisement of a given size is about the same in four large newspaper groups. Random samples of seven newspapers from each group are selected, and the cost of an ad is recorded. The data follow (in dollars). Do you believe that there are differences in the price of an ad across the four groups?

Group A:	57, 65, 50, 45, 70, 62, 48
Group B:	72, 81, 64, 55, 90, 38, 75
Group C:	35, 42, 58, 59, 46, 60, 61
Group D:	73, 85, 92, 68, 82, 94, 66

14–33. Lawyers representing the Beatles filed a $15 million suit in New York against Nike, Inc., over Nike's Air Max shoe commercial set to the Beatles' 1968 hit song "Revolution." As part of all such lawsuits, the plaintiff must prove a financial damage— in this case, that Nike improperly gained from the unlicensed use of the Beatles' song. In proving their case, lawyers for the Beatles had to show that "Revolution," or any Beatles' song, is not just a tune played with the commercial and that, in fact, the use of the song made the Nike commercial more appealing than it would have been if it had featured another song or melody. A statistician was hired to aid in proving this point. The statistician designed a study in which the Air Max commercial was recast using two other randomly chosen songs that were in the public domain and did not require permission, and that were not sung by the Beatles. Then three groups of 12 people each were randomly selected. Each group was shown one of the commercials, and every person's appeal score for the commercial was recorded. Using the following appeal scores, determine whether there is statistical evidence that not all three songs would be equally effective in the commercial. If you do reject the null hypothesis of equal appeal, go the required extra step to prove that the Beatles' "Revolution" does indeed have greater appeal over other songs, and that Nike should pay the Beatles for using it.

"Revolution":	95, 98, 96, 99, 91, 90, 97, 100, 96, 92, 88, 93
Random alternative A:	65, 67, 66, 69, 60, 58, 70, 64, 64, 68, 61, 62
Random alternative B:	59, 57, 55, 63, 59, 44, 49, 48, 46, 60, 47, 45

14–34. A researcher at an accounting firm wants to find out whether the current ratio for three industries is about the same. Random samples of eight firms in industry A, six firms in industry B, and six firms in industry C are available. The current ratios are as follows:

Industry A:	1.38, 1.55, 1.90, 2.00, 1.22, 2.11, 1.98, 1.61
Industry B:	2.33, 2.50, 2.79, 3.01, 1.99, 2.45
Industry C:	1.06, 1.37, 1.09, 1.65, 1.44, 1.11

Conduct the test at $\alpha = 0.05$, and state your conclusion.

14–35. Recently, Japan's asset-rich life insurance companies have been venturing overseas and becoming a force in world financial markets. The insurance companies invest in real estate in New York, Canada, and the U.S. Sun Belt. A real estate–investment broker hired by one Japanese insurance company wanted to find out whether the return on investment in comparable real estate in each of these areas is approximately the same. Random sample data for investments in the three areas follow (figures represent annualized percentage return).

New York:	15, 18, 17, 19, 18, 10, 12, 16
Canada:	10, 9, 8, 11, 7, 13
Sun Belt:	21, 20, 22, 14, 23, 16, 24

Conduct the test. Are returns approximately equal in the three areas?

14–36. The following data are small random samples of rents (in dollars) in five U.S. cities. The New York data are for Manhattan only.

New York (Manhattan):	900, 1,200, 850, 1,320, 1,400, 1,150, 975
Chicago:	625, 640, 775, 1,000, 690, 550, 840, 750
Detroit:	415, 400, 420, 560, 780, 620, 800, 390
Tampa:	410, 310, 320, 280, 500, 385, 440
Orlando:	340, 425, 275, 210, 575, 360

Conduct the Kruskal-Wallis test to determine whether evidence exists that there are differences in the rents in these cities. If differences exist, where are they?

14–37. What assumptions have you used when solving problems 14–31 through 14–36? What assumptions did you not make about the populations in question? Explain.

14–7 The Friedman Test for a Randomized Block Design

Recall the randomized block design, which was discussed in Chapter 9. In this design, each block of units is assigned all k treatments, and our aim is to determine possible differences among treatments or treatment means (in the context of ANOVA). A *block* may be one person who is given all k treatments (asked to try k different products, to rate k different items, etc.). The Kruskal-Wallis test discussed in the previous section is a nonparametric version of the one-way ANOVA with completely randomized design. Similarly, the **Friedman test,** the subject of this section, is a nonparametric version of the randomized block design ANOVA. Sometimes this design is referred to as a two-way ANOVA with one item per cell because it is possible to view the blocks as one factor and the treatment levels as the other. In the randomized block design, however, we are interested in the treatments as a factor and not in the blocks themselves. Like the methods we discussed in preceding sections, the Friedman test is based on ranks. The test may be viewed as an extension of the Wilcoxon signed-rank test or an extension of the sign test to more than two treatments per block. Recall that in each of these tests, there are two treatments assigned to each element in the sample—the observations are paired. In the Friedman test, the observations are more than paired: each block, or person, is assigned to all $k > 2$ treatments.

Since the Friedman test is based on the use of ranks, it is especially useful for testing treatment effects when the observations are in the form of ranks. In fact, in such situations, we cannot use the randomized block design ANOVA because the assumption of a normal distribution cannot hold for very discrete data such as ranks. The Friedman test is a unique test for a situation where data are in the form of ranks within each block. Our example will demonstrate the use of the test in this particular situation. When our data are on an interval scale and not in the form of ranks, but we believe that the assumption of normality may not hold, we use the Friedman test instead of the parametric ANOVA and transform our data to ranks.

> The null and alternative hypotheses of the Friedman test are
>
> H_0: The distributions of the k treatment populations are identical
> H_1: Not all k distributions are identical   (14–29)

The data for the Friedman test are arranged in a table in which the rows are blocks (or units, if each unit is a block). There are n blocks. The columns are treatments, and there are k of them. Let us assume that each block is one person who is assigned to all treatments. The data in this case are arranged as in Table 14–4.

TABLE 14–4   **Data Layout for the Friedman Test**

	Treatment 1	Treatment 2	Treatment 3	. . .	Treatment k
Person 1					
Person 2					
Person 3					
$\vdots$	$\vdots$	$\vdots$	$\vdots$	. . .	$\vdots$
Person n					
Sum of ranks:	R_1	R_2	R_3	. . .	R_k

If the data are not already in the form of ranks within each block, we rank the observations within each block from 1 to k. That is, the smallest observation in the block is given rank 1, the second smallest gets rank 2, and the largest gets rank k. Then we sum all the ranks for every treatment. The sum of all the ranks for treatment 1 is R_1, the sum of the ranks for treatment 2 is R_2, and so on to R_k, the sum of all the ranks given to treatment k.

If the distributions of the k populations are indeed identical, as stated in the null hypothesis, then we expect that the sum of the ranks for each treatment would not differ much from the sum of the ranks of any other treatment. The differences among the sums of the ranks are measured by the Friedman test statistic, denoted by X^2. When this statistic is too large, we reject the null hypothesis and conclude that at least two treatments do not have the same distribution.

The Friedman test statistic is

$$X^2 = \frac{12}{nk(k+1)} \sum_{j=1}^{k} R_j^2 - 3n(k+1) \qquad (14\text{--}30)$$

When the null hypothesis is true, the distribution of X^2 approaches the chi-square distribution with $k - 1$ degrees of freedom as n increases. For small values of k and n, tables of the exact distribution of X^2 under the null hypothesis may be found in nonparametric statistics books. Here we will use the chi-square distribution as our decision rule. We note that for small n, the chi-square approximation is *conservative;* that is, we may not be able to reject the null hypothesis as easily as we would if we use the exact distribution table. Our decision rule is to reject H$_0$ at a given level, α, if X^2 exceeds the critical point of the chi-square distribution with $k - 1$ degrees of freedom and right-tail area α. We now demonstrate the use of the Friedman test with an example.

EXAMPLE 14–10

There is a segment of the population, mostly retired people, who frequently go on low-budget cruises. Many travel agents specialize in this market and maintain mailing lists of people who take frequent cruises. One such travel agent in Fort Lauderdale wanted to find out whether "frequent cruisers" prefer some of the cruise lines in the low-budget range over others. If so, the agent would concentrate on selling tickets on the preferred line(s) rather than on a wider variety of lines. From a mailing list of people who have taken at least one cruise on each of the three cruise lines Carnival, Costa, and Sitmar, the agent selected a random sample of 15 people and asked them to rank their overall experiences with the three lines. The ranks were 1 (best), 2 (second best), and 3 (worst). The results are given in Table 14–5. Are the three cruise lines equally preferred by people in the target population?

Solution

Using the sums of the ranks of the three treatments (the three cruise lines), we compute the Friedman test statistic. From equation 14–30, we get

$$X^2 = \frac{12}{nk(k+1)} (R_1^2 + R_2^2 + R_3^2) - 3n(k+1)$$

$$= \frac{12}{(15)(3)(4)} (31^2 + 21^2 + 38^2) - 3(15)(4) = 9.73$$

TABLE 14–5   Sample Results of Example 14–10

Respondent	Carnival	Costa	Sitmar
1	1	2	3
2	2	1	3
3	1	3	2
4	2	1	3
5	3	1	2
6	3	1	2
7	1	2	3
8	3	1	2
9	2	1	3
10	1	2	3
11	2	1	3
12	3	1	2
13	1	2	3
14	3	1	2
15	3	1	2
	$R_1 = 31$	$R_2 = 21$	$R_3 = 38$

We now compare the computed value of the statistic with values of the right tail of the chi-square distribution with $k - 1 = 2$ degrees of freedom. The critical point for $\alpha = 0.01$ is found from Appendix C, Table 4, to be 9.21. Since 9.73 is greater than 9.21, we conclude that there is evidence that not all three low-budget cruise lines are equally preferred by the frequent cruiser population.

The Template

Figure 14–11 shows the template that can be used to conduct a Friedman test. The data seen in the figure correspond to Example 14–10. The RowSum column in the template can be used to make a quick check of data entry. All the sums must be equal.

The template provides the p-value in cell O10. Since it is less than 1%, we can reject the null hypothesis that all cruise lines are equally preferred at an α of 1%.

PROBLEMS

14–38.   A random sample of 12 consumers are asked to rank their preferences of four new fragrances that a perfume manufacturer wants to introduce to the market in the coming fall. The data are as follows (best liked denoted by 1 and least liked denoted by 4). Do you believe that all four fragrances are equally liked? Explain.

Respondent	Fragrance 1	Fragrance 2	Fragrance 3	Fragrance 4
1	1	2	4	3
2	2	1	3	4
3	1	3	4	2
4	1	2	3	4
5	1	3	4	2
6	1	4	3	2

FIGURE 14–11   The Template for the Friedman Test
[Nonparametric Tests.xls; Sheet: Friedman]

	A	B	C	D	E	F	G	H	I	J	K	L	M	N	O	P
1	Friedman Test						Cruise Lines									
2																
3		R	31	21	38											
4			1	2	3	4	5	6	7	8	9	10	RowSum			
5		1	1	2	3								6		n	15
6		2	2	1	3								6		k	3
7		3	1	3	2								6			
8		4	2	1	3								6		X^2	9.733333
9		5	3	1	2								6			
10		6	3	1	2								6		p-value	0.0077
11		7	1	2	3								6			
12		8	3	1	2								6			
13		9	2	1	3								6			
14		10	1	2	3								6			
15		11	2	1	3								6			
16		12	3	1	2								6			
17		13	1	2	3								6			
18		14	3	1	2								6			
19		15	3	1	2								6			
20																
21																

Respondent	Fragrance 1	Fragrance 2	Fragrance 3	Fragrance 4
7	1	3	4	2
8	2	1	4	3
9	1	3	4	2
10	1	3	2	4
11	1	4	3	2
12	1	3	4	2

14–39.   While considering three managers for a possible promotion, the company president decided to solicit information from employees about the managers' relative effectiveness. Each person in a random sample of 10 employees who had worked with all three managers was asked to rank the managers, where best is denoted by 1, second best by 2, and worst by 3. The data follow. Based on the survey, are all three managers perceived as equally effective? Explain.

Respondent	Manager 1	Manager 2	Manager 3
1	3	2	1
2	3	2	1
3	3	1	2
4	3	2	1
5	2	3	1
6	3	1	2
7	3	2	1
8	3	2	1
9	3	1	2
10	3	1	2

14–40.   In testing to find a cure for a nervous problem, it is not possible to directly quantify the condition of a patient after he or she has been treated with a drug, except to compare the patient's condition with those of other patients with the same illness severity who were treated with other drugs. A pharmaceutical firm conducting clinical trials therefore selects a random sample of 27 patients. The sample is then separated into blocks of three patients each, with the three patients in each block having about the same pretreatment condition. Each person in a block is then randomly assigned to be treated by one of the three drugs under consideration. After the treatment, a physician evaluates each person's condition and ranks the patient in comparison with the others in the same block (with 1 indicating the most improvement and 3 indicating the least improvement). Using the following data, do you believe that all three drugs are equally effective?

Block	Drug A	Drug B	Drug C
1	2	3	1
2	2	3	1
3	2	3	1
4	2	3	1
5	1	3	2
6	2	3	1
7	2	1	3
8	2	3	1
9	1	2	3

14–41. Four different processes for baking cakes commercially are considered. The cakes produced by each process are evaluated in terms of their overall quality. Since the cakes sometimes may not rise, the distribution of quality ratings is different from a normal distribution. When conducting a test of the quality of the four processes, cakes are blocked into groups of four according to the ingredients used. The ratings of the cakes baked by the four processes are as follows. (Ratings are on a scale of 0 to 100.) Are the four processes equally good? Explain.

Block	Process 1	Process 2	Process 3	Process 4
1	87	65	73	20
2	98	60	39	45
3	85	70	50	60
4	90	80	85	50
5	78	40	60	45
6	95	35	70	25
7	70	60	55	40
8	99	70	45	60

14–8   The Spearman Rank Correlation Coefficient

Recall our discussion of correlation in Chapter 10. There we stressed the assumption that the distributions of the two variables in question, X and Y, are normal. In cases where this assumption is not realistic, or in cases where our data are themselves in the form of ranks or are otherwise on an ordinal scale, we have alternative measures of the degree of association between the two variables. The most frequently used nonparametric measure of the correlation between two variables is the *Spearman rank correlation coefficient,* denoted by r_s.

Our data are pairs of n observations on two variables X and Y—pairs of the form (x_i, y_i), where $i = 1, \ldots, n$. To compute the Spearman correlation coefficient, we first rank all the observations of one variable within themselves from smallest to largest. Then we independently rank the values of the second variable from smallest to largest. *The Spearman rank correlation coefficient is the usual (Pearson) correlation coefficient applied to the ranks.* When no ties exist, that is, when there are no two values of X or two values of Y with the same rank, there is an easier computational formula for the Spearman correlation coefficient. The formula follows.

The **Spearman rank correlation coefficient** (assuming no ties) is

$$r_s = 1 - \frac{6 \sum_{i=1}^{n} d_i^2}{n(n^2 - 1)} \qquad (14\text{–}31)$$

where d_i, $i = 1, \ldots, n$, are the differences in the ranks of x_i and y_i: $d_i = R(x_i) - R(y_i)$.

If we do have ties within the X values or the Y values, but the number of ties is small compared with n, equation 14–31 is still useful.

The Spearman correlation coefficient satisfies the usual requirements of correlation measures. It is equal to 1 when the variables X and Y are perfectly positively related, that is, when Y increases whenever X does, and vice versa. It is equal to -1 in the opposite situation, where X increases whenever Y decreases. It is equal to 0 when there is no relation between X and Y. Values between these extremes give a relative indication of the degree of association between X and Y.

As with the parametric Pearson correlation coefficient, there are two possible uses for the Spearman statistic. It may be used as a descriptive statistic giving us an indication of the association between X and Y. We may also use it for *statistical inference.* In the context of inference, we assume that there is a certain correlation in the ranks of the values of the bivariate population of X and Y. This population rank correlation is denoted by ρ_s. We want to test whether $\rho_s = 0$, that is, whether there is an association between the two variables X and Y.

The hypothesis test for association between two variables is

$$\begin{aligned} &H_0\text{: } \rho_s = 0 \\ &H_1\text{: } \rho_s \neq 0 \end{aligned} \qquad (14\text{–}32)$$

This is a two-tailed test for the existence of a relation between X and Y. One-tailed versions of the test are also possible. If we want to test for a positive association between the variables, then the alternative hypothesis is that the parameter ρ_s is strictly greater than zero. If we want to test for a negative association only, then the alternative hypothesis is that ρ_s is strictly less than zero. The test statistic is simply r_s, as defined in equation 14–31.

When the sample size is less than or equal to 30, we use Appendix C, Table 11 (page 864). The table gives critical points for various levels of significance α. For a two-tailed test, we double the α level given in the table and reject the null hypothesis

TABLE 14–6 Data on the MMI and S&P 100 Indexes (Example 14–11)

MMI	S&P 100
220	151
218	150
216	148
217	149
215	147
213	146
219	152
236	165
237	162
235	161

if r_s is either greater than or equal to the table value C or less than or equal to $-C$. In a right-tailed test, we reject only if r_s is greater than or equal to C; and in a left-tailed test, we reject only if r_s is less than or equal to $-C$. In either one-tailed case, we use the α given in one of the columns in the table (we do not double it).

For larger sample sizes, we use the normal approximation to the distribution of r_s under the null hypothesis. The Z statistic for such a case is as follows.

A large-sample test statistic for association is

$$z = r_s \sqrt{n-1} \qquad (14–33)$$

We demonstrate the computation of Spearman's statistic, and a test of whether the population rank correlation is zero, with Example 14–11.

EXAMPLE 14–11 The S&P 100 Index is an index of 100 stock options traded on the Chicago Board Options Exchange. The MMI is an index of 20 stocks with options traded on the American Stock Exchange. Since options are volatile, the assumption of a normal distribution may not be appropriate, and the Spearman rank correlation coefficient may provide us with information about the association between the two indexes.[4] Using the reported data on the two indexes, given in Table 14–6, compute the r_s statistic, and test the null hypothesis that the MMI and the S&P 100 are not related against the alternative that they are positively correlated.

Solution We rank the MMI values and the S&P 100 values and compute the 10 differences: $d_i = \text{rank}(\text{MMI}_i) - \text{rank}(\text{S\&P100}_i)$. This is shown in Table 14–7. The order of the values in the table corresponds to their order in Table 14–6.

---

[4]*Volatility* means that there are jumps to very small and very large values. This gives the distribution long tails and makes it different from the normal distribution. For *stock returns*, however, the normal assumption is a good one, as mentioned in previous chapters.

TABLE 14–7   **Ranks and Rank Differences for Example 14–11**

Rank(MMI)	Rank(S&P 100)	Difference
7	6	1
5	5	0
3	3	0
4	4	0
2	2	0
1	1	0
6	7	−1
9	10	−1
10	9	1
8	8	0

We now use equation 14–31 and compute r_s:

$$r_s = 1 - \frac{6(d_1^2 + d_2^2 + \cdots + d_{10}^2)}{10(10^2 - 1)} = 1 - \frac{24}{990} = 0.9758$$

The sample correlation is very high.

We now use the r_s statistic in testing the hypotheses:

$$\text{H}_0: \rho_s \leq 0$$
$$\text{H}_1: \rho_s > 0 \tag{14–34}$$

We want to test for the existence of a positive rank correlation between MMI and S&P 100 in the *population* of values of the two indexes. We want to test whether the high sample rank correlation we found is statistically significant. Since this is a right-tailed test, we reject the null hypothesis if r_s is greater than or equal to a point C found in Appendix C, Table 11 (page 864), at a level of α given in the table. We find from the table that for $\alpha = 0.005$ and $n = 10$, the critical point is 0.794. Since $r_s = 0.9758 > 0.794$, we reject the null hypothesis and conclude that the MMI and the S&P 100 are positively correlated. The p-value is less than 0.005.

In closing this section, we note that Spearman's rank correlation coefficient is sometimes referred to as *Spearman's rho* (the Greek letter ρ). There is another commonly used nonparametric measure of correlation. This one was developed by Kendall and is called *Kendall's tau* (the Greek letter τ). Since Kendall's measure is not as simple to compute as the Spearman coefficient of rank correlation, we leave it to texts on nonparametric statistics.

The Template

Figure 14–12 shows the template that can be used for calculating and testing Spearman's rank correlation coefficients. The data entered in columns B and C can be raw data or ranks themselves. The p-values in the range J15:J17 appear only if the sample is large ($n > 30$). Otherwise, the message "Look up the tables for p-value" appears in cell J6.

FIGURE 14–12 The Template for Calculating Spearman's Rank Correlation [Nonparametric Tests.xls; Sheet: Spearman]

	A	B	C	D	E	F	G	H	I	J	K	L
1	Spearman Rank Correlation Coefficient								MMI & S&P 100			
2												
3		MMI	S&P100									
4		*X*	*Y*	Rank *X*	Rank *Y*							
5	1	220	151	7	6		n	10				
6	2	218	150	5	5		ρ_s	0.9758	Look up the tables for p-value.			
7	3	216	148	3	3							
8	4	217	149	4	4							
9	5	215	147	2	2							
10	6	213	146	1	1			For large samples (n > 30)				
11	7	219	152	6	7			Test Statistic				
12	8	236	165	9	10				z			
13	9	237	162	10	9						At an α of	
14	10	235	161	8	8			Null Hypothesis		*p*-value	5%	
15								$H_0: \rho_s = 0$				
16								$H_0: \rho_s >= 0$				
17								$H_0: \rho_s <= 0$				
18												

PROBLEMS

14–42. The director of a management training program wants to test whether there is a positive association between an applicant's score on a test prior to her or his being admitted to the program and the same person's success in the program. The director ranks 15 participants according to their performance on the pretest and separately ranks them according to their performance in the program:

Participant:	1	2	3	4	5	6	7	8	9	10	11	12	13	14	15
Pretest rank:	8	9	4	2	3	10	1	5	6	15	13	14	12	7	11
Performance rank:	7	5	9	6	1	8	2	10	15	14	4	3	11	12	13

Using these data, carry out the test for a positive rank correlation between pretest scores and success in the program.

14–43. The following data are a random sample of consumers' income and expenditure on certain luxury items. Compute the Spearman rank correlation coefficient, and test for the existence of a population correlation.

Income ($1000s/year):	23, 17, 34, 56, 49, 31, 28, 80, 65, 40, 26
Luxury Item Spending ($/month):	10, 50, 120, 225, 90, 60, 55, 340, 170, 25, 80

14–44. Recently the European Community (EC) decided to lower its subsidies to makers of pasta. In deciding by what amount to reduce total subsidies, experiments were carried out for determining the possible reduction in exports, mainly to the United States, that would result from the subsidy reduction. Over a small range of values, economists wanted to test whether there is a positive correlation between level of subsidy and level of exports. A computer simulation of the economic variables involved in the pasta exports market was carried out. The results are as follows. Assuming that the simulation is an accurate description of reality and that the values obtained may be viewed as a random sample of the populations of possible outcomes, state whether you believe that a positive rank correlation exists between subsidy level and exports level over the short range of values studied.

Subsidy (millions of dollars/year): 5.1 5.3 5.2 4.9 4.8 4.7 4.5 5.0 4.6 4.4 5.4
Exports (millions of dollars/year): 22 30 35 29 27 36 40 39 42 45 21

14–45.   An advertising research analyst wanted to test whether there is any relationship between a magazine advertisement's color intensity and the ad's appeal. Ten ads of varying degrees of color intensity, but identical in other ways, were shown to randomly selected groups of respondents. The respondents rated each ad for its general appeal. The respondents were segmented in such a way that each group viewed a different ad, and every group's responses were aggregated. The results were ranked as follows.

Color intensity: 8 7 2 1 3 4 10 6 5 9
Appeal score: 1 3 4 2 5 8 7 6 9 10

Is there a rank correlation between color intensity and appeal?

14–9   A Chi-Square Test for Goodness of Fit

In this section and the next two, we describe tests that make use of the chi-square distribution. The data used in these tests are *enumerative:* The data are counts, or frequencies. Our actual observations may be on a nominal (or higher) scale of measurement. Because many real-world situations in business and other areas allow for the collection of count data (e.g., the number of people in a sample who fall into different categories of age, sex, income, and job classification), chi-square analysis is very common and very useful. The tests are easy to carry out and are versatile: we can employ them in a wide variety of situations. The tests presented in this and the next two sections are among the most useful statistical techniques of analyzing data. Quite often, in fact, a computer program designed merely to count the number of items falling in some categories automatically prints out a chi-square value. The user then has to consider the question: What statistical test is implied by the chi-square statistic in this particular situation? Among their other purposes, these sections should help you answer this question.

There is a common principle in all the chi-square tests we will discuss. The principle is summarized in the following steps:

Steps in a chi-square analysis:

1. We hypothesize about a population by stating the null and alternative hypotheses.
2. We compute frequencies of occurrence of certain events that we expect under the null hypothesis. These give us the *expected* counts of data points in different cells.
3. We note the *observed* counts of data points falling in the different cells.
4. We consider the difference between the observed and the expected. This difference leads us to a computed value of the chi-square statistic. The formula of the statistic is given as equation 14–35.
5. We compare the value of the statistic with critical points of the chi-square distribution and make a decision.

The analysis in this section and the next two involves tables of data counts. The chi-square statistic has the same form in the applications in all three sections. The statistic is equal to the *squared difference between the observed count and the expected count in each cell, divided by the expected count, summed over all cells.* If our data table has k cells, let the observed count in cell i be O_i and the expected count (expected under H$_0$) be E_i. The definition is for all cells $i = 1, 2, \ldots, k$.

The chi-square statistic is

$$X^2 = \sum_{i=1}^{k} \frac{(O_i - E_i)^2}{E_i}$$
(14–35)

As the total sample size increases, for a given number of cells k, the distribution of the statistic X^2 in equation 14–35 approaches the chi-square distribution. The degrees of freedom of the chi-square distribution are determined separately in each situation.

Remember the binomial experiment, where the number of *successes* (items falling in a particular category) is a random variable. The probability of a success is a fixed number p. Recall from the beginning of Chapter 4 that as the number of trials n increases, the distribution of the number of binomial successes approaches a normal distribution. In the situations in this and the next two sections, the number of items falling in any of *several* categories is a random variable, and as the number of trials increases, the observed number in any cell O_i approaches a normal random variable. Remember also that the sum of several squared standard normal random variables has a chi-square distribution. The terms summed in equation 14–35 are standardized random variables that are squared. Each one of these variables approaches a normal random variable. The sum, therefore, approaches a chi-square distribution as the sample size n gets large.

> A **goodness-of-fit test** is a statistical test of how well our data support an assumption about the distribution of a population or random variable of interest. The test determines how well an assumed distribution fits the data.

For example, we often make an assumption of a normal population. It may be of interest to test how well a normal distribution fits a given data set. Shortly we will see how to carry out a test of the normal distribution assumption.

We start our discussion of goodness-of-fit tests with a simpler test, and a very useful one—a test of goodness of fit in the case of a **multinomial distribution**. The multinomial distribution is a generalization of the binomial distribution to more than two possibilities (success versus failure). In the multinomial situation, we have $k > 2$ possible categories for the data. A data point can fall into only one of the k categories, and the probability that the point will fall in category i (where $i = 1, 2, \ldots, k$) is constant and equal to p_i. The sum of all k probabilities p_i is 1.

Given five categories, for example, such as five age groups, a respondent can fall into only one of the (nonoverlapping) groups. If the probabilities that the respondent will fall into any of the k groups are given by the five parameters p_1, p_2, p_3, p_4, and p_5, then the multinomial distribution with these parameters and n, the number of people in a random sample, specifies the probability of any combination of cell counts. For example, if $n = 100$ people, the multinomial distribution gives us the probability that 10 people will fall in category 1; 15 in category 2; 12 in category 3; 50 in category 4; and the remaining 13 in category 5. The distribution gives us the probabilities of *all possible counts* of 100 people (or items) distributed into five cells.

When we have a situation such as this, we may use the multinomial distribution to test how well our data fit the assumption of k fixed probabilities $p_1, \ldots, p_k$ of falling into k cells. However, working with the multinomial distribution is difficult, and the chi-square distribution is a very good alternative when sample size considerations allow its use.

A Goodness-of-Fit Test for the Multinomial Distribution

The null and the alternative hypotheses for the multinomial distribution are

H_0: The probabilities of occurrence of events $E_1, E_2, \ldots, E_k$ are given
    by the specified probabilities $p_1, p_2, \ldots, p_k$ (14–36)

H_1: The probabilities of the k events are not the p_i stated in the null
    hypothesis

The test statistic is as given in equation 14–35. For large enough n (a rule for how large is "enough" will be given shortly), the distribution of the statistic may be approximated by a chi-square distribution with $k - 1$ degrees of freedom. We demonstrate the test with Example 14–12.

EXAMPLE 14–12

Raymond Weil is about to come out with a new watch and wants to find out whether people have special preferences for the color of the watchband, or whether all four colors under consideration are equally preferred. A random sample of 80 prospective watch buyers is selected. Each person is shown the watch with four different band colors and asked to state his or her preference. The results—the *observed counts*—are given in Table 14–8.

Solution

The null and alternative hypotheses, equation 14–36, take the following specific form:

H_0: The four band colors are equally preferred; that is, the probabilities of choosing any of the four colors are equal: $p_1 = p_2 = p_3 = p_4 = 0.25$

H_1: Not all four colors are equally preferred (the probabilities of choosing the four colors are not all equal)

To compute the value of our test statistic (equation 14–35), we need to find the *expected* counts in all four cells (in this example, each cell corresponds to a color).

Recall that for a binomial random variable, the mean—the *expected value*—is equal to the number of trials n times the probability of success in a single trial p. Here, in the multinomial experiment, we have k cells, each with probability p_i, where $i = 1, 2, \ldots, k$. For each cell, we have a binomial experiment with probability p_i and number of trials n. The expected number in each cell is therefore equal to n times p_i.

TABLE 14–8 **Watchband Color Preferences**

	Tan	Brown	Maroon	Black	Total
	12	40	8	20	80

> The expected count in cell i is
>
> $$E_i = np_i \qquad\qquad (14\text{–}37)$$

In this example, the number of trials is the number of people in the random sample: $n = 80$. Under the null hypothesis, the expected number of people who will choose color i is equal to $E_i = np_i$. Furthermore, since all the probabilities in this case are equal to 0.25, we have the following:

$$E_1 = E_2 = E_3 = E_4 = (80)(0.25) = 20$$

When the null hypothesis is true, and the probability that any person will choose any one of the four colors is equal to 0.25, we may not observe 20 people in every cell. In fact, observing *exactly* 20 people in each of the four cells is an event with a small probability. However, the number of people we observe in each cell should not be too far from the expected number, 20. Just how far is "too far" is determined by the chi-square distribution. We use the expected counts and the observed counts in computing the value of the chi-square test statistic. From equation 14–35, we get the following:

$$X^2 = \sum_{i=1}^{k} \frac{(O_i - E_i)^2}{E_i} = \frac{(12 - 20)^2}{20} + \frac{(40 - 20)^2}{20} + \frac{(8 - 20)^2}{20} + \frac{(20 - 20)^2}{20}$$

$$= \frac{64}{20} + \frac{400}{20} + \frac{144}{20} + 0 = 3.2 + 20 + 7.2 + 0 = 30.4$$

We now conduct the test by comparing the computed value of our statistic, $X^2 = 30.4$, with critical points of the chi-square distribution with $k - 1 = 4 - 1 = 3$ degrees of freedom. From Appendix C, Table 4, we find that the critical point for a chi-square random variable with 3 degrees of freedom and right-hand-tail area $\alpha = 0.01$ is 11.3. (*Note that all the chi-square tests in this chapter are carried out only on the right-hand tail of the distribution.*) Since the computed value is much greater than the critical point at $\alpha = 0.01$, we conclude that there is evidence to reject the null hypothesis that all four colors are equally likely to be chosen. Some colors are probably preferable to others. Our *p*-value is very small.

The Template

The template for conducting chi-square tests for goodness of fit is shown in Figure 14–13. The data in the template correspond to Example 14–12. The results agree with the hand calculations. In addition, the template reveals that the *p*-value is almost zero.

Unequal Probabilities

The test for multinomial probabilities does not always entail equal probabilities, as was the case in our example. The probabilities may very well be different. All we need to do is to specify the probabilities in the null hypothesis and then use the hypothesized probabilities in computing the expected cell counts (using equation 14–37). Then we use the expected counts along with the observed counts in computing the value of the chi-square statistic.

Under what conditions can we assume that, under the null hypothesis, the distribution of the test statistic in equation 14–37 is well-approximated by a chi-square

FIGURE 14–13   **The Template for Goodness of Fit**
                     [Chi-Square Tests.xls; Sheet: Goodness-of-Fit]

	A	B	C	D	E	F	G	H	I	J	K	L
1	Chi-Square Test for Goodness-of-Fit							Watchband color				
2												
3												
4						Frequency data						
5		Tan	Brown	Maroon	Black							
6	Actual	12	40	8	20							
7	Expected	20	20	20	20							
8												
9	k	4										
10	df	3										
11												
12	Test Statistic											
13		χ^2	30.4									
14												
15	p-value	0.0000										
16												

distribution? This important question has no exact answer. As the sample size n increases, the approximation gets better and better. On the other hand, there is also a dependence on the cell k. If the expected number of counts in some cells is too small, the approximation may not be valid. We will give a good rule of thumb that specifies the minimum expected count in each cell needed for the chi-square approximation to be valid. The rule is conservative in the sense that other rules have been given that allow smaller expected counts under certain conditions. If we follow the rule given here, we will usually be safe using the chi-square distribution.

> The chi-square distribution may be used as long as the expected count in every cell is at least 5.0.

Suppose that while conducting an analysis, we find that for one or more cells, the expected number of items is less than 5. We may still continue our analysis if we can *combine cells* so that the expected number has a total of at least 5. For example, suppose that our null hypothesis is that the distribution of ages in a certain population is as follows: 20% are between the ages of 0 to 15, 10% are in the age group of 16 to 25, 10% are in the age group of 26 to 35, 20% are in the age group of 36 to 45, 30% are in the age group of 45 to 60, and 10% are age 61 or over. If we number the age group cells consecutively from 1 to 6, then the null hypothesis is H_0: $p_1 = 0.20$, $p_2 = 0.10$, $p_3 = 0.10$, $p_4 = 0.20$, $p_5 = 0.30$, $p_6 = 0.10$.

Now suppose that we gather a random sample of $n = 40$ people from this population and use this group to test for goodness of fit of the multinomial assumption in the null hypothesis. What are our expected cell counts? In the 0–15 cell, the expected number of people is $np_1 = (40)(0.20) = 8$, which is fine. But for the next age group, 16 to 25, we find that the expected number is $np_2 = (40)(0.10) = 4$, which is less than 5. If we want to continue the analysis, we may combine age groups that have small expected counts with other age groups. We may combine the 16–25 age group with the 26–35 age group, which also has a low expected count. Or we may combine the 16–25 group with the 0–15 group, and the 26–35 group with the 36–45 group—whichever makes more sense in terms of the interpretation of the analysis. We also need to combine the 61-and-over group with the 45–60 age group. Once we make sure that all expected counts are at least 5, we may use the chi-square distribution. Instead of combining groups, we may choose to increase the sample size.

We will now discuss the determination of the number of degrees of freedom denoted by df. The total sample size is $n = 80$ in Table 14–8 of Example 14–12. The total count acts similarly to the way $\bar{x}$ does when we use it in computing the sample standard deviation. *The total count reduces the number of degrees of freedom by 1.* Why? Because knowing the total allows us not to know directly *any one* of the cell counts. If we knew, for example, the counts in the cells corresponding to tan, black, and maroon but did not know the count in the brown cell, we could still figure out the count for this cell by subtracting the sum of the three cell counts we do know from the total of 80. Thus, when we know the total, 1 degree of freedom is lost from the category cells. Out of four cells in this example, any three are free to move. Out of k cells, since we know their total, only $k - 1$ are free to move: df $= k - 1$.

Next we note another fact that will be important in our next example.

> If we have to use the data for estimating the parameters of the probability distribution stated in the null hypothesis, then for every parameter we estimate from the data, we lose an additional degree of freedom.

The chi-square goodness-of-fit test may be applied to testing any hypothesis about the distribution of a population or a random variable. As mentioned earlier, the test may be applied in particular to testing how well an assumption of a normal distribution is supported by a given data set. The standard normal distribution table, Appendix C, Table 2, gives us the probability that a standard normal random variable will be between any two given values. Through the transformation $X = \mu + \sigma Z$, we may then find boundaries in terms of the original variable X for any given probabilities of occurrence. These boundaries can be used in forming cells with known probabilities and, hence, known expected counts for a given sample size. This analysis, however, assumes that we know μ and σ, the mean and the standard deviation of the population or variable in question.

When μ and σ are *not* known and when the null and alternative hypotheses are stated as

H_0: The population (or random variable) has a normal distribution
H_1: The population (or random variable) is not normally distributed
$(14–38)$

there is no mention in the statement of the hypotheses of what the mean or standard deviation may be, and we need to estimate them directly from our data. When this happens, we lose a degree of freedom for each parameter estimated from the data (unless we use another data set for the estimation). We estimate μ by $\bar{X}$ and σ by S, as usual. The degrees of freedom of the chi-square statistic are df $= k - 2 - 1 = k - 3$ (instead of $k - 1$, as before). We will now demonstrate the test for a normal distribution with Example 14–13.

EXAMPLE 14–13   An analyst working for a department store chain wants to test the assumption that the amount of money spent by a customer in any store is approximately normally distributed. It is important to test this assumption because the analyst plans to conduct an analysis of variance to determine whether average sales per customer are equal at several stores in the same chain (as we recall, the normal-distribution assumption is

FIGURE 14–14 **Intervals and Their Standard Normal Probabilities**

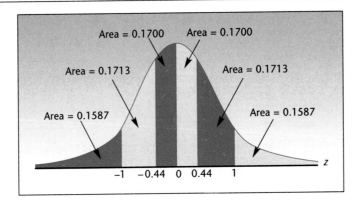

required for ANOVA). A random sample of 100 shoppers at one of the department stores reveals that the average spending is $\bar{x} = \$125$ and the standard deviation is $s = \$40$. These are sample estimates of the *population* mean and standard deviation. (The breakdown of the data into cells is included in the solution.)

Solution

We begin by defining boundaries with known probabilities for the standard normal random variable Z. We know that the probability that the value of Z will be between -1 and $+1$ is about 0.68. We also know that the probability that Z will be between -2 and $+2$ is about 0.95, and we know other such probabilities. We may use Appendix C, Table 2, to find more exact probabilities. Let us use the table and define several nonoverlapping intervals for Z with known probabilities. We will form intervals of about the same probability. Figure 14–14 shows one possible partition of the standard normal distribution to intervals and their probabilities, obtained from Table 2. You may use any partition you desire.

The partition was obtained as follows. We know that the area under the curve between 0 and 1 is 0.3413 (from Table 2). Looking for an area of about half that size, 0.1700, we find that the appropriate point is $z = 0.44$. A similar relationship exists on the negative side of the number line. Thus, using just the values 0.44 and 1 and their negatives, we get a complete partition of the Z scale into the six intervals: $-\infty$ to -1, with associated probability of 0.1587; -1 to -0.44, with probability 0.1713; -0.44 to 0, with probability 0.1700; 0 to 0.44, with probability 0.1700; 0.44 to 1, with probability 0.1713; and, finally, 1 to ∞, with probability 0.1587. Breakdowns into other intervals may also be used.

Now we transform the Z scale values to interval boundaries for the original problem. Taking $\bar{x}$ and s as if they were the mean and the standard deviation of the *population*, we use the transformation $X = \mu + \sigma Z$ with $\bar{x} = 125$ and $s = 40$ substituted for the unknown parameters. The Z value boundaries we just obtained are substituted into the transformation, giving us the following cell boundaries:

$$x_1 = 125 + (-1)(40) = 85$$
$$x_2 = 125 + (-0.44)(40) = 107.4$$
$$x_3 = 125 + (0)(40) = 125$$

TABLE 14–9   Cells and Their Expected Counts

0–$84.99	$85.00–$107.39	$107.4–$124.99	$125–$142.59	$142.6–$164.99	$165 and above	
15.87	17.13	17.00	17.00	17.13	15.87	Total = 100

TABLE 14–10   Observed Cell Counts

0–$84.99	$85.00–$107.39	$107.4–$124.99	$125–$142.59	$142.6–$164.99	$165 and above	
14	20	16	19	16	15	Total = 100

$$x_4 = 125 + (0.44)(40) = 142.6$$
$$x_5 = 125 + (1)(40) = 165$$

The cells and their expected counts are given in Table 14–9. Cell boundaries are broken at the nearest cent. Recall that the expected count in each cell is equal to the cell probability times the sample size $E_i = np_i$. In this example, the p_i are obtained from the normal table and are, in order, 0.1587, 0.1713, 0.1700, 0.1700, 0.1713, and 0.1587. Multiplying these probabilities by $n = 100$ gives us the expected counts. (Note that the theoretical boundaries of $-\infty$ and $+\infty$ have no practical meaning; therefore, the lowest bound is replaced by 0 and the highest bound by *and above*.) Note that all expected cell counts are above 5, and, therefore, the chi-square distribution is an adequate approximation to the distribution of the test statistic X^2 in equation 14–35 under the null hypothesis.

Table 14–10 gives the observed counts of the sales amounts falling in each of the cells. The table was obtained by the analyst by looking at each data point in the sample and classifying the amount into one of the chosen categories.

To facilitate the computation of the chi-square statistic, we arrange the observed and expected cell counts in a single table and show the computations necessary for obtaining the value of the test statistic. This has been done in Table 14–11. The sum of all the entries in the last column in the table is the value of the chi-square statistic. The appropriate distribution has $k - 3 = 6 - 3 = 3$ degrees of freedom. We now consult the chi-square table, Appendix C, Table 4, and we find that the computed statistic value $X^2 = 1.12$ falls in the nonrejection region for any level of α in the table. There is therefore no statistical evidence that the population is not normally distributed.

The Template

The template for testing the normal distribution of a given data set is shown in Figure 14–15. Since the calculated mean and standard deviation of the data set are used to define the class intervals, the degrees of freedom $= k - 3 = 3$.

The chi-square goodness-of-fit test may be applied to testing the fit of any hypothesized distribution. In general, we use an appropriate probability table for obtaining probabilities of intervals of values. The intervals define our data cells. Using the

TABLE 14–11   Computing the Value of the Chi-Square Statistic for Example 14–13

Cell	i	O_i	E_i	$O_i - E_i$	$(O_i - E_i)^2$	$(O_i - E_i)^2/E_i$
0–$84.99	1	14	15.87	−1.87	3.50	0.22
$85.00–$107.39	2	20	17.13	2.87	8.24	0.48
$107.40–$124.99	3	16	17.00	−1.00	1.00	0.06
$125.00–$142.59	4	19	17.00	2.00	4.00	0.24
$142.60–$164.99	5	16	17.13	−1.13	1.28	0.07
$165.00 and above	6	15	15.87	−0.87	0.76	0.05
						1.12

FIGURE 14–15   The Template for Testing Normal Distributions
[Chi-Square Tests.xls; Sheet: Normal Fit]

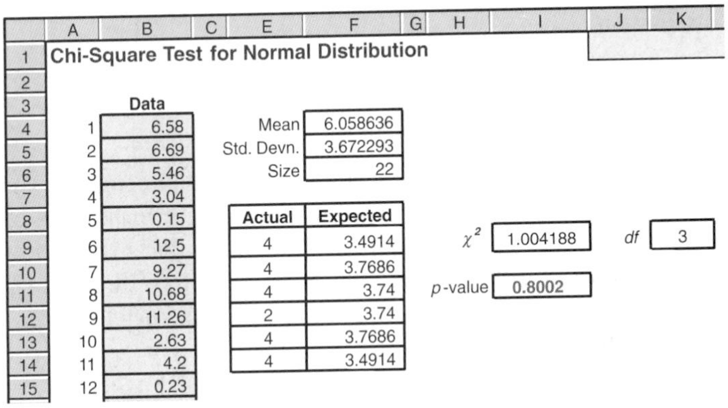

sample size, we then find the expected count in each cell. We compare the expected counts with the observed counts and compute the value of the chi-square test statistic.

The chi-square statistic is useful in other areas as well. In the next section, we will describe the use of the chi-square statistic in the analysis of contingency tables—an analysis of whether two principles of classification are contingent on each other or independent of each other. The following section extends the contingency table analysis to a test of homogeneity of several populations.

PROBLEMS

14–46.   A company is considering five possible names for its new product. Before choosing a name, the firm decides to test whether all five names are equally appealing. A random sample of 100 people is chosen, and each person is asked to state her or his choice of the best name among the five possibilities. The numbers of people who chose each one of the names are as follows.

Name:	A	B	C	D	E
Number of choices:	4	12	34	40	10

Conduct the test.

14–47.   A study reports an analysis of 35 key product categories. At the time of the study, 72.9% of the products sold were of a national brand, 23% were private label, and 4.1% were generic. Suppose that you want to test whether these percentages are still valid for the market today. You collect a random sample of 1,000 products in the 35 product categories studied, and you find the following: 610 products are of a national brand, 290 are private label, and 100 are generic. Conduct the test, and state your conclusions.

14–48.   An industry analyst wants to test the null hypothesis that the market share figures for firms in the industry are as follows:

Firm:	A	B	C	D	E	Other
Share:	1%	9%	14%	26%	0.5%	49.5%

The analysis is to be based on a classification of 100 randomly chosen products.
   a. Can you use the chi-square statistic to conduct this test?
   b. If you cannot use the chi-square statistic, can you make changes that will allow you to use it?
   c. Suppose that the 100 products in the sample are classified as follows: firm A, 0; firm B, 0; firm C, 26; firm D, 15; firm E, 2; other, 57. Conduct the analysis (if you can), and state your conclusion.

14–49.   Returns on an investment have been known to be normally distributed with mean 11% (annualized rate) and standard deviation 2%. A brokerage firm wants to test the null hypothesis that this statement is true and collects the following returns data in percent (assume a random sample): 8, 9, 9.5, 9.5, 8.6, 13, 14.5, 12, 12.4, 19, 9, 10, 10, 11.7, 15, 10.1, 12.7, 17, 8, 9.9, 11, 12.5, 12.8, 10.6, 8.8, 9.4, 10, 12.3, 12.9, 7. Conduct the analysis and state your conclusion.

14–50.   Using the data provided in problem 14–49, test the null hypothesis that returns on the investment are normally distributed, but with *unknown* mean and standard deviation. That is, test only for the validity of the normal-distribution assumption. How is this test different from the one in problem 14–49?

14–10   Contingency Table Analysis— A Chi-Square Test for Independence

Recall the important concept of *independence* of events, which we discussed in Chapter 2. Two events A and B are independent if the probability of their joint occurrence is equal to the product of their marginal (i.e., separate) probabilities. This was given as:

A and B are independent if $P(A \cap B) = P(A)P(B)$

   In this section, we will develop a statistical test that will help us determine whether two classification criteria, such as gender and job performance, are independent of each other. The technique will make use of **contingency tables**—tables with cells corresponding to cross-classifications of attributes or events. In market research studies, such tables are referred to as *cross-tabs*. The basis for our analysis will be the property of independent events just stated.
   The contingency tables may have several rows and several columns. The rows correspond to levels of one classification category, and the columns correspond to another. We will denote the number of rows by r, and the number of columns by c. The

FIGURE 14–16   **Layout of a Contingency Table**

Second Classification Category	First Classification Category						Total
	1	2	3	4	5	6	
1	O_{11}	O_{12}	O_{13}	O_{14}	O_{15}	O_{16}	R_1
2	O_{21}	O_{22}	O_{23}	O_{24}	O_{25}	O_{26}	R_2
3	O_{31}	O_{32}	O_{33}	O_{34}	O_{35}	O_{36}	R_3
4	O_{41}	O_{42}	O_{43}	O_{44}	O_{45}	O_{46}	R_4
5	O_{51}	O_{52}	O_{53}	O_{54}	O_{55}	O_{56}	R_5
Total	C_1	C_2	C_3	C_4	C_5	C_6	n

total sample size is n, as before. The count of the elements in cell (i, j), that is, the cell in row i and column j (where $i = 1, 2, \ldots, r$ and $j = 1, 2, \ldots, c$), is denoted by O_{ij}. The total count for row i is R_i, and the total count for column j is C_j. The general form of a contingency table is shown in Figure 14–16. The table is demonstrated for $r = 5$ and $c = 6$. Note that n is also the sum of all r row totals and the sum of all c column totals.

Let us now state the null and alternative hypotheses.

The hypothesis test for independence is

H_0: The two classification variables are independent of each other   (14–39)

H_1: The two classification variables are not independent

The principle of our analysis is the same as that used in the previous section. The chi-square test statistic for this set of hypotheses is the one we used before, given in equation 14–35. The only difference is that the summation extends over all cells in the table: the c columns and the r rows (in the previous application, goodness-of-fit tests, we only had one row). We will rewrite the statistic to make it clearer:

The chi-square test statistic for independence is

$$X^2 = \sum_{i=1}^{r} \sum_{j=1}^{c} \frac{(O_{ij} - E_{ij})^2}{E_{ij}}$$ (14–40)

The double summation in equation 14–39 means summation over all rows and all columns.

The degrees of freedom of the chi-square statistic are

$$df = (r - 1)(c - 1)$$ (14–41)

Now all we need to do is to find the expected cell counts E_{ij}. Here is where we use the assumption that the two classification variables are independent. Remember that the philosophy of hypothesis testing is to assume that H_0 is true and to use this assumption in determining the distribution of the test statistic. Then we try to show that the result is unlikely under H_0 and thus reject the null hypothesis.

Assuming that the two classification variables are independent, let us derive the expected counts in all cells. Look at a particular cell in row i and column j. Recall from equation 14–37 that the expected number of items in a cell is equal to the sample size times the probability of the occurrence of the event signified by the particular cell. In the context of an $r \times c$ contingency table, the probability associated with cell (i, j) is the probability of occurrence of event i *and* event j. Thus, the expected count in cell (i, j) is $E_{ij} = nP(i \cap j)$. If we assume independence of the two classification variables, then event i and event j are independent events, and by the law of independence of events, $P(i \cap j) = P(i)P(j)$.

From the row totals, we can estimate the probability of event i as R_i/n. Similarly, we estimate the probability of event j by C_j/n. Substituting these estimates of the marginal probabilities, we get the following expression for the expected count in cell (i, j): $E_{ij} = n(R_i/n)(C_j/n) = R_iC_j/n$.

The expected count in cell (i, j) is

$$E_{ij} = \frac{R_iC_j}{n} \tag{14–42}$$

Equation 14–42 allows us to compute the expected cell counts. These, along with the observed cell counts, are used in computing the value of the chi-square statistic, which leads us to a decision about the null hypothesis of independence.

We will now illustrate the analysis with two examples. The first example is an illustration of an analysis of the simplest contingency table, a 2×2 table. In such tables, the two rows correspond to the occurrence versus nonoccurrence of one event, and the two columns correspond to the occurrence or nonoccurrence of another event.

EXAMPLE 14–14   An article in *Business Week* reports profits and losses of firms by industry.[5] A random sample of 100 firms is selected, and for each firm in the sample, we record whether the company made money or lost money, and whether the firm is a service company. The data are summarized in the 2×2 contingency table, Table 14–12. Using the information in the table, determine whether you believe that the two events "the company made a profit this year" and "the company is in the service industry" are independent.

Solution   Table 14–12 is the table of observed counts. We now use its marginal totals R_1, R_2, C_1, and C_2 as well as the sample size n in creating a table of expected counts. Using equation 14–42, we get

---

[5]"The Business Week 1000," *Business Week*, March 27, 1995, pp. 102–65.

TABLE 14–12   Contingency Table of Profit/Loss versus Industry Type

	Industry Type		
	Service	Nonservice	Total
Profit	42	18	60
Loss	6	34	40
Total	48	52	100

TABLE 14–13   Expected Counts (with the observed counts shown in parentheses) for Example 14–14

	Service	Nonservice
Profit	28.8	31.2
	(42)	(18)
Loss	19.2	20.8
	(6)	(34)

$$E_{11} = R_1 C_1/n = (60)(48)/100 = 28.8$$
$$E_{12} = R_1 C_2/n = (60)(52)/100 = 31.2$$
$$E_{21} = R_2 C_1/n = (40)(48)/100 = 19.2$$
$$E_{22} = R_2 C_2/n = (40)(52)/100 = 20.8$$

We now arrange these values in a table of expected counts, Table 14–13. Using the values shown in the table, we now compute the chi-square test statistic of equation 14–40:

$$X^2 = \frac{(42 - 28.8)^2}{28.8} + \frac{(18 - 31.2)^2}{31.2} + \frac{(6 - 19.2)^2}{19.2} + \frac{(34 - 20.8)^2}{20.8} = 29.09$$

To conduct the test, we compare the computed value of the statistic with critical points of the chi-square distribution with $(r - 1)(c - 1) = (2 - 1)(2 - 1) = 1$ degree of freedom. From Appendix C, Table 4, we find that the critical point for $\alpha = 0.01$ is 6.63, and since our computed value of the X^2 statistic is much greater than the critical point, we reject the null hypothesis and conclude that the two qualities, profit/loss and industry type, are probably not independent.

In the analysis of 2 × 2 contingency tables, our chi-square statistic has *1 degree of freedom*. In such cases, it is often recommended that the value of the statistic be "corrected" so that its discrete distribution will be better approximated by the *continuous* chi-square distribution. The correction is called the **Yates correction** and entails subtracting the number 1/2 from the absolute value of the difference between the observed and the expected counts before squaring them as required by equation 14–40. The Yates-corrected form of the statistic is as follows.

FIGURE 14–17 **The Template for Testing Independence [Chi-Square Tests.xls; Sheet: Independence]**

Chi-Square Test for Independence

Frequencies Data

	1	2	3	4	5	6	7	8	9	10	Total
1	42	18									60
2	6	34									40
3											0
4											0
5											0
6											0
7											0
8											0
9											0
10											0
Total	48	52	0	0	0	0	0	0	0	0	100

Yates Correction	0.5
#Rows	2
#Cols	2
df	1

Test Statistic

χ^2	26.9247
p-value	0.0000

$$\text{Yates-corrected } X^2 = \sum_{i=1}^{r}\sum_{j=1}^{c}\frac{(|O_{ij} - E_{ij}| - 0.5)^2}{E_{ij}} \qquad (14\text{–}43)$$

For our example, the corrected value of the chi-square statistic is found as

Yates-corrected X^2

$$= \frac{(13.2 - 0.5)^2}{28.8} + \frac{(13.2 - 0.5)^2}{31.2} + \frac{(13.2 - 0.5)^2}{19.2} + \frac{(13.2 - 0.5)^2}{20.8}$$
$$= 26.92$$

As we see, the correction yields a smaller computed value. This value still leads to a strong rejection of the null hypothesis of independence. In many cases, the correction will not significantly change the results of the analysis. We will not emphasize the correction in the applications in this book.

The Template

The template for testing independence using the chi-square distribution is shown in Figure 14–17. The data correspond to Example 14–14.

EXAMPLE 14–15 To better identify its target market, Alfa Romeo conducted a market research study. A random sample of 669 respondents was chosen, and each was asked to select one of four qualities that best described him or her as a driver. The four possible self-descriptive qualities were *defensive, aggressive, enjoying,* and *prestigious.* Each respondent was then asked to choose one of three Alfa Romeo models as her or his choice of the most suitable car. The three models were Alfasud, Giulia, and Spider. The purpose of the study was to determine whether a relationship existed between a driver's self-image and choice of an Alfa Romeo model. The response data are given in Table 14–14.

Solution Figure 14–18 shows the template solution to Example 14–15. The *p*-value of 0.0019 is less than 1% and therefore we reject the null hypothesis that the choice of Alfa Romeo model and self-image are independent.

TABLE 14–14 **The Observed Counts: Alfa Romeo Study**

Alfa Romeo Model	Self-Image: Defensive	Aggressive	Enjoying	Prestigious	Total
Alfasud	22	21	34	56	133
Giulia	39	45	42	68	194
Spider	77	89	96	80	342
Total	138	155	172	204	669

FIGURE 14–18 **Template Solution to Example 14–15. [Chi-Square Tests.xls; Sheet: Independence]**

Chi-Square Test for Independence — Alfa Romeo

Frequencies Data

	1	2	3	4	5	6	7	8	9	10	Total
1	22	21	34	56							133
2	39	45	42	68							194
3	77	89	96	80							342
4											0
5											0
6											0
7											0
8											0
9											0
10											0
Total	138	155	172	204	0	0	0	0	0	0	669

Yates Correction	0
#Rows	3
#Cols	4
df	6
Test Statistic χ^2	20.867
p-value	0.0019

PROBLEMS

14–51. An article reports that smaller firms seem to be hiring more than large ones as the economy picks up its pace. The table below gives numbers of employees hired and those laid off, out of a random sample of 1,032, broken down by firm size. Is there evidence that hiring practices are dependent on firm size?

	Small Firm	Medium-Size Firm	Large Firm	Total
Number hired	210	290	325	825
Number laid off	32	95	80	207
Total	242	385	405	1,032

14–52. An analyst in the soft drink industry wants to conduct a statistical test to determine whether there is a relationship between a person's preference for one of the four brands—Coke, Pepsi, 7Up, and Dr Pepper—and whether the person drinks regular or diet drinks. A random sample of 330 people is selected, and their responses are as follows.

	Soft Drink Preference				
	Coke	Pepsi	7Up	Dr Pepper	Total
Diet	55	32	47	21	155
Regular	60	43	35	37	175
Total	115	75	82	58	330

Conduct the test, and state your conclusion.

14–53. *Fortune* reports on the competition among car rental companies. The table below gives the number of cars, out of a random sample of 100 rental cars, belonging to each of the listed firms in 1979 and in 1990. Is there evidence of a change in the market shares of the car rental firms?

	Hertz	Avis	National	Budget	Other	Total
1979	39	26	18	14	3	100
1990	29	25	16	19	11	100

Source: "Rental Firms for Market Share," *Fortune*, April 22, 1991, p. 17.

14–54. An article on the stock market describes stock-buying behavior by institutions and individuals, as well as foreign and domestic buyers.[6] The following table describes recent purchases of U.S. stocks by individual or institution as well as domestic or foreign. Is there evidence of a dependence of institutional buying on whether the buyer is foreign or domestic?

	Domestic	Foreign
Individual	25	32
Institution	30	13

14–55. A study was conducted to determine whether a relationship existed between certain shareholder characteristics and the level of risk associated with the shareholders' investment portfolios. As part of the analysis, portfolio risk (measured by the portfolio beta) was divided into three categories: low-risk, medium-risk, and high-risk; and the portfolios were cross-tabulated according to the three risk levels and seven family-income levels. The results of the analysis, conducted using a random sample of 180 investors, are shown in the following contingency table. Test for the existence of a relationship between income and investment risk taking. [Be careful here! (Why?)]

	Portfolio Risk Level			
Income Level ($)	Low	Medium	High	Total
0 to 19,999	5	4	1	10
20,000 to 24,999	6	3	0	9
25,000 to 29,999	22	30	11	63
30,000 to 34,999	11	20	20	51
35,000 to 39,999	8	10	4	22
40,000 to 44,999	2	0	10	12
45,000 and above	1	1	11	13
Total	55	68	57	180

[6]S. Kuhn, "What's Driving the Dow?" *Fortune,* June 12, 1995, pp. 54–56.

14-56. When new paperback novels are promoted at bookstores, a display is often arranged with copies of the same book with differently colored covers. A publishing house wanted to find out whether there is a dependence between the place where the book is sold and the color of its cover. For one of its latest novels, the publisher sent displays and a supply of copies of the novel to large bookstores in five major cities. The resulting sales of the novel for each city–color combination are as follows. Numbers are in thousands of copies sold over a 3-month period.

	Color				
City	Red	Blue	Green	Yellow	Total
New York	21	27	40	15	103
Washington	14	18	28	8	68
Boston	11	13	21	7	52
Chicago	3	33	30	9	75
Los Angeles	30	11	34	10	85
Total	79	102	153	49	383

a. Assume that the data are random samples for each particular color–city combination and that the inference may apply to all novels. Conduct the overall test for independence of color and location.

b. Before the analysis, the publisher stated a special interest in the issue of whether there is any dependence between the red versus blue preference and the two cities Chicago versus Los Angeles. Conduct the test. Explain.

14-11 A Chi-Square Test for Equality of Proportions

Contingency tables and the chi-square statistic are also useful in another kind of analysis. Sometimes we are interested in whether the proportion of some characteristic is equal in several populations. An insurance company, for example, may be interested in finding out whether the proportion of people who submit claims for automobile accidents is about the same for the three age groups 25 and under, over 25 and under 50, and 50 and over. In a sense, the question of whether the proportions are equal is a question of whether the three age populations are *homogeneous* with respect to accident claims. Therefore, tests of equality of proportions across several populations are also called *tests of homogeneity.*

The analysis is carried out in exactly the same way as in the previous application. We arrange the data in cells corresponding to population-characteristic combinations, and for each cell, we compute the expected count based on its row and column totals. The chi-square statistic is computed exactly as before. Two things are different in this analysis. First, we identify our populations of interest before the analysis and sample directly from these populations. Contrast this with the previous application, where we sampled from one population and then cross-classified according to two criteria. Second, because we identify populations and sample from them directly, the sizes of the samples from the different populations of interest are fixed. This is called a *chi-square analysis with fixed marginal totals.* This fact, however, does not affect the analysis.

We will demonstrate the analysis with the insurance company example just mentioned. The null and alternative hypotheses are

TABLE 14–15   **Data for the Insurance Company Example**

	Age Group			
	25 and under	Over 25 and under 50	50 and over	Total
Claim	40	35	60	135
No claim	60	65	40	165
Total	100	100	100	300

There are fixed sample sizes for all three populations.

TABLE 14–16   **Expected Counts for the Insurance Company Example**

	25 and under	Over 25 and under 50	50 and over	Total
Claim	45	45	45	135
No claim	55	55	55	165
Total	100	100	100	300

H_0: The proportion of claims is the same for all three age groups (i.e., the age groups are homogeneous with respect to claim proportions)

H_1: The proportion of claims is not the same across age groups (the age groups are not homogeneous)

$$(14\text{–}44)$$

Suppose that random samples, selected from company records for the three age categories, are classified according to *claim* versus *no claim* and are counted. The data are presented in Table 14–15.

To carry out the test, we first calculate the expected counts in all the cells. The expected cell counts are obtained, as before, by using equation 14–42. The expected count in each cell is equal to the row total times the column total, divided by the total sample size (the pooled sample size from all populations). The reason for the formula in this new context is that if the proportion of items in the class of interest (here, the proportion of people who submit a claim) is equal across all populations, as stated in the null hypothesis, then *pooling* this proportion across populations gives us the expected proportion in the cells for the class. Thus, the expected proportion in the claim class is estimated by the total in the claim class divided by the grand total, or $R_1/n = 135/300 = 0.45$. If we multiply this pooled proportion by the total number in the sample from the population of interest (say, the sample of people 25 and under), this should give us the *expected* count in the cell *claim—25 and under*. We get $E_{11} = C_1(R_1/n) = (C_1 R_1)/n$. This is exactly as prescribed by equation 14–42 in the test for independence. Here we get $E_{11} = (100)(0.45) = 45$. This is the expected count under the null hypothesis. We compute the expected counts for all other cells in the table in a similar manner. Table 14–16 is the table of expected counts in this example.

Note that since we have used equal sample sizes (100 from each age population), the expected count is equal in all cells corresponding to the same class. The proportions are expected to be equal under the null hypothesis. Since these proportions are multiplied by the same sample size, the counts are also equal.

We are now ready to compute the value of the chi-square test statistic. From equation 14–40, we get

$$X^2 = \sum_{\text{all cells}} \frac{(O - E)^2}{E} = \frac{(40 - 45)^2}{45} + \frac{(35 - 45)^2}{45} + \frac{(60 - 45)^2}{45}$$
$$+ \frac{(60 - 55)^2}{55} + \frac{(65 - 55)^2}{55} + \frac{(40 - 55)^2}{55} = 14.14$$

The degrees of freedom are obtained as usual. We have three rows and two columns, so the degrees of freedom are $(3 - 1)(2 - 1) = 2$. Alternatively, cross out any one row and any one column in Table 14–15 or 14–16 (ignoring the *Total* row and column). This leaves you with two cells, giving df $= 2$.

Comparing the computed value of the statistic with critical points of the chi-square distribution with 2 degrees of freedom, we find that the null hypothesis may be rejected and that the *p*-value is less than 0.01. (Check this, using Appendix C, Table 4.) We conclude that the proportions of people who submit claims to the insurance company are not the same across the age groups studied.

In general, when we compare c populations (or r populations, if they are arranged as the rows of the table rather than the columns), the hypotheses in equation 14–44 may be written as

$$H_0: p_1 = p_2 = \cdots = p_c \tag{14–45}$$
$$H_1: \text{Not all } p_i, \; i = 1, \ldots, c, \text{ are equal}$$

where p_i $(i = 1, \ldots, c)$, is the proportion in population i of the characteristic of interest. The test of equation 14–45 is a generalization to c populations of the test of equality of two population proportions discussed in Chapter 8. In fact, when $c = 2$, the test is identical to the simple test for equality of two population proportions. In our present context, the two-population test for proportion may be carried out using a 2×2 contingency table. The results of such a test would be identical to the results of a test using the method of Chapter 8 (a Z test).

The test presented in this section may also be applied to several proportions within each population. That is, instead of just testing for the proportion of *claim* versus *no claim*, we could be testing a more general hypothesis about the proportions of *several* different types of claims: no claim, claim under $1,000, claim of $1,000 to $5,000, and claim over $5,000. Here the null hypothesis would be that the proportion of each type of claim is equal across all populations. (This does not mean that the proportions of all types of claims are equal within a population.) The alternative hypothesis would be that not all proportions are equal across all populations under study. The analysis is done using an $r \times c$ contingency table (instead of the $2 \times c$ table we used in the preceding example). The test statistic is the same, and the degrees of freedom are as before: $(r - 1) \times (c - 1)$. We now discuss another extension of the test presented in this section.

The Median Test

The hypotheses for the median test are

$$H_0: \text{The } c \text{ populations have the same median} \tag{14–46}$$
$$H_1: \text{Not all } c \text{ populations have the same median}$$

TABLE 14–17   Family Incomes ($1,000s per year)

Region A	Region B	Region C
22	31	28
29	37	42
36	26	21
40	25	47
35	20	18
50	43	23
38	27	51
25	41	16
62	57	30
16	32	48

Using the c random samples from the populations of interest, we determine the grand median, that is, the median of all our data points regardless of which population they are from. Then we divide each sample into two sets. One set contains all points that are greater than the grand median, and the second set contains all points in the sample that are less than or equal to the grand median. We construct a $2 \times c$ contingency table in which the cells in the top row contain the counts of all points above the median for all c samples. The second row contains cells with the counts of the data points in each sample that are less than or equal to the grand median. Then we conduct the usual chi-square analysis of the contingency table. If we reject H_0, then we may conclude that there is evidence that not all c population medians are equal. We now demonstrate the median test with Example 14–16.

EXAMPLE 14–16   An economist wants to test the null hypothesis that median family incomes in three rural areas are approximately equal. Random samples of family incomes in the three regions (in thousands of dollars per year) are given in Table 14–17.

Solution   For simplicity, we chose an equal sample size of 10 in each population. This is not necessary; the sample sizes may be different. There is a total of 30 observations, and the median is therefore the average of the 15th and the 16th observations. Since the 15th observation (counting from smallest to largest) is 31 and the 16th is 32, the grand median is 31.5. Table 14–18 shows the counts of the sample points in each sample that are above the grand median and those that are less than or equal to the grand median. The table also shows the expected cell counts (in parentheses). Note that all expected counts are 5—the minimum required for the chi-square test. We now compute the value of the chi-square statistic.

$$X^2 = \frac{1}{5} \left[(4-5)^2 + (5-5)^2 + (6-5)^2 + (6-5)^2 + (5-5)^2 + (4-5)^2 \right]$$

$$= \frac{4}{5} = 0.8$$

TABLE 14-18   Observed and Expected Counts for Example 14-16

	Region A	Region B	Region C	Total
Less than or equal to	4	5	6	15
	(5)	(5)	(5)	
Above grand median	6	5	4	15
	(5)	(5)	(5)	—
Total	10	10	10	30

Comparing this value with critical points of the chi-square distribution with 2 degrees of freedom, we conclude that there is no evidence to reject the null hypothesis. The p-value is greater than 0.20.

Note that the median test is a weak test. Other tests could have resulted in the rejection of the null hypothesis (try them). We presented the test as an illustration of the wide variety of possible uses of the chi-square statistic. Other uses may be found in advanced books. We note that if the test had led to rejection, then other tests would probably have done so, too. Sometimes this test is easier to carry out and may lead to a quick answer (when we reject the null hypothesis).

PROBLEMS

14-57. An advertiser runs a commercial on national television and wants to determine whether the proportion of people exposed to the commercial is equal throughout the country. A random sample of 100 people is selected at each of five locations, and the number of people in each location who have seen the commercial at least once during the week is recorded. The numbers are as follows: location A, 32 people; location B, 59 people; location C, 78 people; location D, 40 people; and location E, 10 people. Do you believe that the proportion of people exposed to the commercial is equal across the five locations?

14-58. An accountant wants to test the hypothesis that the proportion of incorrect transactions at four client accounts is about the same. A random sample of 80 transactions of one client reveals that 21 are incorrect; for the second client, the sample proportion is 25 out of 100; for the third client, the proportion is 30 out of 90 sampled; and for the fourth, 40 are incorrect out of a sample of 110. Conduct the test at $\alpha = 0.05$.

14-59. An article in *The Wall Street Journal* raises the question of whether experienced fund managers perform about the same as less experienced ones in a bull market.[7] A sample of 128 fund managers with at least 8 years of experience revealed that 54 of them beat the bull market, 43 of them did about as well as the market, and 31 did worse than the market. In a sample of 95 less experienced fund managers, 65 were found to beat the bull market, 21 did about the same as the market, and 9 did more poorly than the market. Test the null hypothesis that experienced fund managers do as well as inexperienced ones versus the alternative that differences in performance exist between the two groups.

[7]R. McGough, "Finding Thrills in Boring Index Funds," *The Wall Street Journal*, June 27, 1995, p. c1.

14–60. A quality control engineer wants to test the null hypothesis that the proportion of defective components in three large shipments is approximately equal. A random sample of 100 components is selected from each shipment. The first sample reveals 25 defective items, the second has 15, and the third has 8. Conduct the test at $\alpha = 0.05$.

14–61. As markets become more and more international, many firms invest in research aimed at determining the maximum possible extent of sales in foreign markets. A U.S. manufacturer of coffee makers wants to find out whether the company's market share and the market shares of two main competitors are about the same in three European countries to which all three companies export their products. The results of a market survey are summarized in the following table. The data are random samples of 150 consumers in each country. Conduct the test of equality of population proportions across the three countries.

	Country			
	France	England	Spain	Total
Company	55	38	24	117
First competitor	28	30	21	79
Second competitor	20	18	31	69
Other	47	64	74	185
Total	150	150	150	450

14–62. New production methods stressing teamwork have recently been instituted at car manufacturing plants in Detroit. Three teamwork production methods are to be compared to see if they are equally effective. Since large deviations often occur in the numbers produced daily, it is desired to test for equality of medians (rather than means). Samples of daily production volume for the three methods are as follows. Assume that these are random samples from the populations of daily production volume. Use the median test to help determine whether the three methods are equally effective.

Method A: 5, 7, 19, 8, 10, 16, 14, 9, 22, 4, 7, 8, 15, 18, 7
Method B: 8, 12, 15, 28, 5, 14, 19, 16, 23, 19, 25, 17, 20
Method C: 14, 28, 13, 10, 8, 29, 30, 26, 17, 13, 10, 31, 27, 20

14–12   Summary and Review of Terms

This chapter was devoted to **nonparametric tests** (summarized in Table 14–19). Interpreted loosely, the term refers to statistical tests in situations where stringent assumptions about the populations of interest may not be warranted. Most notably, the very common assumption of a normal distribution—required for the parametric t and F tests—is not necessary for the application of nonparametric methods. The methods often use less of the information in the data and thus tend to be less powerful than parametric methods, when the assumptions of the parametric methods are met. The nonparametric methods include methods for handling categorical data, and here the analysis entails use of a limiting chi-square distribution for our test statistic. **Chi-square analysis** is often discussed separately from nonparametric methods, although the analysis is indeed "nonparametric," as it usually involves no specific reference to population *parameters* such as μ and σ. The other nonparametric methods (ones that

TABLE 14–19   **Summary of Nonparametric Tests**

Situation	Nonparametric Test(s)	Corresponding Parametric Test
Single-sample test for location	Sign test Wilcoxon test (more powerful)	Single-sample *t* test
Goodness of fit	Chi-square test	
Randomness	Runs test	
Paired-differences test	Sign test Wilcoxon test (more powerful)	Paired-data *t* test
Test for difference of two independent samples	Wald-Wolfowitz (weaker) Mann-Whitney (more powerful) Median test (weaker)	Two-sample *t* test
Test for difference of more than two independent samples	Kruskal-Wallis test Median test (weaker)	ANOVA
Test for difference of more than two samples, blocked	Friedman test	Randomized block-design ANOVA
Test for trend in single series	Cox and Stuart test	Time series tests
Correlation	Cox and Stuart test Spearman's statistic and test Chi-square test for independence	Pearson's product-moment
Equality of several population proportions	Chi-square test	

require no assumptions about the distribution of the population) are often called *distribution-free* methods.

Besides chi-square analyses of **goodness of fit, independence,** and tests for **equality of proportions,** the methods we discussed included many based on **ranks.** These included a **rank correlation coefficient** due to Spearman; a test analogous to the parametric paired-sample *t* test—the **Wilcoxon signed-rank test;** a ranks-based ANOVA—the **Kruskal-Wallis test;** and a method for investigating two independent samples analogous to the parametric two-sample *t* test, called the **Mann-Whitney test.** We also discussed a test for randomness—the **runs test;** a paired-difference test called the **sign test,** which uses less information than the Wilcoxon signed-rank test; and several other methods.

ADDITIONAL PROBLEMS

14–63.   The following data are daily price quotations of two stocks:

Stock A: 12.50, 12.75, 12.50, 13.00, 13.25, 13.00, 13.50, 14.25, 14.00
Stock B: 35.25, 36.00, 37.25, 37.25, 36.50, 36.50, 36.00, 36.00, 36.25

Is there a correlation between the two stocks? Explain.

14–64.   The Hyatt Gold Passport is a card designed to allow frequent guests at Hyatt hotels to enjoy privileges similar to the ones enjoyed by frequent air travelers. When the program was initiated, a random sample of 15 Hyatt Gold Passport members were asked to rate the program on a scale of 0 to 100 and also to rate (on the same scale) an airline frequent-flier card that all of them had. The results are as follows.

Hyatt card: 98, 99, 87, 56, 79, 89, 86, 90, 95, 99, 76, 88, 90, 95
Airline card: 84, 62, 90, 77, 80, 98, 65, 97, 58, 74, 80, 90, 85, 70

Is the Hyatt Gold Passport better liked than the airline frequent-flier card by holders of both cards? Explain.

14–65. Two telecommunication systems are to be compared. A random sample of 14 users of one system independently rate the system on a scale of 0 to 100. An independent random sample of 12 users of the other system rate their system on the same scale. The data are as follows.

System A: 65, 67, 83, 39, 45, 20, 95, 64, 99, 98, 76, 78, 82, 90
System B: 45, 57, 76, 54, 60, 72, 34, 50, 63, 39, 44, 70

Based on these data, are the two telecommunication systems equally liked? Explain.

14–66. What is the distinction between *distribution-free* methods and *nonparametric* methods?

14–67. The following data are salaries, in thousands of dollars per year, of a random sample of employees in a certain industry:

28.5, 29, 29.7, 31, 28, 32.5, 32.6, 33, 31.8, 37.4, 24.6, 38.1, 28.8, 30.1, 32.5, 34.4, 26.7, 28.1, 39.2, 40.3, 33.5, 33.8, 38.2, 23.0, 22.9, 29.9, 31.2, 33.0, 31.7, 37, 35, 32.1, 37, 29.9, 22, 23.8, 27, 35, 37.5, 40.1, 22.1, 26.3, 30.0, 40.1

Do you believe that salaries in this industry are normally distributed? Explain.

14–68. In a chi-square analysis, the expected count in one of the cells is 2.1. Can you conduct the analysis? If not, what can be done?

14–69. New credit card machines use two receipts, to be signed by the payer. This has recently caused confusion as many customers forget to sign the copy they leave with the establishment.[8] If 6 out of 17 randomly selected patrons forgot to sign their slips, test the hypothesis that a full one-half of the customers do so, using $\alpha = 0.05$, against a left-tailed alternative.

14–70. The rankings of teams in the National Football League in four performance measures for the year 2000 are given below.

Team	Ranking by Total Yards/ Game	First Downs/ Game	3rd-Down Efficiency	Points/ Game
AMERICAN CONFERENCE				
Baltimore	16	18	14	14
Buffalo	9	12	20	20
Cincinnati	29	26	25	30
Cleveland	31	31	31	31
Denver	2	1	4	2
Indianapolis	3	3	2	4
Jacksonville	7	8	7	8
Kansas City	8	6	21	9
Miami	26	28	23	16
New England	22	19	24	25
New York Jets	12	13	17	17

---

[8] From "Annoyances," *Fortune*, February 16, 1998, p. 48.

	Ranking by			
Team	Total Yards/ Game	First Downs/ Game	3rd-Down Efficiency	Points/ Game
AMERICAN CONFERENCE				
Oakland	6	4	5	3
Pittsburgh	18	20	19	18
San Diego	28	29	27	26
Seattle	19	21	10	19
Tennessee	14	16	8	13
NATIONAL CONFERENCE				
Arizona	24	27	18	29
Atlanta	30	25	30	27
Carolina	20	15	16	21
Chicago	23	30	29	28
Dallas	25	22	22	23
Detroit	27	24	28	22
Green Bay	15	9	11	11
Minnesota	5	7	3	5
New Orleans	10	10	6	10
New York Giants	13	11	13	15
Philadelphia	17	17	12	12
San Francisco	4	5	9	6
St. Louis	1	2	1	1
Tampa Bay	21	23	26	7
Washington	11	14	15	24

a. Which of the first three rankings do you think will be the most correlated with the last ranking (by points/game)?

b. Calculate the Spearman rank correlation coefficient between each of the first three rankings and the ranking by points/game.

c. Test each of the three correlations for significance.

CASE 16 The Nine Nations of North America

In a fascinating article in the *Journal of Marketing* (April 1986), "The Nine Nations of North America and the Value Basis of Geographic Segmentation," Professor Lynn Kahle explores the possible marketing implications of Joel Garreau's idea of the nine nations.

Garreau traveled extensively throughout North America, studying people, customs, traditions, and ways of life. This research led Garreau to the conclusion that state boundaries or the Census Bureau's divisions of the United States into regions are not very indicative of the cultural and social boundaries that really exist on the continent. Instead, Garreau suggested in his best-selling book *The Nine Nations of North America* (New York: Avon, 1981) that the real boundaries divide the entire North American continent into nine separate, homogeneous regions, which he called "nations." Each nation, according to Garreau, is inhabited by people who share the same traditions, values,

EXHIBIT 1   The Nine Nations

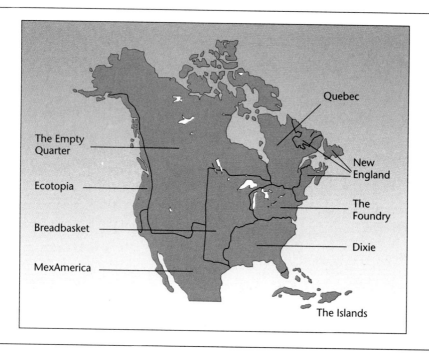

hopes, and world outlook and are different from the people of the other nations. The nine nations cross national boundaries of the United States, Canada, and the Caribbean. Garreau named his nations very descriptively, as follows: New England, Quebec, The Foundry, Dixie, The Islands, Empty Quarter, Breadbasket, MexAmerica, and Ecotopia. Exhibit 1 shows the boundaries of these nations.

Geographic segmentation is a very important concept in marketing. Thus, Garreau's novel idea promised potential gains in marketing. Professor Kahle suggested a statistical test of whether Garreau's division of the country (without the nation of Quebec, which lies entirely outside the United States) could be found valid with respect to marketing-related values. Such a division could then replace currently used geographic segmentation methods.

Two currently used segmentation schemes studied by Kahle were the quadrants and the Census Bureau regions. Kahle used a random sample of 2,235 people across the country and collected responses pertaining to eight self-assessed personal attributes: self-respect, security, warm relationships with others, sense of accomplishment, self-fulfillment, being well respected, sense of belonging, and fun–enjoyment–excitement. Kahle showed that these self-assessment attributes were directly related to marketing variables. The attributes determine, for example, the magazines a person is likely to read and the television programs he or she is likely to watch.

Kahle's results, using the nine-nations division (without Quebec), the quadrants division, and the Census division of the country, are presented in Exhibits 2 through 4. These tables are reprinted by permission from Kahle (1986). (Values reported in the exhibits are percentages.)

Carefully analyze the results presented in the exhibits. Is the nine-nations segmentation a useful alternative to the quadrants or the Census Bureau divisions of the country? Explain.

EXHIBIT 2   Distribution of Values across the Nine Nations

Value	New England	The Foundry	Dixie	The Islands	Bread-basket	Mex-America	Empty Quarter	Eco-topia	N
Self-respect	22.5%	20.5%	22.5%	25.0%	17.9%	22.7%	35.3%	18.0%	471
Security	21.7	19.6	23.3	15.6	20.2	17.3	17.6	19.6	461
Warm relationships with others	14.2	16.7	13.8	9.4	20.5	18.0	5.9	18.5	362
Sense of accomplishment	14.2	11.7	10.0	9.4	12.4	11.3	8.8	12.2	254
Self-fulfillment	9.2	9.9	8.4	3.1	7.5	16.0	5.9	12.7	214
Being well respected	8.3	8.7	11.0	15.6	10.1	2.7	2.9	4.2	196
Sense of belonging	5.0	8.4	7.5	12.5	7.8	6.7	17.6	7.9	177
Fun–enjoyment–excitement	5.0	4.5	3.5	9.4	3.6	5.3	5.9	6.9	100
Total	100.0	100.0	100.0	100.0	100.0	100.0	100.0	100.0	2,235
N	120	750	653	32	307	150	34	189	

EXHIBIT 3   Distribution of Values across Quadrants of the United States

Value	East	Midwest	South	West	N
Self-respect	19.7%	19.1%	23.4%	21.6%	471
Security	18.9	21.6	22.0	18.4	461
Warm relationships with others	16.0	17.8	14.5	17.1	362
Sense of accomplishment	13.2	12.5	9.2	11.4	254
Self-fulfillment	9.5	9.0	8.1	13.5	214
Being well respected	8.0	9.1	11.6	3.6	196
Sense of belonging	8.4	7.3	8.0	8.3	117
Fun–enjoyment–excitement	6.3	3.3	3.4	6.2	100
Total	100.0	100.0	100.0	100.0	2,235
N	476	634	740	385	

EXHIBIT 4   Distribution of Values across Census Regions of the United States

Value	New England	Middle Atlantic	South Atlantic	East South Central	East North Central	West North Central	West South Central	Mountain	Pacific	N
Self-respect	22.6%	18.6%	23.1%	23.4%	20.2%	16.7%	23.8%	29.2%	19.8%	471
Security	21.2	18.0	18.3	26.9	22.1	20.6	23.8	18.1	18.5	461
Warm relationships with others	13.9	16.8	15.7	11.4	16.0	21.6	14.9	15.3	17.6	362
Sense of accomplishment	13.9	13.0	10.7	9.6	11.4	14.7	6.8	8.3	12.1	254
Self-fulfillment	8.0	10.0	10.1	7.8	9.3	8.3	5.5	6.9	15.0	214
Being well respected	8.8	7.7	9.8	12.0	10.0	7.4	14.0	4.2	3.5	196
Sense of belonging	7.3	8.8	9.2	7.8	7.4	6.9	6.4	13.9	7.0	177
Fun–enjoyment–excitement	4.4	7.1	3.3	1.2	3.5	3.9	4.7	4.2	6.4	100
a Total	100.0	100.0	100.0	100.0	100.0	100.0	100.0	100.0	100.0	2,235
N	137	339	338	167	430	204	235	72	313	

15 Bayesian Statistics and Decision Analysis

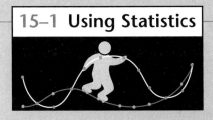

Only two hours after the polls closed in the Spanish general election on October 28, 1982, statistician José M. Bernardo made a stunning prediction about the outcome of the election. Bernardo accurately forecast that the Spanish Socialist party would win a landslide victory and be returned to power for the first time since the Spanish Civil War in the late 1930s. Furthermore, Bernardo predicted (to within a fraction of 1%) the number of votes that each political party would receive. He also estimated the number of seats (to within a single seat) that each party would hold in the new parliament. Bernardo made these amazing predictions hours before any other attempt was made to forecast the outcome of the election.

How did Bernardo do it? Instead of using the usual statistical approach to inference, an approach based solely on random sampling and usually referred to as the *classical* or *frequentist* approach, Bernardo chose to follow a different statistical philosophy. He chose to use the **Bayesian** approach.[1]

The Bayesian approach allows the statistician to use *prior information* about a particular problem, in addition to the information obtained from sampling. This approach is called *Bayesian* because the mathematical link between the probabilities associated with data results and the probabilities associated with the prior information is Bayes' theorem, which was introduced in Chapter 2. The theorem allows us to combine the prior information with the results of our sampling, giving us *posterior* (postsampling) information. A schematic comparison of the classical and the Bayesian approaches is shown in Figure 15–1.

In the case of the Spanish election, Bernardo used the results of the 1979 general election as his prior information. He used the 1979 results in identifying polling stations that seemed to be representative of the political behavior of Spanish voters as a whole, and he sampled from these stations. The undecided votes were distributed using information about proportions obtained in the 1979 election. The use of prior information led to more precise statistical conclusions than would have been obtained by pure random sampling. It is the *mixing* (using Bayes' theorem) of the prior information with sampling information from the 1982 election that gave Bernardo his stunningly precise conclusions.

The Bayesian philosophy does not necessarily lead to conclusions that are more accurate than those obtained by using the frequentist approach. If the prior information we have is accurate, then using it in conjunction with sample information leads to more accurate results than would be obtained without prior information. If, on the other hand, the prior information is inaccurate, then using it in conjunction with our sampling results leads to a worse outcome than would be obtained by using frequentist statistical inference. It is the very use of prior knowledge in a statistical analysis that often brings the entire Bayesian methodology under attack.

When prior information is a direct result of previous statistical surveys, or when prior information reflects no knowledge about the problem at hand (in which case the prior probabilities are called *noninformative*), the Bayesian analysis is purely objective, and few people would argue with its validity. Sometimes, however, the prior information reflects the personal opinions of the individual doing the analysis—or possibly

---

[1]A description of Professor Bernardo's study may be found in his article: J. M. Bernardo, "Monitoring the 1982 Spanish Socialist Victory: A Bayesian Analysis," *Journal of the American Statistical Association* 79 (September 1984), pp. 510–15.

FIGURE 15–1   A Comparison of Bayesian and Classical Approaches

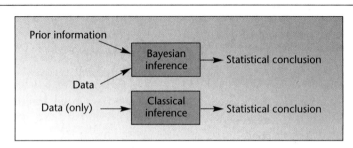

those of an expert who has knowledge of the particular problem at hand. In such cases, where the prior information is of a subjective nature, one may criticize the results of the analysis.

One way to classify statisticians is according to whether they are Bayesian or non-Bayesian (i.e., frequentist). The Bayesian group used to be a minority, but in recent years its numbers have grown. Even though differences between the two groups exist, it can be shown that when noninformative prior probabilities are used, the Bayesian results parallel the frequentist statistical results. This fact lends credibility to the Bayesian approach. If we are careful with the use of any prior information, we may avoid criticism and produce good results via the Bayesian methodology. Bernardo, for example, used no subjective elements in his analysis of the Spanish election: His prior knowledge—his prior probabilities—all came from the known results of the 1979 election. There is, of course, the question: Why would one believe that any results obtained in 1979 should have any relevance to the 1982 election? We may assume that Bernardo's answer to this question was subjective. If indeed it was, he certainly made a good bet.

In the next two sections, we give some basic elements of *Bayesian statistics*. These sections extend the idea of Bayes' theorem, first to discrete random variables and then to continuous ones. Section 15–4 discusses some aspects of subjective probabilities and how they can be elicited from a person who has knowledge of the situation at hand.

There is another important area, not entirely in the realm of statistics, that makes use of Bayes' theorem as well as subjective probabilities. This is the area of **decision analysis.** Decision analysis is a methodology developed in the 1960s, and it quantifies the elements of a decision-making process in an effort to determine the optimal decision.

15–2   Bayes' Theorem and Discrete Probability Models

In Section 2–7, we introduced Bayes' theorem. The theorem was presented in terms of *events*. The theorem was shown to transform *prior probabilities* of the occurrence of certain events into *posterior probabilities* of occurrence of the same events. Recall Example 2–10. In that example, we started with a prior probability that a randomly chosen person has a certain illness, given by $P(I) = 0.001$. Through the information that the person tested positive for the illness, and the reliability of the test, known to be $P(Z \mid I) = 0.92$ and $P(Z \mid \bar{I}) = 0.04$, we obtained through Bayes' theorem (equation 2–21) the posterior probability that the person was sick:

$$P(\text{I} \mid \text{Z}) = \frac{P(\text{Z} \mid \text{I})P(\text{I})}{P(\text{Z} \mid \text{I})P(\text{I}) + P(\text{Z} \mid \bar{\text{I}})P(\bar{\text{I}})} = 0.0225$$

The fact that the person had a positive reaction to the test may be considered our data. The conditional probabilities $P(\text{Z} \mid \text{I})$ and $P(\text{Z} \mid \bar{\text{I}})$ help incorporate the data in the computation. We will now extend these probabilities to include more than just an event and its complement, as was done in this example, or one of three events, as was the case in Example 2–10. Our extension will cover a whole set of values and their prior probabilities. The *conditional* probabilities, when extended over the entire set of values of a random variable, are called the *likelihood function.*

The **likelihood function** is the set of conditional probabilities $P(x \mid \theta)$ for given data x, considered a function of an unknown population parameter θ.

Using the likelihood function and the prior probabilities $P(\theta)$ of the values of the parameter in question, we define Bayes' theorem for discrete random variables in the following form:

> **Bayes' theorem** for a discrete random variable is
>
> $$P(\theta \mid x) = \frac{P(x \mid \theta)P(\theta)}{\sum_i P(x \mid \theta_i)P(\theta_i)} \qquad (15\text{–}1)$$
>
> where θ is an unknown population parameter to be estimated from the data. The summation in the denominator is over all possible values of the parameter of interest θ_i, and x stands for our particular data set.

In Bayesian statistics, we assume that population parameters such as the mean, the variance, or the population proportion are *random variables* rather than fixed (but unknown) quantities, as in the classical approach.

We assume that the parameter of interest is a random variable; thus, we may specify our prior information about the parameter as a **prior probability distribution** of the parameter. Then we obtain our data, and from them we get the likelihood function, that is, a measure of how likely we are to obtain our particular data, given different values of the parameter specified in the parameter's prior probability distribution. This information is transformed via Bayes' theorem, equation 15–1, to a **posterior probability distribution** of the value of the parameter in question. The posterior distribution includes the prior information as well as the data results. The posterior distribution can then be used in statistical inference. Such inference may include computing confidence intervals. Bayesian confidence intervals are often called **credible sets** of given posterior probability.

The following example illustrates the use of Bayes' theorem when the population parameter of interest is the population proportion p.

A market research analyst is interested in estimating the proportion of people in a certain area who use a product made by her client. That is, the analyst is interested in estimating her client's market share. The analyst denotes the parameter in question—the

EXAMPLE 15–1

TABLE 15–1
Prior Probabilities of Market Share S

S	P(S)
0.1	0.05
0.2	0.15
0.3	0.20
0.4	0.30
0.5	0.20
0.6	0.10
	1.00

true (population) market share of her client—by S. From previous studies of a similar nature, and from other sources of information about the industry, the analyst constructs the table of prior probabilities of the possible values of the market share S. This is the analyst's prior probability distribution of S. It contains different values of the parameter in question and the analyst's degree of belief that the parameter is equal to any of the values, given as a probability. The prior probability distribution is presented in Table 15–1.

As seen from the prior probabilities table, the analyst does not believe that her client's market share could be above 0.6 (60% of the market). For example, she may know that a competitor controls 40% of the market, so values above 60% are impossible as her client's share. Similarly, she may know for certain that her client's market share is at least 10%. The assumption that S may equal one of six discrete values is a restrictive approximation. In the next section, we will explore a continuous space of values.

The analyst now gathers a random sample of 20 people and finds out that 4 out of the 20 in the sample do use her client's product. The analyst wishes to use Bayes' theorem to combine her prior distribution of market share with the data results to obtain a posterior distribution of market share. Recall that in the classical approach, all that can be used is the sample estimate of the market share, which is $\hat{p} = x/n = 4/20 = 0.2$ and may be used in the construction of a confidence interval or a hypothesis test.

Solution   Using Bayes' theorem for discrete random variables (equation 15–1), the analyst updates her prior information to incorporate the data results. This is done in a tabular format and is shown in Table 15–2. As required by equation 15–1, the conditional probabilities $P(x \mid S)$ are evaluated. These conditional probabilities are our likelihood function. To evaluate these probabilities, we ask the following questions:

1. How likely are we to obtain the data results we have, that is, 4 successes out of 20 trials, if the probability of success in a single trial (the true *population proportion*) is equal to 0.1?
2. How likely are we to obtain the results we have if the population proportion is 0.2?
3. How likely are we to obtain these results when the population proportion is 0.3?

TABLE 15–2   Prior Distribution, Likelihood, and Posterior Distribution of Market Share (Example 15–1)

S	P(S)	P(x | S)	P(S)P(x | S)	P(S | x)
0.1	0.05	0.0898	0.00449	0.06007
0.2	0.15	0.2182	0.03273	0.43786
0.3	0.20	0.1304	0.02608	0.34890
0.4	0.30	0.0350	0.01050	0.14047
0.5	0.20	0.0046	0.00092	0.01230
0.6	0.10	0.0003	0.00003	0.00040
	1.00		0.07475	1.00000

4. How likely are we to obtain these results when the population proportion is 0.4?

5. How likely are we to obtain these results when the population proportion is 0.5?

6. How likely are we to obtain these results when the population proportion is 0.6?

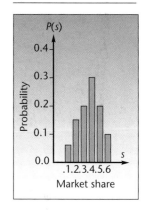

The answers to these six questions are obtained from a table of the binomial distribution (Appendix C, Table 1) and written in the appropriate places in the third column of Table 15–2. The fourth column is the product, for each value of S, of the prior probability of S and its likelihood. The sum of the entries in the fourth column is equal to the denominator in equation 15–1. When each entry in column 4 is divided by the sum of that column, we get the posterior probabilities, which are written in column 5. This procedure corresponds to an application of equation 15–1 for each one of the possible values of the population proportion S.

By comparing the values in column 2 of Table 15–2 with the values in column 5, we see how the prior probabilities of different possible market share values changed by the incorporation, via Bayes' theorem, of the information in the data (i.e., the fact that 4 people in a sample of 20 were found to be product users). The influence of the prior beliefs about the actual market share is evident in the posterior distribution. This is illustrated in Figure 15–2, which shows the prior probability distribution of S, and Figure 15–3, which shows the posterior probability distribution of S.

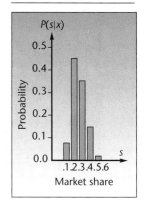

FIGURE 15–3
Posterior Distribution of Market Share (Example 15–1)

As the two figures show, starting with a prior distribution that is spread in a somewhat symmetric fashion over the six possible values of S, we end up, after the incorporation of data results, with a posterior distribution that is concentrated over the three values 0.2, 0.3, and 0.4, with the remaining values having small probabilities. The total posterior probability of the three values 0.2, 0.3, and 0.4 is equal to 0.92723 (from summing Table 15–2 entries). The three adjacent values are thus a set of highest posterior probability and can be taken as a credible set of values for S with posterior probability close to the standard 95% confidence level. Recall that with discrete random variables, it is hard to get values corresponding to exact, prespecified levels such as 95%, and we are fortunate in this case to be close to 95%. We may state as our conclusion that we are about 93% confident that the market share is anywhere between 0.2 and 0.4. Our result is a Bayesian conclusion, which may be stated in terms of a probability; it includes both the data results and our prior information. (As a comparison, compute an approximate classical confidence interval based on the sampling result.)

One of the great advantages of the Bayesian approach is the possibility of carrying out the analysis in a sequential fashion. Information obtained from one sampling study can be used as the prior information set when new information becomes available. The second survey results are considered the data set, and the two sources are combined by use of Bayes' theorem. The resulting posterior distribution may then be used as the prior distribution when new data become available, and so on.

We now illustrate the sequential property by continuing Example 15–1. Suppose that the analyst is able to obtain a *second* sample after her analysis of the first sample is completed. She obtains a sample of 16 people and finds that there are 3 users of the product of interest in this sample. The analyst now wants to combine this new sampling information with what she already knows about the market share. To do this,

TABLE 15–3 **Prior Distribution, Likelihood, and Posterior Distribution of Market Share for Second Sampling**

S	P(S)	P(x | S)	P(S)P(x | S)	P(S | x)
0.1	0.06007	0.1423	0.0085480	0.049074
0.2	0.43786	0.2463	0.1078449	0.619138
0.3	0.34890	0.1465	0.0511138	0.293444
0.4	0.14047	0.0468	0.0065740	0.037741
0.5	0.01230	0.0085	0.0001046	0.000601
0.6	0.00040	0.0008	0.0000003	0.000002
	1.00000		0.1741856	1.000000

the analyst considers her last posterior distribution, from column 5 of Table 15–2, as her new prior distribution when the new data come in. Note that the last posterior distribution contains *all* the analyst's information about market share before the incorporation of the new data, because it includes both her prior information and the results of the first sampling. Table 15–3 shows how this information is transformed into a new posterior probability distribution by incorporating the new sample results. The likelihood function is again obtained by consulting Appendix C, Table 1. We look for the binomial probabilities of obtaining 3 successes in 16 trials, using the given values of $S(0.1, 0.2, \ldots, 0.6)$, each in turn taken as the binomial parameter p.

The new posterior distribution of S is shown in Figure 15–4. Note that the highest posterior probability after the second sampling is given to the value $S = 0.2$, the posterior probability being 0.6191. With every additional sampling, the posterior distribution will get more peaked at values indicated by the data. The posterior distribution keeps moving toward data-derived results, and the effects of the prior distribution become less and less important. This fact becomes clear as we compare the distributions shown in Figures 15–2, 15–3, and 15–4. This property of Bayesian analysis is reassuring. It allows the data to speak for themselves, thus moving away from prior beliefs if these beliefs are away from reality. In the presence of limited data, Bayesian analysis allows us to compensate for the small data set by allowing us to use previous information—obtained either by prior sampling or by other means.

Incidentally, what would have happened if our analyst had decided to combine the results of the two surveys before considering them in conjunction with her prior information? That is, what would have happened if the analyst had decided to consider the two samples as one, where the total number of trials is $20 + 16 = 36$ and the total number of successes is $4 + 3 = 7$ users? Surprisingly, the posterior probability distribution for the combined sample incorporated with the prior distribution would have been exactly the same as the posterior distribution presented in Table 15–3. This fact demonstrates how well the Bayesian approach handles successive pieces of information. It does not matter when or how information is incorporated in the model—the posterior distribution will contain *all* information available at any given time.

In the next section, we discuss Bayesian statistics in the context of continuous probability distributions. In particular, we develop the normal probability model for Bayesian analysis. As will be seen in the next section, the normal distribution is particularly amenable to Bayesian analysis. If our prior distribution is normal and the

FIGURE 15–4
Second Posterior Distribution of Market Share

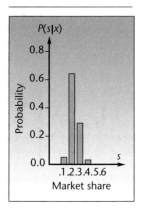

FIGURE 15–5 The Template for Bayesian Revision—Binomial
[Bayesian Revision.xls; Sheet: Binomial]

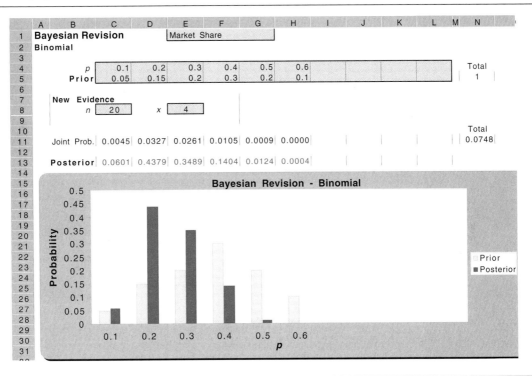

likelihood function is normal, then the posterior distribution is also normal. We will develop two simple formulas: one for the posterior mean and one for the posterior variance (and standard deviation) in terms of the prior distribution parameters and the likelihood function.

The Template

Figure 15–5 shows the template for revising binomial probabilities. The data seen in the figure correspond to Example 15–1.

To get the new posterior distribution, copy the posterior probabilities in the range C13:L13 and use **Paste Special** (values) command to paste them into the range C5:L5. They thus become the new prior probabilities. Add the new evidence of $n = 16$ and $x = 3$. The new posterior probabilities appear in row 13, as seen in Figure 15–6.

PROBLEMS

15–1. In 1995, Wells Fargo Bank of San Francisco launched a special credit card. The card was designed to reward people who pay their bills on time by allowing them to pay a lower-than-usual interest rate. Some research went into designing the new program. The director of the bank's regular credit card systems was consulted, and she gave her prior probability distribution of the proportion of cardholders who would qualify for the new program. Then a random sample of 20 cardholders was

FIGURE 15–6   The New Posterior Probabilities for Example 15–1
[Bayesian Revision.xls; Sheet: Binomial]

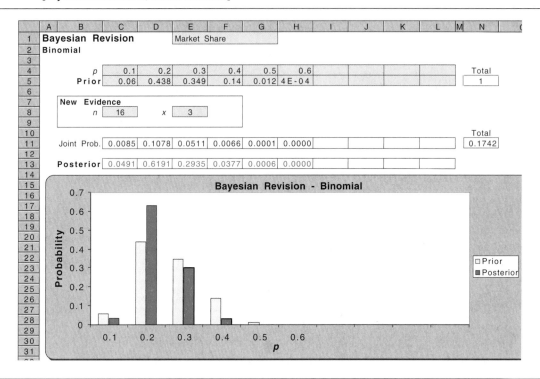

selected and tracked over several months. It was found that 6 of them paid all their credit card bills on time. Using this information and the information in the following table—the director's prior distribution of the proportion of all cardholders who pay their bills on time—construct the posterior probability distribution for this parameter. Also give a credible set of highest posterior probability close to 95% for the parameter in question. Plot both the prior and the posterior distributions of the parameter.

Proportion	Probability
0.1	0.2
0.2	0.3
0.3	0.1
0.4	0.1
0.5	0.1
0.6	0.1
0.7	0.1

15–2. In the situation of problem 15–1, suppose that a second random sample of cardholders was selected, and it was found that 7 out of the 17 people in the sample paid their bills on time. Construct the new posterior distribution containing information from the prior distribution and both samplings. Again, give a credible set of highest posterior probability close to 95%, and plot the posterior distribution.

15–3. For Example 15–1 suppose that a third sample is obtained. Three out of 10 people in the sample are product users. Update the probabilities of market share after the third sampling, and produce the new posterior distribution.

15–4. The company that markets the Yugoslav car Yugo in the United States needed to estimate the proportion of the market that the car would capture once it was introduced. Based on experience with Korean, French, and other foreign cars marketed in the country, one of the company's executives gave his prior probability distribution of the proportion of the foreign-car market that would go to Yugo. The prior probability distribution of this proportion (M = market share) is as follows:

M	P(M)
0.05	0.3
0.15	0.5
0.20	0.1
0.25	0.1

A random sample of 18 potential buyers of new foreign cars revealed that 1 of them would buy a Yugo. Compute the posterior probability distribution of M.

15–5. Recent years have seen a sharp decline in the Alaska king crab fishery. One problem identified as a potential cause of the decline has been the prevalence of a deadly parasite believed to infect a large proportion of the adult king crab population. A fisheries management agency monitoring crab catches needed to estimate the proportion of the adult crab population infected by the parasite. The agency's biologists constructed the following prior probability distribution for the proportion of infected adult crabs (denoted by R):

R	P(R)
0.25	0.1
0.30	0.2
0.35	0.2
0.40	0.3
0.45	0.1
0.50	0.1

A random sample of 10 adult king crabs was collected, and it was found that 3 of them were infected with the parasite. Construct the posterior probability distribution of the proportion of infected adult crabs, and plot it.

15–6. To continue problem 15–5, a second random sample of 12 adult crabs was collected, and it revealed that 4 individual crabs had been infected. Revise your probability distribution, and plot it. Give a credible set of highest posterior probability close to 95%.

15–7. For problem 15–5, suppose the biologists believed the proportion of infected crabs in the population was equally likely to be anywhere from 10% to 90%. Using the discrete points 0.1, 0.2, etc., construct a uniform prior distribution for the proportion of infected crabs, and compute the posterior distribution after the results of the sampling in problem 15–5.

15–8. An airline is interested in the proportion of flights that are full during a given season. The airline uses data from past experience and constructs the following prior distribution of the proportion of flights that are full to capacity:

S	P(S)
0.70	0.1
0.75	0.2
0.80	0.3
0.85	0.2
0.90	0.1
0.95	0.1
	1.0

A sample of 20 flights shows that 17 of these flights are full. Update the probability distribution to obtain a posterior distribution for the proportion of full flights.

15–9.   In the situation of problem 15–8, another sample of 20 flights reveals that 18 of them are full. Obtain the second posterior distribution of the proportion of full flights. Graph the prior distribution of problem 15–8, as well as the first posterior and the second posterior distributions. How did the distribution of the proportion in question change as more information became available?

15–10.   A quality control engineer has the following prior probability distribution for the proportion of defective items in a production process:

x	P(x)
0.05	0.1
0.10	0.2
0.15	0.4
0.20	0.2
0.25	0.1
	1.0

The engineer collects a random sample of 15 items and finds that 2 are defective. Compute the posterior distribution for the proportion of defective items, and give a credible set for the population proportion of defective items with posterior probability close to 95%.

15–3   Bayes' Theorem and Continuous Probability Distributions

We will now extend the results of the preceding section to the case of continuous probability models. Recall that a continuous random variable has a probability density function, denoted by $f(x)$. The function $f(x)$ is nonnegative, and the total area under the curve of $f(x)$ must equal 1.00. Recall that the probability of an event is defined as the area under the curve of $f(x)$ over the interval or intervals corresponding to the event.

> We define $f(\theta)$ as the **prior probability density** of the parameter θ. We define $f(x \mid \theta)$ as the conditional density of the data x, given the value of θ. This is the likelihood function.

The **joint density** of θ and x is obtained as the product:

$$f(\theta, x) = f(x \mid \theta)f(\theta) \qquad (15\text{--}2)$$

Using these functions, we may now write Bayes' theorem for continuous probability distributions. The theorem gives us the **posterior density** of the parameter θ, *given* the data *x*.

Bayes' theorem for continuous distributions[2] is

$$f(\theta \mid x) = \frac{f(x \mid \theta)\, f(\theta)}{\text{total area under } f(\theta, x)} \qquad (15\text{–}3)$$

Equation 15–3 is the analog for continuous random variables of equation 15–1. We may use the equation for updating a prior probability density function of a parameter θ once data *x* are available. In general, computing the posterior density is a complicated operation. However, in the case of a normal prior distribution and a normal data-generating process (or large samples, leading to central-limit conditions), the posterior distribution is also a normal distribution. The parameters of the posterior distribution are easy to calculate, as will be shown next.

The Normal Probability Model

Suppose that you want to estimate the population mean μ of a normal population that has a *known* standard deviation σ. Also suppose that you have some prior beliefs about the population in question. Namely, you view the population mean as a random variable with a normal (prior) distribution with mean *M′ and standard deviation* σ′.

If you draw a random sample of size *n* from the normal population in question and obtain a sample mean *M,* then the posterior distribution for the population mean μ is a *normal distribution with mean M″ and standard deviation* σ″ obtained, respectively, from equations 15–4 and 15–5.

The posterior mean and variance of the normal distribution of the population mean μ are

$$M'' = \frac{(1/\sigma'^2)M' + (n/\sigma^2)M}{1/\sigma'^2 + n/\sigma^2} \qquad (15\text{–}4)$$

$$\sigma''^2 = \frac{1}{1/\sigma'^2 + n/\sigma^2} \qquad (15\text{–}5)$$

The two equations are very useful in many applications. We are fortunate that the normal distribution family is *closed*; that is, when the prior distribution of a parameter is normal and the population (or process) is normal, the posterior distribution of the parameter in question is also normal. Be sure that you understand the distinction among the various quantities involved in the computations—especially the distinction between σ^2 and σ'^2. The quantity σ^2 is the variance of the population, and σ'^2 is the prior variance of the population mean μ. We demonstrate the methodology with Example 15–2.

---

[2]For the reader with knowledge of calculus, we note that Bayes' theorem is written as $f(\theta \mid x) = f(x \mid \theta)f(\theta) / [\int_{-\infty}^{\infty} f(x \mid \theta)f(\theta)\, d\theta]$.

EXAMPLE 15–2

A stockbroker is interested in the return on investment for a particular stock. Since Bayesian analysis is especially suited for the incorporation of opinion or prior knowledge with data, the stockbroker wishes to use a Bayesian model. The stockbroker quantifies his beliefs about the *average return* on the stock by a normal probability distribution with mean 15 (percentage return per year) and a standard deviation of 8. Since it is relatively large, compared with the mean, the stockbroker's prior standard deviation of μ reflects a state of relatively little prior knowledge about the stock in question. However, the prior distribution allows the broker to incorporate into the analysis some of his limited knowledge about the stock. The broker collects a sample of 10 monthly observations on the stock and computes the annualized average percentage return. He gets a mean $M = 11.54$ (percent) and a standard deviation $s = 6.8$. Assuming that the population standard deviation is equal to 6.8 and that returns are normally distributed, what is the posterior distribution of average stock returns?

Solution

We know that the posterior distribution is normal, with mean and variance given by equations 15–4 and 15–5, respectively. We have

$$M'' = \frac{(1/64)15 + (10/46.24)11.54}{1/64 + 10/46.24} = 11.77$$

$$\sigma'' = \sqrt{\frac{1}{1/64 + 10/46.24}} = 2.077$$

Note how simple it is to update probabilities when you start with a normal prior distribution and a normal population. Incidentally, the assumption of a normal population is very appropriate in our case, as the theory of finance demonstrates that stock returns are well approximated by the normal curve. If the population standard deviation is unknown, the sample standard deviation provides a reasonable estimate.

Figure 15–7 shows the stockbroker's prior distribution, the normal likelihood function (normalized to have a unit area), and the posterior density of the average return on the stock of interest. Note that the prior distribution is relatively flat—this is due to the relatively large standard deviation. The standard deviation is a measure of uncertainty, and here it reflects the fact that the broker does not know much about the stock. Prior distributions such as the one used here are called **diffuse priors.** They convey little a priori knowledge about the process in question. A relatively flat prior normal distribution conveys some information but lets the data tell us more.

Credible Sets

Unlike the discrete case, it is possible in the continuous case to construct credible sets for parameters with an exact, prespecified probability level. These are easy to construct. In Example 15–2, the stockbroker may construct a 95% *highest-posterior-density* (HPD) credible set for the average return on the stock directly from the posterior density. The posterior distribution is normal, with mean 11.77 and standard deviation 2.077. Therefore, the 95% HPD credible set for μ is simply

$$M'' \pm 1.96\sigma'' = 11.77 \pm 1.96(2.077)$$
$$= [7.699, 15.841]$$

Thus, the stockbroker may conclude there is a 0.95 probability that the average return on the stock is anywhere from 7.699% to 15.841% per year.

FIGURE 15–7   **Prior Distribution, Likelihood Function, and Posterior Distribution of Average Return μ (Example 15–2)**

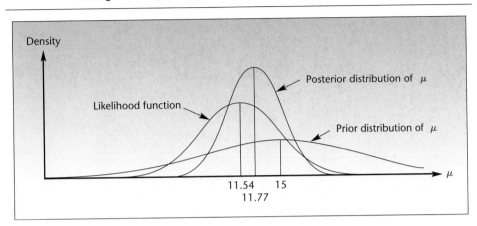

Recall that in the classical approach, we would have to rely only on the data and would not be able to use prior knowledge. As a conclusion, we would have to say: "Ninety-five percent of the intervals constructed in this manner will contain the parameter of interest." In the Bayesian approach, we are free to make *probability* statements as conclusions. Incidentally, the idea of attaching a probability to a result extends to the Bayesian way of testing hypotheses. A Bayesian statistician can give a posterior probability to the null hypothesis. Contrast this with the classical *p*-value, as defined in Chapter 7.

Suppose the stockbroker believed differently. Suppose that he believed that returns on the stock had a mean of 15 and a standard deviation of 4. In this case, the broker admits less uncertainty in his knowledge about average stock returns. The sampling results are the same, so the likelihood is unchanged. However, the posterior distribution does change as it now incorporates the data (through the likelihood) with a prior distribution that is not diffuse, as in the last case, but more peaked over its mean of 15. In our present case, the broker has a stronger belief that the average return is around 15% per year, as indicated by a normal distribution more peaked around its mean. Using equations 15–4 and 15–5, we obtain the posterior mean and standard deviation:

$$M'' = \frac{(1/16)15 + (10/46.24)11.54}{1/16 + 10/46.24} = 12.32$$

$$\sigma'' = \sqrt{\frac{1}{1/16 + 10/46.24}} = 1.89$$

As can be seen, the fact that the broker felt more confident about the average return's being around 15% (as manifested by the smaller standard deviation of his prior probability distribution) caused the posterior mean to be closer to 15% than it was when the same data were used with a more diffuse prior (the new mean is 12.32, compared with 11.77, obtained earlier). The prior distribution, the likelihood function, and the posterior distribution of the mean return on the stock are shown in Figure 15–8. Compare this figure with Figure 15–7, which corresponds to the earlier case with a more diffuse prior.

FIGURE 15–8   **Prior Distribution, Likelihood Function, and Posterior Distribution of Average Return Using a More Peaked Prior Distribution**

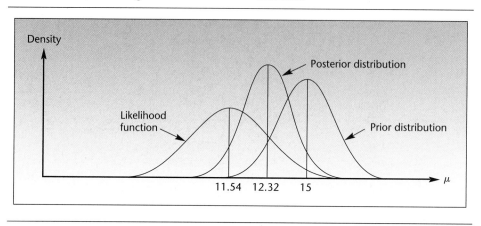

FIGURE 15–9   **The Template for Revising Beliefs about a Normal Mean [Bayesian Revision.xls; Sheet: Normal]**

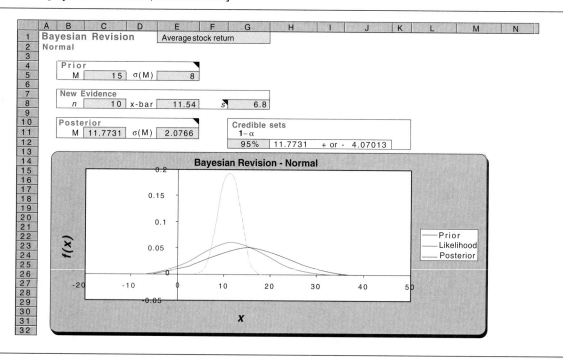

The Template

Figure 15–9 shows the template for revising beliefs about the mean of a normal distribution. The data in the figure correspond to Example 15–2.

Changing the 8 in cell E4 to 4 solves the less diffuse case. See Figure 15–10.

FIGURE 15–10   A Second Case of Revising the Normal Mean
[Bayesian Revision.xls; Sheet: Normal]

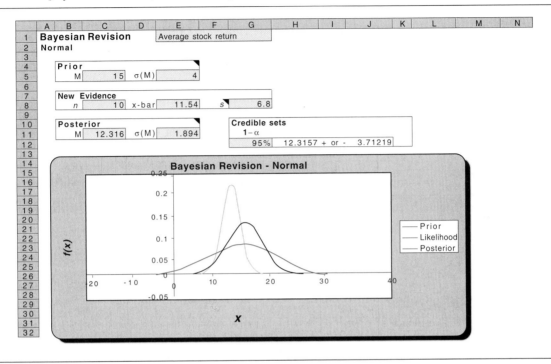

15–11. The partner-manager of a franchise of Au Bon Pain, Inc., the French bakery-restaurant chain, believes that the *average* daily revenue of her business may be viewed as a random variable (she adheres to the Bayesian philosophy) with mean $8,500 and standard deviation $1,000. Her prior probability distribution of average daily revenue is normal. A random sample of 35 days reveals a mean daily revenue of $9,210 and a standard deviation of $365. Construct the posterior distribution of average daily revenue. Assume a normal distribution of daily revenues.

15–12. Video Dog is an ingenious invention. It allows you to enjoy the pleasures of owning a pet without having to feed it, take it for walks, or worry about fleas and ticks. Before marketing the idea, the manufacturer wanted to estimate average monthly sales that would result in an area with a given population. From experience with similar products (probably the Pet Rock), the manufacturer assessed a normal prior probability distribution for average monthly sales in the area, with mean 15,000 and standard deviation 4,000. A random sample of 12 monthly sales in an area of about the same population and in the same geographic region gave a mean of 9,867 and a standard deviation of 1,055. Give a 95% highest-posterior-density credible set for the average monthly sales the manufacturer may expect in a region of the given size. Assume sales are normally distributed.

15-13. Claude Vrinat, owner of Taillevent—one of Europe's most highly acclaimed restaurants—is reported to regularly sample the tastes of his patrons.[3] From experience, Vrinat believes that the average rating (on a scale of 0 to 100) that his clients give his *foie gras de canard* may be viewed as normally distributed with mean 94 and standard deviation 2. A random sample of 10 diners gives an average rating of 96 and standard deviation of 1. What should be Vrinat's posterior distribution of average rating of *foie gras de canard*, assuming ratings are normally distributed?

15-14. In the context of problem 15-13, a second random sample of 15 diners is asked to rate the *foie gras*, giving a mean rating of 95 and standard deviation of 1. Incorporate the new information to give a posterior distribution that accounts for both samplings and the prior distribution. Give a 95% HPD credible set for mean rating.

15-15. The operator of a film developing business believes that the average time it takes to develop a roll of film of a particular type is uncertain and may be described by a normal distribution with mean 22 minutes and standard deviation 2 minutes. Using this information as the prior distribution, and the fact that a random sample of 14 rolls had an average development time of 24 minutes per roll and a standard deviation of 3 minutes, construct the posterior distribution of average development time. Assume normality.

15-16. To continue problem 15-15, a second random sample of 20 rolls of film gave a mean development time of 25 minutes and a standard deviation of 2 minutes. Construct the second posterior distribution, and use it in constructing a 99% HPD credible set for average development time.

15-17. In an effort to predict Alaska's oil-related state revenues, a Delphi session is regularly held where experts in the field give their expectations of the average future price of crude oil over the next year. The views of five prominent experts who participated in the last Delphi session may be stated as normal prior distributions with means and standard deviations given in the following table. To protect their identities (the Delphi sessions are closed to the public), we will denote them by the letters A through E. Data are in dollars per barrel.

Expert	Mean	Standard Deviation
A	23	4
B	19	7
C	25	1
D	20	9
E	27	3

Compare the views of the five experts, using this information. What can you say about the different experts' degrees of belief in their own respective knowledge? One of the experts is the governor of Alaska, who, due to the nature of the post, devotes little time to following oil prices. All other experts have varying degrees of experience with price analysis; one of them is the ARCO expert who assesses oil prices on a daily basis. Looking only at the reported prior standard deviations, who is likely to be the governor, and who is likely to be the ARCO expert? Now suppose that at the end of the year it was found that the average daily price of crude oil was $18 per barrel. Who should be most surprised (and embarrassed), and why?

---

[3]Patricia Wells, "Three-Star Galaxy," in *Beloved Cities: Europe*, ed. A. Rosenthal and A. Gelb (New York: Viking Penguin, 1985), pp. 240–55.

15–4   The Evaluation of Subjective Probabilities

Since Bayesian analysis makes extensive use of people's subjective beliefs in the form of prior probabilities, it is only natural that the field should include methods for the elicitation of personal probabilities. We begin by presenting some simple ideas on how to identify a normal prior probability distribution and give a rough estimate of its mean and standard deviation.

Assessing a Normal Prior Distribution

As you well know by now, the normal probability model is useful in a wide variety of applications. Furthermore, since we know probabilities associated with the normal distribution, results can easily be obtained if we do make the assumption of normality. How can we estimate a decision maker's subjective normal probability distribution? For example, how did the stockbroker of Example 15–2 decide that his prior distribution of average returns was normal with mean 15 and standard deviation 8?

The normal distribution appears naturally and as an approximation in many situations due to the central limit theorem. Therefore, in many instances, it makes sense to assume a normal distribution. In other cases, it frequently happens that we have a distribution that is not normal but still is *symmetric* with a *single mode*. In such cases, it may still make sense to assume a normal distribution as an approximation because this distribution is easily estimated as a subjective distribution, and the resulting inaccuracies will not be great. In cases where the distribution is *skewed*, however, the normal approximation will not be adequate.

Once we determine that the normal distribution is appropriate for describing our personal beliefs about the situation at hand, we need to estimate the mean and the standard deviation of the distribution. For a symmetric distribution with one mode, the mean is equal to the median and to the mode. Therefore, we may ask the decision maker whose subjective probability we are trying to assess what he or she believes to be the center of the distribution. We may also ask for the most likely value. We may ask for the average, or we may ask for the point that splits the distribution into two equal parts. All these questions would lead us to the central value, which we take to be the mean of the subjective distribution. It is useful to ask the person whose probabilities we are trying to elicit several of these questions, so that we have a few checks on the answer. Any discrepancies in the answers may lead to possible violations of our assumption of the symmetry of the distribution or its unimodality (having only one mode), which would obviate the normal approximation. Presumably, questions such as these lead the stockbroker of Example 15–2 to determine that the mean of his prior distribution for average returns is 15%.

How do we estimate the standard deviation of a subjective distribution? Recall the simple rules of thumb for the normal probability model:

Approximately 68% of the distribution lies within 1 standard deviation of the mean.

Approximately 95% of the distribution lies within 2 standard deviations of the mean.

These rules lead us to the following questions for the decision maker whose probabilities we are trying to assess: "Give me two values of the distribution in question such that you are 95% sure that the variable in question is between the two values," or equivalently, "Give me two values such that 95% of the distribution lies between them." We may also ask for two values such that 68% of the distribution is between these values.

For 95% sureness, assuming symmetry, we know that the two values we obtain as answers are each 2 standard deviations away from the mean. In the case of the stockbroker, he must have felt there was a 0.95 chance that the average return on the stock was anywhere from −1% to 31%. The two points −1 and 31 are 2 × 8 units on either side of the mean of 15. Hence, the standard deviation is 8. The stockbroker could also have said that he was 68% sure that the average return was anywhere from 7% to 23% (each of these two values is 1 standard deviation away from the mean of 15). Using 95% bounds is more useful than 68% limits because people are more likely to think in terms of 95% sureness. Be sure you understand the difference between this method of obtaining bounds on values of a population (or random variable) and the construction of confidence intervals (or credible sets) for population *parameters*.

15–5 Decision Analysis: An Overview

Some years ago, the state of Massachusetts had to solve a serious problem: an alarming number of road fatalities caused by icy roads in winter. The state department of transportation wanted to solve the problem by salting the roads to reduce ice buildup. The introduction of large amounts of salt into the environment, however, would eventually cause an increase in the sodium content of drinking water, thus increasing the risk of heart problems in the general population.

This is the kind of problem that can be solved by *decision analysis*. There is a decision to be made: to salt or not to salt. With each of the two possible actions, we may associate a final outcome, and each outcome has a probability of occurrence. An additional number of deaths from heart disease would result if roads were salted. The number of deaths is uncertain, but its probability may be assessed. On the other hand, a number of highway deaths would be prevented if salt were used. Here again, the number is uncertain and governed by some probability law. In decision analysis we seek the best decision in a given situation. Although it is unpleasant to think of deaths, the best (optimal) decision here is the decision that would minimize the expected total number of deaths. *Expected* means averaged using the different probabilities as weights.

The area of decision analysis is independent of most of the material in this book. To be able to perform decision analysis, you need to have a rudimentary understanding of probability and of expected values. Some problems make use of additional information, obtained either by sampling or by other means. In such cases, we may have an idea about the *reliability* of our information—which may be stated as a probability—and the information is incorporated in the analysis by use of Bayes' theorem.

When a company is interested in introducing a new product, decision analysis offers an excellent aid in coming to a final decision. When one company considers a merger with another, decision analysis may be used as a way of evaluating all possible outcomes of the move and deciding whether to go ahead based on the best expected outcome. Decision analysis can help you decide which investment or combination of investments to choose. It could help you choose a job or career. It could help you decide whether to pursue an MBA degree.

We emphasize the use of decision analysis as an aid in corporate decision making. Since it is often difficult to quantify the aspects of human decision making, it is important to understand that decision analysis should not be the only criterion for making a decision. A stockbroker's hunch, for example, may be a much better indication of the best investment decision than a formal mathematical analysis, which may very well miss some important variables.

Decision analysis, as described in this book, has several elements.

The elements of a decision analysis:

1. Actions
2. Chance occurrences
3. Probabilities
4. Final outcomes
5. Additional information
6. Decision

Actions

By an *action*, we mean anything that the decision maker can do. An action is something you, the decision maker, can control. You may choose to take an action, or you may choose not to take it. Often, there are several choices for action: You may buy one of several different products, travel one of several possible routes, etc. Many decision problems are sequential in nature: You choose one action from among several possibilities; later, you are again in a position to take an action. You may keep taking actions until a final outcome is reached; you keep playing the game until the game is over. Finally, you have reached some final outcome—you have gained a certain amount or lost an amount, achieved a goal or failed.

Chance Occurrences

Even if the decision problem is essentially nonsequential (you take an action, something happens, and that is it), we may gain a better understanding of the problem if we view the problem as sequential. We assume that the decision maker takes an action, and afterward "chance takes an action." The action of chance is the chance occurrence. When you decide to buy ABC stock, you have taken an action. When the stock falls 3 points the next day, chance has taken an action.

Probabilities

All actions of chance are governed by probabilities, or at least we view them that way because we cannot predict chance occurrences. The probabilities are obtained by some method. Often, the probabilities of chance occurrences are the decision maker's (or consulted expert's) subjective probabilities. Thus, the chief executive officer of a firm bidding for another firm will assign certain probabilities to the various outcomes that may result from the attempted merger.

In other cases, the probabilities of chance occurrences are more objective. If we use sampling results as an aid in the analysis (see the section on additional information, which follows), then statistical theory gives us measures of the reliability of results and, hence, probabilities of relevant chance occurrences.

Final Outcomes

We assume that the decision problem is of finite duration. After you, the decision maker, have taken an action or a sequence of actions, and after chance has taken action or a sequence of actions, there is a final outcome. An outcome may be viewed as a *payoff* or *reward*, or it may be viewed as a *loss*. We will look at outcomes as rewards (positive or negative). A payoff is an amount of money (or other measure of benefit, called a *utility*) that you receive at the end of the game—at the end of the decision problem.

Additional Information

Each time chance takes over, a random occurrence takes place. We may have some prior information that allows us to assess the probability of any chance occurrence. Often, however, we may be able to purchase additional information. We may consult an expert, at a cost, or we may sample from the population of interest (assuming there is such a population) for a price. The costs of obtaining additional information are subtracted from our final payoff. Therefore, buying new information is, in itself, an action that we may choose to take or not. Deciding whether to obtain such information is part of the entire decision process. We must weigh the benefit of the additional information against its cost.

Decision

The action, or sequential set of actions, we decide to take is called our *decision*. The decision obtained through a useful analysis is that set of actions that maximizes our expected final-outcome payoff. The decision will often give us a set of *alternative actions* in addition to the optimal set of actions. In a decision to introduce a new product, suppose that the result of the decision analysis indicates that we should proceed with the introduction of the product without any market testing—that is, without any sampling. Suppose, however, that a higher official in the company requires us to test the product even though we may not want to do so. A comprehensive solution to the decision problem would provide us not only with the optimal action (market the product), but also with information on how to proceed in the best possible way when we are forced to take some suboptimal actions along the way. The complete solution to the decision problem would thus include information on how to treat the results of the market test. If the results are unfavorable, the optimal action at this point may be not to go ahead with introducing the product. The solution to the decision problem—the *decision*—gives us all information on how to proceed at any given stage or circumstance.

As you see, we have stressed a sequential approach to decision making. At the very least, a decision analysis consists of two stages: The decision maker takes an action out of several possible ones, and then chance takes an action. This sequential approach to decision making is very well modeled and visualized by what is called a **decision tree.**

A decision tree is a set of nodes and branches. At a *decision node*, the decision maker takes an action; the action is the choice of a branch to be followed. The branch leads to a *chance node*, where chance determines the outcome; that is, chance chooses the branch to be followed. Then either the final outcome is reached (the branch ends), or the decision maker gets to take another action, and so on. We mark a decision node by a square and a chance node by a circle. These are connected by the branches of the decision tree. An example of a decision tree is shown in Figure 15–11. The decision tree shown in the figure is a simple one: It consists of only four branches, one decision node, and one chance node. In addition, there is no product-testing option. As we go on, we will see more complicated decision trees, and we will explore related topics.

FIGURE 15–11 An Example of a Decision Tree for New-Product Introduction

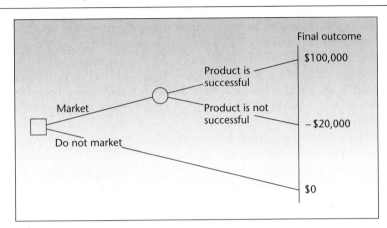

PROBLEMS

15–18. What are the uses of decision analysis?

15–19. What are the limitations of decision analysis?

15–20. List the elements of a decision problem, and explain how they interrelate.

15–21. What is the role of probabilities in a decision problem, and how do these probabilities arise?

15–22. What is a decision tree?

15–6 Decision Trees

As mentioned in the last section, a decision tree is a useful aid in carrying out a decision analysis because it allows us to visualize the decision problem. If nothing else, the tree gives us a good perspective on our decision problem: It lets us see when we, as decision makers, are in control and when we are not. To handle the instances when we are not in control, we use probabilities. These probabilities—assuming they are assessed in some accurate, logical, and consistent way—are our educated guesses as to what will happen when we are not in control.

The aforementioned use of decision trees in clarifying our perspective on a decision problem may not seem terribly important, say, compared with a quantitatively rigorous solution to a problem involving exact numbers. However, this use of decision trees is actually more important than it seems. After you have seen how to use a decision tree in computing the expected payoff at each chance node and the choice of the optimal action at each decision node, and after you have tried several decision problems, you will find that the trees have an added advantage. You will find that just drawing the decision tree helps you better understand the decision problem you need to solve. Then, even if the probabilities and payoffs are not accurately assessed, making you doubt the exact optimality of the solution you have obtained, you will still have gained a better understanding of your decision problem. This in itself should help you find a good solution.

In a sense, a decision tree is a good psychological tool. People are often confused about decisions. They are not always perfectly aware of what they can do and what they cannot do, and they often lack an understanding of uncertainty and how it affects the outcomes of their decisions. This is especially true of large-scale decision problems, where there are several possible actions at several different points, each followed by chance outcomes, leading to a distant final outcome. In such cases, drawing a decision tree is an indispensable way of gaining familiarity with all aspects of the decision problem. The tree shows which actions affect which, and how the actions interrelate with chance outcomes. The tree shows how combinations of actions and chance outcomes lead to possible final outcomes and payoffs.

Having said all this, let us see how decision problems are transformed to visual decision trees and how these trees are analyzed. Let us see how decision trees can lead us to optimal solutions to decision problems. We will start with the simple new-product introduction example shown in the decision tree in Figure 15–11. Going step by step, we will show how that simple tree was constructed. The same technique is used in constructing more complicated trees, with many branches and nodes.

The Payoff Table

The first step in the solution of any decision problem is to prepare the *payoff table* (also called the *payoff matrix*). The payoff table is a table of the possible payoffs we would receive if we took certain actions and certain chance occurrences followed. Generally, what takes place will be called *state of nature*, and what we do will be called the *decision*. This leads us to a table that is very similar to Table 7–1. There we dealt with hypothesis testing, and the state of nature was whether the null hypothesis was true; our decision was either to reject or not reject the null hypothesis. In that context, we could have associated the result "not reject H_0 when H_0 is true" with some payoff (because a correct decision was made); the outcome "not reject H_0 when H_0 is false" could have been associated with another (negative) payoff, and similarly for the other two possible outcomes. In the context of decision analysis, we might view the hypothesis testing as a sequential process. We make a decision (to reject or not to reject H_0), and then "chance" takes over and makes H_0 either true or false.

Let us now write the payoff table for the new-product introduction problem. Here we assume that if we do not introduce the product, there is nothing gained and nothing lost. This assumes we have not invested anything in developing the product, and it assumes no opportunity loss. If we do not introduce the new product, our payoff is zero.

If our action is to introduce the product, two things may happen: The product may be successful, or it may not. If the product is successful, our payoff will be $100,000; and if it is not successful, we will lose $20,000, so our payoff will be −$20,000. The payoff table for this simple problem is Table 15–4. In real-world situations, we may assess more possible outcomes: finely divided *degrees* of success. For example, the product may be extremely successful—with payoff $150,000; very successful—payoff $120,000; successful—payoff $100,000; somewhat successful—payoff $80,000; barely successful—payoff $40,000; breakeven—payoff $0; unsuccessful—payoff −$20,000; or disastrous—payoff −$50,000. Table 15–4 can be easily extended to cover these expanded states of nature. Instead of two columns, we would have eight, and we would still have two rows corresponding to the two possible actions.

The values in Table 15–4 give rise to the decision tree that was shown in Figure 15–11. We take an action: If we do not market the product, the payoff is zero, as shown by the arc from the decision node to the final outcome of zero. If we choose to

TABLE 15–4   **Payoff Table: New-Product Introduction**

	Product Is	
Action	**Successful**	**Not Successful**
Market the product	+$100,000	−$20,000
Do not market the product	0	0

FIGURE 15–12   **Decision Tree for New-Product Introduction**

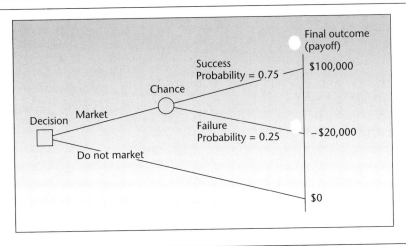

market the product, chance either will take us to success and a payoff of $100,000 or will lead us to failure and a loss of $20,000. We now need to deal with chance. We do so by assigning probabilities to the two possible states of nature, that is, to the two possible actions of chance. Here, some elicitation of personal probabilities is done. Suppose that our marketing manager concludes that the probability of success of the new product is 0.75. The probability of failure then must be $1 - 0.75 = 0.25$. Let us write these probabilities on the appropriate branches of our decision tree. The tree, with payoffs and probabilities, is shown in Figure 15–12.

We now have all the elements of the decision tree, and we are ready to solve the decision problem.

> The solution of decision tree problems is achieved by working backward from the final outcomes.

The method we use is called **averaging out and folding back.** Working backward from the final outcomes, we *average out all chance occurrences.* This means that we find the *expected value* at each chance node. At each chance node (each circle in the tree), we write the expected monetary value of all branches leading out of the node; we *fold back* the tree. At each decision node (each square in the tree), we *choose the action that maximizes our (expected) payoff.* That is, we look at all branches emanating from the decision node, and we choose the branch leading to the highest monetary value. Other branches may be *clipped*; they are not optimal. The problem is solved once we reach the beginning: the first decision node.

Let us solve the decision problem of the new-product introduction. We start at the final outcomes. There are three such outcomes, as seen in Figure 15–12. The outcome with payoff $0 emanates directly from the decision node; we leave it for now. The other two payoffs, $100,000 and −$20,000, both emanate from a chance node. We therefore average them out—using their respective probabilities—and fold back to the chance node. To do this, we find the expected monetary value at the chance node (the circle in Figure 15–12). Recall the definition of the expected value of a random variable, given as equation 3–4.

The expected value of X, denoted $E(X)$, is

$$E(X) = \sum_{\text{all } x} xP(x)$$

The outcome as you leave the chance node is a random variable with two possible values: 100,000 and −20,000. The probability of outcome 100,000 is 0.75, and the probability of outcome −20,000 is 0.25. To find the expected value at the chance node, we apply equation 3–4:

$$E(\text{outcome at chance node}) = (100{,}000)(0.75) + (-20{,}000)(0.25)$$
$$= 70{,}000$$

Thus, the expected value associated with the chance node is +$70,000; we write this value next to the circle in our decision tree. We can now look at the decision node (the square), since we have folded back and reached the first node in the tree. We know the (expected) monetary values associated with the two branches emanating from this node. Recall that at decision nodes we do not average. Rather, we choose the best branch to be followed and clip the other branches, as they are not optimal. Thus, at the decision node, we compare the two values +$70,000 and $0. Since 70,000 is greater than 0, the expected monetary outcome of the decision to market the new product is greater than the monetary outcome of the decision not to market the new product. We follow the rule of choosing the decision that maximizes the expected payoff, so we choose to market the product. (We clip the branch corresponding to "not market" and put a little arrow by the branch corresponding to "market.") In Section 15–8, where we discuss *utility*, we will see an alternative to the "maximum expected monetary value" rule, which takes into account our attitudes toward risk rather than simply aims for the highest *average* payoff, as we have done here. The solution of the decision tree is shown in Figure 15–13.

We follow the arrow and make the decision to market the new product. Then chance takes over, and the product either becomes successful (an event which, a priori, we believe to have a 0.75 probability of occurring) or does not become successful. On average—that is, if we make decisions such as this one very many times—we should expect to make $70,000.

Let us now consider the extended market possibilities mentioned earlier. Suppose that the outcomes and their probabilities in the case of extended possibilities are as given in Table 15–5. In this new example, the payoff is more realistic: It has many possible states. Our payoff is a random variable. The expected value of this random variable is computed, as usual, by multiplying the values by their probabilities and adding (equation 3–4). This can easily be done by adding a column for Payoff ×

FIGURE 15–13 Solution of the New-Product Introduction Decision Tree

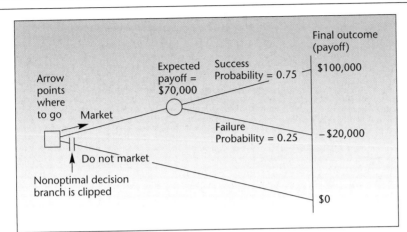

TABLE 15–5 Possible Outcomes and Their Probabilities

Outcome	Payoff	Probability
Extremely successful	$150,000	0.1
Very successful	120,000	0.2
Successful	100,000	0.3
Somewhat successful	80,000	0.1
Barely successful	40,000	0.1
Breakeven	0	0.1
Unsuccessful	−20,000	0.05
Disastrous	−50,000	0.05

Probability to Table 15–5 and adding all entries in the column. This gives us E(pay-off) = $77,500 (verify this). The decision tree for this example—with many branches emanating from the chance node—is shown in Figure 15–14. The optimal decision in this case is, again, to market the product.

We have seen how to analyze a decision problem by using a decision tree. Let us now look at an example. In Example 15–3, chance takes over after either action we take, and the problem involves more than one action. We will take an action; then a chance occurrence will take place. Then we will again decide on an action, after which chance will again take over, leading us to a final outcome.

EXAMPLE 15–3

In 1995, Digital Equipment Corporation arranged to get the Cunard Lines ship *Queen Elizabeth 2* (QE2) for use as a floating hotel for the company's annual convention. The meeting took place in September and lasted nine days. In agreeing to lease the QE2, Cunard had to make a decision. If the cruise ship were leased to Digital, Cunard would get a flat fee and an additional percentage of profits from the gala convention, which could attract as many as 50,000 people. Cunard analysts therefore estimated

FIGURE 15–14   **Extended-Possibilities Decision Tree for New-Product Introduction**

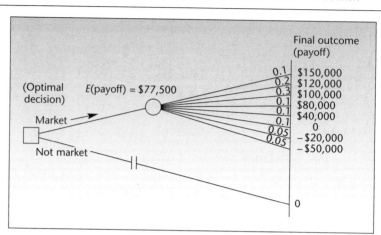

that if the ship were leased, there would be a 0.50 probability that the company would make $700,000 for the nine days; a 0.30 probability that profits from the venture would be about $800,000; a 0.15 probability that profits would be about $900,000; and a 0.05 probability that profits would be as high as $1 million. If the ship were not leased to Digital, the vessel would be used for its usual Atlantic crossing voyage, also lasting nine days. If this happened, there would be a 0.90 probability that profits would be $750,000 and a 0.10 probability that profits would be about $780,000. The tighter distribution of profits on the voyage was due to the fact that Cunard analysts knew much about the company's usual business of Atlantic crossings but knew relatively little about the proposed venture.

Cunard had one additional option. If the ship were leased to Digital, and it became clear within the first few days of the convention that Cunard's profits from the venture were going to be in the range of only $700,000, the steamship company could choose to promote the convention on its own by offering participants discounts on QE2 cruises. The company's analysts believed that if this action were chosen, there would be a 0.60 probability that profits would increase to about $740,000 and a 0.40 probability that the promotion would fail, lowering profits to $680,000 due to the cost of the promotional campaign and the discounts offered. What should Cunard have done?

Solution   Let us analyze all the components of this decision problem. There are two possible actions, one of which must be chosen: to lease or not to lease. We can start constructing our tree by drawing the square denoting this decision node and showing the two appropriate branches leading out of it.

Once we make our choice, chance takes over. If we choose to lease, chance will lead us to one of four possible outcomes. We show these possibilities by attaching a circle node at the end of the lease action branch, with four branches emanating from it. If we choose not to lease, chance again takes over, leading us to two possible outcomes. This is shown by a chance node attached at the end of the not-lease action branch, with two branches leading out of it and into the possible final outcome payoffs of $750,000 and $780,000.

FIGURE 15–15 **Decision Tree for the Cunard Lease Example**

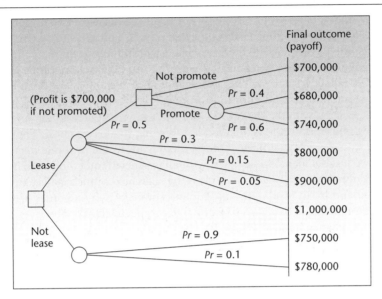

We now go back to the chance occurrences following the lease decision. At the end of the branch corresponding to an outcome of $700,000, we attach another decision node corresponding to the promotion option. This decision node has two branches leaving it: One goes to the final outcome of $700,000, corresponding to nonpromotion of the convention; and the other, the one corresponding to promotion, leads to a chance node, which in turn leads to two possible final outcomes: a profit of $740,000 and a profit of $680,000. All other chance outcomes following the lease decision lead directly to final outcomes. These outcomes are profits of $800,000, $900,000, and $1 million. At each chance branch, we note its probability. The chance outcomes of the lease action have probabilities 0.5, 0.3, 0.15, and 0.05 (in order of increasing monetary outcome). The probabilities of the outcomes following the not-lease action are 0.9 and 0.1, respectively. Finally, the probabilities corresponding to the chance outcomes following the promote action are 0.4 and 0.6, again in order of increasing profit.

Our decision tree for the problem is shown in Figure 15–15. Having read the preceding description of the details of the tree, you will surely agree that "A picture is worth 1,000 words."

We now solve the decision tree. Our method of solution, as you recall, is averaging out and folding back. We start at the end; that is, we look at the final-outcome payoffs and work backward from these values. At every chance node, we average the payoffs, using their respective probabilities. This gives us the expected monetary value associated with the chance node. At every decision node, we choose the action with the highest expected payoff and clip the branches corresponding to all other, nonoptimal actions. Once we reach the first decision node in the tree, we are done and will have obtained a complete solution to the decision problem.

Let us start with the closest chance node to the final outcomes—the one corresponding to the possible outcomes of the promote action. The expected payoff at this chance node is obtained as

$$E(\text{payoff}) = (680,000)(0.4) + (740,000)(0.6) = \$716,000$$

We now move back to the promote/not-promote decision node. Here we must choose the action that maximizes the expected payoff. This is done by comparing the two payoffs: the payoff of $700,000 associated with the not-promote action and the expected payoff of $716,000 associated with the promote action. Since the *expected* value of $716,000 is greater, we choose to promote. We show this with an arrow, and we clip the nonoptimal action not to promote. The expected value of $716,000 now becomes associated with the decision node, and we write it next to the node.

We now fold back to the chance node following the lease action. There are four branches leading out of that node. One of them leads to the promote decision node, which, as we just decided, is associated with an (expected) outcome of $716,000. The probability of reaching the decision node is 0.5. The next branch leads to an outcome of $800,000 and has a probability of 0.3; and the next two outcomes are $900,000 and $1 million, with probabilities 0.15 and 0.05, respectively. We now average out the payoffs at this chance node as follows:

$$E(\text{payoff}) = (716,000)(0.5) + (800,000)(0.3) + (900,000)(0.15) \\ + (1,000,000)(0.05) = \$783,000$$

This expected monetary value is now written next to the chance node.

Let us now look at the last chance node, the one corresponding to outcomes associated with the not-lease action. Here, we have two possible outcomes: a payoff of $750,000, with probability 0.9; and a payoff of $780,000, with probability 0.1. We now find the expected monetary value of the chance node:

$$E(\text{payoff}) = (750,000)(0.9) + (780,000)(0.1) = \$753,000$$

We are now finally at the first decision node of the entire tree. Here we must choose the action that maximizes the expected payoff. The choice is done by comparing the expected payoff associated with the lease action, $783,000, and the expected payoff associated with not leasing, $753,000. Since the higher expected payoff is that associated with leasing, the decision is to lease. This is shown by an arrow on the tree and by clipping the not-lease action as nonoptimal. The stages of the solution are shown in Figure 15–16.

We now have a final solution to the decision problem: We choose to *lease* our ship to Digital Equipment Corporation. Then, if we should find out that our profit from the lease were in the range of only $700,000, our action would be to *promote* the convention. Note that in this decision problem, the decision consists of a pair of actions. The decision tells us what to do in any eventuality.

If the tree had more than just two decision nodes, the final solution to the problem would have consisted of a set of optimal actions at all decision nodes. The solution would have told us what to do at any decision node to maximize the expected payoff, given that we arrive at that decision node. Note, again, that our solution is optimal only in an *expected monetary value* sense. If Cunard is very conservative and does not want to take any risk of getting a profit lower than the minimum of $750,000 it is assured of receiving from an Atlantic crossing, then, clearly, the decision to lease would not be optimal because it does admit such a risk. If the steamship company has some risk aversion but would accept some risk of lower profits, then the analysis should be done using utilities rather than pure monetary values. The use of utility in a way that accounts for people's attitudes toward risk is discussed in Section 15–8.

FIGURE 15–16   **Solution of the Cunard Leasing Problem**

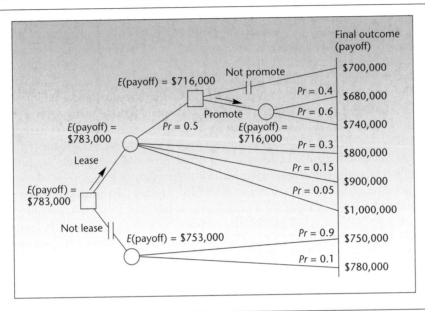

15–23. Spuds MacKenzie, the stocky, bow-legged, white English bull terrier featured in Bud Light beer commercials, was probably one of the greatest advertising discoveries of all time. Few people know, however, that Spuds began appearing for Anheuser-Busch, the makers of Bud Light, in 1983 but played only a minor role in Bud Light advertising. Spuds's career really started 4 years (almost 30 dog-years!) later, at the Super Bowl in January 1987, when Anheuser-Busch paid CBS *$20,000 a second* to put Spuds on the air.

What went into that fateful decision? Let us say that Anheuser-Busch's marketing director felt that there would be a 0.35 probability that a series of three 60-second commercials during the Super Bowl could boost sales volume of Bud Light beer by $8 million during the following 3 months. The director also believed that there was a 0.65 probability that the resulting excess sales would be about $4 million. Considering also the cost of airing the company's commercials, carry out the decision analysis of whether to advertise on television featuring the then-unknown Spuds MacKenzie. Construct a complete decision tree and solve it.

15–24. Boeing has drawn up plans for a new fuel-saving jet, the 7J7, that is designed to compete with the European-built Airbus 320. Boeing is currently conducting research to assess the feasibility of building the 7J7. Company experts believe there is a 0.20 chance that sales of the new plane would bring in gross revenues of $400 million over a period of 2 years; a 0.30 chance that sales of the plane during this period would bring $300 million; and a 0.50 chance that sales would bring in only $200 million. Were Boeing not to manufacture the plane and instead use its resources to make other planes that are currently being used by airlines all over the world, there is a 0.75 chance that sales of these other planes during the period in question will bring in gross revenues of $250 million, and a 0.25 chance that sales will bring in $300 million. Draw a decision tree for this problem, and solve it. What should Boeing do?

15–25. Atari Corporation is facing a decision about whether to acquire all the outstanding stock of Federated Stores, Inc., a stereo and electronic equipment retailer. Atari's financial analysts believe that there is a 0.40 probability that ownership of Federated Stores would bring Atari profits of $55 million over the next 3 years, a 0.45 probability of $70 million in profits over the period in question, and a 0.15 probability of $80 million in profits. Atari is considering two alternative plans. One plan is to start its own chain of retail outlets. Analysts believe that this plan has a 0.10 probability of failing, producing a loss of $10 million over the next 3 years; it has a 0.30 probability of bringing in profits of $30 million, a 0.40 probability of producing profits of $75 million over the same period, and a 0.20 probability of bringing in $100 million in profits. The second plan is to divert the efforts to manufacturing a new computer. This plan is believed to have a 0.5 probability of bringing in profits of $150 million over the next 3 years, a 0.3 probability of bringing in only $50 million, and a 0.2 probability of causing a loss of $20 million over the period in question. Construct and solve the decision tree. Advise Atari about what to do and why.

15–26. Predicting the styles that will prevail in a coming year is one of the most important and difficult problems in the fashion industry. A fashion designer must work on designs for the coming fall long before he or she can find out for certain what styles are going to be "in." A well-known designer believes that there is a 0.20 chance that short dresses and skirts will be popular in the coming fall; a 0.35 chance that popular styles will be of medium length; and a 0.45 chance that long dresses and skirts will dominate fall fashions. The designer must now choose the styles on which to concentrate. If she chooses one style and another turns out to be more popular, profits will be lower than if the new style were guessed correctly. The following table shows what the designer believes she would make, in hundreds of thousands of dollars, for any given combination of her choice of style and the one that prevails in the new season.

	Prevailing Style		
Designer's Choice	Short	Medium	Long
Short	8	3	1
Medium	1	9	2
Long	4	3	10

Construct a decision tree, and determine what style the designer should choose to maximize her expected profits.

15–27. For problem 15–26, suppose that if the designer starts working on long designs, she can change them to medium designs after the prevailing style for the season becomes known—although she must then pay a price for this change because of delays in delivery to manufacturers. In particular, if the designer chooses long and the prevailing style is medium, and she then chooses to change to medium, there is a 0.30 chance that her profits will be $200,000 and a 0.70 chance that her profits will be $600,000. No other change from one style to another is possible. Incorporate this information in your decision tree of problem 15–26, and solve the new tree. Give a complete solution in the form of a pair of decisions under given circumstances that maximize the designer's expected profits.

15–28. Commodity futures provide an opportunity for buyers and suppliers of commodities such as wheat to arrange in advance sales of a commodity, with delivery and payment taking place at a specified time in the future. The price is decided at the time the order is placed, and the buyer is asked to deposit an amount less than the value of the order, but enough to protect the seller from loss in case the buyer should decide not to meet the obligation.

An investor is considering investing $15,000 in wheat futures and believes that there is a 0.10 probability that he will lose $5,000 by the expiration of the contract, a 0.20 probability that he will make $2,000, a 0.25 probability that he will make $3,000, a 0.15 probability he will make $4,000, a 0.15 probability he will make $5,000, a 0.10 probability he will make $6,000, and a 0.05 probability that he will make $7,000. If the investor should find out that he is going to lose $5,000, he can pull out of his contract, losing $3,500 for certain and an additional $3,000 with probability 0.20 (the latter amount deposited with a brokerage firm as a guarantee). Draw the decision tree for this problem, and solve it. What should the investor do?

15–29. For problem 15–28, suppose that the investor is considering another investment as an alternative to wheat. He is considering investing his $15,000 in a limited partnership for the same duration of time as the futures contract. This alternative has a 0.50 chance of earning $5,000 and a 0.50 chance of earning nothing. Add this information to your decision tree of problem 15–28, and solve it.

15–7 Handling Additional Information Using Bayes' Theorem

In any kind of decision problem, it is very natural to ask: Can I gain additional information about the situation? Any additional information will help in making a decision under uncertainty. The more we know, the better able we are to make decisions that are likely to maximize our payoffs. If our information is perfect, that is, if we can find out exactly what chance will do, then there really is no randomness, and the situation is perfectly determined. In such cases, there is no need for decision analysis because we can determine the exact action that will maximize the actual (rather than the expected) payoff. Here we are concerned with making decisions under uncertainty, and we assume that when additional information is available, such information is probabilistic in nature. Our information has a certain degree of reliability. The reliability is stated as a set of conditional probabilities.

If we are considering the introduction of a new product into the market, it is wise to try and gain information about the prospects for success of the new product by sampling potential consumers and soliciting their views about the product. (Is this not what statistics is all about?) Results obtained from random sampling are always probabilistic; the probabilities originate in the sampling distributions of our statistics. The reliability of survey results may be stated as a set of *conditional probabilities* in the following way. Given that the market is ripe for our product and that it is in fact going to be successful, there is a certain probability that the sampling results will tell us so. Conversely, given that the market will not accept the new product, there is a certain probability that the random sample of people we select will be representative enough of the population in question to tell us so.

To show the use of the conditional probabilities, let S denote the event that the product will be a success and $F = \bar{S}$ (the complement of S) be the event that the product will fail. Let IS be the event that the sample indicates that the product will be a success, and IF the event that the sample indicates that the product will fail. The reliability of the sampling results may be stated by the conditional probabilities $P(\text{IS} \mid \text{S})$,

$P(\text{IS} \mid \text{F})$, $P(\text{IF} \mid \text{S})$, and $P(\text{IF} \mid \text{F})$. [Each pair of conditional probabilities with the same condition has a sum of 1.00. Thus, $P(\text{IS} \mid \text{S}) + P(\text{IF} \mid \text{S}) = 1$, and $P(\text{IS} \mid \text{F}) + P(\text{IF} \mid \text{F}) = 1$. So we need to be given only two of the four conditional probabilities.]

Once we have sampled, we know the sample outcome: either the event IS (the sample telling us that the product will be successful) or the event IF (the sample telling us that our product will not be a success). What we need is the probability that the product will be successful given that the sample told us so, or the probability that the product will fail, if that is what the sample told us. In symbols, what we need is $P(\text{S} \mid \text{IS})$ and $P(\text{F} \mid \text{IS})$ (its complement), or the pair of probabilities $P(\text{S} \mid \text{IF})$ and $P(\text{F} \mid \text{IF})$. We have $P(\text{IS} \mid \text{S})$, and we need $P(\text{S} \mid \text{IS})$. The conditions in the two probabilities are reversed. Remember that Bayes' theorem reverses the conditionality of events. This is why decision analysis is usually associated with Bayesian theory. In order to transform information about the reliability of additional information in a decision problem to usable information about the likelihood of states of nature, we need to use Bayes' theorem.

To restate the theorem in this context, suppose that the sample told us that the product will be a success. The (posterior) probability that the product will indeed be a success is given by

$$P(\text{S} \mid \text{IS}) = \frac{P(\text{IS} \mid \text{S})\, P(\text{S})}{P(\text{IS} \mid \text{S})P(\text{S}) + P(\text{IS} \mid \text{F})P(\text{F})} \tag{15–6}$$

The probabilities $P(\text{S})$ and $P(\text{F})$ are our *prior* probabilities of the two possible outcomes: successful product versus unsuccessful product. Knowing these prior probabilities and knowing the reliability of survey results, here in the form of $P(\text{IS} \mid \text{S})$ and $P(\text{IS} \mid \text{F})$, allows us to compute the posterior, updated probability that the product will be successful given that the sample told us that it will be successful.

How is all this used in a decision tree? We extend our decision problem to include two possible actions: to test or not to test, that is, to obtain additional information or not to obtain such information. The decision of whether to test must be made before we make any other decision. In the case of a new-product introduction decision problem, our decision tree must be augmented to include the possibility of testing or not testing before we decide whether to market our new product. We will assume that the test costs $5,000. Our new decision tree is shown in Figure 15–17.

As shown in Figure 15–17, we first decide whether to test. If we test, we get a test result. The result is a *chance outcome*—the test indicates success, or the test indicates failure (event IS or event IF). If the test indicates success, then we may choose to market or not to market. The same happens if the test indicates failure: We may still choose to market, or we may choose not to market. If the test is worthwhile, it is not logical to market once the test tells us the product will fail. But the point of the decision analysis is that we do not know whether the test is worthwhile; this is one of the things that we need to decide. Therefore, we allow for the possibility of marketing even if the test tells us not to market, as well as allowing for all other possible combinations of decisions.

Determining the Payoffs

Recall that if the product is successful, we make $100,000, and if it is not successful, we lose $20,000. The test is assumed to cost $5,000. Thus, we must subtract $5,000 from all final-outcome payoffs that are reached via testing. If we test, we spend $5,000. If we then market and the product is successful, we make $100,000, but we

FIGURE 15–17   New-Product Decision Tree with Testing

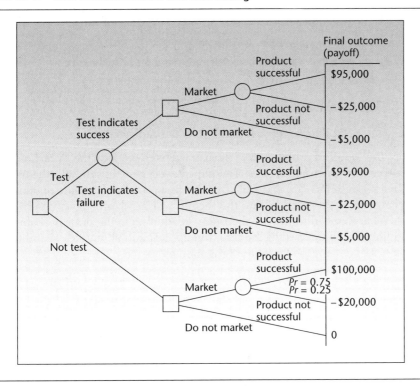

must deduct the $5,000 we had to pay for the test, leaving us a net profit of $95,000. Similarly, we must add the $5,000 cost of the market test to the possible loss of $20,000. This brings the payoff that corresponds to product failure to −$25,000.

Determining the Probabilities

We have now reached the crucial step of determining the probabilities associated with the different branches of our decision tree. As shown in Figure 15–17, we know only two probabilities: the probability of a successful product without any testing and the probability of an unsuccessful product without any testing. (These are our old probabilities from the decision tree of Figure 15–12.) These probabilities are $P(S) = 0.75$ and $P(F) = 0.25$. The two probabilities are also our *prior* probabilities—the probabilities before any sampling or testing is undertaken. As such, we will use them in conjunction with Bayes' theorem for determining the posterior probabilities of success and of failure, and the probabilities of the two possible test results $P(IS)$ and $P(IF)$. The latter are the total probabilities of IS and IF, respectively, and are obtained from the *denominator* in equation 15–6 and its analog, using the event IF. These are sometimes called *predictive probabilities* because they predict the test results.

For the decision tree in Figure 15–17, we have *two* probabilities, and we need to fill in the other six. First, let us look at the particular branches and define our probabilities. The probabilities of the two upper branches of the chance node immediately preceding the payoffs correspond to the two sequences

Test → Test indicates success → Market → Product is successful and

$$\text{Test} \rightarrow \text{Test indicates success} \rightarrow \text{Market} \rightarrow \text{Product is not successful}$$

These are the two sequences of events leading to the payoffs $95,000 and −$25,000, respectively. The probabilities we seek for the two final branches are

$$P(\text{Product is successful} \mid \text{Test has indicated success})$$

and

$$P(\text{Product is not successful} \mid \text{Test has indicated success})$$

These are the required probabilities because we have reached the branches success and no-success via the route: Test → Test indicates success. In symbols, the two probabilities we seek are $P(S \mid IS)$ and $P(F \mid IS)$. The first probability will be obtained from Bayes' theorem, equation 15–6, and the second will be obtained as $P(F \mid IS) = 1 - P(S \mid IS)$. What we need for Bayes' theorem—in addition to the prior probabilities—is the conditional probabilities that contain the information about the reliability of the market test. Let us suppose that these probabilities are

$$P(IS \mid S) = 0.9 \qquad P(IF \mid S) = 0.1 \qquad P(IF \mid F) = 0.85 \qquad P(IS \mid F) = 0.15$$

Thus, when the product is indeed going to be successful, the test has a 0.90 chance of telling us so. Ten percent of the time, however, when the product is going to be successful, the test erroneously indicates failure. When the product is not going to be successful, the test so indicates with probability 0.85 and fails to do so with probability 0.15. This information is assumed to be known to us at the time we consider whether to test.

Applying Bayes' theorem, equation 15–6, we get

$$P(S \mid IS) = \frac{P(IS \mid S)\,P(S)}{P(IS \mid S)P(S) + P(IS \mid F)P(F)} = \frac{(0.9)(0.75)}{(0.9)(0.75) + (0.15)(0.25)}$$

$$= \frac{0.675}{0.7125} = 0.9474$$

The denominator in the equation, 0.7125, is an important number. Recall from Section 2–7 that this is the *total probability* of the conditioning event; it is the probability of IS. We therefore have

$$P(S \mid IS) = 0.9474 \qquad \text{and} \qquad P(IS) = 0.7125$$

These two probabilities give rise to two more probabilities (namely, those of their complements): $P(F \mid IS) = 1 - 0.9474 = 0.0526$ and $P(IF) = 1 - P(IS) = 1 - 0.7125 = 0.2875$.

Using Bayes' theorem and its denominator, we have found that the probability that the test will indicate success is 0.7125, and the probability that it will indicate failure is 0.2875. Once the test indicates success, there is a probability of 0.9474 that the product will indeed be successful and a probability of 0.0526 that it will not be successful. This gives us four more probabilities to attach to branches of the decision tree. Now all we need are the last two probabilities, $P(S \mid IF)$ and $P(F \mid IF)$. These are obtained via an analog of equation 15–6 for when the test indicates failure. It is given as equation 15–7.

$$P(S \mid IF) = \frac{P(IF \mid S)P(S)}{P(IF \mid S)P(S) + P(IF \mid F)P(F)} \qquad (15\text{–}7)$$

FIGURE 15–18   New-Product Decision Tree with Probabilities

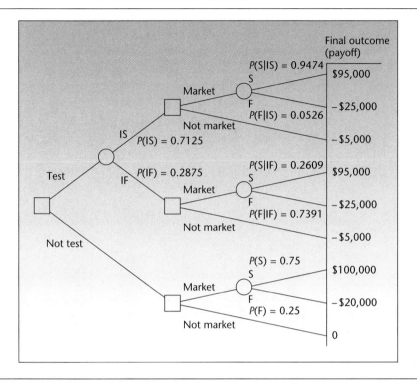

The denominator of equation 15–7 is, by the law of total probability, simply the probability of the event IF, and we have just solved for it: $P(\text{IF}) = 0.2875$. The numerator is equal to $(0.1)(0.75) = 0.075$. We thus get $P(\text{S} \mid \text{IF}) = 0.075/0.2875 = 0.2609$. The last probability we need is $P(\text{F} \mid \text{IF}) = 1 - P(\text{S} \mid \text{IF}) = 1 - 0.2609 = 0.7391$.

We will now enter all these probabilities into our decision tree. The complete tree with all probabilities and payoffs is shown in Figure 15–18. (To save space in the figure, events are denoted by their symbols: S, F, IS, etc.)

We are finally in a position to solve the decision problem by averaging out and folding back our tree. Let us start by averaging out the three chance nodes closest to the final outcomes:

$$E(\text{payoff}) = (0.9474)(95,000) + (0.0526)(-25,000)$$
$$= \$88,688 \quad \text{(top chance node)}$$
$$E(\text{payoff}) = (0.2609)(95,000) + (0.7391)(-25,000)$$
$$= \$6,308 \quad \text{(middle chance node)}$$
$$E(\text{payoff}) = (0.75)(100,000) + (0.25)(-20,000)$$
$$= \$70,000 \quad \text{(bottom chance node)}$$

We can now fold back and look for the optimal actions at each of the three preceding decision nodes. Again, starting from top to bottom, we first compare $88,688 with −$5,000 and conclude that—once the test indicates success—we should market the product. Then, comparing $6,308 with −$5,000, we conclude that even if the test

FIGURE 15–18   New-Product Decision Tree with Probabilities

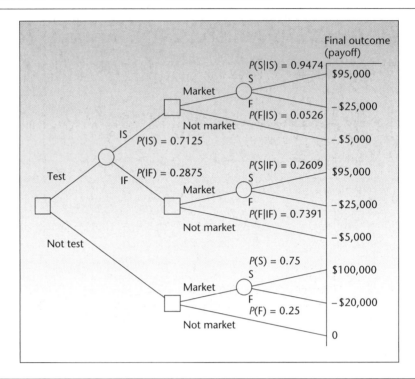

says that the product will fail, we are still better off if we go ahead and market the product (remember, all our conclusions are based on the expected monetary value and have no allowance for risk aversion). The third comparison again tells us to market the product, because $70,000 is greater than $0.

We are now at the chance node corresponding to the outcome of the test. At this point, we need to average out $88,688 and $6,308, with probabilities 0.7125 and 0.2875, respectively. This gives us

$$E(\text{payoff}) = (0.7125)(88,688) + (0.2875)(6,308) = \$65,003.75$$

Finally, we are at the very first decision node, and here we need to compare $65,003.75 with $70,000. Since $70,000 is greater, our optimal decision is not to test and to go right ahead and market the new product. If we must, for some reason, test the product, then we should go ahead and market it regardless of the outcome of the test, if we want to maximize our expected monetary payoff. Note that our solution is, of course, strongly dependent on the numbers we have used. If these numbers were different—for example, if the prior probability of success were not as high as it is—the optimal solution could very well have been to test first and then follow the result of the test. Our solution to this problem is shown in Figure 15–19.

We now demonstrate the entire procedure of decision analysis with additional information by Example 15–4. To simplify the calculations, which were explained earlier on a conceptual level using equations, we will use tables.

FIGURE 15–19   New-Product Introduction: Expected Values and Optimal Decision

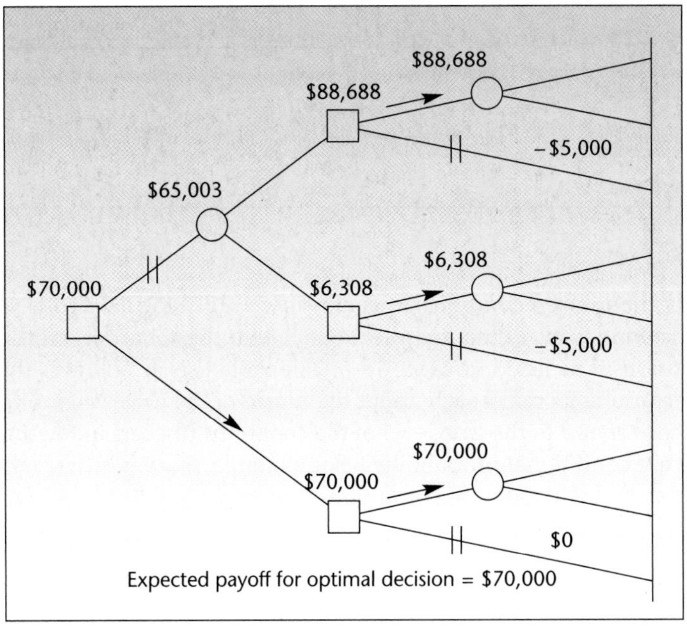

TABLE 15–6   Information for Example 15–4

Profit from Investment	Level of Economic Activity	Probability
$ 3 million	Low	0.20
6 million	Medium	0.50
12 million	High	0.30

Insurance companies need to invest large amounts of money in opportunities that **EXAMPLE 15–4**
provide high yields and are long-term. One type of investment that has recently at-
tracted some insurance companies is real estate.

   Aetna Life and Casualty Company is considering an investment in real estate in
central Florida. The investment is for a period of 10 years, and company analysts believe that the investment will lead to returns that depend on future levels of economic
activity in the area. In particular, the analysts believe that the invested amount would
bring the profits listed in Table 15–6, depending on the listed levels of economic activity and their given (prior) probabilities. The alternative to this investment plan—
one that the company has used in the past—is a particular investment that has a 0.5
probability of yielding a profit of $4 million and a 0.5 probability of yielding $7 million over the period in question.

   The company may also seek some expert advice on economic conditions in central Florida. For an amount that would be equivalent to $1 million 10 years from now
(when invested at a risk-free rate), the company could hire an economic consulting

TABLE 15–7   **Reliability of the Consulting Firm**

	Consultants' Conclusion		
True Future State of Economy	**High**	**Medium**	**Low**
Low	0.05	0.05	0.90
Medium	0.15	0.80	0.05
High	0.85	0.10	0.05

firm to study the future economic prospects in central Florida. From past dealings with the consulting firm, Aetna analysts believe that the reliability of the consulting firm's conclusions is as listed in Table 15–7. The table lists as columns the three conclusions the consultants may reach about the future of the area's economy. The rows of the table correspond to the true level of the economy 10 years in the future, and the table entries are conditional probabilities. For example, if the future level of the economy is going to be high, then the consultants' statement will be "high" with probability 0.85. What should Aetna do?

Solution   First, we construct the decision tree, including all the known information. The decision tree is shown in Figure 15–20. Now we need to use the prior probabilities in Table 15–6 and the conditional probabilities in Table 15–7 in computing both the posterior probabilities of the different payoffs from the investment given the three possible consultants' conclusions, and the predictive probabilities of the three consultants' conclusions. This is done in Tables 15–8, 15–9, and 15–10. Note that the probabilities of the outcomes of the alternative investment do not change with the consultants' conclusions (the consultants' conclusions pertain only to the central Florida investment prospects, not to the alternative investment).

These tables represent a way of using Bayes' theorem in an efficient manner. Each table gives the three posterior probabilities and the predictive probability for a particular consultant's statement. The structure of the tables is the same as that of Table 15–2, for example. Let us define our events in shortened form: H indicates that the level of economic activity will be high; L and M are defined similarly. We let *H* be the event that the consultants will predict a high level of economic activity. We similarly define *L* and *M*. Using this notation, the following is a breakdown of Table 15–8.

The prior probabilities are just the probabilities of events H, L, and M, as given in Table 15–6. Next we consider the event that the consultants predict a low economy: event L. The next column in Table 15–8 consists of the conditional probabilities $P(L \mid L)$, $P(L \mid M)$, and $P(L \mid H)$. These probabilities come from the last column of Table 15–7. The joint probabilities column in Table 15–8 consists of the products of the entries in the first two probabilities columns. The sum of the entries in this column is the denominator in Bayes' theorem: It is the total (or predictive) probability of the event L. Finally, dividing each entry in the joint probabilities column by the sum of that column [i.e., by $P(L)$] gives us the posterior probabilities: $P(L \mid L)$, $P(M \mid L)$, and $P(H \mid L)$. Tables 15–9 and 15–10 are interpreted in the same way, for events M and H, respectively.

Now that we have all the required probabilities, we can enter them in the tree. We can then average out at the chance nodes and fold back the tree. At each decision node, we choose the action that maximizes the expected payoff. The tree, with all its

FIGURE 15–20 **Decision Tree for Example 15–4**

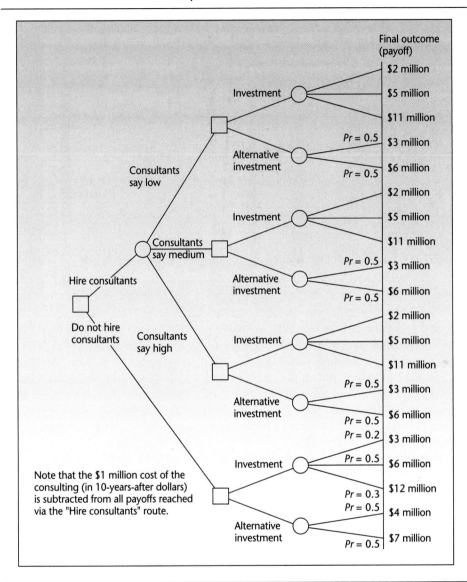

TABLE 15–8 **Events and Their Probabilities: Consultants Say "Low"**

Event	Prior	Conditional	Joint	Posterior
Low	0.20	0.90	0.180	0.818
Medium	0.50	0.05	0.025	0.114
High	0.30	0.05	0.015	0.068
		P(Consultants say "low") = 0.220		1.000

TABLE 15–9   Events and Their Probabilities: Consultants Say "Medium"

Event	Prior	Conditional	Joint	Posterior
Low	0.20	0.05	0.01	0.023
Medium	0.50	0.80	0.40	0.909
High	0.30	0.10	0.03	0.068
		P(Consultants say medium) = 0.44		1.000

TABLE 15–10   Events and Their Probabilities: Consultants Say "High"

Event	Prior	Conditional	Joint	Posterior
Low	0.20	0.05	0.010	0.029
Medium	0.50	0.15	0.075	0.221
High	0.30	0.85	0.255	0.750
		P(Consultants say high) = 0.340		1.000

probabilities, final-outcome payoffs, expected payoffs at the chance nodes, and indicators of the optimal action at each decision node, is shown in Figure 15–21. The figure is a complete solution to this problem.

The final solution is not to hire the consultants and to invest in the central Florida project. If we have to consult, then we should choose the alternative investment if the consultants predict a low level of economic activity for the region, and invest in the central Florida project if they predict either a medium or a high level of economic activity. The expected value of the investment project is $7.2 million.

A template solution to this problem is given in Section 15–10.

PROBLEMS

15–30.   Explain why Bayes' theorem is necessary for handling additional information in a decision problem.

15–31.   Explain the meaning of the term *predictive probability*.

15–32.   For Example 15–4, suppose that hiring the economic consultants costs only $100,000 (in 10-years-after dollars). Redo the analysis. What is the optimal decision? Explain.

15–33.   For problem 15–23, suppose that before deciding whether to advertise on television, Anheuser-Busch has a choice of testing the commercial. Suppose that the test costs $200,000 and has the following reliability. If we call sales of $8 million the high state and $4 million in sales the low state, then if the true state is high, the test has a 0.96 chance of detecting this and a 0.04 chance of erroneously indicating "low." If the true state is low, then the test has a 0.93 chance of determining this and a 0.07 chance of erroneously concluding "high." Use this information, and redo the problem. What is the optimal decision?

FIGURE 15–21   Solution to Aetna Decision Problem

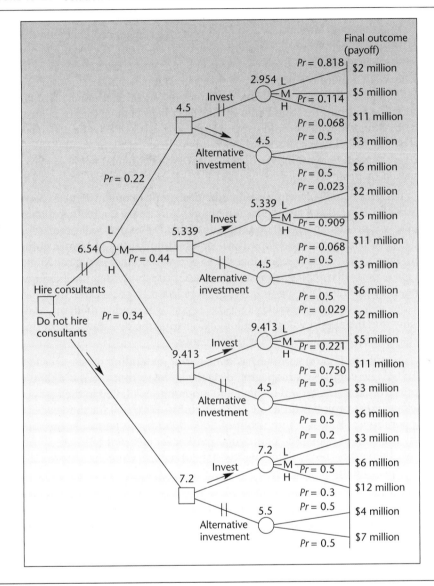

15–34. One of the most powerful people in Hollywood is not an actor, director, or producer. It is Richard Soames, an insurance director for the London-based Film Finances Ltd. Soames is a leading provider of movie completion bond guarantees. The guarantees are like insurance policies that pay the extra costs when films go over budget or are not completed on time. Suppose that Soames is considering insuring the production of a movie and feels there is a 0.65 chance that his company will make $80,000 on the deal (i.e., the production will be on time and not exceed budget). He believes there is a 0.35 chance that the movie will exceed budget and his company will lose $120,000, which would have to be paid to complete production. Soames could pay a movie industry expert $5,000 for an evaluation of the project's success.

He believes that the expert's conclusions are true 90% of the time. What should Soames do?

15–35.   Many airlines flying overseas have recently considered changing the kinds of goods they sell at their in-flight duty-free services. Swissair, for example, is considering selling Swiss watches instead of the usual liquor and cigarettes. A Swissair executive believes that there is a 0.60 chance that passengers would prefer these goods to the usual items and that revenues from in-flight sales would increase by $40,000 over a period of several months. She believes there is a 0.40 chance that revenues would decrease by $20,000, which would happen should people not buy the watches and instead desire the usual items. Testing the new idea on actual flights would cost $6,000, and the results would have a 0.85 probability of correctly detecting the state of nature. What should Swissair do?

15–36.   For problem 15–26, suppose that the designer can obtain some expert advice for a cost of $30,000. If the fashion is going to be short, there is a 0.90 probability that the expert will predict short, a 0.05 probability that the expert will predict medium, and a 0.05 probability that the expert will predict long. If the fashion is going to be medium, there is a 0.10 probability that the expert will predict short, a 0.75 probability that the expert will predict medium, and a 0.15 probability that the expert will predict long. If the fashion is going to be long, there is a 0.10 probability that the expert will predict short, a 0.10 probability that the expert will predict medium, and a 0.80 probability that the expert will predict long. Construct the decision tree for this problem. What is the optimal decision for the designer?

15–37.   A cable television company is considering extending its services to a rural community. The company's managing director believes that there is a 0.50 chance that profits from the service will be high and amount to $760,000 in the first year, and a 0.50 chance that profits will be low and amount to $400,000 for the year. An alternative operation promises a sure profit of $500,000 for the period in question. The company may test the potential of the rural market for a cost of $25,000. The test has a 90% reliability of correctly detecting the state of nature. Construct the decision tree and determine the optimal decision.

15–38.   An investor is considering two brokerage firms. One is a discount broker offering no investment advice, but charging only $50 for the amount the investor intends to invest. The other is a full-service broker who charges $200 for the amount of the intended investment. If the investor chooses the discount broker, there is a 0.45 chance of a $500 profit (before charges) over the period of the investment, a 0.35 chance of making only $200, and 0.20 chance of losing $100. If the investor chooses the full-service broker, then there is a 0.60 chance that the investment will earn $500, a 0.35 chance that it will earn $200, and a 0.05 chance that it will lose $100. What is the best investment advice in this case?

15–8   Utility

Often we have to make decisions where the rewards are not easily quantifiable. The reputation of a company, for example, is not easily measured in terms of dollars and cents. Such rewards as job satisfaction, pride, and a sense of well-being also fall into this category. Although you may feel that a stroll on the beach is "priceless," it is sometimes possible to order such things on a scale of values by gauging them against the amount of money you would require for giving them up. When such scaling is possible, the value system used is called a **utility.**

If a decision affecting a firm involves rewards or losses that either are nonmonetary or—more commonly—represent a mixture of dollars and other benefits such as reputation, long-term market share, and customer satisfaction, then we need to convert all the benefits to a single scale. The scale, often measured in dollars or other units, is a *utility scale*. Once utilities are assessed, the analysis proceeds as before, with the utility units acting as dollars and cents. If the utility function was correctly evaluated, results of the decision analysis may be meaningful.

The concept of utility is derived not only from seemingly nonquantifiable rewards. Utility is a part of the very way we deal with money. For most people, the value of $1,000 is not constant. For example, suppose you were offered $1,000 to wash someone's dirty dishes. Would you do it? Probably yes. Now suppose that you were given $1 million and then asked if you would do the dishes for $1,000. Most people would refuse because the value of an additional $1,000 seems insignificant once you have $1 million (or more), as compared with the value of $1,000 if you do not have $1 million.

The value you attach to money—the *utility* of money—is not a straight-line function, but a curve. Such a curve is shown in Figure 15–22. Looking at the figure, we see that the utility (the value) of one additional dollar, as measured on the vertical axis, *decreases* as our wealth (measured on the horizontal axis) increases. The type of function shown in Figure 15–22 is well suited for modeling a situation where an additional amount of money $x has different worth to us depending on our wealth, that is, where the utility of each additional dollar decreases as we acquire more money. This utility function is the utility curve of a *risk-averse* individual. Indeed, utility can be used to model people's attitudes toward risk. Let us see why and how.

Suppose that you are offered the following choice. You can get $5,000 for certain, or you could get a lottery ticket where you have a 0.5 chance of winning $20,000 and a 0.5 chance of losing $2,000. Which would you choose? The expected payoff from the lottery is $E \text{(payoff)} = (0.5)(20,000) + (0.5)(-2,000) = \$9,000$. This is almost twice

FIGURE 15–22 A Utility-of-Money Curve

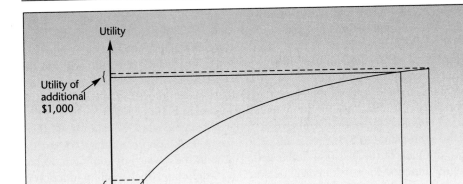

FIGURE 15–23 **Utility of a Risk Avoider**

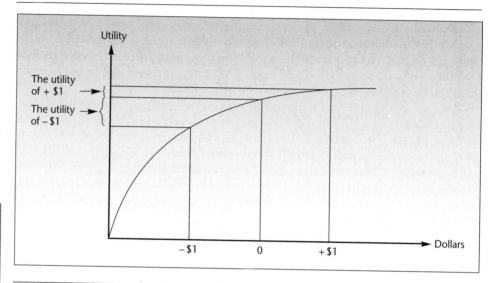

FIGURE 15–24
Utility of a Risk Taker

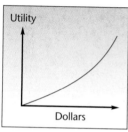

FIGURE 15–25
Utility of a Risk-Neutral Person

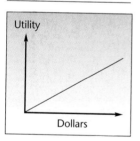

FIGURE 15–26
A Mixed Utility

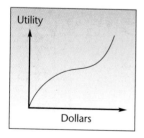

the amount you could get with probability 1.0, the $5,000. Expected monetary payoff would tell us to choose the lottery. However, few people would really do so; most would choose the $5,000, a sure amount. This shows us that a possible loss of $2,000 is not worth the possible gain of $20,000, even if the expected payoff is large. Such behavior is typical of a risk-averse individual. For such a person, the reward of a possible gain of $1 is not worth the "pain" of a possible loss of the same amount.

Risk aversion is modeled by a utility function (the value-of-money function) such as the one shown in Figure 15–22. Again, let us look at such a function. Figure 15–23 shows how, for a risk-averse individual, the utility of $1 earned ($1 to the right of zero) is less than the value of a dollar lost ($1 to the left of zero).

Not everyone is risk-averse, especially if we consider companies rather than individuals. The utility functions in Figures 15–22 and 15–23 are *concave* functions: They are functions with a decreasing slope, and these are characteristic of a risk-averse person. For a *risk-seeking* person, the utility is a *convex* function: a function with an increasing slope. Such a function is shown in Figure 15–24. Look at the curve in the figure, and convince yourself that an added dollar is worth more to the risk taker than the pain of a lost dollar (use the same technique used in Figure 15–23).

For a *risk-neutral* person, a dollar is a dollar no matter what. Such an individual gives the same value to $1 whether he or she has $10 million or nothing. For such a person, the pain of the loss of a dollar is the same as the reward of gaining a dollar. The utility function for a risk-neutral person is a straight line. Such a utility function is shown in Figure 15–25. Again, convince yourself that the utility of +$1 is equal (in absolute value) to the utility of −$1 for such a person. Figure 15–26 shows a mixed utility function. The individual is a risk avoider when his or her wealth is small and a risk taker when his or her wealth is great. We now present a method that may be used for assessing an individual's utility function.

A Method of Assessing Utility

One way of assessing the utility curve of an individual is to do the following:

1. Identify the maximum payoff in a decision problem, and assign it the utility 1: $U(\text{maximum value}) = 1$.
2. Identify the minimum payoff in a decision problem, and assign it the value 0: $U(\text{minimum value}) = 0$.
3. Conduct the following game to determine the utility of any intermediate value in the decision problem (in this chosen scale of numbers). Ask the person whose utility you are trying to assess to determine the probability p such that he or she expresses *indifference* between the two choices: receive the payoff R with certainty or have probability p of receiving the maximum value and probability $1 - p$ of receiving the minimum value. The determined p is the utility of the value R. This is done for all values R for which we want to assess the utility.

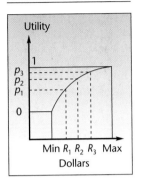

FIGURE 15–27
Assessment of a Utility Function

The assessment of a utility function is demonstrated in Figure 15–27. The utility curve passes through all the points (R_i, p_i) $(i = 1, 2 \ldots)$ for which the utility was assessed. Let us look at an example.

Suppose that an investor is considering decisions that lead to the following possible payoffs: \$1,500, \$4,300, \$22,000, \$31,000, and \$56,000 (the investments have different levels of risk). We now try to assess the investor's utility function.

EXAMPLE 15–5

Starting with step 1, we identify the minimum payoff as \$1,500. This value is assigned the utility of 0. The maximum payoff is \$56,000, and we assign the utility 1 to that figure. We now ask the investor a series of questions that should lead us to the determination of the utilities of the intermediate payoff values. Let us suppose the investor states that he is indifferent between receiving \$4,300 for certain and receiving \$56,000 with probability 0.2 and \$1,500 with probability 0.8. This means that the utility of the payoff \$4,300 is 0.2. We now continue to the next payoff, of \$22,000. Suppose that the investor is indifferent between receiving \$22,000 with certainty and \$56,000 with probability 0.7 and \$1,500 with probability 0.3. The investor's utility of \$22,000 is therefore 0.7. Finally, the investor indicates indifference between a certain payoff of \$31,000 and receiving \$56,000 with probability 0.8 and \$1,500 with probability 0.2. The utility of \$31,000 is thus equal to 0.8. We now plot the corresponding pairs (payoff, utility) and run a rough curve through them.

The curve—the utility function of the investor—is shown in Figure 15–28. Whatever the decision problem facing the investor, the *utilities* rather than the actual payoffs are the values to be used in the analysis. The analysis is based on maximizing the investor's *expected utility* rather than the expected monetary outcome.

Solution

FIGURE 15–28
Investor's Utility

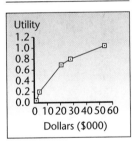

Note that utility is not unique. Many possible scales of values may be used to represent a person's attitude toward risk, as long as the general shape of the curve, for a given individual, remains the same—convex, concave, or linear—and with the same relative curvature. In practice, the assessment of utilities may not always be a feasible procedure, as it requires the decision maker to play the hypothetical game of assessing the indifference probabilities.

PROBLEMS

15–39. What is a utility function?

15–40. What are the advantages of using a utility function?

15–41. What are the characteristics of the utility function of a risk-averse individual? Of a risk taker?

15–42. What can you say about the risk attitude of the investor in Example 15–5?

15–43. Choose a few hypothetical monetary payoffs, and determine your own utility function. From the resulting curve, draw a conclusion about your attitude toward risk.

15–9   The Value of Information

In decision-making problems, the question often arises as to the *value* of information: How much should we be willing to pay for additional information about the situation at hand? The first step in answering this question is to find out how much we should be willing to pay for perfect information, that is, how much we should pay for a crystal ball that would tell us exactly what will happen—what the exact state of nature will be. If we can determine the value of perfect information, this will give us an upper bound on the value of any (imperfect) information. If we are willing to pay D dollars to know exactly what will happen, then we should be willing to pay an amount no greater than D for information that is less reliable. Since sample information is imperfect (in fact, it is probabilistic, as we well know from our discussion of sampling), the value of sample information is less than the value of perfect information. It will equal the value of perfect information only if the entire population is sampled.

Let us see how the upper bound on the value of information is obtained. Since we do not know what the perfect information is, we can only compute the *expected value of perfect information* in a given decision-making situation. The expected value is a *mean* computed using the prior probabilities of the various states of nature. It assumes, however, that at any given point when we actually take an action, we know its exact outcome. Before we (hypothetically) buy the perfect information, we do not know what the state of nature will be, and therefore we must average payoffs using our prior probabilities.

> The **expected value of perfect information (EVPI)** is
>
> EVPI = The expected monetary value of the decision situation when perfect information is available, minus the expected value of the decision situation when no additional information is available

This definition of the expected value of perfect information is logical: It says that the (expected) maximum amount we should be willing to pay for perfect information is equal to the difference between our expected payoff from the decision situation when we have the information and our expected payoff from the decision situation without the information. The expected value of information is equal to what we stand to gain from this information. We will demonstrate the computation of the expected value of perfect information with an example.

EXAMPLE 15–6   An article in the *Journal of Marketing Research* gives an example of decision making in the airline industry. The situation involves a price war that ensues when one airline determines the fare it will set for a particular route. Profits depend on the fare that will be set by a competing airline for the same route. Competitive situations such as this one are modeled using game theory. In this example, however, we will look at the competitor's action as a chance occurrence and consider the problem within the realm of decision analysis.

Table 15–11 shows the payoffs (in millions of dollars) to the airline over a given period of time, for a given fare set by the airline and by its competitor. We assume that there is a certain probability that the competitor will choose the low ($200) price and a certain probability that the competitor will choose the high price. Suppose that the probability of the low price is 0.6 and that the probability of the high price is 0.4.

Solution

The decision tree for this situation is given in Figure 15–29. Solving the tree, we find that if we set our price at $200, the expected payoff is equal to $E(\text{payoff}) = (0.6)(8) + (0.4)(9) = \$8.4$ million. If we set our price at $300, then our expected payoff is $E(\text{payoff}) = (0.6)(4) + (0.4)(10) = \$6.4$ million. The optimal action is, therefore, to set our price at $200. This is shown with an arrow in Figure 15–29.

Now we may ask the question of whether it may be worthwhile to obtain more information. Obtaining new information in this case may entail hiring a consultant who is knowledgeable about the operating philosophy of the competing airline. We may seek other ways of obtaining information; we may, for example, make an analysis of the competitor's past pricing behavior. The important question is: What do we stand to gain from the new information? Suppose that we know exactly what our competitor plans to do. If we know that the competitor plans to set the price at $200, then our optimal action is to set ours at $200 as well; this is seen by comparing the two

TABLE 15–11   **Airline Payoffs (in millions of dollars)**

Airline's Fare (Action)	Competitor's Fare (State of Nature)	
	$200	$300
$200	8	9
$300	4	10

FIGURE 15–29   **Decision Tree for Example 15–6**

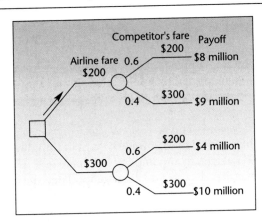

amounts in the first payoff column of Table 15–11, the column corresponding to the competitor's setting the price at $200. We see that our maximum payoff is then $8 million, obtained by choosing $200 as our own price as well. If the competitor chooses to set its price at $300, then our optimal action is to set our price at $300 as well and obtain a payoff of $10 million.

We know that without any additional information, the optimal decision is to set the price at $200, obtaining an *expected* payoff of $8.4 million (the expected value of the decision situation without any information). What is the expected payoff with *perfect* information? We do not know what the perfect information may be, but we assume that the prior probabilities we have are a true reflection of the long-run proportion of the time our competitor sets either price. Therefore, 60% of the time our competitor sets the low price, and 40% of the time the competitor sets the high price. If we had perfect information, we would know—at the time—how high to set the price. If we knew that the competitor planned to set the price at $200, we would do the same because this would give us the maximum payoff, $8 million. Conversely, when our perfect information tells us that the competitor is setting the price at $300, we again follow suit and gain a maximum payoff, $10 million. Analyzing the situation now, we do not know what the competitor will do (we do not have the perfect information), but we do know the probabilities. We therefore average the maximum payoff in each case, that is, the payoff that would be obtained under perfect information, using our probabilities. This gives us the expected payoff under perfect information. We get E(payoff under perfect information) = (Maximum payoff if the competitor chooses $200) × (Probability that the competitor will choose $200) + (Maximum payoff if the competitor chooses $300) × (Probability that the competitor will choose $300) = (8)(0.6) + (10)(0.4) = $8.8 million. If we could get perfect information, we could expect (on the average) to make a profit of $8.8 million. Without perfect information, we expect to make $8.4 million (the optimal decision without any additional information).

We now use the definition of the expected value of perfect information:

$$\text{EVPI} = E(\text{payoff under perfect information}) - E(\text{payoff without information})$$

Applying the rule in this case, we get EVPI = 8.8 − 8.4 = $0.4 million, or simply $400,000. Therefore, $400,000 is the maximum amount of money we should be willing to pay for additional information about our competitor's price intentions. This is the amount of money we should be willing to pay to know for certain what our competitor plans to do. We should pay less than this amount for all information that is not as reliable.

What about sampling—when sampling is possible? (In the airlines example, it probably is not possible to sample.) The expected value of sample information is equal to the expected value of perfect information, minus the expected cost of sampling errors. The expected cost of sampling errors is obtained from the probabilities of errors—known from sampling theory—and the resulting loss of payoff due to making less-than-optimal decisions. The *expected net gain* from sampling is equal to the expected value of sampling information, minus the cost of sampling. As the sample size increases, the expected net gain from sampling first increases, as our new information is valuable and improves our decision-making ability. Then the expected net gain decreases because we are paying for the information at a constant rate, while the information content in each additional data point becomes less and less important as we get more and more data. (A sample of 1,100 does not contain much more information

FIGURE 15–30   **Expected Net Gain from Sampling (in Dollars) as a Function of the Sample Size**

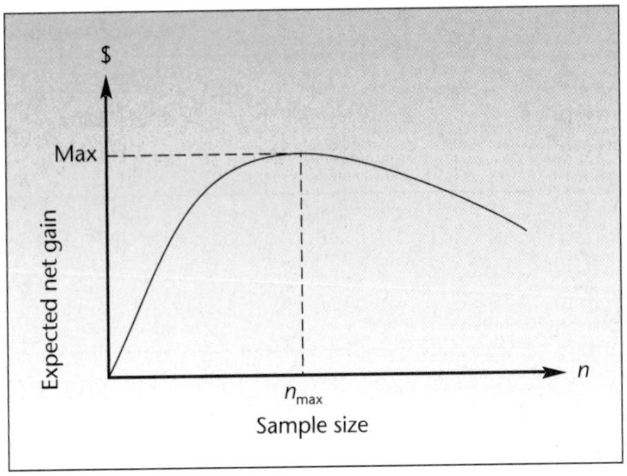

than a sample of 1,000. However, the same 100 data points may be very valuable if they are all we have.)

At some particular sample size n, we maximize our expected net gain from sampling. Determining the optimal sample size to be used in a decision situation is a difficult problem, and we will not say more about it here. We will, however, show the relationship between sample size and the expected net gain from sampling in Figure 15–30. The interested reader is referred to advanced works on the subject.[4]

15–44.   Explain the value of additional information within the context of decision making.

15–45.   Explain how we compute the expected value of perfect information, and why it is computed that way.

15–46.   Compute the expected value of perfect information for the situation in problem 15–26.

15–47.   For the situation in problem 15–26, suppose the designer is offered expert opinion about the new fall styles for a price of $300,000. Should she buy the advice? Explain why or why not.

15–48.   What is the expected value of perfect information in the situation of problem 15–23? Explain.

15–10   Using the Computer

Most statistical computer packages do not have extensive Bayesian statistics or decision analysis capabilities. There are, however, several commercial computer programs that do decision analysis. It is also possible to write your own computer program for solving a decision tree.

---

[4]From C. Barenghi, A. Aczel, and R. Best, "Determining the Optimal Sample Size for Decision Making," *Journal of Statistical Computation and Simulation,* Spring 1986, pp. 135–45. © 1986 Taylor & Francis Ltd. Reprinted with permission.

TABLE 15–12   **The Payoff Table for the Aetna Decision Problem (Example 15–4)**

	State of the Economy		
	Low	**Medium**	**High**
Real estate investment	$ 3 million	$ 6 million	$ 12 million
Alternative investment	5.5 million	5.5 million	5.5 million

FIGURE 15–31   **The Data Sheet for Decision Analysis**
[Decision Analysis.xls; Sheet: Data]

	A	B	C	D	E	F	G	H	I	J	K	L	M
1		Data		Aetna Decision									
2													
3		Payoff Table		Low	Med	High							
4				s 1	s 2	s 3	s 4	s 5	s 6	s 7	s 8	s 9	s 10
5		Invest	d 1	3	6	12							
6		Alternative	d 2	5.5	5.5	5.5							
7			d 3										
8			d 4										
9			d 5										
10			d 6										
11			d 7										
12			d 8										
13			d 9										
14			d 10										
16		Probability		0.2	0.5	0.3							
17													
18		Conditional Probabilities of Additional Information											
19				s 1	s 2	s 3	s 4	s 5	s 6	s 7	s 8	s 9	s 10
20		low	P(I1I.)	0.9	0.05	0.05							
21		med	P(I2I.)	0.05	0.8	0.1							
22		high	P(I3I.)	0.05	0.15	0.85							
23			P(I4I.)										
24			P(I5I.)										

The Template

The decision analysis template can be used only if the problem is representable by a payoff table.[5] In the Aetna decision example, it is possible to represent the problem by Table 15–12. Note that the returns from the alternative investment have been reduced to a constant *expected value* of $5.5 million to fit the payoff table format. If this manipulation of the payoff were not possible, then we could not use the template.

The template consists of three sheets, Data, Results, and Calculation. The Data sheet is shown in Figure 15–31. On this sheet the payoffs are entered at the top and conditional probabilities of additional information (such as a consultant's information) at the bottom. The probabilities of all the states entered in row 16 must add up to 1, or an error message will appear in row 17. Similarly, every column of the conditional probabilities in the bottom must add up to 1.

Once the data are entered properly, we can read off the results on the Results page shown in Figure 15–32. The optimal decision for each possible information from

---

[5]Theoretically speaking, all the problems can be represented in a payoff table format. But the payoff table can get too large and cumbersome for cases where the set of consequences is different for different decision alternatives.

FIGURE 15–32   **The Results**
                [Decision Analysis.xls; Sheet: Results]

	A	B	C	D	E	F	G	H	I
1	**Results**			Aetna Decision					
2									
3		Optimal decisions with additional information							
4									
5			**Best Decision Under**						
6			**No Info.**	**low**	**med**	**high**			
7		**d ***	d1	d2	d1	d1			
8		**EV**	7.2	5.5	6.34091	10.4118			
9		**Prob.**		0.22	0.44	0.34			
10									
11				**EVPI**	0.5				
12				**EVSI**	0.34		EV with SI	7.54	
13				**Efficiency of SI**	68.00%				
14									

the consultant appears in row 7. The corresponding expected payoff (EV) appears in row 8. The marginal probability of obtaining each possible information appears in row 9. The maximum expected payoff achieved by correctly following the optimal decision under each information is $7.54 million, which appears in cell H12. It has been labeled as "EV with SI," where SI stands for sample information. The sample information in this example is the consultant's prediction of the state of the economy.

In cell E11, we get the expected value of perfect information (EVPI), and in cell E12, we get the expected value of sample information (EVSI). This EVSI is the expected value of the consultant's information. Recall that the consultant's fee is $1 million. Since EVSI is only $0.34 million, going by expected values, Aetna should not hire the consultant. Note also that the best decision without any information is d1 (seen in cell C7), which stands for real estate investment. Thus Aetna should simply invest in real estate without hiring the consultant. It can then expect a payoff of $7.2 million (seen in cell C8). All these results agree with what we already saw using manual calculations.

The **efficiency** of the sample information is defined as (EVSI/EVPI) × 100%, and it appears in cell E13.

The third sheet, Calculation, is where all the calculations are carried out. The user need not look at this sheet at all, and it is not shown here.

15–11   Summary and Review of Terms

In this chapter, we presented two related topics: **Bayesian statistics** and **decision analysis.** We saw that the Bayesian statistical methods are extensions of Bayes' theorem to discrete and continuous random variables. We saw how the Bayesian approach allows the statistician to use, along with the data, **prior information** about the situation at hand. The prior information is stated in terms of a **prior probability distribution** of population parameters. We saw that the Bayesian approach is less restrictive in that it allows us to consider an unknown parameter as a random variable. In this context, as more sampling information about a parameter becomes available to us, we can update our prior probability distribution of the parameter, thus creating a **posterior probability distribution.** The posterior distribution may then serve as a prior distribution when more data become available. We also may use the posterior distribution in computing a **credible set** for a population parameter, in the discrete case, and a **highest-posterior-density (HPD) set** in the continuous case. We discussed the possible dangers in using the Bayesian approach and the fact that we must be careful in our choice of prior distributions.

We saw how decision analysis may be used to find the decision that maximizes our **expected payoff** from an uncertain situation. We discussed personal or **subjective probabilities** and saw how these can be assessed and used in a Bayesian statistics problem or in a decision problem. We saw that a **decision tree** is a good method of solving problems of decision making under uncertainty. We learned how to use the method of **averaging out and folding back,** which leads to the determination of the **optimal decision**—the decision that maximizes the expected monetary payoff. We saw how to assess the usefulness of obtaining additional information within the context of the decision problem, using a decision tree and Bayes' theorem. Finally, we saw how to incorporate people's attitudes toward risk into our analysis and how these attitudes lead to a **utility function.** We saw that this leads to solutions to decision problems that maximize the **expected utility** rather than the expected monetary payoff. We also discussed the **expected value of perfect information** and saw how this value serves as an upper bound for the amount of money we are willing to pay for additional information about a decision-making situation.

ADDITIONAL PROBLEMS

15–49. A quality control engineer believes that the proportion of defective items in a production process is a random variable with a probability distribution that is approximated as follows:

x	P(x)
0.1	0.1
0.2	0.3
0.3	0.2
0.4	0.2
0.5	0.1
0.6	0.1

The engineer collects a random sample of items and finds that 5 out of the 16 items in the sample are defective. Find the engineer's posterior probability distribution of the proportion of defective items.

15–50. To continue problem 15–49, determine a credible set for the proportion of defective items with probability close to 0.95. Interpret the meaning of the credible set.

15–51. For problem 15–49, suppose that the engineer collects a second sample of 20 items and finds that 5 items are defective. Update the probability distribution of the population proportion you computed in problem 15–49 to incorporate the new information.

15–52. What are the main differences between the Bayesian approach to statistics and the classical (frequentist) approach? Discuss these differences.

15–53. What is the added advantage of the normal probability distribution in the context of Bayesian statistics?

15–54.   The average life of a battery is believed to be normally distributed with an unknown mean μ. The mean is viewed as a random variable with expected value of 45 hours and standard deviation of 5 hours. The population standard deviation is believed to be 10 hours. A random sample of 100 batteries gives a sample mean of 102 hours. Find the posterior probability distribution of the population mean μ.

15–55.   For problem 15–54, give a highest-posterior-density credible set of probability 0.95 for the population mean.

15–56.   For problem 15–54, a second sample of 60 batteries gives a sample mean of 101.5 hours. Update the distribution of the population mean, and give a new HPD credible set of probability 0.95 for μ.

15–57.   What is a payoff table? What is a decision tree? Can a payoff table be used in decision making without a decision tree?

15–58.   What is a subjective probability, and what are its limitations?

15–59.   Discuss the advantages and the limitations of the assessment of personal probabilities.

15–60.   Why is Bayesian statistics controversial? Try to argue for, and then against, the Bayesian methodology.

15–61.   Suppose that I am indifferent about the following two choices: a sure \$3,000 payoff, and a payoff of \$5,000 with probability 0.2 and \$500 with probability 0.8. Am I a risk taker or a risk-averse individual (within the range \$500 to \$5,000)? Explain.

15–62.   An investment is believed to earn \$2,000 with probability 0.2, \$2,500 with probability 0.3, and \$3,000 with probability 0.5. An alternative investment may earn \$0 with probability 0.1, \$3,000 with probability 0.2, \$4,000 with probability 0.5, and \$7,000 with probability 0.2. Construct a decision tree for this problem, and determine the investment with the highest expected monetary outcome. What are the limitations of the analysis?

15–63.   A company is considering merging with a smaller firm in a related industry. The company's chief executive officer believes that the merger has a 0.55 probability of success. If the merger is successful, the company stands to gain in the next 2 years \$5 million with probability 0.2; \$6 million with probability 0.3; \$7 million with probability 0.3; and \$8 million with probability 0.2. If the attempted merger should fail, the company stands to lose \$2 million (due to loss of public goodwill) over the next 2 years with probability 0.5 and to lose \$3 million over this period with probability 0.5. Should the merger be attempted? Explain.

15–64.   For problem 15–63, suppose that the chief executive officer may hire a consulting firm for a fee of \$725,000. The consulting firm will advise the CEO about the possibility of success of the merger. This consulting firm is known to have correctly predicted the outcomes of 89% of all successful mergers and the outcomes of 97% of all unsuccessful ones. What is the optimal decision?

15–65.   What is the expected value of perfect information about the success or failure of the merger in problem 15–63?

15–66.   A company is interested in hiring a new manager. There are two candidates for the position. Candidate A successfully managed two out of three firms that he headed and improved profits by an average of 10%. In his third previous job, the candidate caused a profit loss of 4%. Candidate B successfully managed four out of five firms she previously headed. This candidate increased profits by an average of 6% at the firms she successfully managed, and caused a profit loss of 5% at the firm she managed unsuccessfully. Which candidate should be hired, and why?

CASE 17   Pizzas 'R' Us

izzas 'R' Us is a national restaurant chain with close to 100 restaurants across the United States. It is continuously in the process of finding and evaluating possible locations for new restaurants. For any potential site, Pizzas 'R' Us needs to decide the size of the restaurant to build at that site—small, medium, or large—or whether to build none. For a particular prospective site, the accounting department estimates the present value (PV) of possible annual profit, in thousands of dollars, for each size as follows:

Size	Demand		
	Low	Medium	High
Small	48	32	30
Medium	−64	212	78
Large	−100	12	350
No restaurant	0	0	0

The prior probabilities are estimated for low demand at 0.42 and for medium demand at 0.36.

1. What is the best decision if the PV of expected profits is to be maximized?
2. What is the EVPI?

Since the EVPI is large, Pizzas 'R' Us decides to gather additional information. It has two potential market researchers, Alice Miller and Becky Anderson, both of whom it has contracted many times in the past. The company has a database of the demand levels predicted by these researchers in the past and the corresponding actual demand levels realized. The cross-tabulation of these records are as follows:

Alice Miller

Pre-dicted	Actual		
	Low	Medium	High
Low	9	4	2
Medium	8	12	4
High	4	3	13

Becky Anderson

Pre-dicted	Actual		
	Low	Medium	High
Low	12	2	1
Medium	2	15	2
High	0	3	20

3. Alice Miller charges $12,500 for the research and Becky Anderson $48,000. With which one should the company contract for additional information? Why?

16 Sampling Methods

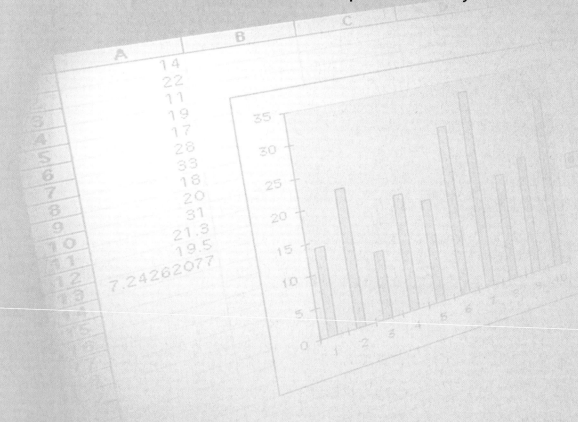

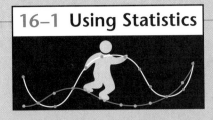

Throughout this book, we have always assumed that information is obtained through random sampling. The method we have used until now is called simple random sampling. In simple random sampling, we assume that our sample is randomly chosen from the entire population of interest, and that every set of n elements in the population has an equal chance of being selected as our sample.

We assume that a randomization device is always available to the experimenter. We also assume that the entire population of interest is known, in the sense that it is possible to draw a random sample from the entire population where every element has an equal chance of being included in our sample. Randomly choosing our sample from the entire population in question is our insurance against sampling bias. This was demonstrated in Chapter 5 with the story of the *Literary Digest*.

But do these conditions always hold in real sampling situations? Also, are there any easier ways–more efficient, more economical ways–of drawing random samples? Are there any situations where, instead of randomly drawing every single sample point, we may randomize less frequently and still obtain an adequately random sample for use in statistical inference? Consider the situation where our population is made up of several groups and the elements within each group are similar to one another, but different from the elements in other groups (e.g., sampling an economic variable in a city where some people live in rich neighborhoods and thus form one group, while others live in poorer neighborhoods, forming other groups). Is there a way of using the homogeneity within the groups to make our sampling more efficient? The answers to these questions, as well as many other questions related to sampling methodology, are the subject of this chapter.

In the next section, we will thoroughly explore the idea of random sampling and discuss its practical limitations. We will see that biases may occur if samples are chosen without randomization. We will also discuss criteria for the prevention of such selection biases. In the following sections, we will present more efficient and involved methods of drawing random samples that are appropriate in different situations.

16–2 Nonprobability Sampling and Bias

The advantage of random sampling is that the probabilities that the sample estimator will be within a given number of units from the population parameter it estimates are known. Sampling methods that do not use samples with known probabilities of selection are known as *nonprobability sampling methods*. In such sampling methods, we have no objective way of evaluating how far away from the population parameter our estimate may be. In addition, when we do not select our sample randomly out of the entire population of interest, our sampling results may be biased. That is, the average value of the estimate in repeated sampling is not equal to the parameter of interest. Put simply, our sample may not be a true representative of the population of interest.

EXAMPLE 16–1
A market research firm wants to estimate the proportion of consumers who might be interested in purchasing the Spanish sherry Jerez if this product were available at liquor stores in this country. How should information for this study be obtained?

Solution   The population relevant to our case is not a clear-cut one. Before embarking on our study, we need to define our population more precisely. Do we mean all consumers in the United States? Do we mean all families of consumers? Do we mean only people of drinking age? Perhaps we should consider our population to be only those people who, at least occasionally, consume similar drinks. These are important questions to answer before we begin the sampling survey. The population must be defined in accordance with the purpose of the study. In the case of a proposed new product such as Jerez, we are interested in the product's potential market share. We are interested in estimating the proportion of the market for alcoholic beverages that will go to Jerez once it is introduced. Therefore, we define our population as all people who, at least occasionally, consume alcoholic beverages.

Now we need to know how to obtain a random sample from this population. To obtain a random sample out of the whole population of people who drink alcoholic beverages at least occasionally, we must have a *frame*. That is, we need a list of all such people, from which we can randomly choose as many people as we need for our sample. In reality, of course, no such list is available. Therefore, we must obtain our sample in some other way. Market researchers send field workers to places where consumers may be found, usually shopping malls. There shoppers are randomly selected and prescreened to ascertain that they are in the population of interest—in this case, whether they are people who at least occasionally consume alcoholic beverages. Then the selected people are given a taste test of the new product and asked to fill out a questionnaire about their response to the product and their future purchase intent. This method of obtaining a random sample works as long as the interviewers do not choose people in a nonrandom fashion, for example, choosing certain types of people because of their appearance. If this should happen, a bias may be introduced if the variable favoring selection is somehow related to interest in the product.

Another point we must consider is the requirement that people selected at the shopping mall constitute a representative sample from the entire population in which we are interested. We must consider the possibility that potential buyers of Jerez may not be found in shopping malls, and if their proportion outside the malls is different from what it is in the malls where surveys take place, a bias will be introduced. We must consider the location of shopping malls and must ascertain that we are not favoring some segments of the population over others. Preferably, several shopping malls, located in different areas, should be chosen.

We should randomize the selection of people or elements in our sample. However, if our sample is not chosen in a purely random way, it may still suffice for our purposes as long as it behaves as a purely random sample and no biases are introduced. In designing the study, we should collect a few random samples at different locations, chosen at different times and handled by different field workers, to minimize the chances of a bias. The results of different samples validate the assumption that we are indeed getting a representative sample.

16–3   Stratified Random Sampling

In some cases, a population may be viewed as comprising different groups where elements in each group are similar to one another in some way. In such cases, we may gain sampling precision (i.e., reduce the variance of our estimators) as well as reduce the costs of the survey by treating the different groups separately. If we consider these groups, or *strata,* as separate subpopulations and draw a separate random sample from each stratum and combine the results, our sampling method is called *stratified random sampling.*

> In **stratified random sampling,** we assume that the population of N units may be divided into m groups with N_i units in group i, $i = 1, \ldots, m$. The m strata are nonoverlapping and together they make up the total population: $N_1 + N_2 + \cdots + N_m = N$.

We define the true *weight* of stratum i as $W_i = N_i/N$. That is, the weight of stratum i is equal to the proportion of the size of stratum i in the whole population. Our total sample, of size n, is divided into subsamples from each of the strata. We sample n_i items in stratum i, and $n_1 + n_2 + \cdots + n_m = n$. We define the sampling fraction in stratum i as $f_i = n_i/N_i$.

The true mean of the entire population is μ, and the true mean in stratum i is μ_i. The variance of stratum i is σ_i^2, and the variance of the entire population is σ^2. The sample mean in stratum i is $\overline{X}_i$, and the combined estimator, the sample mean in stratified random sampling, $\overline{X}_{st}$ is defined as follows:

> The **estimator of the population mean** in stratified random sampling is
>
> $$\overline{X}_{st} = \sum_{i=1}^{m} W_i\overline{X}_i \qquad (16\text{–}1)$$

In simple random sampling with no stratification, the stratified estimator in equation 16–1 is, in general, *not equal* to the simple estimator of the population mean. The reason is that the estimator in equation 16–1 uses the true weights of the strata W_i. The simple random sampling estimator of the population mean is $\overline{X} = (\Sigma_{\text{all data}} X)/n = (\Sigma_{i=1}^{m} n_i\overline{X}_i)/n$. This is equal to $\overline{X}_{st}$ only if we have $n_i/n = N_i/N$ for each stratum, that is, if the proportion of the sample taken from each stratum is equal to the proportion of each stratum in the entire population. Such a stratification is called stratification with *proportional allocation.*

Following are some important properties of the stratified estimator of the population mean:

1. If the estimator of the mean in each stratum $\overline{X}_i$ is *unbiased,* then the stratified estimator of the mean $\overline{X}_{st}$ is an unbiased estimator of the population mean μ.
2. If the samples in the different strata are drawn *independently* of one another, then the variance of the stratified estimator of the population mean $\overline{X}_{st}$ is given by

$$V(\overline{X}_{st}) = \sum_{i=1}^{m} W_i^2 V(\overline{X}_i) \qquad (16\text{–}2)$$

   where $V(\overline{X}_i)$ is the variance of the sample mean in stratum i.
3. If sampling in all strata is *random,* then the variance of the estimator, given in equation 16–2, is further equal to

$$V(\overline{X}_{st}) = \sum_{i=1}^{m} W_i^2 \left(\frac{\sigma_i^2}{n_i}\right)(1 - f_i) \qquad (16\text{--}3)$$

When the sampling fractions f_i are small and may be ignored, we get

$$V(\overline{X}_{st}) = \sum_{i=1}^{m} W_i^2 \frac{\sigma_i^2}{n_i} \qquad (16\text{--}4)$$

4. If the sample allocation is proportional $[n_i = n(N_i/N)$ for all $i]$, then

$$V(\overline{X}_{st}) = \frac{1 - f}{n} \sum_{i=1}^{m} W_i \sigma_i^2 \qquad (16\text{--}5)$$

which reduces to $(1/n) \sum_{i=1}^{m} W_i \sigma_i^2$ when the sampling fraction is small. Note that f is the sampling fraction, that is, the size of the sample divided by the population size.

In addition, if the population variances in all the strata are equal, then

$$V(\overline{X}_{st}) = \frac{\sigma^2}{n} \qquad (16\text{--}6)$$

when the sampling fraction is small.

Practical Applications

In practice, the true population variances in the different strata are usually not known. When the variances are not known, we estimate them from our data. An unbiased estimator of σ_i^2, the population variance in stratum i, is given by

$$S_i^2 = \sum_{\text{data in stratum } i} \frac{(X - \overline{X}_i)^2}{n_i - 1} \qquad (16\text{--}7)$$

The estimator in equation 16–7 is the usual unbiased sample estimator of the population variance in each stratum as a separate population. A particular estimate of the variance in stratum i will be denoted by s_i^2. If sampling in each stratum is random, then an unbiased estimator of the variance of the sample estimator of the population mean is

$$S^2(\overline{X}_{st}) = \sum_{i=1}^{m} \left(\frac{W_i^2 S_i^2}{n_i}\right)(1 - f_i) \qquad (16\text{--}8)$$

Any of the preceding formulas apply in the special situations where they can be used with the estimated variances substituted for the population variances.

Confidence Intervals

We now give a confidence interval for the population mean μ obtained from stratified random sampling.

A $(1 - \alpha)$ 100% confidence interval for the population mean μ using stratified sampling is

$$\overline{x}_{st} \pm z_{\alpha/2} \, s(\overline{X}_{st}) \qquad (16\text{--}9)$$

where $s(\overline{X}_{st})$ is the square root of the estimate of the variance of $\overline{X}_{st}$ given in equation 16–8.

When the sample sizes in some of the strata are small and the population variances are unknown, but the populations are at least approximately normal, we use the t distribution instead of Z. We denote the degrees of freedom of the t distribution by df. The exact value of df is difficult to determine, but it lies somewhere between the smallest $n_i - 1$ (the degrees of freedom associated with the sample from stratum i) and the sum of the degrees of freedom associated with the samples from all strata $\sum_{i=1}^{m}(n_i - 1)$. An approximation for the effective number of degrees of freedom is given by

$$\text{Effective df} = \frac{\left[\sum_{i=1}^{m} N_i(N_i - n_i)s_i^2/n_i \right]^2}{\sum_{i=1}^{m} [N_i(N_i - n_i)/n_i]^2 s_i^4/(n_i - 1)} \tag{16-10}$$

We demonstrate the application of the theory of stratified random sampling presented so far by the following example.

Once a year, *Fortune* magazine publishes the Fortune Service 500, a list of the largest service companies in the United States. The 500 firms belong to six major industry groups. The industry groups and the number of firms in each group are listed in Table 16–1.

EXAMPLE 16–2

The 500 firms are considered a complete population: the population of the top 500 service companies in the United States. An economist who is interested in this population wants to estimate the mean net income of all firms in the index. However, obtaining the data for all 500 firms in the index is difficult, time-consuming, or costly. Therefore, the economist wants to gather a random sample of the firms, compute a quick average of the net income for the firms in the sample, and use it to estimate the mean net income for the entire population of 500 firms.

The economist believes that firms in the same industry group share common characteristics related to net income. Therefore, the six groups are treated as different strata, and a random sample is drawn from each stratum. The weights of each of the strata are known exactly as computed from the strata sizes in Table 16–1. Using the definition of the population weights $W_i = N_i/N$, we get the following weights:

Solution

$$W_1 = N_1/N = 100/500 = 0.2$$
$$W_2 = N_2/N = 100/500 = 0.2$$

TABLE 16–1 **The Fortune Service 500**

Group	Number of Firms
1. Diversified service companies	100
2. Commercial banking companies	100
3. Financial service companies (including savings and insurance)	150
4. Retailing companies	50
5. Transportation companies	50
6. Utilities	50
	500

$$W_3 = N_3/N = 150/500 = 0.3$$
$$W_4 = N_4/N = 50/500 = 0.1$$
$$W_5 = N_5/N = 50/500 = 0.1$$
$$W_6 = N_6/N = 50/500 = 0.1$$

The economist decides to select a random sample of 100 of the 500 firms listed in *Fortune*. The economist chooses to use a proportional allocation of the total sample to the six strata (another method of allocation will be presented shortly). With proportional allocation, the total sample of 100 must be allocated to the different strata in proportion to the computed strata weights. Thus, for each i, $i = 1, \ldots, 6$, we compute n_i as $n_i = nW_i$. This gives the following sample sizes:

$$n_1 = 20 \qquad n_2 = 20 \qquad n_3 = 30 \qquad n_4 = 10 \qquad n_5 = 10 \qquad n_6 = 10$$

We will assume that the net income values in the different strata are approximately normally distributed and that the estimated strata variances (to be estimated from the data) are the true strata variances σ_i^2, so that the normal distribution may be used.

The economist draws the random samples and computes the sample means and variances. The results, in millions of dollars (for the means) and in millions of dollars squared (for the variances), are given in Table 16–2, along with the sample sizes in the different strata and the strata weights. From the table, and with the aid of equation 16–1 for the mean and equation 16–5 for the variance of the sample mean (with the estimated sample variances substituted for the population variances in the different strata), we will now compute the stratified sample mean and the estimated variance of the stratified sample mean.

$$\bar{x}_{st} = \sum_{i=1}^{6} W_i \bar{x}_i = (0.2)(52.7) + (0.2)(112.6) + (0.3)(85.6) + (0.1)(12.6)$$
$$+ (0.1)(8.9) + (0.1)(52.3)$$
$$= \$66.12 \text{ million}$$

and

$$s(\bar{X}_{st}) = \sqrt{\frac{1-f}{n} \sum_{i=1}^{6} W_i s_i^2}$$
$$= \sqrt{\frac{0.8}{100}[(0.2)(97,650) + (0.2)(64,300) + (0.3)(76,990) + (0.1)(18,320) +}$$
$$(0.1)(9,037) + (0.1)(83,500)]$$
$$= 23.08$$

(Our sampling fraction is $f = 100/500 = 0.2$.) Our unbiased point estimate of the average net income of all firms in the Fortune Service 500 is $66.12 million.

Using equation 16–9, we now compute a 95% confidence interval for μ, the mean net income of all firms in the index. We have the following:

$$\bar{x}_{st} \pm z_{\alpha/2} \, s(\bar{X}_{st}) = 66.12 \pm (1.96)(23.08) = [20.88, 111.36]$$

Thus, the economist may be 95% confident that the average net income for all firms in the Fortune Service 500 is anywhere from 20.88 to 111.36 million dollars. Inciden-

TABLE 16–2 **Sampling Results for Example 16–2**

Stratum	Mean	Variance	n_i	W_i
1	52.7	97,650	20	0.2
2	112.6	64,300	20	0.2
3	85.6	76,990	30	0.3
4	12.6	18,320	10	0.1
5	8.9	9,037	10	0.1
6	52.3	83,500	10	0.1

FIGURE 16–1 **The Template for Estimating Means by Stratified Sampling [Stratified Sampling.xls; Sheet: Mean]**

Stratified Sampling for Estimating μ — Fortune Service 500

Stratum	*N*	*n*	*x*-bar	s^2
1	100	20	52.7	97650
2	100	20	112.6	64300
3	150	30	85.6	76990
4	50	10	12.6	18320
5	50	10	8.9	9037
6	50	10	52.3	83500
7				
8				
9				
10				
11				
12				
13				
14				
15				
16				
17				
18				
19				
20				
Total	500	100		

X-bar	V(*X*-bar)	S(*X*-bar)	*df*
66.12	532.5816	23.07773	80

1–α	(1-α) CI for *X*-bar				
95%	66.12	+ or -	45.9262	or	20.193

tally, the true population mean net income for all 500 firms in the index is $\mu = \$61.496$ million.

The Template

Figure 16–1 shows the template that can be used to estimate population means by stratified sampling in the example.

Stratified Sampling for the Population Proportion

The theory of stratified sampling extends in a natural way to sampling for the population proportion p. Let the sample proportion in stratum i be $\hat{P}_i = X_i/n_i$, where X_i is the number of successes in a sample of size n_i. Then the stratified estimator of the population proportion p is the following.

The stratified estimator of the population proportion p is

$$\hat{P}_{st} = \sum_{i=1}^{m} W_i \hat{P}_i \qquad (16\text{--}11)$$

where the weights W_i are defined as in the case of sampling for the population mean: $W_i = N_i/N$

The following is an approximate expression for the variance of the estimator of the population proportion $\hat{P}_{st}$, for use with large samples.

The approximate variance of $\hat{P}_{st}$ is

$$V(\hat{P}_{st}) = \sum_{i=1}^{m} W_i^2 \frac{\hat{P}_i \hat{Q}_i}{n_i} \qquad (16\text{--}12)$$

where $\hat{Q}_i = 1 - \hat{P}_i$

When finite-population correction factors f_i must be considered, the following expression is appropriate for the variance of $\hat{P}_{st}$:

$$V(\hat{P}_{st}) = \frac{1}{N^2} \sum_{i=1}^{m} N_i^2 (N_i - n_i) \frac{\hat{P}_i \hat{Q}_i}{(N_i - 1)n_i} \qquad (16\text{--}13)$$

When proportional allocation is used, an approximate expression is

$$V(\hat{P}_{st}) = \frac{1 - f}{n} \sum_{i=1}^{m} W_i \hat{P}_i \hat{Q}_i \qquad (16\text{--}14)$$

Let us now return to Example 16–1, sampling for the proportion of people who might be interested in purchasing the Spanish sherry Jerez. Suppose that the market researchers believe that preferences for imported wines differ between consumers in metropolitan areas and those in other areas. The area of interest for the survey covers a few states in the Northeast, where it is known that 65% of the people live in metropolitan areas and 35% live in nonmetropolitan areas. A sample of 130 people randomly chosen at shopping malls in metropolitan areas shows that 28 are interested in Jerez, while a random sample of 70 people selected at malls outside the metropolitan areas shows that 18 are interested in the sherry.

Let us use these results in constructing a 90% confidence interval for the proportion of people in the entire population who are interested in the product. From equation 16–11, using two strata with weights 0.65 and 0.35, we get

$$\hat{p}_{st} = \sum_{i=1}^{2} W_i \hat{p}_i = (0.65)\frac{28}{130} + (0.35)\frac{18}{70} = 0.23$$

Our allocation is proportional because $n_1 = 130$, $n_2 = 70$, and $n = 130 + 70 = 200$, so that $n_1/n = 0.65 = W_1$ and $n_2/n = 0.35 = W_2$. In addition, the sample sizes of 130 and 70 represent tiny fractions of the two strata; hence, no finite-population correction factor is required. The equation for the estimated variance of the sample estimator of the proportion is therefore equation 16–14 without the finite-population correction:

$$V(\hat{P}_{st}) = \frac{1}{n} \sum_{i=1}^{2} W_i \hat{p}_i \hat{q}_i = \frac{1}{200}[(0.65)(0.215)(0.785) + (0.35)(0.257)(0.743)]$$

$$= 0.0008825$$

The standard error of $\hat{P}_{st}$ is therefore $\sqrt{V(\hat{P}_{st})} = \sqrt{0.0008825} = 0.0297$. Thus, our 90% confidence interval for the population proportion of people interested in Jerez is

$$\hat{p}_{st} \pm z_{\alpha/2} \, s(\hat{P}_{st}) = 0.23 \pm (1.645)(0.0297) = [0.181, 0.279]$$

The stratified point estimate of the percentage of people in the proposed market area for Jerez who may be interested in the product, if it is introduced, is 23%. A 90% confidence interval for the population percentage is 18.1% to 27.9%.

The Template

Figure 16–2 shows the template that can be used to estimate population proportions by stratified sampling. The data in the figure correspond to the Jerez example.

What Do We Do When the Population Strata Weights Are Unknown?

Here and in the following subsections, we will explore some optional, advanced aspects of stratified sampling. When the true strata weights $W_i = N_i/N$ are unknown—that is, when we do not know what percentage of the whole population belongs to each stratum—we may still use stratified random sampling. In such cases, we use estimates of the true weights, denoted by w_i. The consequence of using estimated weights instead of the true weights is the introduction of a bias into our sampling results. The interesting thing about this kind of bias is that it is not eliminated as the sample size increases. When errors in the stratum weights exist, our results are always biased; the greater the errors, the greater the bias. These errors also cause the standard error of the sample mean $s(\overline{X}_{st})$ to underestimate the true standard deviation. Consequently,

FIGURE 16–2 **The Template for Estimating Proportions by Stratified Sampling**
[Stratified Sampling.xls; Sheet: Proportion]

	A	B	C	E	G	H	I	J	K	L	M	N	O	P	Q	R
1		Stratified Sampling for Estimating Proportion						Jerez								
2																
3		Stratum	*N*	*n*	*x*	*p*-hat										
4		1	650000	130	28	0.21538										
5		2	350000	70	18	0.25714		*P*-hat	V(*P*-hat)	S(*P*-hat)						
6		3						0.2300	0.000883	0.0297						
7		4														
8		5														
9		6						1−α			(1-α) CI for *P*					
10		7						90%	0.2300	+ or −	0.0489	or	0.1811	to 0.2789		
11		8														
12		9														
13		10														
14		11														
15		12														
16		13														
17		14														
18		15														
19		16														
20		17														
21		18														
22		19														
23		20														
24		Total	1000000	200												

confidence intervals for the population parameter of interest tend to be narrower than they should be.

How Many Strata Should We Use?

The number of strata to be used is an important question to consider when you are designing any survey that uses stratified sampling. In many cases, there is a natural breakdown of the population into a given number of strata. In other cases, there may be no clear, unique way of separating the population into groups. For example, if age is to be used as a stratifying variable, there are many ways of breaking the variable and forming strata. There are two guidance rules for constructing strata. The rules are presented below.

Rules for Constructing Strata

1. The number of strata should preferably be less than or equal to 6.
2. Choose the strata so that Cum $\sqrt{f(x)}$ is approximately constant for all strata [where Cum $\sqrt{f(x)}$ is the cumulative square root of the frequency of X, the variable of interest].

The first rule is clear. The second rule says that in the absence of other guidelines for breaking down a population into strata, we partition the variable used for stratification into categories so that the cumulative square root of the frequency function of the variable is approximately equal for all strata. We illustrate this rule in the hypothetical case of Table 16–3. As can be seen in this simplified example, the combined age groups 20–30, 31–35, and 36–45 all have a sum of $\sqrt{f}$ equal to 5; hence, these groups make good strata with respect to age as a stratifying variable according to rule 2.

Postsampling Stratification

At times, we conduct a survey using simple random sampling with no stratification, and after obtaining our results, we note that the data may be broken into categories of similar elements. Can we now use the techniques of stratified random sampling and enjoy its benefits in terms of reduced variances of the estimators? Surprisingly, the answer is yes. In fact, if the subsamples in each of our strata contain at least 20 elements, and if our estimated weights of the different strata w_i (computed from the data as n_i/n, or from more accurate information) are close to the true population strata weights W_i, then our stratified estimator will be almost as good as that of stratified random sampling with proportional allocation. This procedure is called *poststratification*. We close this section with a discussion of an alternative to proportional allocation of the sample in stratified random sampling, called *optimum allocation*.

TABLE 16–3 **Constructing Strata by Age**

Age	Frequency f	$\sqrt{f}$	Cum $\sqrt{f}$
20–25	1	1	
26–30	16	4	5
31–35	25	5	5
36–40	4	2	
41–45	9	3	5

Optimum Allocation

With optimum allocation, we select the sample sizes to be allocated to each of the strata so as to minimize one of two criteria. Either we minimize the cost of the survey for a given value of the variance of our estimator, or we minimize the variance of our estimator for a given cost of taking the survey.

We assume a cost function of the form

$$C = C_0 + \sum_{i=1}^{m} C_i n_i \qquad (16\text{–}15)$$

where C is the total cost of the survey, C_0 is the fixed cost of setting up the survey, and C_i is the cost per item sampled in stratum i. Clearly, the total cost of the survey is the sum of the fixed cost and the costs of sampling in all the strata (where the cost of sampling in stratum i is equal to the sample size n_i times the cost per item sampled C_i).

Under the assumption of a cost function given in equation 16–15, the optimum allocation that will minimize our total cost for a fixed variance of the estimator, or minimize the variance of the estimator for a fixed total cost, is as follows.

Optimum allocation:

$$\frac{n_i}{n} = \frac{W_i \sigma_i / \sqrt{C_i}}{\sum_{i=1}^{m} W_i \sigma_i / \sqrt{C_i}} \qquad (16\text{–}16)$$

Equation 16–16 has an intuitive appeal. It says that for a given stratum, we should take a larger sample if the stratum is *more variable internally* (greater σ_i), if the *relative size of the stratum is larger* (greater W_i), or if *sampling in the stratum is cheaper* (smaller C_i).

If the cost per unit sampled is the same in all the strata (i.e., if $C_i = c$ for all i), then the optimum allocation for a fixed total cost is the same as the optimum allocation for fixed sample size, and we have what is called the *Neyman allocation* (after J. Neyman, although this allocation was actually discovered earlier by A. A. Tschuprow in 1923).

The Neyman allocation:

$$\frac{n_i}{n} = \frac{W_i \sigma_i}{\sum_{i=1}^{m} W_i \sigma_i} \qquad (16\text{–}17)$$

Suppose that we want to allocate a total sample of size 1,000 to three strata, where stratum 1 has weight 0.4, standard deviation 1, and cost per sampled item of 4 cents; stratum 2 has weight 0.5, standard deviation 2, and cost per item of 9 cents; and stratum 3 has weight 0.1, standard deviation 3, and cost per item of 16 cents. How should we allocate this sample if optimum allocation is to be used? We have

$$\sum_{i=1}^{3} \frac{W_i \sigma_i}{\sqrt{C_i}} = \frac{(0.4)(1)}{\sqrt{4}} + \frac{(0.5)(2)}{\sqrt{9}} + \frac{(0.1)(3)}{\sqrt{16}} = 0.608$$

From equation 16–16, we get

$$\frac{n_1}{n} = \frac{W_1\sigma_1/\sqrt{C_1}}{\displaystyle\sum_{i=1}^{3}(W_i\sigma_i/\sqrt{C_i})} = \frac{(0.4)(1)/\sqrt{4}}{0.608} = 0.329$$

$$\frac{n_2}{n} = \frac{W_2\sigma_2/\sqrt{C_2}}{\displaystyle\sum_{i=1}^{3}(W_i\sigma_i/\sqrt{C_i})} = \frac{(0.5)(2)/\sqrt{9}}{0.608} = 0.548$$

$$\frac{n_3}{n} = \frac{W_3\sigma_3/\sqrt{C_3}}{\displaystyle\sum_{i=1}^{3}(W_i\sigma_i/\sqrt{C_i})} = \frac{(0.1)(3)/\sqrt{16}}{0.608} = 0.123$$

The optimum allocation in this case is 329 items from stratum 1; 548 items from stratum 2; and 123 items from stratum 3 (making a total of 1,000 sample items, as specified).

Let us now compare this allocation with proportional allocation. With a sample of size 1,000 and a proportional allocation, we would allocate our sample only by the stratum weights, which are 0.4, 0.5, and 0.1, respectively. Therefore, our allocation will be 400 from stratum 1; 500 from stratum 2; and 100 from stratum 3. The optimum allocation is different, as it incorporates the cost and variance considerations. Here, the difference between the two sets of sample sizes is not large.

Suppose, in this example, that the costs of sampling from the three strata are the same. In this case, we can use the Neyman allocation and get, from equation 16–17,

$$\frac{n_1}{n} = \frac{W_1\sigma_1}{\displaystyle\sum_{i=1}^{3}W_i\sigma_i} = \frac{(0.4)(1)}{1.7} = 0.235$$

$$\frac{n_2}{n} = \frac{W_2\sigma_2}{\displaystyle\sum_{i=1}^{3}W_i\sigma_i} = \frac{(0.5)(2)}{1.7} = 0.588$$

$$\frac{n_3}{n} = \frac{W_3\sigma_3}{\displaystyle\sum_{i=1}^{3}W_i\sigma_i} = \frac{(0.1)(3)}{1.7} = 0.176$$

Thus, the Neyman allocation gives a sample of size 235 to stratum 1; 588 to stratum 2; and 176 to stratum 3. Note that these subsamples add only to 999, due to rounding error. The last sample point may be allocated to any of the strata.

In general, stratified random sampling gives more precise results than those obtained from simple random sampling: The standard errors of our estimators from stratified random sampling are usually smaller than those of simple random sampling. Furthermore, in stratified random sampling, an optimum allocation will produce more precise results than a proportional allocation if some strata are more expensive to sample than others or if the variances within strata are different from one another.

The Template

Figure 16–3 shows the template that can be used to estimate population proportions by stratified sampling. The data in the figure correspond to the example we have been discussing.

FIGURE 16–3 **The Template for Optimal Allocation for Stratified Sampling [Stratified Sampling.xls; Sheet: Allocation]**

	A	B	C	D	E	H	I	J	K	L
1	Optimum Allocation									
2										
3				Planned Total Sample Size			1000			
5		Stratum	*W*	σ	*C*		*n*			
6		1	0.4	1	$ 4.00		329		Fixed Cost	
7		2	0.5	2	$ 9.00		548		Variable Cost	$ 8,216.00
8		3	0.1	3	$ 16.00		123		Total Cost	$ 8,216.00
9		4								
10		5								
11		6								
12		7								
13		8								
14		9								
15		10								
16		11								
17		12								
18		13								
19		14								
20		15								
21		16								
22		17								
23		18								
24		19								
25		20								
26		Total	1		Actual total		1000			

The same template can be used for Neyman allocation. In column E, enter the same cost *C*, say $1, for all the strata.

PROBLEMS

16–1. A securities analyst wants to estimate the average percentage of institutional holding of all publicly traded stocks in the United States. The analyst believes that stocks traded on the three major exchanges have different characteristics and therefore decides to use stratified random sampling. The three strata are the New York Stock Exchange (NYSE), the American Exchange (AMEX), and the Over the Counter (OTC) exchange. The weights of the three strata, as measured by the number of stocks listed in each exchange, divided by the total number of stocks, are NYSE, 0.44; AMEX, 0.15; OTC, 0.41. A total random sample of 200 stocks is selected, with proportional allocation. It is found that the average percentage of institutional holdings of the subsample selected from the issues of the NYSE is 46%, and the standard deviation is 8%. The corresponding results for the AMEX are 9% average institutional holdings and a standard deviation of 4%, and the corresponding results for the OTC stocks are 29% average institutional holdings and a standard deviation of 16%.

a. Give a stratified estimate of the mean percentage of institutional holdings per stock.

b. Give the standard error of the estimate in *a.*

c. Give a 95% confidence interval for the mean percentage of institutional holdings.

d. Explain the advantages of using stratified random sampling in this case. Compare with simple random sampling.

16–2. A company has 2,100 employees belonging to the following groups: production, 1,200; marketing, 600; management, 100; other, 200. The company president wants to obtain an estimate of the views of all employees about a certain impending executive decision. The president knows that the management employees' views are most variable, along with employees in the "other" category, while the marketing and production people have rather uniform views within their groups. The production people are the most costly to sample, because of the time required to find them at their different jobs, and the management people are easiest to sample.

    a. Suppose that a total sample of 100 employees is required. What are the sample sizes in the different strata under proportional allocation?

    b. Discuss how you would design an optimum allocation in this case.

16–3. Last year, consumers increasingly bought fleece (industry jargon for hot-selling jogging suits, which now rival jeans as the uniform for casual attire). A New York designer of jogging suits is interested in the new trend and wants to estimate the amount spent per person on jogging suits during the year. The designer knows that people who belong to health and fitness clubs will have different buying behavior than people who do not. Furthermore, the designer finds that, within the proposed study area, 18% of the population are members of health and fitness clubs. A random sample of 300 people is selected, and the sample is proportionally allocated to the two strata: members of health clubs and nonmembers of health clubs. It is found that among members, the average amount spent is $152.43 and the standard deviation is $25.77, while among the nonmembers, the average amount spent is $15.33 and the standard deviation is $5.11.

    a. What is the stratified estimate of the mean?

    b. What is the standard error of the estimator?

    c. Give a 90% confidence interval for the population mean μ.

    d. Discuss one possible problem with the data. (*Hint:* Can the data be considered normally distributed? Why?)

16–4. A financial analyst is interested in estimating the average amount of a foreign loan by U.S. banks. The analyst believes that the amount of a loan may be different depending on the bank, or, more precisely, on the extent of the bank's involvement in foreign loans. The analyst obtains the following data on the percentage of profits of U.S. banks from loans to Mexico and proposes to use these data in the construction of stratum weights. The strata are the different banks: First Chicago, 33%; Manufacturers Hanover, 27%; Bankers Trust, 21%; Chemical Bank, 19%; Wells Fargo Bank, 19%; Citicorp, 16%; Mellon Bank, 16%; Chase Manhattan, 15%; Morgan Guarantee Trust, 9%.

    a. Construct the stratum weights for proportional allocation.

    b. Discuss two possible problems with this study.

16–4 Cluster Sampling

Let us consider the case where we have no frame (i.e., no list of all the elements in the population) and the elements are *clustered* in larger units. Each unit, or cluster, contains several elements of the population. In this case, we may choose to use the method of **cluster sampling.** This may also be the case when the population is large and spread over a geographic area in which smaller subregions are easily sampled and where a simple random sample or a stratified random sample may not be carried out as easily.

    Suppose that the population is composed of M clusters and there is a list of all M clusters from which a random sample of m clusters is selected. There are two possibilities. First, we may sample *every element* in every one of the m selected clusters. In this case, our sampling method is called *single-stage cluster sampling*. Second, we may select a random sample of m clusters and then select a random sample of n elements from each of the selected clusters. In this case, our sampling method is called *two-stage cluster sampling*.

The Relation with Stratified Sampling

In stratified sampling, we sample elements from every one of our strata, and this assures us of full representation of all segments of the population in the sample. In cluster sampling, we sample only some of the clusters, and although elements within any cluster may tend to be homogeneous, as is the case with strata, not all the clusters are represented in the sample; this leads to lowered precision of the cluster sampling method. In stratified random sampling, we use the fact that the population may be broken into subgroups. This usually leads to a smaller variance of our estimators. In cluster sampling, however, the method is used mainly because of ease of implementation or reduction in sampling costs, and the estimates do not usually lead to more precise results.

Single-Stage Cluster Sampling for the Population Mean

Let $n_1, n_2, \ldots, n_m$ be the number of elements in each of the m sampled clusters. Let $\overline{X}_1, \overline{X}_2, \ldots, \overline{X}_m$ be the means of the sampled clusters. The cluster sampling unbiased estimator of the population mean μ is given as follows.

The cluster sampling estimator of μ is

$$\overline{X}_{cl} = \frac{\displaystyle\sum_{i=1}^{m} n_i \overline{X}_i}{\displaystyle\sum_{i=1}^{m} n_i} \qquad (16\text{–}18)$$

An estimator of the variance of the estimator of μ in equation 16–18 is

$$s^2(\overline{X}_{cl}) = \frac{M - m}{Mm\bar{n}^2} \cdot \frac{\displaystyle\sum_{i=1}^{m} n_i^2(\overline{X}_i - \overline{X}_{cl})^2}{m - 1} \qquad (16\text{–}19)$$

where $\bar{n} = (\sum_{i=1}^{m} n_i)/m$ is the average number of units in the sampled clusters.

Single-Stage Cluster Sampling for the Population Proportion

The cluster sampling estimator of the population proportion p is

$$\hat{P}_{cl} = \frac{\displaystyle\sum_{i=1}^{m} n_i \hat{P}_i}{\displaystyle\sum_{i=1}^{m} n_i} \qquad (16\text{–}20)$$

where the $\hat{P}_i$ are the proportions of interest within the sampled clusters

The estimated variance of the estimator in equation 16–20 is given by

$$s^2(\hat{P}_{cl}) = \frac{M - m}{Mm\bar{n}^2} \frac{\displaystyle\sum_{i=1}^{m} n_i^2(\hat{P}_i - \hat{P}_{cl})^2}{m - 1} \tag{16–21}$$

We now demonstrate the use of cluster sampling for the population mean with the following example.

EXAMPLE 16–3   J. B. Hunt Transport Company is especially interested in lowering fuel costs in order to survive in the tough world of deregulated trucking. Recently, the company introduced new measures to reduce fuel costs for all its trucks. Suppose that company trucks are based in 110 centers throughout the country and that the company's management wants to estimate the average amount of fuel saved per truck for the week following the institution of the new measures. For reasons of lower cost and administrative ease, management decides to use single-stage cluster sampling, select a random sample of 20 trucking centers, and measure the weekly fuel saving for each of the trucks in the selected centers (each center is a cluster). The average fuel savings per truck, in gallons, for each of the 20 selected centers are as follows (the number of trucks in each center is given in parentheses): 21 (8), 22 (8), 11 (9), 34 (10), 28 (7), 25 (8), 18 (10), 24 (12), 19 (11), 20 (6), 30 (8), 26 (9), 12 (9), 17 (8), 13 (10), 29 (8), 24 (8), 26 (10), 18 (10), 22 (11). From these data, compute an estimate of the average amount of fuel saved per truck for all Hunt's trucks over the week in question. Also give a 95% confidence interval for this parameter.

Solution   From equation 16–18, we get

$$\bar{x}_{cl} = \frac{\displaystyle\sum_{i=1}^{m} n_i\bar{x}_i}{\displaystyle\sum_{i=1}^{m} n_i} = [21(8) + 22(8) + 11(9) + 34(10) + 28(7) + 25(8) + 18(10) + 24(12) + 19(11) + 20(6) + 30(8) + 26(9) + 12(9) + 17(8) + 13(10) + 29(8) + 24(8) + 26(10) + 18(10) + 22(11)]/(8 + 8 + 9 + 10 + 7 + 8 + 10 + 12 + 11 + 6 + 8 + 9 + 9 + 8 + 10 + 8 + 8 + 10 + 10 + 11)$$

$$= 21.83$$

From equation 16–19, we find that the estimated variance of our sample estimator of the mean is

$$s^2(\overline{X}_{cl}) = \frac{M - m}{Mm\bar{n}^2} \frac{\displaystyle\sum_{i=1}^{20} n_i^2(\bar{x}_i - \bar{x}_{cl})^2}{m - 1}$$

$$= \frac{110 - 20}{(110)(20)(9)^2}[8^2(21 - 21.83)^2 + 8^2(22 - 21.83)^2 + 9^2(11 - 21.83)^2 + \cdots + 11^2(22 - 21.83)^2]/19$$

$$= 1.587$$

Using the preceding information, we construct a 95% confidence interval for μ as follows:

$$\bar{x}_{cl} \pm 1.96s(\overline{X}_{cl}) = 21.83 \pm 1.96\sqrt{1.587} = [19.36, 24.30]$$

Thus, based on the sampling results, Hunt's management may be 95% confident that average fuel savings per truck for all trucks over the week in question is anywhere from 19.36 to 24.30 gallons.

The Template

The template for estimating a population mean by cluster sampling is shown in Figure 16–4. The data in the figure correspond to Example 16–3.

The template for estimating a population proportion by cluster sampling is shown in Figure 16–5.

Two-Stage Cluster Sampling

When clusters are very large or when elements within each cluster tend to be similar, we may gain little information by sampling every element within the selected clusters. In such cases, it may be more economical to select more clusters and to sample only some of the elements within the chosen clusters. The formulas for the estimators and their variances in the case of two-stage cluster sampling are more complicated and may be found in advanced books on sampling methodology.

PROBLEMS

16–5. There are 602 aerobics and fitness centers in Japan (up from 170 five years ago). Adidas, the European maker of sports shoes and apparel, is very interested in this fast-growing potential market for its products. As part of a marketing survey, Adidas wants to estimate the average income of all members of Japanese fitness centers. (Members of one such club pay \$62 to join and another \$62 per month. Adidas believes that the

FIGURE 16–4 The Template for Estimating Means by Cluster Sampling
[Cluster Sampling.xls; Sheet: Mean]

	A	B	C	D	E	F	G	H	I	J	K	L	M
1	Cluster	Sampling					J. B. Hunt Transport Co.						
2													
3		Cluster	n	x-bar		M							
4		1	8	21		110							
5		2	8	22									
6		3	9	11		m		n-bar					
7		4	10	34		20		9					
8		5	7	28									
9		6	8	25		X-bar	V(X-bar)	s(X-bar)					
10		7	10	18		21.8333	1.58691	1.25973					
11		8	12	24									
12		9	11	19		1−α	(1-α) CI for X-bar						
13		10	6	20		95%	21.833	+ or -	2.469	or	19.364	to	24.302
14		11	8	30									
15		12	9	26									
16		13	9	12									
17		14	8	17									
18		15	10	13									
19		16	8	29									
20		17	8	24									
21		18	10	26									
22		19	10	18									
23		20	11	22									
24		Total	180										

FIGURE 16–5   The Template for Estimating Proportions by Cluster Sampling
[Cluster Sampling.xls; Sheet: Proportion]

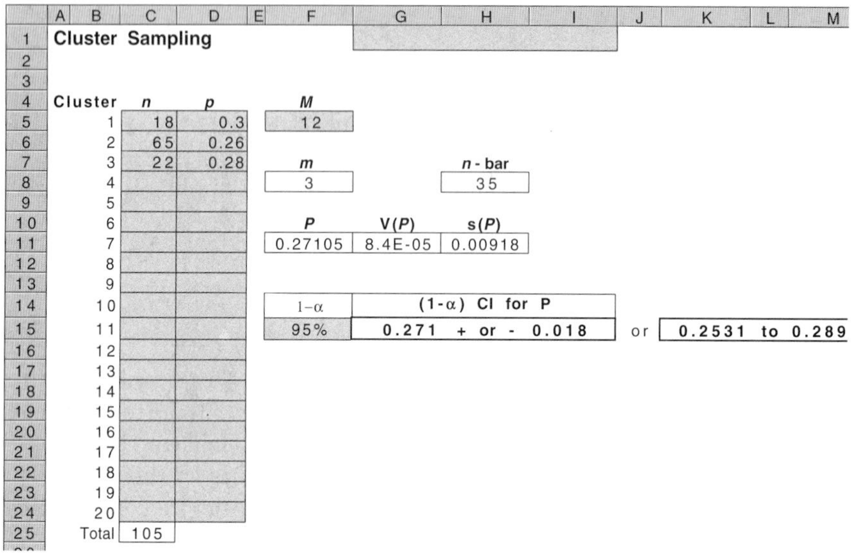

average income of all fitness club members in Japan may be higher than that of the general population, for which census data exist.) Since travel and administrative costs for conducting a simple random sample of all members of fitness clubs throughout Japan would be prohibitive, Adidas decided to conduct a cluster sampling survey. Five clubs were chosen at random out of the entire collection of 602 clubs, and all members of the five clubs were interviewed. The following are the average incomes (in U.S. dollars) for the members of each of the five clubs (the number of members in each club is given in parentheses): $37,237 (560), $41,338 (435), $28,800 (890), $35,498 (711), $47,446 (230). Give the cluster sampling estimate of the population mean income for all fitness club members in Japan. Also give a 90% confidence interval for the population mean. Are there any limitations to the methodology in this case?

16–6.   Israel's kibbutzim are by now well diversified beyond their agrarian roots, producing everything from lollipops to plastic pipe. These 282 widely scattered communes of several hundred members maintain hundreds of factories and other production facilities. An economist wants to estimate the average annual revenues of all kibbutz production facilities. Since each kibbutz has several production units, and since travel and other costs are high, the economist wants to consider a sample of 15 randomly chosen kibbutzim and find the annual revenues of all production units in the selected kibbutzim. From these data, the economist hopes to estimate the average annual revenue per production unit in all 282 kibbutzim. The sample results are as follows:

Kibbutz	Number of Production Units	Total Kibbutz Annual Revenues (in millions of dollars)
1	4	4.5
2	2	2.8
3	6	8.9
4	2	1.2

Kibbutz	Number of Production Units	Total Kibbutz Annual Revenues (in millions of dollars)
5	5	7.0
6	3	2.2
7	2	2.3
8	1	0.8
9	8	12.5
10	4	6.2
11	3	5.5
12	3	6.2
13	2	3.8
14	5	9.0
15	2	1.4

From these data, compute the cluster sampling estimate of the mean annual revenue of all kibbutzim production units, and give a 95% confidence interval for the mean.

16–7. Under what conditions would you use cluster sampling? Explain the differences among cluster sampling, simple random sampling, and stratified random sampling. Under what conditions would you use two-stage cluster sampling? Explain the difference between single-stage and two-stage cluster sampling. What are the limitations of cluster sampling?

16–8. Recently a survey was conducted to assess the quality of investment brokers. A random sample of 6 brokerage houses was selected from a total of 27 brokerage houses. Each of the brokers in the selected brokerage houses was evaluated by an independent panel of industry experts as "highly qualified" (HQ) or was given an evaluation below this rating. The designers of the survey wanted to estimate the proportion of all brokers in the entire industry who would be considered highly qualified. The survey results are in the following table.

Brokerage House	Total Number of Brokers	Number of HQ Brokers
1	120	80
2	150	75
3	200	100
4	100	65
5	88	45
6	260	200

Use the cluster sampling estimator of the population proportion to estimate the proportion of all highly qualified brokers in the investment industry. Also give a 99% confidence interval for the population proportion you estimated.

16–9. Forty-two cruise ships come to Alaska's Glacier Bay every year. The state tourist board wants to estimate the average cruise passenger's satisfaction from this experience, rated on a scale of 0 to 100. Since the ships' arrivals are evenly spread throughout the season, simple random sampling is costly and time-consuming. Therefore, the agency decides to send its volunteers to board the first five ships of the season, consider them as clusters, and randomly choose 50 passengers in each ship for interviewing.

    a. Is the method employed single-stage cluster sampling? Explain.

    b. Is the method employed two-stage cluster sampling? Explain.

    c. Suppose that each of the ships has exactly 50 passengers. Is the proposed method single-stage cluster sampling?

d. The 42 ships belong to 12 cruise ship companies. Each company has its own characteristics in terms of price, luxury, services, and type of passengers. Suggest an alternative sampling method, and explain its benefits.

16–5   Systematic Sampling

Sometimes a population is arranged in some order: files in a cabinet, crops in a field, goods in a warehouse, etc. In such cases it may be easier to draw our random sample in a *systematic* way rather than generate a simple random sample that would entail looking for particular items within the population. To select a **systematic sample** of n elements from a population of N elements, we divide the N elements in the population into n groups of k elements and then use the following rule:

> We randomly select the first element out of the first k elements in the population, and then we select every kth unit afterward until we have a sample of n elements.

For example, suppose $k = 20$, and we need a sample of $n = 30$ items. We randomly select the first item from the integers 1 to 20. If the random number selected is 11, then our systematic sample will contain the elements 11, 11 + 20 = 31, 31 + 20 = 51, . . . , and so on until we have 30 elements in our sample.

A variant of this rule, which solves the problems that may be encountered when k is not an integer multiple of the sample size n (their product being N), is to let k be the nearest integer to N/n. We now regard the N elements as being arranged in a circle (with the last element preceding the first element). We randomly select the first element from all N population members and then select every kth item until we have n items in our sample.

The Advantages of Systematic Sampling

In addition to the ease of drawing samples in a systematic way–for example, by simply measuring distances with a ruler in a file cabinet and sampling every fixed number of inches–the method has some statistical advantages as well. First, when $k = N/n$, the sample estimator of the population mean is unbiased. Second, systematic sampling is usually more precise than simple random sampling because it actually stratifies the population into n strata, each stratum containing k elements. Therefore, systematic sampling is approximately as precise as stratified random sampling with one unit per stratum. The difference between the two methods is that the systematic sample is spread more evenly over the entire population than a stratified sample, because in stratified sampling the samples in the strata are drawn separately. This adds precision in some cases. Systematic sampling is also related to cluster sampling in that it amounts to selecting one cluster out of a population of k clusters.

Estimation of the Population Mean in Systematic Sampling

> The systematic sampling estimator of the population mean μ is
>
> $$\overline{X}_{sy} = \frac{\sum_{i=1}^{n} X_i}{n} \qquad (16\text{–}22)$$

The estimator is, of course, the same as the simple random sampling estimator of the population mean based on a sample of size n. The variance of the estimator in equation 16–22 is difficult to estimate from the results of a single sample. The estimation requires some assumptions about the order of the population. The estimated variances of $\bar{X}_{sy}$ in different situations are given below.

1. When the population values are assumed to be in no particular order with respect to the variable of interest, the estimated variance of the estimator of the mean is the same as in the case of simple random sampling

$$s^2(\bar{X}_{sy}) = \frac{N - n}{Nn} S^2 \qquad (16\text{–}23)$$

   where S^2 is the usual sample variance, and the first term accounts for finite-population correction as well as division by n.

2. When the mean is constant within each stratum of k elements but different from stratum to stratum, the estimated variance of the sample mean is

$$s^2(\bar{X}_{sy}) = \frac{N - n}{Nn} \frac{\displaystyle\sum_{i=1}^{n} (X_i - X_{i+k})^2}{2(n - 1)} \qquad (16\text{–}24)$$

3. When the population is assumed to be either increasing or decreasing linearly in the variable of interest, and when the sample size is large, the appropriate estimator of the variance of our estimator of the mean is

$$s^2(\bar{X}_{sy}) = \frac{N - n}{Nn} \frac{\displaystyle\sum_{i=1}^{n-2} (X_i - 2X_{i+k} + X_{i+2k})^2}{6(n - 2)} \qquad (16\text{–}25)$$

There are formulas that apply in more complicated situations as well.
   We demonstrate the use of systematic sampling with the following example.

EXAMPLE 16–4

An investor obtains a copy of *The Wall Street Journal* and wants to get a quick estimate of how the New York Stock Exchange has performed since the previous day. The investor knows that there are about 2,100 stocks listed on the NYSE and wants to look at a quick sample of 100 stocks and determine the average price change for the sample. The investor thus decides on an "every 21st" systematic sampling scheme. The investor uses a ruler and finds that this means that a stock should be selected about every 1.5 inches along the listings columns in the *Journal*. The first stock is randomly selected from among the first 21 stocks listed on the NYSE by using a random-number generator in a calculator. The selected stock is the seventh from the top, which happens to be ANR. For the day in question, the price change for ANR is -0.25. The next stock to be included in the sample is the one in position $7 + 21 = 28$th from the top. The stock is Aflpb, which on this date had a price change of 0 from the previous day. As mentioned, the selection is not done by counting the stocks, but by the faster method of successively measuring 1.5 inches down the column from each selected stock. The resulting sample of 100 stocks gives a sample mean of $\bar{x}_{sy} = +0.5$ and $s^2 = 0.36$. Give a 95% confidence interval for the average price change of all stocks listed on the NYSE.

Solution We have absolutely no reason to believe that the order in which the NYSE stocks are listed in *The Wall Street Journal* (i.e., alphabetically) has any relationship to the stocks' price changes. Therefore, the appropriate equation for the estimated variance of $\overline{X}_{sy}$ is equation 16–23. Using this equation, we get

$$s^2(\overline{X}_{sy}) = \frac{N - n}{Nn} s^2 = \frac{2{,}100 - 100}{210{,}000} 0.36 = 0.0034$$

A 95% confidence interval for μ, the average price change on this day for all stocks on the NYSE, is therefore

$$\overline{x}_{sy} \pm 1.96 s(\overline{X}_{sy}) = 0.5 \pm (1.96)(\sqrt{0.0034}) = [0.386, 0.614]$$

The investor may be 95% sure that the average stock on the NYSE gained anywhere from \$0.386 to \$0.614.

When sampling for the population proportion, use the same equations as the ones used for simple random sampling if it may be assumed that no inherent order exists in the population. Otherwise use variance estimators given in advanced texts.

The Template

The template for estimating a population mean by systematic sampling is shown in Figure 16–6. The data in the figure correspond to Example 16–4.

PROBLEMS

16–10. A tire manufacturer maintains strict quality control of its tire production. This entails frequent sampling of tires from large stocks shipped to retailers. Samples of tires are selected and run continuously until they are worn out, and the average number of miles "driven" in the laboratory is noted. Suppose a warehouse contains 11,000 tires arranged in a certain order. The company wants to select a systematic sample of 50 tires to be tested in the laboratory. Use randomization to determine the first item to be sampled, and give the rule for obtaining the rest of the sample in this case.

FIGURE 16–6 **The Template for Estimating Population Means by Systematic Sampling [Systematic Sampling.xls; Sheet: Sheet 1]**

	A	B	C	D	E	F	G	H	I	J
1	Systematic Sampling				Average price change					
2										
3		*N*	*n*	*x*-bar	*s*²					
4		2100	100	0.5	0.36					
5										
6				X-bar	0.5					
7				V(X-bar)	0.003429	Assuming that the population is in no particular order.				
8				S(X-bar)	0.058554					
9										
10		1−α		(1-α) CI for X-bar						
11		95%		0.5	+ or −	0.11476	or	0.38524	to 0.61476	
12										

16–11. A large discount store gives its sales personnel bonuses based on their average sale amount. Since each salesperson makes hundreds of sales each month, the store management decided to base average sale amount for each salesperson on a random sample of the person's sales. Since records of sales are kept in books, it is convenient to use systematic sampling. Suppose a salesperson has made 855 sales over the month, and management wants to choose a sample of 30 sales for estimation of the average amount of all sales. Suggest a way of doing this so that no problems would result due to the fact that 855 is not an integer multiple of 30. Give the first element you choose to select, and explain how the rest of the sample is obtained.

16–12. An accountant always audits the second account and every fourth account thereafter when sampling a client's accounts.

  a. Does the accountant use systematic sampling? Explain.

  b. Explain the problems that may be encountered when this sampling scheme is used.

16–13. Beer sales in a tavern are cyclical over the week, with large volume during weekend nights, lower volume during the beginning of the week, and somewhat higher volume at midweek. Explain the possible problems that could arise, and the conditions under which they might arise, if systematic sampling were used to estimate beer sales volume per night.

16–14. A population is composed of 100 items arranged in some order. It is known that every stratum of 10 items in the order of arrangement tends to be similar in its values. An "every 10th" systematic sample is selected. The first item, randomly chosen, is the 6th item, and its value is 20. The following items in the sample are, of course, the 16th, the 26th, etc. The values of all items in the systematic sample are as follows: 20, 25, 27, 34, 28, 22, 28, 21, 37, 31. Give a 90% confidence interval for the population mean.

16–15. Explain the relationship between the method employed in problem 16–14 and the method of stratified random sampling. Explain the differences between the two methods.

16–6 Nonresponse

Nonresponse to sample surveys is one of the most serious problems that occur in practical applications of sampling methodology. The problem is one of loss of information. For example, suppose that a survey questionnaire dealing with some issue is mailed to a randomly chosen sample of 500 people and that only 300 people respond to the survey. The question is: What can you say about the 200 people who did not respond? This is a very important question, and there is no immediate answer to it, precisely because the people did not respond; we know nothing about them. Suppose that the questionnaire asks for a yes or no answer to a particular public issue over which people have differing views, and we want to estimate the proportion of people who would respond yes. People may have such strong views about the issue that those who would respond no may refuse to respond altogether. In this case, the 200 nonrespondents to our survey will contain a higher proportion of "no" answers than the 300 responses we have. But, again, we would not know about this. The result will be a bias. How can we compensate for such a possible bias?

We may want to consider the population as made up of two *strata:* the respondents' stratum and the nonrespondents' stratum. In the original survey, we managed to sample only the respondents' stratum, and this caused the bias. What we need to do is to obtain a random sample from the nonrespondents' stratum. This is easier said than done. Still, there are ways we can at least reduce the bias and get some idea about the proportion of "yes" answers in the nonresponse stratum. This entails *callbacks:* returning to the nonrespondents and asking them again. In some mail questionnaires, it is common to send several requests for response, and these reduce the uncertainty. There may, however, be hard-core refusers who just do not want to answer the questionnaire. It is likely that these people have very distinct views about the issue in question, and if you leave them out, there will be a significant bias in your conclusions. In such a situation, it may be useful to gather a small random sample of the hard-core refusers and offer them some monetary reward for their answers. In cases where people may find the question embarrassing or may worry about revealing their personal views, there is the possibility of using a random-response mechanism whereby the respondent randomly answers one of two questions, one is the sensitive question, and the other is an innocuous question of no relevance. The interviewer does not know which question any particular respondent answered but does know the probability of answering the sensitive question. This still allows for computation of the aggregated response to the sensitive question while protecting any given respondent's privacy.

16–7   Summary and Review of Terms

In this chapter, we considered some advanced sampling methods that allow for better precision than simple random sampling, or for lowered costs and easier survey implementation. We concentrated on **stratified random sampling,** the most important and useful of the advanced methods and one that offers statistical advantages of improved precision. We then discussed **cluster sampling** and **systematic sampling,** two methods that are used primarily for their ease of implementation and reduced sampling costs. We mentioned a few other advanced methods, which are described in books devoted to sampling methodology. We discussed the problem of **nonresponse.**

ADDITIONAL PROBLEMS

16–16.   Bloomingdale's in New York has the following departments on its mezzanine level: Stendahl, Ralph Lauren, Beauty Spot, and Lauder Prescriptives. The mezzanine level is managed separately from the other levels, and during the store's postholiday sale, the level manager wanted to estimate the average sales amount per customer throughout the sale. The following table gives the relative weights of the different departments (known from previous operation of the store), as well as the sample means and variances of the different strata for a total sample of 1,000 customers, proportionally allocated to the four strata. Give a 95% confidence interval for the average sale (in dollars) per customer for the entire level over the period of the postholiday sale.

Stratum	Weight	Sample Mean	Sample Variance
Stendahl	0.25	65.00	123.00
Ralph Lauren	0.35	87.00	211.80
Beauty Spot	0.15	52.00	88.85
Lauder Prescriptives	0.25	38.50	100.40

Note: We assume that shoppers visit the mezzanine level to purchase from only one of its departments. Since the brands and designers are competitors, and since shoppers are known to have a strong brand loyalty in this market, the assumption seems reasonable.

16–17. A state department of transportation is interested in sampling commuters to determine certain of their characteristics. The department arranges for its field workers to board buses at random as well as stop private vehicles at intersections and ask the commuters to fill out a short questionnaire. Is this method cluster sampling? Explain.

16–18. Use systematic sampling to estimate the average performance of all stocks in one of the listed stock exchanges on a given day. Compare your results with those reported in the media for the day in question.

16–19. An economist wants to estimate average annual profits for all businesses in a given community and proposes to draw a systematic sample of all businesses listed in the local Yellow Pages. Comment on the proposed methodology. What potential problem do you foresee?

16–20. In an "every 23rd" systematic sampling scheme, the first item was randomly chosen to be the 17th element. Give the numbers of 6 sample items out of a population of 120.

16–21. A quality control sampling scheme was carried out by Sony for estimating the percentage of defective radios in a large shipment of 1,000 containers with 100 radios in each container. Twelve containers were chosen at random, and every radio in them was checked. The numbers of defective radios in each of the containers are 8, 10, 4, 3, 11, 6, 9, 10, 2, 7, 6, 12. Give a 95% confidence interval for the proportion of defective radios in the entire shipment.

16–22. Suppose that the radios in the 1,000 containers of problem 16–21 were produced in five different factories, each factory known to have different internal production controls. Each container is marked with a number denoting the factory where the radios were made. Suggest an appropriate sampling method in this case, and discuss its advantages.

16–23. The makers of Taster's Choice instant coffee want to estimate the proportion of underfilled jars of a given size. The jars are in 14 warehouses around the country, and each warehouse contains crates of cases of jars of coffee. Suggest a sampling method, and discuss it.

16–24. Cadbury, Inc., is interested in estimating people's responses to a new chocolate. The company believes that people in different age groups differ in their preferences for chocolate. The company believes that in the region of interest, 25% of the population are children, 55% are young adults, and 20% are older people. A proportional allocation of a total random sample of size 1,000 is undertaken, and people's responses on a scale of 0 to 100 are solicited. The results are as follows. For the children, $\bar{x} = 90$ and $s = 5$; for the young adults, $\bar{x} = 82$ and $s = 11$; and for the older people, $\bar{x} = 88$ and $s = 6$. Give a 95% confidence interval for the population average rating for the new chocolate.

16–25. For problem 16–24, suppose that it costs twice as much money to sample a child as the younger and older adults, where costs are the same per sampled person. Use the information in problem 16–24 (the weights and standard deviations) to determine an optimal allocation of the total sample.

16–26. Refer to the situation in problem 16–24. Suppose that the following relative age frequencies in the population are known:

Age Group	Frequency
Under 10	0.10
10 to 15	0.10
16 to 18	0.05
19 to 22	0.05
23 to 25	0.15
26 to 30	0.15
31 to 35	0.10
36 to 40	0.10
41 to 45	0.05
46 to 50	0.05
51 to 55	0.05
56 and over	0.05

Define strata to be used in the survey.

16–27.   Name two sampling methods that are useful when there is information about a variable related to the variable of interest.

16–28.   Suppose that a study was undertaken using a simple random sample from a particular population. When the results of the study became available, it was apparent that the population could be viewed as consisting of several strata. What can be done now?

16–29.   For problem 16–28, suppose that the population is viewed as comprising 18 strata. Is using this number of strata advantageous? Are there any alternative solutions?

16–30.   Discuss and compare the three sampling methods: cluster sampling, stratified sampling, and systematic sampling.

16–31.   The following table reports return on capital for insurance companies. Consider the data a population of U.S. insurance companies, and select a random sample of firms to estimate mean return on capital. Do the sampling two ways: first, take a systematic sample considering the entire list as a uniform, ordered population; and second, use stratified random sampling, the strata being the types of insurance company. Compare your results.

Company Diversified	Return on Capital Latest 12 Mos. %	Company Life & Health	Return on Capital Latest 12 Mos. %	Company Property & Casualty	Return on Capital Latest 12 Mos. %
Marsh & McLennan Cos	25.4	Conseco	13.7	20th Century Inds	25.1
Loews	13.8	First Capital Holding	10.7	Geico	20.4
American Intl Group	14.6	Torchmark	18.4	Argonaut Group	17.1
General Re	17.4	Capital Holding	9.9	Hartford Steam Boiler	19.1
Safeco	11.6	American Family	8.5	Progressive	10.1
Leucadia National	27.0	Kentucky Central Life	6.3	WR Berkley	11.5
CNA Financial	8.6	Provident Life & Acc	13.1	Mercury General	28.3
Aon	12.8	NWNL	8.4	Selective Insurance	13.6
Kemper	1.4	UNUM	13.2	Hanover Insurance	6.3
Cincinnati Financial	10.7	Liberty Corp	10.1	St Paul Cos	16.9
Reliance Group	23.1	Jefferson-Pilot	9.5	Chubb	14.3
Alexander & Alexander	9.9	USLife	6.7	Ohio Casualty	9.3
Zenith National Ins	9.8	American Natl Ins	5.2	First American Finl	4.8
Old Republic Intl	13.4	Monarch Capital	0	Berkshire Hathaway	7.3
Transamerica	7.5	Washington National	1.2	ITT	10.2
Uslico	7.7	Broad	8.0	USF&G	6.6
Aetna Life & Cas	8.0	First Executive	0	Xerox	5.3
American General	8.2	ICH	0	Orion Capital	10.8
Lincoln National	9.2			Fremont General	12.6
Sears, Roebuck	7.2			Foremost Corp of Amer	0
Independent Insurance	7.8			Continental Corp	6.6
Cigna	7.0			Alleghany	9.0
Travelers	0				
American Bankers	8.3				
Unitrin	6.1				

CASE 18   The Boston Redevelopment Authority

The Boston Redevelopment Authority is mandated with the task of improving and developing urban areas in Boston. One of the Authority's main concerns is the development of the community of Roxbury. This community has undergone many changes in recent years, and much interest is given to its future development.

Currently, only 2% of the total land in this community is used in industry, and 9% is used commercially. As part of its efforts to develop the community, the Boston Redevelopment Authority is interested in determining the attitudes of the residents of Roxbury toward the development of more business and industry in their region. The authority therefore plans to sample residents of the community to determine their views and use the sample or samples to infer about the views of all residents of the community. Roxbury is divided into 11 planning subdistricts. The population density is believed to be uniform across all 11 subdistricts, and the population of each subdistrict is approximately proportional to the subdistrict's size. There is no known list of all the people in Roxbury. A map of the community is shown in Exhibit 1. Advise the Boston Redevelopment Authority on designing the survey.

EXHIBIT 1   Roxbury Planning Subdistrict

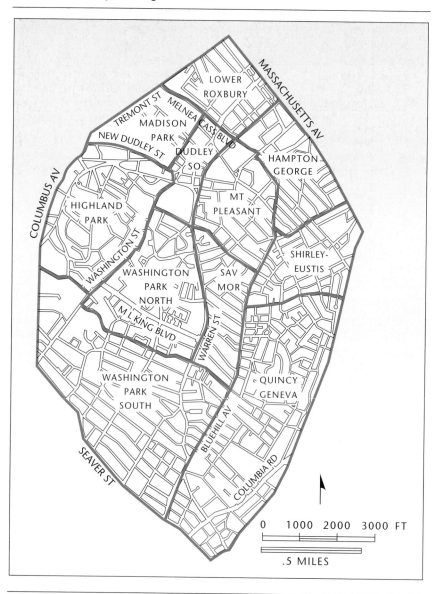

Appendixes

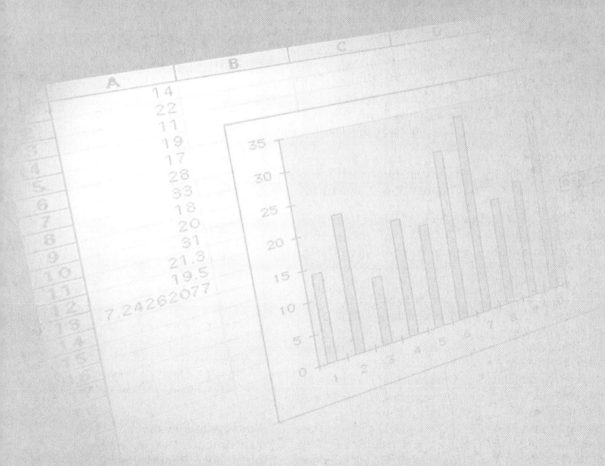

APPENDIX A   References

Books on Data Analysis (Chapter 1):

Chambers, J. M.; W. S. Cleveland; B. Kleiner; and P. A. Tukey. *Graphical Methods for Data Analysis*. Boston: Duxbury Press, 1983. An interesting approach to graphical techniques and EDA using computer-intensive methods. The book requires no mathematical training.

Tukey, J. W. *Exploratory Data Analysis*. Reading, Mass.: Addison-Wesley Publishing, 1977. This is the original EDA book. Some material in our Chapter 1 is based on this text.

Books Primarily about Probability and Random Variables (Chapters 2, 3, 4):

Chung, K. L. *Probability Theory with Stochastic Processes*. New York: Springer-Verlag, 1979. This is a lucidly written book. The approach to the theory of probability is similar to the one used in our text.

Feller, William. *An Introduction to Probability Theory and Its Applications*. Vol. 1, 3rd ed.; vol. 2, 2nd ed. New York: John Wiley & Sons, 1968, 1971. This is a classic textbook in probability theory. Volume 1 should be understandable to a reader of our text. Volume 2, which deals with continuous probability models, is more difficult and requires considerable mathematical ability.

Loève, Michel. *Probability Theory*. New York: Springer-Verlag, 1994. This is a mathematically demanding classic text in probability (an understanding of mathematical analysis is required).

Ross, Sheldon M. *A First Course in Probability*. 3rd ed. New York: Macmillan, 1988. An intuitive introduction to probability that requires a knowledge of calculus.

Ross, Sheldon M. *Introduction to Probability Models*. 4th ed. New York: Academic Press, 1989. A very intuitive introduction to probability theory that is consistent with the development in our text.

Statistical Theory and Sampling (Chapters 5, 6, 7, 8, and 16):

Cochran, William G. *Sampling Techniques*. 3rd ed. New York: John Wiley & Sons, 1977. This is a classic text on sampling methodology. Much of the material in our Chapter 16 draws on the results in this book.

Cox, D. R., and D. V. Hinkley. *Theoretical Statistics*. London: Chapman and Hall, 1974. A thorough discussion of the theory of statistics.

Fisher, Sir Ronald A. *The Design of Experiments*. 7th ed. Edinburgh: Oliver and Boyd, 1960. A classic treatise on statistical inference.

Fisher, Sir Ronald A. *Statistical Methods for Research Workers*.

Edinburgh: Oliver and Boyd, 1941.

Hogg, R. V., and A. T. Craig. *Introduction to Mathematical Statistics*. 4th ed. New York: Macmillan, 1978. A good introduction to mathematical statistics that requires an understanding of calculus.

Kendall, M. G., and A. Stuart. *The Advanced Theory of Statistics*. Vol. 1, 2nd ed.; vols. 2, 3. London: Charles W. Griffin, 1963, 1961, 1966.

Mood, A. M.; F. A. Graybill; and D. C. Boes. *Introduction to the Theory of Statistics*. 3rd ed. New York: McGraw-Hill, 1974.

Rao, C. R. *Linear Statistical Inference and Its Applications*. 2nd ed. New York: John Wiley & Sons, 1973. This is a classic book on statistical inference that provides in-depth coverage of topics ranging from probability to analysis of variance, regression analysis, and multivariate methods. This book contains theoretical results that are the basis of statistical inference. The book requires advanced mathematical ability.

Books Primarily about Experimental Design, Analysis of Variance, Regression Analysis, and Econometrics (Chapters 9, 10, and 11):

Chatterjee, S., and B. Price. *Regression Analysis by Example*.

2nd ed. New York: John Wiley & Sons, 1991.

Cochran, W. G., and G. M. Cox. *Experimental Designs*. 2nd ed. New York: John Wiley & Sons, 1957.

Draper, N. R., and H. Smith. *Applied Regression Analysis*. 3rd ed. New York: John Wiley & Sons, 1998. A thorough text on regression analysis that requires an understanding of matrix algebra.

Johnston, J. *Econometric Methods*. 3rd ed. New York: McGraw-Hill, 1984. A good, comprehensive introduction to econometric models and regression analysis at a somewhat higher level than that of our text.

Judge, G. R.; C. Hill; W. Griffiths; H. Lutkepohl; and T. Lee. *Introduction to the Theory and Practice of Econometrics*. 2nd ed. New York: John Wiley & Sons, 1985.

Montgomery, D. C., and E. A. Peck. *Introduction to Linear Regression Analysis*. 2nd ed. New York: John Wiley & Sons, 1992. A very readable book on regression analysis that is recommended for further reading after our Chapter 11.

Neter, J.; W. Wasserman; and M. H. Kutner. *Applied Linear Regression Models*. 3rd ed. New York: McGraw-Hill/Irwin, 1996. A good introduction to regression analysis.

Neter, J.; W. Wasserman; and M. H. Kutner. *Applied Linear*

Statistical Models. 4th ed. New York: McGraw-Hill/Irwin, 1996. A good introduction to regression and analysis of variance that requires no advanced mathematics.

Scheffé, H. *The Analysis of Variance*. New York: John Wiley & Sons, 1959. This is a classic text on analysis of variance that requires advanced mathematical ability.

Seber, G. A. F. *Linear Regression Analysis*. New York: John Wiley and Sons, 1977. An advanced book on regression analysis. Some of the results in this book are used in our Chapter 11.

Snedecor, George W., and William G. Cochran. *Statistical Methods*. 7th ed. Ames: Iowa State University Press, 1980. This well-known book is an excellent introduction to analysis of variance and experimental design, as well as regression analysis. The book is very readable and requires no advanced mathematics.

Weisberg, S. *Applied Linear Regression*. 2nd ed. New York: John Wiley & Sons, 1985. A good introduction to regression analysis.

Books on Forecasting (Chapter 12):

Abraham, B., and J. Ledolter. *Statistical Methods for Forecasting*. New York: John Wiley & Sons, 1983. This is an excellent book on forecasting methods.

Armstrong, S. *Long-Range Forecasting*. 2nd ed. New York: John Wiley & Sons, 1985.

Granger, C. W. J., and P. Newbold. *Forecasting Economic Time Series*. 2nd ed. New York: Academic Press, 1986. A good introduction to forecasting models.

Books on Quality Control (Chapter 13):

Duncan, A. J. *Quality Control and Industrial Statistics*. 5th ed. New York: McGraw-Hill/Irwin, 1986.

Gitlow, H.; A. Oppenheim; and R. Oppenheim. *Quality Management*. New York: McGraw-Hill/Irwin, 1994.

Ott, E. R., and E. G. Schilling. *Process Quality Control*. 2nd ed. New York: McGraw-Hill, 1990.

Ryan, T. P. *Statistical Methods for Quality Improvement*. New York: John Wiley & Sons, 1989. Much of the material in our Chapter 13 is inspired by the approach in this book.

Books on Nonparametric Methods (Chapter 14):

Conover, W. J. *Practical Nonparametric Statistics*. 2nd ed. New York: John Wiley & Sons, 1980. This is an excellent, readable textbook covering a wide range of nonparametric methods. Much of the material in our Chapter 14 is based on results in this book.

Hollander, M., and D. A. Wolfe. *Nonparametric Statistical Methods*. New York: John Wiley & Sons, 1973.

Siegel, S. *Nonparametric Statistics for the Behavioral Sciences*. 2nd ed. New York: McGraw-Hill, 1988.

Books on Subjective Probability, Bayesian Statistics, and Decision Analysis (Chapter 15 and Chapter 2):

Berger, James O. *Statistical Decision Theory and Bayesian Analysis.* 2nd ed. New York: Springer-Verlag, 1985. A comprehensive book on Bayesian methods at an advanced level.

de Finetti, Bruno. *Probability, Induction, and Statistics.* New York: John Wiley & Sons, 1972. This excellent book on subjective probability and the Bayesian philosophy is the source of the de Finetti game in our Chapter 15. The book is readable at about the level of our text.

de Finetti, Bruno. *Theory of Probability.* Vols. 1 and 2. New York: John Wiley & Sons, 1974, 1975. An excellent introduction to subjective probability and the Bayesian approach by one of its pioneers.

DeGroot, M. H. *Optimal Statistical Decisions.* New York: McGraw-Hill, 1970.

Good, I. J. *Good Thinking: The Foundations of Probability and Its Applications.* Minneapolis: University of Minnesota Press, 1983.

Jeffreys, Sir Harold. *Theory of Probability.* 3rd rev. ed. London: Oxford University Press, 1983. First published in 1939, this book truly came before its time. The book explains the Bayesian philosophy of science and its application in probability and statistics. It is readable and thought-provoking and is highly recommended for anyone with an interest in the ideas underlying Bayesian inference.

APPENDIX B   Answers to Most Odd-Numbered Problems

Chapter 1

1–1.
1. quantitative/ratio
2. qualitative/nominal
3. quantitative/ratio
4. qualitative/nominal
5. quantitative/ratio
6. quantitative/interval
7. quantitative/ratio
8. quantitative/ratio
9. quantitative/ratio
10. quantitative/ratio
11. quantitative/ordinal

1–3. Weakest to strongest: nominal, ordinal, interval, ratio.

1–5. Ordinal

1–7. Non-random sample; frame is random sample

1–11. Ordinal

1–13. LQ = 121
MQ = 128
UQ = 133.5
10th percentile = 114.8
15th percentile = 118.1
65th percentile = 131.1
IQR = 12.5

1–15. Median = 70
20th percentile = 45
30th percentile = 53.8
60th percentile = 76.8
90th percentile = 89.4

1–17. Median = 51
LQ = 31.5
UQ = 162.75
IQR = 131.25
45th percentile = 42.2

1–19. mean = 126.64
median = 128
modes = 128, 134, 136

1–21. mean = 66.955
median = 70
mode = 45

1–23. mean = 199.875
median = 51
mode = none

1–25.

Investments	Mean	Median
Lg & mid cap stock	20.06	20.5
Sm cap stock	12.46	12.2
Foreign stock	11.8	10.8
Emerging mkts	17.2	19.4
Intermediate bonds	3.33	2.9
Short-term bonds	2.5	2.55

1–27. mean = 592.93
median = 566
outlier: 940
s.d. = 117.03

1–29. Variance, standard deviation

1–31. range = 27
var = 57.74
s.d. = 7.5986

1–33. range = 60
var = 321.38
s.d. = 17.927

1–35. range = 1186
var = 110,287.45
s.d. = 332.096

1–37. Chebyshev holds; data not mound-shaped, empirical rule does not apply

1–39. Chebyshev holds; data not mound-shaped, empirical rule does not apply

1–45. mean = 49.4
median = 39

1–47. 5 | 5688
6 | 0123677789
7 | 00222333455667889
8 | 224

1–49. Stem-and-leaf is similar to a histogram but it retains the individual data points; box plot is useful in identifying outliers and the shape of the distribution of the data.

1–51. Data are concentrated about the median; 2 outliers

1–53. mean = 126.84
var = 137
s.d. = 11.7
mode = 127

1–55. Can use stem-and-leaf or box plots to identify outliers; outliers need to be evaluated instead of just eliminated.

1–57.

Mine A:	Mine B:
3	2457	2	3489
4	12355689	3	24578
5	123	4	034789
6	0	5	0129

–out values–
7 | 36
8 | 5

1–63. mean = 504.688
s.d. = 94.547

1–65. range = 346
90th percentile = 632.7
LQ = 419.25
MQ = 501.5
UQ = 585.75

1–67. 1 | 2456789
2 | 02355
3 | 24
4 | 01

1–69. 1 | 012
–out values–
1 | 9
2 | 11222334556677889
3 | 02457
–out values–
6 | 2

1–71. mean = 12.38
s.d. = 3.199

1–73. mean = 33.271
s.d. = 16.945
var = 287.15
LQ = 25.41
MQ = 26.71

UQ = 35
1–75. *a.* IQR = 3.5
　　b. right-skewed
　　c. 9.5
　　d. will not affect the
　　　plot
1–77. mean (billion of tons)
　　= 1.439
　　mean (per capital) =
　　9.98
　　dividing billions of
　　tons by rate per capita
　　for U.S. we get a
　　population estimate of
　　256 million, which is
　　close to the actual.
1–79. mean = 7781.5
　　var = 105.87
　　s.d. = 10.289
1–81. mean = 17.587
　　var = 0.2172
　　s.d. = 0.466
1–83. mean = 4.8394
　　med = 4.86
　　s.d. = 0.08
1–85. *a.* VARP = 3.5 +
　　　offset^2
　　b. VARP = 3.5

Chapter 2

2–1. Objective and
　　subjective
2–3. The sample space is
　　the set of all possible
　　outcomes of an
　　experiment.
2–5. $G \cup F$: the baby is
　　either a girl, or is over
　　5 pounds (of either
　　sex). $G \cap F$: the baby
　　is a girl over 5 pounds.
2–7. 0.417
2–9. $S \cup B$: purchase stock
　　or bonds, or both.
　　$S \cap B$: purchase stock
　　and bonds.
2–11. 3rd sample space is
　　correct
2–13. *a.* 1/52
　　b. 1/52

c. 1/51
2–15. 0.85 is a typical "very
　　likely" probability.
2–17. The team is very likely
　　to win.
2–19. *a.* Mutually exclusive
　　b. 0.035
　　c. 0.985
2–21. 0.49
2–23. 0.7909
2–25. 0.500
2–27. 0.34
2–29. 0.60
2–31. *a.* 0.1002
　　b. 0.2065
　　c. 0.59
　　d. 0.144
　　e. 0.451
　　f. 0.571
　　g. 0.168
　　h. 0.454
　　i. 0.569
2–33. *a.* 0.484
　　b. 0.455
　　c. 0.138
　　d. 0.199
　　e. 0.285
　　f. 0.634
　　g. 0.801
2–35. *a.* 5/25 = 0.20
　　b. 5/13 = 0.38
2–37. 0.8143
2–39. 0.72675
2–41. 0.99055
2–43. not independent
2–45. not independent
2–47. 0.3686
2–49. 0.0001, Yes
2–51. 0.0039, 0.684
2–53. 362,880
2–55. 120
2–57. 0.00275
2–59. 0.0000924
2–61. 0.86
2–63. 0.78
2–65. 0.9944
2–67. 0.2857
2–69. 0.8824
2–71. 0.0248
2–73. 0.6

2–75. 0.20
2–77. 0.60
2–79. not independent
2–81. 0.388
2–83. 0.59049, 0.40951
2–85. 0.132, not random
2–87. 0.6667
2–89. *a.* 0.255
　　b. 0.8235
2–91. 0.5987
2–93. 0.5825
2–95. Practically speaking,
　　the probabilities
　　involved vary only
　　slightly, and their role
　　in the outcome of the
　　game should be more
　　or less unnoticeable
　　when averaged over a
　　span of games.
2–97. 1/2
2–99. 0.767
2–101. *a.* 0.202
　　b. 1.00
　　c. 0.0

Chapter 3

3–1. *a.* $\Sigma\, P(x) = 1.0$
　　b.

x	F(x)
0	0.3
1	0.5
2	0.7
3	0.8
4	0.9
5	1.0

　　c. 0.3
3–3. *a.* $\Sigma\, P(x) = 1.0$
　　b.

x	F(x)
0	0.10
10	0.30
20	0.65
30	0.85
40	0.95
50	1.00

　　c. 0.35
3–5.

X	P(x)	F(x)
2	1/36	1/36
3	2/36	3/36
4	3/36	6/36
5	4/36	10/36
6	5/36	15/36

X	P(x)	F(x)
7	6/36	21/36
8	5/36	26/36
9	4/36	30/36
10	3/36	33/36
11	2/36	35/36
12	1/36	36/36

Most likely sum is 7

3–7. a. 0.55

b.

x	F(x)
2	0.20
3	0.40
4	0.70
5	0.80
6	0.90
7	0.95
8	1.00

c. 0.9

d. 0.5

3–9. a. $\Sigma\, P(x) = 1.0$

b. 0.50

c.

x	F(x)
9	0.05
10	0.20
11	0.50
12	0.70
13	0.85
14	0.95
15	1.00

3–11. $E(X) = 1.8$
$E(X^2) = 6$
$V(X) = 2.76$
$SD(X) = 1.661$

3–13. $E(X) = 21.5$
$E(X^2) = 625$
$V(X) = 162.75$

3–15. E(sum of 2 dice) = 7

x	P(x)	xP(x)
2	1/36	2/36
3	2/36	6/36
4	3/36	12/36
5	4/36	20/36
6	5/36	30/36
7	6/36	42/36
8	5/36	40/36
9	4/36	36/36
10	3/36	30/36
11	2/36	22/36
12	1/36	12/36
		252/36 = 7

3–17. $E(X) = 4.05$
$E(X^2) = 19.15$
$V(X) = 2.7475$
$SD(X) = 1.6576$

3–19. Three standard deviations
$8/9 = 1 - 1/3^2$ $k = 3$

3–21. a. 2000
b. Yes $P(X > 0) = .6$
c. $E(X) = +800$
d. good measure of risk is the standard deviation
$E(X^2) = 2,800,000$
$V(X) = 2,160,000$
$SD(X) = 1,469.69$

3–23. $E(aX + b) = 25,000 + 5000\, E(X) = 45,250$

3–25. Penalty $= X^2$
$E(X^2) = 12.39$

3–27. Variance is a measure of the spread or uncertainty of the random variable.

3–29. $V(\text{Cost}) = V(aX + b) = a^2 V(X) = 68,687,500$
$SD(\text{Cost}) = 8,287.79$

3–31. 3.11

3–33. X is binomial if sales calls are independent.

3–35. X is not binomial because members of the same family are related and not independent of each other.

3–37. a. slightly skewed; becomes more symmetric as n increases.
b. Symmetric if $p = 0.5$. Left-skewed if $p > 0.5$; right-skewed if $p < 0.5$

3–39. a. 0.8889
b. 11
c. 0.55

3–41. a. 0.9981
b. 0.889, 0.935
c. 4, 5

d. increase the reliability of each engine

3–43. a. mean = 6.25, var = 6.77083
b. 61.80%
c. 11
d. $p = 0.7272$

3–45. a. mean = 2.857, var = 5.306
b. 82.15%
c. 7
d. $p = 0.5269$

3–47. a. 0.5000
b. Add 4 more women or remove 3 men.

3–49. a. 0.8430
b. 6
c. $\mu = 1.972$

3–51. a. MTBF = 5.74 days
b. $P(x \le 1) = 0.1599$
c. 0.1599
d. 0.4185

3–55. As s or p increases skewness decreases.

3–57. a. 0.7807
b. 0.2858

3–59. a. 0.7627
b. 0.2373

3–61. a. 0.2119
b. 0.2716
c. 0.1762

3–63. a. $P(x = 5) = 0.2061$
b. $P(x = 4) = 0.2252$
c. $P(x = 3) = 0.2501$
d. $P(x = 2) = 0.2903$
e. $P(x = 1) = 0.3006$

3–65. $P(x \ge 2) = 0.5134$

3–67. a. $\Sigma\, P(x) = 1.0$

b.

x	F(x)
0	.05
1	.10
2	.20
3	.35
4	.55
5	.70
6	.85
7	.95
8	1.00

c. $P(3 \le x \le 7) = 0.65$

d. $P(X \le 5) = 0.70$

e. $E(X) = 4.25$

f. $E(X^2) = 22.25$
   $V(X) = 4.1875$
   $SD(X) = 2.0463$

g. [0.1574, 8.3426] vs
   $P(1 \le X \le 8) = 0.95$

3–69. a. $\Sigma\, P(x) = 1.0$

b.

x	F(x)
0	.10
1	.30
2	.60
3	.75
4	.90
5	.95
6	1.00

c. 0.35, 0.40, 0.10

d. 0.20

e. 0.0225

f. $E(X) = 2.4$
   $SD(X) = 1.562$

3–71. a. $E(X) = 17.56875$
   profit = \$31.875

b. $SD(X) = 0.3149$
   a measure of risk

c. the assumption of
   stationary and
   independence of the
   stock prices

3–73. a. Each of the 10
   choices is made
   independently of
   the others

b. $P(X \ge 2) = 0.456$

3–75. a. The distribution is
   binomial if the cars
   are independent of
   each other.

b.

x	P(x)
0	.5987
1	.3152
2	.0746
3	.0105
4	.0009
5	.0001

c. 0.0861

d. ½ a car

3–77. $N/n < 10$

3–79. b. 1.00

c. 0.75

3–81. 0.9945

3–83. a. 11.229

b. 20/20 : +1.02
   G.B.: −0.33
   M.C.: −1.48

3–85. 0.133; 10

3–87. 0.3935

3–89. $P(X \ge 5) = 0.0489$

3–91. 0.9999; 0.0064; 0.6242

Chapter 4

4–1. 0.6826; 0.95; 0.9802

4–3. 0.1805

4–5. 0.0215

4–7. 0.9901

4–9. a very small number,
   close to 0

4–11. 0.9544

4–13. Not likely, $P = 0.00003$

4–15. $z = 0.48$

4–17. $z = 1.175$

4–19. $z = \pm 1.96$

4–21. 0.0164

4–23. 0.927

4–25. 0.003

4–27. 0.8609; 0.2107; 0.6306

4–29. 0.0931

4–31. a. 0.1357

b. 0.0951

4–33. 0.8644

4–35. 0.3759; 0.0135; 0.8766

4–37. 0.8167

4–39. 15.67

4–41. 76.35; 99.65

4–43. −46.15

4–45. 832.6; 435.5

4–47. [18,130.2; 35,887.8]

4–49. 119.58

4–51. 1606.896

4–53. 0.99998

4–55. 0.3804

4–57. 0.2743

4–59. 0.7642

4–61. 0.1587; 0.9772

4–63. 791,580

4–65. [7.02, 8.98]

4–67. 1555.52, [1372.64,
   3223.36]

4–69. 8.856 KW

4–71. more than 0.26%

4–73. $u = 64.31$, s.d. $= 5.49$

4–75. 6015.6

4–77. 0.0228

4–79. a. $N(248, 5.3852^2)$

b. 0.6448

c. 0.0687

4–81. a. 0.0873

b. 0.4148

c. 16,764.55

d. 0.0051

Chapter 5

5–1. Parameters are
   numerical measures of
   populations. Sample
   statistics are numerical
   measures of samples.
   An estimator is a
   sample statistic used
   for estimating a
   population parameter.

5–3. 5/12 = 0.41667

5–5. a. 1.690385

b. 0.2475

5–11. The probability
   distribution of a
   sample statistic; useful
   in determining the
   accuracy of estimation
   results.

5–13. $E(\overline{X}) = 125$
   $SE(\overline{X}) = 8.944$

5–15. When the population
   distribution is
   unknown.

5–17. Binomial. Cannot use
   normal approximation
   since $np = 1.2$

5–19. 0.075

5–21. 0.8426

5–23. 0.2308

5–25. 0.0907

5–27. 0.0497

5–29. 0.0190

5–31. A consistent estimator
   means as $n \to \infty$ the
   probability of getting
   close to the parameter
   increases. A generous
   budget affords a large

sample size, making this probability high.

5-33. Advantage: uses all information in the data. Disadvantage: may be too sensitive to the influence of outliers.

5-37. *a.* mean = 43.667,
SSD = 358,
MSD = 44.75
b. use Means: 40.75, 49.667, 40.5
c. SSD = 195.917,
MSD = 32.6528
d. SSD = 719,
MSD = 89.875

5-39. Yes, we can solve the equation for the one unknown amount.

5-41. $E(\overline{X}) = 1065$
$V(\overline{X}) = 2500$

5-43. $E(\overline{X}) = 53$
$SE(\overline{X}) = 0.5$

5-45. $E(\hat{p}) = 0.2$
$SE(\hat{p}) = 0.04216$

5-47. 0.6388

5-49. 0.9544

5-51. *a.* 8128.08
b. 0.012

5-55. The sample median is unbiased. The sample mean is more efficient and is sufficient. Must assume normality for using the sample median to estimate μ; it is more resistant to outliers.

5-57. 0.1727

5-59. *a.* 0.9044
b. 1.00 approximately

5-61. 0.9503

5-63. No minimum ($n = 1$ is enough for normality).

5-65. This estimator is consistent, and is more efficient than $\overline{X}$, because $\sigma^2/n^2 < \sigma^2/n$.

5-67. Relative minimum sample sizes:

$n_a < n_b < n_d < n_e < n_c$

5-69. Use a computer simulation, draw repeated samples, determine the empirical distribution.

5-71. $P(Z < -5) =$ 0.0000003   Not probable

5-73. 0.923

Chapter 6

6-5. [86,978.12, 92,368.12]

6-7. [31.098, 32.902] m.p.g.

6-9. [9.045, 9.555] percent

6-11. Yes; all values in the 99% confidence interval are above 245.

6-13. 95% C.I.: [136.99, 156.51]
90% C.I.: [138.56, 154.94]
99% C.I.: [133.93, 159.57]

6-15. [17.4, 22.6] percent

6-17. 8.393

6-19. 95% C.I.: [15,684.37, 17,375.63]
99% C.I.: [15,418.6, 17,641.4]

6-21. [27.93, 33.19] thousand miles

6-23. [72.599, 89.881]

6-25. [29.87, 59.33] dollars

6-27. [2.344, 2.856] days

6-29. [93.75, 108.71] sales

6-31. [15.86, 17.14] dollars

6-33. [5.44, 7.96] years

6-35. [55.85, 67.48] containers

6-37. [9.764, 10.380]

6-39. $\overline{X} = 15.4375$; [15.032, 15.843]

6-41. [0.4658, 0.7695]

6-43. [0.5694, 0.6306]

6-45. [0.3213, 0.3787]

6-47. [0.0375, 0.2702]

6-49. [0.5357, 0.6228]

6-51. [0.7138, 0.7486]

6-53. [61.11, 197.04]

6-55. [19.25, 74.92]

6-57. [1268.03, 1676.68]

6-59. 271

6-61. 39

6-63. 131

6-65. 865

6-67. [21.507, 25.493]

6-69. [0.6211, 0.7989]

6-71. 11.478

6-73. [1.0841, 1.3159]

6-75. [0.6974, 0.7746]

6-77. [$23,246.92, $26,753.08]

6-79. [1.383, 3.817]

6-81. [0.508, 0.692]

6-83. [0.902, 0.918]

6-85. [0.0632, 0.1088]

6-87. 75% C.I.: [7.76, 8.24]

6-89. 95% CI: [0.9859, 0.9981]
99% CI: [0.9837, 1.0003]

Chapter 7

7-1. H_0: $p = 0.8$
H_1: $p \neq 0.8$

7-3. H_0: $\mu \leq 12$
H_1: $\mu > 12$

7-5. H_0: $\mu \leq \$1.78$
H_1: $\mu > \$1.78$

7-11. *a.* left-tailed H_1:
$\mu < 10$
b. right-tailed H_1:
$p > 0.5$
c. left-tailed H_1:
$\mu < 100$
d. right-tailed H_1:
$\mu > 20$
e. two-tailed H_1:
$p \neq 0.22$
f. right-tailed H_1:
$\mu > 50$
g. two-tailed H_1:
$\sigma^2 \neq 140$

7-13. *a.* to the left tail
b. to the right tail
c. either to the left or to the right tail

7-15. *a.* *p*-value will decrease
b. *p*-value increases

c. p-value decreases

7-17. $z = 1.936$,
Do not reject H_0
(p-value = 0.0528)

7-19. $z = 4.68$, Reject H_0

7-21. $t_{(15)} = 1.55$, Do not
reject H_0 (p-value >
0.10)

7-23. $z = -3.269$, Reject H_0
(p-value = 0.0011)

7-25. $t_{(24)} = 2.25$, Do not
reject H_0 at $\alpha = 0.01$,
reject at $\alpha = 0.05$

7-27. $z = 0.472$, Do not
reject H_0 (p-value =
0.637)

7-29. $z = 1.539$, Do not
reject H_0 (p-value =
0.1238)

7-31. $z = 1.622$, Do not reject
H_0 (p-value = 0.1048)

7-33. $t_{(23)} = 2.939$, Reject H_0

7-35. $z = 16.0$, Reject H_0

7-37. $z = -20$, Reject H_0

7-39. $z = -1.304$, Do not
reject H_0 (p-value =
0.0962)

7-41. $z = 9.643$, Reject H_0

7-43. a. cars: $z = -1.25$, Do
not reject H_0
b. trucks: $z = -3.115$,
Reject H_0

7-45. a. start: $z = 2.789$,
Reject H_0 (p-value
= 0.0026)
b. braking: $z = 1.00$,
Do not reject H_0
(p-value = 0.1587)

7-47. power = 0.9092

7-51. $z = -2.711$, Reject H_0

7-53. two-tailed test; $z =$
2.07, Reject H_0
(p-value = 0.384)

7-55. $z = 4.249$, Reject H_0
(p-value = 0.00001)

7-65. $z = -4.86$, Reject H_0,
power = 0.5214

7-67. $\chi^2_{(24)} = 26.923$, Do not
reject H_0 (p-value >
0.10)

7-69. $t_{(20)} = 1.06$, Do not
reject H_0 at $\alpha = 0.10$
(p-value = 0.15)

7-71. $t_{(13)} = -3.74$, Reject H_0
(p-value < 0.005)

7-73. $z = 2.53$, Reject H_0
(p-value = 0.0057)

7-75. Do not reject H_0 at
0.05 level of
significance

7-77. Do not reject H_0 at
0.05 level of
significance

7-79. Do not reject H_0 at
0.05 level of
significance

7-81. Do not reject H_0 at 0.05
level of significance

7-83. b. power = 0.4968
c. No.

7-85. b. power = 0.6779
c. Yes.

Chapter 8

8-1. $t_{(24)} = 3.11$ Reject H_0,
p-value < 0.01

8-3. $t_{(11)} = 2.034$ Cannot
reject H_0.

8-5. $t_{(14)} = 1.469$ Cannot
reject H_0.

8-7. Power = $P(Z > 1.55)$
= 0.0606

8-9. $z = -3.13$ Reject H_0

8-11. $z = 3.3$ Reject H_0

8-13. $z = 4.24$ Reject H_0

8-15. a. One-tailed: H_0:
$\mu_1 - \mu_2 \leq 0$
b. $z = 1.53$
c. At $\alpha = .05$ Do not
reject H_0.
d. 0.063
e. $t_{(19)} = 0.846$ Do
not reject H_0.

8-17. [2.416, 2.664] percent

8-19. $t_{(26)} = 1.132$ Do not
reject H_0. p-value >
0.10

8-21. $z = -8.2496$ Strongly
reject H_0. p-value very
small

8-23. $t_{(13)} = 1.164$ Strongly
reject H_0. p-value >
0.10

8-25. $z = 2.785$ Reject H_0.
p-value = 0.0026

8-27. $t_{(28)} = 5.136$ Strongly
reject H_0.

8-29. $z = 2.835$ Reject H_0.
p-value = 0.0023

8-31. $z = -0.228$ Do not
reject H_0.

8-33. [0.0419, 0.0781]

8-35. $z = 1.601$ Do not
reject H_0 at $\alpha = .05$.
p-value = 0.0547

8-37. $z = 5.664$ Reject H_0.
p-value is very small

8-39. $z = 5.33$ Strongly
reject H_0. p-value is
very small

8-41. $F_{(17,11)} = 2.16$ Do not
reject H_0 at $\alpha = 0.10$.

8-43. $F_{(27,20)} = 1.838$ At
$\alpha = 0.10$, cannot reject
H_0 [0.652, 4.837].

8-45. $F_{(24,24)} = 1.538$ Do
not reject H_0.

8-47. Independent random
sampling from the
populations and
normal population
distributions

8-49. [−3.4777, 0.6803]

8-51. [−0.465, 9.737]

8-53. [−812.9, 1,372.9]

8-55. $z = 1.447$ Do not
reject H_0. p-value =
0.1478

8-57. $t_{(22)} = 2.719$ Reject H_0.
(0.01 < p-value < 0.02)

8-59. $z = 23.479$ Reject
H_0. p-value is very
small.

8-61. $t_{(26)} = 2.479$ Reject
H_0. p-value = 0.02

8-63. $t_{(29)} = 1.08$ Do not
reject H_0.

8-65. Since $s_1^2 < s_2^2$, do not
reject H_0.

8–67. $t_{(15)} = -0.9751$ Do not reject H_0.

8–69. [0.0366, 0.1474]

8–71. $t_{(46)} = 0.866$ Do not reject H_0.

8–73. $z = 7.071$ Reject H_0. p-value is very small

8–75. Do not reject H_0.

8–77. Reject H_0.

Chapter 9

9–1. H_0: All 4 means are equal
H_1: All 4 are different; or 2 equal, 2 different; or 3 equal, 1 different; or 2 equal, other 2 equal but different from first 2.

9–3. Series of paired t tests are *dependent* on each other. No control over the probability of a type I error.

9–5. $F_{(3,176)} = 12.53$ Reject H_0.

9–7. The sum of all the deviations from a mean is equal to 0.

9–11. Both MSTR and MSE are *sample statistics* given to natural variation about their own means.

9–19.

Source	df	SS
Treatment	3	0.1152
Error	28	0.7315
Total	31	0.8467

MS	F
0.0384	1.47
0.0261	

Critical point $F_{(3,28)}$ for $\alpha = 0.01$ is 4.568; cannot reject H_0.

9–21.

Source	df	SS
Treatment	2	91.043
Error	38	140.529
Total	40	231.571

MS	F
45.521	12.31
3.698	

Critical point $F_{(2,38)}$ for $\alpha = 0.01$ is 3.24 Reject H_0.

9–23. $F_{(7,792)} = 108.53$ Reject H_0.

9–25. $T = 4.738$. The mean for squares is significantly greater than those for circles and for triangles; circles and triangles show no significant difference.

9–27. $T = 0.22$ No differences are significant, as expected.

9–29. $T = 3.85$ Mean for Zenith is significantly greater than all others; Sylvania, Philco, Sears, and RCA are next, not significantly different from one another; then Panasonic, different from the preceding; and then GE and Magnavox, not significantly different from each other.

9–31. No; the 3 prototypes were not randomly chosen from a population.

9–33. Fly all 3 planes on the same route every time.

9–35. Otherwise not a random sample from a population of treatments, and inference is thus not valid for the entire "population."

9–37. If the locations and the artists are chosen randomly, we have a random effects model.

9–41. Nearest is $F_{(4,150)} = 3.44$ Reject H_0 of no interaction (p-value < 0.01).

9–43.

Source	SS	df
Network	145	2
Newstime	160	2
Interaction	240	4
Error	6200	441
Total	6745	449

MS	F
72.5	5.16
80	5.69
60	4.27
14.06	

All are significant at $\alpha = 0.01$. There are interactions. There are Network main effects averaged over Newstime levels. There are Newstime main effects averaged over Network levels.

9–45. *a.* Explained is treatment = Factor A + Factor B + (AB)

b. $a = 3$

c. $b = 2$

d. $N = 150$

e. $n = 25$ There are no exercise-price main effects.

g. There are time-of-expiration main effects at 0.05 but not at 0.01.

h. There are no interactions.

i. Some evidence for time-of-expiration main effects; no evidence for exercise-price main effects or interaction effects.

j. For time-of-expiration main effects, 0.01 <

p-value < 0.05. For
the other two tests,
the p-values are
very high.
k. Could use a t test
for time-of-
expiration effects:
$t^2_{(144)} = F_{(1,144)}$

9–47. Advantages: reduced
experimental errors
and great economy of
sample size.
Disadvantages:
restrictive, because it
requires that number
of treatments =
number of rows =
number of columns.

9–49. Could use a
randomized blocking
design.

9–51. Yes; have people of the
same occupation/
age/demographics use
sweaters of the 3 kinds
under study. Each
group of 3 people is a
block.

9–53. Group the executives
into blocks according
to some choice of
common
characteristics such as
age, sex, or years
employed at current
firm; these blocks
would then form a
third variable beyond
Location and Type to
use in a 3-way
ANOVA.

9–55. $F_{(2,198)} = 25.84$ Reject
H_0 (p-value very
small).

9–57. $F_{(7,152)} = 14.67$ Reject
H_0 (p-value very
small).

9–59.

Source	SS	df
Software	77,645	2
Computer	54,521	3
Interaction	88,699	6
Error	434,557	708
Total	655,422	719

MS	F
38,822.5	63.25
18,173.667	29.60
14,783.167	24.09
613.78	

Both main effects and
the interactions are
highly significant.

9–61.

Source	SS	df
Pet	22,245	3
Location	34,551	3
Interaction	31,778	9
Error	554,398	144
Total	642,972	159

MS	F
7,415	1.93
11,517	2.99
3,530.89	0.92
3,849.99	

No interactions; no pet
main effects. There are
location main effects at
$\alpha = 0.05$

9–63. b. $F_{(2,58)} = 11.47$;
Reject H_0.

9–65. $F_{(2,98)} = 0.14958$;
Do not reject H_0.

9–67. Rents are equal on
average; no evidence
of differences among
the four cities.

Chapter 10

10–1. A set of mathematical
formulas and
assumptions that
describe some real-
world situation.

10–3. 1. a straight-line
relationship
between X and Y
2. the values of X are
fixed

3. the regression errors,
ϵ, are identically
normally distributed
random variables,
uncorrelated with
each other through
time.

10–5. It is the population
regression line.

10–7. 1. It captures the
randomness in the
process.
2. It makes the result
(Y) a random
variable.
3. It captures the
effects on Y of other
unknown
components not
accounted by the
regression model.

10–9. The line is the best
unbiased linear
estimator of the true
regression line. Least-
squares line is obtained
by minimizing the sum
of the squared
deviations of the data
points about the line.

10–11. $b_0 = 1113.35$
$b_1 = -14.8045$ The
fourth data set may be
an outlier. Sample is
too small.

10–13. $b_0 = -3.057$
$b_1 = 0.187$

10–15. $b_0 = 126.82$
$b_1 = -0.0284$ No.
There appears to be a
strong negative
relationship between
the two variables, but
causation cannot be
determined.

10–17. 1. off-line debit vs
on-line debit:
$b_0 = -255.94$
$b_1 = 1.5518$

2. no indication of causation; third variable: growth in the economy as measured by changes in per capita income.

10–19. [1.1158, 1.3949]

10–21. $s(b_0) = 94.54$
$s(b_1) = 12.76$

10–23. $s(b_0) = 0.971$
$s(b_1) = 0.016$ Estimate of error variance is MSE = 0.991

10–25. s^2 gives information about the variation of the data points about the computed regression line.

10–27. $r = -0.502$

10–29. $t_{(5)} = 0.601$ Do not reject H_0.

10–31. $t_{(8)} = 5.11$ Reject H_0.

10–35. $z = 1.297$ Do not reject H_0.

10–37. $t_{(16)} = 1.0727$ Do not reject H_0.

10–39. $t_{(11)} = 11.69$ Strongly reject H_0.

10–41. $t_{(58)} = 5.90$ Reject H_0.

10–43. $t_{(211)} = 0.0565$ Do not reject H_0.

10–45. No surprise since there is no linear relationship between the two variables.

10–47. 25.16% of the variation in Y is explained by the regression relationship.

10–49. $r^2 = 0.067$

10–51. $r^2 = 0.873$

10–53. $r^2 = 0.835$

10–57. $F_{(1,11)} = 129.525$
$t_{(11)} = 11.381$ $t^2 = F$
$(11.381)^2 = 129.525$

10–59. $F_{(1,17)} = 85.90$ Very strongly reject H_0.

10–61. $F_{(1,20)} = 0.3845$
Do not reject H_0.

10–63. *a.* Heteroscedasticity
b. No apparent inadequacy
c. Data display curvature, not a linear relationship.

10–65. *a.* no serious inadequacy.
b. Yes. A deviation from the normal-distribution assumption is apparent.

10–69. 6551.35
P.I.: [5854.4, 7248.3]

10–71. [5605.75, 7496.95]

10–73. C.I. [869.49, 1209.17]

10–75. P.I.: [2.243, 26.975]

10–77. *a.* $Y = 2.779337X - 0.284157$. When $X = 10$, $Y = 27.5092$
b. $Y = 2.741537X$. When $X = 10$, $Y = 27.41537$
c. $Y = 2.825566X - 1.12783$. When $X = 10$, $Y = 27.12783$
d. $Y = 2X + 4.236$. When $X = 10$, $Y = 24.236$

10–79. $b_0 = 26.657$
$b_1 = -0.2843$
$r^2 = 0.777$
$F = 20.91$ Strongly reject H_0.

10–81. Profits = 11.1 + 0.206 × Time
$r^2 = 0.68$ $F = 12.77$
A linear relationship does exist.

10–83. Takeover value = 3.377 + 1.359 Price
$r^2 = 0.923$ $F = 154.85$
Strong relationship exists.

Chapter 11

11–5. 8 equations

11–7. $b_0 = -1.134$

$b_1 = 0.048$
$b_2 = 10.897$

11–11. $n - 13$

11–21. $R^2 = 0.9174$, a good regression, $\bar{R}^2 = 0.8983$

11–23. $\bar{R}^2 = 0.8907$ Do not include the new variable.

11–25. *a.* Assume $n = 50$;
Regression is:
Return = 0.484 − 0.030 (Siz rnk) − 0.017 (Prc rnk)
b. $R^2 = 0.130$ 13% of the variation is due to the two independent variables.
c. Adjusted R^2 is quite low; try regressing on size alone.

11–27. $F_{(6,243)} = 128.19$
Reject H_0. $R^2 = 0.7599$
$\bar{R}^2 = 0.754$

11–31. $\beta_2 = [3.052, 8.148]$
$\beta_3 = [-3.135, 23.835]$
$\beta_4 = [-1.842, 8.742]$
$\beta_5 = [-4.995, -3.505]$

11–33. Yes

11–35. Lend seems insignificant because of collinearity with M_1 or price.

11–37. Autocorrelation of the regression errors.

11–39. $b_0 = 31.569$
$b_1 = 0.1759$
$b_2 = 0.2963$
$R^2 = 0.435$
$F_{(2,14)} = 5.39$
There is a good regression present that explains 43.5% of the variation in Total Return to Investors.

11–41. *a.* Residuals appear to be normally distributed.

b. Residuals are not normally distributed.

11-47. Creates a bias. There is no reason to force the regression surface to go through the origin.

11-51. 363.78

11-53. 0.341. 0.085

11-55. The estimators are the same although their standard errors are different.

11-59. Two-way ANOVA

11-61. Early investment is not statistically significant (or may be collinear with another variable). Rerun the regression without it. The dummy variables are both significant. Investment is significant.

11-63. The STEPWISE routine chooses Price and M_1 *Price as the best set of explanatory variables. Exports $= -1.39 + 0.0229$Price $+ 0.00248 M_1$ *Price. t statistics: $-2.36, 4.57, 9.08$, respectively. $R^2 = 0.822$

11-67. Quadratic regression (should get a negative estimated x^2 coefficient)

11-69. Linearizing a model; finding a more parsimonious model than is possible without a transformation; stabilizing the variance.

11-71. The transformation log Y

11-73. A logarithmic model

11-77. No

11-79. Taking reciprocals of both sides of the equation.

11-81. No. They minimize the sum of the squared deviations relevant to the estimated, transformed model.

11-83.

Earn		Prod	Prom
Prod	.867		
Prom	.882	.638	
Book	.547	.402	.319

Multicollinearity does not seem to be serious.

11-85. Sample correlation is 0.740

11-89. Not true. Predictions may be good when carried out within the same region of the multicollinearity as used in the estimation procedure.

11-91. X_2 and X_3 are probably collinear.

11-93. Drop some of the other variables one at a time and see what happens to the suspected sign of the estimate.

11-97. 1. The test checks only for first-order autocorrelation.
2. The test may not be conclusive.
3. The usual limitations of a statistical test owing to the two possible types of errors.

11-99. DW $= 2.13$ At the 0.10 level, no evidence of a first-order autocorrelation.

11-103. $F_{(r, n-(k+1))} = 0.0275$ Cannot reject H_0.

11-105. The STEPWISE procedure selects all three variables. $R^2 = 0.9667$

11-107. Because a variable may lose explanatory power and become insignificant once other variables are added to the model.

11-109. No. There may be several different "best" models.

11-111. All variables except power distance and collectivism in section B are significant.

11-113. After removing the 1991 data as an outlier, a plot of the six remaining near collisions numbers versus flights looks like a cubic curve. $NC = 3.32F^3 - 54.12F^2 + 291.73F - 518.52$

Chapter 12

12-3. $y(1998) = 198.182$ $y(1999) = 210.748$

12-5. No, because of the seasonality.

12-11. Forecast using trend and seasonal index $= 403.82$

12-13. Forecast using trend and seasonal index $= 1,582.61$

12-15. Forecast using trend and seasonal index $= 19.24$

12-17. The $w = 0.8$ forecasts follow the raw data much more closely.

12-19. Forecast $= 17.14$

12-23.

Year	Old CPI	New CPI
1950	72.1	24.9
1951	77.8	26.9
1952	79.5	27.5
1953	80.1	27.7

12-25. A simple price index reflects changes in a single price variable of time, relative to a single base time.

12-27. *a.* 1988

b. Divide each index number by 163/100

c. It fell, from 145% of the 1988 output down to 133% of that output.

12-29. Sales = 4.23987 − 0.03870 Month. Forecast for July 1997 (month #19) = 3.5046

12-33. Exponential Model: $Y_t = 4.475(1.14)^t$. Forecast = 62.343

12-35. Change in GDBP = 6.357 + 0.2 Year. Forecast for 1998 is 7.96

Chapter 13

13-3. 1. Natural, random variation
2. variation due to assignable causes

13-9. *a.* 77.62%
b. Omissions, Quantity Entry, Part Numbers (90.70%)

13-15. Random sampling, so that the observations are independent.

13-17. Process is in control.

13-21. Process is in control.

13-23. Process is in control.

13-25. Process is out of control (9th sample).

13-27. All points are well within the p chart limits; process is in control.

13-29. All points are well within the p chart limits; process is in control.

13-31. The tenth sample barely exceeds the UCL = 8.953; otherwise in control.

13-33. All points within c chart limits; process is in control.

13-37. The 20th observation far exceeds the UCL = 8.92/100; the last nine observations are all on one side of the center line $\bar{P}$ = 3.45/100

13-39. Last group's mean is below the LCL = 2.136

Chapter 14

14-1. $\Sigma + = 10$ Cannot reject H_0 (p-value = 0.18).

14-3. $z = 1.46$ Cannot reject H_0.

14-5. $T = 9$ Cannot reject H_0 (p-value = .593).

14-7. $z = 2.145$ Reject H_0.

14-9. $z = -3.756$ Reject H_0.

14-11. $z = -3.756$ Reject H_0.

14-13. $U = 3.5$ Reject H_0.

14-17. $U = 12$ Reject H_0.

14-21. $T = 10$ Do not reject H_0.

14-23. Sign Test

14-25. Wilcoxon Signed-Rank Test

14-27. $T = 27 > 17$ Do not reject H_0.

14-29. p-value < 0.001 Reject H_0.

14-31. $H = 12.5$ Reject H_0.

14-33. $H = 29.61$ Reject H_0. $C_{KW} = 11.68$

14-35. $H = 13.01$ Reject H_0.

14-39. The three managers are not equally effective.

14-41. No, the 4 baking processes are not equally good.

14-43. $\chi^2 = 119.97$ Reject H_0.

14-45. $\chi^2 = 0.586$ Do not reject H_0.

14-47. $\chi^2 = 12.193$ Reject H_0.

14-49. $\chi^2 = 6.94$ Do not reject H_0 at $\alpha = 0.05$

14-51. $\chi^2 = 50.991$ Reject H_0.

14-53. $\chi^2 = 109.56$ Reject H_0.

14-55. $\chi^2 = 16.15$ Reject H_0.

14-57. $\chi^2 = 24.36$ Reject H_0.

14-61. $\Sigma - = 32.5$
$\Sigma + = 72.5$
2-tailed p-value > 0.10

Chapter 15

15-1. 0.02531, 0.46544, 0.27247, 0.17691, 0.05262, 0.00697, 0.00028. Credible set is [0.2, 0.4].

15-3. 0.0126, 0.5829, 0.3658, 0.0384, 0.0003, 0.0000

15-5. 0.1129, 0.2407, 0.2275, 0.2909, 0.0751, 0.0529

15-7. 0.0633, 0.2216, 0.2928, 0.2364, 0.1286, 0.0465, 0.0099, 0.0009, 0.0000

15-9. 0.0071, 0.0638, 0.3001, 0.3962, 0.1929, 0.0399

15-11. Normal with mean 9,207.3 and standard deviation 61.58

15-13. Normal with mean 95.95 and standard deviation 0.312

15-15. Normal with mean 23.72 and standard deviation 0.7442

15-17. Governor: D (largest s.d.) ARCO expert: C (smallest s.d.) Most embarrassed: C

15-23. Expected profit = $1.8 million

15-25. Expected profit is $86 million

15-27. Optimal decision is *long*–change if possible. Expected profit is $698,000

15-29. Optimal decision is invest in wheat futures. Expected value is $3,040

15-33. Don't test; E(payoff) = $1.64 million. If you

15-35. Don't test. E(payoff) = $16,000. If you must test, sell watches anyway. E(payoff w/test) = $9999.

must test, advertise anyway.

15-37. Test and follow the test's recommendation. E(payoff) = $587,000

15-39. A utility function is a value-of-money function of an individual.

15-47. EVPI = $290,000. Buy information if it is perfect.

15-49. 0.01142, 0.30043, 0.35004, 0.27066, 0.05561, 0.01184

15-51. 0.0026, 0.3844, 0.4589, 0.1480, 0.0060, 0.0001

15-55. 95% HPD region = [97.888, 101.732]

15-61. 1,400 < 3,000: a risk taker

15-63. Merge; E(payoff) = $2.45 million

15-65. EVPI = $1.125 million

15-67. Hire candidate A

Chapter 16

16-1. *a.* $\bar{x}_{st}$ = 33.48%
b. S.D. = 0.823%
c. [31.87, 35.09]

16-3. *a.* $\bar{x}_{st}$ = $40.01
b. S.D. = 0.6854
c. [38.88, 41.14]
d. Data has many zero values

16-5. $35,604.5   C.I. = [30,969.19, 40,239.87]

16-9. *a–c.* All no. Clusters need to be randomly chosen.
d. Consider the companies as strata, ships as clusters. Randomly draw clusters from the strata.

16-11. Arrange elements in a circle, randomly choose a number from 1 to 28, and then add 28 to element number

until you have a sample of 30 sales.

16-13. If k is a multiple of 7, we would sample on the same day for different weeks and bias the results due to weekly sales cycles.

16-17. Yes, with each vehicle a cluster.

16-19. OK unless a nonnegligible fraction of the businesses in the community are unlisted in the Yellow Pages.

16-21. [0.055, 0.091]

16-25. Sample about 109 children, 744 young adults, 147 older people.

16-27. Regression and ratio estimators

16-29. No; benefits are not substantial when the number of strata is much greater than 6. Combine some.

APPENDIX C   Statistical Tables

TABLE 1   Cumulative Binomial Distribution

$$F(x) = P(X \le x) = \sum_{i=0}^{x} \binom{n}{i} p^i (1-p)^{n-i}$$

Example: if $p = 0.10$, $n = 5$, and $x = 2$, then $F(x) = 0.991$

n	x	.01	.05	.10	.20	.30	.40	.50	.60	.70	.80	.90	.95	.99
5	0	.951	.774	.590	.328	.168	.078	.031	.010	.002	.000	.000	.000	.000
	1	.999	.977	.919	.737	.528	.337	.187	.087	.031	.007	.000	.000	.000
	2	1.000	.999	.991	.942	.837	.683	.500	.317	.163	.058	.009	.001	.000
	3	1.000	1.000	1.000	.993	.969	.913	.813	.663	.472	.263	.081	.023	.001
	4	1.000	1.000	1.000	1.000	.998	.990	.969	.922	.832	.672	.410	.226	.049
6	0	.941	.735	.531	.262	.118	.047	.016	.004	.001	.000	.000	.000	.000
	1	.999	.967	.886	.655	.420	.233	.109	.041	.011	.002	.000	.000	.000
	2	1.000	.998	.984	.901	.744	.544	.344	.179	.070	.017	.001	.000	.000
	3	1.000	1.000	.999	.983	.930	.821	.656	.456	.256	.099	.016	.002	.000
	4	1.000	1.000	1.000	.998	.989	.959	.891	.767	.580	.345	.114	.033	.001
	5	1.000	1.000	1.000	1.000	.999	.996	.984	.953	.882	.738	.469	.265	.059
7	0	.932	.698	.478	.210	.082	.028	.008	.002	.000	.000	.000	.000	.000
	1	.998	.956	.850	.577	.329	.159	.063	.019	.004	.000	.000	.000	.000
	2	1.000	.996	.974	.852	.647	.420	.227	.096	.029	.005	.000	.000	.000
	3	1.000	1.000	.997	.967	.874	.710	.500	.290	.126	.033	.003	.000	.000
	4	1.000	1.000	1.000	.995	.971	.904	.773	.580	.353	.148	.026	.004	.000
	5	1.000	1.000	1.000	1.000	.996	.981	.937	.841	.671	.423	.150	.044	.002
	6	1.000	1.000	1.000	1.000	1.000	.998	.992	.972	.918	.790	.522	.302	.068
8	0	.923	.663	.430	.168	.058	.017	.004	.001	.000	.000	.000	.000	.000
	1	.997	.943	.813	.503	.255	.106	.035	.009	.001	.000	.000	.000	.000
	2	1.000	.994	.962	.797	.552	.315	.145	.050	.011	.001	.000	.000	.000
	3	1.000	1.000	.995	.944	.806	.594	.363	.174	.058	.010	.000	.000	.000
	4	1.000	1.000	1.000	.990	.942	.826	.637	.406	.194	.056	.005	.000	.000
	5	1.000	1.000	1.000	.999	.989	.950	.855	.685	.448	.203	.038	.006	.000
	6	1.000	1.000	1.000	1.000	.999	.991	.965	.894	.745	.497	.187	.057	.003
	7	1.000	1.000	1.000	1.000	1.000	.999	.996	.983	.942	.832	.570	.337	.077
9	0	.914	.630	.387	.134	.040	.010	.002	.000	.000	.000	.000	.000	.000
	1	.997	.929	.775	.436	.196	.071	.020	.004	.000	.000	.000	.000	.000
	2	1.000	.992	.947	.738	.463	.232	.090	.025	.004	.000	.000	.000	.000
	3	1.000	.999	.992	.914	.730	.483	.254	.099	.025	.003	.000	.000	.000
	4	1.000	1.000	.999	.980	.901	.733	.500	.267	.099	.020	.001	.000	.000
	5	1.000	1.000	1.000	.997	.975	.901	.746	.517	.270	.086	.008	.001	.000
	6	1.000	1.000	1.000	1.000	.996	.975	.910	.768	.537	.262	.053	.008	.000
	7	1.000	1.000	1.000	1.000	1.000	.996	.980	.929	.804	.564	.225	.071	.003
	8	1.000	1.000	1.000	1.000	1.000	1.000	.998	.990	.960	.866	.613	.370	.086
10	0	.904	.599	.349	.107	.028	.006	.001	.000	.000	.000	.000	.000	.000
	1	.996	.914	.736	.376	.149	.046	.011	.002	.000	.000	.000	.000	.000
	2	1.000	.988	.930	.678	.383	.167	.055	.012	.002	.000	.000	.000	.000
	3	1.000	.999	.987	.879	.650	.382	.172	.055	.011	.001	.000	.000	.000
	4	1.000	1.000	.998	.967	.850	.633	.377	.166	.047	.006	.000	.000	.000
	5	1.000	1.000	1.000	.994	.953	.834	.623	.367	.150	.033	.002	.000	.000
	6	1.000	1.000	1.000	.999	.989	.945	.828	.618	.350	.121	.013	.001	.000
	7	1.000	1.000	1.000	1.000	.998	.988	.945	.833	.617	.322	.070	.012	.000
	8	1.000	1.000	1.000	1.000	1.000	.998	.989	.954	.851	.624	.264	.086	.004
	9	1.000	1.000	1.000	1.000	1.000	1.000	.999	.994	.972	.893	.651	.401	.096

TABLE 1 *(continued)*   **Cumulative Binomial Distribution**

							p							
n	x	.01	.05	.10	.20	.30	.40	.50	.60	.70	.80	.90	.95	.99
15	0	.860	.463	.206	.035	.005	.000	.000	.000	.000	.000	.000	.000	.000
	1	.990	.829	.549	.167	.035	.005	.000	.000	.000	.000	.000	.000	.000
	2	1.000	.964	.816	.398	.127	.027	.004	.000	.000	.000	.000	.000	.000
	3	1.000	.995	.944	.648	.297	.091	.018	.002	.000	.000	.000	.000	.000
	4	1.000	.999	.987	.836	.515	.217	.059	.009	.001	.000	.000	.000	.000
	5	1.000	1.000	.998	.939	.722	.403	.151	.034	.004	.000	.000	.000	.000
	6	1.000	1.000	1.000	.982	.869	.610	.304	.095	.015	.001	.000	.000	.000
	7	1.000	1.000	1.000	.996	.950	.787	.500	.213	.050	.004	.000	.000	.000
	8	1.000	1.000	1.000	.999	.985	.905	.696	.390	.131	.018	.000	.000	.000
	9	1.000	1.000	1.000	1.000	.996	.966	.849	.597	.278	.061	.002	.000	.000
	10	1.000	1.000	1.000	1.000	.999	.991	.941	.783	.485	.164	.013	.001	.000
	11	1.000	1.000	1.000	1.000	1.000	.998	.982	.909	.703	.352	.056	.005	.000
	12	1.000	1.000	1.000	1.000	1.000	1.000	.996	.973	.873	.602	.184	.036	.000
	13	1.000	1.000	1.000	1.000	1.000	1.000	1.000	.995	.965	.833	.451	.171	.010
	14	1.000	1.000	1.000	1.000	1.000	1.000	1.000	1.000	.995	.965	.794	.537	.140
20	0	.818	.358	.122	.012	.001	.000	.000	.000	.000	.000	.000	.000	.000
	1	.983	.736	.392	.069	.008	.001	.000	.000	.000	.000	.000	.000	.000
	2	.999	.925	.677	.206	.035	.004	.000	.000	.000	.000	.000	.000	.000
	3	1.000	.984	.867	.411	.107	.016	.001	.000	.000	.000	.000	.000	.000
	4	1.000	.997	.957	.630	.238	.051	.006	.000	.000	.000	.000	.000	.000
	5	1.000	1.000	.989	.804	.416	.126	.021	.002	.000	.000	.000	.000	.000
	6	1.000	1.000	.998	.913	.608	.250	.058	.006	.000	.000	.000	.000	.000
	7	1.000	1.000	1.000	.968	.772	.416	.132	.021	.001	.000	.000	.000	.000
	8	1.000	1.000	1.000	.990	.887	.596	.252	.057	.005	.000	.000	.000	.000
	9	1.000	1.000	1.000	.997	.952	.755	.412	.128	.017	.001	.000	.000	.000
	10	1.000	1.000	1.000	.999	.983	.872	.588	.245	.048	.003	.000	.000	.000
	11	1.000	1.000	1.000	1.000	.995	.943	.748	.404	.113	.010	.000	.000	.000
	12	1.000	1.000	1.000	1.000	.999	.979	.868	.584	.228	.032	.000	.000	.000
	13	1.000	1.000	1.000	1.000	1.000	.994	.942	.750	.392	.087	.002	.000	.000
	14	1.000	1.000	1.000	1.000	1.000	.998	.979	.874	.584	.196	.011	.000	.000
	15	1.000	1.000	1.000	1.000	1.000	1.000	.994	.949	.762	.370	.043	.003	.000
	16	1.000	1.000	1.000	1.000	1.000	1.000	.999	.984	.893	.589	.133	.016	.000
	17	1.000	1.000	1.000	1.000	1.000	1.000	1.000	.996	.965	.794	.323	.075	.001
	18	1.000	1.000	1.000	1.000	1.000	1.000	1.000	.999	.992	.931	.608	.264	.017
	19	1.000	1.000	1.000	1.000	1.000	1.000	1.000	1.000	.999	.988	.878	.642	.182
25	0	.778	.277	.072	.004	.000	.000	.000	.000	.000	.000	.000	.000	.000
	1	.974	.642	.271	.027	.002	.000	.000	.000	.000	.000	.000	.000	.000
	2	.998	.873	.537	.098	.009	.000	.000	.000	.000	.000	.000	.000	.000
	3	1.000	.966	.764	.234	.033	.002	.000	.000	.000	.000	.000	.000	.000
	4	1.000	.993	.902	.421	.090	.009	.000	.000	.000	.000	.000	.000	.000
	5	1.000	.999	.967	.617	.193	.029	.002	.000	.000	.000	.000	.000	.000
	6	1.000	1.000	.991	.780	.341	.074	.007	.000	.000	.000	.000	.000	.000
	7	1.000	1.000	.998	.891	.512	.154	.022	.001	.000	.000	.000	.000	.000
	8	1.000	1.000	1.000	.953	.677	.274	.054	.004	.000	.000	.000	.000	.000
	9	1.000	1.000	1.000	.983	.811	.425	.115	.013	.000	.000	.000	.000	.000

TABLE 1 *(concluded)*   **Cumulative Binomial Distribution**

								p						
n	*x*	.01	.05	.10	.20	.30	.40	.50	.60	.70	.80	.90	.95	.99
	10	1.000	1.000	1.000	.994	.902	.586	.212	.034	.002	.000	.000	.000	.000
	11	1.000	1.000	1.000	.998	.956	.732	.345	.078	.006	.000	.000	.000	.000
	12	1.000	1.000	1.000	1.000	.983	.846	.500	.154	.017	.000	.000	.000	.000
	13	1.000	1.000	1.000	1.000	.994	.922	.655	.268	.044	.002	.000	.000	.000
	14	1.000	1.000	1.000	1.000	.998	.966	.788	.414	.098	.006	.000	.000	.000
	15	1.000	1.000	1.000	1.000	1.000	.987	.885	.575	.189	.017	.000	.000	.000
	16	1.000	1.000	1.000	1.000	1.000	.996	.946	.726	.323	.047	.000	.000	.000
	17	1.000	1.000	1.000	1.000	1.000	.999	.978	.846	.488	.109	.002	.000	.000
	18	1.000	1.000	1.000	1.000	1.000	1.000	.993	.926	.659	.220	.009	.000	.000
	19	1.000	1.000	1.000	1.000	1.000	1.000	.998	.971	.807	.383	.033	.001	.000
	20	1.000	1.000	1.000	1.000	1.000	1.000	1.000	.991	.910	.579	.098	.007	.000
	21	1.000	1.000	1.000	1.000	1.000	1.000	1.000	.998	.967	.766	.236	.034	.000
	22	1.000	1.000	1.000	1.000	1.000	1.000	1.000	1.000	.991	.902	.463	.127	.002
	23	1.000	1.000	1.000	1.000	1.000	1.000	1.000	1.000	.998	.973	.729	.358	.026
	24	1.000	1.000	1.000	1.000	1.000	1.000	1.000	1.000	1.000	.996	.928	.723	.222

TABLE 2 Areas of the Standard Normal Distribution

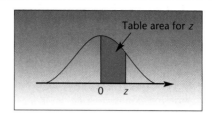

Table area for z

0   z

The table areas are probabilities that the standard normal random variable is between 0 and z.

Second Decimal Place in z

z	0.00	0.01	0.02	0.03	0.04	0.05	0.06	0.07	0.08	0.09
0.0	0.0000	0.0040	0.0080	0.0120	0.0160	0.0199	0.0239	0.0279	0.0319	0.0359
0.1	0.0398	0.0438	0.0478	0.0517	0.0557	0.0596	0.0636	0.0675	0.0714	0.0753
0.2	0.0793	0.0832	0.0871	0.0910	0.0948	0.0987	0.1026	0.1064	0.1103	0.1141
0.3	0.1179	0.1217	0.1255	0.1293	0.1331	0.1368	0.1406	0.1443	0.1480	0.1517
0.4	0.1554	0.1591	0.1628	0.1664	0.1700	0.1736	0.1772	0.1808	0.1844	0.1879
0.5	0.1915	0.1950	0.1985	0.2019	0.2054	0.2088	0.2123	0.2157	0.2190	0.2224
0.6	0.2257	0.2291	0.2324	0.2357	0.2389	0.2422	0.2454	0.2486	0.2517	0.2549
0.7	0.2580	0.2611	0.2642	0.2673	0.2704	0.2734	0.2764	0.2794	0.2823	0.2852
0.8	0.2881	0.2910	0.2939	0.2967	0.2995	0.3023	0.3051	0.3078	0.3106	0.3133
0.9	0.3159	0.3186	0.3212	0.3238	0.3264	0.3289	0.3315	0.3340	0.3365	0.3389
1.0	0.3413	0.3438	0.3461	0.3485	0.3508	0.3531	0.3554	0.3577	0.3599	0.3621
1.1	0.3643	0.3665	0.3686	0.3708	0.3729	0.3749	0.3770	0.3790	0.3810	0.3830
1.2	0.3849	0.3869	0.3888	0.3907	0.3925	0.3944	0.3962	0.3980	0.3997	0.4015
1.3	0.4032	0.4049	0.4066	0.4082	0.4099	0.4115	0.4131	0.4147	0.4162	0.4177
1.4	0.4192	0.4207	0.4222	0.4236	0.4251	0.4265	0.4279	0.4292	0.4306	0.4319
1.5	0.4332	0.4345	0.4357	0.4370	0.4382	0.4394	0.4406	0.4418	0.4429	0.4441
1.6	0.4452	0.4463	0.4474	0.4484	0.4495	0.4505	0.4515	0.4525	0.4535	0.4545
1.7	0.4554	0.4564	0.4573	0.4582	0.4591	0.4599	0.4608	0.4616	0.4625	0.4633
1.8	0.4641	0.4649	0.4656	0.4664	0.4671	0.4678	0.4686	0.4693	0.4699	0.4706
1.9	0.4713	0.4719	0.4726	0.4732	0.4738	0.4744	0.4750	0.4756	0.4761	0.4767
2.0	0.4772	0.4778	0.4783	0.4788	0.4793	0.4798	0.4803	0.4808	0.4812	0.4817
2.1	0.4821	0.4826	0.4830	0.4834	0.4838	0.4842	0.4846	0.4850	0.4854	0.4857
2.2	0.4861	0.4864	0.4868	0.4871	0.4875	0.4878	0.4881	0.4884	0.4887	0.4890
2.3	0.4893	0.4896	0.4898	0.4901	0.4904	0.4906	0.4909	0.4911	0.4913	0.4916
2.4	0.4918	0.4920	0.4922	0.4925	0.4927	0.4929	0.4931	0.4932	0.4934	0.4936
2.5	0.4938	0.4940	0.4941	0.4943	0.4945	0.4946	0.4948	0.4949	0.4951	0.4952
2.6	0.4953	0.4955	0.4956	0.4957	0.4959	0.4960	0.4961	0.4962	0.4963	0.4964
2.7	0.4965	0.4966	0.4967	0.4968	0.4969	0.4970	0.4971	0.4972	0.4973	0.4974
2.8	0.4974	0.4975	0.4976	0.4977	0.4977	0.4978	0.4979	0.4979	0.4980	0.4981
2.9	0.4981	0.4982	0.4982	0.4983	0.4984	0.4984	0.4985	0.4985	0.4986	0.4986
3.0	0.4987	0.4987	0.4987	0.4988	0.4988	0.4989	0.4989	0.4989	0.4990	0.4990
3.1	0.4990	0.4991	0.4991	0.4991	0.4992	0.4992	0.4992	0.4992	0.4993	0.4993
3.2	0.4993	0.4993	0.4994	0.4994	0.4994	0.4994	0.4994	0.4995	0.4995	0.4995
3.3	0.4995	0.4995	0.4995	0.4996	0.4996	0.4996	0.4996	0.4996	0.4996	0.4997
3.4	0.4997	0.4997	0.4997	0.4997	0.4997	0.4997	0.4997	0.4997	0.4997	0.4998
3.5	0.4998									
4.0	0.49997									
4.5	0.499997									
5.0	0.4999997									
6.0	0.49999999									

TABLE 3   Critical Values of the t Distribution

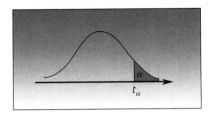

Degrees of Freedom	$t_{.100}$	$t_{.050}$	$t_{.025}$	$t_{.010}$	$t_{.005}$
1	3.078	6.314	12.706	31.821	63.657
2	1.886	2.920	4.303	6.965	9.925
3	1.638	2.353	3.182	4.541	5.841
4	1.533	2.132	2.776	3.747	4.604
5	1.476	2.015	2.571	3.365	4.032
6	1.440	1.943	2.447	3.143	3.707
7	1.415	1.895	2.365	2.998	3.499
8	1.397	1.860	2.306	2.896	3.355
9	1.383	1.833	2.262	2.821	3.250
10	1.372	1.812	2.228	2.764	3.169
11	1.363	1.796	2.201	2.718	3.106
12	1.356	1.782	2.179	2.681	3.055
13	1.350	1.771	2.160	2.650	3.012
14	1.345	1.761	2.145	2.624	2.977
15	1.341	1.753	2.131	2.602	2.947
16	1.337	1.746	2.120	2.583	2.921
17	1.333	1.740	2.110	2.567	2.898
18	1.330	1.734	2.101	2.552	2.878
19	1.328	1.729	2.093	2.539	2.861
20	1.325	1.725	2.086	2.528	2.845
21	1.323	1.721	2.080	2.518	2.831
22	1.321	1.717	2.074	2.508	2.819
23	1.319	1.714	2.069	2.500	2.807
24	1.318	1.711	2.064	2.492	2.797
25	1.316	1.708	2.060	2.485	2.787
26	1.315	1.706	2.056	2.479	2.779
27	1.314	1.703	2.052	2.473	2.771
28	1.313	1.701	2.048	2.467	2.763
29	1.311	1.699	2.045	2.462	2.756
30	1.310	1.697	2.042	2.457	2.750
40	1.303	1.684	2.021	2.423	2.704
60	1.296	1.671	2.000	2.390	2.660
120	1.289	1.658	1.980	2.358	2.617
∞	1.282	1.645	1.960	2.326	2.576

Source: M. Merrington, "Table of Percentage Points of the t-Distribution," *Biometrika* 32 (1941), p. 300.
Reproduced by permission of the *Biometrika* trustees.

TABLE 4   Critical Values of the Chi-Square Distribution

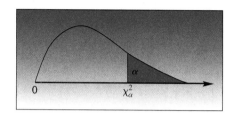

Degrees of Freedom	$\chi^2_{.995}$	$\chi^2_{.990}$	$\chi^2_{.975}$	$\chi^2_{.950}$	$\chi^2_{.900}$
1	0.0000393	0.0001571	0.0009821	0.0039321	0.0157908
2	0.0100251	0.0201007	0.0506356	0.102587	0.210720
3	0.0717212	0.114832	0.215795	0.351846	0.584375
4	0.206990	0.297110	0.484419	0.710721	1.063623
5	0.411740	0.554300	0.831211	1.145476	1.61031
6	0.675727	0.872085	1.237347	1.63539	2.20413
7	0.989265	1.239043	1.68987	2.16735	2.83311
8	1.344419	1.646482	2.17973	2.73264	3.48954
9	1.734926	2.087912	2.70039	3.32511	4.16816
10	2.15585	2.55821	3.24697	3.94030	4.86518
11	2.60321	3.05347	3.81575	4.57481	5.57779
12	3.07382	3.57056	4.40379	5.22603	6.30380
13	3.56503	4.10691	5.00874	5.89186	7.04150
14	4.07468	4.66043	5.62872	6.57063	7.78953
15	4.60094	5.22935	6.26214	7.26094	8.54675
16	5.14224	5.81221	6.90766	7.96164	9.31223
17	5.69724	6.40776	7.56418	8.67176	10.0852
18	6.26481	7.01491	8.23075	9.39046	10.8649
19	6.84398	7.63273	8.90655	10.1170	11.6509
20	7.43386	8.26040	9.59083	10.8508	12.4426
21	8.03366	8.89720	10.28293	11.5913	13.2396
22	8.64272	9.54249	10.9823	12.3380	14.0415
23	9.26042	10.19567	11.6885	13.0905	14.8479
24	9.88623	10.8564	12.4011	13.8484	15.6587
25	10.5197	11.5240	13.1197	14.6114	16.4734
26	11.1603	12.1981	13.8439	15.3791	17.2919
27	11.8076	12.8786	14.5733	16.1513	18.1138
28	12.4613	13.5648	15.3079	16.9279	18.9392
29	13.1211	14.2565	16.0471	17.7083	19.7677
30	13.7867	14.9535	16.7908	18.4926	20.5992
40	20.7065	22.1643	24.4331	26.5093	29.0505
50	27.9907	29.7067	32.3574	34.7642	37.6886
60	35.5346	37.4848	40.4817	43.1879	46.4589
70	43.2752	45.4418	48.7576	51.7393	55.3290
80	51.1720	53.5400	57.1532	60.3915	64.2778
90	59.1963	61.7541	65.6466	69.1260	73.2912
100	67.3276	70.0648	74.2219	77.9295	82.3581

TABLE 4 *(concluded)*   Critical Values of the Chi-Square Distribution

Degrees of Freedom	$\chi^2_{.100}$	$\chi^2_{.050}$	$\chi^2_{.025}$	$\chi^2_{.010}$	$\chi^2_{.005}$
1	2.70554	3.84146	5.02389	6.63490	7.87944
2	4.60517	5.99147	7.37776	9.21034	10.5966
3	6.25139	7.81473	9.34840	11.3449	12.8381
4	7.77944	9.48773	11.1433	13.2767	14.8602
5	9.23635	11.0705	12.8325	15.0863	16.7496
6	10.6446	12.5916	14.4494	16.8119	18.5476
7	12.0170	14.0671	16.0128	18.4753	20.2777
8	13.3616	15.5073	17.5346	20.0902	21.9550
9	14.6837	16.9190	19.0228	21.6660	23.5893
10	15.9871	18.3070	20.4831	23.2093	25.1882
11	17.2750	19.6751	21.9200	24.7250	26.7569
12	18.5494	21.0261	23.3367	26.2170	28.2995
13	19.8119	22.3621	24.7356	27.6883	29.8194
14	21.0642	23.6848	26.1190	29.1413	31.3193
15	22.3072	24.9958	27.4884	30.5779	32.8013
16	23.5418	26.2962	28.8454	31.9999	34.2672
17	24.7690	27.5871	30.1910	33.4087	35.7185
18	25.9894	28.8693	31.5264	34.8053	37.1564
19	27.2036	30.1435	32.8523	36.1908	38.5822
20	28.4120	31.4104	34.1696	37.5662	39.9968
21	29.6151	32.6705	35.4789	38.9321	41.4010
22	30.8133	33.9244	36.7807	40.2894	42.7956
23	32.0069	35.1725	38.0757	41.6384	44.1813
24	33.1963	36.4151	39.3641	42.9798	45.5585
25	34.3816	37.6525	40.6465	44.3141	46.9278
26	35.5631	38.8852	41.9232	45.6417	48.2899
27	36.7412	40.1133	43.1944	46.9630	49.6449
28	37.9159	41.3372	44.4607	48.2782	50.9933
29	39.0875	42.5569	45.7222	49.5879	52.3356
30	40.2560	43.7729	46.9792	50.8922	53.6720
40	51.8050	55.7585	59.3417	63.6907	66.7659
50	63.1671	67.5048	71.4202	76.1539	79.4900
60	74.3970	79.0819	83.2976	88.3794	91.9517
70	85.5271	90.5312	95.0231	100.425	104.215
80	96.5782	101.879	106.629	112.329	116.321
90	107.565	113.145	118.136	124.116	128.299
100	118.498	124.342	129.561	135.807	140.169

Source: C. M. Thompson, "Tables of the Percentage Points of the χ^2-Distribution," *Biometrika* 32 (1941), pp. 188–89. Reproduced by permission of the *Biometrika* Trustees.

TABLE 5   Critical Values of the F Distribution for α = 0.10

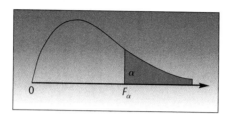

Denominator Degrees of Freedom (k_2)	Numerator Degrees of Freedom (k_1)								
	1	2	3	4	5	6	7	8	9
1	39.86	49.50	53.59	55.83	57.24	58.20	58.91	59.44	59.86
2	8.53	9.00	9.16	9.24	9.29	9.33	9.35	9.37	9.38
3	5.54	5.46	5.39	5.34	5.31	5.28	5.27	5.25	5.24
4	4.54	4.32	4.19	4.11	4.05	4.01	3.98	3.95	3.94
5	4.06	3.78	3.62	3.52	3.45	3.40	3.37	3.34	3.32
6	3.78	3.46	3.29	3.18	3.11	3.05	3.01	2.98	2.96
7	3.59	3.26	3.07	2.96	2.88	2.83	2.78	2.75	2.72
8	3.46	3.11	2.92	2.81	2.73	2.67	2.62	2.59	2.56
9	3.36	3.01	2.81	2.69	2.61	2.55	2.51	2.47	2.44
10	3.29	2.92	2.73	2.61	2.52	2.46	2.41	2.38	2.35
11	3.23	2.86	2.66	2.54	2.45	2.39	2.34	2.30	2.27
12	3.18	2.81	2.61	2.48	2.39	2.33	2.28	2.24	2.21
13	3.14	2.76	2.56	2.43	2.35	2.28	2.23	2.20	2.16
14	3.10	2.73	2.52	2.39	2.31	2.24	2.19	2.15	2.12
15	3.07	2.70	2.49	2.36	2.27	2.21	2.16	2.12	2.09
16	3.05	2.67	2.46	2.33	2.24	2.18	2.13	2.09	2.06
17	3.03	2.64	2.44	2.31	2.22	2.15	2.10	2.06	2.03
18	3.01	2.62	2.42	2.29	2.20	2.13	2.08	2.04	2.00
19	2.99	2.61	2.40	2.27	2.18	2.11	2.06	2.02	1.98
20	2.97	2.59	2.38	2.25	2.16	2.09	2.04	2.00	1.96
21	2.96	2.57	2.36	2.23	2.14	2.08	2.02	1.98	1.95
22	2.95	2.56	2.35	2.22	2.13	2.06	2.01	1.97	1.93
23	2.94	2.55	2.34	2.21	2.11	2.05	1.99	1.95	1.92
24	2.93	2.54	2.33	2.19	2.10	2.04	1.98	1.94	1.91
25	2.92	2.53	2.32	2.18	2.09	2.02	1.97	1.93	1.89
26	2.91	2.52	2.31	2.17	2.08	2.01	1.96	1.92	1.88
27	2.90	2.51	2.30	2.17	2.07	2.00	1.95	1.91	1.87
28	2.89	2.50	2.29	2.16	2.06	2.00	1.94	1.90	1.87
29	2.89	2.50	2.28	2.15	2.06	1.99	1.93	1.89	1.86
30	2.88	2.49	2.28	2.14	2.05	1.98	1.93	1.88	1.85
40	2.84	2.44	2.23	2.09	2.00	1.93	1.87	1.83	1.79
60	2.79	2.39	2.18	2.04	1.95	1.87	1.82	1.77	1.74
120	2.75	2.35	2.13	1.99	1.90	1.82	1.77	1.72	1.68
∞	2.71	2.30	2.08	1.94	1.85	1.77	1.72	1.67	1.63

TABLE 5 *(continued)* **Critical Values of the *F* Distribution for $\alpha = 0.05$**

Denominator Degrees of Freedom (k_2)	Numerator Degrees of Freedom (k_1)									
	10	12	15	20	24	30	40	60	120	∞
1	241.9	243.9	245.9	248.0	249.1	250.1	251.1	252.2	253.3	254.3
2	19.40	19.41	19.43	19.45	19.45	19.46	19.47	19.48	19.49	19.50
3	8.79	8.74	8.70	8.66	8.64	8.62	8.59	8.57	8.55	8.53
4	5.96	5.91	5.86	5.80	5.77	5.75	5.72	5.69	5.66	5.63
5	4.74	4.68	4.62	4.56	4.53	4.50	4.46	4.43	4.40	4.36
6	4.06	4.00	3.94	3.87	3.84	3.81	3.77	3.74	3.70	3.67
7	3.64	3.57	3.51	3.44	3.41	3.38	3.34	3.30	3.27	3.23
8	3.35	3.28	3.22	3.15	3.12	3.08	3.04	3.01	2.97	2.93
9	3.14	3.07	3.01	2.94	2.90	2.86	2.83	2.79	2.75	2.71
10	2.98	2.91	2.85	2.77	2.74	2.70	2.66	2.62	2.58	2.54
11	2.85	2.79	2.72	2.65	2.61	2.57	2.53	2.49	2.45	2.40
12	2.75	2.69	2.62	2.54	2.51	2.47	2.43	2.38	2.34	2.30
13	2.67	2.60	2.53	2.46	2.42	2.38	2.34	2.30	2.25	2.21
14	2.60	2.53	2.46	2.39	2.35	2.31	2.27	2.22	2.18	2.13
15	2.54	2.48	2.40	2.33	2.29	2.25	2.20	2.16	2.11	2.07
16	2.49	2.42	2.35	2.28	2.24	2.19	2.15	2.11	2.06	2.01
17	2.45	2.38	2.31	2.23	2.19	2.15	2.10	2.06	2.01	1.96
18	2.41	2.34	2.27	2.19	2.15	2.11	2.06	2.02	1.97	1.92
19	2.38	2.31	2.23	2.16	2.11	2.07	2.03	1.98	1.93	1.88
20	2.35	2.28	2.20	2.12	2.08	2.04	1.99	1.95	1.90	1.84
21	2.32	2.25	2.18	2.10	2.05	2.01	1.96	1.92	1.87	1.81
22	2.30	2.23	2.15	2.07	2.03	1.98	1.94	1.89	1.84	1.78
23	2.27	2.20	2.13	2.05	2.01	1.96	1.91	1.86	1.81	1.76
24	2.25	2.18	2.11	2.03	1.98	1.94	1.89	1.84	1.79	1.73
25	2.24	2.16	2.09	2.01	1.96	1.92	1.87	1.82	1.77	1.71
26	2.22	2.15	2.07	1.99	1.95	1.90	1.85	1.80	1.75	1.69
27	2.20	2.13	2.06	1.97	1.93	1.88	1.84	1.79	1.73	1.67
28	2.19	2.12	2.04	1.96	1.91	1.87	1.82	1.77	1.71	1.65
29	2.18	2.10	2.03	1.94	1.90	1.85	1.81	1.75	1.70	1.64
30	2.16	2.09	2.01	1.93	1.89	1.84	1.79	1.74	1.68	1.62
40	2.08	2.00	1.92	1.84	1.79	1.74	1.69	1.64	1.58	1.51
60	1.99	1.92	1.84	1.75	1.70	1.65	1.59	1.53	1.47	1.39
120	1.91	1.83	1.75	1.66	1.61	1.55	1.50	1.43	1.35	1.25
∞	1.83	1.75	1.67	1.57	1.52	1.46	1.39	1.32	1.22	1.00

TABLE 5 *(continued)* **Critical Values of the *F* Distribution for $\alpha = 0.025$**

Denominator Degrees of Freedom (k_2)	Numerator Degrees of Freedom (k_1)								
	1	2	3	4	5	6	7	8	9
1	647.8	799.5	864.2	899.6	921.8	937.1	948.2	956.7	963.3
2	38.51	39.00	39.17	39.25	39.30	39.33	39.36	39.37	39.39
3	17.44	16.04	15.44	15.10	14.88	14.73	14.62	14.54	14.47
4	12.22	10.65	9.98	9.60	9.36	9.20	9.07	8.98	8.90
5	10.01	8.43	7.76	7.39	7.15	6.98	6.85	6.76	6.68
6	8.81	7.26	6.60	6.23	5.99	5.82	5.70	5.60	5.52
7	8.07	6.54	5.89	5.52	5.29	5.12	4.99	4.90	4.82
8	7.57	6.06	5.42	5.05	4.82	4.65	4.53	4.43	4.36
9	7.21	5.71	5.08	4.72	4.48	4.32	4.20	4.10	4.03
10	6.94	5.46	4.83	4.47	4.24	4.07	3.95	3.85	3.78
11	6.72	5.26	4.63	4.28	4.04	3.88	3.76	3.66	3.59
12	6.55	5.10	4.47	4.12	3.89	3.73	3.61	3.51	3.44
13	6.41	4.97	4.35	4.00	3.77	3.60	3.48	3.39	3.31
14	6.30	4.86	4.24	3.89	3.66	3.50	3.38	3.29	3.21
15	6.20	4.77	4.15	3.80	3.58	3.41	3.29	3.20	3.12
16	6.12	4.69	4.08	3.73	3.50	3.34	3.22	3.12	3.05
17	6.04	4.62	4.01	3.66	3.44	3.28	3.16	3.06	2.98
18	5.98	4.56	3.95	3.61	3.38	3.22	3.10	3.01	2.93
19	5.92	4.51	3.90	3.56	3.33	3.17	3.05	2.96	2.88
20	5.87	4.46	3.86	3.51	3.29	3.13	3.01	2.91	2.84
21	5.83	4.42	3.82	3.48	3.25	3.09	2.97	2.87	2.80
22	5.79	4.38	3.78	3.44	3.22	3.05	2.93	2.84	2.76
23	5.75	4.35	3.75	3.41	3.18	3.02	2.90	2.81	2.73
24	5.72	4.32	3.72	3.38	3.15	2.99	2.87	2.78	2.70
25	5.69	4.29	3.69	3.35	3.13	2.97	2.85	2.75	2.68
26	5.66	4.27	3.67	3.33	3.10	2.94	2.82	2.73	2.65
27	5.63	4.24	3.65	3.31	3.08	2.92	2.80	2.71	2.63
28	5.61	4.22	3.63	3.29	3.06	2.90	2.78	2.69	2.61
29	5.59	4.20	3.61	3.27	3.04	2.88	2.76	2.67	2.59
30	5.57	4.18	3.59	3.25	3.03	2.87	2.75	2.65	2.57
40	5.42	4.05	3.46	3.13	2.90	2.74	2.62	2.53	2.45
60	5.29	3.93	3.34	3.01	2.79	2.63	2.51	2.41	2.33
120	5.15	3.80	3.23	2.89	2.67	2.52	2.39	2.30	2.22
∞	5.02	3.69	3.12	2.79	2.57	2.41	2.29	2.19	2.11

TABLE 5 *(continued)* Critical Values of the *F* Distribution for $\alpha = 0.025$

Denominator Degrees of Freedom (k_2)	Numerator Degrees of Freedom (k_1)									
	10	12	15	20	24	30	40	60	120	∞
1	968.6	976.7	984.9	993.1	997.2	1001	1006	1010	1014	1018
2	39.40	39.41	39.43	39.45	39.46	39.46	39.47	39.48	39.49	39.50
3	14.42	14.34	14.25	14.17	14.12	14.08	14.04	13.99	13.95	13.90
4	8.84	8.75	8.66	8.56	8.51	8.46	8.41	8.36	8.31	8.26
5	6.62	6.52	6.43	6.33	6.28	6.23	6.18	6.12	6.07	6.02
6	5.46	5.37	5.27	5.17	5.12	5.07	5.01	4.96	4.90	4.85
7	4.76	4.67	4.57	4.47	4.42	4.36	4.31	4.25	4.20	4.14
8	4.30	4.20	4.10	4.00	3.95	3.89	3.84	3.78	3.73	3.67
9	3.96	3.87	3.77	3.67	3.61	3.56	3.51	3.45	3.39	3.33
10	3.72	3.62	3.52	3.42	3.37	3.31	3.26	3.20	3.14	3.08
11	3.53	3.43	3.33	3.23	3.17	3.12	3.06	3.00	2.94	2.88
12	3.37	3.28	3.18	3.07	3.02	2.96	2.91	2.85	2.79	2.72
13	3.25	3.15	3.05	2.95	2.89	2.84	2.78	2.72	2.66	2.60
14	3.15	3.05	2.95	2.84	2.79	2.73	2.67	2.61	2.55	2.49
15	3.06	2.96	2.86	2.76	2.70	2.64	2.59	2.52	2.46	2.40
16	2.99	2.89	2.79	2.68	2.63	2.57	2.51	2.45	2.38	2.32
17	2.92	2.82	2.72	2.62	2.56	2.50	2.44	2.38	2.32	2.25
18	2.87	2.77	2.67	2.56	2.50	2.44	2.38	2.32	2.26	2.19
19	2.82	2.72	2.62	2.51	2.45	2.39	2.33	2.27	2.20	2.13
20	2.77	2.68	2.57	2.46	2.41	2.35	2.29	2.22	2.16	2.09
21	2.73	2.64	2.53	2.42	2.37	2.31	2.25	2.18	2.11	2.04
22	2.70	2.60	2.50	2.39	2.33	2.27	2.21	2.14	2.08	2.00
23	2.67	2.57	2.47	2.36	2.30	2.24	2.18	2.11	2.04	1.97
24	2.64	2.54	2.44	2.33	2.27	2.21	2.15	2.08	2.01	1.94
25	2.61	2.51	2.41	2.30	2.24	2.18	2.12	2.05	1.98	1.91
26	2.59	2.49	2.39	2.28	2.22	2.16	2.09	2.03	1.95	1.88
27	2.57	2.47	2.36	2.25	2.19	2.13	2.07	2.00	1.93	1.85
28	2.55	2.45	2.34	2.23	2.17	2.11	2.05	1.98	1.91	1.83
29	2.53	2.43	2.32	2.21	2.15	2.09	2.03	1.96	1.89	1.81
30	2.51	2.41	2.31	2.20	2.14	2.07	2.01	1.94	1.87	1.79
40	2.39	2.29	2.18	2.07	2.01	1.94	1.88	1.80	1.72	1.64
60	2.27	2.17	2.06	1.94	1.88	1.82	1.74	1.67	1.58	1.48
120	2.16	2.05	1.94	1.82	1.76	1.69	1.61	1.53	1.43	1.31
∞	2.05	1.94	1.83	1.71	1.64	1.57	1.48	1.39	1.27	1.00

TABLE 5 *(continued)*   Critical Values of the F Distribution for $\alpha = 0.01$

Denominator Degrees of Freedom (k_2)	Numerator Degrees of Freedom (k_1)								
	1	2	3	4	5	6	7	8	9
1	4,052	4,999.5	5,403	5,625	5,764	5,859	5,928	5,982	6,022
2	98.50	99.00	99.17	99.25	99.30	99.33	99.36	99.37	99.39
3	34.12	30.82	29.46	28.71	28.24	27.91	27.67	27.49	27.35
4	21.20	18.00	16.69	15.98	15.52	15.21	14.98	14.80	14.66
5	16.26	13.27	12.06	11.39	10.97	10.67	10.46	10.29	10.16
6	13.75	10.92	9.78	9.15	8.75	8.47	8.26	8.10	7.98
7	12.25	9.55	8.45	7.85	7.46	7.19	6.99	6.84	6.72
8	11.26	8.65	7.59	7.01	6.63	6.37	6.18	6.03	5.91
9	10.56	8.02	6.99	6.42	6.06	5.80	5.61	5.47	5.35
10	10.04	7.56	6.55	5.99	5.64	5.39	5.20	5.06	4.94
11	9.65	7.21	6.22	5.67	5.32	5.07	4.89	4.74	4.63
12	9.33	6.93	5.95	5.41	5.06	4.82	4.64	4.50	4.39
13	9.07	6.70	5.74	5.21	4.86	4.62	4.44	4.30	4.19
14	8.86	6.51	5.56	5.04	4.69	4.46	4.28	4.14	4.03
15	8.68	6.36	5.42	4.89	4.56	4.32	4.14	4.00	3.89
16	8.53	6.23	5.29	4.77	4.44	4.20	4.03	3.89	3.78
17	8.40	6.11	5.18	4.67	4.34	4.10	3.93	3.79	3.68
18	8.29	6.01	5.09	4.58	4.25	4.01	3.84	3.71	3.60
19	8.18	5.93	5.01	4.50	4.17	3.94	3.77	3.63	3.52
20	8.10	5.85	4.94	4.43	4.10	3.87	3.70	3.56	3.46
21	8.02	5.78	4.87	4.37	4.04	3.81	3.64	3.51	3.40
22	7.95	5.72	4.82	4.31	3.99	3.76	3.59	3.45	3.35
23	7.88	5.66	4.76	4.26	3.94	3.71	3.54	3.41	3.30
24	7.82	5.61	4.72	4.22	3.90	3.67	3.50	3.36	3.26
25	7.77	5.57	4.68	4.18	3.85	3.63	3.46	3.32	3.22
26	7.72	5.53	4.64	4.14	3.82	3.59	3.42	3.29	3.18
27	7.68	5.49	4.60	4.11	3.78	3.56	3.39	3.26	3.15
28	7.64	5.45	4.57	4.07	3.75	3.53	3.36	3.23	3.12
29	7.60	5.42	4.54	4.04	3.73	3.50	3.33	3.20	3.09
30	7.56	5.39	4.51	4.02	3.70	3.47	3.30	3.17	3.07
40	7.31	5.18	4.31	3.83	3.51	3.29	3.12	2.99	2.89
60	7.08	4.98	4.13	3.65	3.34	3.12	2.95	2.82	2.72
120	6.85	4.79	3.95	3.48	3.17	2.96	2.79	2.66	2.56
∞	6.63	4.61	3.78	3.32	3.02	2.80	2.64	2.51	2.41

TABLE 5 *(concluded)*   Critical Values of the *F* Distribution for $\alpha = 0.01$

Denominator Degrees of Freedom (k_2)	Numerator Degrees of Freedom (k_1)									
	10	12	15	20	24	30	40	60	120	∞
1	6,056	6,106	6,157	6,209	6,235	6,261	6,287	6,313	6,339	6,366
2	99.40	99.42	99.43	99.45	99.46	99.47	99.47	99.48	99.49	99.50
3	27.23	27.05	26.87	26.69	26.60	26.50	26.41	26.32	26.22	26.13
4	14.55	14.37	14.20	14.02	13.93	13.84	13.75	13.65	13.56	13.46
5	10.05	9.89	9.72	9.55	9.47	9.38	9.29	9.20	9.11	9.02
6	7.87	7.72	7.56	7.40	7.31	7.23	7.14	7.06	6.97	6.88
7	6.62	6.47	6.31	6.16	6.07	5.99	5.91	5.82	5.74	5.65
8	5.81	5.67	5.52	5.36	5.28	5.20	5.12	5.03	4.95	4.86
9	5.26	5.11	4.96	4.81	4.73	4.65	4.57	4.48	4.40	4.31
10	4.85	4.71	4.56	4.41	4.33	4.25	4.17	4.08	4.00	3.91
11	4.54	4.40	4.25	4.10	4.02	3.94	3.86	3.78	3.69	3.60
12	4.30	4.16	4.01	3.86	3.78	3.70	3.62	3.54	3.45	3.36
13	4.10	3.96	3.82	3.66	3.59	3.51	3.43	3.34	3.25	3.17
14	3.94	3.80	3.66	3.51	3.43	3.35	3.27	3.18	3.09	3.00
15	3.80	3.67	3.52	3.37	3.29	3.21	3.13	3.05	2.96	2.87
16	3.69	3.55	3.41	3.26	3.18	3.10	3.02	2.93	2.84	2.75
17	3.59	3.46	3.31	3.16	3.08	3.00	2.92	2.83	2.75	2.65
18	3.51	3.37	3.23	3.08	3.00	2.92	2.84	2.75	2.66	2.57
19	3.43	3.30	3.15	3.00	2.92	2.84	2.76	2.67	2.58	2.49
20	3.37	3.23	3.09	2.94	2.86	2.78	2.69	2.61	2.52	2.42
21	3.31	3.17	3.03	2.88	2.80	2.72	2.64	2.55	2.46	2.36
22	3.26	3.12	2.98	2.83	2.75	2.67	2.58	2.50	2.40	2.31
23	3.21	3.07	2.93	2.78	2.70	2.62	2.54	2.45	2.35	2.26
24	3.17	3.03	2.89	2.74	2.66	2.58	2.49	2.40	2.31	2.21
25	3.13	2.99	2.85	2.70	2.62	2.54	2.45	2.36	2.27	2.17
26	3.09	2.96	2.81	2.66	2.58	2.50	2.42	2.33	2.23	2.13
27	3.06	2.93	2.78	2.63	2.55	2.47	2.38	2.29	2.20	2.10
28	3.03	2.90	2.75	2.60	2.52	2.44	2.35	2.26	2.17	2.06
29	3.00	2.87	2.73	2.57	2.49	2.41	2.33	2.23	2.14	2.03
30	2.98	2.84	2.70	2.55	2.47	2.39	2.30	2.21	2.11	2.01
40	2.80	2.66	2.52	2.37	2.29	2.20	2.11	2.02	1.92	1.80
60	2.63	2.50	2.35	2.20	2.12	2.03	1.94	1.84	1.73	1.60
120	2.47	2.34	2.19	2.03	1.95	1.86	1.76	1.66	1.53	1.38
∞	2.32	2.18	2.04	1.88	1.79	1.70	1.59	1.47	1.32	1.00

Source: M. Merrington and C. M. Thompson, "Tables of Percentage Points of the Inverted Beta (*F*)-Distribution," *Biometrika* 33 (1943), pp. 73–88. Reproduced by permission of the *Biometrika* Trustees.

TABLE 5A The *F* Distribution for $\alpha = 0.05$ and $\alpha = 0.01$ for Many Possible Degrees of Freedom

Denominator Degrees of Freedom (k_2)	Numerator Degrees of Freedom (k_1)																							
	1	2	3	4	5	6	7	8	9	10	11	12	14	16	20	24	30	40	50	75	100	200	500	∞
1	161	200	216	225	230	234	237	239	241	242	243	244	245	246	248	249	250	251	252	253	253	254	254	254
	4,052	4,999	5,403	5,625	5,764	5,859	5,928	5,981	6,022	6,056	6,082	6,106	6,142	6,169	6,208	6,234	6,261	6,286	6,302	6,323	6,334	6,352	6,361	6,366
2	18.51	19.00	19.16	19.25	19.30	19.33	19.36	19.37	19.38	19.39	19.40	19.41	19.42	19.43	19.44	19.45	19.46	19.47	19.47	19.48	19.49	19.49	19.50	19.50
	98.49	99.00	99.17	99.25	99.30	99.33	99.36	99.37	99.39	99.40	99.41	99.42	99.43	99.44	99.45	99.46	99.47	99.48	99.48	99.49	99.49	99.49	99.50	99.50
3	10.13	9.55	9.28	9.12	9.01	8.94	8.88	8.84	8.81	8.78	8.76	8.74	8.71	8.69	8.66	8.64	8.62	8.60	8.58	8.57	8.56	8.54	8.54	8.53
	34.12	30.82	29.46	28.71	28.24	27.91	27.67	27.49	27.34	27.23	27.13	27.05	26.92	26.83	26.69	26.60	26.50	26.41	26.35	26.27	26.23	26.18	26.14	26.12
4	7.71	6.94	6.59	6.39	6.26	6.16	6.09	6.04	6.00	5.96	5.93	5.91	5.87	5.84	5.80	5.77	5.74	5.71	5.70	5.68	5.66	5.65	5.64	5.63
	21.20	18.00	16.69	15.98	15.52	15.21	14.98	14.80	14.66	14.54	14.45	14.37	14.24	14.15	14.02	13.93	13.83	13.74	13.69	13.61	13.57	13.52	13.48	13.46
5	6.61	5.79	5.41	5.19	5.05	4.95	4.88	4.82	4.78	4.74	4.70	4.68	4.64	4.60	4.56	4.53	4.50	4.46	4.44	4.42	4.40	4.38	4.37	4.36
	16.26	13.27	12.06	11.39	10.97	10.67	10.45	10.29	10.15	10.05	9.96	9.89	9.77	9.68	9.55	9.47	9.38	9.29	9.24	9.17	9.13	9.07	9.04	9.02
6	5.99	5.14	4.76	4.53	4.39	4.28	4.21	4.15	4.10	4.06	4.03	4.00	3.96	3.92	3.87	3.84	3.81	3.77	3.75	3.72	3.71	3.69	3.68	3.67
	13.74	10.92	9.78	9.15	8.75	8.47	8.26	8.10	7.98	7.87	7.79	7.72	7.60	7.52	7.39	7.31	7.23	7.14	7.09	7.02	6.99	6.94	6.90	6.88
7	5.59	4.74	4.35	4.12	3.97	3.87	3.79	3.73	3.68	3.63	3.60	3.57	3.52	3.49	3.44	3.41	3.38	3.34	3.32	3.29	3.28	3.25	3.24	3.23
	12.25	9.55	8.45	7.85	7.46	7.19	7.00	6.84	6.71	6.62	6.54	6.47	6.35	6.27	6.15	6.07	5.98	5.90	5.85	5.78	5.75	5.70	5.67	5.65
8	5.32	4.46	4.07	3.84	3.69	3.58	3.50	3.44	3.39	3.34	3.31	3.28	3.23	3.20	3.15	3.12	3.08	3.05	3.03	3.00	2.98	2.96	2.94	2.93
	11.26	8.65	7.59	7.01	6.63	6.37	6.19	6.03	5.91	5.82	5.74	5.67	5.56	5.48	5.36	5.28	5.20	5.11	5.06	5.00	4.96	4.91	4.88	4.86
9	5.12	4.26	3.86	3.63	3.48	3.37	3.29	3.23	3.18	3.13	3.10	3.07	3.02	2.98	2.93	2.90	2.86	2.82	2.80	2.77	2.76	2.73	2.72	2.71
	10.56	8.02	6.99	6.42	6.06	5.80	5.62	5.47	5.35	5.26	5.18	5.11	5.00	4.92	4.80	4.73	4.64	4.56	4.51	4.45	4.41	4.36	4.33	4.31
10	4.96	4.10	3.71	3.48	3.33	3.22	3.14	3.07	3.02	2.97	2.94	2.91	2.86	2.82	2.77	2.74	2.70	2.67	2.64	2.61	2.59	2.56	2.55	2.54
	10.04	7.56	6.55	5.99	5.64	5.39	5.21	5.06	4.95	4.85	4.78	4.71	4.60	4.52	4.41	4.33	4.25	4.17	4.12	4.05	4.01	3.96	3.93	3.91
11	4.84	3.98	3.59	3.36	3.20	3.09	3.01	2.95	2.90	2.86	2.82	2.79	2.74	2.70	2.65	2.61	2.57	2.53	2.50	2.47	2.45	2.42	2.41	2.40
	9.65	7.20	6.22	5.67	5.32	5.07	4.88	4.74	4.63	4.54	4.46	4.40	4.29	4.21	4.10	4.02	3.94	3.86	3.80	3.74	3.70	3.66	3.62	3.60
12	4.75	3.88	3.49	3.26	3.11	3.00	2.92	2.85	2.80	2.76	2.72	2.69	2.64	2.60	2.54	2.50	2.46	2.42	2.40	2.36	2.35	2.32	2.31	2.30
	9.33	6.93	5.95	5.41	5.06	4.82	4.65	4.50	4.39	4.30	4.22	4.16	4.05	3.98	3.86	3.78	3.70	3.61	3.56	3.49	3.46	3.41	3.38	3.36
13	4.67	3.80	3.41	3.18	3.02	2.92	2.84	2.77	2.72	2.67	2.63	2.60	2.55	2.51	2.46	2.42	2.38	2.34	2.32	2.28	2.26	2.24	2.22	2.21
	9.07	6.70	5.74	5.20	4.86	4.62	4.44	4.30	4.19	4.10	4.02	3.96	3.85	3.78	3.67	3.59	3.51	3.42	3.37	3.30	3.27	3.21	3.18	3.16

Numerator Degrees of Freedom (k_1)

Denominator Degrees of Freedom (k_2)	1	2	3	4	5	6	7	8	9	10	11	12	14	16	20	24	30	40	50	75	100	200	500	∞
14	4.60	3.74	3.34	3.11	2.96	2.85	2.77	2.70	2.65	2.60	2.56	2.53	2.48	2.44	2.39	2.35	2.31	2.27	2.24	2.21	2.19	2.16	2.14	2.13
	8.86	**6.51**	**5.56**	**5.03**	**4.69**	**4.46**	**4.28**	**4.14**	**4.03**	**3.94**	**3.86**	**3.80**	**3.70**	**3.62**	**3.51**	**3.43**	**3.34**	**3.26**	**3.21**	**3.14**	**3.11**	**3.06**	**3.02**	**3.00**
15	4.54	3.68	3.29	3.06	2.90	2.79	2.70	2.64	2.59	2.55	2.51	2.48	2.43	2.39	2.33	2.29	2.25	2.21	2.18	2.15	2.12	2.10	2.08	2.07
	8.68	**6.36**	**5.42**	**4.89**	**4.56**	**4.32**	**4.14**	**4.00**	**3.89**	**3.80**	**3.73**	**3.67**	**3.56**	**3.48**	**3.36**	**3.29**	**3.20**	**3.12**	**3.07**	**3.00**	**2.97**	**2.92**	**2.89**	**2.87**
16	4.49	3.63	3.24	3.01	2.85	2.74	2.66	2.59	2.54	2.49	2.45	2.42	2.37	2.33	2.28	2.24	2.20	2.16	2.13	2.09	2.07	2.04	2.02	2.01
	8.53	**6.23**	**5.29**	**4.77**	**4.44**	**4.20**	**4.03**	**3.89**	**3.78**	**3.69**	**3.61**	**3.55**	**3.45**	**3.37**	**3.25**	**3.18**	**3.10**	**3.01**	**2.96**	**2.88**	**2.86**	**2.80**	**2.77**	**2.75**
17	4.45	3.59	3.20	2.96	2.81	2.70	2.62	2.55	2.50	2.45	2.41	2.38	2.33	2.29	2.23	2.19	2.15	2.11	2.08	2.04	2.02	1.99	1.97	1.96
	8.40	**6.11**	**5.18**	**4.67**	**4.34**	**4.10**	**3.93**	**3.79**	**3.68**	**3.59**	**3.52**	**3.45**	**3.35**	**3.27**	**3.16**	**3.08**	**3.00**	**2.92**	**2.86**	**2.79**	**2.76**	**2.70**	**2.67**	**2.65**
18	4.41	3.55	3.16	2.93	2.77	2.66	2.58	2.51	2.46	2.41	2.37	2.34	2.29	2.25	2.19	2.15	2.11	2.07	2.04	2.00	1.98	1.95	1.93	1.92
	8.28	**6.01**	**5.09**	**4.58**	**4.25**	**4.01**	**3.85**	**3.71**	**3.60**	**3.51**	**3.44**	**3.37**	**3.27**	**3.19**	**3.07**	**3.00**	**2.91**	**2.83**	**2.78**	**2.71**	**2.68**	**2.62**	**2.59**	**2.57**
19	4.38	3.52	3.13	2.90	2.74	2.63	2.55	2.48	2.43	2.38	2.34	2.31	2.26	2.21	2.15	2.11	2.07	2.02	2.00	1.96	1.94	1.91	1.90	1.88
	8.18	**5.93**	**5.01**	**4.50**	**4.17**	**3.94**	**3.77**	**3.63**	**3.52**	**3.43**	**3.36**	**3.30**	**3.19**	**3.12**	**3.00**	**2.92**	**2.84**	**2.76**	**2.70**	**2.63**	**2.60**	**2.54**	**2.51**	**2.49**
20	4.35	3.49	3.10	2.87	2.71	2.60	2.52	2.45	2.40	2.35	2.31	2.28	2.23	2.18	2.12	2.08	2.04	1.99	1.96	1.92	1.90	1.87	1.85	1.84
	8.10	**5.85**	**4.94**	**4.43**	**4.10**	**3.87**	**3.71**	**3.56**	**3.45**	**3.37**	**3.30**	**3.23**	**3.13**	**3.05**	**2.94**	**2.86**	**2.77**	**2.69**	**2.63**	**2.56**	**2.53**	**2.47**	**2.44**	**2.42**
21	4.32	3.47	3.07	2.84	2.68	2.57	2.49	2.42	2.37	2.32	2.28	2.25	2.20	2.15	2.09	2.05	2.00	1.96	1.93	1.89	1.87	1.84	1.82	1.81
	8.02	**5.78**	**4.87**	**4.37**	**4.04**	**3.81**	**3.65**	**3.51**	**3.40**	**3.31**	**3.24**	**3.17**	**3.07**	**2.99**	**2.88**	**2.80**	**2.72**	**2.63**	**2.58**	**2.51**	**2.47**	**2.42**	**2.38**	**2.36**
22	4.30	3.44	3.05	2.82	2.66	2.55	2.47	2.40	2.35	2.30	2.26	2.23	2.18	2.13	2.07	2.03	1.98	1.93	1.91	1.87	1.84	1.81	1.80	1.78
	7.94	**5.72**	**4.82**	**4.31**	**3.99**	**3.76**	**3.59**	**3.45**	**3.35**	**3.26**	**3.18**	**3.12**	**3.02**	**2.94**	**2.83**	**2.75**	**2.67**	**2.58**	**2.53**	**2.46**	**2.42**	**2.37**	**2.33**	**2.31**
23	4.28	3.42	3.03	2.80	2.64	2.53	2.45	2.38	2.32	2.28	2.24	2.20	2.14	2.10	2.04	2.00	1.96	1.91	1.88	1.84	1.82	1.79	1.77	1.76
	7.88	**5.66**	**4.76**	**4.26**	**3.94**	**3.71**	**3.54**	**3.41**	**3.30**	**3.21**	**3.14**	**3.07**	**2.97**	**2.89**	**2.78**	**2.70**	**2.62**	**2.53**	**2.48**	**2.41**	**2.37**	**2.32**	**2.28**	**2.26**
24	4.26	3.40	3.01	2.78	2.62	2.51	2.43	2.36	2.30	2.26	2.22	2.18	2.13	2.09	2.02	1.98	1.94	1.89	1.86	1.82	1.80	1.76	1.74	1.73
	7.82	**5.61**	**4.72**	**4.22**	**3.90**	**3.67**	**3.50**	**3.36**	**3.25**	**3.17**	**3.09**	**3.03**	**2.93**	**2.85**	**2.74**	**2.66**	**2.58**	**2.49**	**2.44**	**2.36**	**2.33**	**2.27**	**2.23**	**2.21**
25	4.24	3.38	2.99	2.76	2.60	2.49	2.41	2.34	2.28	2.24	2.20	2.16	2.11	2.06	2.00	1.96	1.92	1.87	1.84	1.80	1.77	1.74	1.72	1.71
	7.77	**5.57**	**4.68**	**4.18**	**3.86**	**3.63**	**3.46**	**3.32**	**3.21**	**3.13**	**3.05**	**2.99**	**2.89**	**2.81**	**2.70**	**2.62**	**2.54**	**2.45**	**2.40**	**2.32**	**2.29**	**2.23**	**2.19**	**2.17**
26	4.22	3.37	2.98	2.74	2.59	2.47	2.39	2.32	2.27	2.22	2.18	2.15	2.10	2.05	1.99	1.95	1.90	1.85	1.82	1.78	1.76	1.72	1.70	1.69
	7.72	**5.53**	**4.64**	**4.14**	**3.82**	**3.59**	**3.42**	**3.29**	**3.17**	**3.09**	**3.02**	**2.96**	**2.86**	**2.77**	**2.66**	**2.58**	**2.50**	**2.41**	**2.36**	**2.28**	**2.25**	**2.19**	**2.15**	**2.13**

TABLE 5A (continued)   The F Distribution for α = 0.05 and α = 0.01 for Many Possible Degrees of Freedom

Numerator Degrees of Freedom (k_1)

Denominator Degrees of Freedom (k_2)	1	2	3	4	5	6	7	8	9	10	11	12	14	16	20	24	30	40	50	75	100	200	500	∞
27	4.21 7.68	3.35 5.49	2.96 4.60	2.73 4.11	2.57 3.79	2.46 3.56	2.37 3.39	2.30 3.26	2.25 3.14	2.20 3.06	2.16 2.98	2.13 2.93	2.08 2.83	2.03 2.74	1.97 2.63	1.93 2.55	1.88 2.47	1.84 2.38	1.80 2.33	1.76 2.25	1.74 2.21	1.71 2.16	1.68 2.12	1.67 2.10
28	4.20 7.64	3.34 5.45	2.95 4.57	2.71 4.07	2.56 3.76	2.44 3.53	2.36 3.36	2.29 3.23	2.24 3.11	2.19 3.03	2.15 2.95	2.12 2.90	2.06 2.80	2.02 2.71	1.96 2.60	1.91 2.52	1.87 2.44	1.81 2.35	1.78 2.30	1.75 2.22	1.72 2.18	1.69 2.13	1.67 2.09	1.65 2.06
29	4.18 7.60	3.33 5.42	2.93 4.54	2.70 4.04	2.54 3.73	2.43 3.50	2.35 3.33	2.28 3.20	2.22 3.08	2.18 3.00	2.14 2.92	2.10 2.87	2.05 2.77	2.00 2.68	1.94 2.57	1.90 2.49	1.85 2.41	1.80 2.32	1.77 2.27	1.73 2.19	1.71 2.15	1.68 2.10	1.65 2.06	1.64 2.03
30	4.17 7.56	3.32 5.39	2.92 4.51	2.69 4.02	2.53 3.70	2.42 3.47	2.34 3.30	2.27 3.17	2.21 3.06	2.16 2.98	2.12 2.90	2.09 2.84	2.04 2.74	1.99 2.66	1.93 2.55	1.89 2.47	1.84 2.38	1.79 2.29	1.76 2.24	1.72 2.16	1.69 2.13	1.66 2.07	1.64 2.03	1.62 2.01
32	4.15 7.50	3.30 5.34	2.90 4.46	2.67 3.97	2.51 3.66	2.40 3.42	2.32 3.25	2.25 3.12	2.19 3.01	2.14 2.94	2.10 2.86	2.07 2.80	2.02 2.70	1.97 2.62	1.91 2.51	1.86 2.42	1.82 2.34	1.76 2.25	1.74 2.20	1.69 2.12	1.67 2.08	1.64 2.02	1.61 1.98	1.59 1.96
34	4.13 7.44	3.28 5.29	2.88 4.42	2.65 3.93	2.49 3.61	2.38 3.38	2.30 3.21	2.23 3.08	2.17 2.97	2.12 2.89	2.08 2.82	2.05 2.76	2.00 2.66	1.95 2.58	1.89 2.47	1.84 2.38	1.80 2.30	1.74 2.21	1.71 2.15	1.67 2.08	1.64 2.04	1.61 1.98	1.59 1.94	1.57 1.91
36	4.11 7.39	3.26 5.25	2.86 4.38	2.63 3.89	2.48 3.58	2.36 3.35	2.28 3.18	2.21 3.04	2.15 2.94	2.10 2.86	2.06 2.78	2.03 2.72	1.98 2.62	1.93 2.54	1.87 2.43	1.82 2.35	1.78 2.26	1.72 2.17	1.69 2.12	1.65 2.04	1.62 2.00	1.59 1.94	1.56 1.90	1.55 1.87
38	4.10 7.35	3.25 5.21	2.85 4.34	2.62 3.86	2.46 3.54	2.35 3.32	2.26 3.15	2.19 3.02	2.14 2.91	2.09 2.82	2.05 2.75	2.02 2.69	1.96 2.59	1.92 2.51	1.85 2.40	1.80 2.32	1.76 2.22	1.71 2.14	1.67 2.08	1.63 2.00	1.60 1.97	1.57 1.90	1.54 1.86	1.53 1.84
40	4.08 7.31	3.23 5.18	2.84 4.31	2.61 3.83	2.45 3.51	2.34 3.29	2.25 3.12	2.18 2.99	2.12 2.88	2.07 2.80	2.04 2.73	2.00 2.66	1.95 2.56	1.90 2.49	1.84 2.37	1.79 2.29	1.74 2.20	1.69 2.11	1.66 2.05	1.61 1.97	1.59 1.94	1.55 1.88	1.53 1.84	1.51 1.81
42	4.07 7.27	3.22 5.15	2.83 4.29	2.59 3.80	2.44 3.49	2.32 3.26	2.24 3.10	2.17 2.96	2.11 2.86	2.06 2.77	2.02 2.70	1.99 2.64	1.94 2.54	1.89 2.46	1.82 2.35	1.78 2.26	1.73 2.17	1.68 2.08	1.64 2.02	1.60 1.94	1.57 1.91	1.54 1.85	1.51 1.80	1.49 1.78
44	4.06 7.24	3.21 5.12	2.82 4.26	2.58 3.78	2.43 3.46	2.31 3.24	2.23 3.07	2.16 2.94	2.10 2.84	2.05 2.75	2.01 2.68	1.98 2.62	1.92 2.52	1.88 2.44	1.81 2.32	1.76 2.24	1.72 2.15	1.66 2.06	1.63 2.00	1.58 1.92	1.56 1.88	1.52 1.82	1.50 1.78	1.48 1.75
46	4.05 7.21	3.20 5.10	2.81 4.24	2.57 3.76	2.42 3.44	2.30 3.22	2.22 3.05	2.14 2.92	2.09 2.82	2.04 2.73	2.00 2.66	1.97 2.60	1.91 2.50	1.87 2.42	1.80 2.30	1.75 2.22	1.71 2.13	1.65 2.04	1.62 1.98	1.57 1.90	1.54 1.86	1.51 1.80	1.48 1.76	1.46 1.72
48	4.04 7.19	3.19 5.08	2.80 4.22	2.56 3.74	2.41 3.42	2.30 3.20	2.21 3.04	2.14 2.90	2.08 2.80	2.03 2.71	1.99 2.64	1.96 2.58	1.90 2.48	1.86 2.40	1.79 2.28	1.74 2.20	1.70 2.11	1.64 2.02	1.61 1.96	1.56 1.88	1.53 1.84	1.50 1.78	1.47 1.73	1.45 1.70

Numerator Degrees of Freedom (k_1)

Denominator Degrees of Freedom (k_2)	1	2	3	4	5	6	7	8	9	10	11	12	14	16	20	24	30	40	50	75	100	200	500	∞
50	4.03 / 7.17	3.18 / 5.06	2.79 / 4.20	2.56 / 3.72	2.40 / 3.41	2.29 / 3.18	2.20 / 3.02	2.13 / 2.88	2.07 / 2.78	2.02 / 2.70	1.98 / 2.62	1.95 / 2.56	1.90 / 2.46	1.85 / 2.39	1.78 / 2.26	1.74 / 2.18	1.69 / 2.10	1.63 / 2.00	1.60 / 1.94	1.55 / 1.86	1.52 / 1.82	1.48 / 1.76	1.46 / 1.71	1.44 / 1.68
55	4.02 / 7.12	3.17 / 5.01	2.78 / 4.16	2.54 / 3.68	2.38 / 3.37	2.27 / 3.15	2.18 / 2.98	2.11 / 2.85	2.05 / 2.75	2.00 / 2.66	1.97 / 2.59	1.93 / 2.53	1.88 / 2.43	1.83 / 2.35	1.76 / 2.23	1.72 / 2.15	1.67 / 2.06	1.61 / 1.96	1.58 / 1.90	1.52 / 1.82	1.50 / 1.78	1.46 / 1.71	1.43 / 1.66	1.41 / 1.64
60	4.00 / 7.08	3.15 / 4.98	2.76 / 4.13	2.52 / 3.65	2.37 / 3.34	2.25 / 3.12	2.17 / 2.95	2.10 / 2.82	2.04 / 2.72	1.99 / 2.63	1.95 / 2.56	1.92 / 2.50	1.86 / 2.40	1.81 / 2.32	1.75 / 2.20	1.70 / 2.12	1.65 / 2.03	1.59 / 1.93	1.56 / 1.87	1.50 / 1.79	1.48 / 1.74	1.44 / 1.68	1.41 / 1.63	1.39 / 1.60
65	3.99 / 7.04	3.14 / 4.95	2.75 / 4.10	2.51 / 3.62	2.36 / 3.31	2.24 / 3.09	2.15 / 2.93	2.08 / 2.79	2.02 / 2.70	1.98 / 2.61	1.94 / 2.54	1.90 / 2.47	1.85 / 2.37	1.80 / 2.30	1.73 / 2.18	1.68 / 2.09	1.63 / 2.00	1.57 / 1.90	1.54 / 1.84	1.49 / 1.76	1.46 / 1.71	1.42 / 1.64	1.39 / 1.60	1.37 / 1.56
70	3.98 / 7.01	3.13 / 4.92	2.74 / 4.08	2.50 / 3.60	2.35 / 3.29	2.23 / 3.07	2.14 / 2.91	2.07 / 2.77	2.01 / 2.67	1.97 / 2.59	1.93 / 2.51	1.89 / 2.45	1.84 / 2.35	1.79 / 2.28	1.72 / 2.15	1.67 / 2.07	1.62 / 1.98	1.56 / 1.88	1.53 / 1.82	1.47 / 1.74	1.45 / 1.69	1.40 / 1.62	1.37 / 1.56	1.35 / 1.53
80	3.96 / 6.96	3.11 / 4.88	2.72 / 4.04	2.48 / 3.56	2.33 / 3.25	2.21 / 3.04	2.12 / 2.87	2.05 / 2.74	1.99 / 2.64	1.95 / 2.55	1.91 / 2.48	1.88 / 2.41	1.82 / 2.32	1.77 / 2.24	1.70 / 2.11	1.65 / 2.03	1.60 / 1.94	1.54 / 1.84	1.51 / 1.78	1.45 / 1.70	1.42 / 1.65	1.38 / 1.57	1.35 / 1.52	1.32 / 1.49
100	3.94 / 6.90	3.09 / 4.82	2.70 / 3.98	2.46 / 3.51	2.30 / 3.20	2.19 / 2.99	2.10 / 2.82	2.03 / 2.69	1.97 / 2.59	1.92 / 2.51	1.88 / 2.43	1.85 / 2.36	1.79 / 2.26	1.75 / 2.19	1.68 / 2.06	1.63 / 1.98	1.57 / 1.89	1.51 / 1.79	1.48 / 1.73	1.42 / 1.64	1.39 / 1.59	1.34 / 1.51	1.30 / 1.46	1.28 / 1.43
125	3.92 / 6.84	3.07 / 4.78	2.68 / 3.94	2.44 / 3.47	2.29 / 3.17	2.17 / 2.95	2.08 / 2.79	2.01 / 2.65	1.95 / 2.56	1.90 / 2.47	1.86 / 2.40	1.83 / 2.33	1.77 / 2.23	1.72 / 2.15	1.65 / 2.03	1.60 / 1.94	1.55 / 1.85	1.49 / 1.75	1.45 / 1.68	1.39 / 1.59	1.36 / 1.54	1.31 / 1.46	1.27 / 1.40	1.25 / 1.37
150	3.91 / 6.81	3.06 / 4.75	2.67 / 3.91	2.43 / 3.44	2.27 / 3.14	2.16 / 2.92	2.07 / 2.76	2.00 / 2.62	1.94 / 2.53	1.89 / 2.44	1.85 / 2.37	1.82 / 2.30	1.76 / 2.20	1.71 / 2.12	1.64 / 2.00	1.59 / 1.91	1.54 / 1.83	1.47 / 1.72	1.44 / 1.66	1.37 / 1.56	1.34 / 1.51	1.29 / 1.43	1.25 / 1.37	1.22 / 1.33
200	3.89 / 6.76	3.04 / 4.71	2.65 / 3.88	2.41 / 3.41	2.26 / 3.11	2.14 / 2.90	2.05 / 2.73	1.98 / 2.60	1.92 / 2.50	1.87 / 2.41	1.83 / 2.34	1.80 / 2.28	1.74 / 2.17	1.69 / 2.09	1.62 / 1.97	1.57 / 1.88	1.52 / 1.79	1.45 / 1.69	1.42 / 1.62	1.35 / 1.53	1.32 / 1.48	1.26 / 1.39	1.22 / 1.33	1.19 / 1.28
400	3.86 / 6.70	3.02 / 4.66	2.62 / 3.83	2.39 / 3.36	2.23 / 3.06	2.12 / 2.85	2.03 / 2.69	1.96 / 2.55	1.90 / 2.46	1.85 / 2.37	1.81 / 2.29	1.78 / 2.23	1.72 / 2.12	1.67 / 2.04	1.60 / 1.92	1.54 / 1.84	1.49 / 1.74	1.42 / 1.64	1.38 / 1.57	1.32 / 1.47	1.28 / 1.42	1.22 / 1.32	1.16 / 1.24	1.13 / 1.19
1,000	3.85 / 6.66	3.00 / 4.62	2.61 / 3.80	2.38 / 3.34	2.22 / 3.04	2.10 / 2.82	2.02 / 2.66	1.95 / 2.53	1.89 / 2.43	1.84 / 2.34	1.80 / 2.26	1.76 / 2.20	1.70 / 2.09	1.65 / 2.01	1.58 / 1.89	1.53 / 1.81	1.47 / 1.71	1.41 / 1.61	1.36 / 1.54	1.30 / 1.44	1.26 / 1.38	1.19 / 1.28	1.13 / 1.19	1.08 / 1.11
∞	3.84 / 6.63	2.99 / 4.60	2.60 / 3.78	2.37 / 3.32	2.21 / 3.02	2.09 / 2.80	2.01 / 2.64	1.94 / 2.51	1.88 / 2.41	1.83 / 2.32	1.79 / 2.24	1.75 / 2.18	1.69 / 2.07	1.64 / 1.99	1.57 / 1.87	1.52 / 1.79	1.46 / 1.69	1.40 / 1.59	1.35 / 1.52	1.28 / 1.41	1.24 / 1.36	1.17 / 1.25	1.11 / 1.15	1.00 / 1.00

Reprinted by permission from *Statistical Methods*, 7th ed., by George W. Snedecor and William G. Cochran, © 1980 by the Iowa State University Press, Ames, Iowa, 50010

TABLE 6   Critical Values of the Studentized Range Distribution for $\alpha = 0.05$

r

$n-r$	2	3	4	5	6	7	8	9	10	11	12	13	14	15	16	17	18	19	20
1	18.0	27.0	32.8	37.1	40.4	43.1	45.4	47.4	49.1	50.6	52.0	53.2	54.3	55.4	56.3	57.2	58.0	58.8	59.6
2	6.08	8.33	9.80	10.9	11.7	12.4	13.0	13.5	14.0	14.4	14.7	15.1	15.4	15.7	15.9	16.1	16.4	16.6	16.8
3	4.50	5.91	6.82	7.50	8.04	8.48	8.85	9.18	9.46	9.72	9.95	10.2	10.3	10.5	10.7	10.8	11.0	11.1	11.2
4	3.93	5.04	5.76	6.29	6.71	7.05	7.35	7.60	7.83	8.03	8.21	8.37	8.52	8.66	8.79	8.91	9.03	9.13	9.23
5	3.64	4.60	5.22	5.67	6.03	6.33	6.58	6.80	6.99	7.17	7.32	7.47	7.60	7.72	7.83	7.93	8.03	8.12	8.21
6	3.46	4.34	4.90	5.30	5.63	5.90	6.12	6.32	6.49	6.65	6.79	6.92	7.03	7.14	7.24	7.34	7.43	7.51	7.59
7	3.34	4.16	4.68	5.06	5.36	5.61	5.82	6.00	6.16	6.30	6.43	6.55	6.66	6.76	6.85	6.94	7.02	7.10	7.17
8	3.26	4.04	4.53	4.89	5.17	5.40	5.60	5.77	5.92	6.05	6.18	6.29	6.39	6.48	6.57	6.65	6.73	6.80	6.87
9	3.20	3.95	4.41	4.76	5.02	5.24	5.43	5.59	5.74	5.87	5.98	6.09	6.19	6.28	6.36	6.44	6.51	6.58	6.64
10	3.15	3.88	4.33	4.65	4.91	5.12	5.30	5.46	5.60	5.72	5.83	5.93	6.03	6.11	6.19	6.27	6.34	6.40	6.47
11	3.11	3.82	4.26	4.57	4.82	5.03	5.20	5.35	5.49	5.61	5.71	5.81	5.90	5.98	6.06	6.13	6.20	6.27	6.33
12	3.08	3.77	4.20	4.51	4.75	4.95	5.12	5.27	5.39	5.51	5.61	5.71	5.80	5.88	5.95	6.02	6.09	6.15	6.21
13	3.06	3.73	4.15	4.45	4.69	4.88	5.05	5.19	5.32	5.43	5.53	5.63	5.71	5.79	5.86	5.93	5.99	6.05	6.11
14	3.03	3.70	4.11	4.41	4.64	4.83	4.99	5.13	5.25	5.36	5.46	5.55	5.64	5.71	5.79	5.85	5.91	5.97	6.03
15	3.01	3.67	4.08	4.37	4.59	4.78	4.94	5.08	5.20	5.31	5.40	5.49	5.57	5.65	5.72	5.78	5.85	5.90	5.96
16	3.00	3.65	4.05	4.33	4.56	4.74	4.90	5.03	5.15	5.26	5.35	5.44	5.52	5.59	5.66	5.73	5.79	5.84	5.90
17	2.98	3.63	4.02	4.30	4.52	4.70	4.86	4.99	5.11	5.21	5.31	5.39	5.47	5.54	5.61	5.67	5.73	5.79	5.84
18	2.97	3.61	4.00	4.28	4.49	4.67	4.82	4.96	5.07	5.17	5.27	5.35	5.43	5.50	5.57	5.63	5.69	5.74	5.79
19	2.96	3.59	3.98	4.25	4.47	4.65	4.79	4.92	5.04	5.14	5.23	5.31	5.39	5.46	5.53	5.59	5.65	5.70	5.75
20	2.95	3.58	3.96	4.23	4.45	4.62	4.77	4.90	5.01	5.11	5.20	5.28	5.36	5.43	5.49	5.55	5.61	5.66	5.71
24	2.92	3.53	3.90	4.17	4.37	4.54	4.68	4.81	4.92	5.01	5.10	5.18	5.25	5.32	5.38	5.44	5.49	5.55	5.59
30	2.89	3.49	3.85	4.10	4.30	4.46	4.60	4.72	4.82	4.92	5.00	5.08	5.15	5.21	5.27	5.33	5.38	5.43	5.47
40	2.86	3.44	3.79	4.04	4.23	4.39	4.52	4.63	4.73	4.82	4.90	4.98	5.04	5.11	5.16	5.22	5.27	5.31	5.36
60	2.83	3.40	3.74	3.98	4.16	4.31	4.44	4.55	4.65	4.73	4.81	4.88	4.94	5.00	5.06	5.11	5.15	5.20	5.24
120	2.80	3.36	3.68	3.92	4.10	4.24	4.36	4.47	4.56	4.64	4.71	4.78	4.84	4.90	4.95	5.00	5.04	5.09	5.13
∞	2.77	3.31	3.63	3.86	4.03	4.17	4.29	4.39	4.47	4.55	4.62	4.68	4.74	4.80	4.85	4.89	4.93	4.97	5.01

TABLE 6 (concluded) **Critical Values of the Studentized Range Distribution for α = 0.01**

r

$n-r$	2	3	4	5	6	7	8	9	10	11	12	13	14	15	16	17	18	19	20
1	90.0	135	164	186	202	216	227	237	246	253	260	266	272	277	282	286	290	294	298
2	14.0	19.0	22.3	24.7	26.6	28.2	29.5	30.7	31.7	32.6	33.4	34.1	34.8	35.4	36.0	36.5	37.0	37.5	37.9
3	8.26	10.6	12.2	13.3	14.2	15.0	15.6	16.2	16.7	17.1	17.5	17.9	18.2	18.5	18.8	19.1	19.3	19.5	19.8
4	6.51	8.12	9.17	9.96	10.6	11.1	11.5	11.9	12.3	12.6	12.8	13.1	13.3	13.6	13.7	13.9	14.1	14.2	14.4
5	5.70	6.97	7.80	8.42	8.91	9.32	9.67	9.97	10.2	10.5	10.7	10.9	11.1	11.2	11.4	11.6	11.7	11.8	11.9
6	5.24	6.33	7.03	7.56	7.97	8.32	8.61	8.87	9.10	9.30	9.49	9.65	9.81	9.95	10.1	10.2	10.3	10.4	10.5
7	4.95	5.92	6.54	7.01	7.37	7.68	7.94	8.17	8.37	8.55	8.71	8.86	9.00	9.12	9.24	9.35	9.46	9.55	9.65
8	4.74	5.63	6.20	6.63	6.96	7.24	7.47	7.68	7.87	8.03	8.18	8.31	8.44	8.55	8.66	8.76	8.85	8.94	9.03
9	4.60	5.43	5.96	6.35	6.66	6.91	7.13	7.32	7.49	7.65	7.78	7.91	8.03	8.13	8.23	8.32	8.41	8.49	8.57
10	4.48	5.27	5.77	6.14	6.43	6.67	6.87	7.05	7.21	7.36	7.48	7.60	7.71	7.81	7.91	7.99	8.07	8.15	8.22
11	4.39	5.14	5.62	5.97	6.25	6.48	6.67	6.84	6.99	7.13	7.25	7.36	7.46	7.56	7.65	7.73	7.81	7.88	7.95
12	4.32	5.04	5.50	5.84	6.10	6.32	6.51	6.67	6.81	6.94	7.06	7.17	7.26	7.36	7.44	7.52	7.59	7.66	7.73
13	4.26	4.96	5.40	5.73	5.98	6.19	6.37	6.53	6.67	6.79	6.90	7.01	7.10	7.19	7.27	7.34	7.42	7.48	7.55
14	4.21	4.89	5.32	5.63	5.88	6.08	6.26	6.41	6.54	6.66	6.77	6.87	6.96	7.05	7.12	7.20	7.27	7.33	7.39
15	4.17	4.83	5.25	5.56	5.80	5.99	6.16	6.31	6.44	6.55	6.66	6.76	6.84	6.93	7.00	7.07	7.14	7.20	7.26
16	4.13	4.78	5.19	5.49	5.72	5.92	6.08	6.22	6.35	6.46	6.56	6.66	6.74	6.82	6.90	6.97	7.03	7.09	7.15
17	4.10	4.74	5.14	5.43	5.66	5.85	6.01	6.15	6.27	6.38	6.48	6.57	6.66	6.73	6.80	6.87	6.94	7.00	7.05
18	4.07	4.70	5.09	5.38	5.60	5.79	5.94	6.08	6.20	6.31	6.41	6.50	6.58	6.65	6.72	6.79	6.85	6.91	6.96
19	4.05	4.67	5.05	5.33	5.55	5.73	5.89	6.02	6.14	6.25	6.34	6.43	6.51	6.58	6.65	6.72	6.78	6.84	6.89
20	4.02	4.64	5.02	5.29	5.51	5.69	5.84	5.97	6.09	6.19	6.29	6.37	6.45	6.52	6.59	6.65	6.71	6.76	6.82
24	3.96	4.54	4.91	5.17	5.37	5.54	5.69	5.81	5.92	6.02	6.11	6.19	6.26	6.33	6.39	6.45	6.51	6.56	6.61
30	3.89	4.45	4.80	5.05	5.24	5.40	5.54	5.65	5.76	5.85	5.93	6.01	6.08	6.14	6.20	6.26	6.31	6.36	6.41
40	3.82	4.37	4.70	4.93	5.11	5.27	5.39	5.50	5.60	5.69	5.77	5.84	5.90	5.96	6.02	6.07	6.12	6.17	6.21
60	3.76	4.28	4.60	4.82	4.99	5.13	5.25	5.36	5.45	5.53	5.60	5.67	5.73	5.79	5.84	5.89	5.93	5.98	6.02
120	3.70	4.20	4.50	4.71	4.87	5.01	5.12	5.21	5.30	5.38	5.44	5.51	5.56	5.61	5.66	5.71	5.75	5.79	5.83
∞	3.64	4.12	4.40	4.60	4.76	4.88	4.99	5.08	5.16	5.23	5.29	5.35	5.40	5.45	5.49	5.54	5.57	5.61	5.65

E. S. Pearson and H. O. Hartley, eds., *Biometrika Tables for Statisticians*, vol. 1, 3rd ed. (Cambridge University Press, 1966). Reprinted by permission by the *Biometrika* Trustees.

TABLE 7 Critical Values of the Durbin-Watson Test Statistic for $\alpha = 0.05$

	$k = 1$		$k = 2$		$k = 3$		$k = 4$		$k = 5$	
n	d_L	d_U	d_L	d_U	d_L	d_U	d_L	d_U	d_L	d_U
15	1.08	1.36	0.95	1.54	0.82	1.75	0.69	1.97	0.56	2.21
16	1.10	1.37	0.98	1.54	0.86	1.73	0.74	1.93	0.62	2.15
17	1.13	1.38	1.02	1.54	0.90	1.71	0.78	1.90	0.67	2.10
18	1.16	1.39	1.05	1.53	0.93	1.69	0.82	1.87	0.71	2.06
19	1.18	1.40	1.08	1.53	0.97	1.68	0.86	1.85	0.75	2.02
20	1.20	1.41	1.10	1.54	1.00	1.68	0.90	1.83	0.79	1.99
21	1.22	1.42	1.13	1.54	1.03	1.67	0.93	1.81	0.83	1.96
22	1.24	1.43	1.15	1.54	1.05	1.66	0.96	1.80	0.86	1.94
23	1.26	1.44	1.17	1.54	1.08	1.66	0.99	1.79	0.90	1.92
24	1.27	1.45	1.19	1.55	1.10	1.66	1.01	1.78	0.93	1.90
25	1.29	1.45	1.21	1.55	1.12	1.66	1.04	1.77	0.95	1.89
26	1.30	1.46	1.22	1.55	1.14	1.65	1.06	1.76	0.98	1.88
27	1.32	1.47	1.24	1.56	1.16	1.65	1.08	1.76	1.01	1.86
28	1.33	1.48	1.26	1.56	1.18	1.65	1.10	1.75	1.03	1.85
29	1.34	1.48	1.27	1.56	1.20	1.65	1.12	1.74	1.05	1.84
30	1.35	1.49	1.28	1.57	1.21	1.65	1.14	1.74	1.07	1.83
31	1.36	1.50	1.30	1.57	1.23	1.65	1.16	1.74	1.09	1.83
32	1.37	1.50	1.31	1.57	1.24	1.65	1.18	1.73	1.11	1.82
33	1.38	1.51	1.32	1.58	1.26	1.65	1.19	1.73	1.13	1.81
34	1.39	1.51	1.33	1.58	1.27	1.65	1.21	1.73	1.15	1.81
35	1.40	1.52	1.34	1.58	1.28	1.65	1.22	1.73	1.16	1.80
36	1.41	1.52	1.35	1.59	1.29	1.65	1.24	1.73	1.18	1.80
37	1.42	1.53	1.36	1.59	1.31	1.66	1.25	1.72	1.19	1.80
38	1.43	1.54	1.37	1.59	1.32	1.66	1.26	1.72	1.21	1.79
39	1.43	1.54	1.38	1.60	1.33	1.66	1.27	1.72	1.22	1.79
40	1.44	1.54	1.39	1.60	1.34	1.66	1.29	1.72	1.23	1.79
45	1.48	1.57	1.43	1.62	1.38	1.67	1.34	1.72	1.29	1.78
50	1.50	1.59	1.46	1.63	1.42	1.67	1.38	1.72	1.34	1.77
55	1.53	1.60	1.49	1.64	1.45	1.68	1.41	1.72	1.38	1.77
60	1.55	1.62	1.51	1.65	1.48	1.69	1.44	1.73	1.41	1.77
65	1.57	1.63	1.54	1.66	1.50	1.70	1.47	1.73	1.44	1.77
70	1.58	1.64	1.55	1.67	1.52	1.70	1.49	1.74	1.46	1.77
75	1.60	1.65	1.57	1.68	1.54	1.71	1.51	1.74	1.49	1.77
80	1.61	1.66	1.59	1.69	1.56	1.72	1.53	1.74	1.51	1.77
85	1.62	1.67	1.60	1.70	1.57	1.72	1.55	1.75	1.52	1.77
90	1.63	1.68	1.61	1.70	1.59	1.73	1.57	1.75	1.54	1.78
95	1.64	1.69	1.62	1.71	1.60	1.73	1.58	1.75	1.56	1.78
100	1.65	1.69	1.63	1.72	1.61	1.74	1.59	1.76	1.57	1.78

TABLE 7 *(concluded)*   **Critical Values of the Durbin-Watson Test Statistic for $\alpha = 0.01$**

	$k = 1$		$k = 2$		$k = 3$		$k = 4$		$k = 5$	
n	d_L	d_U	d_L	d_U	d_L	d_U	d_L	d_U	d_L	d_U
15	0.81	1.07	0.70	1.25	0.59	1.46	0.49	1.70	0.39	1.96
16	0.84	1.09	0.74	1.25	0.63	1.44	0.53	1.66	0.44	1.90
17	0.87	1.10	0.77	1.25	0.67	1.43	0.57	1.63	0.48	1.85
18	0.90	1.12	0.80	1.26	0.71	1.42	0.61	1.60	0.52	1.80
19	0.93	1.13	0.83	1.26	0.74	1.41	0.65	1.58	0.56	1.77
20	0.95	1.15	0.86	1.27	0.77	1.41	0.68	1.57	0.60	1.74
21	0.97	1.16	0.89	1.27	0.80	1.41	0.72	1.55	0.63	1.71
22	1.00	1.17	0.91	1.28	0.83	1.40	0.75	1.54	0.66	1.69
23	1.02	1.19	0.94	1.29	0.86	1.40	0.77	1.53	0.70	1.67
24	1.05	1.20	0.96	1.30	0.88	1.41	0.80	1.53	0.72	1.66
25	1.05	1.21	0.98	1.30	0.90	1.41	0.83	1.52	0.75	1.65
26	1.07	1.22	1.00	1.31	0.93	1.41	0.85	1.52	0.78	1.64
27	1.09	1.23	1.02	1.32	0.95	1.41	0.88	1.51	0.81	1.63
28	1.10	1.24	1.04	1.32	0.97	1.41	0.90	1.51	0.83	1.62
29	1.12	1.25	1.05	1.33	0.99	1.42	0.92	1.51	0.85	1.61
30	1.13	1.26	1.07	1.34	1.01	1.42	0.94	1.51	0.88	1.61
31	1.15	1.27	1.08	1.34	1.02	1.42	0.96	1.51	0.90	1.60
32	1.16	1.28	1.10	1.35	1.04	1.43	0.98	1.51	0.92	1.60
33	1.17	1.29	1.11	1.36	1.05	1.43	1.00	1.51	0.94	1.59
34	1.18	1.30	1.13	1.36	1.07	1.43	1.01	1.51	0.95	1.59
35	1.19	1.31	1.14	1.37	1.08	1.44	1.03	1.51	0.97	1.59
36	1.21	1.32	1.15	1.38	1.10	1.44	1.04	1.51	0.99	1.59
37	1.22	1.32	1.16	1.38	1.11	1.45	1.06	1.51	1.00	1.59
38	1.23	1.33	1.18	1.39	1.12	1.45	1.07	1.52	1.02	1.58
39	1.24	1.34	1.19	1.39	1.14	1.45	1.09	1.52	1.03	1.58
40	1.25	1.34	1.20	1.40	1.15	1.46	1.10	1.52	1.05	1.58
45	1.29	1.38	1.24	1.42	1.20	1.48	1.16	1.53	1.11	1.58
50	1.32	1.40	1.28	1.45	1.24	1.49	1.20	1.54	1.16	1.59
55	1.36	1.43	1.32	1.47	1.28	1.51	1.25	1.55	1.21	1.59
60	1.38	1.45	1.35	1.48	1.32	1.52	1.28	1.56	1.25	1.60
65	1.41	1.47	1.38	1.50	1.35	1.53	1.31	1.57	1.28	1.61
70	1.43	1.49	1.40	1.52	1.37	1.55	1.34	1.58	1.31	1.61
75	1.45	1.50	1.42	1.53	1.39	1.56	1.37	1.59	1.34	1.62
80	1.47	1.52	1.44	1.54	1.42	1.57	1.39	1.60	1.36	1.62
85	1.48	1.53	1.46	1.55	1.43	1.58	1.41	1.60	1.39	1.63
90	1.50	1.54	1.47	1.56	1.45	1.59	1.43	1.61	1.41	1.64
95	1.51	1.55	1.49	1.57	1.47	1.60	1.45	1.62	1.42	1.64
100	1.52	1.56	1.50	1.58	1.48	1.60	1.46	1.63	1.44	1.65

TABLE 8 Cumulative Distribution Function: $F(r)$ for the Total Number of Runs R in Samples of Sizes n_1 and n_2

(n_1, n_2)	Number of Runs, r								
	2	3	4	5	6	7	8	9	10
(2, 3)	0.200	0.500	0.900	1.000					
(2, 4)	0.133	0.400	0.800	1.000					
(2, 5)	0.095	0.333	0.714	1.000					
(2, 6)	0.071	0.286	0.643	1.000					
(2, 7)	0.056	0.250	0.583	1.000					
(2, 8)	0.044	0.222	0.533	1.000					
(2, 9)	0.036	0.200	0.491	1.000					
(2, 10)	0.030	0.182	0.455	1.000					
(3, 3)	0.100	0.300	0.700	0.900	1.000				
(3, 4)	0.057	0.200	0.543	0.800	0.971	1.000			
(3, 5)	0.036	0.143	0.429	0.714	0.929	1.000			
(3, 6)	0.024	0.107	0.345	0.643	0.881	1.000			
(3, 7)	0.017	0.083	0.283	0.583	0.833	1.000			
(3, 8)	0.012	0.067	0.236	0.533	0.788	1.000			
(3, 9)	0.009	0.055	0.200	0.491	0.745	1.000			
(3, 10)	0.007	0.045	0.171	0.455	0.706	1.000			
(4, 4)	0.029	0.114	0.371	0.629	0.886	0.971	1.000		
(4, 5)	0.016	0.071	0.262	0.500	0.786	0.929	0.992	1.000	
(4, 6)	0.010	0.048	0.190	0.405	0.690	0.881	0.976	1.000	
(4, 7)	0.006	0.033	0.142	0.333	0.606	0.833	0.954	1.000	
(4, 8)	0.004	0.024	0.109	0.279	0.533	0.788	0.929	1.000	
(4, 9)	0.003	0.018	0.085	0.236	0.471	0.745	0.902	1.000	
(4, 10)	0.002	0.014	0.068	0.203	0.419	0.706	0.874	1.000	
(5, 5)	0.008	0.040	0.167	0.357	0.643	0.833	0.960	0.992	1.000
(5, 6)	0.004	0.024	0.110	0.262	0.522	0.738	0.911	0.976	0.998
(5, 7)	0.003	0.015	0.076	0.197	0.424	0.652	0.854	0.955	0.992
(5, 8)	0.002	0.010	0.054	0.152	0.347	0.576	0.793	0.929	0.984
(5, 9)	0.001	0.007	0.039	0.119	0.287	0.510	0.734	0.902	0.972
(5, 10)	0.001	0.005	0.029	0.095	0.239	0.455	0.678	0.874	0.958
(6, 6)	0.002	0.013	0.067	0.175	0.392	0.608	0.825	0.933	0.987
(6, 7)	0.001	0.008	0.043	0.121	0.296	0.500	0.733	0.879	0.966
(6, 8)	0.001	0.005	0.028	0.086	0.226	0.413	0.646	0.821	0.937
(6, 9)	0.000	0.003	0.019	0.063	0.175	0.343	0.566	0.762	0.902
(6, 10)	0.000	0.002	0.013	0.047	0.137	0.288	0.497	0.706	0.864
(7, 7)	0.001	0.004	0.025	0.078	0.209	0.383	0.617	0.791	0.922
(7, 8)	0.000	0.002	0.015	0.051	0.149	0.296	0.514	0.704	0.867
(7, 9)	0.000	0.001	0.010	0.035	0.108	0.231	0.427	0.622	0.806
(7, 10)	0.000	0.001	0.006	0.024	0.080	0.182	0.355	0.549	0.743
(8, 8)	0.000	0.001	0.009	0.032	0.100	0.214	0.405	0.595	0.786
(8, 9)	0.000	0.001	0.005	0.020	0.069	0.157	0.319	0.500	0.702
(8, 10)	0.000	0.000	0.003	0.013	0.048	0.117	0.251	0.419	0.621
(9, 9)	0.000	0.000	0.003	0.012	0.044	0.109	0.238	0.399	0.601
(9, 10)	0.000	0.000	0.002	0.008	0.029	0.077	0.179	0.319	0.510
(10, 10)	0.000	0.000	0.001	0.004	0.019	0.051	0.128	0.242	0.414

TABLE 8 *(concluded)* **Cumulative Distribution Function:** $F(r)$ **for the Total Number of Runs** R **in Samples of Sizes** n_1 **and** n_2

					Number of Runs, r					
(n_1, n_2)	2	3	4	5	6	7	8	9	10	
(2, 3)										
(2, 4)										
(2, 5)										
(2, 6)										
(2, 7)										
(2, 8)										
(2, 9)										
(2, 10)										
(3, 3)										
(3, 4)										
(3, 5)										
(3, 6)										
(3, 7)										
(3, 8)										
(3, 9)										
(3, 10)										
(4, 4)										
(4, 5)										
(4, 6)										
(4, 7)										
(4, 8)										
(4, 9)										
(4, 10)										
(5, 5)										
(5, 6)	1.000									
(5, 7)	1.000									
(5, 8)	1.000									
(5, 9)	1.000									
(5, 10)	1.000									
(6, 6)	0.998	1.000								
(6, 7)	0.992	0.999	1.000							
(6, 8)	0.984	0.998	1.000							
(6, 9)	0.972	0.994	1.000							
(6, 10)	0.958	0.990	1.000							
(7, 7)	0.975	0.996	0.999	1.000						
(7, 8)	0.949	0.988	0.998	1.000	1.000					
(7, 9)	0.916	0.975	0.994	0.999	1.000					
(7, 10)	0.879	0.957	0.990	0.998	1.000					
(8, 8)	0.900	0.968	0.991	0.999	1.000	1.000				
(8, 9)	0.843	0.939	0.980	0.996	0.999	1.000	1.000			
(8, 10)	0.782	0.903	0.964	0.990	0.998	1.000	1.000			
(9, 9)	0.762	0.891	0.956	0.988	0.997	1.000	1.000	1.000		
(9, 10)	0.681	0.834	0.923	0.974	0.992	0.999	1.000	1.000	1.000	
(10, 10)	0.586	0.758	0.872	0.949	0.981	0.996	0.999	1.000	1.000	1.000

Reproduced from F. Swed and C. Eisenhart, "Tables for Testing Randomness of Grouping in a Sequence of Alternatives," *Annals of Mathematical Statistics* 14 (1943) by permission of the authors and of the Editor, *Annals of Mathematical Statistics*.

TABLE 9 Cumulative Distribution Function of the Mann-Whitney U Statistic: $F(u)$ for $n_1 \leq n_2$ and $3 \leq n_2 \leq 10$

$n_2 = 3$

u	n_1		
	1	**2**	**3**
0	0.25	0.10	0.05
1	0.50	0.20	0.10
2		0.40	0.20
3		0.60	0.35
4			0.50

$n_2 = 4$

u	n_1			
	1	**2**	**3**	**4**
0	0.2000	0.0667	0.0286	0.0143
1	0.4000	0.1333	0.0571	0.0286
2	0.6000	0.2667	0.1143	0.0571
3		0.4000	0.2000	0.1000
4		0.6000	0.3143	0.1714
5			0.4286	0.2429
6			0.5714	0.3429
7				0.4429
8				0.5571

$n_2 = 5$

u	n_1				
	1	**2**	**3**	**4**	**5**
0	0.1667	0.0476	0.0179	0.0079	0.0040
1	0.3333	0.0952	0.0357	0.0159	0.0079
2	0.5000	0.1905	0.0714	0.0317	0.0159
3		0.2857	0.1250	0.0556	0.0278
4		0.4286	0.1964	0.0952	0.0476
5		0.5714	0.2857	0.1429	0.0754
6			0.3929	0.2063	0.1111
7			0.5000	0.2778	0.1548
8				0.3651	0.2103
9				0.4524	0.2738
10				0.5476	0.3452
11					0.4206
12					0.5000

TABLE 9 *(continued)*   **Cumulative Distribution Function of the Mann-Whitney U Statistic: $F(u)$ for $n_1 \leq n_2$ and $3 \leq n_2 \leq 10$**

$$n_2 = 6$$

$$n_1$$

u	1	2	3	4	5	6
0	0.1429	0.0357	0.0119	0.0048	0.0022	0.0011
1	0.2857	0.0714	0.0238	0.0095	0.0043	0.0022
2	0.4286	0.1429	0.0476	0.0190	0.0087	0.0043
3	0.5714	0.2143	0.0833	0.0333	0.0152	0.0076
4		0.3214	0.1310	0.0571	0.0260	0.0130
5		0.4286	0.1905	0.0857	0.0411	0.0206
6		0.5714	0.2738	0.1286	0.0628	0.0325
7			0.3571	0.1762	0.0887	0.0465
8			0.4524	0.2381	0.1234	0.0660
9			0.5476	0.3048	0.1645	0.0898
10				0.3810	0.2143	0.1201
11				0.4571	0.2684	0.1548
12				0.5429	0.3312	0.1970
13					0.3961	0.2424
14					0.4654	0.2944
15					0.5346	0.3496
16						0.4091
17						0.4686
18						0.5314

$$n_2 = 7$$

$$n_1$$

u	1	2	3	4	5	6	7
0	0.1250	0.0278	0.0083	0.0030	0.0013	0.0006	0.0003
1	0.2500	0.0556	0.0167	0.0061	0.0025	0.0012	0.0006
2	0.3750	0.1111	0.0333	0.0121	0.0051	0.0023	0.0012
3	0.5000	0.1667	0.0583	0.0212	0.0088	0.0041	0.0020
4		0.2500	0.0917	0.0364	0.0152	0.0070	0.0035
5		0.3333	0.1333	0.0545	0.0240	0.0111	0.0055
6		0.4444	0.1917	0.0818	0.0366	0.0175	0.0087
7		0.5556	0.2583	0.1152	0.0530	0.0256	0.0131
8			0.3333	0.1576	0.0745	0.0367	0.0189
9			0.4167	0.2061	0.1010	0.0507	0.0265
10			0.5000	0.2636	0.1338	0.0688	0.0364
11				0.3242	0.1717	0.0903	0.0487
12				0.3939	0.2159	0.1171	0.0641
13				0.4636	0.2652	0.1474	0.0825
14				0.5364	0.3194	0.1830	0.1043
15					0.3775	0.2226	0.1297
16					0.4381	0.2669	0.1588
17					0.5000	0.3141	0.1914
18						0.3654	0.2279
19						0.4178	0.2675
20						0.4726	0.3100
21						0.5274	0.3552
22							0.4024
23							0.4508
24							0.5000

TABLE 9 *(continued)*   Cumulative Distribution Function of the Mann-Whitney U Statistic: $F(u)$ for $n_1 \le n_2$ and $3 \le n_2 \le 10$

$n_2 = 8$

n_1

u	1	2	3	4	5	6	7	8
0	0.1111	0.0222	0.0061	0.0020	0.0008	0.0003	0.0002	0.0001
1	0.2222	0.0444	0.0121	0.0040	0.0016	0.0007	0.0003	0.0002
2	0.3333	0.0889	0.0242	0.0081	0.0031	0.0013	0.0006	0.0003
3	0.4444	0.1333	0.0424	0.0141	0.0054	0.0023	0.0011	0.0005
4	0.5556	0.2000	0.0667	0.0242	0.0093	0.0040	0.0019	0.0009
5		0.2667	0.0970	0.0364	0.0148	0.0063	0.0030	0.0015
6		0.3556	0.1394	0.0545	0.0225	0.0100	0.0047	0.0023
7		0.4444	0.1879	0.0768	0.0326	0.0147	0.0070	0.0035
8		0.5556	0.2485	0.1071	0.0466	0.0213	0.0103	0.0052
9			0.3152	0.1414	0.0637	0.0296	0.0145	0.0074
10			0.3879	0.1838	0.0855	0.0406	0.0200	0.0103
11			0.4606	0.2303	0.1111	0.0539	0.0270	0.0141
12			0.5394	0.2848	0.1422	0.0709	0.0361	0.0190
13				0.3414	0.1772	0.0906	0.0469	0.0249
14				0.4040	0.2176	0.1142	0.0603	0.0325
15				0.4667	0.2618	0.1412	0.0760	0.0415
16				0.5333	0.3108	0.1725	0.0946	0.0524
17					0.3621	0.2068	0.1159	0.0652
18					0.4165	0.2454	0.1405	0.0803
19					0.4716	0.2864	0.1678	0.0974
20					0.5284	0.3310	0.1984	0.1172
21						0.3773	0.2317	0.1393
22						0.4259	0.2679	0.1641
23						0.4749	0.3063	0.1911
24						0.5251	0.3472	0.2209
25							0.3894	0.2527
26							0.4333	0.2869
27							0.4775	0.3227
28							0.5225	0.3605
29								0.3992
30								0.4392
31								0.4796
32								0.5204

TABLE 9 (continued) Cumulative Distribution Function of the Mann-Whitney U Statistic: $F(u)$ for $n_1 \leq n_2$ and $3 \leq n_2 \leq 10$

$n_2 = 9$

n_1

u	1	2	3	4	5	6	7	8	9
0	0.1000	0.0182	0.0045	0.0014	0.0005	0.0002	0.0001	0.0000	0.0000
1	0.2000	0.0364	0.0091	0.0028	0.0010	0.0004	0.0002	0.0001	0.0000
2	0.3000	0.0727	0.0182	0.0056	0.0020	0.0008	0.0003	0.0002	0.0001
3	0.4000	0.1091	0.0318	0.0098	0.0035	0.0014	0.0006	0.0003	0.0001
4	0.5000	0.1636	0.0500	0.0168	0.0060	0.0024	0.0010	0.0005	0.0002
5		0.2182	0.0727	0.0252	0.0095	0.0038	0.0017	0.0008	0.0004
6		0.2909	0.1045	0.0378	0.0145	0.0060	0.0026	0.0012	0.0006
7		0.3636	0.1409	0.0531	0.0210	0.0088	0.0039	0.0019	0.0009
8		0.4545	0.1864	0.0741	0.0300	0.0128	0.0058	0.0028	0.0014
9		0.5455	0.2409	0.0993	0.0415	0.0180	0.0082	0.0039	0.0020
10			0.3000	0.1301	0.0599	0.0248	0.0115	0.0056	0.0028
11			0.3636	0.1650	0.0734	0.0332	0.0156	0.0076	0.0039
12			0.4318	0.2070	0.0949	0.0440	0.0209	0.0103	0.0053
13			0.5000	0.2517	0.1199	0.0567	0.0274	0.0137	0.0071
14				0.3021	0.1489	0.0723	0.0356	0.0180	0.0094
15				0.3552	0.1818	0.0905	0.0454	0.0232	0.0122
16				0.4126	0.2188	0.1119	0.0571	0.0296	0.0157
17				0.4699	0.2592	0.1361	0.0708	0.0372	0.0200
18				0.5301	0.3032	0.1638	0.0869	0.0464	0.0252
19					0.3497	0.1924	0.1052	0.0570	0.0313
20					0.3986	0.2280	0.1261	0.0694	0.0385
21					0.4491	0.2643	0.1496	0.0836	0.0470
22					0.5000	0.3035	0.1755	0.0998	0.0567
23						0.3445	0.2039	0.1179	0.0680
24						0.3878	0.2349	0.1383	0.0807
25						0.4320	0.2680	0.1606	0.0951
26						0.4773	0.3032	0.1852	0.1112
27						0.5227	0.3403	0.2117	0.1290
28							0.3788	0.2404	0.1487
29							0.4185	0.2707	0.1701
30							0.4591	0.3029	0.1933
31							0.5000	0.3365	0.2181
32								0.3715	0.2447
33								0.4074	0.2729
34								0.4442	0.3024
35								0.4813	0.3332
36								0.5187	0.3652
37									0.3981
38									0.4317
39									0.4657
40									0.5000

TABLE 9 *(concluded)*　Cumulative Distribution Function of the Mann-Whitney U Statistic: $F(u)$ for $n_1 \leq n_2$ and $3 \leq n_2 \leq 10$

$n_2 = 10$

n_1

u	1	2	3	4	5	6	7	8	9	10
0	0.0909	0.0152	0.0035	0.0010	0.0003	0.0001	0.0001	0.0000	0.0000	0.0000
1	0.1818	0.0303	0.0070	0.0020	0.0007	0.0002	0.0001	0.0000	0.0000	0.0000
2	0.2727	0.0606	0.0140	0.0040	0.0013	0.0005	0.0002	0.0001	0.0000	0.0000
3	0.3636	0.0909	0.0245	0.0070	0.0023	0.0009	0.0004	0.0002	0.0001	0.0000
4	0.4545	0.1364	0.0385	0.0120	0.0040	0.0015	0.0006	0.0003	0.0001	0.0001
5	0.5455	0.1818	0.0559	0.0180	0.0063	0.0024	0.0010	0.0004	0.0002	0.0001
6		0.2424	0.0804	0.0270	0.0097	0.0037	0.0015	0.0007	0.0003	0.0002
7		0.3030	0.1084	0.0380	0.0140	0.0055	0.0023	0.0010	0.0005	0.0002
8		0.3788	0.1434	0.0529	0.0200	0.0080	0.0034	0.0015	0.0007	0.0004
9		0.4545	0.1853	0.0709	0.0276	0.0112	0.0048	0.0022	0.0011	0.0005
10		0.5455	0.2343	0.0939	0.0376	0.0156	0.0068	0.0031	0.0015	0.0008
11			0.2867	0.1199	0.0496	0.0210	0.0093	0.0043	0.0021	0.0010
12			0.3462	0.1518	0.0646	0.0280	0.0125	0.0058	0.0028	0.0014
13			0.4056	0.1868	0.0823	0.0363	0.0165	0.0078	0.0038	0.0019
14			0.4685	0.2268	0.1032	0.0467	0.0215	0.0103	0.0051	0.0026
15			0.5315	0.2697	0.1272	0.0589	0.0277	0.0133	0.0066	0.0034
16				0.3177	0.1548	0.0736	0.0351	0.0171	0.0086	0.0045
17				0.3666	0.1855	0.0903	0.0439	0.0217	0.0110	0.0057
18				0.4196	0.2198	0.1099	0.0544	0.0273	0.0140	0.0073
19				0.4725	0.2567	0.1317	0.0665	0.0338	0.0175	0.0093
20				0.5275	0.2970	0.1566	0.0806	0.0416	0.0217	0.0116
21					0.3393	0.1838	0.0966	0.0506	0.0267	0.0144
22					0.3839	0.2139	0.1148	0.0610	0.0326	0.0177
23					0.4296	0.2461	0.1349	0.0729	0.0394	0.0216
24					0.4765	0.2811	0.1574	0.0864	0.0474	0.0262
25					0.5235	0.3177	0.1819	0.1015	0.0564	0.0315
26						0.3564	0.2087	0.1185	0.0667	0.0376
27						0.3962	0.2374	0.1371	0.0782	0.0446
28						0.4374	0.2681	0.1577	0.0912	0.0526
29						0.4789	0.3004	0.1800	0.1055	0.0615
30						0.5211	0.3345	0.2041	0.1214	0.0716
31							0.3698	0.2299	0.1388	0.0827
32							0.4063	0.2574	0.1577	0.0952
33							0.4434	0.2863	0.1781	0.1088
34							0.4811	0.3167	0.2001	0.1237
35							0.5189	0.3482	0.2235	0.1399
36								0.3809	0.2483	0.1575
37								0.4143	0.2745	0.1763
38								0.4484	0.3019	0.1965
39								0.4827	0.3304	0.2179
40								0.5173	0.3598	0.2406
41									0.3901	0.2644
42									0.4211	0.2894
43									0.4524	0.3153
44									0.4841	0.3421
45									0.5159	0.3697
46										0.3980
47										0.4267
48										0.4559
49										0.4853
50										0.5147

TABLE 10 Critical Values of the Wilcoxon *T* Statistic

One-Tailed	Two-Tailed	n = 5	n = 6	n = 7	n = 8	n = 9	n = 10
P = 0.05	P = 0.10	1	2	4	6	8	11
P = 0.025	P = 0.05		1	2	4	6	8
P = 0.01	P = 0.02			0	2	3	5
P = 0.005	P = 0.01				0	2	3

One-Tailed	Two-Tailed	n = 11	n = 12	n = 13	n = 14	n = 15	n = 16
P = 0.05	P = 0.10	14	17	21	26	30	36
P = 0.025	P = 0.05	11	14	17	21	25	30
P = 0.01	P = 0.02	7	10	13	16	20	24
P = 0.005	P = 0.01	5	7	10	13	16	19

One-Tailed	Two-Tailed	n = 17	n = 18	n = 19	n = 20	n = 21	n = 22
P = 0.05	P = 0.10	41	47	54	60	68	75
P = 0.025	P = 0.05	35	40	46	52	59	66
P = 0.01	P = 0.02	28	33	38	43	49	56
P = 0.005	P = 0.01	23	28	32	37	43	49

One-Tailed	Two-Tailed	n = 23	n = 24	n = 25	n = 26	n = 27	n = 28
P = 0.05	P = 0.10	83	92	101	110	120	130
P = 0.025	P = 0.05	73	81	90	98	107	117
P = 0.01	P = 0.02	62	69	77	85	93	102
P = 0.005	P = 0.01	55	68	68	76	84	92

One-Tailed	Two-Tailed	n = 29	n = 30	n = 31	n = 32	n = 33	n = 34
P = 0.05	P = 0.10	141	152	163	175	188	201
P = 0.025	P = 0.05	127	137	148	159	171	183
P = 0.01	P = 0.02	111	120	130	141	151	162
P = 0.005	P = 0.01	100	109	118	128	138	149

One-Tailed	Two-Tailed	n = 35	n = 36	n = 37	n = 38	n = 39
P = 0.05	P = 0.10	214	228	242	256	271
P = 0.025	P = 0.05	195	208	222	235	250
P = 0.01	P = 0.02	174	186	198	211	224
P = 0.005	P = 0.01	160	171	183	195	208

One-Tailed	Two-Tailed	n = 40	n = 41	n = 42	n = 43	n = 44	n = 45
P = 0.05	P = 0.10	287	303	319	336	353	371
P = 0.025	P = 0.05	264	279	295	311	327	344
P = 0.01	P = 0.02	238	252	267	281	297	313
P = 0.005	P = 0.01	221	234	248	262	277	292

One-Tailed	Two-Tailed	n = 46	n = 47	n = 48	n = 49	n = 50
P = 0.05	P = 0.10	389	408	427	446	466
P = 0.025	P = 0.05	361	379	397	415	434
P = 0.01	P = 0.02	329	345	362	380	398
P = 0.005	P = 0.01	307	323	339	356	373

Reproduced from F. Wilcoxon and R. A. Wilcox, *Some Rapid Approximate Statistical Procedures* (1964), p. 28, with the permission of American Cyanamid Company.

TABLE 11   Critical Values of Spearman's Rank Correlation Coefficient

n	$\alpha = 0.05$	$\alpha = 0.025$	$\alpha = 0.01$	$\alpha = 0.005$
5	0.900	—	—	—
6	0.829	0.886	0.943	—
7	0.714	0.786	0.893	—
8	0.643	0.738	0.833	0.881
9	0.600	0.683	0.783	0.833
10	0.564	0.648	0.745	0.794
11	0.523	0.623	0.736	0.818
12	0.497	0.591	0.703	0.780
13	0.475	0.566	0.673	0.745
14	0.457	0.545	0.646	0.716
15	0.441	0.525	0.623	0.689
16	0.425	0.507	0.601	0.666
17	0.412	0.490	0.582	0.645
18	0.399	0.476	0.564	0.625
19	0.388	0.462	0.549	0.608
20	0.377	0.450	0.534	0.591
21	0.368	0.438	0.521	0.576
22	0.359	0.428	0.508	0.562
23	0.351	0.418	0.496	0.549
24	0.343	0.409	0.485	0.537
25	0.336	0.400	0.475	0.526
26	0.329	0.392	0.465	0.515
27	0.323	0.385	0.456	0.505
28	0.317	0.377	0.448	0.496
29	0.311	0.370	0.440	0.487
30	0.305	0.364	0.432	0.478

Reproduced by permission from E. G. Olds, "Distribution of Sums of Squares of Rank Differences for Small Samples," *Annals of Mathematical Statistics* 9 (1938).

TABLE 12 Poisson Probability Distribution

This table gives values of

$$P(x) = \frac{\mu^x e^{-\mu}}{x!}$$

					μ					
x	.005	.01	.02	.03	.04	.05	.06	.07	.08	.09
0	.9950	.9900	.9802	.9704	.9608	.9512	.9418	.9324	.9231	.9139
1	.0050	.0099	.0192	.0291	.0384	.0476	.0565	.0653	.0738	.0823
2	.0000	.0000	.0002	.0004	.0008	.0012	.0017	.0023	.0030	.0037
3	.0000	.0000	.0000	.0000	.0000	.0000	.0000	.0001	.0001	.0001

					μ					
x	0.1	0.2	0.3	0.4	0.5	0.6	0.7	0.8	0.9	1.0
0	.9048	.8187	.7408	.6703	.6065	.5488	.4966	.4493	.4066	.3679
1	.0905	.1637	.2222	.2681	.3033	.3293	.3476	.3595	.3659	.3679
2	.0045	.0164	.0333	.0536	.0758	.0988	.1217	.1438	.1647	.1839
3	.0002	.0011	.0033	.0072	.0126	.0198	.0284	.0383	.0494	.0613
4	.0000	.0001	.0002	.0007	.0016	.0030	.0050	.0077	.0111	.0153
5	.0000	.0000	.0000	.0001	.0002	.0004	.0007	.0012	.0020	.0031
6	.0000	.0000	.0000	.0000	.0000	.0000	.0001	.0002	.0003	.0005
7	.0000	.0000	.0000	.0000	.0000	.0000	.0000	.0000	.0000	.0001

					μ					
x	1.1	1.2	1.3	1.4	1.5	1.6	1.7	1.8	1.9	2.0
0	.3329	.3012	.2725	.2466	.2231	.2019	.1827	.1653	.1496	.1353
1	.3662	.3614	.3543	.3452	.3347	.3230	.3106	.2975	.2842	.2707
2	.2014	.2169	.2303	.2417	.2510	.2584	.2640	.2678	.2700	.2707
3	.0738	.0867	.0998	.1128	.1255	.1378	.1496	.1607	.1710	.1804
4	.0203	.0260	.0324	.0395	.0471	.0551	.0636	.0723	.0812	.0902
5	.0045	.0062	.0084	.0111	.0141	.0176	.0216	.0260	.0309	.0361
6	.0008	.0012	.0018	.0026	.0035	.0047	.0061	.0078	.0098	.0120
7	.0001	.0002	.0003	.0005	.0008	.0011	.0015	.0020	.0027	.0034
8	.0000	.0000	.0001	.0001	.0001	.0002	.0003	.0005	.0006	.0009
9	.0000	.0000	.0000	.0000	.0000	.0000	.0001	.0001	.0001	.0002

					μ					
x	2.1	2.2	2.3	2.4	2.5	2.6	2.7	2.8	2.9	3.0
0	.1225	.1108	.1003	.0907	.0821	.0743	.0672	.0608	.0550	.0498
1	.2572	.2438	.2306	.2177	.2052	.1931	.1815	.1703	.1596	.1494
2	.2700	.2681	.2652	.2613	.2565	.2510	.2450	.2384	.2314	.2240
3	.1890	.1966	.2033	.2090	.2138	.2176	.2205	.2225	.2237	.2240
4	.0992	.1082	.1169	.1254	.1336	.1414	.1488	.1557	.1622	.1680
5	.0417	.0476	.0538	.0602	.0668	.0735	.0804	.0872	.0940	.1008
6	.0146	.0174	.0206	.0241	.0278	.0319	.0362	.0407	.0455	.0504
7	.0044	.0055	.0068	.0083	.0099	.0118	.0139	.0163	.0188	.0216
8	.0011	.0015	.0019	.0025	.0031	.0038	.0047	.0057	.0068	.0081
9	.0003	.0004	.0005	.0007	.0009	.0011	.0014	.0018	.0022	.0027
10	.0001	.0001	.0001	.0002	.0002	.0003	.0004	.0005	.0006	.0008
11	.0000	.0000	.0000	.0000	.0000	.0001	.0001	.0001	.0002	.0002
12	.0000	.0000	.0000	.0000	.0000	.0000	.0000	.0000	.0000	.0001

TABLE 12 *(continued)* **Poisson Probability Distribution**

					μ					
x	3.1	3.2	3.3	3.4	3.5	3.6	3.7	3.8	3.9	4.0
0	.0450	.0408	.0369	.0334	.0302	.0273	.0247	.0224	.0202	.0183
1	.1397	.1304	.1217	.1135	.1057	.0984	.0915	.0850	.0789	.0733
2	.2165	.2087	.2008	.1929	.1850	.1771	.1692	.1615	.1539	.1465
3	.2237	.2226	.2209	.2186	.2158	.2125	.2087	.2046	.2001	.1954
4	.1734	.1781	.1823	.1858	.1888	.1912	.1931	.1944	.1951	.1954
5	.1075	.1140	.1203	.1264	.1322	.1377	.1429	.1477	.1522	.1563
6	.0555	.0608	.0662	.0716	.0771	.0826	.0881	.0936	.0989	.1042
7	.0246	.0278	.0312	.0348	.0385	.0425	.0466	.0508	.0551	.0595
8	.0095	.0111	.0129	.0148	.0169	.0191	.0215	.0241	.0269	.0298
9	.0033	.0040	.0047	.0056	.0066	.0076	.0089	.0102	.0116	.0132
10	.0010	.0013	.0016	.0019	.0023	.0028	.0033	.0039	.0045	.0053
11	.0003	.0004	.0005	.0006	.0007	.0009	.0011	.0013	.0016	.0019
12	.0001	.0001	.0001	.0002	.0002	.0003	.0003	.0004	.0005	.0006
13	.0000	.0000	.0000	.0000	.0001	.0001	.0001	.0001	.0002	.0002
14	.0000	.0000	.0000	.0000	.0000	.0000	.0000	.0000	.0000	.0001

					μ					
x	4.1	4.2	4.3	4.4	4.5	4.6	4.7	4.8	4.9	5.0
0	.0166	.0150	.0136	.0123	.0111	.0101	.0091	.0082	.0074	.0067
1	.0679	.0630	.0583	.0540	.0500	.0462	.0427	.0395	.0365	.0337
2	.1393	.1323	.1254	.1188	.1125	.1063	.1005	.0948	.0894	.0842
3	.1904	.1852	.1798	.1743	.1687	.1631	.1574	.1517	.1460	.1404
4	.1951	.1944	.1933	.1917	.1898	.1875	.1849	.1820	.1789	.1755
5	.1600	.1633	.1662	.1687	.1708	.1725	.1738	.1747	.1753	.1755
6	.1093	.1143	.1191	.1237	.1281	.1323	.1362	.1398	.1432	.1462
7	.0640	.0686	.0732	.0778	.0824	.0869	.0914	.0959	.1002	.1044
8	.0328	.0360	.0393	.0428	.0463	.0500	.0537	.0575	.0614	.0653
9	.0150	.0168	.0188	.0209	.0232	.0255	.0280	.0307	.0334	.0363
10	.0061	.0071	.0081	.0092	.0104	.0118	.0132	.0147	.0164	.0181
11	.0023	.0027	.0032	.0037	.0043	.0049	.0056	.0064	.0073	.0082
12	.0008	.0009	.0011	.0014	.0016	.0019	.0022	.0026	.0030	.0034
13	.0002	.0003	.0004	.0005	.0006	.0007	.0008	.0009	.0011	.0013
14	.0001	.0001	.0001	.0001	.0002	.0002	.0003	.0003	.0004	.0005
15	.0000	.0000	.0000	.0000	.0001	.0001	.0001	.0001	.0001	.0002

					μ					
x	5.1	5.2	5.3	5.4	5.5	5.6	5.7	5.8	5.9	6.0
0	.0061	.0055	.0050	.0045	.0041	.0037	.0033	.0030	.0027	.0025
1	.0311	.0287	.0265	.0244	.0225	.0207	.0191	.0176	.0162	.0149
2	.0793	.0746	.0701	.0659	.0618	.0580	.0544	.0509	.0477	.0446
3	.1348	.1293	.1239	.1185	.1133	.1082	.1033	.0985	.0938	.0892
4	.1719	.1681	.1641	.1600	.1558	.1515	.1472	.1428	.1383	.1339
5	.1753	.1748	.1740	.1728	.1714	.1697	.1678	.1656	.1632	.1606
6	.1490	.1515	.1537	.1555	.1571	.1584	.1594	.1601	.1605	.1606
7	.1086	.1125	.1163	.1200	.1234	.1267	.1298	.1326	.1353	.1377
8	.0692	.0731	.0771	.0810	.0849	.0887	.0925	.0962	.0998	.1033
9	.0392	.0423	.0454	.0486	.0519	.0552	.0586	.0620	.0654	.0688
10	.0200	.0220	.0241	.0262	.0285	.0309	.0334	.0359	.0386	.0413
11	.0093	.0104	.0116	.0129	.0143	.0157	.0173	.0190	.0207	.0225
12	.0039	.0045	.0051	.0058	.0065	.0073	.0082	.0092	.0102	.0113
13	.0015	.0018	.0021	.0024	.0028	.0032	.0036	.0041	.0046	.0052
14	.0006	.0007	.0008	.0009	.0011	.0013	.0015	.0017	.0019	.0022
15	.0002	.0002	.0003	.0003	.0004	.0005	.0006	.0007	.0008	.0009
16	.0001	.0001	.0001	.0001	.0001	.0002	.0002	.0002	.0003	.0003
17	.0000	.0000	.0000	.0000	.0000	.0001	.0001	.0001	.0001	.0001

TABLE 12 *(concluded)* **Poisson Probability Distribution**

					μ					
x	6.1	6.2	6.3	6.4	6.5	6.6	6.7	6.8	6.9	7.0
0	.0022	.0020	.0019	.0017	.0015	.0014	.0012	.0011	.0010	.0009
1	.0137	.0126	.0116	.0106	.0098	.0090	.0082	.0076	.0070	.0064
2	.0417	.0390	.0364	.0340	.0318	.0296	.0276	.0258	.0240	.0223
3	.0848	.0806	.0765	.0726	.0688	.0652	.0617	.0584	.0552	.0521
4	.1294	.1249	.1205	.1162	.1118	.1076	.1034	.0992	.0952	.0912
5	.1579	.1549	.1519	.1487	.1454	.1420	.1385	.1349	.1314	.1277
6	.1605	.1601	.1595	.1586	.1575	.1562	.1546	.1529	.1511	.1490
7	.1399	.1418	.1435	.1450	.1462	.1472	.1480	.1486	.1489	.1490
8	.1066	.1099	.1130	.1160	.1188	.1215	.1240	.1263	.1284	.1304
9	.0723	.0757	.0791	.0825	.0858	.0891	.0923	.0954	.0985	.1014
10	.0441	.0469	.0498	.0528	.0558	.0588	.0618	.0649	.0679	.0710
11	.0245	.0265	.0285	.0307	.0330	.0353	.0377	.0401	.0426	.0452
12	.0124	.0137	.0150	.0164	.0179	.0194	.0210	.0227	.0245	.0264
13	.0058	.0065	.0073	.0081	.0089	.0098	.0108	.0119	.0130	.0142
14	.0025	.0029	.0033	.0037	.0041	.0046	.0052	.0058	.0064	.0071
15	.0010	.0012	.0014	.0016	.0018	.0020	.0023	.0026	.0029	.0033
16	.0004	.0005	.0005	.0006	.0007	.0008	.0010	.0011	.0013	.0014
17	.0001	.0002	.0002	.0002	.0003	.0003	.0004	.0004	.0005	.0006
18	.0000	.0001	.0001	.0001	.0001	.0001	.0001	.0002	.0002	.0002
19	.0000	.0000	.0000	.0000	.0000	.0000	.0000	.0001	.0001	.0001

					μ					
x	7.1	7.2	7.3	7.4	7.5	7.6	7.7	7.8	7.9	8.0
0	.0008	.0007	.0007	.0006	.0006	.0005	.0005	.0004	.0004	.0003
1	.0059	.0054	.0049	.0045	.0041	.0038	.0035	.0032	.0029	.0027
2	.0208	.0194	.0180	.0167	.0156	.0145	.0134	.0125	.0116	.0107
3	.0492	.0464	.0438	.0413	.0389	.0366	.0345	.0324	.0305	.0286
4	.0874	.0836	.0799	.0764	.0729	.0696	.0663	.0632	.0602	.0573
5	.1241	.1204	.1167	.1130	.1094	.1057	.1021	.0986	.0951	.0916
6	.1468	.1445	.1420	.1394	.1367	.1339	.1311	.1282	.1252	.1221
7	.1489	.1486	.1481	.1474	.1465	.1454	.1442	.1428	.1413	.1396
8	.1321	.1337	.1351	.1363	.1373	.1382	.1388	.1392	.1395	.1396
9	.1042	.1070	.1096	.1121	.1144	.1167	.1187	.1207	.1224	.1241
10	.0740	.0770	.0800	.0829	.0858	.0887	.0914	.0941	.0967	.0993
11	.0478	.0504	.0531	.0558	.0585	.0613	.0640	.0667	.0695	.0722
12	.0283	.0303	.0323	.0344	.0366	.0388	.0411	.0434	.0457	.0481
13	.0154	.0168	.0181	.0196	.0211	.0227	.0243	.0260	.0278	.0296
14	.0078	.0086	.0095	.0104	.0113	.0123	.0134	.0145	.0157	.0169
15	.0037	.0041	.0046	.0051	.0057	.0062	.0069	.0075	.0083	.0090
16	.0016	.0019	.0021	.0024	.0026	.0030	.0033	.0037	.0041	.0045
17	.0007	.0008	.0009	.0010	.0012	.0013	.0015	.0017	.0019	.0021
18	.0003	.0003	.0004	.0004	.0005	.0006	.0006	.0007	.0008	.0009
19	.0001	.0001	.0001	.0002	.0002	.0002	.0003	.0003	.0003	.0004
20	.0000	.0000	.0001	.0001	.0001	.0001	.0001	.0001	.0001	.0002
21	.0000	.0000	.0000	.0000	.0000	.0000	.0000	.0000	.0001	.0001

TABLE 13   **Control Chart Constants**

n	For Estimating Sigma		For $\bar{X}$ Chart		For $\bar{X}$ Chart (Standard Given)	For R Chart		For R Chart (Standard Given)		For s Chart (Standard Given)			
	c_4	d_2	A_2	A_3	A	D_3	D_4	D_1	D_2	B_3	B_4	B_5	B_6
2	0.7979	1.128	1.880	2.659	2.121	0	3.267	0	3.686	0	3.267	0	2.606
3	0.8862	1.693	1.023	1.954	1.732	0	2.575	0	4.358	0	2.568	0	2.276
4	0.9213	2.059	0.729	1.628	1.500	0	2.282	0	4.698	0	2.266	0	2.088
5	0.9400	2.326	0.577	1.427	1.342	0	2.115	0	4.918	0	2.089	0	1.964
6	0.9515	2.534	0.483	1.287	1.225	0	2.004	0	5.078	0.030	1.970	0.029	1.874
7	0.9594	2.704	0.419	1.182	1.134	0.076	1.924	0.205	5.203	0.118	1.882	0.113	1.806
8	0.9650	2.847	0.373	1.099	1.061	0.136	1.864	0.387	5.307	0.185	1.815	0.179	1.751
9	0.9693	2.970	0.337	1.032	1.000	0.184	1.816	0.546	5.394	0.239	1.761	0.232	1.707
10	0.9727	3.078	0.308	0.975	0.949	0.223	1.777	0.687	5.469	0.284	1.716	0.276	1.669
15	0.9823	3.472	0.223	0.789	0.775	0.348	1.652	1.207	5.737	0.428	1.572	0.421	1.544
20	0.9869	3.735	0.180	0.680	0.671	0.414	1.586	1.548	5.922	0.510	1.490	0.504	1.470
25	0.9896	3.931	0.153	0.606	0.600	0.459	1.541	1.804	6.058	0.565	1.435	0.559	1.420

T. P. Ryan, *Statistical Methods for Quality Improvement* © 1989 New York: John Wiley & Sons. This material is used by permission of John Wiley & Sons, Inc.

TABLE 14   Random Numbers

1559	9068	9290	8303	8508	8954	1051	6677	6415	0342
5550	6245	7313	0117	7652	5069	6354	7668	1096	5780
4735	6214	8037	1385	1882	0828	2957	0530	9210	0177
5333	1313	3063	1134	8676	6241	9960	5304	1582	6198
8495	2956	1121	8484	2920	7934	0670	5263	0968	0069
1947	3353	1197	7363	9003	9313	3434	4261	0066	2714
4785	6325	1868	5020	9100	0823	7379	7391	1250	5501
9972	9163	5833	0100	5758	3696	6496	6297	5653	7782
0472	4629	2007	4464	3312	8728	1193	2497	4219	5339
4727	6994	1175	5622	2341	8562	5192	1471	7206	2027
3658	3226	5981	9025	1080	1437	6721	7331	0792	5383
6906	9758	0244	0259	4609	1269	5957	7556	1975	7898
3793	6916	0132	8873	8987	4975	4814	2098	6683	0901
3376	5966	1614	4025	0721	1537	6695	6090	8083	5450
6126	0224	7169	3596	1593	5097	7286	2686	1796	1150
0466	7566	1320	8777	8470	5448	9575	4669	1402	3905
9908	9832	8185	8835	0384	3699	1272	1181	8627	1968
7594	3636	1224	6808	1184	3404	6752	4391	2016	6167
5715	9301	5847	3524	0077	6674	8061	5438	6508	9673
7932	4739	4567	6797	4540	8488	3639	9777	1621	7244
6311	2025	5250	6099	6718	7539	9681	3204	9637	1091
0476	1624	3470	1600	0675	3261	7749	4195	2660	2150
5317	3903	6098	9438	3482	5505	5167	9993	8191	8488
7474	8876	1918	9828	2061	6664	0391	9170	2776	4025
7460	6800	1987	2758	0737	6880	1500	5763	2061	9373
1002	1494	9972	3877	6104	4006	0477	0669	8557	0513
5449	6891	9047	6297	1075	7762	8091	7153	8881	3367
9453	0809	7151	9982	0411	1120	6129	5090	2053	7570
0471	2725	7588	6573	0546	0110	6132	1224	3124	6563
5469	2668	1996	2249	3857	6637	8010	1701	3141	6147
2782	9603	1877	4159	9809	2570	4544	0544	2660	6737
3129	7217	5020	3788	0853	9465	2186	3945	1696	2286
7092	9885	3714	8557	7804	9524	6228	7774	6674	2775
9566	0501	8352	1062	0634	2401	0379	1697	7153	6208
5863	7000	1714	9276	7218	6922	1032	4838	1954	1680
5881	9151	2321	3147	6755	2510	5759	6947	7102	0097
6416	9939	9569	0439	1705	4680	9881	7071	9596	8758
9568	3012	6316	9065	0710	2158	1639	9149	4848	8634
0452	9538	5730	1893	1186	9245	6558	9562	8534	9321
8762	5920	8989	4777	2169	7073	7082	9495	1594	8600
0194	0270	7601	0342	3897	4133	7650	9228	5558	3597
3306	5478	2797	1605	4996	0023	9780	9429	3937	7573
7198	3079	2171	6972	0928	6599	9328	0597	5948	5753
8350	4846	1309	0612	4584	4988	4642	4430	9481	9048
7449	4279	4224	1018	2496	2091	9750	6086	1955	9860
6126	5399	0852	5491	6557	4946	9918	1541	7894	1843
1851	7940	9908	3860	1536	8011	4314	7269	7047	0382
7698	4218	2726	5130	3132	1722	8592	9662	4795	7718
0810	0118	4979	0458	1059	5739	7919	4557	0245	4861
6647	7149	1409	6809	3313	0082	9024	7477	7320	5822
3867	7111	5549	9439	3427	9793	3071	6651	4267	8099
1172	7278	7527	2492	6211	9457	5120	4903	1023	5745
6701	1668	5067	0413	7961	7825	9261	8572	0634	1140
8244	0620	8736	2649	1429	6253	4181	8120	6500	8127
8009	4031	7884	2215	2382	1931	1252	8088	2490	9122
1947	8315	9755	7187	4074	4743	6669	6060	2319	0635
9562	4821	8050	0106	2782	4665	9436	4973	4879	8900
0729	9026	9631	8096	8906	5713	3212	8854	3435	4206
6904	2569	3251	0079	8838	8738	8503	6333	0952	1641

T. P. Ryan, *Statistical Methods for Quality Improvement* © 1989 New York: John Wiley & Sons. This material is used by permission of John Wiley & Sons, Inc.